# Osteoimmunology

# Osteoimmunology: Interactions of the Immune and Skeletal Systems

Edited by

### Joseph Lorenzo, MD
PROFESSOR OF MEDICINE IN THE DIVISION OF ENDOCRINOLOGY AND METABOLISM AND DIRECTOR OF BONE BIOLOGY RESEARCH AT THE UNIVERSITY OF CONNECTICUT HEALTH CENTER, FARMINGTON, CT, USA

### Yongwon Choi, PhD
LEONARD JARETT PROFESSOR OF PATHOLOGY AND LABORATORY MEDICINE, UNIVERSITY OF PENNSYLVANIA SCHOOL OF MEDICINE, PHILADELPHIA, PA, USA

### Mark Horowitz, PhD
PROFESSOR OF ORTHOPEDICS AND REHABILITATION AT THE YALE UNIVERSITY SCHOOL OF MEDICINE, NEW HAVEN, CT, USA

### Hiroshi Takayanagi, MD PhD
PROFESSOR IN THE DEPARTMENT OF CELL SIGNALLING, TOKYO MEDICAL AND DENTAL UNIVERSITY, TOKYO, JAPAN

AMSTERDAM • BOSTON • HEIDELBERG • LONDON • NEW YORK • OXFORD
PARIS • SAN DIEGO SAN FRANCISCO • SINGAPORE • SYDNEY • TOKYO

Academic Press is an imprint of Elsevier

Academic Press is an imprint of Elsevier
32 Jamestown Road, London NW1 7BY, UK
30 Corporate Drive, Suite 400, Burlington, MA 01803, USA
525 B Street, Suite 1800, San Diego, CA 92101-4495, USA

First edition 2011

Copyright © 2011 Elsevier Inc. All rights reserved

No part of this publication may be reproduced, stored in a retrieval system or transmitted in any form or by any means electronic, mechanical, photocopying, recording or otherwise without the prior written permission of the publisher

Permissions may be sought directly from Elsevier's Science & Technology Rights Department in Oxford, UK: phone (+ 44) (0) 1865 843830; fax (+44) (0) 1865 853333; email: permissions@elsevier.com. Alternatively, visit the Science and Technology Books website at www.elsevierdirect.com/rights for further information

**Notice**
No responsibility is assumed by the publisher for any injury and/or damage to persons or property as a matter of products liability, negligence or otherwise, or from any use or operation of any methods, products, instructions or ideas contained in the material herein. Because of rapid advances in the medical sciences, in particular, independent verification of diagnoses and drug dosages should be made

> Medicine is an ever-changing field. Standard safety precautions must be followed, but as new research and clinical experience broaden our knowledge, changes in treatment and drug therapy may become necessary or appropriate. Readers are advised to check the most current product information provided by the manufacturer of each drug to be administered to verify the recommended dose, the method and duration of administrations, and contraindications. It is the responsibility of the treating physician, relying on experience and knowledge of the patient, to determine dosages and the best treatment for each individual patient. Neither the publisher nor the authors assume any liability for any injury and/or damage to persons or property arising from this publication.

**British Library Cataloguing-in-Publication Data**
A catalogue record for this book is available from the British Library

**Library of Congress Cataloging-in-Publication Data**
A catalog record for this book is available from the Library of Congress

ISBN: 978-0-12-375670-1

> For information on all Academic Press publications
> visit our website at www.elsevierdirect.com

Typeset by TNQ Books and Journals

Printed and bound in United States of America

10 11 12 13   10 9 8 7 6 5 4 3 2 1

Working together to grow
libraries in developing countries

www.elsevier.com | www.bookaid.org | www.sabre.org

ELSEVIER   BOOK AID International   Sabre Foundation

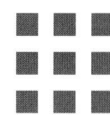

# Contents

Contributors    vii
Foreword    xi
Preface    xiii

1. Overview: The Developing Field of Osteoimmunology    1
   *Joseph Lorenzo, Yongwon Choi, Mark Horowitz, and Hiroshi Takayanagi*

2. Origins of Osteoclasts    7
   *Deborah L. Galson and G. David Roodman*

3. The Adaptive Immune Response    43
   *David G. Hesslein, Hector L. Aguila, and Mark C. Horowitz*

4. The Role of Bone Cells in Establishing the Hematopoietic Stem Cell Niche    81
   *Joy Y. Wu and Henry M. Kronenberg*

5. Osteoblasts and their Signaling Pathways: New Frontiers for Linkage to the Immune System    101
   *Jane B. Lian, Ellen M. Gravallese, and Gary S. Stein*

6. The Osteoclast: The Pioneer of Osteoimmunology    141
   *Roberta Faccio, Yongwon Choi, Steven L. Teitelbaum, and Hiroshi Takayanagi*

7. The Effects of Immune Cell Products (Cytokines and Hematopoietic Cell Growth Factors) on Bone Cells    187
   *Joseph Lorenzo*

8. Interactions Among Osteoblasts, Osteoclasts, and Other Cells in Bone    227
   *T. John Martin, Natalie A. Sims, and Julian M. W. Quinn*

v

9. The Role of the Immune System in the Development of Osteoporosis ............ 269
   *Ulrike I. Mödder, B. Lawrence Riggs, and Sundeep Khosla*

10. The Role of the Immune System in the Bone Loss of Inflammatory Arthritis ............ 301
    *Steven R. Goldring and Georg Schett*

11. Inflammatory Bowel Disease and Bone ............ 325
    *Francisco A. Sylvester and Anthony T. Vella*

12. The Role of the Immune System in Fracture Healing ............ 343
    *Brandon M. Steen, Louis C. Gerstenfeld, and Thomas A. Einhorn*

13. The Role of the Immune System and Bone Cells in Acute and Chronic Osteomyelitis ............ 369
    *Brendan F. Boyce, Lianping Xing, and Edward M. Schwarz*

14. The Role of the Immune System in Hematologic Malignancies that Affect Bone ............ 391
    *Jessica A. Fowler, Claire M. Edwards, and Gregory R. Mundy*

15. Osteoimmunology in the Oral Cavity (Periodontal Disease, Lesions of Endodontic Origin and Orthodontic Tooth Movement) ............ 411
    *Dana T. Graves, Rayyan A. Kayal, Thomas Oates, and Gustavo P. Garlet*

Index ............ 443

Color Plate Section

# Contributors

**Hector L. Aguila**  Department of Immunology, University of Connecticut Health Center, Farmington, CT, USA

**Brendan F. Boyce**  Center for Musculoskeletal Research, University of Rochester Medical Center, Rochester, NY, USA and Department of Pathology and Laboratory Medicine, University of Rochester Medical Center, Rochester, NY, USA

**Yongwon Choi**  Department of Pathology and Laboratory Medicine, University of Pennsylvania School of Medicine, Philadelphia, PA, USA

**Claire M. Edwards**  Vanderbilt Center for Bone Biology, Department of Cancer Biology, Vanderbilt University, Nashville, TN, USA

**Thomas A. Einhorn**  Department of Orthopaedic Surgery, Boston University School of Medicine, Boston, MA, USA

**Roberta Faccio**  Department of Orthopedics, Washington University, St. Louis, MO, USA

**Jessica A. Fowler**  Vanderbilt Center for Bone Biology, Department of Cancer Biology, Vanderbilt University, Nashville, TN, USA

**Deborah L. Galson**  The Center for Bone Biology of UPMC, Departments of Medicine and of Microbiology and Molecular Genetics, University of Pittsburgh School of Medicine, VA Pittsburgh Healthcare System, Pittsburgh, PA, USA

**Gustavo P. Garlet**  Department of Biological Sciences, School of Dentistry of Bauru, Sao Paulo University, Bauru, Brazil

**Louis C. Gerstenfeld**  Department of Orthopaedic Surgery, Boston University School of Medicine, Boston, MA, USA

**Steven R. Goldring**  Departments of Orthopedics and Rheumatology, Hospital for Special Surgery, Weill Medical College of Cornell University, New York, USA

**Ellen M. Gravallese**  Department of Medicine, Division of Rheumatology, University of Massachusetts Medical School, Worcester, MA, USA

**Dana T. Graves**  Department of Periodontics, UMDNJ, Newark, NJ, USA

**David G. Hesslein**  Department of Microbiology and Immunology, University of California San Francisco, San Francisco, CA, USA

**Mark C. Horowitz**  Department of Orthopedics and Rehabilitation, Yale University School of Medicine, New Haven, CT, USA

**Rayyan A. Kayal**   Division of Periodontics, Department of Oral Basic Science, Faculty of Dentistry, King Abdulaziz University, Jeddah, Saudi Arabia

**Sundeep Khosla**   Endocrine Research Unit, College of Medicine, Mayo Clinic, Rochester, MN, USA

**Henry M. Kronenberg**   Endocrine Unit, Massachusetts General Hospital, Boston, MA, USA

**Jane B. Lian**   Departments of Cell Biology, Orthopedics and Physical Rehabilitation, University of Massachusetts Medical School, Worcester, MA

**Joseph Lorenzo**   University of Connecticut Health Center, Farmington, CT, USA

**T. John Martin**   St. Vincent's Institute of Medical Research and University of Melbourne Department of Medicine, Fitzroy, Victoria, Australia

**Ulrike I. Mödder**   Endocrine Research Unit, College of Medicine, Mayo Clinic, Rochester, MN, USA

**Gregory R. Mundy**[¥]   Vanderbilt Center for Bone Biology, Department of Medicine/Clinical Pharmacology, Vanderbilt University, Nashville, TN, USA

**Thomas Oates**   Department of Periodontics, UTHSC, San Antonio, TX, USA

**Julian M. W. Quinn**   Prince Henry's Institute, Monash Medical Centre, Clayton, Victoria, Australia

**B. Lawrence Riggs**   Endocrine Research Unit, College of Medicine, Mayo Clinic, Rochester, MN, USA

**G. David Roodman**   The Center for Bone Biology of UPMC, Department of Medicine, University of Pittsburgh School of Medicine, VA Pittsburgh Healthcare System, Pittsburgh, PA, USA

**Georg Schett**   Department of Internal Medicine III and Institute for Clinical Immunology, University of Erlangen-Nuremberg, Erlangen, Germany

**Edward M. Schwarz**   Center for Musculoskeletal Research, University of Rochester Medical Center, Rochester, NY, USA and Department of Orthopaedics and Rehabilitation, University of Rochester Medical Center, Rochester, NY, USA

**Natalie A. Sims**   St. Vincent's Institute of Medical Research and University of Melbourne Department of Medicine, Fitzroy, Victoria, Australia

**Brandon Steen**   Department of Orthopaedic Surgery, Boston University School of Medicine, Boston, MA, USA

**Gary Stein**   Departments of Cell Biology and Cancer Center, University of Massachusetts Medical School, Worcester, MA, USA

**Francisco A. Sylvester**   University of Connecticut School of Medicine, Hartford, CT, USA

**Hiroshi Takayanagi**  Department of Cell Signalling, Tokyo Medical and Dental University, Tokyo, Japan

**Steven L. Teitelbaum**  Department of Pathology and Immunology, Washington University, St. Louis, MO, USA

**Anthony T. Vella**  University of Connecticut School of Medicine, Farmington, CT, USA

**Joy Y. Wu**  Endocrine Unit, Massachusetts General Hospital, Boston, MA, USA

**Lianping Xing**  Center for Musculoskeletal Research, University of Rochester Medical Center, Rochester, NY, USA and Department of Pathology and Laboratory Medicine, University of Rochester Medical Center, Rochester, NY, USA

# Foreword

## A Promising Bridge Between Immune and Skeletal Systems

The idea that the immune and skeletal systems may have some close interactions has only recently been appreciated. The first observation that immune cells could influence the activity of bone cells came from the finding that supernatants from phytohemagglutinin (PHA)-stimulated peripheral blood monocytes of healthy humans contained a factor called osteoclast- activating factor (OAF), which has now been identified as IL-1 (1985). Since then, TNF (1986) and IL-6 (1990) were also found to have the activity of inducing bone resorption. Osteoclasts, the principal cells responsible for bone resorption, are differentiated from hematopoietic precursors in response to M-CSF and RANKL. Production of proinflammatory cytokines such as IL-1, TNF and IL-6 enhances responses of osteoclasts to RANKL.

From these experimental backgrounds, the field of osteoimmunology was proposed by the Editors of the present book, Joseph Lorenzo, Yongwon Choi, Mark Horowitz and Hiroshi Takayanagi in 2000. They also established the International Conference on Osteoimmunology in 2006, which has been held three times in Greece. These four scientists are indeed pioneers in the field of osteoimmunology. This book contains chapters by outstanding experts. The chapters span the breadth and depth of our current knowledge of osteoimmunology from cell and molecular biology to clinical problems.

According to the U.S. Surgeon General Reports on Bone Health and Osteoporosis, one in two Americans over the age of 50 will be at risk for fractures from osteoporosis and low bone mass by 2020. These bone health concerns become more prominent as people live longer and expect to remain active as they age. Many of the pathologic processes of skeletal and immune systems are major targets for therapeutic intervention. However, the search for novel treatments for these conditions is often pursued in the absence of a scientific understanding of the molecular and cellular pathways that underlie these processes. More recently, RANKL was reported to be a critical factor controlling the differentiation of antigen-sampling microfold (M) cells situated over Peyer's patches from RANK-expressing intestinal epithelial precursor cells (2009). This book will provide readers recent advances in osteoimmunology from basic science to diseases of bone and the immune system.

Tatsuo Suda

MEMBER OF THE JAPAN ACADEMY, ASSOCIATE MEMBER, SCIENCE COUNCIL OF JAPAN, EMERUTUS PROFESSOR OF SHOWA UNIVERSITY, VISITING PROFESSOR, RESEARCH CENTER FOR GENOMIC MEDICINE, SAITAMA MEDICAL UNIVERSITY, JAPAN

# Preface

The editors welcome readers to the first book on the topic of osteoimmunology. The importance of the interactions of bone and immune cells was only really appreciated less than 40 years ago. In addition, the term osteoimmunology was first used in an editorial in Nature by Yongwon Choi just 10 years ago. Hence, as disciplines go, osteoimmunology is relatively young. However, over its short existence, it has seen great progress. Perhaps its most important discovery was the identification of RANKL as the master regulator of osteoclasts. This TNF superfamily member was first isolated and cloned because of its ability to regulate the interactions between T-lymphocytes and antigen presenting dendritic cells. However, after its original description, it was soon found to be the critical signal for bone resorption. In addition to RANKL, a large number of cytokines and immune cells are now known to influence bone cell function and bone mass. This has become important for understanding how bone loss develops in diseases like inflammatory arthritis, inflammatory bowel disease, periodontal disease and after organ or bone marrow transplant. Conversely, bone is now known to provide important signals to the hematopoietic and immune system. Hematopoietic stem cells colonize the marrow and initiate the production of all blood and immune cells. It is well appreciated that these cells exist in the bone marrow because they receive critical signals from bone. In addition, the bone marrow is the site where memory immune cells reside. These T and B-lymphocytes are central for the development of immunity from repeat infection. It is highly likely that bone cell-derived signals also regulate the various "niches" that support these important functions.

This book is designed to bring the reader a broad overview of the latest knowledge about these interactions in a wide variety of areas. It is hoped that both experienced investigators and those just learning about this field will find the information useful as a reference for their own studies in this area. The editors and the authors have tried to be as comprehensive as possible and to provide a detailed list of references in this area where the reader can find additional information about this topic.

# Overview: The Developing Field of Osteoimmunology

Joseph Lorenzo[1], Yongwon Choi[2], Mark Horowitz[3], Hiroshi Takayanagi[4]

[1]UNIVERSITY OF CONNECTICUT, HEALTH CENTER, FARMINGTON, CT, USA
[2]UNIVERSITY OF PENNSYLVANIA, SCHOOL OF MEDICINE, DEPARTMENT OF PATHOLOGY AND LABORATORY MEDICINE, PHILADELPHIA, PA, USA
[3]DEPARTMENT OF ORTHOPAEDICS AND REHABILITATION, YALE UNIVERSITY SCHOOL OF MEDICINE, NEW HAVEN, CT, USA
[4]TOKYO MEDICAL AND DENTAL UNIVERSITY, TOKYO, JAPAN

It has been almost 40 years since the first observations that cells of the immune system could influence the functions of bone cells. Since that time, significant strides have been made in our understanding of the interactions between hematopoietic, immune, and bone cells, which is now known as the field of "osteoimmunology".

In this introductory chapter, we will briefly establish some of the key features of osteoimmunology, which are described in greater detail in the subsequent chapters of this book.

Bone cells derive from two lineages. Osteoclasts, which are responsible for bone resorption, are large, multinucleated cells that are uniquely capable of removing both the organic and mineral components of bone. Osteoclasts share a common origin with cells of the myeloid dendritic cell and macrophage lineages and, as such, respond to and produce many of the cytokines that regulate macrophage and dendritic cell function. The discovery of a tumor necrosis factor (TNF) family member, receptor activator of NF-κB ligand (RANKL), on activated T cells and its subsequent identification as one of the key differentiation and survival factors for osteoclasts, provided critical evidence for a potential link between normal immune responses and bone metabolism.

Bone is formed by osteoblasts, which originate from mesenchymal stem cells (MSC). Osteoblast-lineage cells carry out at least three major functions: (1) they secrete bone matrix, which mineralizes over time to form new bone; (2) they regulate osteoclast differentiation; (3) they support hematopoietic cell growth and differentiation. It is now well accepted that MSC can differentiate into a variety of lineages including osteoblasts, adipocytes, muscle cells, and hematopoiesis-supporting stromal cells [1]. Osteoblast-lineage cells, which are sometimes referred to as stromal cells, produce a variety of cytokines that are critical for hematopoietic cell differentiation.

The first observation that immune cells could influence the activity of bone cells came from the finding that supernatants from phytohemagglutinin-stimulated peripheral blood monocytes of normal humans contained factors that stimulated bone resorption [2]. This activity was named osteoclast-activating factor (OAF). When it was eventually purified and sequenced, the principal stimulator of bone resorption in these crude OAF preparations was identified as the cytokine interleukin-1 (IL-1) [3]. In addition to its ability to stimulate osteoclast formation and resorbing activity, IL-1 is a mediator of a variety of inflammatory responses and a potent stimulus of prostaglandin synthesis, which independently increases bone resorption. It also is an inhibitor of osteoblast activity and bone formation.

Subsequent to the identification of IL-1 as a bone resorption stimulus, tumor necrosis factor (TNF) [4] and interleukin-6 (IL-6) [5] were also found to have this activity. Like IL-1, these cytokines are critical mediators of inflammatory responses. It has now been demonstrated that a long list of cytokines can have both positive and negative effects on bone mass and bone cell activity.

Production of cytokines by immune cells has been linked to human diseases that involve bone. Perhaps the most extensive studies have been of the role of cytokines in the development of the bone loss and lytic lesions that occur in inflammatory arthritis, inflammatory bowel disease, and periodontal disease [6–8]. Here, production of RANKL from a variety of cell types mediates osteolysis by stimulating osteoclastic activity. In addition, production of proinflammatory cytokines such as IL-1, TNF, and IL-6 enhances the response of osteoclasts to RANKL.

Estrogen withdrawal after menopause is also associated with a rapid and sustained increase in the rate of bone loss. This phenomenon seems to result from an increase in bone resorption, which is not met by an equivalent increase in bone formation. It was initially demonstrated that conditioned medium from cultured peripheral blood monocytes from osteoporotic women with rapid bone turnover contained more IL-1 activity than did conditioned medium from the cells of women with slow bone turnover or normal controls [9]. In rodents, treatment with inhibitors of IL-1 and TNF prevented the bone loss that occurred with ovariectomy. In addition, ovariectomy was not associated with bone loss in mice that were genetically prevented from responding to IL-1 [10] and TNF [11] or unable to produce IL-6 [12, 13]. These findings strongly link the bone loss of estrogen withdrawal to effects of estrogen on the production or activity of proinflammatory cytokines. Most recently it was shown that inhibitors of IL-1 and TNF reduced the rate of bone resorption in post-menopausal women [14].

The role of cytokines in the bone disease that occurs with malignancy has also been studied extensively [15]. In hematological malignancies such as lymphomas or multiple myeloma, which are associated with increased osteoclast formation and activity, a variety of cytokines have been implicated as mediating the bone loss that can occur in these conditions. Unlike the bone disease of solid tumors, which is typically mediated by parathyroid-hormone-related protein (PTHrP), hematological malignancies are often characterized by an uncoupling of resorption from formation and the development of purely lytic bone lesions.

The immune system is also involved in normal fracture healing and the response of bone to infections (osteomyelitis). Understanding the interactions of bone and immune cells during these events is best accomplished by an osteoimmunologic approach, which integrates an appreciation of the crosstalk between these two organ systems [16, 17].

The question of whether the immune system influences normal skeletal development and function is not well answered. Ontogenically, skeletal development precedes early immune system development. Therefore, it is unlikely that the immune system influences early skeletal formation. However, bone homeostasis and remodeling occur throughout life. Anatomically, bone marrow spaces are loosely compartmentalized, which allows immune and bone cells to interact and influence each other. Hence, bone homeostasis is often regulated by immune responses, particularly when the immune system has been activated or becomes pathologic.

It is not difficult to imagine that crosstalk occurs throughout life between activated lymphocytes and bone cells because all mammals are constantly challenged by a variety of infectious agents, which produce some level of sustained low-grade immune system activation. Furthermore, as we age, there is an accumulation in the bone marrow of memory T cells, which can express RANKL on their surface [18, 19]. The role that these cells play in skeletal homeostasis is unknown. However, it is conceivable that they might influence bone turnover and be responsible for some of the changes that occur in the skeleton with aging.

Immunologists and hematologists are well aware that, at least in adult mammals, the development of the immune system depends on the normal function of hematopoietic stem cells (HSCs), which are now known to reside in close association with bone cells. It is not surprising to learn that the development of the immune system in the bone marrow is dependent on the production of a facilitative microenvironment by bone cells. This fact was made clearer by data demonstrating that osteoblast-lineage cells provide key factors that regulate HSC development. There is also accumulating evidence that bone continues to play a role in adaptive immunity, beyond its influence on lymphocyte development. It is now known that long-lived memory T and B cells return to specialized niches in the bone marrow. These cells are capable of circulating throughout the organism. However, the questions of why they remain in specific areas of bone marrow and what factors draw them there remain unanswered. It is likely that the answers to these questions will come from experiments that are designed in the context of osteoimmunology by investigators who have knowledge of both the immune system and bone.

We are quite honored to have obtained 14 outstanding contributions for this book. The chapters of this book span the breadth and depth of our current knowledge of osteoimmunology. In this volume, the contributions are organized according to their scientific messages, though these connections are not absolute.

The initial chapters deal with the development of osteoblasts, osteoclasts, hematopoietic stem cells, T and B lymphocytes, and communications between these cellular elements. There is also a detailed chapter on the signaling pathways by which RANKL influences osteoclast development and function. Subsequent chapters explore the effects that estrogen has on bone and the immune system and the development of pathologic conditions, which involve osteoimmunology, like osteoporosis and the bone loss of inflammatory arthritis, inflammatory bowel disease, periodontal disease, and hematologic malignancies. The book concludes with chapters on the role that immune and bone cell interactions have in osteomyelitis and fracture healing.

After reading this book, one will hopefully appreciate the intricate interaction between the immune system and bone. However, despite the progress that has already been made towards understanding the cross-regulation between bone and the immune system, the biological implications of such interactions are only beginning to be identified. The fields of immunology and bone biology have matured sufficiently so that key cellular and molecular mechanisms governing the homeostasis of the individual systems are extensively described. Hence, progress toward understanding osteoimmunologic networks will likely be greatly facilitated by creating an environment conducive to its study. It is hoped that this endeavor will lead to better treatments for human diseases involving both systems.

Many of the pathologic processes of the skeletal and immune systems are major targets for therapeutic intervention. However, the search for novel treatments for these conditions is often pursued in the absence of a solid scientific understanding of the molecular and cellular pathways that underlie these processes. According to the U.S. Surgeon General Report on Bone Health and Osteoporosis, by 2020 one in two Americans over the age of 50 will be at risk for fractures from osteoporosis or low bone mass. These health concerns become more prominent as people live longer and expect to remain active as they age. Future interventions to prevent and treat bone diseases will require a high degree of specificity, especially because these therapies are often tailored for a segment of the population that is already suffering from or is vulnerable to other age-related ailments. These issues place osteoimmunology in a position of unique clinical significance and make its study highly relevant.

# References

[1] Rosen CJ, Ackert-Bicknell C, Rodriguez JP, Pino AM. Marrow fat and the bone microenvironment: developmental, functional, and pathological implications (Translated from English). Crit Rev Eukaryot Gene Expr 2009;19(2):109–24 (in English).

[2] Horton JE, Raisz LG, Simmons HA, Oppenheim JJ, Mergenhagen SE. Bone resorbing activity in supernatant fluid from cultured human peripheral blood leukocytes. Science 1972;177(51):793–5.

[3] Dewhirst FE, Stashenko PP, Mole JE, Tsurumachi T. Purification and partial sequence of human osteoclast-activating factor: identity with interleukin 1 beta. J Immunol 1985;135:2562–8.

[4] Bertolini DR, Nedwin GE, Bringman TS, Smith DD, Mundy GR. Stimulation of bone resorption and inhibition of bone formation in vitro by human tumour necrosis factors. Nature 1986;319:516–18.

[5] Ishimi Y, Miyaura C, Jin CH, et al. IL-6 is produced by osteoblasts and induces bone resorption. J Immunol 1990;145:3297–303.

[6] Schett G. Osteoimmunology in rheumatic diseases. Arthritis Res Ther 2009;11(1):210.

[7] Sylvester FA. IBD and skeletal health: children are not small adults! Inflamm Bowel Dis 2005; 11(11):1020–3.

[8] Taubman MA, Valverde P, Han X, Kawai T. Immune response: the key to bone resorption in periodontal disease. Journal of Periodontology 2005;76(11-s):2033–41.

[9] Pacifici R, Rifas L, Teitelbaum S, et al. Spontaneous release of interleukin 1 from human blood monocytes reflects bone formation in idiopathic osteoporosis. Proc Natl Acad Sci USA 1987;84: 4616–20.

[10] Lorenzo JA, Naprta A, Rao Y, et al. Mice lacking the type I interleukin-1 receptor do not lose bone mass after ovariectomy. Endocrinology 1998;139(6):3022–5.

[11] Ammann P, Rizzoli R, Bonjour JP, et al. Transgenic mice expressing soluble tumor necrosis factor-receptor are protected against bone loss caused by estrogen deficiency. J Clin Invest 1997; 99(7):1699–703.

[12] Jilka RL, Hangoc G, Girasole G, et al. Increased osteoclast development after estrogen loss: mediation by interleukin-6. Science 1992;257:88–91.

[13] Poli V, Balena R, Fattori E, et al. Interleukin-6 deficient mice are protected from bone loss caused by estrogen depletion. EMBO Journal 1994;13:1189–96.

[14] Charatcharoenwitthaya N, Khosla S, Atkinson EJ, McCready LK, Riggs BL. Effect of blockade of TNF-alpha and interleukin-1 action on bone resorption in early postmenopausal women. J Bone Miner Res 2007;22(5):724–9.

[15] Mundy GR. Metastasis to bone: causes, consequences and therapeutic opportunities. Nat Rev Cancer 2002;2(8).

[16] Einhorn TA. The science of fracture healing. J Orthop Trauma 2005;19(Suppl. 10):S4–6.

[17] Marriott I. Osteoblast responses to bacterial pathogens: a previously unappreciated role for bone-forming cells in host defense and disease progression. Immunol Res 2004;30(3):291–308.

[18] Effros RB. Replicative senescence of CD8 T cells: effect on human ageing (Translated from English). Exp Gerontol 2004;39(4):517–24 (in English).

[19] Josien R, Wong BR, Li HL, Steinman RM, Choi YW. TRANCE, a TNF family member, is differentially expressed on T cell subsets and induces cytokine production in dendritic cells. Journal of Immunology 1999;162(5):2562–8.

# 2

# Origins of Osteoclasts

Deborah L. Galson[1], G. David Roodman[2]

[1]THE CENTER FOR BONE BIOLOGY OF UPMC, DEPARTMENTS OF MEDICINE AND OF MICROBIOLOGY AND MOLECULAR GENETICS, UNIVERSITY OF PITTSBURGH SCHOOL OF MEDICINE, VA PITTSBURGH HEALTHCARE SYSTEM, PITTSBURGH, PA, USA,
[2]THE CENTER FOR BONE BIOLOGY OF UPMC, DEPARTMENT OF MEDICINE, UNIVERSITY OF PITTSBURGH SCHOOL OF MEDICINE, VA PITTSBURGH HEALTHCARE SYSTEM, PITTSBURGH, PA, USA

## CHAPTER OUTLINE

**Introduction: osteoclasts** ............ 8
**Hematopoietic origins of osteoclasts – insights from osteopetrosis** ............ 11
    Hematopoietic cells ............ 11
    Osteoclast precursors are in the monocyte–macrophage lineage ............ 11
**Generation of osteoclasts** ............ 17
    External signals/receptors ............ 18
        *M-CSF/c-fms (Csf1R)* ............ 18
        *RANKL/RANK/OPG* ............ 18
    Signal transduction molecules ............ 19
        *TRAF6* ............ 20
        *DAP12/FcRγ and their co-receptors TREM-2, PIR-A, OSCAR, and SIRPβ1* ............ 20
        *CaMKIV and calcineurin* ............ 21
    Transcription factors ............ 21
        *Spi1/PU.1* ............ 21
        *NF-κB/IKKα (IKK1), IKKβ (IKK2), IKKγ (IKK3, NEMO)* ............ 22
        *PPARγ* ............ 23
        *c-Fos* ............ 23
        *NFATc1* ............ 24
        *MITF* ............ 24
    Regulation of pre-osteoclast fusion ............ 24
        *CD47/TSP1/SIRPα (MFR)* ............ 24
        *DC-STAMP* ............ 26
        *Multi-subunit V-ATPase (oc/oc (Atp6i) and Atp6v0d2)* ............ 26
        *CD44* ............ 27
        *ADAM8/α$_9$β$_1$-integrin* ............ 27

|   |   |
|---|---|
| *CTR* | 28 |

**More than one type of osteoclast?** .................................................................................. 28

    Distinctive morphological and biochemical characteristics of mature osteoclasts
    that suggest the presence of osteoclast subtypes ............................................................ 29

        *Carbonic anhydrase II (CAII)* ................................................................................ 29

        *Anion exchangers (Ae2 and Slc4a4)* ...................................................................... 30

        *Cathepsin K (CatK)* ................................................................................................. 30

        *Matrix metalloproteinase (MMP9)* ....................................................................... 30

    Different osteoclasts in different bone sites: trabecular vs. cortical, long bones
    vs. jaw and calvarial sites ................................................................................................ 31

**Other OCL precursors** ............................................................................................................ 32

    Other macrophage lineage cells ..................................................................................... 32

        *Dendritic cells* ........................................................................................................ 32

        *Alveolar macrophages* ........................................................................................... 33

    B cells ............................................................................................................................... 33

    Multiple myeloma cells .................................................................................................... 33

**Conclusion** ............................................................................................................................. 34

**References** ............................................................................................................................ 34

# Introduction: osteoclasts

Bone remodeling is an essential process, which creates the marrow space utilized for hematopoietic stem cell differentiation, shapes and sculpts the bones during growth, enables tooth eruption, is critical for maintaining bone quality and strength, and is part of the system generating calcium homeostasis. Excessive bone remodeling is a feature of a number of pathological states, such as rheumatoid arthritis, hypercalcemia of malignancy, Paget's disease, and osteoporosis, while defective bone remodeling is seen in osteopetrosis. Mammalian bone undergoes continuous remodeling to remove old bone and stress-induced microfractures, which involves a process of bone resorption at selected sites followed by bone formation at the previous site of resorption. Without remodeling the skeleton would eventually collapse. Humans remodel their skeleton at different rates depending upon the bone location and the number of additional factors, including mechanical forces, autocrine and paracrine hormone status, and immunological influences. However, on average, during normal bone remodeling in the adult *human skeleton*, 5–10% of the existing bone is replaced every *year*.

    Osteoclasts are large multinucleated giant cells that contains between 3 and 100 nuclei per cell, but usually contain 10–20 nuclei per cell [1] and are highly motile. They form by fusion of mononuclear precursors and become adherent to bone. Osteoclasts are polarized cells that have a basolateral domain that doesn't face bone and a resorptive surface that forms a characteristic sealing zone and F-actin ring at sites of bone contact (Figure 2-1). The integrin $\alpha_v\beta_3$ (also known as vitronectin receptor (VNR) and

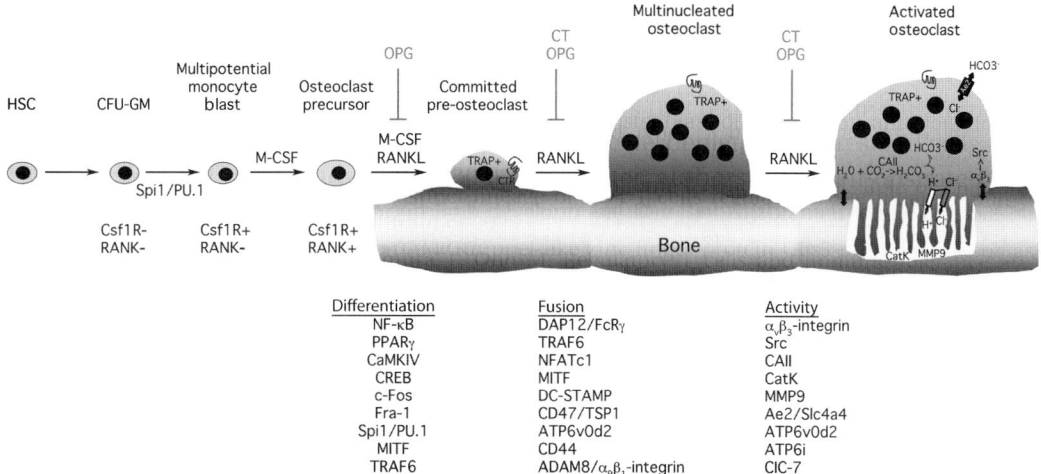

FIGURE 2-1 **Stages of osteoclast differentiation.** In basal osteoclastogenesis, membrane-bound forms of M-CSF and RANKL produced and presented largely by osteoblasts induce the differentiation, activation, and survival of osteoclasts from osteoclast precursors through a series of steps. Depicted are some of the key transcription factors regulating the steps of osteoclastogenesis (Spi1/PU.1, MITF, NF-κB, PPARγ, CREB, c-Fos, Fra-1, NFATc1), signal transduction molecules (TRAF6, CaMKIV, DAP12/FcRγ, Src), fusion regulators (transmembrane proteins DC-STAMP, CTR, CD47, and CD44, the CD47 ligand TSP1, the vacuolar-ATPase subunit ATP6v0d2, and the disintegrin and metalloproteinase ADAM8 and its cognate receptor $\alpha_9\beta_1$-integrin), and genes important for osteoclast function ($\alpha_v\beta_3$-integrin, CAII, proteinases CatK and MMP9, anion exchangers Ae2 and Slc4a4, vacuolar-ATPase subunit ATP6i, the chloride channel ClC-7). Only a few of the regulatory cytokines are denoted (M-CSF, RANKL, OPG, and CT). Please refer to color plate section.

$CD51^+CD61^+$) mediates the attachment of the sealing zone to the bone surface by binding Arg-Gly-Asp (RGD)-containing extracellular matrix proteins such as osteopontin. Activation of Src kinase by $\alpha_v\beta_3$ is required to form the actin ring structure and create a sealing zone. Inside the sealing zone, the resorptive surface of the osteoclast forms a unique and specialized ruffled border at the interface with bone from which proteolytic enzymes and hydrogen ions are released to degrade and resorb both the mineral and organic components of the bone matrix [2]. This feature distinguishes osteoclasts from macrophage polykaryons. When osteoclasts are plated on bone surfaces, a characteristic "resorption pit" is formed below the cell within the sealing zone. These resorption lacunae are never seen in the absence of osteoclasts and are not produced by macrophages or macrophage polykaryons. The capability to efficiently excavate bone is a unique function of osteoclasts and requires many specialized systems as well as exquisite regulation to maintain healthy bone.

The acidification process involves a key enzymatic reaction and a set of transports in and out of the cell to acidify the extracellular space between the osteoclast and the bone and maintain the $pH_i$ of the osteoclast (reviewed in [3]). Carbonic anhydrase II (CAII) in the cytoplasm of the osteoclast forms carbonic acid from carbon dioxide and water, which then dissociates to form bicarbonate ($HCO_3^-$) and a proton ($H^+$). The protons are then pumped out of the ruffled membrane by vacuolar proton pump ($H^+$-ATPase) into the extracellular space next to the bone to generate an acidic environment (~pH 4.5) that dissolves the mineral in the bone matrix. Chloride channel 7 (CIC-7) encoded by the *CLCN7* gene is coupled to the vacuolar $H^+$-ATPase (V-ATPase) and pumps $Cl^-$ out of the cell into the resorption lacunae in order to balance the charge of ions across the membrane. Additionally, in the basolateral membrane there is a $Cl^-/HCO_3^-$ anion exchanger (AE2 or SLC4A2), which passively transports the excess bicarbonate generated by CAII out of the cell in exchange for $Cl^-$ entering the cell, thereby maintaining the pH of the cell and providing $Cl^-$ for the chloride channel on the resorptive surface.

Osteoclasts are located on the endosteal surface of Haversian tunnels running through cortical bone and trabeculae thicker than 200 μm and on the periosteal surface beneath the periosteum. Osteoclasts are generally rare cells (only 2–3 per $μm^3$), except at sites of increased bone turnover such as the metaphysis of growing bones. Although potential precursors are found in the peripheral blood and spleen as well as within the bone marrow, mature osteoclasts are rarely observed off the bone surface, except in some particular disease states such as giant cell tumors [4].

Osteoclasts have several unique biochemical features that are useful tools in their identification and have important functional roles as well. These include, but are not limited to, expression of the calcitonin receptor (CTR), $β_3$ integrin, CAII, V-ATPase subunit ATP6i, cathepsin K (CatK), matrix metalloproteinase 9 (MMP9). A key biochemical feature often used to identify osteoclasts is the presence of binuclear iron protein tartrate-resistant acid phosphatase (TRAP; type 5 acid phosphatase, *Acp5*). TRAP appears very early during osteoclast differentiation, and continues to increase throughout. Although by RT-PCR analysis, the mRNA for TRAP is expressed in other tissues, such as gut, kidney, and lung, it is most highly expressed in bone and can serve as a marker enzyme for osteoclasts [5, 6]. However, activated human macrophages, but not murine macrophages, can also express TRAP. While increased TRAP expression is observed in pathological conditions in which osteoclast activity is increased, such as osteoporosis and hypercalcemia of malignancy, there are also pathological conditions in which TRAP expression is increased in non-osteoclastic cells, including splenic macrophages in Gaucher disease [7] and hairy cell leukemia cells, which derive from mature B cells [8]. Additionally, TRAP is expressed in only the most mature mononuclear osteoclast precursors, just prior to their fusion into multinuclear cells, so cell surface markers need to be established that define the osteoclast precursors. This chapter will discuss the efforts to identify the cell surface markers of the osteoclast precursor.

Analysis of gene deficiencies and mutations that cause osteopetrosis provide insight into the origins of osteoclasts and the molecular mechanisms that regulate their differentiation and function. In general, these split into two types of mechanisms that result in osteopetrosis. In one group are gene alterations that cause osteopetrosis due to effects on osteoclast differentiation that result in decreased or absent multinucleated osteoclasts. In the other group are gene alterations that cause osteopetrosis in which osteoclast numbers are approximately normal or even elevated, but resorption is impaired. This chapter will focus more on those that affect differentiation, and also on those that suggest that there are multiple types of osteoclasts.

## Hematopoietic origins of osteoclasts – insights from osteopetrosis

### Hematopoietic cells

The hematopoietic origins of osteoclasts were first demonstrated by the elegant experiments of Walker using parabiosis to rescue osteopetrotic gray-lethal (*gl/gl*) and microphthalmic (*mi/mi*) mice [9]. This was complemented by experiments using bone marrow or spleen cell reconstitution of lethally irradiated mice to restore normal bone resorption by transfer of cells from normal mice into the osteopetrotic *gl/gl* and *mi/mi* [10], and conversely, to induce osteopetrosis in normal mice by transfer of spleen cells from *gl/gl* or *mi/mi* mice into lethally irradiated normal mice [11]. Parabiotic union between a rat with monocytes labeled with thorotrast and an unlabeled lethally irradiated rat revealed that the osteoclasts formed in the irradiated rat were derived from the non-irradiated partner [12]. Correction of osteopetrosis in humans by bone marrow transplantation [13, 14] further confirmed that the osteoclast precursor is present in the hematopoietic tissue. Hattersley and Chambers [15] showed that when either bone marrow hematopoietic cells or a factor-dependent mouse multipotential hematopoietic cell line (FDCP-mix A4) were cultured on bone slices with $1,25(OH)_2D_3$ and either live or killed, fixed bone marrow stromal cells, they were both able to generate bone-resorbing osteoclasts. These results demonstrated that although stromal cells play a required role in supporting osteoclast differentiation, they are not the osteoclast progenitors.

### Osteoclast precursors are in the monocyte–macrophage lineage

Considerable evidence supports the concept that osteoclast precursors derive from multipotent precursors of the monocyte–macrophage lineage. Young [16] showed using [$^3$H]-thymidine that osteoclasts form by fusion of mononuclear cells rather than by mitotic division. Early evidence which suggested that osteoclasts were derived from monocyte precursors included histological studies that revealed that mononuclear cells with a low nuclear–cytoplasmic ratio and an abundance of ribosomes invade sites of bone resorption (for a review see [17]). The key evidence that osteoclasts could form by fusion of peripheral blood monocytes was provided by the studies of Tinkler et al. [18] who injected [$^3$H]-thymidine-labeled peripheral blood monocytes into syngeneic hosts treated with $1,25(OH)_2D_3$ and found that the labeled nuclei from these cells were present in the host osteoclasts. Confirmation of these findings was presented by direct observation of fusion between chicken peripheral blood monocytes and purified osteoclasts in vitro [19]. Further support for the relatedness between osteoclasts and monocyte–macrophages is the finding that a number of the same antigens are expressed on osteoclast precursors and/or osteoclasts as well as on monocyte–macrophages and/or macrophage polykaryons. These include CD11b ($\alpha_M$ integrin, a subunit of Mac1, also known as complement receptor 3 (CR3)), Csf1R (colony-stimulating factor 1 receptor, also known as c-Fms, M-CSF R, and CD115), CD68 (also called macrosialin), and Kn22. Flow cytometric analysis of postmitotic committed osteoclast precursors undergoing osteoclast differentiation revealed that they expressed macrophage-associated phenotypes such as non-specific esterase, Mac1, Mac2, Gr1, but not F4/80 [20]. Additionally, the cells were negative for the B-cell marker B220 and the T-cell marker CD3e, and had a myelomonocytic appearance by Wright-Giemsa staining. However, as already noted for F4/80 expression, there are also some differences in the surface antigens expressed by macrophages and osteoclasts. These include loss of CD11b (Mac1) with osteoclast differentiation, and gain of expression of the 121F antigen (related to the magnesium iron superoxide dismutase), calcitonin receptor (CTR), and $\alpha_v\beta_3$ integrin [17].

Although osteoclasts are derived from the monocyte–macrophage lineage, their precise origin remains unclear. Monocytes develop from hematopoietic stem cells in the bone marrow through a series of intermediate multipotential stages and lineage commitment decisions that successively restrict their developmental potential. The pluripotent hematopoietic stem cell gives rise to a myeloid progenitor cell, which can further differentiate to megakaryocytes, granulocytes, monocyte–macrophages, myeloid dendritic cells (mDC), and osteoclasts. The current paradigm suggests that monocytes go through a common myeloid progenitor (CMP), the granulocyte/macrophage progenitor (GMP), and the macrophage/DC progenitor (MDP) stages. There is some disagreement about what constitutes the point for divergence of the osteoclast lineage and there seems to be plasticity in the system as regards the differentiation of the various specialized resident tissue macrophages, such as osteoclasts (bone), mDC (immune system), alveolar macrophages (lung), Langerhans cells (epidermis), microglia (brain), histiocytes (connective tissue), and Kupffer cells (liver). These various tissue macrophages require continuous renewal and three alternatives have been proposed that are not mutually exclusive and may operate in parallel to regenerate these subsets: (1) self-renewal of differentiated cells in the peripheral tissues; (2) proliferation of bone-marrow-derived precursors in the peripheral tissues; and (3) continuous extravasation and differentiation of circulating blood precursors. However, although there is evidence that monocyte–macrophages, osteoclasts, and DC all derive from myelomonocytic precursors, the precise lineage of the osteoclast and its relationship to other hematopoietic cells remains uncertain.

Monocytes are defined as blood mononuclear cells with bean-shaped nuclei that express CD11b, CD11c ($\alpha_X$ integrin, a subunit of CR4), and CD14 (LPS receptor subunit) in humans, and CD11b plus F4/80 in mice and lack markers for pDC, B, T, NK cells. However, the monocytes within this population are morphologically and physiologically heterogeneous with distinct subsets. The functional subset that contains circulating osteoclast precursors is still being defined. Furthermore, how this subset relates to the osteoclast precursor that resides within the bone marrow is also not well understood. The question is unresolved as to whether osteoclast precursors involved in normal bone remodeling ever need to leave the marrow, that is, do they need to spend time outside the marrow to mature in some undefined manner and then be recruited back? Are circulating osteoclast precursors recruited to sites of inflammation different from osteoclast precursors involved in normal bone remodeling? In addition, there is some evidence that osteoclasts at different bone sites are not identical – these differences could arise because they are induced by signals from the unique microenvironments in each location or unique subsets of osteoclast precursors could selectively migrate to each location.

Kurihara and coworkers [21] have shown using spleen cells from 5-fluorouracil-treated mice (to expand the hematopoietic precursor pool) that the earliest identifiable hematopoietic precursor that can form osteoclasts in vitro when cultured in the appropriate cytokine milieu is the multipotent hematopoietic precursor, colony-forming unit (CFU)-blast which can differentiate into all the hematopoietic lineages. Within this population are the CFU-GM, the granulocyte–macrophage progenitor cells. This cell proliferates when stimulated by GM-CSF (7–10 days) in semi-solid media to yield colonies of cells that form osteoclasts at very high efficiency when cultured with interleukin (IL)-3, GM-CSF, and $1,25(OH)_2D_3$ (21) or RANK ligand (RANKL), macrophage CSF (M-CSF, also known as CSF-1), and dexamethasone [22] (M-CSF and RANKL, and their respective receptors Csf1R and RANK play important roles in inducing osteoclast formation and are discussed in more detail below). In contrast, more committed macrophage CFU (CFU-M)-derived cells form few osteoclasts under these conditions. These observations suggest that osteoclasts differentiate from early myelomonocytic progenitors rather than more differentiated monocyte–macrophage progenitors. In humans, G-CSF administration can mobilize $CD34^+$ (binds L-selectin) cells from the marrow into peripheral blood [23]. This $CD34^+$ cell population was $Stro-1^-$ (indicating that there were no stromal cells) and formed osteoclasts when cultured with GM-CSF, IL-1, and IL-3 in vitro.

The work of Arai et al. [24] revealed that in the mouse bone marrow c-Kit$^+$(CD117$^+$) CD11b$^{dull}$ cells contained the osteoclast progenitors as opposed to the c-Kit$^+$CD11b$^{hi}$ population, which were more mature and contained macrophages and granulocytes (Figure 2-2). Furthermore, the c-Kit$^+$CD11b$^{dull}$ population could be divided into Csf1R$^+$ and Csf1R$^-$ subsets with different properties. Compared to the Csf1R$^+$ subset, the Csf1R$^-$ subset contained cells that were more multipotent and immature, and took longer to become osteoclasts. Incubation with stem cell factor (SCF; the ligand for c-Kit) induced the Csf1R$^-$ population to change into Csf1R$^+$ cells after 2 days (early osteoclast precursors), which could then be induced to differentiate into osteoclasts more quickly. Exposure of the c-Kit$^+$CD11b$^{dull}$Csf1R$^+$ cells to M-CSF induced the expression of RANK (found on osteoclast precursors). Addition of M-CSF and RANKL to the Csf1R$^+$RANK$^+$ cells quickly induced them to become osteoclasts; however, if addition of RANKL was delayed, the M-CSF induced the progenitors to become macrophages (the default pathway). Therefore the c-Kit$^+$CD11b$^{dull}$ Csf1R$^+$RANK$^+$ cell was a bipotential precursor until exposed to RANKL.

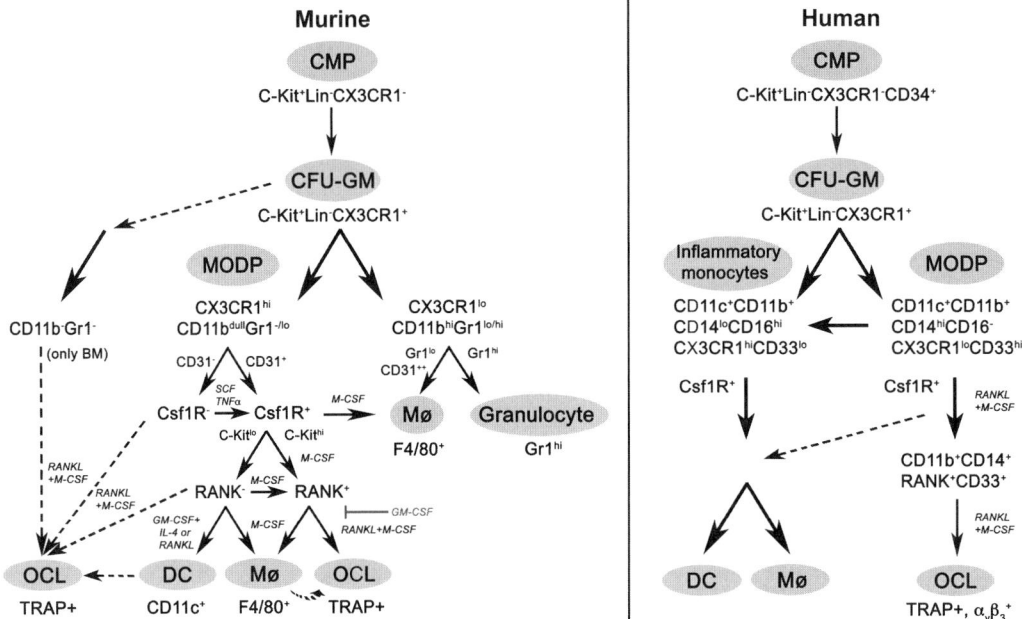

**FIGURE 2-2 Murine and human osteoclast progenitor populations.** A schematic representing a synopsis of the murine and human cell surface markers differentiating different osteoclast precursor pathways starting from the common myeloid progenitor (CMP) and the granulocyte–monocyte precursor (CFU-GM) as discussed in the text. Not depicted in the diagram is the branching of the erythroid and megakaryocyte lineages from the CMP or the steps before the CMP. The multipotential precursors for macrophages, osteoclasts, mDCs are represented as MODP. Only a few key cytokines are noted. The CD11b$^-$Gr1$^-$ population in mice can become osteoclasts only if harvested from the bone marrow (BM), and not if harvested from the peripheral blood or spleen, suggesting that there is a difference between those two populations. Note that the CX3CR1 levels on mouse and human progenitors define different subpopulations.

Mouse Flt3$^+$ bone marrow cells expanded in the presence of Flt3$^+$-ligand were found to alter their differentiation capacity in a time-dependent manner in response to the appropriate cytokines, becoming macrophages (all times), osteoclasts (day 6), mDCs (day 8), microglia (day 11). These results suggest that these cell types share common progenitors [25] with macrophage fate always the default pathway.

Recent work from de Vries et al. [26] analyzed the osteoclastogenic potential of different stages of mouse myeloid development isolated from bone marrow: early blasts (CD31$^{hi}$Ly6C$^-$), myeloid blasts (CD31$^+$Ly6C$^+$), and monocytes (CD31$^-$Ly6C$^{hi}$). The myeloid blasts responded most quickly to M-CSF plus RANKL and developed into multinucleated (MNC) osteoclasts within 4 days, whereas the other two populations took 8 days to reach maximal MNC osteoclasts. In mice, blood monocytes newly released from the bone marrow are exclusively Ly6C$^{hi}$ (Gr1$^{hi}$), and the level of Ly6C is downregulated in the circulation [27]. Jacquin et al. [28] showed that mouse bone-marrow-derived osteoclasts form with highest efficiency from CD3$^-$CD45R$^-$CD11b$^{-/lo}$Csf1R$^+$ cells and that these cells could be further subdivided into c-Kit$^{hi}$ (rapidly formed osteoclasts when cultured with M-CSF and RANKL) and c-Kit$^{-/lo}$ (formed osteoclasts more slowly in vitro). Yao et al. [29] found that both CD11b$^-$/Gr1$^-$ and CD11b$^+$Gr1$^{-/lo}$, but not CD11b$^+$Gr1$^{hi}$ cells, isolated from the marrow responded to M-CSF plus RANKL to form osteoclasts. However, when these three populations were isolated from the blood, only CD11b$^+$Gr1$^{-/lo}$ could differentiate into osteoclasts. The Gr-1 antigen is constitutively expressed on granulocytes (the CD11b$^+$Gr1$^{hi}$ cells), and transiently expressed by monocytic cells during differentiation into macrophages. It is likely that the marrow CD11b$^-$Gr1$^-$ give rise to the CD11b$^+$Gr1$^{-/lo}$ cells. These latter cells are still multipotent and can give rise to macrophages and DCs as well as osteoclasts and contain both early (Csf1R$^-$RANK$^-$; Csf1R$^+$RANK$^-$) and late (Csf1R$^+$RANK$^+$) osteoclast precursors. M-CSF and sRANKL induce osteoclastogenesis, while GM-CSF with sRANKL (or GM-CSF plus IL-4) induces dendritic cell differentiation from single common precursors (c-Kit$^+$Csf1R$^+$RANK$^-$ cells) that can also form macrophages [30]. M-CSF and GM-CSF appear to antagonize each other in the regulation of osteoclast vs. dendritic cell differentiation. However, the effect of GM-CSF on osteoclastogenesis is biphasic as it supports the generation of CFU-GM, which are efficient osteoclast precursors, but later inhibits human osteoclastogenesis while promoting the formation of CD1a$^+$/TRAP$^-$ DC clusters [31]. Tumor necrosis factor (TNF)α can also induce the Csf1R$^-$/RANK$^-$ cells to express Csf1R resulting in increased osteoclast progenitors in the bone marrow and the blood. The CD11b$^+$Gr1$^{-/lo}$Csf1R$^+$ cells can then be induced by M-CSF to express RANK.

Shalhoub et al. [32] used a fluorescent form of RANKL to identify osteoclast precursors in human peripheral blood mononuclear cells (PBMCs). They excluded T and B cells (which express RANK) by analyzing the CD14$^+$ population (10–15% of PBMC). They found that all CD14$^+$RANK$^+$ cells were also positive for CD33 (SEGLIC3; contains an ITIM), CD61 (β$_3$ integrin), CD11b, CD38 (cyclic ADP ribose hydrolase), CD45 (PTPRC), and CD54 (ICAM-1), but not CD34 or CD56 (NCAM). The expression of β$_3$ integrin suggests that these cells had already become committed pre-osteoclasts as this gene is induced late in osteoclast differentiation.

Recently, Geissmann et al. [33] proposed that CD11b$^+$F4/80$^+$ murine blood monocytes can be divided into two functional subsets according to expression patterns of an array of surface proteins. One subset is short-lived inflammatory monocytes that are actively recruited to inflamed tissues and produce inflammatory cytokines. These cells are CX3CR1$^{lo}$Ly6C/G$^+$(Gr1$^+$)CCR2$^+$CD62L$^+$VLA2$^+$VLA4$^+$LFA1$^+$CD31$^{++}$ (large granular cells). The other subset is comprised of long-lived resident monocytes that give rise to specialized cell types, including osteoclasts and are CX3CR1$^{hi}$Ly6C/G$^-$(Gr1$^-$) CCR2$^-$CD62L$^-$VLA2$^-$VLA4$^+$LFA1$^{++}$CD31$^+$ (small cells). Both subsets are non-cycling in the blood. Utilizing an adoptive transfer model and mice expressing enhanced green fluorescent protein (EGFP) because the EGFP was knocked into one allele of the CX3CR1 gene (heterozygous CX3CR1-EGFP), they showed that CX3CR1$^{lo}$Gr1$^+$ represents immediate circulating precursors for antigen-presenting DC and CD11c$^-$ myeloid cells in inflammatory conditions and CX3CR1$^{hi}$Gr1$^-$ serves as a precursor for resident myeloid cells in non-inflamed tissues. CX3CR1 binds CX3CL1 (fractalkine), which can trigger adhesion or chemotaxis, depending upon whether the CX3CL1 is membrane-associated or soluble. Further use of this model allowed them to demonstrate that single-cell clones of mouse bone marrow progenitors retained the capacity to differentiate into monocytes, several macrophages subsets, and resident spleen DCs in vivo and in vitro [34].

A recent very elegant study by Ishii et al. [35] using intravital two-photon imaging of calvaria bone tissues and using both heterozygous CX3CR1-EGFP and Csf1R-EGFP mice revealed that all the TRAP-positive osteoclasts on the bone surface had expressed both CX3CR1 and Csf1R as in each of the mouse strains these cells were EGFP$^+$. As expected, the TRAP$^+$ cells were a minor percentage of the EGFP$^+$ cells in the marrow space. FACS analysis of blood monocytes from these animals showed twice as many cells were EGFP$^+$ from Csf1R-EGFP mice as compared to CX3CR1-EGFP mice. Interestingly, more than half of both EGFP$^+$ populations were also RANK$^+$ (67% CX3CR1-EGFP; 51% Csf1R-EGFP). In view of the low percentage TRAP$^+$EGFP$^+$ cells in the marrow, most of these RANK$^+$EGFP$^+$ cells in the circulation do not get recruited to form osteoclasts and negative regulation of RANKL signaling may dominate in these circulating cells.

Koizumi et al. [36] recently reported that immunohistochemical staining for CX3CR1$^+$ cells in human bone revealed that immature pre-osteoclasts (1–3 nuclei) in close localization with CX3CL1-positive osteoblasts on the bone are CX3CR1-positive, whereas mature multinucleated and strongly CatK-positive osteoclasts were CX3CR1-negative. Similarly, they found that mouse osteoclast precursors derived from either mouse splenocytes or the mouse RAW264.7 osteoclast precursor cell line were positive for CX3CR1, and that its expression was down-regulated during differentiation. Blocking the CX3CL1–CX3CR1 interaction with anti-CX3CL1 antibody strongly inhibited osteoblast-induced differentiation of osteoclasts in vitro and the numbers of mature osteoclasts actively resorbing the bone in vivo. However, it did not reduce the number of TRAP-positive pre-osteoclasts in the bone, suggesting that the CX3CL1–CX3CR1 interaction is not necessary for osteoclast precursor migration into the bone.

Human peripheral blood monocytes consist of two major subsets, CD16$^+$ and CD16$^-$ (CD16 is FcγRIIIa/Fcγ), comprising 5–10% and 90–95% of the monocytes, respectively. The CD16$^+$ and CD16$^-$ monocyte subsets show functional differences in migration, cytokine production and differentiation into macrophages or dendritic cells. The two main subsets of human blood monocytes also differ in CX3CR1 expression [33]. In human peripheral blood two monocyte populations negative for the T, B, pDC, and NK markers were evident: CX3CR1$^{lo}$CD14$^{hi}$CD16$^-$CD11b$^{hi}$CD11c$^{hi}$CCR2$^+$CD62L$^+$ (CX3CR1$^{lo}$CD14$^{++}$CD16$^-$ monocytes) that resemble murine CX3CR1$^{lo}$Gr1$^+$ monocytes in shape and CX3CR1$^{hi}$CD14$^{lo}$CD16$^{hi}$CD11b$^+$CD11c$^+$CCR2$^-$CD62L$^-$ (CX3CR1$^{hi}$CD14$^+$CD16$^+$ monocytes) that resemble murine CX3CR1$^{hi}$/Gr1$^-$ in shape. Like the mouse CX3CR1$^{lo}$Gr1$^+$ monocytes, the human CX3CR1$^{lo}$CD14$^+$ monocytes express several chemokine receptors that the CX3CR1$^{hi}$ cells do not express. This correlation between mouse and human subtypes would predict that the human CX3CR1$^{hi}$CD16$^+$ subset should comprise the resident monocyte precursors that would give rise to osteoclasts. However, Komano et al. [37] compared the CD16$^-$ and the CD16$^+$ monocyte subsets from human peripheral blood for their capacity to form osteoclasts with RANKL plus M-CSF treatment. They found that the two subsets of monocytes responded differently to these cytokines. Only the CD16$^-$ subset differentiated into osteoclasts that expressed $\beta_3$-integin mRNA and the $\alpha_v\beta_3$ heterodimer. The CD16$^+$ subset responded to RANKL by increasing production of TNF$\alpha$ and IL-6. The notable difference between the molecular responses to RANKL of the two subsets is that RANKL induced ERK and p38 MAPK phosphorylation and increased NFATc1 mRNA in CD16$^-$ cells, but not in CD16$^+$ cells. Additionally, the CD33$^{hi}$ monocytes, which are largely CD16$^-$, produced lots of osteoclasts, whereas the CD33$^{lo}$ monocytes, which are largely CD16$^+$, did not. Selection on the basis of CD33 avoids the possibility that the lack of osteoclastogenesis by CD16$^+$ cells is a result of the anti-CD16 immune complex triggering signals via the ITAM in the Fc$\gamma$ chain.

In addition, in contrast to the murine CX3CR1$^{lo}$Gr1$^+$ monocytes that were recruited into inflammatory sites [33], the CX3Cr1$^{hi}$CD16$^+$ monocytes are expanded in many inflammatory diseases, exhibit preferential migration across the endothelial layer in response to the chemokines fractalkine (CX3CL1), and produce TNF$\alpha$ and IL-1 in response to LPS suggesting that these are the human inflammatory monocytes. The CD14$^+$CD16$^+$ monocytes are more mature than CD14$^{hi}$CD16$^-$ [38].

A study following repopulation of mouse blood monocytes after toxic liposome administration suggests that the Ly6C$^{hi}$ (Gr1$^{hi}$) cells mature into the Ly6C$^{lo}$ (Gr1$^{lo}$) monocytes [27]. However, other studies have concluded that these are separate populations. At present, it is not clear whether mouse monocyte subsets, Gr1$^{hi}$ and Gr1$^{lo}$, represent human monocyte subsets, CD16$^-$ and CD16$^+$ monocytes, respectively, and whether the biologic functions of mouse monocyte subtypes are analogous to those of human monocytes. Human and mouse monocytes have other differences. Mouse monocytes do not express MHC class II antigens or CD11c (both associated with DC cells), whereas human monocytes do express both and also have some ability to be antigen-presenting cells.

The results discussed above suggest that osteoclast precursors can be derived from a multiplicity of cells within the monocyte–macrophage lineage and that multiple differentiation branch points may be utilized under differing circumstances. It remains unclear whether circulating cells form bone-resorbing osteoclasts or whether they are derived from bone marrow precursors. In particular, does recruitment of circulating cells to form osteoclasts only occur during an inflammatory state? Further, these results also indicate that there are differences between mouse and human osteoclast precursors. Hence, caution must be used in extrapolating results from mouse to human systems.

## Generation of osteoclasts

The multipotential progenitor may have several forms as discussed. The committed osteoclast precursors have been exposed to RANKL, but have not ceased dividing yet, whereas the pre-osteoclasts are post-mitotic (Figure 2-1). They can initiate and participate in cell fusion to generate mature multinucleated osteoclasts, which upon activation become polarized cells that resorb bone. The pre-osteoclasts can also be activated to resorb bone, but they are not very effective at resorption. Osteoclasts have a limited lifespan and eventually die via apoptosis. Osteoclasts in normal mammalian bone are rare cells and in histological sections are always observed in the vicinity of bone. Since monocytes are not rare and are found in all the soft tissues as well as in the marrow and peripheral blood, there must be strong negative regulation against the generation of committed osteoclast precursors and/or of their further differentiation in the absence of bone. Further bone must provide inhibitors of both the negative signals as well as positive signals enhancing the recruitment and differentiation of osteoclast precursors.

## External signals/receptors

### M-CSF/c-fms (Csf1R)

Lineage commitment of osteoclasts is governed by a number of growth factors and cytokines. Early osteoclast precursors are proliferative cells, which respond to hematopoietic growth factors, such as IL-3, GM-CSF, and M-CSF. While M-CSF (encoded by the *Csf1* gene) induces differentiation of macrophages and prevents apoptosis, the important role of M-CSF in the generation of osteoclasts has been demonstrated in rodent models in which mutations in the *Csf1* gene result in severe osteopetrosis (for a review of osteopetrosis see [3]). The *op/op* mouse lacks M-CSF due to a point mutation in the *Csf1* gene that results in a stop codon and production of a truncated M-CSF protein, and both osteoclasts and mature macrophages are absent at birth in this mouse [39]. The *op/op* mice was not cured by transplantation of normal bone marrow cells, indicating that the defect in *op/op* mice is associated with an abnormal hematopoietic microenvironment rather than with an intrinsic defect in hematopoietic progenitors. In contrast, mice deficient (by targeted deletion) in the gene encoding the receptor for M-CSF, *Csf1r*, have an intrinsic defect in osteoclastogenesis and showed an even more extreme osteopetrosis [40], suggesting the possibility of an additional ligand for Csf1R. IL-34 has recently been identified as a novel ligand for Csf1R [41]. Mice deficient in either M-CSF (*op/op* mice) [42] or its receptor *Csf1R* [40] are born osteopetrotic but have an age-related recovery of osteoclast production, due largely to the actions of other growth factors, such as either vascular endothelial growth factor (VEGF) [43] or GM-CSF plus IL-3 [44]. What remains unclear is why these factors fail to rescue osteoclast production during earlier development. Is it because the nature of the osteoclast progenitor is altered with age, making it less dependent upon M-CSF?

### RANKL/RANK/OPG

Takahashi et al. [45] developed a co-culture system with mouse spleen cells and osteoblastic cells from fetal mouse calvariae that, when stimulated by 1,25(OH)$_2$D$_3$, generated multinucleated cells expressing an osteoclast phenotype, demonstrating a role for stromal/osteoblastic cells in inducing hematopoietic precursors to form osteoclasts. This process required cell–cell contact as no osteoclasts were formed if a membrane separated the spleen cells from the osteoblastic cells. The major breakthrough in osteoclast biology was the discovery of the critical role that the cytokine RANKL (receptor activator of nuclear factor-κB ligand, osteoprotegerin ligand, TRANCE), encoded by the *Tnfsf11* gene, has in regulating osteoclastogenesis [46, 47]. RANKL is a member of the TNF superfamily whose expression is induced in bone marrow stromal cells by factors that regulate bone marrow stromal cell support of osteoclastogenesis. Examples include 1,25(OH)$_2$D$_3$, TNFα, parathyroid hormone (PTH), IL-11, and prostaglandin E2 (PGE2) [17]. RANKL is predominantly expressed as a transmembrane protein on the surface of bone marrow stromal cells, and can be cleaved from the surface by MMP14 as an inactive molecule [48] or as an active soluble protein by TNFα converting enzyme-like proteins [49]; although it is not clear that the latter occurs in vivo. RANKL is also expressed by other cells, such as activated T lymphocytes, which also express membrane-bound RANKL and can secrete a soluble form of RANKL. It interacts with its cognate receptor RANK (TNFRSF-11A), a type 1 transmembrane protein that is a member of the TNF receptor family. It functions as a homotrimer, on osteoclast precursors and osteoclasts to promote differentiation of osteoclast precursors and activation of mature osteoclasts to resorb bone. Addition of RANKL along with M-CSF, to normal osteoclast progenitors in vitro in the absence of bone marrow stromal cells is sufficient to induce osteoclastogenesis and resorption. Further, both RANKL$^{-/-}$ mice [46] and RANK$^{-/-}$ mice [50] exhibit a dramatic phenotype. They are severely osteopetrotic (no osteoclasts), and therefore lack tooth eruption, along with defects in immune cell differentiation. Interestingly, RANKL-deficient mice have normal dendritic cell and monocyte development.

Osteoprotegerin (OPG; TNFRSF-11B) is a soluble decoy receptor for RANKL that is a secreted member of the TNFR family that lacks a transmembrane domain and is structurally distinct from RANK [51]. OPG, which is also produced by bone marrow stromal cells, inhibits osteoclast differentiation by binding RANKL with high affinity, thereby preventing RANKL from binding its cognate receptor RANK. OPG$^{-/-}$ mice demonstrate severe osteoporosis, which is a result of the unopposed activity of endogenous RANKL, leading to excessive osteoclast differentiation and activity and also arterial calcification [52, 53]. In contrast, transgenic animals engineered to over-express OPG develop osteopetrosis because the excess OPG interacts with RANKL and decreases the RANKL available to bind RANK on the osteoclast precursors, thereby decreasing the number of osteoclasts formed [54]. OPG is also expressed by B-lymphocytes and DCs. Factors that regulate bone remodeling often do so by affecting the balance of RANKL and OPG synthesis. In diseases with pathological bone remodeling, such as hypercalcemia of malignancy, this ratio has been shown to be abnormal.

## Signal transduction molecules

M-CSF and RANKL exert their effects on osteoclast precursors throughout the differentiation/activation process by interacting with their respective receptors, Csf1R and RANK, and activating cascades of intracellular signals. Upon M-CSF binding, Csf1R, a transmembrane receptor tyrosine kinase, becomes autophosphorylated at seven tyrosine residues within its cytoplasmic tail, which then serve as recruiting sites for Src homology 2 (SH2)-containing signal transduction molecules, Src, PI3-kinase, and Grb2. RANKL interaction with RANK activates the various TRAF signal transduction molecules that bind to specific sites within the cytoplasmic domain of RANK. TRAF6 activates the NF-κB pathways and the mitogen-activated protein kinases (MAPKs), including ERK, JNK1, and p38, as well as PI3-kinase and Src.

### TRAF6

Interaction of RANKL with RANK initiates signal cascades in waves that orchestrate the complicated steps of osteoclast differentiation and activation. Unlike many TNFR family members, RANK does not contain a Death Domain and does not induce apoptosis. Instead, RANK recruits TNFR-associated factor (TRAF) proteins to TRAF interaction motifs (TIMs) within its cytoplasmic domain. The recruited TRAFs mediate the downstream signaling cascades. The cytoplasmic domain of RANK is unique in that there are at least three independent TIM regions. Although several TRAF family members are recruited to bind RANK, TRAF6 binds to a unique set of TIM sites and is critical for RANK signaling. TRAF2, TRAF3, and TRAF5 bind to two regions near the C-terminus, whereas TRAF6 binds to a highly specific membrane-proximal TIM. Deficiency of either TRAF2 [55] or TRAF5 [56] in osteoclast progenitors had only a minor effect on RANKL-induced osteoclastogenesis, although the TRAF2 deficiency had a large effect on the ability of TNFα to support osteoclastogenesis. TRAF6 is the major adapter molecule linking RANK to osteoclastogenesis. Two groups independently showed that TRAF6$^{-/-}$ mice developed severe osteopetrosis due to impaired bone resorption although the type of osteoclast defect was different in the two reports [57, 58]. In one case, dysfunctional osteoclasts were formed [57], but in the other osteoclast differentiation was strongly affected [58]. Further analysis of the role of TRAF6 by many has supported an important role for TRAF6 in RANKL-stimulated osteoclast differentiation (see review [59]).

### DAP12/FcRγ and their co-receptors TREM-2, PIR-A, OSCAR, and SIRPβ1

The immunoreceptor tyrosine-based activation motif (ITAM) is a highly conserved region in the cytoplasmic domain of signaling chains of adapter proteins and receptors and is a critical mediator of intracellular signals. ITAM signaling is required for the differentiation and function of B and T cells in adaptive immunity and regulates the function of innate immune cells, including natural killer cells, and myeloid cells such as macrophages, neutrophils, and dendritic cells. Recent studies have demonstrated that ITAM adapter proteins are involved in the formation and function of osteoclasts. They signal to activate Syk kinase and PLCγ2, which initiates $Ca^{2+}$ oscillations that can result in activation of the key transcription factor, NFATc1, controlling differentiation of pre-osteoclasts and multinucleation. Mice deficient in the ITAM adapter protein, DNAX-activating protein (DAP) 12, are osteopetrotic with the presence in bone of numerous inactive multinucleated osteoclasts. However, in culture, the DAP12$^{-/-}$ precursors differentiated with M-CSF and RANKL were unable to multinucleate, although this could be partially rescued if stromal cells were present. The other ITAM adapter protein present in osteoclasts is Fcε receptor γ-chain (FcRγ), but mice with the loss of this gene alone do not have a bone phenotype. Mice deficient in both of the ITAM adapter proteins, DAP12 and FcRγ, are osteopetrotic, owing to impaired osteoclast formation and bone resorption [60–62]. These ITAM-containing adapters associate with activating receptors for which the ligands are unknown: DAP12 interacts with triggering receptors expressed on myeloid cells-2 (TREM-2) and signal regulatory protein β1 (SIRPβ1), whereas FcRγ interacts with paired immunoglobulin-like receptor-A (PIR-A) and osteoclast-associated receptor (OSCAR).

Although the single ITAM gene (DAP12 and FcRγ) deficiencies and WT showed ovariectomized (OVX)-induced vertebral body trabecular bone loss, the DAP12/FcRγ double knockout did not. None of the ITAM gene deficiencies were protected from OVX-induced bone loss in the femur and tibia relative to basal bone volumes, but they all protected the cortical bone, as only WT developed significant OVX-induced cortical bone loss. These results indicate that although ITAM adapter signaling is critical for normal bone remodeling, estrogen deficiency induces an ITAM adapter-independent bypass mechanism allowing for enhanced osteoclastogenesis and activation in specific bony microenvironments [63].

### CaMKIV and calcineurin

The calcium/calmodulin-dependent protein kinase type IV (CaMKIV) encoded by the *Camk4* gene has been shown to be regulated by ITAM signaling and important for both physiological and pathological osteoclastogenesis by analysis of *Camk4* knockout mice [64]. Proliferation of precursor cells was unaffected by the deletion, but the number of osteoclasts was reduced both in vivo and in M-CSF plus RANKL-treated cultures. Although the *Camk4*$^{-/-}$ osteoclast precursors were able to respond to RANKL and activate MAPKs, the level of c-Fos mRNA and protein was not elevated, nor was the transcription factor CREB phosphorylated. CREB phosphorylation was shown to be critical for c-Fos gene induction, and thereby for NFATc1 expression. The calcium/calmodulin-dependent protein phosphatase calcineurin also plays a critical role in regulating osteoclastogenesis. Upon activation, calcineurin dephosphorylates nuclear factor of activated T cells c1 ((NFATc1); described below), thereby promoting its translocation into the nucleus [59].

## Transcription factors

### Spi1/PU.1

Loss or mutation of lineage regulating transcription factors has yielded insight into the lineage derivation and stages of osteoclast differentiation. Spi1/PU.1 (*Spi1* gene) is a member of the ETS domain transcription factors that has a key role in regulating the production of B cells, pDC, and all the myelomonocyte–macrophage lineages, including mDC. Mice with targeted deletion of the *Spi1* gene fail to generate monocyte progenitors that express the receptors for GM-CSF, G-CSF, and M-CSF and have severe osteopetrosis due to the complete lack of osteoclasts [65]. This defect is intrinsic to the osteoclast progenitor as bone marrow transplantation reverses the osteopetrosis in Spi1/PU.1-deficient mice. Spi1/PU.1 expression increases as the cells differentiated from monocyte to osteoclast, similar to what has been reported for DCs. Spi1/PU.1 has been demonstrated to interact with the microphthalmia transcription factor (MITF) to regulate TRAP gene expression [66]. Since mice deficient in either M-CSF [42] or its receptor *Csf1r* [40] are born osteopetrotic but have an age-related recovery of osteoclast production, due largely to the actions of other growth factors, there must be a cell-autonomous function of Spi1/PU.1 in the generation of the monocyte–macrophage lineage independent of its role in regulating expression of *Csf1r*. In support of this, *Csf1r* expression (by transduction) cannot rescue macrophage differentiation in Spi1/PU.1-deficient cells [67], indicating that, in the absence of Spi1/PU.1, *Csf1r* signaling is not sufficient to drive macrophage differentiation.

### NF-κB/IKKα (IKK1), IKKβ (IKK2), IKKγ (IKK3, NEMO)

NF-κB is a composite transcription factor, made of varied homo- and heterodimers of the Rel family members, p50 (*Nfkb1*), p52 (*Nfkb2*), p65 (RelA), RelB, and c-Rel. The p105 and p100 Rel proteins have long C-terminal domains that contain multiple copies of ankyrin repeats, which act to inhibit these molecules. These are processed in the proteasome by limited cleavage to generate active DNA-binding proteins p50 (from p105) and p52 (from p100). Targeted disruption of either of the *Nfkb1* or *Nfkb2* genes did not affect osteoclasts; however, the double knockout of these genes generated osteopetrotic mice with few osteoclasts, but elevated CD11b$^+$RANK$^+$ osteoclast precursors in the spleen, although macrophage function was also impaired [68, 69]. This defect could be rescued by bone marrow transplantation, indicating that this was an osteoclast precursor-intrinsic defect that blocked differentiation. Ectopic expression of cFos or NFATc1 rescued the ability of RANKL or TNFα to induce osteoclastogenesis, indicating that both are downstream of NF-κB [70]. Recently, Xing et al. [71] have shown that although p50 and p52 NF-κB expression is essential for RANK expressing progenitors to differentiate into osteoclasts in response to RANKL, NF-κB is not required for formation of RANK expressing osteoclast progenitors. NF-κB also plays a critical role in the expression of a variety of cytokines involved in osteoclast differentiation including IL-1, TNFα, IL-6, GM-CSF, and RANKL.

NF-κB proteins reside in the cytoplasm in a complex with a member of the inhibitor family of κB (IκB), often IκBα. Upon cytokine stimulation, activation of the IκB kinase (IKK) complex triggers phosphorylation of IκBs and targets them for ubiquitin-mediated proteasomal degradation, thereby allowing activation of NF-κB and translocation into the nucleus. The classical NF-κB activation pathway utilizes an IKK complex composed of

IKKα, IKKβ, and IKKγ (also known as NEMO). RANKL has been reported to regulate stimulated, but not basal, osteoclast formation by a pathway involving NF-κB inducing kinase (NIK) plus a dimer of IKKα and NF-κB p100 processing to p52 in the alternative NF-κB activation pathway [72]. This pathway activates the p52:RelB NF-κB. Targeted disruption of either the *NIK* or *IKKα* gene does not affect basal osteoclastogenesis in vivo, but does impair in vitro osteoclastogenesis. Further RelB$^{-/-}$ mice fail to form osteoclasts in culture. Most interestingly, lack of NIK or RelB prevented tumor-induced osteolysis using a B16 melanoma tumor model.

### PPARγ

The nuclear receptor peroxisome proliferator-activated receptor γ (PPARγ), known for its lineage-regulating role in adipogenesis, in which it activates adipogenesis and represses osteoblastogenesis, also has a role in osteoclastogenesis. Lack of PPARγ function by mutation or by inhibitors in osteoclast precursors intrinsically decreases RANKL-stimulated osteoclast formation due to impaired c-Fos induction [73], a key step in generating an osteoclast, as described below.

### c-Fos

Another critical transcription factor for OCL differentiation is c-Fos, a component of the dimeric transcription factor AP-1. Mice lacking c-Fos develop osteopetrosis and have no tooth eruption due to a block in osteoclast differentiation that can be rescued by bone marrow transplantation [74]. Ectopic c-Fos expression can also overcome the block in osteoclast differentiation. Interestingly, in contrast to the PU.1-deficient mice that lack monocyte–macrophages, the c-Fos knockout mice have elevated numbers of monocyte–macrophages [74]. This suggests that the level of c-Fos expression regulates the switch between monocyte–macrophage and osteoclast differentiation. Furthermore, signals that down-regulate c-Fos and Fra-1, such as GM-CSF [30] or interferon-β [75], block osteoclastogenesis. As noted above, the induction of c-Fos expression is dependent on NF-κB [70]. It is also dependent on PPARγ [73] and CREB, which is activated by the $Ca^{2+}$/CaMKIV [64].

A key target of c-Fos in osteoclast precursors is the transcription factor NFATc1, and ectopic expression of NFATc1 in bone marrow monocytes rescues osteoclastogenesis in cells lacking c-Fos [76]. However, Fra-1 (also known as Fosl) is thought to be a target of c-Fos, which is supported by the observation that Fra-1 expression can rescue the osteoclastogenesis defect seen in c-Fos-deficient mice [77] with equal efficiency as c-Fos. However, when Owens and coworkers [78] transduced immortalized bipotential osteoclast precursors with c-Fos or Fra-1 using retroviral vectors, they found that over-expression of Fra-1, but not c-Fos, generated a large increase in the ability of the cells to form osteoclasts. Fra-1 transduction into primary spleen cells did not have this effect. These results indicate that c-Fos is necessary but not sufficient to stimulate osteoclastogenesis in the absence of a cytokine signal.

### NFATc1

The $Ca^{2+}$-calcineurin regulated NFATc1 is the master regulator of osteoclastogenesis and highly stimulated by RANKL (reviewed in [59]). It is a direct downstream target of c-Fos and retroviral transduction of NFATc1 into c-Fos-deficient BM cells can rescue osteoclastogenesis and bone resorption induced in vitro; although not all osteoclastic gene expression is regained [76]. NFATc1 induces many target genes responsible for osteoclast differentiation and function, including itself, OSCAR, TRAP, CatK, Src, $β_3$ integrin, Atp6v0d2, DC-STAMP, and CTR. RANKL-stimulated $Ca^{2+}$ oscillations necessary to activate NFATc1 are only observed 24–48 h post stimulation, indicating the need to induce the pathway capable of generating $Ca^{2+}$ oscillations sufficient to activate calcineurin and thereby activate NFATc1. The nature of this pathway is not entirely clear yet. However, co-stimulatory signals via the ITAM-containing Dap12/FcRγ and activation of PLCγ2 have been shown to play a role. Furthermore, the tyrosine kinases Tec and Btk have been demonstrated to be involved in linking the RANK and ITAM signal to activate PLCγ2 [79].

### MITF

The microphthalmic (*mi/mi*) mouse has a mutation within the gene encoding the microphthalmia transcription factor (MITF) – it is a 3-bp deletion causing the loss of an arginine in the N-terminal DNA-binding domain. The mutation results in severe osteopetrosis due to an intrinsic defect in the osteoclast as the *mi/mi* osteoclast precursors do not fuse. The mice are also deaf and blind with a white coat due to impairment of melanocyte development. M-CSF induction of ERK activation and, consequently, MITF phosphorylation is one of the keys to regulation of osteoclast fusion [80].

## Regulation of pre-osteoclast fusion

A number of molecules have been proposed to have a role in directing osteoclast fusion events, including the macrophage-fusion receptor (MFR; also called protein tyrosine phosphatase, non-receptor type substrate 1, SIRPα or CD172a) [81], thrombospondin-1 (TSP1), CD47 – a receptor for MFR and TSP1, the dendritic cell specific transmembrane protein (DC-STAMP), and tetraspanins.

### CD47/TSP1/SIRPα (MFR)

CD47 (also called integrin-associated protein) is the "receptor" for TSP1 and the "ligand" for SHP-1-recruiting inhibitory immunoreceptor signal regulatory protein (SIRPα), also known as MFR and CD172a. Both CD47 and SIRPα belong to the immunoglobulin superfamily and both conduct signaling inwards. TSP1 and TSP2 constitute a small family of secreted, modular glycoproteins that form homotrimers and have considerable sequence homology [82]. CD47 also associates within the same membrane with $α_vβ_3$ and other integrins and can augment integrin function. TSP1-CD47 has been reported to inhibit NO signaling in several cell types. As NO signaling inhibits osteoclast differentiation, the role of TSP1–CD47 interaction may be to stimulate osteoclastogenesis by modulating NO. The CD47-deficient mouse had osteopetrosis due to osteoclast dysfunction not a lack of osteoclasts in vivo [83]. However, $CD47^{-/-}$ osteoclast formation in vitro was impaired. This defect in $CD47^{-/-}$ formation and resorption could be largely rescued by high-dose RANKL both in vitro and in vivo via injection. $CD47^{-/-}$ cells had elevated iNOS and inhibition of NOS also rescued the osteoclast defect in $CD47^{-/-}$ cells in vitro. $CD47^{-/-}$ bone was more resistant to tumor-induced osteolysis when injected with tumor cells that did not make RANKL (B16-FL) and also had decreased tumor burden. Although CD47 signaling may act by augmenting $\alpha_v\beta_3$ activity, the rescue of $CD47^{-/-}$ by high-dose RANKL and the inhibition of NO generation suggests the possibility that CD47 might be preventing NO inhibition of Src by nitrosylation.

Recently, Kukreja [84] reported that anti-TSP1 blockade inhibited osteoclast formation from normal human $CD14^+$ PBMC precursors and from myeloma-triggered DC (discussed further below). They also found that injection of anti-TSP1 mAb but not isotype control mAb led to clear inhibition of PTH-induced hypercalcemia in mice. This suggests that the TSP1 signal is important for osteoclast fusion and activity. However, with regards to skeletal development, mice lacking TSP1 (encoded by the *Thbs1* gene) are born with curvature of the spine, and they display minor abnormalities in trabecular bone [82]. It is unclear whether this represents a difference in the role of TSP1 in humans vs. mice, or a difference in its role between basal and stimulated osteoclastogenesis. Of note, CD47 is not the only receptor for TSP1. TSP1 is a complex multidomain protein that also binds scavenger receptor CD36, $\alpha_v\beta_1$ and $\alpha_v\beta_3$ integrins, along with heparin sulfate proteoglycans such as syndecan (CD138), glucosaminoglycans, reelin receptors (ApoER2 and VLDLR), cathepsin G, MMP, elastase, and several other proteases [82]. TSP1 participates in the differentiation of Th17 cells through its ability to activate latent TGF-β by dissociating latency-associated protein (LAP) from the latent TGF-β complex [85]. Carron et al. [86] demonstrated that a peptide present in TSP1 that can interact with CD36 stimulates resorption by avian osteoclasts in a manner similar to the intact TSP1 molecule. However, it was recently reported using bone marrow monocytes from $CD36^{-/-}$ mice that while CD36 is required for macrophage giant cell fusion induced by IL-4 and GM-CSF, it is dispensable for osteoclast fusion. However, it is not known if there is compensation from another TSP1-interacting protein such as CD47 in the osteoclast and whether a double knockout of CD36 and CD47 would reveal a more severe phenotype than either knockout alone.

Physical interaction between the cell surface receptors CD47 and SIRPα was reported to regulate cell migration, phagocytosis, cytokine production, and macrophage fusion. Functional blocking antibodies to either CD47 or SIRPα strongly reduced formation of multinucleated TRAP+ osteoclasts in cultures of murine hematopoietic cells stimulated in vitro by M-CSF and RANKL. ITAM signaling in the immune system is counterbalanced by receptors bearing ITIM domains that recruit SH2-protein tyrosine phosphatases (SH2-PTPases), such as SHP-1 and SHP-2. The *motheaten* (*me/me*) mice have a loss of function mutation in SHP-1 and develop severe osteoporosis associated with increased osteoclast differentiation and resorption. SIRPα contains an ITIM domain that becomes phosphorylated upon binding CD47 and can mediate recruitment of SHP-1 and SHP-2. Mice with a mutated SIRPα that lacks the ability to signal have reduced cortical bone mass (in contrast to CD47-deficient mice), but formed normal numbers of multinucleated osteoclasts (in contrast to SHP-1-deficient mice) with normal expression of major differentiation markers TRAP, CatK, CTR, and DC-STAMP and no change in the average nuclei/osteoclast. However, the SIRPα-mutant osteoclasts had enhanced actin ring formation and increased bone resorption activity in vitro (in contrast to CD47$^{-/-}$) with greater total resorption area and pit size, indicating that the bone-resorbing capacity per cell is increased, suggesting that SIRPα is a negative regulator of osteoclast activity [87]. These results suggest that osteoclast formation is supported by CD47–SIRPα interaction on the CD47 side, but osteoclast function is inhibited on the SIRP1α side by its recruitment of SHP-1. In addition, CD47 interaction with TSP1 may affect the balance of these signals, perhaps by stimulating CD47 signaling while blocking SIRPα negative regulation. Interestingly, the in vivo effect of SIRPα mutant caused a decrease in cortical thickness, but no effect on trabecular bone mass, suggesting functional heterogeneity between osteoclasts in cortical bone and those in trabecular bone.

In summary, the exact role of TSP1, CD47, and SIRPα in basal osteoclastogenesis as well as how those roles may be different in pathological osteoclastogenesis remains unclear.

### DC-STAMP

DC-STAMP is a seven-transmembrane-spanning receptor with no homology to any other known protein or multimembrane-spanning receptor. DC-STAMP is expressed both in immature and mature dendritic cells (DCs), and its mRNA levels fall upon activation of DCs with CD40 ligand (CD40L). DC-STAMP has been demonstrated to be an NFATc1 direct target gene and is involved in the fusion of both osteoclasts and macrophage giant cells [88, 89]. Mice deficient in DC-STAMP had mild osteopetrosis with mononucleated osteoclasts due to lack of fusion rather than a defect in differentiation, and therefore resorption was inefficient. Use of bone marrow monocytes from mice with GFP replacing DC-STAMP mixed with unlabeled WT bone marrow cells revealed that the WT cells could fuse with the GFP+DC-STAMP cells confirming that only one partner in osteoclast fusion needs to be expressing DC-STAMP [88]. The ligand(s) for DC-STAMP is still unknown and the requirement for its expression is unclear, including the issue of whether osteoclasts and macrophages express different DC-STAMP ligands. Furthermore, the molecular mechanism by which DC-STAMP exerts its effects in osteoclasts remains unknown.

### Multi-subunit V-ATPase (oc/oc (Atp6i) and Atp6v0d2)

The structure of the multisubunit V-ATPase complex in the osteoclast $V_0$ domain conducts the proton transport and $V_1$ domain is involved in the ATP hydrolysis. Each domain has multiple different subunits ($V_0$ 5 subunits; $V_1$ 8 subunits), some of which have multiple isoforms. One of the isoforms for $V_0$ subunit a (termed a3, ATP6i, or Tcirg1) is predominantly expressed in osteoclasts and disruption of the *Atp6i* gene in mice (*oc/oc*) results in a severe osteopetrosis phenotype due to loss of extracellular acidification [90] and mutations of this gene cause types of human osteopetrosis [91, 92]. On the other hand, disruption of the gene encoding the $V_0$ subunit d2 (*Atp6v0d2*), an NFATc1 direct target gene, demonstrated that it is necessary for fusion [93]. Comparing knockdown of the *Atp6v0d2* early vs. late during osteoclast differentiation revealed that it was both required early to regulate cell fusion, and later for extracellular acidification as an essential component of the osteoclast proton pump [94]. Notably, the late knockdown of *Atp6v0d2* did not affect cell fusion.

## CD44

CD44 is a type I transmembrane protein and member of the cartilage link protein family. Several CD44 ligands have been identified. CD44 is a major cell surface receptor for hyaluronan, a component of the extracellular matrix, and it also interacts with osteopontin (OPN), collagen, and laminin. Kania et al. [95] determined that CD44, a cell surface glycoprotein known to function as an adhesion receptor, also is involved in osteoclast differentiation. They demonstrated that antibodies against CD44 inhibited osteoclast formation in mouse bone marrow co-cultures stimulated with $1,25(OD)_2D_3$, but did not inhibit the bone resorption activity of mature osteoclasts. However, de Vries et al. [96] reported that $CD44^{-/-}$ mice do not have an osteopetrotic phenotype in vivo. Further, in vitro analysis of bone marrow cells from $CD44^{-/-}$ and WT mice demonstrated that they formed equal numbers of osteoclasts on bone and the resorption capacity was also similar. This group did not find that CD44-blocking antibodies altered osteoclastogenesis induced by RANKL plus M-CSF. This discrepancy with the previous report may be due to different pathways utilized during osteoclastogenesis. It is possible that in vivo there are compensating signals for the loss of CD44.

## ADAM8/$\alpha_9\beta_1$-integrin

Choi et al. [97] reported that ADAM8, a member of the ADAM (a disintegrin and metalloproteinase) family is an autocrine regulator of osteoclastogenesis. ADAM8 is the only member of the ADAM family that is increased in expression in mature osteoclasts compared to osteoclast precursors. Soluble ADAM8 induced formation of bone-resorbing osteoclasts and acted at the later stages of osteoclast differentiation and pre-osteoclast fusion. Structure–function studies demonstrated that the disintegrin domain of ADAM8 mediated its effects on osteoclastogenesis. This group more recently reported that the receptor for ADAM8 is $\alpha_9\beta_1$ integrin [98]. The $\alpha_9\beta_1$ integrin is expressed on both murine and human osteoclast precursors at a very early stage and increases as the cells differentiate. Neutralizing antibodies to $\alpha_9$ integrin blocked human osteoclast formation in vitro, and osteoclasts from $\alpha_9$-deficient mice are defective. The $\alpha_9$-deficient osteoclasts are small and contracted, resorb bone poorly, and are similar to osteoclasts that lack $\beta_3$ integrin. They can make actin rings, but did not form lamellipodia and had impaired mobility. Further, the $\alpha_9$-deficient mice are mildly osteopetrotic, largely due to decreased osteoclast mobility and functionality. We have recently shown that mice lacking ADAM8 do not have a bone phenotype under basal conditions, but in contrast to wild-type littermates, they fail to increase osteoclast formation in response to TNF$\alpha$ injected over the calvaria (unpublished results).

*CTR*

The calcitonin receptor (CTR) is a seven-transmembrane class II G-protein-coupled receptor for the 32-amino-acid calcitonin, and is highly expressed on the osteoclast plasma membrane. CTR exists as several different isoforms resulting from alternative splicing of a single gene that are different in different species [99]. Numerous studies have demonstrated that it is a specific marker of osteoclast differentiation and the degree of surface expression increases as osteoclasts undergo fusion and activation. It is one of the best differentiation markers for distinguishing mammalian osteoclasts from macrophage polykaryons. However, it is not a marker for avian osteoclasts, as they do not express it. The CTR is coupled to multiple signal transduction pathways via $G_s$, $G_q$, and $G_i$ to adenylyl cyclase-cAMP-PKA, MAP kinases, and phospholipases A2, C, and D generating increased cytosolic $Ca^{+2}$ and PKC activation, which contribute to different responses (reviewed in [100]). Interaction of CTR with calcitonin can inhibit multinucleation of TRAP+ mononucleated pre-osteoclasts without altering the RNA expression of RANK, Cfs1R, or a wide array of transcription factors and genes important for differentiation and activation, including the genes discussed above that have been suggested to have roles regulating osteoclast fusion [101]. The ability of calcitonin–CTR interaction to block multinucleation was mimicked by both activating the cAMP–PKA pathway, and by activating the exchange protein directly activated by the cAMP (Epac) pathway. Calcitonin also rapidly induces loss of the ruffled border, quiescence, and cell retraction of mature activated osteoclasts resulting in inhibition of bone resorption both in vivo and in vitro. However, global deletion of the cytoplasmic C-terminus of CTR using a Cre/loxP system showed that CTR has a modest physiological role in the regulation of bone and calcium homeostasis in the basal state in mice, which was accentuated after $1,25(OH)_2D_3$-induced hypercalcemia. Woodrow et al. [102] have shown that the normal loss of bone mineral density that occurs during lactation is increased in CT/CGRP KO mice compared to littermate controls and this is normalized by administration of salmon CT to the CT/CRGP KO mice. This supports the concept that CTR plays an important role in protecting the skeleton in times of calcium stress such as induced hypercalcemia, pregnancy, lactation, and in states of high bone turnover.

# More than one type of osteoclast?

Not all bone sites are the same. Cortical bone tissue is the hard outer layer of bones and accounts for 80% of the total bone mass of an adult skeleton. Trabecular bone fills the

**Table 2-1** Gene knockouts with different osteoclast phenotypes in different bone regions.

| Knockout mice | Calvarial | Long bones | | Vertebral |
| --- | --- | --- | --- | --- |
| | | Trabecular | Cortical | |
| DAP12 or FcRγ [63] | ND | No protection from OVX-induced bone loss (like WT) | Protected from OVX-induced bone loss | No protection from OVX-induced bone loss (like WT) |
| DAP12+FcRγ [63] | ND | No protection from OVX-induced bone loss (like WT) | Protected from OVX-induced bone loss | Protected from OVX-induced bone loss |
| CAII [104] | ND | Osteopetrotic | Normal function | ND |
| Ae2ab [106] | Normal Function | Osteoporotic | Osteopetrotic | ND |
| CatK [107, 108] | Normal | Osteopetrotic | Osteopetrotic | Osteopetrotic |
| MMP9 [110] | Normal | Temporary osteopetrosis at 3 wks | Normal | Normal |
| SIRPα mutant that lacks signaling [87] | ND | Normal | Osteoporotic | ND |

ND, not determined.

interior of the bone and is only 20% of the bone mass but has much a greater surface area than cortical bone. In addition, bone formation is by different mechanisms at different sites, i.e. endochondral ossification in long bones and intramembraneous ossification in the skull. In growing long bones, resorption of cartilage occurs at the epiphyseal/metaphyseal border, an area containing both non-mineralized and mineralized cartilage. This raises the question of whether all osteoclasts are the same as they have to degrade different types of matrices within different bone microenvironments at different bone sites. Some resorption abnormalities are systemic in nature; however, others can be restricted to certain parts of the skeleton, such as occurs in Paget's disease, which gives rise to focal lesions, or as seen in several of the mouse knockouts that affect osteoclasts discussed in this chapter (Table 2-1).

## Distinctive morphological and biochemical characteristics of mature osteoclasts that suggest the presence of osteoclast subtypes

### Carbonic anhydrase II (CAII)

Disorders of acidification result in poor bone resorption, but little or no defects in differentiation. The CAII (*Ca2* gene) is highly expressed in resorbing osteoclasts and CAII deficiency leads to an intermediate form of osteopetrosis in humans for reasons that are not clear, along with renal tubular acidosis and cerebral calcification [103]. CAII-deficient mice were reported to have a similar phenotype with osteopetrosis evident in the trabecular, but not cortical bone [104]. These mice also had increased trabecular bone volume, whereas their bone formation rate was decreased. Furthermore, osteoclast number was significantly increased in CAII-deficient mice but osteoblast number was similar to that of WT. This implies that CAII deficiency impairs osteoclast resorptive function. Antisense constructs to CAII block osteoclastic bone resorption both in isolated osteoclasts and in bone organ culture [105].

### Anion exchangers (Ae2 and Slc4a4)

The long bones of $Ae2_{a,b}^{-/-}$ mice (deficient in the main isoforms Ae2a, Ae2b1, and Ae2b2) had normal numbers of multinucleated osteoclasts, but they lacked a ruffled border and displayed impaired bone resorption activity due to failure to maintain the $pH_i$, resulting in osteopetrosis [106]. Strikingly, osteoclasts in the $Ae2_{a,b}^{-/-}$ calvaria were functional due to the selective expression of a sodium-bicarbonate cotransporter (possibly *Slc4a4*) and their teeth erupted suggesting that osteoclasts in the jaw were also functional. This difference between osteoclasts of the long bones and of the calvaria strongly suggests the existence of osteoclast subtypes.

### Cathepsin K (CatK)

Cathepsin K (CatK; also termed cathepsin O, encoded by the *Ctsk* gene) is a cysteine protease active at low pH that plays a critical role in osteoclastic bone resorption, being largely responsible for cleaving and removing the organic matrix of the bone (type I collagen fibers). The rare human disease pycnodysostosis results from mutations in the *Ctsk* gene that result in decreased CatK enzymatic activity. The bones in these patients are characterized by defective remodeling, leading to poor bone structure and hypomineralization. The lack of CatK caused high levels of collagen fibers to remain in the resorption pit, as well as the accumulation of non-digested collagen fibers in vesicles inside the osteoclast. MMP activity was able to compensate for the lack of CatK to some degree. CatK-deficient mice have a mild osteopetrosis and a high number of osteoclasts with impaired function [107, 108]. Interestingly, these mice also had elevated bone formation rates.

### Matrix metalloproteinase (MMP9)

In adult bone, matrix metalloproteinase 9 (MMP9, also known as 92 kDa gelatinase or type IV collagenase) is predominantly expressed by osteoclasts and committed osteoclast precursors. Bone resorption can be specifically reduced by chemical inhibition of MMP9, which results in inhibition of osteoclast migration [109]. Similar effects are seen with MMP9 antisense oligos treatment or by MMP9 gene knockout [110]. $MMP9^{-/-}$ mice display delayed endochondral ossification and vessel invasion and osteoclastic recruitment into the primary ossification center, accompanied by lengthened growth plates. MMP9 is specifically required for the invasion of osteoclasts into the discontinuously mineralized hypertrophic cartilage that fills the core of the diaphysis. However, this phenotype resolves, and adults have normal-appearing bones that are slightly shorter (10%) than WT animals. Likewise, lack of MMP9 inhibits vessel invasion and healing of long bone fractures. In adult bone, MMP9 is predominantly expressed by osteoclasts and committed osteoclast precursors, but can be expressed at very low levels by other cell types such as osteoblasts and hypertrophic chondrocytes during the development of neutrophils and macrophages during fracture repair. Osteoclast expression of MMP9 has a key role in the ability of osteoclasts to migrate and induce angiogenesis [111].

## Different osteoclasts in different bone sites: trabecular vs. cortical, long bones vs. jaw and calvarial sites

Data in the literature suggest that site-specific differences exist in the skeleton with respect to digestion of bone by osteoclasts (Table 2-1). Bone resorption by calvarial osteoclasts (intramembranous bone) was demonstrated to differ from resorption by long bone osteoclasts (endochondral bone) [109]. Osteoclastic resorption of calvarial bone depends on the activity of both cathepsins B and K and MMPs, whereas long bone resorption depends on cathepsins, but not on the activity of MMPs. MMP inhibitors did not affect resorption by osteoclasts isolated from long bones, whereas resorption by calvarial osteoclasts was inhibited when each was cultured on slices of bovine skull or cortical bone. The question is whether these differences in osteoclast behavior occur because osteoclasts originate from different sets of progenitors or because they are a product of the effect of the local microenvironment. Additionally, Jemtland et al. [112] found that during embryonic bone development at ED17, when the primary ossification center is formed, MMP9+ cells were found at the chondro-osseous margin between the growth plate chondrocytes and bone as well as some cells in the marrow cavity. The latter cells scattered through the marrow cavity on the surfaces of trabecular bones were positive as well for TRAP and Csf1R mRNA and were bi- or multinucleated. At later stages, all three mRNAs were co-expressed in cells lining the cortical bone surfaces as well as increasingly at the chondro-osseous junction. MMP9 may have a unique role in the degradation of the cartilage matrix in the chondro-osseous junction.

The multinucleated cells that resorb cartilage, chondroclasts, are osteoclast-related cells. The chondroclasts are more rounded non-polarized cells that form fewer ruffled membranes and express more intracellular TRAP than the polarized osteoclasts present in the lower metaphysis, which degrade bone [113]. Chondroclast invasion of non-mineralized cartilage depends on MMP9 [109]. The CAII-deficient mice demonstrate differences between osteoclasts in cortical and trabecular bone as only the trabecular bone areas were osteopetrotic [104]. Examples of differences between osteoclasts in calvarial bone and long bones include differences in the roles of the proteases cathepsin K and MMP9. Osteoclasts in long bones have much higher levels of CatK than osteoclasts in calvaria, suggesting that CatK plays a bigger role in resorption of long bones than of calvaria. In fact, CatK$^{-/-}$ mice did not have any abnormalities in calvarial bones while there were resorption defects in long bones. In contrast, MMP inhibitors did not affect osteoclast resorption of long bones, but did inhibit resorption of calvarial bones. The role of MMP9 in the long bone is primarily osteoclast migration. In addition, $Ae2_{a,b}^{-/-}$ mice had selective osteopetrosis of the long bones, whereas osteoclast resorption in calvaria and the jaw was normal. The osteoclast phenotypical heterogeneity in different bone regions suggests that there are osteoclast subtypes that may be due to recruitment of different osteoclast precursors to the particular microenvironments. Alternatively, signals from their microenvironments that regulate osteoclast function may be responsible for these differences.

Thus, the osteoclast contains high levels of several enzymes that distinguish it from other bone cells, and possibly contribute to the definition of different osteoclast subtypes.

## Other OCL precursors

A number of studies have suggested that under certain circumstances other cell types can become osteoclasts. This is a different issue than coming from a common progenitor, although it is hard to be definitive about the possible presence of multipotent progenitors within a mature cell population.

### Other macrophage lineage cells

#### Dendritic cells

A number of studies have indicated that DCs can form osteoclasts through novel pathways. Inflammatory cytokines can induce the formation of osteoclasts from human DCs in vitro [114], as well as from murine DCs in vitro and in the bone marrow microenvironment in vivo [115–117]. Conventional DC (cDC) from mice spleen defined as $CD11c^{hi}MHC-II^+$ were shown to be $Ly6C^-F4/80^-B220^-$ (most were also positive for CD80, CD86, and CD40, but could be further matured with LPS), therefore these are not monocytes or macrophages. RANKL plus M-CSF efficiently induced the $CD4^-CD8^-$ splenic cDC subset to form osteoclasts, and this was inhibited by GM-CSF [117]. Most importantly, it was found that intraperitoneal injection of normal FACS-purified DCs partially reverted the osteopetrotic phenotype of oc/oc mice by restoring bone resorption activity in vivo and generating marrow cells that could be recovered and induced to form bone-resorbing osteoclasts in vitro. The requirement of a $CD11c^+$ cell for the rescue was demonstrated by the lack of rescue if $CD11c^+$ cells, which were engineered to be specifically sensitive to killing by diphtheria toxin (DT), were injected and then treated with DT. These results suggest that differentiation of OCLs from DCs represent an alternative OCL differentiation pathway induced by high levels of osteoclastogenic factors.

Recently, Kukreja et al. [84] have also shown that via cell–cell contact myeloma cells can induce human immature DCs ($CD14^-$,$HLA-DR^+CD80^{lo}CD86^{lo}CD83^-$), but not $CD14^+$ monocytes or macrophages, to fuse and become osteoclasts via upregulation of

TSP1 in DCs. TSP1 is one of two ligands for CD47 (the other is SIRPα). Blockade of CD47–TSP1 interactions blocked osteoclast formation both from myeloma-induced DCs (generated from CD14⁻ PBMC with GM-CSF plus IL-4) and from RANKL plus M-CSF-induced monocytes (CD14⁺ PBMC). The capacity of myeloma cells to induce DC to form osteoclasts was inhibited by down-regulation of CD47 expression in the myeloma cell.

*Alveolar macrophages*

Another case of transdifferentiation of a tissue macrophage into an osteoclast involves alveolar macrophages. It has been reported that alveolar macrophages also differentiated into osteoclasts in co-culture with stromal cells [20, 118].

## B cells

Several studies have proposed that B220⁺ B cells in the marrow contain a bipotential precursor that can generate both B-lymphocytes and osteoclasts [119–121]. Katavic et al. [122] reported that murine bone marrow cells FACS-selected for expression of the B lymphocyte marker CD45R (also known as B220) could form osteoclasts in culture. However, this was later shown to be due to a 1–2% contamination with CD45R⁻ cells, and that further FACS purification to more than 99.9% pure CD45R⁺ cells eliminated the ability of this population to form osteoclasts in vitro [28]. The nature of the CD45R⁻ cells that contributed to osteoclast formation is not clear. They could be highly proliferative osteoclast precursors or they may be osteoclast precursors that can recruit CD45R⁺ cells to undergo heterotypic fusion into multinucleated osteoclasts. This remains to be determined.

## Multiple myeloma cells

Bone-resorbing osteoclasts from myeloma patients have been shown to contain both normal nuclei and nuclei with translocated chromosomes of myeloma B-cell clone origin [123]. The myeloma-originating nuclei were transcriptionally active within the osteoclast and as high as 30% of the nuclei within the osteoclasts were of myeloma origin. The osteoclast–myeloma hybrids were found in the region of bone containing myeloma cells and were more active than normal osteoclasts. Osteoclast–myeloma hybrids were also formed in co-cultures. The myeloma nuclei were tracked by either labeling them with BrDU or via mixing male myeloma cells with female osteoclasts, and assaying for the presence of the Y chromosome. As noted above myeloma cells express CD47 and also CD44, which are both thought to have some role in osteoclast fusion. Further analysis has indicated that only pre-osteoclasts that were already fusogenic could incorporate myeloma cell nuclei, suggesting that the osteoclast needs to be the instigator of the fusion [124]. Further, the myeloma cells that participated in forming osteoclast–myeloma hybrids were predominantly a subset with a propensity to adhere.

## Conclusion

With the many varied osteoclast progenitors in the circulation as well as present in the bone marrow and the spleen, it is intriguing that osteoclasts are only found on the bone. This is in spite of circulating levels of osteoclastogenic cytokines, particularly in inflammatory states. The developmental OCL precursor involved in normal bone remodeling may not be the same as that which is utilized in particular inflammatory states, such as myeloma of the bone, rheumatoid arthritis, postmenopausal osteoporosis, or periodontitis. Furthermore, the signals required to induce OCL differentiation may differ as well [112]. A number of gene deficiencies result in no effects on basal osteoclastogenesis, but impair stimulated osteoclastogenesis and often in vitro osteoclastogenesis (which might be thought of as a type of stimulated osteoclastogenesis). Examples of this are osteoclast progenitors with deletions of ADAM8, DAP12, RelB (NF-κB), NIK, IKKα and p62 (sequestosome1). The last, p62, is an adapter protein that is involved in RANKL signal transduction, but as with the other genes noted here, loss of p62 only affects osteoclastogenesis induced in vivo by PTHrP and in vitro [125]. While many of the cytokines known to support osteoclastogenesis work through indirect action on supporting cells, such as stromal cells or T-cells, inducing their production of RANKL, some have direct actions on osteoclast precursors and on osteoclasts. There is an array of cytokines that have direct positive and negative effects on osteoclastogenesis, discussed elsewhere in this book, that may affect or recruit particular types of osteoclast progenitors in certain pathological states.

## References

[1] Lucht U. Osteoclasts – ultrastructure and function, in the reticuloendothelial system. A comprehensive treatise. In: Carr I, Daems W, editors. Morphology. New York: Plenum Press; 1980. p. 705–33.

[2] Holtrop ME, King GJ. The ultrastructure of the osteoclast and its functional implications. Clin Orthop Relat Res 1977 Mar-Apr;123:177–96.

[3] Tolar J, Teitelbaum SL, Orchard PJ. Osteopetrosis. N Engl J Med 2004 Dec 30;351(27):2839–49.

[4] Darling JM, Goldring SR, Harada Y, Handel ML, Glowacki J, Gravallese EM. Multinucleated cells in pigmented villonodular synovitis and giant cell tumor of tendon sheath express features of osteoclasts. Am J Pathol 1997 Apr;150(4):1383–93.

[5] Reddy SV, Kuzhandaivelu N, Acosta LG, Roodman GD. Characterization of the 5′-flanking region of the human tartrate-resistant acid phosphatase (TRAP) gene. Bone 1995 May;16(5):587–93.

[6] Reddy SV, Scarcez T, Windle JJ, Leach RJ, Hundley JE, Chirgwin JM, et al. Cloning and characterization of the 5′-flanking region of the mouse tartrate-resistant acid phosphatase gene. J Bone Miner Res 1993 Oct;8(10):1263–70.

[7] Robinson DB, Glew RH. A tartrate-resistant acid phosphatase from Gaucher spleen. Purification and properties. J Biol Chem 1980 Jun 25;255(12):5864–70.

[8] Krause JR, Dekker A. Hairy cell leukemia (leukemic reticuloendotheliosis) in serous effusions. Acta Cytol 1978 Mar-Apr;22(2):80–2.

[9] Walker DG. Osteopetrosis cured by temporary parabiosis. Science 1973 May 25;180(88):875.

[10] Walker DG. Bone resorption restored in osteopetrotic mice by transplants of normal bone marrow and spleen cells. Science 1975 Nov 21;190(4216):784–5.

[11] Walker DG. Spleen cells transmit osteopetrosis in mice. Science 1975 Nov 21;190(4216):785–7.

[12] Gothlin G, Ericsson JL. On the histogenesis of the cells in fracture callus. Electron microscopic autoradiographic observations in parabiotic rats and studies on labeled monocytes. Virchows Arch B Cell Pathol 1973 Mar 30;12(4):318–29.

[13] Coccia PF, Krivit W, Cervenka J, Clawson C, Kersey JH, Kim TH, et al. Successful bone-marrow transplantation for infantile malignant osteopetrosis. N Engl J Med 1980 Mar 27;302(13):701–8.

[14] Sorell M, Kapoor N, Kirkpatrick D, Rosen JF, Chaganti RS, Lopez C, et al. Marrow transplantation for juvenile osteopetrosis. Am J Med 1981 Jun;70(6):1280–7.

[15] Hattersley G, Chambers TJ. Generation of osteoclasts from hemopoietic cells and a multipotential cell line in vitro. J Cell Physiol 1989 Sep;140(3):478–82.

[16] Young RW. Cell proliferation and specialization during endochondral osteogenesis in young rats. J Cell Biol 1962 Sep;14:357–70.

[17] Roodman GD. Regulation of osteoclast differentiation. Ann N Y Acad Sci 2006 Apr;1068:100–9.

[18] Tinkler SM, Linder JE, Williams DM, Johnson NW. Formation of osteoclasts from blood monocytes during 1 alpha-OH Vit D-stimulated bone resorption in mice. J Anat 1981 Oct;133(Pt 3):389–96.

[19] Zambonin Zallone A, Teti A, Primavera MV. Monocytes from circulating blood fuse in vitro with purified osteoclasts in primary culture. J Cell Sci 1984 Mar;66:335–42.

[20] Tsurukai T, Takahashi N, Jimi E, Nakamura I, Udagawa N, Nogimori K, et al. Isolation and characterization of osteoclast precursors that differentiate into osteoclasts on calvarial cells within a short period of time. J Cell Physiol 1998 Oct;177(1):26–35.

[21] Kurihara N, Suda T, Miura Y, Nakauchi H, Kodama H, Hiura K, et al. Generation of osteoclasts from isolated hematopoietic progenitor cells. Blood 1989 Sep;74(4):1295–302.

[22] Menaa C, Kurihara N, Roodman GD. CFU-GM-derived cells form osteoclasts at a very high efficiency. Biochem Biophys Res Commun 2000 Jan 27;267(3):943–6.

[23] Matayoshi A, Brown C, DiPersio JF, Haug J, Abu-Amer Y, Liapis H, et al. Human blood-mobilized hematopoietic precursors differentiate into osteoclasts in the absence of stromal cells. Proc Natl Acad Sci U S A 1996 Oct 1;93(20):10785–90.

[24] Arai F, Miyamoto T, Ohneda O, Inada T, Sudo T, Brasel K, et al. Commitment and differentiation of osteoclast precursor cells by the sequential expression of c-Fms and receptor activator of nuclear factor kappaB (RANK) receptors. J Exp Med 1999 Dec 20;190(12):1741–54.

[25] Servet-Delprat C, Arnaud S, Jurdic P, Nataf S, Grasset MF, Soulas C, et al. Flt3+ macrophage precursors commit sequentially to osteoclasts, dendritic cells and microglia. BMC Immunol 2002 Oct 24;3:15.

[26] de Vries TJ, Schoenmaker T, Hooibrink B, Leenen PJ, Everts V. Myeloid blasts are the mouse bone marrow cells prone to differentiate into osteoclasts. J Leukoc Biol 2009 Jun;85(6):919–27.

[27] Sunderkotter C, Nikolic T, Dillon MJ, Van Rooijen N, Stehling M, Drevets DA, et al. Subpopulations of mouse blood monocytes differ in maturation stage and inflammatory response. J Immunol 2004 Apr 1;172(7):4410–17.

[28] Jacquin C, Gran DE, Lee SK, Lorenzo JA, Aguila HL. Identification of multiple osteoclast precursor populations in murine bone marrow. J Bone Miner Res 2006 Jan;21(1):67–77.

[29] Yao Z, Li P, Zhang Q, Schwarz EM, Keng P, Arbini A, et al. Tumor necrosis factor-alpha increases circulating osteoclast precursor numbers by promoting their proliferation and differentiation in the bone marrow through up-regulation of c-Fms expression. J Biol Chem 2006 Apr 28;281(17):11846–55.

[30] Miyamoto T, Ohneda O, Arai F, Iwamoto K, Okada S, Takagi K, et al. Bifurcation of osteoclasts and dendritic cells from common progenitors. Blood 2001 Oct 15;98(8):2544–54.

[31] Hodge JM, Kirkland MA, Aitken CJ, Waugh CM, Myers DE, Lopez CM, et al. Osteoclastic potential of human CFU-GM: biphasic effect of GM-CSF. J Bone Miner Res 2004 Feb;19(2):190–9.

[32] Shalhoub V, Elliott G, Chiu L, Manoukian R, Kelley M, Hawkins N, et al. Characterization of osteoclast precursors in human blood. Br J Haematol 2000 Nov;111(2):501–12.

[33] Geissmann F, Jung S, Littman DR. Blood monocytes consist of two principal subsets with distinct migratory properties. Immunity 2003 Jul;19(1):71–82.

[34] Fogg DK, Sibon C, Miled C, Jung S, Aucouturier P, Littman DR, et al. A clonogenic bone marrow progenitor specific for macrophages and dendritic cells. Science 2006 Jan 6;311(5757):83–7.

[35] Ishii M, Egen JG, Klauschen F, Meier-Schellersheim M, Saeki Y, Vacher J, et al. Sphingosine-1-phosphate mobilizes osteoclast precursors and regulates bone homeostasis. Nature 2009 Mar 26;458(7237):524–8.

[36] Koizumi K, Saitoh Y, Minami T, Takeno N, Tsuneyama K, Miyahara T, et al. Role of CX3CL1/fractalkine in osteoclast differentiation and bone resorption. J Immunol 2009 Dec 15;183(12):7825–31.

[37] Komano Y, Nanki T, Hayashida K, Taniguchi K, Miyasaka N. Identification of a human peripheral blood monocyte subset that differentiates into osteoclasts. Arthritis Res Ther 2006;8(5):R152.

[38] Ziegler-Heitbrock L. The CD14+ CD16+ blood monocytes: their role in infection and inflammation. J Leukoc Biol 2007 Mar;81(3):584–92.

[39] Yoshida H, Hayashi S, Kunisada T, Ogawa M, Nishikawa S, Okamura H, et al. The murine mutation osteopetrosis is in the coding region of the macrophage colony stimulating factor gene. Nature 1990 May 31;345(6274):442–4.

[40] Dai XM, Ryan GR, Hapel AJ, Dominguez MG, Russell RG, Kapp S, et al. Targeted disruption of the mouse colony-stimulating factor 1 receptor gene results in osteopetrosis, mononuclear phagocyte deficiency, increased primitive progenitor cell frequencies, and reproductive defects. Blood 2002 Jan 1;99(1):111–20.

[41] Lin H, Lee E, Hestir K, Leo C, Huang M, Bosch E, et al. Discovery of a cytokine and its receptor by functional screening of the extracellular proteome. Science 2008 May 9;320(5877):807–11.

[42] Begg SK, Bertoncello I. The hematopoietic deficiencies in osteopetrotic (op/op) mice are not permanent, but progressively correct with age. Exp Hematol 1993 Apr;21(4):493–5.

[43] Niida S, Kaku M, Amano H, Yoshida H, Kataoka H, Nishikawa S, et al. Vascular endothelial growth factor can substitute for macrophage colony-stimulating factor in the support of osteoclastic bone resorption. J Exp Med 1999 Jul 19;190(2):293–8.

[44] Myint YY, Miyakawa K, Naito M, Shultz LD, Oike Y, Yamamura K, et al. Granulocyte/macrophage colony-stimulating factor and interleukin-3 correct osteopetrosis in mice with osteopetrosis mutation. Am J Pathol 1999 Feb;154(2):553–66.

[45] Takahashi N, Akatsu T, Udagawa N, Sasaki T, Yamaguchi A, Moseley JM, et al. Osteoblastic cells are involved in osteoclast formation. Endocrinology 1988 Nov;123(5):2600–2.

[46] Kong YY, Yoshida H, Sarosi I, Tan HL, Timms E, Capparelli C, et al. OPGL is a key regulator of osteoclastogenesis, lymphocyte development and lymph-node organogenesis. Nature 1999 Jan 28;397(6717):315–23.

[47] Fuller K, Wong B, Fox S, Choi Y, Chambers TJ. TRANCE is necessary and sufficient for osteoblast-mediated activation of bone resorption in osteoclasts. J Exp Med 1998 Sep 7;188(5):997–1001.

[48] Hikita A, Yana I, Wakeyama H, Nakamura M, Kadono Y, Oshima Y, et al. Negative regulation of osteoclastogenesis by ectodomain shedding of receptor activator of NF-kappaB ligand. J Biol Chem 2006 Dec 1;281(48):36846–55.

[49] Lum L, Wong BR, Josien R, Becherer JD, Erdjument-Bromage H, Schlondorff J, et al. Evidence for a role of a tumor necrosis factor-alpha (TNF-alpha)-converting enzyme-like protease in shedding of TRANCE, a TNF family member involved in osteoclastogenesis and dendritic cell survival. J Biol Chem 1999 May 7;274(19):13613–18.

[50] Dougall WC, Glaccum M, Charrier K, Rohrbach K, Brasel K, De Smedt T, et al. RANK is essential for osteoclast and lymph node development. Genes Dev 1999 Sep 15;13(18):2412–24.

[51] Yasuda H, Shima N, Nakagawa N, Mochizuki SI, Yano K, Fujise N, et al. Identity of Osteoclastogenesis inhibitory factor (OCIF) and osteoprotegerin (OPG): a mechanism by which OPG/OCIF inhibits osteoclastogenesis in vitro. Endocrinology 1998 Mar;139(3):1329–37.

[52] Bucay N, Sarosi I, Dunstan CR, Morony S, Tarpley J, Capparelli C, et al. Osteoprotegerin-deficient mice develop early onset osteoporosis and arterial calcification. Genes Dev 1998 May 1;12(9):1260–8.

[53] Mizuno A, Amizuka N, Irie K, Murakami A, Fujise N, Kanno T, et al. Severe osteoporosis in mice lacking osteoclastogenesis inhibitory factor/osteoprotegerin. Biochem Biophys Res Commun 1998 Jun 29;247(3):610–15.

[54] Simonet WS, Lacey DL, Dunstan CR, Kelley M, Chang MS, Luthy R, et al. Osteoprotegerin: a novel secreted protein involved in the regulation of bone density. Cell 1997 Apr 18;89(2):309–19.

[55] Kanazawa K, Kudo A. TRAF2 is essential for TNF-alpha-induced osteoclastogenesis. J Bone Miner Res 2005 May;20(5):840–7.

[56] Kanazawa K, Azuma Y, Nakano H, Kudo A. TRAF5 functions in both RANKL- and TNFalpha-induced osteoclastogenesis. J Bone Miner Res 2003 Mar;18(3):443–50.

[57] Lomaga MA, Yeh WC, Sarosi I, Duncan GS, Furlonger C, Ho A, et al. TRAF6 deficiency results in osteopetrosis and defective interleukin-1, CD40, and LPS signaling. Genes Dev 1999 Apr 15;13(8):1015–24.

[58] Naito A, Azuma S, Tanaka S, Miyazaki T, Takaki S, Takatsu K, et al. Severe osteopetrosis, defective interleukin-1 signalling and lymph node organogenesis in TRAF6-deficient mice. Genes Cells 1999 Jun;4(6):353–62.

[59] Asagiri M, Takayanagi H. The molecular understanding of osteoclast differentiation. Bone 2007 Feb;40(2):251–64.

[60] Mocsai A, Humphrey MB, Van Ziffle JA, Hu Y, Burghardt A, Spusta SC, et al. The immunomodulatory adapter proteins DAP12 and Fc receptor gamma-chain (FcRgamma) regulate development of functional osteoclasts through the Syk tyrosine kinase. Proc Natl Acad Sci U S A 2004 Apr 20;101(16):6158–63.

[61] Humphrey MB, Ogasawara K, Yao W, Spusta SC, Daws MR, Lane NE, et al. The signaling adapter protein DAP12 regulates multinucleation during osteoclast development. J Bone Miner Res 2004 Feb;19(2):224–34.

[62] Koga T, Inui M, Inoue K, Kim S, Suematsu A, Kobayashi E, et al. Costimulatory signals mediated by the ITAM motif cooperate with RANKL for bone homeostasis. Nature 2004 Apr 15;428(6984):758–63.

[63] Wu Y, Torchia J, Yao W, Lane NE, Lanier LL, Nakamura MC, et al. Bone microenvironment specific roles of ITAM adapter signaling during bone remodeling induced by acute estrogen-deficiency. PLoS One 2007;2(7):e586.

[64] Sato K, Suematsu A, Nakashima T, Takemoto-Kimura S, Aoki K, Morishita Y, et al. Regulation of osteoclast differentiation and function by the CaMK-CREB pathway. Nat Med 2006 Dec;12(12):1410–16.

[65] Tondravi MM, McKercher SR, Anderson K, Erdmann JM, Quiroz M, Maki R, et al. Osteopetrosis in mice lacking haematopoietic transcription factor PU.1. Nature 1997 Mar 6;386(6620):81–4.

[66] Luchin A, Suchting S, Merson T, Rosol TJ, Hume DA, Cassady AI, et al. Genetic and physical interactions between microphthalmia transcription factor and PU.1 are necessary for osteoclast gene expression and differentiation. J Biol Chem 2001 Sep 28;276(39):36703–10.

[67] DeKoter RP, Walsh JC, Singh H. PU.1 regulates both cytokine-dependent proliferation and differentiation of granulocyte/macrophage progenitors. EMBO J 1998 Aug 3;17(15):4456–68.

[68] Franzoso G, Carlson L, Xing L, Poljak L, Shores EW, Brown KD, et al. Requirement for NF-kappaB in osteoclast and B-cell development. Genes Dev 1997 Dec 15;11(24):3482–96.

[69] Iotsova V, Caamano J, Loy J, Yang Y, Lewin A, Bravo R. Osteopetrosis in mice lacking NF-kappaB1 and NF-kappaB2. Nat Med 1997 Nov;3(11):1285–9.

[70] Yamashita T, Yao Z, Li F, Zhang Q, Badell IR, Schwarz EM, et al. NF-kappaB p50 and p52 regulate receptor activator of NF-kappaB ligand (RANKL) and tumor necrosis factor-induced osteoclast precursor differentiation by activating c-Fos and NFATc1. J Biol Chem 2007 Jun 22;282(25):18245–53.

[71] Xing L, Bushnell TP, Carlson L, Tai Z, Tondravi M, Siebenlist U, et al. NF-kappaB p50 and p52 expression is not required for RANK-expressing osteoclast progenitor formation but is essential for RANK- and cytokine-mediated osteoclastogenesis. J Bone Miner Res 2002 Jul;17(7):1200–10.

[72] Vaira S, Johnson T, Hirbe AC, Alhawagri M, Anwisye I, Sammut B, et al. RelB is the NF-kappaB subunit downstream of NIK responsible for osteoclast differentiation. Proc Natl Acad Sci U S A 2008 Mar 11;105(10):3897–902.

[73] Wan Y, Chong LW, Evans RM. PPAR-gamma regulates osteoclastogenesis in mice. Nat Med 2007 Dec;13(12):1496–503.

[74] Grigoriadis AE, Wang ZQ, Cecchini MG, Hofstetter W, Felix R, Fleisch HA, et al. c-Fos: a key regulator of osteoclast-macrophage lineage determination and bone remodeling. Science 1994 Oct 21;266(5184):443–8.

[75] Takayanagi H, Kim S, Matsuo K, Suzuki H, Suzuki T, Sato K, et al. RANKL maintains bone homeostasis through c-Fos-dependent induction of interferon-beta. Nature 2002 Apr 18;416 (6882):744–9.

[76] Matsuo K, Galson DL, Zhao C, Peng L, Laplace C, Wang KZ, et al. Nuclear factor of activated T-cells (NFAT) rescues osteoclastogenesis in precursors lacking c-Fos. J Biol Chem 2004 Jun 18;279 (25):26475–80.

[77] Matsuo K, Owens JM, Tonko M, Elliott C, Chambers TJ, Wagner EF. Fosl1 is a transcriptional target of c-Fos during osteoclast differentiation. Nat Genet 2000 Feb;24(2):184–7.

[78] Owens JM, Matsuo K, Nicholson GC, Wagner EF, Chambers TJ. Fra-1 potentiates osteoclastic differentiation in osteoclast-macrophage precursor cell lines. J Cell Physiol 1999 May;179 (2):170–8.

[79] Shinohara M, Koga T, Okamoto K, Sakaguchi S, Arai K, Yasuda H, et al. Tyrosine kinases Btk and Tec regulate osteoclast differentiation by linking RANK and ITAM signals. Cell 2008 Mar 7;132 (5):794–806.

[80] Weilbaecher KN, Motyckova G, Huber WE, Takemoto CM, Hemesath TJ, Xu Y, et al. Linkage of M-CSF signaling to Mitf, TFE3, and the osteoclast defect in Mitf(mi/mi) mice. Mol Cell 2001 Oct; 8(4):749–58.

[81] Vignery A. Macrophage fusion: the making of osteoclasts and giant cells. J Exp Med 2005 Aug 1;202 (3):337–40.

[82] Alford AI, Hankenson KD. Matricellular proteins: extracellular modulators of bone development, remodeling, and regeneration. Bone 2006 Jun;38(6):749–57.

[83] Uluckan O, Becker SN, Deng H, Zou W, Prior JL, Piwnica-Worms D, et al. CD47 regulates bone mass and tumor metastasis to bone. Cancer Res 2009 Apr 1;69(7):3196–204.

[84] Kukreja A, Radfar S, Sun BH, Insogna K, Dhodapkar MV. Dominant role of CD47-thrombospondin-1 interactions in myeloma-induced fusion of human dendritic cells: implications for bone disease. Blood 2009 Oct 15;114(16):3413–21.

[85] Yang K, Vega JL, Hadzipasic M, Schatzmann Peron JP, Zhu B, Carrier Y, et al. Deficiency of thrombospondin-1 reduces Th17 differentiation and attenuates experimental autoimmune encephalomyelitis. J Autoimmun 2009 Mar;32(2):94–103.

[86] Carron JA, Wagstaff SC, Gallagher JA, Bowler WB. A CD36-binding peptide from thrombospondin-1 can stimulate resorption by osteoclasts in vitro. Biochem Biophys Res Commun 2000 Apr 21;270(3):1124–7.

[87] van Beek EM, de Vries TJ, Mulder L, Schoenmaker T, Hoeben KA, Matozaki T, et al. Inhibitory regulation of osteoclast bone resorption by signal regulatory protein alpha. FASEB J 2009 Dec;23(12):4081–90.

[88] Yagi M, Miyamoto T, Sawatani Y, Iwamoto K, Hosogane N, Fujita N, et al. DC-STAMP is essential for cell-cell fusion in osteoclasts and foreign body giant cells. J Exp Med 2005 Aug 1;202(3):345–51.

[89] Kukita T, Wada N, Kukita A, Kakimoto T, Sandra F, Toh K, et al. RANKL-induced DC-STAMP is essential for osteoclastogenesis. J Exp Med 2004 Oct 4;200(7):941–6.

[90] Li YP, Chen W, Liang Y, Li E, Stashenko P. Atp6i-deficient mice exhibit severe osteopetrosis due to loss of osteoclast-mediated extracellular acidification. Nat Genet 1999 Dec;23(4):447–51.

[91] Kornak U, Schulz A, Friedrich W, Uhlhaas S, Kremens B, Voit T, et al. Mutations in the a3 subunit of the vacuolar H(+)-ATPase cause infantile malignant osteopetrosis. Hum Mol Genet 2000 Aug 12;9(13):2059–63.

[92] Frattini A, Orchard PJ, Sobacchi C, Giliani S, Abinun M, Mattsson JP, et al. Defects in TCIRG1 subunit of the vacuolar proton pump are responsible for a subset of human autosomal recessive osteopetrosis. Nat Genet 2000 Jul;25(3):343–6.

[93] Lee SH, Rho J, Jeong D, Sul JY, Kim T, Kim N, et al. v-ATPase V0 subunit d2-deficient mice exhibit impaired osteoclast fusion and increased bone formation. Nat Med 2006 Dec;12(12):1403–9.

[94] Wu H, Xu G, Li YP. Atp6v0d2 is an essential component of the osteoclast-specific proton pump that mediates extracellular acidification in bone resorption. J Bone Miner Res 2009 May;24(5):871–85.

[95] Kania JR, Kehat-Stadler T, Kupfer SR. CD44 antibodies inhibit osteoclast formation. J Bone Miner Res 1997 Aug;12(8):1155–64.

[96] de Vries TJ, Schoenmaker T, Beertsen W, van der Neut R, Everts V. Effect of CD44 deficiency on in vitro and in vivo osteoclast formation. J Cell Biochem 2005 Apr 1;94(5):954–66.

[97] Choi SJ, Han JH, Roodman GD. ADAM8: a novel osteoclast stimulating factor. J Bone Miner Res 2001 May;16(5):814–22.

[98] Rao H, Lu G, Kajiya H, Garcia-Palacios V, Kurihara N, Anderson J, et al. Alpha9beta1: a novel osteoclast integrin that regulates osteoclast formation and function. J Bone Miner Res 2006 Oct;21(10):1657–65.

[99] Galson DL, Goldring SR. Structure and molecular biology of the calcitonin receptor. In: Bilezikian J, Raisz L, Rodan G, editors. Principles of Bone Biology. 2nd ed. New York, NY: Academic Press; 2002. p. 603–17.

[100] Del Fattore A, Teti A, Rucci N. Osteoclast receptors and signaling. Arch Biochem Biophys 2008 May 15;473(2):147–60.

[101] Granholm S, Lundberg P, Lerner UH. Calcitonin inhibits osteoclast formation in mouse haematopoetic cells independently of transcriptional regulation by receptor activator of NF-{kappa}B and c-Fms. J Endocrinol 2007 Dec;195(3):415–27.

[102] Woodrow JP, Sharpe CJ, Fudge NJ, Hoff AO, Gagel RF, Kovacs CS. Calcitonin plays a critical role in regulating skeletal mineral metabolism during lactation. Endocrinology 2006 Sep;147(9):4010–21.

[103] Sly WS, Hewett-Emmett D, Whyte MP, Yu YS, Tashian RE. Carbonic anhydrase II deficiency identified as the primary defect in the autosomal recessive syndrome of osteopetrosis with renal tubular acidosis and cerebral calcification. Proc Natl Acad Sci U S A 1983 May;80(9):2752–6.

[104] Margolis DS, Szivek JA, Lai LW, Lien YH. Phenotypic characteristics of bone in carbonic anhydrase II-deficient mice. Calcif Tissue Int 2008 Jan;82(1):66–76.

[105] Laitala T, Vaananen HK. Inhibition of bone resorption in vitro by antisense RNA and DNA molecules targeted against carbonic anhydrase II or two subunits of vacuolar H(+)-ATPase. J Clin Invest 1994 Jun;93(6):2311–18.

[106] Jansen ID, Mardones P, Lecanda F, de Vries TJ, Recalde S, Hoeben KA, et al. Ae2(a, b)-deficient mice exhibit osteopetrosis of long bones but not of calvaria. FASEB J 2009 Oct;23(10):3470–81.

[107] Kiviranta R, Morko J, Alatalo SL, NicAmhlaoibh R, Risteli J, Laitala-Leinonen T, et al. Impaired bone resorption in cathepsin K-deficient mice is partially compensated for by enhanced osteoclastogenesis and increased expression of other proteases via an increased RANKL/OPG ratio. Bone 2005 Jan;36(1):159–72.

[108] Pennypacker B, Shea M, Liu Q, Masarachia P, Saftig P, Rodan S, et al. Bone density, strength, and formation in adult cathepsin K (−/−) mice. Bone 2009 Feb;44(2):199–207.

[109] Everts V, Korper W, Jansen DC, Steinfort J, Lammerse I, Heera S, et al. Functional heterogeneity of osteoclasts: matrix metalloproteinases participate in osteoclastic resorption of calvarial bone but not in resorption of long bone. FASEB J 1999 Jul;13(10):1219–30.

[110] Vu TH, Shipley JM, Bergers G, Berger JE, Helms JA, Hanahan D, et al. MMP-9/gelatinase B is a key regulator of growth plate angiogenesis and apoptosis of hypertrophic chondrocytes. Cell 1998 May 1;93(3):411–22.

[111] Cackowski FC, Anderson JL, Patrene KD, Choksi RJ, Shapiro SD, Windle JJ, et al. Osteoclasts are important for bone angiogenesis. Blood 2009 Nov 3.

[112] Jemtland R, Lee K, Segre GV. Heterogeneity among cells that express osteoclast-associated genes in developing bone. Endocrinology 1998 Jan;139(1):340–9.

[113] Nordahl J, Andersson G, Reinholt FP. Chondroclasts and osteoclasts in bones of young rats: comparison of ultrastructural and functional features. Calcif Tissue Int 1998 Nov;63(5):401–8.

[114] Rivollier A, Mazzorana M, Tebib J, Piperno M, Aitsiselmi T, Rabourdin-Combe C, et al. Immature dendritic cell transdifferentiation into osteoclasts: a novel pathway sustained by the rheumatoid arthritis microenvironment. Blood 2004 Dec 15;104(13):4029–37.

[115] Alnaeeli M, Park J, Mahamed D, Penninger JM, Teng YT. Dendritic cells at the osteo-immune interface: implications for inflammation-induced bone loss. J Bone Miner Res 2007 Jun;22(6):775–80.

[116] Speziani C, Rivollier A, Gallois A, Coury F, Mazzorana M, Azocar O, et al. Murine dendritic cell transdifferentiation into osteoclasts is differentially regulated by innate and adaptive cytokines. Eur J Immunol 2007 Mar;37(3):747–57.

[117] Wakkach A, Mansour A, Dacquin R, Coste E, Jurdic P, Carle GF, et al. Bone marrow microenvironment controls the in vivo differentiation of murine dendritic cells into osteoclasts. Blood 2008 Dec 15;112(13):5074–83.

[118] Udagawa N, Takahashi N, Akatsu T, Tanaka H, Sasaki T, Nishihara T, et al. Origin of osteoclasts: mature monocytes and macrophages are capable of differentiating into osteoclasts under a suitable microenvironment prepared by bone marrow-derived stromal cells. Proc Natl Acad Sci U S A 1990 Sep;87(18):7260–4.

[119] Okahashi N, Murase Y, Koseki T, Sato T, Yamato K, Nishihara T. Osteoclast differentiation is associated with transient upregulation of cyclin-dependent kinase inhibitors p21(WAF1/CIP1) and p27(KIP1). J Cell Biochem 2001;80(3):339–45.

[120] Manabe N, Kawaguchi H, Chikuda H, Miyaura C, Inada M, Nagai R, et al. Connection between B lymphocyte and osteoclast differentiation pathways. J Immunol 2001 Sep 1;167(5):2625–31.

[121] Blin-Wakkach C, Wakkach A, Rochet N, Carle GF. Characterization of a novel bipotent hematopoietic progenitor population in normal and osteopetrotic mice. J Bone Miner Res 2004 Jul;19(7):1137–43.

[122] Katavic V, Grcevic D, Lee SK, Kalinowski J, Jastrzebski S, Dougall W, et al. The surface antigen CD45R identifies a population of estrogen-regulated murine marrow cells that contain osteoclast precursors. Bone 2003 Jun;32(6):581–90.

[123] Andersen TL, Boissy P, Sondergaard TE, Kupisiewicz K, Plesner T, Rasmussen T, et al. Osteoclast nuclei of myeloma patients show chromosome translocations specific for the myeloma cell clone: a new type of cancer-host partnership? J Pathol 2007 Jan;211(1):10–17.

[124] Kupisiewicz K, Soe K, Anderson TL, Soendergaard TE, Lund T, Plesner T, et al. Biological prerequisites for hetrotypic fusion between myeloma calls and osteoclasts. Cancer Treat Rev 2008;34(Suppl. 1):S65–6.

[125] Duran A, Serrano M, Leitges M, Flores JM, Picard S, Brown JP, et al. The atypical PKC-interacting protein p62 is an important mediator of RANK-activated osteoclastogenesis. Dev Cell 2004 Feb;6(2):303–9.

# 3

# The Adaptive Immune Response

David G. Hesslein, Hector L. Aguila, Mark C. Horowitz

DEPARTMENT OF MICROBIOLOGY AND IMMUNOLOGY, UNIVERSITY OF CALIFORNIA SAN FRANCISCO, SAN FRANCISCO, CA, USA,
DEPARTMENT OF IMMUNOLOGY, UNIVERSITY OF CONNECTICUT HEALTH CENTER, FARMINGTON, CT, USA,
DEPARTMENT OF ORTHOPAEDICS AND REHABILITATION, YALE UNIVERSITY SCHOOL OF MEDICINE, NEW HAVEN, CT, USA

## CHAPTER OUTLINE

Introduction ............................................................................................................... 43
T-cell development and function ............................................................................ 44
    Introduction ............................................................................................................ 44
    T-cell recognition .................................................................................................... 46
    T-cell functional heterogeneity ............................................................................ 48
    Thymopoiesis ........................................................................................................ 49
    T-cell activation ..................................................................................................... 53
    T-cell memory ....................................................................................................... 57
B-cell development and function ............................................................................ 58
    Introduction ............................................................................................................ 58
    B-cell development ............................................................................................... 59
    Extrinsic requirements ......................................................................................... 62
    IL-7 ......................................................................................................................... 63
    Intrinsic requirements .......................................................................................... 64
    The RANK–RANKL pathway and B-cell differentiation ..................................... 68
    B-cell function ....................................................................................................... 69
    Activated B cells ................................................................................................... 71
Acknowledgments ..................................................................................................... 72
References ................................................................................................................. 72

## Introduction

B cells and T cells comprise the adaptive immune system, which allows an individual to develop a specific response to an infection (antigen) and retain memory of that infection. Memory allows for a faster and more robust response if that same infection occurs again. The adaptive immune response is divided between two components: humoral immunity, which is mediated by antibodies produced by B lymphocytes (B cells) and cell-mediated immunity, which is mediated by T lymphocytes (T cells). T cells can directly kill infected cells or activate other cells, like macrophages, to kill or phagocytize infected cells. T cells are also often needed to "help" B cells mount an antibody response. The other major mechanism of host defense is the innate immune system. In contrast to adaptive immunity, innate immunity is not antigen specific, functions to recognize microorganisms, works very quickly (within hours), and is mediated by other cellular mechanisms. The body, through pattern recognition receptors such as Toll-like receptors, recognizes these organisms. Once recognized, invading microorganisms can be attacked by circulating antibodies, opsonized and killed by complement activation, or phagocytized by macrophages. This chapter will focus on the adaptive immune response and how its components interact with bone.

All of the bloodborne elements arise from pluripotent hematopoietic stem cells (HSCs). These reside in growth and differentiation supportive structures, known as niches, which are found predominantly on endosteal bone surfaces in contact with the bone marrow. HSC are also in contact with cells that are adherent to the endosteal bone surface. These support cells secrete or express on their cell surface cytokines and growth factors that facilitate HSC growth and differentiation [1–3]. Often these cells are referred to as osteoblasts (OBs) or stromal cells, and although they are in the osteoblast lineage their exact stage of differentiation is unknown. However, it is unlikely that they are mature osteoid-secreting osteoblasts.

Long-term reconstituting HSCs (LT-HSC) are multipotential in that they have the capacity for self-renewal as well as producing daughter cells that at each stage of differentiation become more restricted in cell lineage. LT-HSCs give rise to short-term reconstituting HSCs that differentiate to multipotent progenitors (MPPs). It is at this stage that an important separation of the lineages occurs (Figure 3-1). MPPs differentiate into the common myeloid progenitor (CMP), which gives rise to the granulocyte-macrophage progenitor (GMP) and then, in turn, to osteoclasts (OC), macrophages, granulocytes, and dendritic cells [4]. The CMP also produces the megakaryocyte-erythrocyte progenitor in a separate lineage. Alternatively, MPPs can also differentiate into the common lymphoid progenitor (CLP), which gives rise to B, T and natural killer cells. Thus, it is at the MPP stage that myeloid and lymphoid differentiation shares a common progenitor and critical cell fate decisions are made.

# T-cell development and function

## Introduction

T lymphocytes (T cells) are the central cells that define developmental and functional outcomes of adaptive immune responses. Like B lymphocytes (B cells), they respond to environmental antigens with high specificity, inducing clonal expansion and generating an anamnestic (memory) response with the potential to be recalled upon subsequent challenges. T cells and B cells derived from shared progenitors and their receptors share

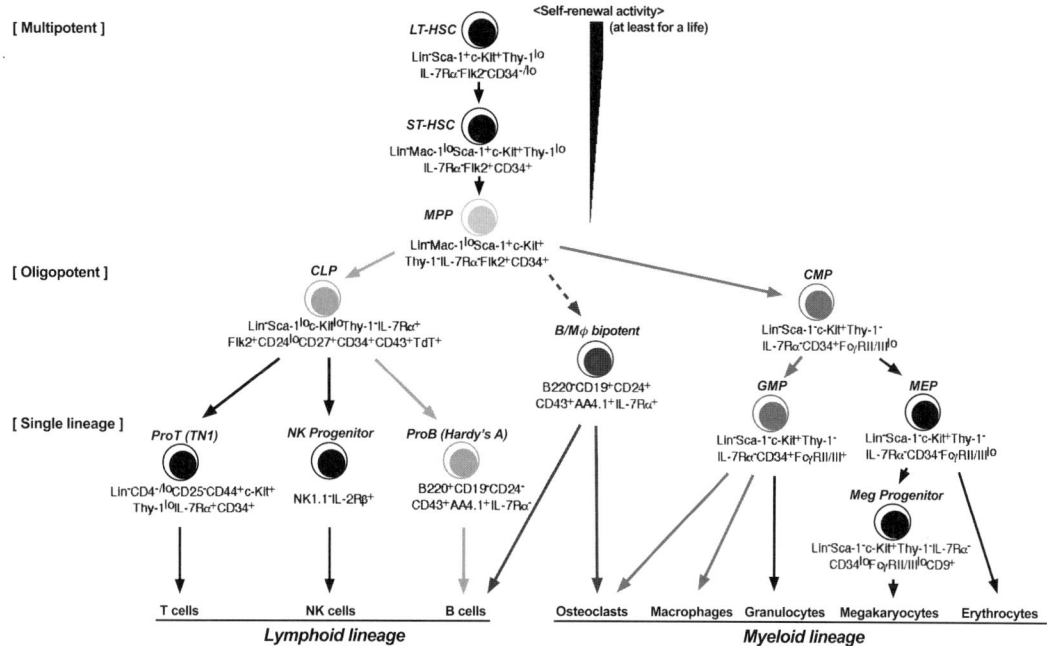

**FIGURE 3-1 Hematopoietic cell differentiation.** All hematopoietic cells arise from hematopoietic stem cells (HSC) that give rise to multipotential progenitor cells (MPP, blue). B cells differentiate from the common lymphoid progenitor (CLP, green) and osteoclasts arise from the common myeloid progenitor (CMP, pink). There appears to exist a distinct lineage derived from the MPP that expresses both macrophage and B-cell characteristics (B/Mφ bipotent), which can differentiate to B cells, macrophages and possibly osteoclasts (purple). *(Adapted from Kondo M, Wagers AJ, Manz MG, Prohaska SS, Scherer DC, Beilhack GF, Shizuru JA, Weissman IL. Ann. Rev. Immunol. 2003; 21:759–806).* Please refer to color plate section.

structural features and rearrange through similar mechanisms. However, the recognition of cognate antigens (antigen) by T cells differs from B cells, in that T-cell receptors (TCRs) do not bind antigens directly, but instead, bind processed peptides that are derived from antigens, which are presented in the context of self-major histocompatibility complex (MHC) molecules. In addition, TCRs are not secreted. Instead, they remain as transmembrane molecules that are tightly associated to co-receptors. They are responsible for signal transduction, the control of cellular functions, and the modulation of cell fate allocation upon antigen binding.

Developmentally, T cells are generated in the thymus, where T-cell precursors acquire cognate receptors and functional reactivity. This occurs through a coordinated series of maturation steps that are tightly controlled by developmental checkpoints. The final result is a population of cells with a diverse repertoire of antigen receptors and potential functional capabilities. Mature T cells are able to recognize foreign antigens in the periphery in a cascade that culminates with elimination of the antigen and the development of a small population of memory T cells. These memory T cells have the ability to mount an efficient and rapid recall response upon secondary challenge with the eliciting antigen. Depending on the T cell type activated, the elimination of the antigen occurs through direct interaction with activated T cells or indirectly through mediators or other cells that are influenced by the responding T-cell population.

The peripheral T-cell compartment is functionally heterogeneous, with at least five different T-cell types: cytotoxic T lymphocytes (CTL), T helper type 1 (Th1) cells, T helper type 2 (Th2) cells, regulatory T cells (Treg), and T helper type 17 (Th17) cells. These differ in features associated with T-cell recognition, co-receptors associated with the TCR, and their ability to secrete cytokines and mediate cell–cell interactions.

From their generation to the establishment of memory, T cells interact with the microenvironment at multiple levels. In their pre-thymic stage they develop in the context of the bone marrow microenvironment. In the thymic phase their maturation and antigen receptor repertoire selection occur through various regions within the thymus. As circulating cells, their homeostatic levels are constantly regulated by interactions within secondary lymphoid organs, which are sites where they are also activated. Finally, at the memory phase it has been proposed that they locate in defined microenvironmental niches that maintain the memory T-cell pool. This versatility makes the T-cell compartment a highly interactive population with the ability to modify other cellular systems in the context of physiological and pathological processes.

## T-cell recognition

The T-cell receptor (TCR) is a heterodimeric molecule assembled by association of an $\alpha$ chain with a $\beta$ chain, or a $\gamma$ chain with a $\delta$ chain. These define two separate and mutually exclusive populations of T cells: the $\alpha\beta$ and the $\gamma\delta$ T-cell compartments. The $\alpha\beta$ compartment includes the majority of the T cells found in an adult individual, with the $\gamma\delta$ compartment accounting for 5% of the total T cells that are mostly associated with mucosal tissues. These polypeptide chains are organized in domains that have structural features of the immunoglobulin (Ig) gene superfamily, with an N-terminus variable region and C-terminus constant region that contains a transmembrane domain [5, 6]. Each of these chains is encoded by a separate gene locus, and as in the case of Ig genes, they are divided in gene segments that upon rearrangement generate the variable and constant regions of the TCR [7, 8]. The TCR$\alpha$ locus is located on chromosome 14 in humans and it is organized in a cluster of approximately 80 V$\alpha$ regions followed by a collection of approximately 60 J$\alpha$ regions and a single C$\alpha$ gene. The TCR$\beta$ regions, located on chromosome 7 in humans, contain approximately 50 V$\beta$ regions followed by two clusters. The first cluster contains a single D$\beta$1 followed by 6 J$\beta$1 segments and one C$\beta$1 gene. The second cluster contains a single D$\beta$2 followed by a J$\beta$2 segment and one C$\beta$2 gene. In an analogous fashion, $\gamma$ and $\delta$ chains are organized in gene segments. TCR $\gamma$ is encoded in chromosome 7 in humans and it is organized in a similar way as the TCR$\beta$ locus with a series of V$\gamma$ regions followed by two C$\gamma$ regions preceded by their own clusters of J$\gamma$ segments. Interestingly, the TCR$\delta$ locus, organized with V$\delta$, D$\delta$, J$\delta$ region segments and a single Cδ gene, is located within the α-chain locus, just before the Jα clusters. All constant region genes are organized in separate exons containing each domain, plus the hinge, transmembrane and cytoplasmic regions [9].

Because of the clonal nature of these receptors, their rearrangement and assembly must be tightly coordinated to achieve functional diversity and allelic exclusion. In a temporal fashion the rearrangement of TCRβ, TCRγ, and TCRδ precede the rearrangement of TCRα. These initial events will determine if a cell will become an αβ or γδ T cell. In this scheme, a cell will develop into an αβ T cell only after successful rearrangement of the D-Jβ segments towards a productive V-D-Jβ rearrangement, before a productive rearrangement of the γ and δ loci will take place. Because TCRδ is located within the TCRα locus, rearrangement of the TCRα will delete the δ genes [10]. All these events are regulated through temporal expression of recombination enzymes and molecules that stabilize newly formed polypeptides, allowing their selection [11].

The TCR antigen-binding site is formed by the association of the variable regions from each chain of the heterodimer. The binding of the TCR to its antigen does not induce signaling directly through the TCR moiety, because the TCR does not have a large cytoplasmic signaling domain. Instead, the TCR associates with a complex of four polypeptides called the CD3 complex [12, 13]. These polypeptides are named: CD3γ, δ, ε, and ζ. The stoichiometry of the TCR–CD3 complex, in general, includes a CD3δε heterodimer, a CD3γε heterodimer, a CD3ζζ homodimer, and the TCR heterodimer. This complex is assembled mostly in the endoplasmic reticulum and subsequently expressed as a trans-membrane receptor complex. Engagement by antigen triggers a signaling cascade initiated through modulations of signaling motifs located in the cytoplasmic regions of CD3γ, δ, ε, and ζ that leads to cellular activation (Figure 3-2) [14].

In contrast to the B-cell receptor, which is similar to an antibody, the TCR does not bind antigens directly. Instead it recognizes peptide fragments that are presented on the surface of cells, non-covalently associated with molecules of the MHC [15, 16]. MHC molecules are members of a polymorphic gene family that is widely distributed on most cells of the body and serve as the signature of immunological identity. Two main types of MHC molecules exist in humans and mice: MHC class I molecules, which are expressed on most cells of the body; and MHC class II molecules, expressed mostly on cells that are involved in the generation and modulation of immune responses like macrophages, dendritic cells, and B cells. Because individual TCR molecules recognize antigenic peptides associated with only one MHC type, T-cell recognition can be divided into receptors restricted by MHC class I and receptors restricted to MHC class II. This difference is highly significant because the type of interactions with MHC determines T-cell developmental and functional pathways. Critical to these recognition events is the expression of two co-receptors: the CD4 and CD8 molecules. These two molecules facilitate and enhance the binding of TCR to MHC class II and MHC class I respectively [17]. In addition, it has been reported that CD4 influences signal transduction initiated by TCR/CD3 complex engagement.

In summary, antigen recognition by T cells is not dependent on a single molecule, but in a multimolecular unit that is composed of the TCR, the CD3 complex, and the CD4 or

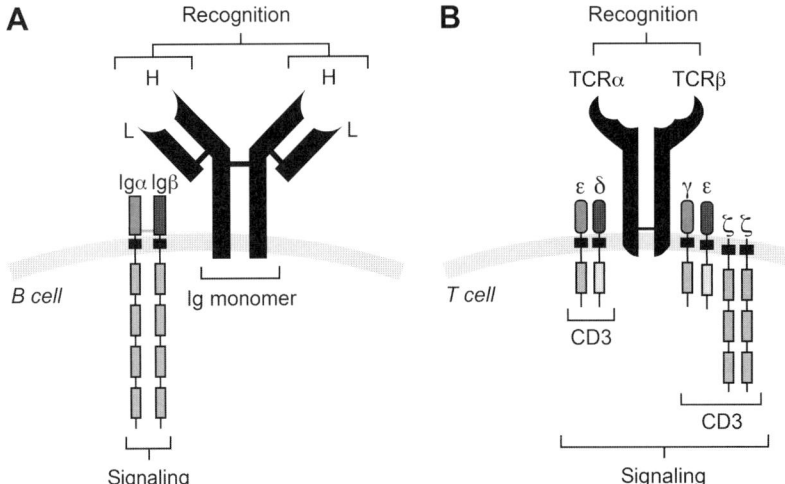

**FIGURE 3-2 Structures of the complete BCR and TCR complexes. (A)** The BCR complex is composed of an Ig monomer associated with an Igα and an Igβ chain. An Ig monomer is composed of two identical heavy (H) chains and two identical light (L) chains. These chains combine to form two identical antigen-binding sites that can reorganize unprocessed antigen. Igα and Igβ transduce intracellular signaling initiated by the binding of antigen to the Ig monomer. **(B)** The TCR is composed of a heterodimeric receptor protein and the CD3 complex. In a majority of T cells, the receptor protein is composed of a TCRα and a TCRβ chain; a minority of T cells bears a receptor protein composed of a TCRγ and a TCRδ chain (not shown). The CD3 complex transduces intracellular signaling initiated by the engagement of the receptor protein. Please refer to color plate section.

CD8 co-receptors. This complex not only defines recognition, but also determines the T-cell sub-types.

## T-cell functional heterogeneity

Based on the expression of CD4 and CD8 co-receptors, T cells can be divided into two major functional compartments: cytotoxic T cells (CTLs) expressing CD8, and T helper cells expressing CD4. In this early classification, CTLs are cells with the ability to kill cellular targets presenting antigens with self-MHC.

In contrast, T helper cells, upon antigen engagement, secrete cytokines, which modulate the activities of other cells in close proximity. In 1986 it was shown that T helper cells could be further divided according to the pattern of cytokines that they produced [18]. This division identified two types of cells: Th1 cells, secreting mostly interferon-γ and IL-2, and Th2 cells, secreting mostly IL-4, IL-13, and IL-5. This differential expression of cytokines by T-cell subsets has profound functional outcomes, which define the immune response to different types of antigens. Th1 responses are elicited mostly against intracellular bacteria and viruses, with activation of monocytic lineage cells like macrophages and dendritic cells. Th2 responses are elicited against parasitic infections with eosinophils and basophils being the effector cells. In addition Th1 and Th2 cells control B-cell responses by influencing the production and secretion of immunoglobulin isotypes.

Recently the identification of T helper cells that did not fit either of the cytokine patterns produced by Th1 or Th2 cells led to the identification of a third subset of T helper cells that secrete predominantly IL-17, IL-21, and IL-22. These cells, termed Th17, are located mostly in mucosal surfaces, are elicited in immune responses against extracellular bacteria and fungi and have the ability to stimulate neutrophils [19, 20].

Finally, regulatory T cells (Tregs) have the capacity to inhibit the activation of T cells and play an important role in controlling autoreactive cells in the periphery. Tregs are heterogeneous and are composed of cells that develop directly in the thymus and cells that acquire their regulatory characteristics in periphery [21].

## Thymopoiesis

T cells develop in the thymus from lymphoid progenitors that are generated in the bone marrow. In the thymus, under the influence of environmental signals, they initiate a coordinated developmental program, which includes: rearrangement of TCR genes, selection of a functional TCR, elimination of self-reactive TCR, acquisition of co-receptors and finally, migration to the periphery [22, 23].

The progenitors that initially seed the thymus are not yet committed to the T-cell lineage as evidenced by the expression of cell surface markers of early hematopoietic progenitors. They have not rearranged their TCR genes, and do not express CD3, CD4, or CD8. Based on the expression of these later molecules, they are called double-negative (CD4-CD8-), or triple-negative (CD4-CD8-CD3-) cells. Gradually, within the thymus, double-negative cells modulate the expression of cell surface markers, defining four sequential developmental stages called DN1, 2, 3, and 4. The initial progression is regulated through signals from the Notch pathway and the IL-7 receptor [24–27]. From DN1 to DN2, cells acquire the expression of CD25, CD24, and CD90, and gradually lose the expression of CD117 and CD44. DN2 defines the stage at which cells become fully committed to the T-cell lineage and is characterized by activation of the recombinase machinery and the initiation of TCR rearrangement at the TCR$\beta$, $\gamma$, and $\delta$ loci. At the DN3 stage, CD117 and CD44 are completely downregulated and rearrangement proceeds with high efficiency. At this stage, cells that have successfully rearranged their TCR$\gamma$ and TCR$\delta$ chains become $\gamma\delta$ T cells, while cells unable to productively rearrange their TCR genes die through apoptosis. Cells that rearrange their TCR$\beta$ chain move to the DN4 stage where TCR$\beta$ polypeptide associates with the CD3 complex and form a surrogate TCR$\alpha$ chain called preT$\alpha$ [28]. This pre-TCR complex localizes in the membrane and initiates a cascade of signaling events that are similar to those observed in mature T cells upon engagement of the TCR. However, this early signaling seems not to be ligand-dependent. Through this process, called $\beta$ selection, there is a rapid proliferation, coupled with extinction of TCR$\beta$ rearrangement and expression of both CD4 and CD8. At the final phase of $\beta$ selection the TCR$\alpha$ locus rearranges and the assembly of the TCR$\alpha\beta$ takes place. The resulting cells, which constitute the bulk of the thymocyte population, are called double-positive (DP) and presumably contain all the possible antigen specificities that can be generated by gene rearrangements.

The correct rearrangement and expression of the TCR complex on the surface of the cell ensures a wide diversity of antigen receptors. However, not all these receptors have the capacity to successfully bind antigens according to the requirements imposed by the nature of TCR recognition. Functional TCR should be able to recognize peptide antigens in the context of self-MHC molecules. Thus, useful specificities must be selected from the pool of TCR in the newly generated DP population. To fulfill this requirement, a sequential selection process occurs, which is dependent on the ability of the TCR to bind putative antigens. DP cells have a finite life span of approximately 3 days, unless they are rescued by signals associated with TCR engagement [29]. These signals are provided through the initial selection process, called positive selection. Positive selection operates through the interaction between newly developed DP T cells and thymic epithelial cells expressing self-MHC on their surface [30]. DP cells bearing TCR are able to bind self-MHC expressed on thymic stromal cells, survive, and mature. Apart from rescuing developing thymocytes from programmed cell death, positive selection also plays a role in setting the program to link the expression of CD4 and CD8 with the recognition properties of the T-cell receptor [31]. Thus, during positive selection cells expressing TCR that are able to bind MHC class I molecules express CD8, while thymocytes expressing TCR that binds MHC class II express CD4. The selection process is dependent on the ability of the cells to couple signaling pathways upon dual recognition of the MHC molecule by the TCR and the co-receptor.

After positive selection, the pool of thymocytes is limited to cells with the ability to bind self-MHC but it has not been further selected to eliminate those T cells that recognize self-antigens and are potentially dangerous to the host. For this reason, a second selection mechanism has evolved to eliminate autoreactive T cells. This process is called negative selection [32]. Developing thymocytes migrate to specialized areas rich in thymic epithelial cells, which present peptides that are derived from endogenous antigen, in the context of self-MHC. Through this interaction, thymocytes, expressing TCRs that bind self-peptides with high affinity are deleted, while T cells with TCRs that bind with low to medium affinities are spared. The molecular mechanisms responsible for this discrimination are not yet clear. The result of this process ensures that expressed TCRs have the potential to bind self-MHC most avidly only when associated with a foreign antigenic peptide [33]. This process is also referred to as central tolerance, and its failure can lead to the development of autoimmunity.

The whole progression of T-cell development and selection is not cell autonomous and for each of the stages and checkpoints active interactions between the developing thymocytes and the thymic microenvironment take place. In addition, temporal development also occurs in a strictly compartmentalized fashion. The thymus is a lobular organ in which each lobule is organized with a central medullar area and a peripheral cortical area surrounded by a subcapsular zone [34, 35]. Circulating bone marrow progenitors enter the thymus via postcapillary venules in the cortical–medullary junction. The cortex is the site where early progenitors commit to the T-cell lineage, while the subcapsular zone is the site where β selection occurs. From there cells move to the outer cortex where they actively proliferate as DP thymocytes. Positive selection occurs in the cortex. Once cells are positively selected, they enter the medulla where the process of negative selection takes place. The migration of cells within different areas of the thymus is controlled by chemo-attractant signals and the differential expression of adhesion molecules. This is regulated in conjunction with the expression of cytokines that guide the expansion and contraction of populations during selection. The most important cytokines in T lymphopoiesis are IL2, IL4, and IL7. These act on receptors, which share a common gamma chain and are referred to as common gamma chain cytokines [36].

Crucial to the progression of development and selection is the interaction between thymocytes and thymic stromal cells. These cells are known generically as thymic epithelial cells (TECs). Two types of TECs are present in the thymic environment, consistent with the compartmentalization of this organ: cortical TECs (cTECs) and medullary TECs (mTECs). There is strong evidence indicating that these two cell types share a common progenitor [37, 38]. However, they are distinct in terms of cell surface phenotype and functionality. cTECs are responsible for positive selection, while mTECs direct negative selection. One important feature of mTECs is that some subsets develop the ability to express multiple proteins that are not commonly expressed in the thymus [39]. This anomalous expression of peripheral tissue antigens is dependent on the transcription factor AIRE (autoimmune regulator) [40]. Interestingly, the T cells that have been recently selected can express RANK ligand that binds to its receptor expressed in a fraction of mTECs inducing the expression of AIRE [41–43].

Once TCRs are selected against auto reactivities they leave the thymus and become part of the pool of peripheral T cells [44]. Mature T cells that have not been exposed to antigen are referred to as naïve. When they leave the thymus they are maintained in the circulation with little proliferation and in adults their number remains quite constant. In the absence of antigenic stimulation naïve T cells have a finite lifespan and thymic output permits the maintenance of T-cell numbers and the preservation of the repertoire available for antigen recognition. The net output of cells from the thymus is not the sole mechanism to account for the maintenance of the peripheral T-cell pool. Active mechanisms are present that control the homeostasis of the naïve T-cell population. It was appreciated early that the constant number of T cells was dependent on space, thus newly arrived thymic emigrants compete for space with pre-existent cells [45, 46].

The two best-studied mechanisms responsible for maintenance of the T-cell pool are the engagement of the TCR by self-MHC and the responsiveness to cytokines, which confer survival without changing the naïve phenotype. These events also occur in the context of specialized microenvironments. Upon entering the circulation naïve T cells recirculate continuously through blood, lymph, and peripheral lymphoid organs. This trafficking is essential for the triggering of effective immune responses as it maximizes the chance of interaction with antigen-presenting cells. Also, this migration is required for the maintenance of the naïve T-cell repertoire.

As a consequence of positive selection in the thymus, T cells retain the ability to interact with self-MHC, and there is strong evidence showing that naïve T cells present a basal level of signaling through the TCR, presumably as a consequence of a continuous interaction with self-MHC. This interaction is weak and does not induce activation of naïve T cells. However, it is important for long-term naïve T-cell survival. One of the key aspects of the maintenance of T cells in a naïve state is the preservation of a functional repertoire, and self-MHC interactions likely contribute to this outcome. It is not entirely clear how this process is regulated.

Early studies on maintenance of T cells in culture indicated that soluble cytokines could be an important factor for the survival of naïve T cells in vivo. Common gamma cytokines like IL-2, IL-4, IL-7, and IL-15 have known activities that support the survival and expansion of different T cell populations [36, 47].

IL-7 is probably the predominant cytokine controlling lymphoid development and survival. IL-7 is produced by multiple stromal sources in primary and secondary lymphoid organs. It signals through a heterodimeric receptor formed by an exclusive IL-7R$\alpha$ subunit and a common $\gamma$ chain ($\gamma$c). The signaling properties of the receptor are mostly mediated by $\gamma$c, which engage two major transcriptional pathways, the Janus kinase (JAK1/JAK3), which signal through STAT5, and the phosphoinositine 3-kinase (PI3K) pathway. These two pathways intersect to promote survival by altering the balance between pro-apoptotic and pro-survival signals and by enhancing metabolic fitness in responding populations. In addition, through activation of defined transcriptional targets, these two pathways influence migration and proliferation. Recent studies have demonstrated that the engagement of self-MHC and response to IL-7 are coordinated through unknown mechanisms [48].

To receive antigenic signals, naive T cells traffic continuously from blood to secondary lymphoid organs. The entrance and retention of T cells in lymph nodes is controlled by the expression of CCL19 and CCL21. These are chemokines that are found in the luminal areas of high endothelial venules. The receptor for these chemokines, CCR7, is expressed on T cells and binding of its ligands controls the migration of T cells through the T-cell zone of secondary lymphoid organs. The T-cell zone consists of a dense network of fibroblastic reticular cells and professional antigen-presenting cells, which provide the right microenvironment for the T cells to sample multiple antigenic epitopes with high efficiency [49, 50]. It has been demonstrated that trafficking through secondary lymphoid organs is required for maintenance of the naïve T-cell pool. This suggests that self-MHC recognition and survival signals are delivered in the T-cell zones. Fibroblastic reticular cells and other stromal components can present MHC class I and antigen-presenting cells of hematopoietic origin can present both MHC class I and class II. Interestingly, fibroblastic reticular cells also express CCL19, CCL21, and IL-7 making them candidates for the delivery of both chemoattractant and survival signals. After being retained within the lymph nodes, T cells exit the nodes via the efferent lymph and recycle through the circulation. The signals responsible for migration out to the periphery are controlled by downregulation of CCR7 and expression of sphingosine 1-phosphate receptor 1 (S1PR1) on the surface of the T cells. The ligand for S1PR1 is the phospholipid, sphingosine 1-phosphate (S1P), which is present in blood and lymph. Thus, a balance between entrance and exit determines the retention rate of T cells within the lymph node and it is during this time that homeostatic interactions and signals occur, which maintain the naïve T-cell population.

## T-cell activation

During recirculation through secondary lymphoid organs naïve T cells sample the environment for antigens that are presented on the surface of cells expressing self-MHC molecules. These interactions determine whether naïve T cells will survive, will die through programmed cell death, or will activate with subsequent proliferation and development of effector functions. Which of these fates occurs depends on the cell presenting the antigen and on the strength of the interaction. A weak interaction with self-peptide and self-MHC complexes is essential for maintenance of the naïve T-cell repertoire, while a strong interaction with self-peptide and self-MHC complexes induces cell death. This latter outcome is an additional mechanism to eliminate self-reactive T cells that have escaped negative selection or central tolerance in the thymus and one of several mechanisms that are responsible for peripheral tolerance.

Antigen activation leading to a T-cell immune response occurs in the T-cell zone of peripheral lymphoid organs. This is where antigen-presenting cells accumulate after incorporating antigens from spleen or local sites of infection. Before migrating to lymph nodes APCs take up antigens, process them, load antigenic peptides on self-MHC molecules, and migrate to lymph nodes. During this progression, APCs mature under the influence of signals delivered by the inflammatory milieu that, in the case of infections, include the presence of pathogen-associated molecules. As a result, APCs upregulate chemokine receptors like CCR7 that senses its ligand CCL21 that is produced in lymph nodes. These signals also induce the expression of co-stimulatory molecules, which are critical for the initiation or priming of an immune response. Not all cells can serve as a professional APC, which is defined by the ability to efficiently present antigen for priming naïve T cells. There are a group of phagocytic cells derived from the monocytic lineage, which have the capacity to upregulate co-stimulatory molecules that are required for priming. The most important professional APCs are dendritic cells and to a lesser extent macrophages.

Recognition of foreign antigens by the TCR occurs during the passage of T cells through the lymph node. This is facilitated by transient interactions between APCs and T cells, a process stabilized by adhesion molecules expressed in the surface of the cells. The predominant adhesion molecules include LFA-1, ICAM-3, and CD2 on the surface of T cells and ICAM-1, ICAM-2, DC SIGN, and CD58 on the surface of the APCs. The relative importance of each of these interactions is not entirely understood, and they seem to have a redundant role in maximizing the chances for sampling of possible cognate antigens.

When a defined TCR binds an antigenic peptide/MHC complex, signaling through the TCR complex occurs. However, this signal alone is not sufficient to activate T cells and secondary signals must be delivered to initiate clonal proliferation. Co-stimulatory molecules expressed in the surface of T cells and mature APCs provide these secondary signals. The best-known pair of co-stimulatory molecules is CD80 and CD86 molecules on APCs. These bind the molecule CD28 that is highly expressed on the surface of T cells [51, 52]. The TCR engagement together with co-stimulation reorganizes the area of interaction in a so-called "immunological synapse". This clusters molecules involved in signaling and prologs the interaction between T cells and APCs [53]. Signaling through TCR engages three major signaling pathways: NFAT, AP-1, and NF-κB, in addition CD28 signaling involves the AP-1 and NF-κB pathways. In addition to inducing proliferation, this signaling cascade initiates the translation of IL-2, CD25, and the α chain of the IL-2 receptor. IL-2 is one of the first cytokines described to have the ability to enhance T-cell proliferation. It binds a receptor composed of three chains: α chain (CD25), β chain (CD122), and the common γ chain (CD135). In basal conditions T cells express the βγ dimer form of the receptor, which confers a low-affinity binding to the cytokine. The addition of CD25 generates a high-affinity receptor, which directs the active proliferation that is triggered during priming.

During the process of priming, several other molecules are expressed on the surface of the T cells, which are important modulators of co-stimulation and further signaling. One important molecule related to CD28 is CTLA-4 (CD152). This molecule binds CD80/CD86 with much higher affinity than CD28. Its primary activity is inhibiting the proliferation of activated cells by downregulating IL-2 production and signaling [54, 55]. There are also a series of TNF family molecules that have co-stimulatory properties in this process. Some of the best studied of these include CD40L/CD154 and 41BB/CD137 in T-cell binding, respectively, the ligands CD40 and 41BBL on APCs. Interactions among these molecules influence both the T cells and the activated APCs [56].

Once T cells are primed they develop functional capacities that allow them to eliminate their activating antigen. The acquisition of functions is determined at the priming stage, and is guided principally through the response to cytokines, which are provided initially by the APCs and later by cells in the antigenic microenvironment. This is a third signal after TCR engagement and co-stimulation in the response of T cells to antigen. Different antigenic stimuli trigger different micoenvironmental conditions that appear to direct APC to take up and process antigen. These differential pressures cause APCs to secrete cytokines that are delivered to naïve T cells during the process of priming. This third signal plays a fundamental role in the generation of CD4 cell types.

Intracellular bacteria, fungi, and viruses favor the development of Th1 CD4 T cells. This is initiated by secretion of IL-12 and IFN-γ by cells of the innate immune system (DCs, macrophages, and NK cells) during the priming process. These cytokines activate the JAK-STAT signaling pathway and through a predominant activation of STAT1, they induce the expression of the lineage-specific transcription factor T-bet. This turns on the IFN-γ gene in T cells and enhances their responsiveness to IL-12 [57–60]. These lineage-committed cells then exit the lymph node by inhibiting expression of adhesion molecules, such as L-selectin (CD62-L). This molecule retains cells in lymph nodes and its downregulation facilitates the emigration of effector T cells [61, 62].

At the site of infection Th1 cells will secrete large amounts of IFN-γ, which activates macrophages, upregulates MHC expression in multiple cell types, activates NK cells, and influences B cells to secrete certain immunoglobulin isotypes (i.e. IgG2a in mice). Apart from IFN-γ, other effector molecules are also produced by Th1 cells, including TNF-α, lymphotoxin, CD40 ligand, and fas ligand.

Th2 cells are elicited mostly in response to parasitic infections. In this case IL-4 is the polarizing cytokine driving the primed T cells towards the Th2 lineage [63, 64]. This engagement activates STAT4 which induces the expression of the lineage-specific transcription factor GATA-3 in the T cell. GATA-3 activates the expression of IL-4 and other Th2-related cytokines like IL-5 and IL-13 [65]. These cytokines have multiple actions at the priming stage and at the effector stage at sites of infection. IL-4 inhibits macrophage activation and stimulates growth of mast cells; at the B-cell level it upregulates the expression of MHC class II and is one of the main signals for the production of IgG1 and IgE antibodies [66]. IL-5 has macrophage-inhibitory properties and is involved in the differentiation and accumulation of eosinophils. At the B-cell level it influences the synthesis of IgA antibodies [67]. IL-13 shares many properties with IL-5 and has been associated with the induction of allergy and asthma [68].

Th1 and Th2 responses regulate each other as IFN-γ inhibits Th2 polarization and IL-4 does the same for Th1 polarization [69]. Most immune responses generate both effector sub-types, with biases towards one or the other. However, the final balance is important as deregulation or exacerbated predominance of one over the other can have profound pathological consequences.

A third CD4 population with the ability to secrete IL-17, but not IFN-γ or IL-4, has been recognized as a separate CD4 effector lineage called Th17. Its generation is dependent on the initial exposure of primed cells to TGF-β and IL-6 [70–72]. The presence of these polarizing cytokines during priming induces the expression of the transcription factor ROR-γt, through the activation of signaling pathways that include the activation of STAT3, IRF-1, and Runx-1 [73, 74]. The activation of ROR-γt stimulates the production of IL-17 in T cells, and also the expression of the receptor for IL-23. Once in periphery, IL-23 engagement induces the production of IL-17, IL-21, and IL-22, and at the same time suppressed IL-10 and IFN-γ production [75–79]. The outcome of the Th17 effector response includes induction of matrix destruction by recruitment of neutrophils, and induction of fibroblasts and epithelial cells to proliferate and to secrete various cytokines. Among the cytokines induced by IL-17 in parenchymal cells is RANK ligand, and recent studies have shown that within effector cells the Th17 cells are the predominant cells with this capacity. Hence, they are likely crucial players in the induction of the bone resorption that is associated to immune responses [80, 81].

Regulatory T cells include several types of cells, some of which are not yet well characterized. There is a central regulatory population that is generated in the thymus during negative selection. These cells are called natural Tregs and share several characteristics. They are CD4 positive, autoreactive, and express the IL-2Rα chain (CD25) and the transcription factor FoxP3 [82, 83]. In vivo they control autoimmune responses and can inhibit specific T-cell responses in vitro. Their mode of action is not completely elucidated but it has been proposed that they act through direct cell-to-cell interaction and the secretion of immunosuppressive cytokines like TGF-β and IL-10 [84]. Adaptive regulatory T cells, in contrast, develop extrathymically. It has been proposed that naïve T cells in a milieu rich in TGF-β and retinoic acid but without IL-6 are induced to activate the expression of FoxP3. These developmental stages are probably driven by APCs presenting commensal bacteria and food antigens rather than pathogenic antigens. The expression of FoxP3 signals the production of immunosuppressive cytokines.

Finally, CD8 T cells do not recognize the functional heterogeneity displayed in the CD4 compartment. Most naïve T cells will develop into cells that are able to destroy the cellular targets that are presenting antigens in the context of MHC class I. Interestingly, to generate fully functional cytotoxic cells, CD8 T cells need to be assisted by CD4 T cells, as direct interaction between naïve CD8 T cells with APCs during priming is not enough to activate them. This is accomplished by upregulation of CD80/CD86 in antigen-presenting cells, which is mediated by CD4 cells interacting with the same APC that bind the CD8 cells [85, 86]. This process also involves the expression of other co-stimulatory molecules and the secretion of IL-2 by the activated CD4 cells. As for CD4 effector cells, once they are in the periphery the delivery of their effector functions is not very dependent on co-stimulation. CD8 T cells mediate their killing through at least three mechanisms: granular exocytosis of cytotoxic molecules, fas-fas ligand-induced programmed cell death, and secretion of TNF-α [87, 88]. From these, granule-mediated cytotoxicity is the predominant mechanism. It involves the assembly of a cytotoxic granule-containing perforin, which is a molecule that is able to create pores in the surface of target cells, and various serine proteases, or granzymes, that can be internalized in the target cells to induce apoptotic responses [89]. After recognition of the infected cell, a tight binding complex initiated by TCR-antigen interaction and multiple adhesion molecules is established. Subsequently, active cytoskeletal processes inside the T cell polarize the granules, containing cytotoxic molecules, to the interphase between the cells where exocytosis liberates their contents delivering killing signals to the target cells [90].

The ultimate goal of an effective immune response is the elimination of the offending antigen. To accomplish this the immune system has evolved complex mechanisms to link specificity with effector functions. In the case of T cells this is accomplished through the generation of distinct functional lineages, defined by the nature of the antigenic stimuli. This inter-relationship coordinates the best combination of effectors necessary to eliminate the antigenic sources. In addition, the activation process has developed ways to generate populations of cells that can be recalled upon subsequent challenges.

## T-cell memory

The activation of naïve T cells induces the clonal proliferation of reactive specificities in conjunction with acquisition of effector functions that ultimately clear the inducing antigen. This response expands the population of specific cells 100–1000-fold and after the antigen is eliminated, the kinetics of the response is followed by a contraction phase in which the majority of the activated cells die, mostly through programmed cell death. However, this contraction does not reach basal levels and long-lived antigen-specific cells are retained with little or no effector capability. These cells, which form the memory T-cell compartment, are maintained in a resting status and upon recall they respond rapidly, expanding the population of specific cells with little co-stimulation [91, 92].

Several markers have been used to distinguish naïve, effector, and memory T cells. Tracking of immune responses has made it possible to define temporally the development of these cell types and these studies have shown that memory T cells are not homogeneous. Based on the differential expression of CD62L and CCR7 and the increased expression of IL-7Rα (CD127) two populations of memory T cells have been identified [93]. The "central memory cells", expressing both molecules, can recirculate from the blood to lymph nodes, and retain their ability to proliferate in secondary lymph nodes. The "effector memory T cells" lack the expression of these molecules, can be rapidly restimulated adopting effector functions, and can migrate to inflamed tissues with high efficiency.

How memory T cells are generated is a matter of current controversy [94]. Alternative models have been proposed in which memory T cells are required to go through the effector phase, generating in a sequential fashion effector memory cells followed by central memory T cells. Another model proposes that the imprinting for central memory T cells occurs during the priming phase and that such cells do not need to go through effector phase, while effector memory T cells develop from a pool of cells programmed to become effector cells. A third model proposes that effector memory T cells develop from central memory T cells generated from the priming stage. Experimental evidence for each of these models has been provided, suggesting that multiple pathways are probably responsible for the generation of the memory T-cell compartment.

The main characteristic of memory cells is that they need to be maintained in a slow replication status, and analogous to the maintenance of the naïve T-cell pool, they should do so through cues imposed by the microenvironment. Signals through the TCR and response to cytokines are believed to be involved. A requirement for the persistence of antigen has been explored and in certain conditions this seems to be involved. However, in the majority of situations, antigens are readily eliminated after the acute response. The role of basal signaling through MHC has also been explored, and it seems that this is a requirement for prolonged maintenance of effective memory recall in vivo [95, 96]. Regarding cytokines, two common gamma cytokines are fundamental for the maintenance of memory: IL-7 and IL-15 [97]. IL-7 is required for both CD4 and CD8 memory and IL-15 only for CD8.

As mentioned before, the maintenance of naïve T cells occurs in association with cells within secondary lymphoid organs. This finding indicates that their maintenance requires a specialized microenvironment that is able to deliver specific signals in an organized fashion. For memory maintenance, there is recent evidence that such a microenvironmental requirement also exists. Memory T-cell niches appear present in bone marrow where the existence of stromal components produces chemokines that attract memory T cells, express adhesion molecules for their retention, and synthesize IL-7 for their maintenance [98]. Because memory T cells need to have a self-renewing potential, they have been compared to stem cells that need a specialized niche for their perpetuation.

# B-cell development and function

## Introduction

B cells are responsible for the generation and production of antibodies or immunoglobulins in the body. Immunoglobulins can bind a wide range of molecules or antigens including native proteins and saccharides. The range of antigens that antibodies can bind is much greater than T-cell receptors as antibodies are not restricted to peptide binding, amino acid length, or binding within the context of MHC (see T-cell section). Each antibody contains two heavy chains and two light chains held together via disulfide bonds. The variable portions of a single heavy and single light chain each contribute to an antigen-binding domain resulting in an immunoglobulin molecule containing two antigen-binding domains. The other end of the heavy-chain proteins encodes the constant region, which can bind receptors on a variety of other immune cells and activate them. The ability of the variable portion of these proteins to bind a wide array of antigens with incredible specificity gives the adaptive immune system its unique power to fight infection and disease. Much of the diversity seen in the variable portions of immunoglobulins (and T-cell receptors) is generated prior to antigen encounter through a process of DNA rearrangement called V(D)J recombination. V(D)J recombination occurs during development in the bone marrow and is controlled so that a mature B-cell expresses an antigen receptor with unique specificity.

Soluble or secreted immunoglobulin or antibody is found in fluid throughout the body. Soluble immunoglobulins protect the host in a variety of ways: (1) Neutralization; whereby important surface molecules on pathogens are prevented from functioning properly by antibody blocking. (2) Opsonization; whereby pathogens are coated by antibodies, recognized via receptors for the immunoglobulin constant domain and "eaten" (phagocytosis) by innate immune cells such as macrophages. (3) Complement activation; whereby antibody binding to antigen can trigger the complement system, which kills some pathogens directly and also enhances opsonization.

The antigen receptor on B cells, known as the B-cell receptor (BCR), contains a membrane-bound immunoglobulin and the integral membrane proteins Igβ and Igα, which are critical for signaling and non-covalently associate with the surface-bound immunoglobulin. Igβ and Igα contain immunoreceptor tyrosine-based activation motifs (ITAMs) that upon receptor cross-linking are phosphorylated via Src kinases, initiating a signaling cascade through a variety of signaling molecules such as Syk, BLNK, and Btk resulting in B-cell activation, proliferation, and differentiation. The B-cell co-receptor, containing CD19, augments the antigen-dependent signal coming from the BCR. Interestingly, ITAM-containing adapter proteins DAP12 and the FcRγ chain play a critical role in osteoclast differentiation [99]. Mice deficient in both DAP12 and FcRγ ITAM-bearing adapters are osteopetrotic with a pronounced defect in osteoclast differentiation.

The mature B cell circulates through blood, lymph, and tissues and becomes activated upon binding to antigen via the B-cell receptor and stimulation by T cells within the secondary lymphoid organs such as spleen or lymph nodes. However, some B cells express B-cell receptors, which bind antigens that do not require T-cell stimulation. Activated B cells will proliferate in germinal centers within lymphoid organs and undergo affinity maturation and class switching. Affinity maturation is the process by which the genes encoding surface immunoglobulin are mutated and improved antigen specificity is selected. Class switching changes the constant region of the immunoglobulin and allows specific hematopoietic effector cells to be recruited and activated. B-cell activation initiates the differentiation program for memory B cells and plasma cells. Memory B cells are dormant and long-lived cells that become reactivated when its BCR is bound by its specific antigen. Memory B cells, as their name suggests, "remember" the previous infection and assist the immune system in responding faster and more potently to the new infection. Plasma cells are the terminally differentiated cells of the B-cell lineage. They reside predominantly within the bone marrow, produce enormous quantities of antibodies, and are the main source of soluble immunoglobulin within the body.

## B-cell development

As the role of the mature B cell centers around the use of the immunoglobulin molecule, B-cell development can be characterized by the generation of a functional B-cell receptor and the accompanying signaling apparatus that is required for the B cell to respond to antigen. B cells share a common lymphoid precursor with T cells, NK cells, and some dendritic cell subsets and primarily develop to functional maturity within the bone marrow in the adult and the liver in the fetus. This text will focus primarily on the adult development and on the standard B cell also known as the B-2 cell. Two other minor subsets of B cells, B-1 and marginal zone B cells, also contribute an innate-like protection via a rapid response. Marginal zone B cells are distinctly located within the spleen, while B1 cells develop primarily from the fetal liver and are found predominantly in peritoneal and pleural cavities [100].

The immunoglobulin gene loci in germline configuration contain component gene segments that will eventually encode for the variable portion of the immunoglobulin. The variable region DNA for the heavy-chain gene (IgH) is assembled from V, D, and J gene segments while the variable region DNA of the light-chain genes (Igκ and Igλ) is encoded by V and J gene segments. Each type of gene segment is most often found in multi-member clusters with each gene segment being very similar but not identical in DNA sequence. Each of these gene sequences is flanked by specific DNA sequences that allow the recombination activating gene proteins, RAG1 and RAG2, to bind and cleave the flanking DNA of two segments. The two cleaved gene segments are then joined together by non-homologous double-strand DNA break repair enzymes. In the joining phase of this reaction, DNA nucleotides can be added or deleted to the ends of the gene segments by TdT and endonucleases before the DNA ends are ligated back together [101].

There is a specific rearrangement order where IgH D to J recombination occurs first followed by V gene segment recombination to the already rearranged DJ gene. Once a functional IgH gene and subsequent protein is made, the light-chain genes can recombine. Igκ recombines first and if a functional Igκ protein is not generated then the Igλ gene locus will recombine. The light-chain loci can recombine repeatedly until a productive rearrangement is achieved, as there are only two different gene segments to join together. The chances of generating a productive immunoglobulin protein are increased as each of these gene loci has the opportunity to undergo gene rearrangement on either of two alleles. The remarkable diversity of the immunoglobulin repertoire allows B cells and antibodies to recognize most pathogens leading to their destruction and removal from the body. Multiple aspects of gene rearrangement contribute to this variability. (1) Combinatorial diversity is generated via the combinations of different types of gene segments. (2) Junctional diversity is generated through the addition and deletion of nucleotides at the joints between recombined gene segments. (3) Lastly, the multitude of combinations of heavy- and light-chain variable regions that pair to create antigen-binding domains adds to the diversity seen in immunoglobulins.

The process of V(D)J recombination is a highly regulated process and is controlled in a lineage-, temporal-, and allele-specific manner [101]. One level of regulation is control of RAG1 and RAG2 expression, which is limited to the lymphoid lineages and thus V(D)J rearrangement is similarly limited [102]. For B cells, high expression of the RAG proteins is restricted to early stages of development when the immunoglobulin genes are being rearranged. Once a functional BCR is generated, a feedback mechanism shuts off RAG expression. A second level of regulation is control of the accessibility of the immunoglobulin loci for the RAG proteins. These unrearranged loci are kept silent in other cell lineages except for a portion of the IgH locus that is "open" or accessible in non-B-cell lymphoid precursors. The chromatin of these loci are opened up through mechanisms linked with gene transcription such as DNA demethylation, histone acetylation, and location within active nuclear territories, which allow the RAG proteins access to their binding sites [101]. However, the precise mechanisms controlling this access are not yet known. There are many cis-acting elements, both promoters and enhancers, found throughout the immunoglobulin loci, often in close proximity to each gene segment, that control recombination to all or specific sections of each locus. Many of these elements are also used to control the expression of the recombined immunoglobulin gene in developing and mature B cells. Deletion of these control elements reduces or eliminates recombination of all or sections of individual loci as well as transcription of the fully rearranged immunoglobulin genes.

B-cell development generates a B cell that can respond when its unique and specific antigen receptor binds antigen. The field has split B-cell development into a variety of stages for ease of study, named A through F using cell surface markers and immunoglobulin rearrangement status [100]. The RAG1 and 2 gene expression and IgH locus accessibility begins in common lymphoid precursor (CLP) before lineage commitment, resulting in some D to J joints found in lymphoid cells other than B cells. After commitment to the lineage, a variety of essential transcription factors (see below) are expressed in fraction A and, often in synergy, induce the accessibility of the IgH locus as well as the expression of numerous genes necessary for B-cell function. These include TdT, Igβ, Igα, CD19, BLNK, and the surrogate light chains, λ5 and VpreB. Fractions A through C constitute the progenitor-B (pro-B) cell stage. IgH D to J recombination is completed in fraction B and V to DJ recombination occurs in fraction C whereupon the IgH protein is tested for competency in the context of the pre-B-cell receptor. The pre-B-cell receptor is similar to the B-cell receptor except that it lacks the immunoglobulin light chains, as they have not yet been rearranged at this stage of development. The surrogate light chains, VpreB and L5, are expressed in early B-cell development and substitute for the immunoglobulin light chains so that a successfully rearranged immunoglobulin heavy chain can be expressed on the cell membrane, acting as an important checkpoint between the pro-B-cell and pre-B-cell stage. If no productive IgH rearrangement occurs on either allele then the developing B cell dies, as is the case in mice lacking the RAG1 or RAG2 genes. Signaling from the pre-B-cell receptor induces: (1) Proliferation of the developing B cell harboring a successful IgH rearrangement. (2) Allelic exclusion, a process that silences the IgH allele not containing the successful IgH rearrangement insuring that the future B cell contains a BCR with a unique specificity. (3) Differentiation to the pre-B-cell stage where the immunoglobulin light-chain gene loci are accessible to the RAG proteins, which usually recombines the Igκ locus prior to the Igλ locus. If none of the Igκ or Igλ alleles productively rearrange to generate a competent light chain, the pre-B cell will die. The ability to repeatedly recombine each of four immunoglobulin light-chain alleles ensures fewer developing B cells die at this stage than at the pro-to-pre-B-cell transition.

Once a complete B-cell receptor is expressed on the surface of the pre-B cell, it is tested for reactivity with self-antigens. If the immature B cell does have strong autoreactivity, the RAG genes are continually expressed and the cell undergoes a process of receptor editing in which light-chain recombination continues. The old light-chain rearrangement may get deleted and replaced by another light chain. If the new resulting BCR is not autoreactive, the pre-B cell will mature. If the BCR remains strongly autoreactive, the developing B cell will undergo programmed cell death. B cells that bind weakly to self-antigens in the bone marrow will be rendered anergic, a condition in which the cell lives, but is completely inactivated. Unfortunately, some B cells are generated that are reactive to self-antigen because the antigen is not "seen" by the developing B cell in the bone marrow and thus the B cell is not deleted. These B cells have the capacity to initiate or play a role in autoimmune disease. Finally, if the immature B cell has no reactivity to self-antigens, RAG gene expression is turned off, the cell continues to differentiate and leaves the bone marrow to enter the bloodstream via the central sinus. B cells finish maturation in the spleen.

## Extrinsic requirements

The microenvironment of the bone marrow provides the signals necessary for the development of lymphocyte progenitors into precursor B cells. These external signals initiate and maintain a B-cell-specific gene expression program. Stromal cells express these extrinsic signals, which include cell adhesion molecules and soluble and membrane-bound cytokines and chemokines.

Hematopoietic stem cells (HSC) and developing B cells are found on the endosteal surface of bones in contact with adherent stromal cells. As the developing B cell matures, it moves towards the center of the marrow space near the central sinus. Stromal cells isolated from the bone marrow are mesenchymal in origin and have the potential to differentiate into mature osteoblasts. HSC and B-cell differentiation is supported by these osteoblast-lineage cells in niches on the endosteal surface [1–3]. While these stromal cells do not look precisely like osteoblasts, which are defined as mature cuboidal cells, on bone surfaces that actively secrete osteoid, the phenotype of these stromal cells is consistent with osteoblast-lineage cells and needs to be more fully characterized [103]. Stromal cells that are often required for in vitro B-cell development culture systems are osteoblast-lineage cells.

The cell surface tyrosine kinase receptors c-kit and Flt3 are expressed on developing common lymphoid precursors (CLPs) and bind to the membrane-bound stem cell factor (SCF) and Flt3 ligand (Flt3L) respectively, which are expressed on stromal cells. Both these receptors and their ligands are important for the development of CLPs and subsequent B-cell development as mice deficient in c-kit or Flt3L are defective in these cell types [104, 105]. Stromal-cell-derived factor 1 (SDF-1) or CXCL12 is also expressed by stromal cells and plays a role in early B-cell development. Production of B-cell precursors from progenitors in vitro requires contact with osteoblasts and expression of CXCL12 (SDF-1) and IL-7, which is induced by parathyroid hormone (PTH) [1, 3]. Selective elimination of osteoblasts by treatment of Col2.3d-TK transgenic mice with gancyclovir also severely depleted pre-pro-B cells from the BM confirming the supportive role of osteoblasts in B-cell development [2]. Pro-B cells express α4β1 integrin, which binds to VCAM-1, expressed by these stromal cells tethering the two cell-types together.

It is now known that signaling through the PTH/PTH-related peptide receptor (PPR) in osteoblastic cells increases trabecular bone and importantly increases HSCs [1]. PTH is known to increase production of CXCL12 and IL-7 by osteoblastic cells in vitro suggesting that downstream signaling through the PPR regulates B-cell development [1, 3]. Mice made deficient in PTH signaling specifically in osteoblasts by ablation of the G protein α subunit $G_s\alpha$ had a striking decrease in trabecular bone and an almost 50% reduction in BM B cells, while other hematopoietic lineages were unaffected [106]. In addition, IL-7 expression was reduced in $G_s\alpha$-deficient osteoblasts, confirming the importance of osteoblast-lineage cells in B-cell growth and differentiation.

## IL-7

The cytokine interleukin seven (IL-7) and signaling via its receptor are required for normal adult progenitor B-cell development. IL-7 is apparently produced by stromal cells and osteoblasts and the IL-7 receptor (IL-7R), composed of the unique IL-7Rα chain and the common γ chain, which is shared by many interleukin receptors, is expressed on developing lymphocytes and is important for their expansion and differentiation before and after lineage commitment [3, 106, 107]. The requirement for IL-7 in in vitro cultures of progenitor B cells illustrates the importance of this pathway in B-cell development. Deletion of IL-7 or IL-7Rα blocks B-cell development with a stronger defect observed in IL-7Rα$^{-/-}$ mice [108, 109]. However, both IL-7 and IL-7Rα-deficient mice possess readily detectable numbers of peripheral B cells, indicating that the block in B lymphopoiesis is not absolute in these animals. These remaining B cells are thought to develop during fetal and neonatal periods. The cytokines, thymic stromal-derived lymphopoietin (TSLP), and Flt3 ligand in addition to IL-7 contribute to in vivo fetal and neonatal B-cell development. TSLP and IL-7 both use IL-7Rα as part of their receptors and subsequently, mice deficient in both IL-7Rα and Flk-2/Flt3 lack B cells completely [110, 111].

Studies have demonstrated that IL-7 also plays an important role in the regulation of bone homeostasis [112, 113]. As stated above, production of IL-7 by osteoblast-lineage cells appears critical for normal B lymphopoiesis [106]. However, the precise nature of how IL-7 affects osteoclasts and osteoblasts is controversial, because it has a variety of actions in different target cells. Systemic administration of IL-7 increased osteoclast formation from human peripheral blood cells by increasing osteoclastogenic cytokine production in T cells [114]. Furthermore, mice with global over-expression of IL-7 had a phenotype of decreased bone mass, with increased osteoclasts and no change in osteoblasts [115]. However, the interpretation of results from in vivo IL-7 treatment studies is complicated by secondary effects of IL-7, which result from the production of bone-resorbing cytokines by T cells in response to activation by this cytokine [114, 116, 117]. Consistent with this conclusion, IL-7 administration did not induce bone resorption or bone loss in T-cell-deficient nude mice [116].

In contrast with previously reported studies, it was found that IL-7 had differential effects on osteoclastogenesis [112, 114, 116, 118]. IL-7 inhibited osteoclast formation in murine bone marrow cells that were cultured for 5 days with M-CSF and RANKL [118]. It was also found that IL-7-deficient mice had markedly increased osteoclast number and decreased trabecular bone mass compared to wild-type controls [119].

Treatment of newborn murine calvaria cultures with IL-7 inhibited bone formation, as did injection of IL-7 above the calvaria of mice in vivo [113]. However, when IL-7 was transgenically over-expressed in osteoblasts, trabecular bone mass was increased compared with wild-type mice [120]. Furthermore, targeted expression of IL-7 in the osteoblasts of IL-7-deficient mice rescued the osteopenic bone phenotype of the IL-7-deficient mice [121].

## Intrinsic requirements

Multiple transcription factors are essential for B-cell development and function [122]. Analysis of mice deficient for these factors results in an order of requirement for these factors in B-cell commitment and differentiation: (1) PU.1, (2) Ikaros, (3) E2A, (4) Ebf1, and (5) Pax5 [123–128]. These factors may play additional roles in B-cell biology beyond the block found in deficient animals. However, these roles may go unseen if the block is prior to the additional function. Lineage and maturation-specific deletion of conditional gene alleles has allowed and will continue to allow the definition of these potential additional roles.

PU.1, a member of the Ets family of transcription factors, is expressed in the myeloid lineages, NK cells, B cells, in early developing thymocytes and the TH2 subset of CD4 T cells. Analysis of PU.1-deficient mice demonstrated this protein to be an important regulator of the development of multiple hematopoietic lineages. $PU.1^{-/-}$ mice, in addition to having no B cells, fail to develop osteoclasts and macrophages [129, 130] (Figure 3-3). Due to a lack of osteoclasts their bones are severely osteopetrotic. These data were central in showing that osteoclast differentiation required PU.1 expression early in lineage development and osteoclasts and macrophages were members of the myeloid lineage. Control of these lineages is dependent on the levels of PU.1 expression; low concentrations of PU.1 protein induce B-cell differentiation while high concentrations inhibit B-cell development and promote macrophage differentiation [131]. The regulation of both myeloid and lymphoid lineages by PU.1 places it upstream of the other factors discussed here. PU.1 regulates these progenitors by controlling the expression of c-fms (M-CSF receptor), an essential differentiation and growth factor for macrophages and osteoclasts, and the IL-7Rα and Flk-2/Flt3 receptor genes, which are important for the generation of CLPs as well as pro-B cells [132, 133]. While PU.1 is critical for B-cell lineage determination, PU.1 is not required for B-cell differentiation after lineage commitment as conditional deletion of PU.1 in committed B cells allows for fairly normal B-cell maturation [134]. PU.1 also controls the expression of a variety of important hematopoietic lineage genes like the signaling phosphatase, CD45, the integrin, CD11b, and the GM-CSF receptor as well as BTK and Ebf (see below) [122, 132–134].

Ikaros is a member of a Kruppel-like zinc finger family of transcription factors that also includes Helios and Aiolos [135, 136]. The Ikaros gene (*Ikzf1*) is widely expressed in hematopoietic lineage cells including self-renewing HSCs and MPPs [137]. Mice deficient in *Ikzf1* have substantially less HSC activity and their MPPs are inhibited from

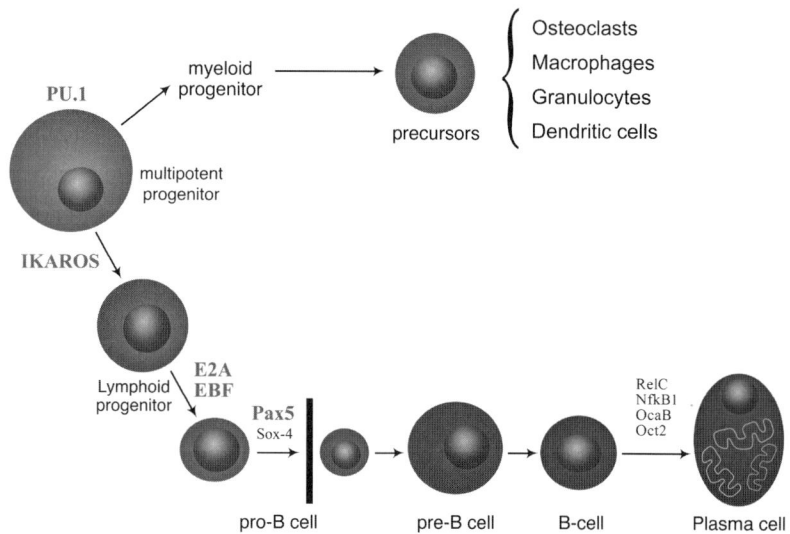

**FIGURE 3-3 Transcriptional regulation of B-cell differentiation.** B-cell differentiation is regulated, in part, by the expression of a series of transcription factors that function in a temporal manner. These transcription factors include PU.1, Ikaros, E2A, Ebf1, and Pax5. Loss of these specific factors precludes the cells from continued maturation, and results in a developmental block of cells at the latest stage of differentiation prior to the arrest. In addition to the absence of B cells, mice deficient in PU.1, Ebf1, and Pax5 have profound changes to their skeletons. No data are available on the bone phenotype in mice deficient in Ikaros or E2A. Please refer to color plate section.

differentiating into the CLP. This results in a complete failure of the B-cell lineage to develop [138, 139]. $Ikzf1^{-/-}$ mice have altered T-cell development in the thymus and functional changes in peripheral T cells [140]. Ikaros also regulates cytokine gene expression [140]. Ikaros-deficient mice have increased myelopoiesis with increased numbers of Mac-1$^+$ cells in the BM and spleen [139]. Ikaros activates or represses gene expression through the recruitment of chromatin remodeling complexes. Hematopoietic progenitors from Ikaros$^{-/-}$ mice are deficient in the expression of Flk2/Flt3 and c-kit, accounting for the loss of early lymphoid differentiation in these mice [122, 134]. Recent evidence suggest that Ikaros plays an essential role in B-cell lineage commitment as well as control of V(D)J recombination [141]. At present, no data are available on the status of the skeleton of $Ikzf1^{-/-}$ mice.

The *E2A* gene is alternatively spliced to encode two, ubiquitously expressed, basic helix-loop-helix proteins, E47 and E12. These proteins, along with HEB and E2-2, are members of a family of transcription factors that bind a DNA motif known as the E-box and play important roles in lymphocyte development [142]. E2A proteins are expressed in HSCs and in different subsets of hematopoietic progenitors. As an example, BM from E2A$^{-/-}$ mice had reduced numbers of long-term HSCs, granulocyte-macrophage progenitors and severe depletion of erythroid progenitors [143]. Although these data suggest E2A$^{-/-}$ mice may have a bone phenotype they remain unexamined. E2A-deficient mice exhibit defects in T-cell development and a complete block very early in

B-cell development, before V(D)J recombination occurs [123, 128]. E2A interacts with the histone acetyltransferases p300/CBP as well as the SAGA histone acetyltransferase complex and may activate transcription by recruiting these remodeling complexes to target sequences. The Id family of proteins antagonizes the function of E2A gene products and other E box family members by binding and inhibiting their ability to bind DNA. Endogenous expression of individual Id proteins prior to lineage commitment or exogenous over-expression can perturb T-cell, B-cell, and NK-cell development and function. Thus, the presence and absence of the Id proteins add another layer of regulation of E2A function. E2A regulates V(D)J recombination by controlling the expression of RAG1 and RAG2 [3]. Additionally, exogenous expression of E2A in the presence of the RAG proteins resulted in V(D)J recombination in non-B-lineage cells, presumably by enhancing gene segment accessibility via E2A-binding sites found in cis-acting elements of the immunoglobulin gene loci that are important for V(D)J recombination [101, 122]. E2A controls expression of TdT and Ebf (see below) as well as essential members of the pre-BCR, λ5, VpreB, and BCR, Igα (mb-1) [134, 144]. At present, no data are available on the status of the skeleton E2A$^{-/-}$ mice.

Ebf1 (O/E-1) is the founding member of the COE (Collier-Olf-EBF) family of helix-loop-helix transcription factors that are evolutionarily conserved from drosophila to human [145]. In addition to Ebf1, mice express at least three additional members of this family, Ebf2 (mMot1/O/E-3), Ebf3 (O/E2), and O/E-4. All known members of this family have been detected in neuronal tissue. Importantly for this volume, Ebf1, Ebf2, and Ebf3 are expressed in adipocytes. We have shown that it is expressed in osteoblast-lineage cells including mesenchymal stem cells and plays important roles in controlling these lineages (see below for further discussion) [125, 146–149]. Ebf1 is expressed in B cells and is essential for B-cell development as mice deficient in Ebf1 contain only early progenitor B cells and lack mature B cells [108, 150]. The low levels of D-J rearrangement and lack of Pax5 expression in Ebf1 deficient mice suggest Ebf1 lies upstream of Pax5 (see below). The role of Ebf1 is to both promote commitment and restrict lymphopoiesis to the B-cell lineage as forced expression of Ebf1 in hematopoietic progenitors resulted in a dramatic increase in B-lineage cells at the expense of other lineages [151]. Similar to E2A, Ebf1 has the capacity to induce V(D)J recombination when expressed ectopically with the RAG genes in non-B-lineage cells. Ebf controls the expression of Pax5 (see below), the surrogate light chains, λ5 and VpreB, and the BCR signaling adaptors, Igα and Igβ [122, 134, 144].

We have reported that $Ebf1^{-/-}$ mice, in stark contrast to $Pax5^{-/-}$ mice, had increases in all bone-formation parameters measured, with marked increases in osteoblast numbers, osteoid volume, and the bone formation rate [152]. In contrast, the number of osteoclasts was similar to controls in young mice but more than doubled in older mice. Mice lacking Ebf1 were found to have substantial increases in bone marrow adipocyte formation, accompanied by a significant loss of both subcutaneous and visceral white adipose tissue (WAT) deposition. Brown adipose tissue (BAT), however, was more modestly affected [153]. The $Ebf1^{-/-}$ mice were mildly hypoglycemic and hypotriglyceridemic. Metabolically, the $Ebf1^{-/-}$ mice had a higher metabolic rate, consumed more food and water, and expended more energy. The circulating levels of insulin were the same as their wild-type (WT) littermates. However, circulating glucagon was elevated in the Ebf1$^{-/-}$ mice, while at the same time markers of gluconeogenesis were decreased. These data demonstrate for the first time that Ebf1 is required for maintaining the regulation of both osteogenesis and adipogenesis and it affects metabolism.

Pax5 is a member of a large family of transcription factors that bind DNA through a characteristic paired domain. Pax5 is expressed in developing and mature B cells but not in plasma cells and other hematopoietic lineages. In Pax5-deficient mice, bone marrow B-cell development is blocked at an early progenitor B-cell stage, containing normal levels of D-to-J$_H$ rearrangements [31]. Pax5 is required for rearrangements to specific V genes and expression of target genes such as Igα, BLNK, and CD19 [122, 126, 154]. Conversely, Pax5 also represses non-B-cell genes such as c-fms (M-CSFR), Flt3, and Notch1 [122, 133, 155, 156]. Pax5 is required for both B-cell-lineage commitment and maintenance. Strikingly, Pax5-deficient pro-B cells express B-cell-specific genes and markers, but can differentiate into many other hematopoietic lineages if cultured under appropriate conditions [157]. Thus, Pax5 activates B-cell gene programs, but importantly, also represses gene expression that is required for differentiation of other lineages.

Loss of Pax5 results in an unanticipated massive decrease in trabecular bone in both the tibia and femur of 15-day-old mice [152]. Bone volume (tibia) was reduced by 67% and osteoid volume was reduced by 55%. This was found to be the result of increases in bone resorption and can be accounted for by a >100% increase in the number of osteoclasts in Pax5$^{-/-}$ bone. These data not only indicate a marked increase in the number of osteoclasts but also suggest that they are functional. The number of osteoblasts in the mutant mice was reduced, although not significantly, to that of controls. This implies that the osteopenia was due, in large part, to the increase in osteoclasts. Thus, Pax5$^{-/-}$ mice have a bone phenotype distinct from PU.1 and Ebf1-deficient mice.

There are positive feedback loops for many of these mentioned factors that add additional layers of regulation of B-cell development. IL-7 signaling through Stat5, E2A, and PU.1 contributes to Ebf1 expression. Additionally, Ebf1 can activate its own expression as well as that of Pax5, which in turn has been shown to activate Ebf1 expression. Thus, there is a reciprocal relationship that serves to mutually reinforce Ebf1 and Pax5 expression and the resulting B-cell differentiation program.

There are several other newly characterized transcription factors worthy of mention that regulate B-cell development: Bcl11α is a Kruppel-related zinc-finger protein expressed in hematopoietic progenitors as well as multiple lineages. B-cell development is blocked in Bcl11α-deficient mice [158]. Mice deficient in FOXp1 or Foxo1, forkhead transcription factors, have been shown to have severe blocks in B-cell development, which is likely due in part to defects in RAG1 and 2 (both transcription factors) and IL-7Rα gene expression (Foxo1) [102, 159]. IRF4 and IRF8, members of the interferon regulatory factor family, control the ability of developing B cells to exit the cell cycle within the pro-B- to pre-B-cell transition [122].

## The RANK–RANKL pathway and B-cell differentiation

It is well accepted that the interaction between RANKL and its cognate receptor RANK is required for the differentiation of osteoclasts from precursors. However, what is often overlooked is the role the genes that encode these proteins and their downstream effector genes play in regulating B-cell development. RANKL-deficient mice have severe osteopetrosis due to the inability of osteoblasts to support osteoclast differentiation. These mice are also one of the few mutants that lack all lymph nodes. The number of splenic IgM$^+$sIgD$^+$ or CD45R/B220$^+$sIgM$^+$ B cells was significantly reduced in RANKL$^{-/-}$ mice as compared to controls [160]. To examine B-cell differentiation chimeric mice were made by injecting fetal liver cells from RANKL$^{-/-}$ or WT mice into sublethally irradiated Rag$^{-/-}$ mice [160]. The BM from Rag$^{-/-}$ mice reconstituted with RANKL$^{-/-}$ cells had normal numbers of CD45R/B220$^+$CD43$^+$ and CD45R/B220$^+$CD25$^-$ pro-B cells but significantly reduced numbers of CD45R/B220$^+$CD43$^-$, CD45R/B220$^+$CD25$^+$ and CD45R/B220$^+$sIgM$^+$ B cells, indicating a conspicuous block in the progression from pro-B cells to pre-B cells in the chimeric mice. Therefore, even in the RANKL sufficient environment of the Rag$^{-/-}$ mice, the RANKL$^{-/-}$ cell had a defect in B lymphopoiesis [160].

The question then arises; do other members of the RANK-RANKL pathway affect B-cell differentiation or activation? RANK-deficient mice are severely osteopetrotic due to a failure of osteoclast precursors to differentiate to mature osteoclasts. The number of cells in the BM of RANK$^{-/-}$ mice is markedly reduced with few hematopoietic colonies due to osteopetrosis and the failure of osteoclasts to develop a bone marrow cavity [161]. As hematopoiesis relocates, the spleens of RANK$^{-/-}$ mice are approximately twice the size of controls but have a greater than 50% reduction in CD45R/B220$^+$ B cells [161]. Mature, CD45R/B220$^+$IgM$^+$ splenic B cells were also reduced by 50% as compared to controls. Like the RANKL-deficient mice RANK-deficient mice lack all lymph nodes [162].

Following RANKL–RANK interaction intracellular signal transduction is mediated, in part, by TNF receptor-associated factor 6 (TRAF6). TRAF6$^{-/-}$ mice are osteopetrotic [161]. Although TRAF6$^{-/-}$ mice have similar numbers of osteoclasts compared to controls, they fail to resorb bone because they do not develop an attachment zone or ruffled border. Interestingly, splenic B cells from TRAF6$^{-/-}$ mice fail to proliferate in response to anti-CD40 or lipopolysaccharide (LPS) stimulation, suggesting a role for TRAF6 in B-cell proliferation to specific signals.

Osteoprotegerin (OPG) is a naturally circulating decoy protein receptor that binds RANKL, reducing RANKL's availability to bind RANK, thus regulating osteoclast differentiation. OPG-deficient mice develop severe osteoporosis with associated fractures due to increased numbers of osteoclasts. OPG is a CD40-regulated gene in B cells and dendritic cells. To analyze the effects of OPG deficiency on B-cell development, pro-B cells (CD45R/B220$^+$CD43$^+$) from OPG$^{-/-}$ mice were sorted by FACS from BM and cultured with IL-7. Pro-B cells from OPG$^{-/-}$ mice proliferated two-fold more after IL-7 stimulation than did WT cells [163]. FACS analysis revealed that CD45R/B220$^+$CD43$^+$IgM$^-$ or CD45R/

B220$^+$CD25$^-$IgM$^-$ pro-B cells were increased in the BM of OPG$^{-/-}$ mice as compared to WT controls. In addition, the number of CD45R/B220$^+$CD19$^+$ B cells in the spleen and lymph nodes was greater in OPG$^{-/-}$ mice than in controls.

These data show that the loss of OPG increases pro- and mature B cells, while the loss of RANKL or RANK decreases these populations of cells. In the RANKL$^{-/-}$ mice B-cell differentiation is arrested between the pro- and pre-B-cell stage of differentiation. The stage of differentiation arrest in the RANK$^{-/-}$ mice is unknown.

## B-cell function

B-cell activation is initiated through binding of antigen by the B-cell receptor. This activation signal is greatly amplified by the B-cell co-receptor composed of CD19, CD21, and CD81 when antigen is complexed with components of the complement system. However, additional activation signals are needed which are usually provided by T cells. The antigen is internalized after binding by the BCR and processed into peptides and bound onto MHC class II molecules within intracellular endocytic compartments. On the surface of the B cell, these peptide:MHC complexes are recognized by an activated helper T cell which upregulates CD40 ligand and secretes cytokines. Ligation of CD40 expressed on the B cell and activation of the B cell via cytokines such as IL-4, IL-5, and IL-6 induces B-cell proliferation and differentiation.

In some circumstances, B-cell activation can occur by T-cell-independent (TI) antigens that have separated into two categories: TI-1 and TI-2. TI-1 antigens are polyclonal B-cell activators at high concentrations and are often Toll-like ligands. TI-2 antigens contain repetitive features that can engage multiple BCRs. The second signal for activation can be delivered by the antigen itself, such as LPS binding to Toll-like receptors or by dendritic cells and macrophages.

Antigen binding by the B cell triggers the activation of adhesion molecules like LFA-1 and traps the B cell within secondary lymphoid tissues where B cells encounter helper T cells. These activated B cells, along with their cognate T cells, migrate to form a germinal center where the B cells undergo clonal expansion and differentiation. B-1 cells and marginal zone B cells are thought to provide a rapid antibody response that allows for protection from a pathogen while the main germinal center reaction generates antibodies with improved specificities.

Affinity maturation occurs within the germinal center through a process called somatic hypermutation. Point mutations are introduced and focused on the DNA encoding the variable region but not the constant region of the B-cell receptor. The generation of these point mutations is dependent on the DNA-editing enzyme, activation-induced cytidine deaminase (AID), and accumulate as an individual B-cell proliferates. These mutations can affect the ability of the BCR to bind antigen. Most mutations are deleterious for antigen affinity while some improve the affinity. The B cells that contain these improved BCRs are selected for and expanded. This cycle of mutation and selection continues over and over.

The constant region domains can also be altered at the DNA level through class switching, which is also dependent on AID. The BCR on a circulating B cell is initially expressed as an IgM or IgD molecule without the need for class switching. Stimuli such as CD40L, IL-4, IL-5, LPS, TGF-β, and IFNγ can induce specific switching of the constant domain to IgG (four types in mouse), IgE, or IgA allowing for the activation variety of innate immune cells such as macrophages, NK cells, mast cells, neutrophils, and eosinophils, depending on the specific constant region.

A subset of these activated B cells from the germinal center will become memory cells, ready to respond faster and better than a naïve B cell the next time an individual encounters the same or a similar pathogen. Other activated B cells will terminally differentiate into plasma cells, which are unable to respond to T-cell help or antigen but possess specific characteristics: (1) reduced surface immunoglobulin expression; (2) lack of MHC class II expression; (3) the inability to undergo class switching or somatic hypermutation; (4) expanded endoplasmic reticulum reflecting; (5) secretion of vast amounts of immunoglobulin.

These changes are reflected in the gene expression program of plasma cells compared to that of mature and activated B cells and is controlled by the switch of transcription factors that control plasma cells compared to mature B cells [164, 165]. As discussed earlier, Pax5 is required for B-cell-lineage commitment as well as the maintenance of B-cell identity. However, reduction of Pax5 expression is required for plasma cell differentiation while the expression of an important plasma cell transcription factor, B-lymphocyte-induced maturation protein 1 (BLIMP1), is upregulated. Pax5, along with BCL6 and MITF, are important for maintaining the mature B-cell phenotype and repressing many genes important for plasma cell function [155, 165]. Conversely, BLIMP1, as well as two other transcription factors, XBP1 and IRF4, are necessary for plasma cell gene expression and the repression of mature B-cell genes. Additionally, these transcription factors regulate each other. Pax5 positively regulates expression of such genes as Igα, CD19, BLNK, and AID expression and represses expression of XBP1 and BLIMP1 and reduces expression of IgH and IgL. Conversely, BLIMP1 represses the expression of Pax5 and CTIIA, the activator of MHCII, resulting in the downregulation of proteins such as BCR and MHCII on plasma cells [164, 165]. BLIMP1 also upregulates plasma-cell-specific genes such as XBP1, a transcription factor essential for plasma cell immunoglobulin secretion. Thus, these different sets of transcription factors control mutually exclusive expression patterns for mature B cells and plasma cells.

Upon differentiation plasma cells reside in inflamed and lymphoid tissues, but most home to survival niches within the bone marrow, which are comprised of stromal cells. These niches provide signals that allow the long-term survival of plasma cells (IgG$^+$CD138$^+$) [164–167]. Plasma cells express CXCR4, which is important for homing to the stromal cells that secrete its ligand CXCL12. Plasma cells express α4β1 integrin, which binds to VCAM-1 expressed by these stromal cells keeping these two cells in close contact. IL-6 is produced by stromal cells upon plasma cell contact. IL-6, B-cell maturation antigen (BCMA, expressed on plasma cells) and the BCMA ligands, BAFF and APRIL, are crucial for plasma cell survival. It is interesting to note that BM stromal cells, which are necessary for progenitor B-cell differentiation and plasma-cell survival, are highly similar if not the same and that a subset of required signals for the beginning and end of the B-cell differentiation pathway are shared [168]. The precise role and nature of these osteoblast-related stromal cells and a possible role of osteoblasts and osteoclasts in B-cell development and plasma-cell survival is an area of future study.

## Activated B cells

Activation of B cells is involved in the development of inflammatory arthritis as well as periodontal disease [169, 170]. Activated B cells can produce RANKL as well as other cytokines that are involved in bone resorption and bone formation [169, 171] (Figure 3-4). In addition, a subpopulation of memory B cells also expresses RANKL [172]. A variety of B-cell-related malignancies produce factors that regulate bone cells. Multiple myeloma is a malignancy of plasma cells, which derive from B cells. It often causes enhanced bone resorption and decreased bone formation [173]. Derangements of the RANK/RANKL/OPG system are frequent in this condition and may be due to production of factors by the tumor that stimulates RANKL production in the bone microenvironment or because of direct production of RANKL by the myeloma cells [174, 175]. In addition, a variety of additional factors including MIP-1α and DKK1 are produced by myeloma cells and directly affect bone cell function [173]. B-cell lymphomas can also produce RANKL and through this mechanism cause bone loss and hypercalcemia [176].

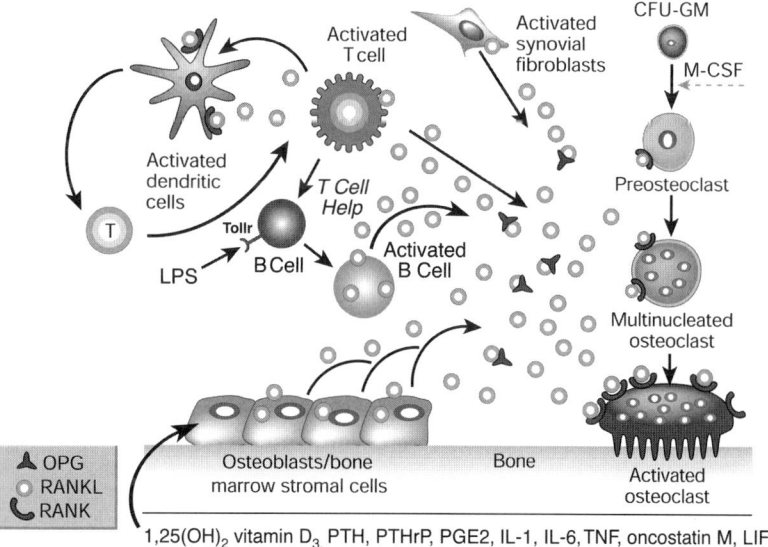

**FIGURE 3-4 Activated B cells induce osteoclast differentiation.** B cells activated by the adaptive immune system (antigen specific) or through the innate immune system (LPS-Toll receptor) result in B cells that secrete or express on their cell surface molecules like RANKL that induce osteoclastogenesis. *(Adapted from Boyle WJ, Simonet WS, Lacey DL. Nature 2003;**423**:337–942).* Please refer to color plate section.

## Acknowledgments

Writing and editing of this review and experimental work was enabled by support from the National Institute of Arthritis and Musculoskeletal and Skin Diseases/National Institutes of Health grants RO1AR047342, RO1AR049190, RO1AR052690 to MCH; the Yale Core Center for Musculoskeletal Disorders P30AR046032; RO1AR048714 and RO1AR052690 to JAL, and the Department of Orthopaedics and Rehabilitation, Yale University School of Medicine, New Haven, CT. David G. T. Hesslein is a Special Fellow of the Leukemia and Lymphoma Society.

## References

[1] Calvi LM, Adams GB, Weibrecht KW, et al. Osteoblastic cells regulate the haematopoietic stem cell niche. Nature 2003;425:841–6.

[2] Visnjic D, Kalajzic Z, Rowe DW, Katavic V, Lorenzo J, Aquila HL. Hematopoiesis is severely altered in mice with an induced osteoblast deficiency. Blood 2004;103:3258–64.

[3] Zhu J, Garrett R, Jung Y, et al. Osteoblasts support B-lymphocyte commitment and differentiation from hematopoietic stem cells. Blood 2007;109:3706–12.

[4] Kondo M, Wagers AJ, Manz MG, et al. Biology of hematopoietic stem cells and progenitors: implications for clinical application. Ann Rev Immunol 2003;21:759–806.

[5] Davis MM. The evolutionary and structural "logic" of antigen receptor diversity. Semin Immunol 2004;16:239–43.

[6] Davis MM, Lyons DS, Altman JD, et al. T cell receptor biochemistry, repertoire selection and general features of TCR and Ig structure. Ciba Foundation Symposium 1997;204:94–100.

[7] Hedrick SM, Cohen DI, Nielsen EA, Davis MM. Isolation of cDNA clones encoding T cell-specific membrane-associated proteins. Nature 1984;308:149–53.

[8] Yanagi Y, Yoshikai Y, Leggett K, Clark SP, Aleksander I, Mak TW. A human T cell-specific cDNA clone encodes a protein having extensive homology to immunoglobulin chains. Nature 1984;308:145–9.

[9] Lefranc MP, Giudicelli V, Ginestoux C, et al. IMGT, the international ImMunoGeneTics information system. Nucleic Acids Res 2009;37:D1006–12.

[10] Chien YH, Iwashima M, Kaplan KB, Elliott JF, Davis MM. A new T-cell receptor gene located within the alpha locus and expressed early in T-cell differentiation. Nature 1987;327:677–82.

[11] Davis MM, Bjorkman PJ. T-cell antigen receptor genes and T-cell recognition. Nature 1988;334:395–402.

[12] Call ME, Pyrdol J, Wiedmann M, Wucherpfennig KW. The organizing principle in the formation of the T cell receptor-CD3 complex. Cell 2002;111:967–79.

[13] Clevers H, Alarcon B, Wileman T, Terhorst C. The T cell receptor/CD3 complex: a dynamic protein ensemble. Annu Rev Immunol 1988;6:629–62.

[14] Guy CS, Vignali DA. Organization of proximal signal initiation at the TCR: CD3 complex. Immunol Rev 2009;232:7–21.

[15] Bjorkman PJ. MHC restriction in three dimensions: a view of T cell receptor/ligand interactions. Cell 1997;89:167–70.

[16] Garcia KC, Teyton L, Wilson IA. Structural basis of T cell recognition. Annu Rev Immunol 1999;17:369–97.

[17] Janeway Jr CA. The T cell receptor as a multicomponent signalling machine: CD4/CD8 coreceptors and CD45 in T cell activation. Annu Rev Immunol 1992;10:645–74.

[18] Mosmann TR, Coffman RL. TH1 and TH2 cells: different patterns of lymphokine secretion lead to different functional properties. Annu Rev Immunol 1989;7:145–73.

[19] Miossec P, Korn T, Kuchroo VK. Interleukin-17 and type 17 helper T cells. N Engl J Med 2009;361:888–98.

[20] Park H, Li Z, Yang XO, et al. A distinct lineage of CD4 T cells regulates tissue inflammation by producing interleukin 17. Nat Immunol 2005;6:1133–41.

[21] Sakaguchi S, Yamaguchi T, Nomura T, Ono M. Regulatory T cells and immune tolerance. Cell 2008;133:775–87.

[22] Ladi E, Yin X, Chtanova T, Robey EA. Thymic microenvironments for T cell differentiation and selection. Nat Immunol 2006;7:338–43.

[23] Miller JF, Osoba D. Current concepts of the immunological function of the thymus. Physiol Rev 1967;47:437–520.

[24] Ciofani M, Zuniga-Pflucker JC. Notch promotes survival of pre-T cells at the beta-selection checkpoint by regulating cellular metabolism. Nat Immunol 2005;6:881–8.

[25] Fry TJ, Mackall CL. The many faces of IL-7: from lymphopoiesis to peripheral T cell maintenance. J Immunol 2005;174:6571–6.

[26] Tanigaki K, Tsuji M, Yamamoto N, et al. Regulation of alphabeta/gammadelta T cell lineage commitment and peripheral T cell responses by Notch/RBP-J signaling. Immunity 2004;20:611–22.

[27] Washburn T, Schweighoffer E, Gridley T, et al. Notch activity influences the alphabeta versus gammadelta T cell lineage decision. Cell 1997;88:833–43.

[28] von Boehmer H, Fehling HJ. Structure and function of the pre-T cell receptor. Annu Rev Immunol 1997;15:433–52.

[29] Droge W. Hypothesis on the origin of the strong alloreactivity. Immunobiology 1979;156:2–12.

[30] Wilkinson RW, Anderson G, Owen JJ, Jenkinson EJ. Positive selection of thymocytes involves sustained interactions with the thymic microenvironment. J Immunol 1995;155:5234–40.

[31] Germain RN. T-cell development and the CD4-CD8 lineage decision. Nat Rev 2002;2:309–22.

[32] Siggs OM, Makaroff LE, Liston A. The why and how of thymocyte negative selection. Curr Opin Immunol 2006;18:175–83.

[33] Teh HS, Motyka B, Teh SJ. Influence of the affinity of selecting ligands on T cell positive and negative selection and the functional maturity of the positively selected T cells. Crit Rev Immunol 1997;17(5-6):399–410.

[34] Petrie HT, Zuniga-Pflucker JC. Zoned out: functional mapping of stromal signaling microenvironments in the thymus. Annu Rev Immunol 2007;25:649–79.

[35] Takahama Y. Journey through the thymus: stromal guides for T-cell development and selection. Nat Rev 2006;6:127–35.

[36] Rochman Y, Spolski R, Leonard WJ. New insights into the regulation of T cells by gamma(c) family cytokines. Nat Rev 2009;9:480–90.

[37] Bleul CC, Corbeaux T, Reuter A, Fisch P, Monting JS, Boehm T. Formation of a functional thymus initiated by a postnatal epithelial progenitor cell. Nature 2006;441:992–6.

[38] Rossi SW, Jenkinson WE, Anderson G, Jenkinson EJ. Clonal analysis reveals a common progenitor for thymic cortical and medullary epithelium. Nature 2006;441:988–91.

[39] Kyewski B, Klein L. A central role for central tolerance. Annu Rev Immunol 2006;24:571–606.

[40] Mathis D, Benoist C. Aire. Annu Rev Immunol 2009;27:287–312.

[41] Akiyama T, Shimo Y, Yanai H, et al. The tumor necrosis factor family receptors RANK and CD40 cooperatively establish the thymic medullary microenvironment and self-tolerance. Immunity 2008;29:423–37.

[42] Rossi SW, Kim MY, Leibbrandt A, et al. RANK signals from CD4(+)3(−) inducer cells regulate development of Aire-expressing epithelial cells in the thymic medulla. J Exp Med 2007;204:1267–72.

[43] White AJ, Withers DR, Parnell SM, et al. Sequential phases in the development of Aire-expressing medullary thymic epithelial cells involve distinct cellular input. Eur J Immunol 2008;38:942–7.

[44] Weinreich MA, Hogquist KA. Thymic emigration: when and how T cells leave home. J Immunol 2008;181:2265–70.

[45] Almeida AR, Rocha B, Freitas AA, Tanchot C. Homeostasis of T cell numbers: from thymus production to peripheral compartmentalization and the indexation of regulatory T cells. Semin Immunol 2005;17:239–49.

[46] Jameson SC. Maintaining the norm: T-cell homeostasis. Nat Rev 2002;2:547–56.

[47] Jiang Q, Li WQ, Aiello FB, et al. Cell biology of IL-7, a key lymphotrophin. Cytokine Growth Factor Rev 2005;16:513–33.

[48] Seddon B, Zamoyska R. TCR signals mediated by Src family kinases are essential for the survival of naive T cells. J Immunol 2002;169:2997–3005.

[49] Cyster JG. Chemokines, sphingosine-1-phosphate, and cell migration in secondary lymphoid organs. Annu Rev Immunol 2005;23:127–59.

[50] Link A, Vogt TK, Favre S, et al. Fibroblastic reticular cells in lymph nodes regulate the homeostasis of naive T cells. Nat Immunol 2007;8:1255–65.

[51] Boden E, Tang Q, Bour-Jordan H, Bluestone JA. The role of CD28 and CTLA4 in the function and homeostasis of CD4+CD25+ regulatory T cells. Novartis Foundation Symposium 2003;252:55–63.

[52] Greenwald RJ, Freeman GJ, Sharpe AH. The B7 family revisited. Annu Rev Immunol 2005;23:515–48.

[53] Fooksman DR, Vardhana S, Vasiliver-Shamis G, et al. Functional anatomy of T cell activation and synapse formation. Annu Rev Immunol 2010;28:79–105.

[54] Carreno BM, Bennett F, Chau TA, et al. CTLA-4 (CD152) can inhibit T cell activation by two different mechanisms depending on its level of cell surface expression. J Immunol 2000;165:1352–6.

[55] Masteller EL, Chuang E, Mullen AC, Reiner SL, Thompson CB. Structural analysis of CTLA-4 function in vivo. J Immunol 2000;164:5319–27.

[56] Watts TH. TNF/TNFR family members in costimulation of T cell responses. Annu Rev Immunol 2005;23:23–68.

[57] Djuretic IM, Levanon D, Negreanu V, Groner Y, Rao A, Ansel KM. Transcription factors T-bet and Runx3 cooperate to activate Ifng and silence Il4 in T helper type 1 cells. Nat Immunol 2007;8:145–53.

[58] Schoenborn JR, Wilson CB. Regulation of interferon-gamma during innate and adaptive immune responses. Adv Immunol 2007;96:41–101.

[59] Szabo SJ, Sullivan BM, Peng SL, Glimcher LH. Molecular mechanisms regulating Th1 immune responses. Annu Rev Immunol 2003;21:713–58.

[60] Thieu VT, Yu Q, Chang HC, et al. Signal transducer and activator of transcription 4 is required for the transcription factor T-bet to promote T helper 1 cell-fate determination. Immunity 2008;29:679–90.

[61] Luster AD, Alon R, von Andrian UH. Immune cell migration in inflammation: present and future therapeutic targets. Nat Immunol 2005;6:1182–90.

[62] Mora JR, von Andrian UH. T-cell homing specificity and plasticity: new concepts and future challenges. Trends in Immunology 2006;27:235–43.

[63] Ansel KM, Djuretic I, Tanasa B, Rao A. Regulation of Th2 differentiation and Il4 locus accessibility. Annu Rev Immunol 2006;24:607–56.

[64] Mowen KA, Glimcher LH. Signaling pathways in Th2 development. Immunol Rev 2004;202:203–22.

[65] Yamashita M, Ukai-Tadenuma M, Miyamoto T, et al. Essential role of GATA3 for the maintenance of type 2 helper T (Th2) cytokine production and chromatin remodeling at the Th2 cytokine gene loci. J Biol Chem 2004;279:26983–90.

[66] Li-Weber M, Krammer PH. Regulation of IL4 gene expression by T cells and therapeutic perspectives. Nat Rev 2003;3:534–43.

[67] Takatsu K, Nakajima H. IL-5 and eosinophilia. Curr Opin Immunol 2008;20:288–94.

[68] Wills-Karp M. Interleukin-13 in asthma pathogenesis. Immunol Rev 2004;202:175–90.

[69] Constant SL, Bottomly K. Induction of Th1 and Th2 CD4+ T cell responses: the alternative approaches. Annu Review Immunol 1997;15:297–322.

[70] Bettelli E, Carrier Y, Gao W, et al. Reciprocal developmental pathways for the generation of pathogenic effector TH17 and regulatory T cells. Nature 2006;441:235–8.

[71] Mangan PR, Harrington LE, O'Quinn DB, et al. Transforming growth factor-beta induces development of the T(H)17 lineage. Nature 2006;441:231–4.

[72] Veldhoen M, Hocking RJ, Atkins CJ, Locksley RM, Stockinger B. TGFbeta in the context of an inflammatory cytokine milieu supports de novo differentiation of IL-17-producing T cells. Immunity 2006;24:179–89.

[73] Yang XO, Pappu BP, Nurieva R, et al. T helper 17 lineage differentiation is programmed by orphan nuclear receptors ROR alpha and ROR gamma. Immunity 2008;28:29–39.

[74] Zhang F, Meng G, Strober W. Interactions among the transcription factors Runx1, RORgammat and Foxp3 regulate the differentiation of interleukin 17-producing T cells. Nat Immunol 2008;9:1297–306.

[75] Langrish CL, Chen Y, Blumenschein WM, Mattson J, Basham B, Sedgwick JD, et al. IL-23 drives a pathogenic T cell population that induces autoimmune inflammation. J Exp Med 2005;201:233–40.

[76] Manel N, Unutmaz D, Littman DR. The differentiation of human T(H)-17 cells requires transforming growth factor-beta and induction of the nuclear receptor RORgammat. Nat Immunol 2008;9:641–9.

[77] Volpe E, Servant N, Zollinger R, et al. A critical function for transforming growth factor-beta, interleukin 23 and proinflammatory cytokines in driving and modulating human T(H)-17 responses. Nat Immunol 2008;9:650–7.

[78] Wilson NJ, Boniface K, Chan JR, McKenzie BS, Blumenschein WM, Mattson JD, et al. Development, cytokine profile and function of human interleukin 17-producing helper T cells. Nat Immunol 2007;8:950–7.

[79] Yang L, Anderson DE, Baecher-Allan C, et al. IL-21 and TGF-beta are required for differentiation of human T(H)17 cells. Nature 2008;454:350–2.

[80] Sato K, Suematsu A, Okamoto K, et al. Th17 functions as an osteoclastogenic helper T cell subset that links T cell activation and bone destruction. J Exp Med 2006;203:2673–82.

[81] Takayanagi H. Osteoimmunology: shared mechanisms and crosstalk between the immune and bone systems. Nat Rev 2007;7:292–304.

[82] Fontenot JD, Rudensky AY. A well adapted regulatory contrivance: regulatory T cell development and the forkhead family transcription factor Foxp3. Nat Immunol 2005;6:331–7.

[83] Sakaguchi S. Naturally arising CD4+ regulatory T cells for immunologic self-tolerance and negative control of immune responses. Annu Rev Immunol 2004;22:531–62.

[84] Vignali DA, Collison LW, Workman CJ. How regulatory T cells work. Nat Rev 2008;8:523–32.

[85] Blazevic V, Trubey CM, Shearer GM. Analysis of the costimulatory requirements for generating human virus-specific in vitro T helper and effector responses. J Clin Immunol 2001;21:293–302.

[86] Weninger W, Manjunath N, von Andrian UH. Migration and differentiation of CD8+ T cells. Immunol Rev 2002;186:221–33.

[87] Barry M, Bleackley RC. Cytotoxic T lymphocytes: all roads lead to death. Nat Rev 2002;2:401–9.

[88] Russell JH, Ley TJ. Lymphocyte-mediated cytotoxicity. Annu Rev Immunol 2002;20:323–70.

[89] Lieberman J. The ABCs of granule-mediated cytotoxicity: new weapons in the arsenal. Nat Rev 2003;3:361–70.

[90] Trambas CM, Griffiths GM. Delivering the kiss of death. Nat Immunol 2003;4:399–403.

[91] Ahmed R, Gray D. Immunological memory and protective immunity: understanding their relation. Science 1996;272:54–60.

[92] Sprent J, Surh CD. T cell memory. Annu Rev Immunol 2002;20:551–79.

[93] Sallusto F, Geginat J, Lanzavecchia A. Central memory and effector memory T cell subsets: function, generation, and maintenance. Annu Rev Immunol 2004;22:745–63.

[94] Ahmed R, Bevan MJ, Reiner SL, Fearon DT. The precursors of memory: models and controversies. Nat Rev 2009;9:662–8.

[95] Robertson JM, MacLeod M, Marsden VS, Kappler JW, Marrack P. Not all CD4+ memory T cells are long lived. Immunol Rev 2006;211:49–57.

[96] Seddon B, Tomlinson P, Zamoyska R. Interleukin 7 and T cell receptor signals regulate homeostasis of CD4 memory cells. Nat Immunol 2003;4:680–6.

[97] Schluns KS, Lefrancois L. Cytokine control of memory T-cell development and survival. Nat Rev 2003;3:269–79.

[98] Tokoyoda K, Zehentmeier S, Hegazy AN, et al. Professional memory CD4+ T lymphocytes preferentially reside and rest in the bone marrow. Immunity 2009;30:721–30.

[99] Humphrey MB, Lanier LL, Nakamura MC. Role of ITAM-containing adapter proteins and their receptors in the immune system and bone. Immunol Rev 2005;208:50–65.

[100] Hardy RR, Hayakawa K. B cell development pathways. Ann Rev Immunol 2001;19:595–621.

[101] Hesslein DG, Schatz DG. Factors and forces controlling V(D)J recombination. Adv Immunol 2001;78:169–232.

[102] Kuo TC, Schlissel MS. Mechanisms controlling expression of the RAG locus during lymphocyte development. Curr Opin Immunol 2009;21:173–8.

[103] Mayack SR, Wagers AJ. Osteolineage niche cells initiate hematopoietic stem cell mobilization. Blood 2008;112:519–31.

[104] Mackarehtschian K, Hardin JD, Moore KA, Boast S, Goff SP, Lemischka IR. Targeted disruption of the flk2/flt3 gene leads to deficiencies in primitive hematopoietic progenitors. Immunity 1995;3:147–61.

[105] Waskow C, Paul S, Haller C, Gassmann M, Rodewald HR. Viable c-Kit(W/W) mutants reveal pivotal role for c-kit in the maintenance of lymphopoiesis. Immunity 2002;17:277–88.

[106] Wu JY, Purton LE, Rodda SJ, et al. Osteoblastic regulation of B lymphopoiesis is mediated by $G_s\alpha$-dependent signaling pathways. Proc Natl Acad Sci U S A 2008;105:16976–81.

[107] Mazzucchelli RM, Durum SK. Interleukin-7 receptor expression: intelligent design. Nat Rev Immunol 2007;7:144–54.

[108] Peschon JJ, Morrissey PJ, Grabstein KH, et al. Early lymphocyte expansion is severely impaired in interleukin 7 receptor-deficient mice. J Exp Med 1994;180:1955–60.

[109] von Freeden-Jeffry U, Vieira P, Lucian LA, McNeil T, Burdach SEG, Murray R. Lymphopenia in interleukin (IL)-7 gene-deleted mice identifies IL-7 as a nonredundant cytokine. J Exp Med 1995;181:1519–26.

[110] Sitnicka E, Brakebusch C, Martensson I, et al. Complementary signaling through flt3 and interleukin-7 receptor alpha is indispensable for fetal and adult B cell genesis. J Exp Med 2003;198:1495–506.

[111] Vobhenrich CAJ, Cumano A, Muller W, Di Santo JP, Vieira P. Thymic stromal-derived lymphopoietin distinguishes fetal from adult B cell development. Nature Immunol 2003;4:773–9.

[112] Miyaura C, Onoe Y, Inada M, et al. Increased B-lymphopoiesis by interleukin 7 induces bone loss in mice with intact ovarian function: similarity to estrogen deficiency. Proc Natl Acad Sci U S A 1997;94:9360–5.

[113] Weitzmann MN, Roggia C, Toraldo G, Weitzmann L, Pacifici R. Increased production of IL-7 uncouples bone formation from bone resorption during estrogen deficiency. J Clin Invest 2002;110:1643–50.

[114] Weitzmann MN, Cenci S, Rifas L, Brown C, Pacifici R. Interleukin-7 stimulates osteoclast formation by up-regulating the T-cell production of soluble osteoclastogenic cytokines. Blood 2000;96: 1873–8.

[115] Salopek D, Grcevic D, Katavic V, Kovacic N, Lukic IK, Marusic A. Increased bone resorption and osteopenia are a part of the lymphoproliferative phenotype of mice with systemic over-expression of interleukin-7 gene driven by MHC class II promoter. Immunol Lett 2008;121:134–9.

[116] Toraldo G, Roggia C, Qian WP, Pacifici R, Weitzmann MN. IL-7 induces bone loss in vivo by induction of receptor activator of nuclear factor kappa B ligand and tumor necrosis factor alpha from T cells. Proc Natl Acad Sci U S A 2003;100:125–30.

[117] Gendron S, Boisvert M, Chetoui N, Aoudjit F. Alpha1beta1 integrin and interleukin-7 receptor up-regulate the expression of RANKL in human T cells and enhance their osteoclastogenic function. Immunology 2008;125:359–69.

[118] Lee SK, Kalinowski JF, Jastrzebski SL, Puddington L, Lorenzo JA. Interleukin-7 is a direct inhibitor of in vitro osteoclastogenesis. Endocrinology 2003;144:3524–31.

[119] Lee SK, Kalinowski JF, Jacquin C, Adams DJ, Gronowicz G, Lorenzo JA. Interleukin-7 influences osteoclast function in vivo but is not a critical factor in ovariectomy-induced bone loss. J Bone Miner Res 2006;21:695–702.

[120] Lee S, Kalinowski JF, Adams DJ, Aguila HL, Lorenzo JA. Osteoblast specific overexpression of human interleukin-7 increases femoral trabecular bone mass in female mice and inhibits in vitro osteoclastogenesis. J Bone Miner Res 2004;19:S410.

[121] Lee S, Kalinowski JF, Adams DJ, Aguila HL, Lorenzo JA. Osteoblast specific overexpression of human interlukin-7 rescues the bone phenotype of interleukin-7 deficient female mice. J Bone Miner Res 2005;20:S48.

[122] Busslinger M. Transcriptional control of early B cell development. Ann Rev Immunol 2004;22: 55–79.

[123] Bain G, Maandag ECR, Izon DJ, et al. E2A proteins are required for proper B cell development and initiation of immunoglobulin gene rearrangements. Cell 1994;79:885–92.

[124] Georgopoulos K, Bigby M, Wang JH, et al. The Ikaros gene is required for the development of all lymphoid lineages. Cell 1994;79:143–56.

[125] Lin HH, Grosschedl R. Failure of B-cell differentiation in mice lacking the transcription factor EBF. Nature 1995;376:263–7.

[126] Nutt SL, Urbanek P, Rolink A, Busslinger M. Essential functions of PAX5 (BSAP) in pro-B cell development – difference between fetal and adult B lymphopoiesis and reduced V-to-DJ recombination at the IgH locus. Genes & Development 1997;11:476–91.

[127] Urbanek P, Wang ZQ, Fetka I, Wagner EF, Busslinger M. Complete block of early B cell differentiation and altered patterning of the posterior midbrain in mice lacking Pax5/BSAP. Cell 1994;79:901–12.

[128] Zhuang Y, Soriano P, Weintraub H. The helix-loop-helix gene E2A is required for B cell formation. Cell 1994;79:875–84.

[129] Scott EW, Simon MC, Anastasi J, Singh H. Requirement of transcription factor PU.1 in the development of multiple hematopoietic lineages. Science 1994;265:1573–7.

[130] Tondravi MM, McKercher SR, Anderson K, et al. Osteopetrosis in mice lacking haematopoietic transcription factor PU.1. Nature 1997;386:81–4.

[131] DeKoter RP, Singh H. Regulation of B lymphocyte and macrophage development by graded expression of PU.1. Science 2000;288:1439–41.

[132] DeKoter RP, Lee HJ, Singh H. PU.1 regulates expression of the interleukin-7 receptor in lymphoid progenitors. Immunity 2002;16:297–309.

[133] DeKoter RP, Walsh JC, Singh H. PU.1 regulates both cytokine-dependent proliferation and differentiation of granulocyte/macrophage progenitors. Embo J 1998;17:4456–68.

[134] Nutt SL, Kee BL. The transcriptional regulation of B cell lineage commitment. Immunity 2007;26:715–25.

[135] Georgopoulos K. Haematopoietic cell-fate decisions, chromatin regulation and Ikaros. Nat Rev Immunol 2002;2:162–74.

[136] Cobb BS, Smale ST. Ikaros-family proteins: in search of molecular functions during lymphocytes development. Curr Top Microbiol Immunol 2005;290:29–47.

[137] Kelley CM, Ikeda T, Koipally J, et al. Helios, a novel dimerization partner of Ikaros expressed in the earliest hematopoietic progenitors. Curr Biol 1998;8:508–15.

[138] Georgopoulos K, Bigby M, Wang JH, et al. The Ikaros gene is required for the development of all lymphoid lineages. Cell 1994;79:143–56.

[139] Wang JH, Nichogiannopoulou A, Wu L, et al. Selective defects in the development of the fetal and adult lymphoid system in mice with an Ikaros null mutation. Immunity 1996;5:537–49.

[140] Georgopoulos K, Winady S, Avitahl N. The role of the Ikaros gene in lymphocyte development and homeostasis. Ann Rev Immunol 1997;15:155–76.

[141] Reynaud D, Demarco IA, Reddy KL, et al. Regulation of B cell fate commitment and immunoglobulin heavy-chain gene rearrangements by Ikaros. Nat Immunol 2008;9:927–36.

[142] Zhuang Y, Cheng P, Weintraub H. B-lymphocyte development is regulated by the combined dosage of three basic helix-loop-helix genes, E2A, E2-2, and HEB. Mol Cell Biol 1996;16:2898–905.

[143] Semerad CL, Mercer EM, Inlay MA, Weissman IL, Murre C. E2A proteins maintain the hematopoietic stem cell pool and promote the maturation of myelolymphoid and myeloerythroid progenitors. Proc Natl Acad Sci U S A 2009;106:1930–5.

[144] Singh H, Medina KL, Pongubala JM. Contingent gene regulatory networks and B cell fate specification. Proc Natl Acad Sci U S A 2005;102:4949–53.

[145] Liberg D, Sigvardsson M, Akerblad P. The EBF/Olf/Collier family of transcription factors: regulators of differentiation in cells originating from all three embryonal germ layers. Mol Cell Biol 2002;22:8389–97.

[146] Hesslein DGT, Fretz JA, Xi Y, et al. Ebf1 dependent control of the osteoblast and adipocyte lineages. Bone 2009;44:537–46.

[147] Akerblad P, Lind U, Liberg D, Bamberg K, Sigvardsson M. Early B-cell factor (O/E-1) is a promoter of adipogenesis and involved in control of genes important for terminal adipocyte differentiation. Mol Cell Biol 2002;22:8015–25.

[148] Hagman J, Belanger C, Travis A, Turck CW, Grosschedl R. Cloning and functional characterization of early B-cell factor, a regulator of lymphocyte-specific gene expression. Genes Dev 1993;7: 760–73.

[149] Kieslinger M, Folberth S, Dobreva G, et al. EBF2 regulates osteoblast-dependent differentiation of osteoclasts. Dev Cell 2005;9:757–67.

[150] Lukin K, Fields S, Hartley J, Hagman J. Early B cell factor: regulator of B lineage specification and commitment. Semin Immunol 2008;20:221–7.

[151] Pongubala JM, Northrup DL, Lancki DW, et al. Transcription factor EBF restricts alternative lineage options and promotes B cell fate commitment independently of Pax5. Nat Immunol 2008;9:203–15.

[152] Horowitz MC, Xi Y, Pflugh DL, et al. Pax5 deficient mice exhibit early onset osteoporosis with increased osteoclast progenitors. J Immunol 2004;173:6583–91.

[153] Fretz JA, Nelson T, Xi Y, Adams DJ, Rosen CJ, Horowitz MC. Altered metabolism and lipodystrophy in the Ebf1-deficient mouse. Endocrinology 2010;151:1611–21.

[154] Hesslein DG, Pflugh DL, Chowdhury D, Bothwell AL, Sen R, Schatz DG. Pax5 is required for recombination of transcribed, acetylated, 5′ IgH V gene segments. Genes Dev 2003;17:37–42.

[155] Delogu A, Schebesta A, Sun Q, Aschenbrenner K, Perlot T, Busslinger M. Gene repression by Pax5 in B cells is essential for blood cell homeostasis and is reversed in plasma cells. Immunity 2006;24:269–81.

[156] Tagoh H, Ingram R, Wilson N, et al. The mechanism of repression of the myeloid-specific c-fms gene by Pax5 during B lineage restriction. Embo J 2006;25:1070–80.

[157] Nutt SL, Heavey B, Rolink AG, Busslinger M. Commitment to the B-lymphoid lineage depends on the transcription factor Pax5. Nature 1999;401:556–62.

[158] Liu P, Keller JR, Ortiz M, et al. Bcl11a is essential for normal lymphoid development. Nat Immunol 2003;4:525–32.

[159] Dengler HS, Baracho GV, Omori SA, et al. Distinct functions for the transcription factor Foxo1 at various stages of B cell differentiation. Nat Immunol 2008;9:1388–98.

[160] Kong Y-Y, Yoshida H, Sarosi I, et al. OPGL is a key regulator of osteoclastogenesis, lymphocyte development and lymph-node organogenesis. Nature 1999;397:315–23.

[161] Lomaga MA, Yeh W-C, Sarosi I, et al. TRAF6 deficiency results in osteopetrosis and defective interleukin-1, CD40, and LPS signaling. Genes & Dev 1999;15:1015–24.

[162] Dougall WC, Glaccum M, Charrier K, et al. RANK is essential for osteoclast and lymph node development. Genes & Dev 1999;13:2412–24.

[163] Yun TJ, Tallquist MD, Aicher A, et al. Osteoprotegerin, a crucial regulator of bone metabolism, also regulates B cell development and function. J Immunol 2001;166:1482–91.

[164] Fairfax KA, Kallies A, Nutt SL, Tarlinton DM. Plasma cell development: from B-cell subsets to long-term survival niches. Semin Immunol 2008;20:49–58.

[165] Shapiro-Shelef M, Calame K. Regulation of plasma-cell development. Nat Rev Immunol 2005;5:230–42.

[166] Moser K, Tokoyoda K, Radbruch A, MacLennan I, Manz RA. Stromal niches, plasma cell differentiation and survival. Curr Opin Immunol 2006;18:265–70.

[167] Tokoyoda K, Zehentmeier S, Chang HD, Radbruch A. Organization and maintenance of immunological memory by stroma niches. Eur J Immunol 2009;39:2095–9.

[168] Tokoyoda K, Egawa T, Sugiyama T, Choi BI, Nagasawa T. Cellular niches controlling B lymphocyte behavior within bone marrow during development. Immunity 2004;20:707–18.

[169] Schett G. Osteoimmunology in rheumatic diseases. Arthritis Res Ther 2009;11:210.

[170] Kawai T, Matsuyama T, Hosokawa Y, et al. B and T lymphocytes are the primary sources of RANKL in the bone resorptive lesion of periodontal disease. Am J Pathol 2006;169:987–98.

[171] Hayer S, Polzer K, Brandl A, et al. B-cell infiltrates induce endosteal bone formation in inflammatory arthritis. J Bone Miner Res 2008;23:1650–60.

[172] Ehrhardt GR, Hijikata A, Kitamura H, Ohara O, Wang JY, Cooper MD. Discriminating gene expression profiles of memory B cell subpopulations. J Exp Med 2008;205:1807–17.

[173] Roodman GD. Pathogenesis of myeloma bone disease. J Cell Biochem 2010;109:283–91.

[174] Pearse RN, Sordillo EM, Yaccoby S, et al. Multiple myeloma disrupts the TRANCE/osteoprotegerin cytokine axis to trigger bone destruction and promote tumor progression. Proc Natl Acad Sci U S A 2001;98:11581–6.

[175] Sezer O, Heider U, Jakob C, et al. Immunocytochemistry reveals RAANKL expression of myeloma cells. Blood 2002;99:4646–7.

[176] Shibata H, Abe M, Hiura K, et al. Malignant B-lymphoid cells with bone lesions express receptor activator of nuclear factor-kappaB ligand and vascular endothelial growth factor to enhance osteoclastogenesis. Clin Cancer Res 2005;11:6109–15.

# 4

# The Role of Bone Cells in Establishing the Hematopoietic Stem Cell Niche

Joy Y. Wu, Henry M. Kronenberg

ENDOCRINE UNIT, MASSACHUSETTS GENERAL HOSPITAL, BOSTON, MA, USA

**CHAPTER OUTLINE**

Introduction ............................................................................................................. 81
The osteoblast lineage is heterogeneous................................................................. 82
Cells of the osteoblast lineage support hematopoietic stem cells ........................... 83
Signaling pathways implicated in osteoblast–HSC communication ......................... 86
    Angiopoietin-1/Tie2................................................................................................ 86
    Thrombopoietin/Mpl .............................................................................................. 86
    Osteopontin ............................................................................................................ 86
    Calcium-sensing receptor ...................................................................................... 87
    N-cadherin ............................................................................................................. 87
    Notch signaling ...................................................................................................... 87
Other signaling pathways in the microenvironment that may involve osteoblasts ... 88
    Wnt signaling ......................................................................................................... 88
    CXCL12/CXCR4 .................................................................................................... 89
    Prostaglandin E2 .................................................................................................... 89
    TGFβ1 .................................................................................................................... 90
Osteoblasts and their precursors also support other hematopoietic lineages.......... 90
Other components of the hematopoietic niche........................................................ 91
The microenvironment and disease ......................................................................... 93
The bone marrow HSC microenvironment is complex............................................. 93
References............................................................................................................... 94

# Introduction

Stem cells have the unique properties of self-renewal and multipotency, and the ability to use stem cells for therapeutic purposes holds tremendous appeal. Therefore, the regulatory mechanisms governing stem cell function are an area of great scientific and public interest. Cues extrinsic to stem cells, emanating from the surrounding microenvironment or niche, are important in the maintenance of stem cell self-renewal and differentiation capacity. The existence of a stem cell niche has been demonstrated for a variety of stem cell populations in diverse organisms [1]. The idea that specific microenvironments control the balance between self-renewal and multipotency has several consequences including: (1) the act of leaving or staying in such an environment itself can be an important regulatory step, and (2) the identification of a specific niche helps focus hypotheses about cellular sources of regulatory molecules.

Among mammalian stem cells, the best characterized are hematopoietic stem cells (HSCs). A single HSC can reconstitute the entire hematopoietic system in vivo, and HSCs have been used with great clinical benefit in bone marrow transplantation [2]. Maintenance of HSC self-renewal and differentiation are dependent upon the presence of a supportive microenvironment, or niche. In adult mammals, the differentiation of hematopoietic cells takes place within the bone marrow cavity, and Schofield first proposed the concept of a bone marrow microenvironment over 30 years ago [3]. The first appearance of hematopoiesis in bones coincides with the onset of mineralization during skeletal development, and in recent decades it has become apparent that within the bone marrow the surrounding non-hematopoietic tissues, or microenvironment, play an active role in regulating hematopoiesis. The bone marrow stromal environment is comprised of a multitude of cellular components, including fibroblasts, endothelial cells, adipocytes, and osteoblasts. In addition, cells of the hematopoietic lineage themselves likely also regulate the microenvironment [4]. This chapter will highlight recent data regarding the role of cells of the osteoblast lineage in the bone marrow hematopoietic microenvironment.

## The osteoblast lineage is heterogeneous

During embryogenesis hematopoiesis first emerges in the yolk sac, then shifts to the aorta, gonads and mesonephros (AGM), and then the liver [5]. The HSC niche migrates to the bone at the time of the onset of mineralization, late in embryogenesis. There it remains throughout the lifespan of the organism unless forced to relocate to organs such as the spleen and liver by pathological processes affecting the bone marrow. The establishment of hematopoiesis in the bone marrow cavity is absolutely dependent upon a normal mineralized skeleton. Mice lacking Runx2/Cbfa1, the earliest known transcription factor defining the osteoblast lineage, are characterized by a failure of osteoblast differentiation with resulting absence of mineralized bone and consequent extramedullary hematopoiesis [6–8]. Conversely, osteopetrotic mice lacking normal osteoclast function have reduced marrow space due to impaired bone resorption and also exhibit extramedullary hematopoiesis [9].

Formation of the skeleton occurs via two processes, intramembranous and endochondral ossification. In endochondral ossification, which occurs predominantly in the axial and appendicular skeleton, mesenchymal condensations arise at sites of future bones, differentiating into cartilage-forming chondrocytes [10]. Chondrocytes proliferate and ultimately differentiate into hypertrophic chondrocytes. Apoptosis of hypertrophic chondrocytes triggers vascular invasion, and simultaneously osteoblast progenitors migrate in and differentiate on the pre-formed cartilage template. In contrast, intramembranous ossification, such as occurs in most of the skull, takes place without a cartilage intermediate. Much of what is currently understood about early stages of osteoblast differentiation has been gleaned from in vitro studies of cells cultured under osteogenic or mineralizing conditions. Once committed to the osteoblast lineage, osteoprogenitors differentiate and express a well-characterized sequence of markers of the osteoblast lineage [11]. Mature, post-mitotic osteoblasts line the bone surface, produce extracellular matrix, and express markers of terminal differentiation such as osteocalcin. From here osteoblasts encounter one of three possible fates – engulfment by mineral matrix to become an osteocyte, death by apoptosis, or quiescence as a lining cell [12]. Thus the cells of the osteoblast lineage are a heterogeneous population, representing various stages of differentiation.

Although mature osteoblasts are easily identified in vivo by their cuboidal morphology and proximity to mineralized bone, the location and behavior of osteogenic precursors are poorly defined. Osteoblast precursors are presumed to be located within the bone marrow. When bone marrow stromal cells are isolated based on adherence to plastic and cultured with β-glycerol phosphate and ascorbic acid, a fraction of cells will differentiate, express osteogenic markers, and ultimately mineralize the surrounding matrix [11]. Furthermore, in patients with primary or secondary hyperparathyroidism [13] or fibrous dysplasia [14], disorders both associated with enhanced signaling via the protein kinase A pathway in osteoblasts, as well as in transgenic mice overexpressing constitutively active parathyroid hormone (PTH)/PTH-related peptide (PTHrP) receptor (PPR) in osteoblasts [15], fibroblastoid stromal cells that express markers of the osteoblast lineage accumulate. Currently our ability to prospectively define and/or isolate populations of osteoblast precursors is limited by the lack of well-characterized cell surface markers, although recent targeting of green fluorescent protein (GFP) to specific stages within the osteoblast lineage promises to facilitate our understanding of the osteoblast lineage [16–18].

# Cells of the osteoblast lineage support hematopoietic stem cells

Far from being passive neighbors in the vicinity of hematopoietic stem cells, cells of the osteoblast lineage are active participants in the support of hematopoiesis. The first indications of an association between the skeletal and hematopoietic systems were largely anatomical. In the 1970s it was noted that the subcutaneous transfer of bone marrow led to the formation first of bone, followed by the appearance of vascularized bone marrow, and that hematopoietic recovery from marrow injury was intimately associated with bone and the marrow stromal network [19, 20]. Within the bone marrow cavity, hematopoietic progenitors with multilineage potential, assayed by formation of splenic colonies during recovery from irradiation (CFU-S), were found to be enriched near the endosteal surface [21, 22]. In contrast, more differentiated hematopoietic cells were located more centrally, closer to the central sinus where they might be expected to more easily enter the circulation in large numbers [23, 24].

Development of the first long-term bone marrow cultures capable of supporting hematopoietic cells in vitro required the presence of adherent stromal cells [25]. Although these stromal cells could express alkaline phosphatase, a marker of cells of the osteoblast lineage, it was not until 20 years later that Taichman and Emerson demonstrated that human trabecular osteoblasts could support hematopoiesis in vitro [26, 27].

The first functional evidence that osteoblasts regulate hematopoiesis in vivo came with the demonstration that targeted expression of thymidine kinase to osteoblasts followed by gancyclovir administration and ablation of osteoblasts was accompanied by a dramatic decrease in bone marrow cellularity, with resultant extramedullary hematopoiesis [28]. This proved reversible, as withdrawal of gancyclovir resulted in rapid hematopoietic recovery that was accompanied by exuberant bone formation. Subsequently, this group showed that following the depletion of osteoblasts there was an abrupt decline in bone marrow cellularity, including a significant reduction in the HSC-enriched Lin$^-$c-Kit$^+$Sca-1$^+$ (LKS) population [29].

That osteoblasts can specifically regulate the HSC niche became apparent when two groups demonstrated that increases in osteoblast number are associated with higher HSC numbers. Conditional inactivation of the bone morphogenetic protein (BMP) receptor type IA (Bmpr1a) resulted in an increase in HSCs, due to an expansion of the quiescent long-term (LT) HSC-enriched fraction [30]. Bmpr1a is not expressed in HSCs, but rather its inactivation resulted in formation of expanded trabecular bone-like regions. In mutant mice LT-HSCs, as defined by long-term retention of BrdU, could be identified in contact with spindle-shaped N-cadherin$^+$ osteoblastic cells on the surface of trabecular bone. Therefore the authors proposed that the increased number of osteoblastic cells due to ectopic bone formation provided an increased niche size for HSCs [30]. Consistent with the finding that increased osteoblast number may augment the HSC niche, targeted expression of the constitutively active PPR to osteoblasts also led to an increase in osteoblast number with a concomitant increase in HSCs [31]. In these initial reports, increased HSC number was accompanied by an increase in osteoblast number, while loss of osteoblasts resulted in a declining HSC population. However, biglycan knockout mice develop osteoporosis with reduced osteoblast numbers by 3 months of age, but there is no alteration in either HSC frequency or function [32]. As such, the number of osteoblasts alone does not appear to determine HSC number in a simple linear fashion, at least with moderate decreases in the osteoblast lineage pool.

Although HSCs are not detectable in mouse limb bone marrow until late in embryogenesis with the onset of mineralization, E14.5 bones transplanted under the renal capsule can develop a bone marrow cavity filled with host-derived marrow and HSCs [33]. Dissociation of fetal bones into single cells demonstrated that the ability to create a hematopoietic-supporting environment was contained within the mesenchymal progenitor fraction of CD105$^+$Thy1$^-$ cells. In contrast, more differentiated CD105$^+$Thy1$^+$cells expressed higher levels of the osteoblast marker osteocalcin, and could form ectopic bone but no hematopoietic marrow. These data suggest that less-differentiated cells of the osteoblast lineage may play a crucial role in formation of the HSC niche, although once the niche is established whether osteoprogenitors themselves directly support HSCs remains to be seen. Interestingly, CD105$^+$Thy1$^-$ cells isolated from intramembranous bones were incapable of forming a hematopoietic microenvironment, suggesting that endochondral ossification is required for an adequate HSC niche [33]. However, others have demonstrated that bone marrow cavities in adult mouse calvariae contain HSCs similar in frequency and repopulating ability to those in limb bones [34]. Perhaps fetal calvariae are delayed in niche-forming potential; alternatively the niche-forming cells in intramembranous bone may not be contained within the CD105$^+$Thy1$^-$ population.

Unlike fetal HSCs that have tremendous proliferative potential, adult HSCs do not have the ability to self-renew indefinitely. Therefore a critical feature of adult HSCs is the maintenance of quiescence. The switch from proliferative fetal liver HSCs to more quiescent bone marrow HSCs occurs approximately 4 weeks after birth [35, 36]. Serial HSC transplantation, the gold standard for assessment of HSC self-renewal, appears to be limited to approximately six rounds for adult HSCs [37]. Mutations that affect HSC self-renewal often result in increased cycling of HSCs, with resultant premature depletion of the HSC pool [38]. In order to ensure adequate repopulation of hematopoietic lineages over the lifespan of an organism, a pool of HSCs must therefore be held in reserve. Indeed, recent studies have found that a population of HSCs exists in a G0 dormant state, dividing on average every 145 days, or approximately five times over the lifetime of the mouse. As expected, these quiescent HSCs have enhanced long-term reconstitution activity [39]. LT-HSCs are located closest to the endosteal surface, while more mature hematopoietic progenitor cells with greater proliferative capacity are found at progressively greater distances from the bone surface [34], suggesting that osteoblasts may play an important role in maintaining HSC quiescence. Of note, several studies have suggested that HSCs are located as single cells within the bone marrow, and some are detected near the endosteum [34, 40, 41]. Although HSCs are rare, the availability of niches is also limited. Increased engraftment occurs with transplantation of HSCs in doses from 10 to 250 cells, but transplantation of greater than 250 cells results in minimal increases in donor chimerism [42]. However, treatment with ACK2, an antibody that blocks c-Kit activity and leads to depletion of endogenous HSCs, dramatically improves engraftment of transplanted HSCs. The factors limiting niche availability are still largely unknown. In irradiated mice as well as in W/Wv mutant mice that exhibit impaired engraftment of endogenous HSCs, transplanted HSCs can lodge closer to the endosteal surface [34, 41]. Moreover, HSCs transplanted into mice, expressing constitutively active PPR in osteoblasts, are also located nearer to osteoblasts [34]. Together these findings highlight the role of osteoblasts in maintaining niches for quiescent HSCs.

# Signaling pathways implicated in osteoblast–HSC communication

Many stromal cell-derived cytokines and growth factors have been identified that play important roles in regulating HSC survival and function, and several are expressed by cells of the osteoblast lineage. Here we review factors that have been implicated in osteoblast–HSC communication. To date, however, definitive proof that these osteoblasts are an essential source of these factors, as demonstrated by osteoblast-specific deletion of candidate factors, is still lacking.

## Angiopoietin-1/Tie2

Signaling via the Tie2 tyrosine kinase receptor plays a crucial role in maintenance of quiescence in HSCs. Hoechst dye-effluxing side population (SP) cells are enriched in HSCs [43]. SP cells are enriched in $Tie2^+$ but not $Tie2^-$ LKS cells, and these cells are resistant to the myelotoxic agent 5-fluorouracil (5-FU) [44]. HSCs lacking Tie2 are able to home to the bone marrow, but then are subsequently lost [45]. The Tie2 ligand, angiopoietin-1 (Ang-1), is expressed by osteoblasts, and following 5-FU treatment $Tie2^+$ cells are detected along the bone surface adjacent to $Ang-1^+$ cells that co-express osteocalcin. When co-cultured with $Tie2^+$ LKS SP cells, Ang-1 increased cobblestone frequency and repopulating activity of cells subsequently transplanted into mice. Thus osteoblasts express Ang-1 in vivo, and Ang-1/Tie2 signaling maintains quiescence in HSCs.

## Thrombopoietin/Mpl

LKS SP cells are enriched in quiescent HSCs, and adhere to osteoblastic cells on the endosteal surface [46]. Expression of the thrombopoietin (TPO) receptor Mpl is increased in quiescent HSCs, and $Mpl^+$ LKS cells are resistant to 5-FU treatment and have greater long-term repopulating activity. Following 5-FU treatment $Mpl^+$ HSCs can be found near the endosteum, and co-express Tie2. A proportion of cells along the bone surface that express alkaline phosphatase, a marker of the osteoblast lineage, also express TPO, and $Mpl^+$ HSCs have been observed in contact with $TPO^+$ osteoblasts. Furthermore, mice lacking either TPO or Mpl have reduced bone marrow HSC numbers. A neutralizing antibody against Mpl (AMM2) significantly increases HSC cycling, and reduces the quiescent HSC pool, while administration of TPO increases the fraction of quiescent LT-HSCs [46]. Consistent with these findings, TPO-deficient mice exhibit progressive loss of HSCs with age, with increased HSC cycling [47]. Of note, TPO is dispensable for expansion of fetal HSCs, but serves a crucial function in maintenance of the adult HSC pool [47].

## Osteopontin

Osteopontin is an integrin-binding glycoprotein expressed at high levels by osteoblasts at the endosteal surface. In the absence of osteopontin there is an increase in bone marrow

HSC number that is dependent on the microenvironment [48, 49]. Osteopontin inhibits proliferation of hematopoietic stem and progenitor cells (HSPCs) in vitro, and indeed in vivo a greater proportion of HSCs are cycling in OPN-deficient mice [48]. Furthermore, treatment with PTH leads to a greater expansion of HSCs in KO mice than in WT mice [49]. These studies demonstrate that osteopontin acts to negatively regulate the size of the HSC fraction. A cleaved fragment of osteopontin accumulates in bone marrow and may contribute to the actions of osteopontin by binding to $\alpha_9\beta_1$ and $\alpha_4\beta_1$ integrins on HSCs and hematopoietic progenitors [50].

## Calcium-sensing receptor

Osteoblasts produce bone by secreting extracellular matrix, which in turn undergoes mineralization in the presence of calcium and phosphate. Particularly near sites of bone resorption by osteoclasts, the local concentration of calcium ions is markedly higher than that found in serum. The calcium-sensing receptor is detected in HSCs, and mice lacking the receptor exhibit marked hypocellularity and decreased frequency of HSCs [51]. Calcium-sensing receptor-deficient HSCs demonstrate reduced repopulating activity, and impaired lodgment at the endosteal surface. Therefore the ability to sense the high calcium ion concentration plays a critical role in the osteoblast HSC niche.

## N-cadherin

In mice lacking Bmpr1a, areas of ectopic bone formation contain more HSCs [30]. Zhang et al. reported that LT-HSCs express N-cadherin, and in turn are in contact with spindle-shaped N-cadherin$^+$ osteoblast cells. However, the role of N-cadherin on HSCs remains controversial. Others have reported that N-cadherin is not detectable in HSCs, and furthermore that HSC repopulating ability is restricted to the N-cadherin-negative fraction of bone marrow [32, 52].

## Notch signaling

As mentioned earlier, transgenic mice with constitutively active PPR targeted to mature osteoblasts have an increase in trabecular bone and osteoblast number, with a concomitant increase in HSC number [31]. Moreover, administration of PTH following irradiation markedly augments engraftment of transplanted HSCs and survival. Since the PPR is not expressed on HSCs, PTH likely acts on osteoblastic cells to augment production of one or more HSC supporting factors. One candidate pathway for mediating the enhanced HSC supporting effects of PTH treatment is the Notch pathway. Binding of the Jagged or Delta transmembrane ligands to the Notch receptor leads to cleavage of the Notch intracellular domain (NICD) by the γ-secretase complex; the NICD then translocates to the nucleus where it initiates transcription of Notch target genes [53]. The Notch ligand Jagged1 is expressed by osteoblasts and upregulated by PTH treatment and constitutive PPR activation in osteoblasts, and HSCs from PPR transgenic mice contained increased levels of

NICD [31]. PTH-dependent enhancement of stromal cell support for hematopoiesis requires direct cell-to-cell contact, and is attenuated by a γ-secretase inhibitor that disrupts Notch signaling [31].

In contrast, ablation of either Jagged1 or Notch1 in the microenvironment with Mx1-Cre does not lead to an HSC defect under basal conditions [54]. Mx1-Cre and Osx-Cre gene targeting can lead to differing phenotypes [55, 56], so the absence of an Mx1-Cre-mediated HSC defect does not rule out a role for osteoblast-derived Notch ligands in regulation of the HSC niche. However, recent studies suggest that Notch signaling is not required in hematopoietic cells, but rather in the microenvironment. More generalized disruption of Notch signaling in HSCs, either by targeted expression of the dominant negative Notch inhibitor Mastermind-like1 or targeted ablation of the requisite DNA-binding factor RBPJ, did not yield defects in HSC reconstituting ability [57]. Thus Notch signaling is not required in a cell-autonomous manner for maintenance of HSCs. Of note, disruption of Notch signaling in the microenvironment by deleting Mind bomb-1 (Mib1), using two independent promoters to drive Cre recombinase, resulted in the development of myeloproliferative disease. Moreover, the myeloproliferative phenotype could be imposed upon wild-type bone marrow cells transplanted into a Mib1-deficient microenvironment [58]. Therefore Notch signaling within the niche is required for normal hematopoiesis, and perhaps is the relevant target for osteoblast-derived Jagged1.

# Other signaling pathways in the microenvironment that may involve osteoblasts

## Wnt signaling

The canonical Wnt signaling pathway plays a crucial role in regulating the differentiation of osteoblasts. Osteoblasts and osteocytes secrete several inhibitors of Wnt signaling, including Dickkopf1 (Dkk1) and sclerostin. These inhibitors may therefore inhibit Wnt signaling not only in osteoblasts, but also in nearby hematopoietic cells. Activation of canonical Wnt signaling leads to nuclear translocation of β-catenin and TCF/LEF-mediated gene transcription. In hematopoietic cells studies of Wnt signaling have yielded somewhat variable results. Activation of Wnt signaling has been reported to enhance HSC proliferation, with varying effects on HSC differentiation [59–62]. However, ablation of β-catenin, the major mediator of canonical Wnt signaling, has no apparent effect on HSC function [63].

Targeted overexpression of Dkk1 to osteoblasts leads to dramatic suppression of Wnt signaling in HSCs [64]. In contrast to previous studies, this reduced Wnt signaling in HSCs led to increased cycling of HSCs. Limiting dilution assays demonstrated an increase in the number of reconstituting HSCs in Dkk1 Tg mice. However, serial transplantation revealed a significant loss of repopulating ability over time in HSCs derived from transgenic mice; even exposure of wild-type bone marrow to a Dkk1-overexpressing microenvironment resulted in an irreversible loss of repopulating ability. Dkk1-exposed HSCs showed a significant decrease in the percentage of cells in the G0 quiescent state, with increased BrdU incorporation. Inhibition of Wnt signaling in HSCs therefore leads to an inability to maintain quiescence. Of note, the overexpression of Dkk1 leads to defects in the microenvironment as well, with a dramatic decrease in trabecular bone volume [65]. Perhaps additional secondary alterations in the niche contribute to the dramatic loss of long-term repopulating ability in these HSCs.

## CXCL12/CXCR4

The chemokine CXCL12 (SDF1) plays a critical role in maintenance of HSCs. Its sole receptor, CXCR4, is expressed on HSCs, and CXCL12 is produced by the stromal microenvironment. CXCL12/CXCR4 signaling is crucial to HSC number, as deletion of CXCR4 leads to a dramatic reduction in HSC number and increased sensitivity to 5-FU treatment [66]. Within the stromal microenvironment CXCL12 expression has been detected in osteoblasts along the bone surface as well as in reticular cells [66, 67]. Moreover, expression of CXCL12 by stromal cells of the osteoblast lineage is enhanced by PTH treatment [68] or targeted expression of the constitutively active PPR in osteoblasts [31]. CXCR4 antagonists rapidly mobilize HSCs into the periphery and are now approved for this purpose in human HSC transplantation [69].

HSCs can be mobilized into the periphery by administration of the cytokine granulocyte-colony stimulating factor (G-CSF), an approach used clinically for bone marrow transplantation. Mobilization of HSCs in response to G-CSF is mediated by cleavage of CXCL12 [70, 71]. G-CSF also suppresses osteoblast activity, with flattening of the usual plump morphology of these cells. The effects of G-CSF on osteoblasts and CXCL12 levels appear to be mediated by the sympathetic nervous system via norepinephrine signaling [72]. In addition to G-CSF, other cytokines capable of mobilizing HSPCs include Flt3L and SCF. Treatment with all three results in loss of endosteal and trabecular osteoblasts [73], and decreased CXCL12 expression in osteoblasts but not stromal cells. Thus reduced CXCL12 expression in osteoblasts is a common mechanism in HSPC mobilization in response to multiple cytokines [73].

## Prostaglandin E2

Prostaglandin E2 (PGE2) is an arachidonic acid derivative generated by the actions of cyclo-oxgenases COX1 and COX2. A chemical genetic screen for regulators of HSC development in zebrafish led to the identification of PGE2 as an important HSC stimulatory factor, while inhibition of COX enzyme activity with indomethacin decreased HSC formation [74]. Similarly, exposure of murine bone marrow to the long-acting PGE2 analog 16,16-dimethyl PGE2 ex vivo increased hematopoietic stem/progenitor cell numbers, with enhanced engraftment. The PGE2 receptors EP2R and EP4R are G protein-coupled receptors that can activate protein kinase A-dependent pathways via the G protein subunit $G_s\alpha$ [75]. Interestingly, HSCs lacking $G_s\alpha$ are unable to engraft in the bone marrow, while ex vivo treatment of HSCs with cholera toxin, which constitutively activates $G_s\alpha$, results in enhanced homing and engraftment [76]. Consistent with these reports, in vivo treatment of mice with PGE2 preferentially expands short-term HSCs, accelerating early engraftment [77].

Whether PGE2 is the relevant endogenous ligand in activating $G_s\alpha$-coupled GPCRs in vivo remains to be determined. If so, osteoblasts could constitute a potential source of PGE2. PGE2 has potent anabolic effects on bone when infused systemically [78, 79], and its expression in calvarial bones is increased by PTH treatment [80]. More recently, matrix-embedded osteocytes, believed to serve as mechanosensors in bone, have been shown to increase PGE2 in response to fluid flow-induced shear stress [81].

## TGFβ1

TGFβ1 is one of the most abundant cytokines in bone matrix, and osteoclast-mediated bone resorption plays an important role in releasing active TGFβ1 to recruit osteoprogenitors to sites of bone remodeling [82]. In vitro, isolated CD34$^-$ LKS cells proliferate rapidly when cultured with stem cell factor (SCF) and TPO, with limited survival. However, in the presence of TGFβ, single HSCs can survive for up to 7 days in culture, and exhibit competitive repopulating activity similar to that of freshly isolated HSCs. Moreover, Smad 2/3, downstream mediators of TGFβ signaling, are highly phosphorylated in CD34$^-$ LKS HSCs, while phosphorylated Smad 2/3 is barely detectable in CD34$^+$ hematopoietic progenitors. TGFβ1 is produced in an inactive latent form, and HSCs cannot self-activate TGFβ1, suggesting that one function of the bone matrix and/or surrounding microenvironment may be to provide active TGFβ to maintain HSC quiescence [83].

TGFβ1 is a potent inhibitor of lipid raft clustering, the formation of cholesterol- and sphingolipid-enriched domains within the plasma membrane [83]. In HSCs, lipid raft clustering may augment cytokine-induced proliferative signals. Lipid raft clustering is absent in freshly isolated quiescent CD34$^-$ LKS HSCs [84], while stimulation of proliferation with SCF and TPO induces lipid raft formation with co-localization of c-Kit and c-Mpl receptors [81]. Lipid raft clustering correlates with Akt activation and exclusion of FOXO3a from the nucleus, so inhibition of lipid raft clustering, with resultant nuclear accumulation of FOXO transcription factors, may play a significant role in maintenance of HSC quiescence by TGFβ1 [83]. Consistent with this model, conditional ablation of FoxO1, FoxO3, and FoxO4 in adult hematopoietic cells leads to increased HSC cycling and impaired long-term repopulation [85].

# Osteoblasts and their precursors also support other hematopoietic lineages

Beyond their role in the HSC niche, accumulating evidence indicates that cells of the osteoblast lineage may provide supportive microenvironments for more differentiated hematopoietic lineages as well. The most abundant evidence exists for osteoblastic support of B lymphocyte development within the bone marrow. Among the factors that are crucial for B-cell development, CXCL12 and IL-7 are expressed by distinct stromal cell populations within the bone marrow [86]. While pre-pro-B cells contact CXCL12$^+$ reticular cells, more differentiated pro-B cells are in contact with a distinct population of IL-7-expressing stromal cells [86]. Cells of the osteoblast lineage can express CXCL12 and IL-7 [31, 68, 87], and the bone marrow stroma contains osteoblast precursors, raising the possibility that osteoblastic cells might serve a supportive role in B-cell development. When cultured in the presence of murine primary osteoblasts, hematopoietic progenitors could progress through all the developmental stages to become B220$^+$IgM$^+$ immature B lymphocytes [68]. Furthermore, osteoblasts secreted IL-7 and CXCL12, and this could be enhanced by PTH treatment. In vivo, ablation of osteoblasts led to a rapid decline in B-cell precursors that preceded the loss of HSCs. That osteoblasts support B lymphocyte development in vivo was further demonstrated by the finding that conditional deletion in osteoprogenitors of $G_s\alpha$, a major mediator of protein kinase A signaling downstream of PPR, leads to severe osteoporosis and a marked decline in B-cell precursors from the pro-B-cell stage and beyond [88]. IL-7 is a critical regulatory factor for the pre-pro-B to pro-B cell transition, and IL-7 mRNA levels were reduced in Gsa-deficient osteoblasts. Exogenous administration of IL-7 significantly restored B lineage development in the bone marrow [88]. Together these studies provide conclusive evidence that cells of the osteoblast lineage provide a niche for more differentiated B-cell precursors in addition to HSCs [89].

## Other components of the hematopoietic niche

In addition to osteoblasts and their precursors, the bone marrow stromal microenvironment is comprised of multiple cellular lineages, including those associated with the vasculature and adipose tissue. Recent studies suggest that these components also exert influence on hematopoietic development. Tissues lacking mineralizing osteoblasts can support hematopoiesis during normal development as well as under pathologic conditions, and in some organisms such as zebrafish, hematopoiesis never occurs in association with bone; therefore, the presence of osteoblasts is not an absolute requirement for hematopoiesis [90]. To date the greatest abundance of data points to a significant role for the vasculature. Using SLAM markers to detect in vivo HSCs by immunohistochemical labeling, the majority of HSCs were located in closer proximity to endothelial sinusoids than to the endosteal surface [40], and endothelial cells can promote HSC survival in culture. In the spleen, where there are no osteoblasts, HSCs are also located close to sinusoids, highlighting that the vasculature can provide a supportive microenvironment [40].

Of note, visualization with in vivo microscopy demonstrated that the vasculature and bone surface are closely apposed, indicating that at least in calvariae, these are not likely to be anatomically distinct niches [34, 41]. However, the functional roles of osteoblastic and vascular cells may differ. Bone marrow sinusoids are enlarged as early as 7 days after irradiation, and infusion of endothelial progenitor cells can accelerate hematopoietic recovery [91, 92]. Sinusoidal endothelial cells (SECs) are disrupted by myeloablative injury, and hematopoietic recovery occurs in association with reorganization of sinusoidal vessels. VEGFR2 is expressed on bone marrow SECs and arterioles, and treatment with a neutralizing antibody against VEGFR2 inhibits this recovery [93]. Conditional ablation of VEGFR2 does not affect steady-state hematopoiesis, but significantly impairs hematopoietic reconstitution in lethally irradiated mice receiving bone marrow transplantation [93]. However, others have demonstrated that hematopoietic recovery after irradiation occurs within 48 hours in association with expansion of the osteoblast lineage, at a time when disrupted vascular structures have yet to regenerate [94].

The specific roles of the vasculature and osteoblast niches with respect to supporting HSC engraftment, mobilization, and homeostasis remain to be clarified. For example, some have suggested that the vascular niche contains primarily self-renewing differentiating HSCs [95]. Hematopoietic stem and progenitor cell populations with the greatest enrichment in quiescent HSCs are located closest to the endosteal surface [34]. Similarly in long bones, hematopoietic progenitors are located at greater distance from the bone surface [23, 24]. Thus perhaps quiescent HSCs are maintained singly close to the endosteal surface, while cycling HSCs have migrated closer to the vasculature, where they can be mobilized more readily.

In human bone, recently identified mesenchymal progenitors are located perivascularly, express CD146, and can form ectopic bone with hematopoietic marrow on transplantation [96]. $CD146^+$ perivascular cells similar to pericytes can be identified in multiple human organs, demonstrate differentiation into multiple mesenchymal lineages, and can generate ectopic bone in vivo upon transplantation [97]. In mice mesenchymal stem cells can be prospectively isolated based on expression of Sca-1 and PDGFRα. These cells also occupy a perivascular location in vivo, express Ang-1 and CXCL12, and, upon transplantation, can contribute to the hematopoietic niche [98]. Whether such cells are analogous to reticular cells within the bone marrow expressing high levels of CXCL12 [67] remains to be determined. These studies raise the intriguing possibility that such osteogenic progenitors represent one of the cellular components of the perivascular niche.

Bone formation by osteoblasts is tightly coupled to bone resorption by osteoclasts, descendants of the monocyte–macrophage lineage, so not surprisingly osteoclasts have also been implicated in the regulation of the HSC niche. In particular, stimulation of osteoclasts by stress or RANKL increases mobilization of HSPCs into the periphery. These effects may be mediated by resorption-induced alterations in the concentrations of niche-regulating factors such as SDF1, SCF, and osteopontin along the endosteum [99].

In contrast, adipocytes, which become increasingly numerous in the bone marrow as a function of age, have a negative impact on hematopoiesis [100]. Adipocyte-containing tail vertebrae have reduced HSC number as compared to thoracic vertebrae that lack adipocytes. Transgenic A-ZIP/F1 fatless mice exhibit enhanced hematopoietic recovery following bone marrow transplantation. Intriguingly, A-ZIP/F1 fatless mice have a dramatic osteogenic response following lethal irradiation [100]. Thus whether the absence of adipocytes enhances hematopoiesis directly or via promotion of osteogenesis remains to be seen.

## The microenvironment and disease

The clinical relevance of improved understanding of the osteoblast HSC niche, for instance enhancing hematopoietic recovery after bone marrow transplantation, has been further strengthened by recent studies demonstrating that the microenvironment can play a significant role in the pathogenesis of hematopoietic diseases. Ablation of RARγ within the microenvironment leads to a myeloproliferative syndrome (MPS) [101]. Similarly, absence of Rb in the microenvironment also predisposes to MPS when Rb is concurrently deleted from hematopoietic cells [56]. With the growing realization that leukemias may arise from leukemic stem cells (LSCs) in a manner analogous to hematopoietic stem cell differentiation, there is increasing interest in the possibility of an LSC niche [102]. In a xenograft model of human AML, leukemic stem cells home to the endosteal region [103]. There the stem cells may be protected from chemotherapy, and be enriched for quiescent cells. Moreover, human $CD34^+$ cord blood cells transformed with MLL-AF9 give rise to either ALL or AML depending upon the recipient, demonstrating that lineage fate of hematopoietic malignancies may be determined by host microenvironment [104]. That the microenvironment might be a target for therapeutic interventions has been suggested by the finding that an antibody against CD44, a receptor for osteopontin, may prevent LSC engraftment in vivo [105, 106].

## The bone marrow HSC microenvironment is complex

The bone marrow stromal environment is comprised of many different cell types. Currently, Mx1-Cre is frequently used in combination with transplanted wild-type marrow to demonstrate an "extrinsic" effect on hematopoiesis. However, Mx1 is expressed in a variety of stromal tissues, and therefore Mx1-cre experiments cannot be used to draw conclusions about which stromal cells are responsible for the effects seen. For example, although Mx1 is clearly expressed in the skeleton [107], ablation of Rb with Mx1-Cre and the early osteoprogenitor-specific Osx-Cre gives very different results [55, 56]. To add to the complexity, within a lineage cells at different stages of differentiation may serve differing functions in support of hematopoiesis. The accumulating evidence suggests that HSCs are closely associated with endosteal, presumably mature, osteoblasts. In contrast, cells of the osteoblast lineage also support B-cell development, at least in part by expression of IL-7, and B-cell precursors are largely found in the marrow space, where osteoprogenitors are expected. Major questions regarding the role of the osteoblast lineage remain. Do osteoprogenitors and mature osteoblasts serve differing functions and support different stages of hematopoietic development? Is direct contact required between hematopoietic cells and the osteoblast lineage? In addition, crucial cytokines may be produced by more than one cell type. In order to gain a better understanding of the relative contributions of the varying components of the bone marrow hematopoietic microenvironment, conditional tissue-specific ablation of candidate supporting cells and factors will be required.

# References

[1] Voog J, Jones DL. Stem cells and the niche: a dynamic duo. Cell Stem Cell 2010;6(2):103–15.

[2] Weissman IL, Anderson DJ, Gage F. Stem and progenitor cells: origins, phenotypes, lineage commitments, and transdifferentiations. Annu Rev Cell Dev Biol 2001;17:387–403.

[3] Schofield R. The relationship between the spleen colony-forming cell and the haemopoietic stem cell. Blood Cells 1978;4(1-2):7–25.

[4] Jung Y, Song J, Shiozawa Y, et al. Hematopoietic stem cells regulate mesenchymal stromal cell induction into osteoblasts thereby participating in the formation of the stem cell niche. Stem Cells 2008;26(8):2042–51.

[5] Cumano A, Godin I. Ontogeny of the hematopoietic system. Annu Rev Immunol 2007;25:745–85.

[6] Deguchi K, Yagi H, Inada M, et al. Excessive extramedullary hematopoiesis in Cbfa1-deficient mice with a congenital lack of bone marrow. Biochem Biophys Res Commun 1999;255(2):352–9.

[7] Komori T, Yagi H, Nomura S, et al. Targeted disruption of Cbfa1 results in a complete lack of bone formation owing to maturational arrest of osteoblasts. Cell 1997;89(5):755–64.

[8] Otto F, Thornell AP, Crompton T, et al. Cbfa1, a candidate gene for cleidocranial dysplasia syndrome, is essential for osteoblast differentiation and bone development. Cell 1997;89(5):765–71.

[9] Yoshida H, Hayashi S, Kunisada T, et al. The murine mutation osteopetrosis is in the coding region of the macrophage colony stimulating factor gene. Nature 1990;345(6274):442–4.

[10] Kronenberg HM. Developmental regulation of the growth plate. Nature 2003;423(6937):332–6.

[11] Aubin JE, Triffit JT. Mesenchymal stem cells and osteoblast differentiation. In: Bilezikiah JP, Raisz LG, Rodan GA, editors. Principles of Bone Biology. 2nd ed. New York: Academic Press; 2002. p. 59–81.

[12] Manolagas SC. Birth and death of bone cells: basic regulatory mechanisms and implications for the pathogenesis and treatment of osteoporosis. Endocr Rev 2000;21(2):115–37.

[13] Pyrah LN, Hodgkinson A, Anderson CK. Primary hyperparathyroidism. Br J Surg 1966;53(4):245–316.

[14] Marie PJ, de Pollak C, Chanson P, et al. Increased proliferation of osteoblastic cells expressing the activating Gs alpha mutation in monostotic and polyostotic fibrous dysplasia. Am J Pathol 1997;150(3):1059–69.

[15] Calvi LM, Sims NA, Hunzelman JL, et al. Activated parathyroid hormone/parathyroid hormone-related protein receptor in osteoblastic cells differentially affects cortical and trabecular bone. J Clin Invest 2001;107(3):277–86.

[16] Bilic-Curcic I, Kronenberg M, Jiang X, et al. Visualizing levels of osteoblast differentiation by a two-color promoter-GFP strategy: Type I collagen-GFPcyan and osteocalcin-GFPtpz. Genesis 2005;43(2):87–98.

[17] Rodda SJ, McMahon AP. Distinct roles for Hedgehog and canonical Wnt signaling in specification, differentiation and maintenance of osteoblast progenitors. Development 2006;133(16):3231–44.

[18] Wang YH, Liu Y, Buhl K, et al. Comparison of the action of transient and continuous PTH on primary osteoblast cultures expressing differentiation stage-specific GFP. J Bone Miner Res 2005;20(1):5–14.

[19] Patt HM, Maloney MA. Bone formation and resorption as a requirement for marrow development. Proc Soc Exp Biol Med 1972;140(1):205–7.

[20] Patt HM, Maloney MA. Bone marrow regeneration after local injury: a review. Exp Hematol 1975; 3(2):135–48.

[21] Gong JK. Endosteal marrow: a rich source of hematopoietic stem cells. Science 1978; 199(4336):1443–5.

[22] Lord BI, Testa NG, Hendry JH. The relative spatial distributions of CFUs and CFUc in the normal mouse femur. Blood 1975;46(1):65–72.

[23] Lambertsen RH, Weiss L. A model of intramedullary hematopoietic microenvironments based on stereologic study of the distribution of endocloned marrow colonies. Blood 1984;63(2):287–97.

[24] Nilsson SK, Johnston HM, Coverdale JA. Spatial localization of transplanted hemopoietic stem cells: inferences for the localization of stem cell niches. Blood 2001;97(8):2293–9.

[25] Dexter TM, Allen TD, Lajtha LG, et al. Stimulation of differentiation and proliferation of haemopoietic cells in vitro. J Cell Physiol 1973;82(3):461–73.

[26] Taichman RS, Emerson SG. Human osteoblasts support hematopoiesis through the production of granulocyte colony-stimulating factor. J Exp Med 1994;179(5):1677–82.

[27] Taichman RS, Reilly MJ, Emerson SG. Human osteoblasts support human hematopoietic progenitor cells in vitro bone marrow cultures. Blood 1996;87(2):518–24.

[28] Visnjic D, Kalajzic I, Gronowicz G, et al. Conditional ablation of the osteoblast lineage in Col2.3-deltatk transgenic mice. J Bone Miner Res 2001;16(12):2222–31.

[29] Visnjic D, Kalajzic Z, Rowe DW, et al. Hematopoiesis is severely altered in mice with an induced osteoblast deficiency. Blood 2004;103(9):3258–64.

[30] Zhang J, Niu C, Ye L, et al. Identification of the haematopoietic stem cell niche and control of the niche size. Nature 2003;425(6960):836–41.

[31] Calvi LM, Adams GB, Weibrecht KW, et al. Osteoblastic cells regulate the haematopoietic stem cell niche. Nature 2003;425(6960):841–6.

[32] Kiel MJ, Radice GL, Morrison SJ. Lack of evidence that hematopoietic stem cells depend on N-cadherin-mediated adhesion to osteoblasts for their maintenance. Cell Stem Cell 2007;1(2):204–17.

[33] Chan CK, Chen CC, Luppen CA, et al. Endochondral ossification is required for haematopoietic stem-cell niche formation. Nature 2009;457(7228):490–4.

[34] Lo Celso C, Fleming HE, Wu JW, et al. Live-animal tracking of individual haematopoietic stem/progenitor cells in their niche. Nature 2009;457(7225):92–6.

[35] Bowie MB, Kent DG, Dykstra B, et al. Identification of a new intrinsically timed developmental checkpoint that reprograms key hematopoietic stem cell properties. Proc Natl Acad Sci U S A 2007;104(14):5878–82.

[36] Bowie MB, McKnight KD, Kent DG, et al. Hematopoietic stem cells proliferate until after birth and show a reversible phase-specific engraftment defect. J Clin Invest 2006;116(10):2808–16.

[37] Harrison DE, Astle CM. Loss of stem cell repopulating ability upon transplantation. Effects of donor age, cell number, and transplantation procedure. J Exp Med 1982;156(6):1767–79.

[38] Orford KW, Scadden DT. Deconstructing stem cell self-renewal: genetic insights into cell-cycle regulation. Nat Rev Genet 2008;9(2):115–28.

[39] Wilson A, Laurenti E, Oser G, et al. Hematopoietic stem cells reversibly switch from dormancy to self-renewal during homeostasis and repair. Cell 2008;135(6):1118–29.

[40] Kiel MJ, Yilmaz OH, Iwashita T, et al. SLAM family receptors distinguish hematopoietic stem and progenitor cells and reveal endothelial niches for stem cells. Cell 2005;121(7):1109–21.

[41] Xie Y, Yin T, Wiegraebe W, et al. Detection of functional haematopoietic stem cell niche using real-time imaging. Nature 2009;457(7225):97–101.

[42] Czechowicz A, Kraft D, Weissman IL, et al. Efficient transplantation via antibody-based clearance of hematopoietic stem cell niches. Science 2007;318(5854):1296–9.

[43] Goodell MA, Brose K, Paradis G, et al. Isolation and functional properties of murine hematopoietic stem cells that are replicating in vivo. J Exp Med 1996;183(4):1797–806.

[44] Arai F, Hirao A, Ohmura M, et al. Tie2/angiopoietin-1 signaling regulates hematopoietic stem cell quiescence in the bone marrow niche. Cell 2004;118(2):149–61.

[45] Puri MC, Bernstein A. Requirement for the TIE family of receptor tyrosine kinases in adult but not fetal hematopoiesis. Proc Natl Acad Sci U S A 2003;100(22):12753–8.

[46] Arai F, Yoshihara H, Hosokawa K, et al. Niche regulation of hematopoietic stem cells in the endosteum. Ann N Y Acad Sci 2009;1176:36–46.

[47] Qian H, Buza-Vidas N, Hyland CD, et al. Critical role of thrombopoietin in maintaining adult quiescent hematopoietic stem cells. Cell Stem Cell 2007;1(6):671–84.

[48] Nilsson SK, Johnston HM, Whitty GA, et al. Osteopontin, a key component of the hematopoietic stem cell niche and regulator of primitive hematopoietic progenitor cells. Blood 2005;106(4):1232–9.

[49] Stier S, Ko Y, Forkert R, et al. Osteopontin is a hematopoietic stem cell niche component that negatively regulates stem cell pool size. J Exp Med 2005;201(11):1781–91.

[50] Grassinger J, Haylock DN, Storan MJ, et al. Thrombin-cleaved osteopontin regulates hemopoietic stem and progenitor cell functions through interactions with alpha9beta1 and alpha4beta1 integrins. Blood 2009;114(1):49–59.

[51] Adams GB, Chabner KT, Alley IR, et al. Stem cell engraftment at the endosteal niche is specified by the calcium-sensing receptor. Nature 2006;439(7076):599–603.

[52] Kiel MJ, Acar M, Radice GL, et al. Hematopoietic stem cells do not depend on N-cadherin to regulate their maintenance. Cell Stem Cell 2009;4(2):170–9.

[53] Schroeter EH, Kisslinger JA, Kopan R. Notch-1 signalling requires ligand-induced proteolytic release of intracellular domain. Nature 1998;393(6683):382–6.

[54] Mancini SJ, Mantei N, Dumortier A, et al. Jagged1-dependent Notch signaling is dispensable for hematopoietic stem cell self-renewal and differentiation. Blood 2005;105(6):2340–2.

[55] Walkley CR, Qudsi R, Sankaran VG, et al. Conditional mouse osteosarcoma, dependent on p53 loss and potentiated by loss of Rb, mimics the human disease. Genes Dev 2008;22(12):1662–76.

[56] Walkley CR, Shea JM, Sims NA, et al. Rb regulates interactions between hematopoietic stem cells and their bone marrow microenvironment. Cell 2007;129(6):1081–95.

[57] Maillard I, Koch U, Dumortier A, et al. Canonical notch signaling is dispensable for the maintenance of adult hematopoietic stem cells. Cell Stem Cell 2008;2(4):356–66.

[58] Kim YW, Koo BK, Jeong HW, et al. Defective Notch activation in microenvironment leads to myeloproliferative disease. Blood 2008;112(12):4628–38.

[59] Kirstetter P, Anderson K, Porse BT, et al. Activation of the canonical Wnt pathway leads to loss of hematopoietic stem cell repopulation and multilineage differentiation block. Nat Immunol 2006;7(10):1048–56.

[60] Reya T, Duncan AW, Ailles L, et al. A role for Wnt signalling in self-renewal of haematopoietic stem cells. Nature 2003;423(6938):409–14.

[61] Scheller M, Huelsken J, Rosenbauer F, et al. Hematopoietic stem cell and multilineage defects generated by constitutive beta-catenin activation. Nat Immunol 2006;7(10):1037–47.

[62] Trowbridge JJ, Xenocostas A, Moon RT, et al. Glycogen synthase kinase-3 is an in vivo regulator of hematopoietic stem cell repopulation. Nat Med 2006;12(1):89–98.

[63] Cobas M, Wilson A, Ernst B, et al. Beta-catenin is dispensable for hematopoiesis and lymphopoiesis. J Exp Med 2004;199(2):221–9.

[64] Fleming HE, Janzen V, Lo Celso C, et al. Wnt signaling in the niche enforces hematopoietic stem cell quiescence and is necessary to preserve self-renewal in vivo. Cell Stem Cell 2008;2(3):274–83.

[65] Guo J, Liu M, Yang D, et al. Suppression of Wnt signaling by Dkk1 attenuates PTH-mediated stromal cell response and new bone formation. Cell Metabolism 2010;11(2):161–71.

[66] Sugiyama T, Kohara H, Noda M, et al. Maintenance of the hematopoietic stem cell pool by CXCL12-CXCR4 chemokine signaling in bone marrow stromal cell niches. Immunity 2006;25(6):977–88.

[67] Nagasawa T. Microenvironmental niches in the bone marrow required for B-cell development. Nat Rev Immunol 2006;6(2):107–16.

[68] Zhu J, Garrett R, Jung Y, et al. Osteoblasts support B-lymphocyte commitment and differentiation from hematopoietic stem cells. Blood 2007;109(9):3706–12.

[69] Broxmeyer HE, Orschell CM, Clapp DW, et al. Rapid mobilization of murine and human hematopoietic stem and progenitor cells with AMD3100, a CXCR4 antagonist. J Exp Med 2005;201(8):1307–18.

[70] Levesque JP, Hendy J, Takamatsu Y, et al. Disruption of the CXCR4/CXCL12 chemotactic interaction during hematopoietic stem cell mobilization induced by GCSF or cyclophosphamide. J Clin Invest 2003;111(2):187–96.

[71] Petit I, Szyper-Kravitz M, Nagler A, et al. G-CSF induces stem cell mobilization by decreasing bone marrow SDF-1 and up-regulating CXCR4. Nat Immunol 2002;3(7):687–94.

[72] Katayama Y, Battista M, Kao WM, et al. Signals from the sympathetic nervous system regulate hematopoietic stem cell egress from bone marrow. Cell 2006;124(2):407–21.

[73] Christopher MJ, Liu F, Hilton MJ, et al. Suppression of CXCL12 production by bone marrow osteoblasts is a common and critical pathway for cytokine-induced mobilization. Blood 2009;114(7):1331–9.

[74] North TE, Goessling W, Walkley CR, et al. Prostaglandin E2 regulates vertebrate haematopoietic stem cell homeostasis. Nature 2007;447(7147):1007–11.

[75] Li X, Okada Y, Pilbeam CC, et al. Knockout of the murine prostaglandin EP2 receptor impairs osteoclastogenesis in vitro. Endocrinology 2000;141(6):2054–61.

[76] Adams GB, Alley IR, Chung UI, et al. Haematopoietic stem cells depend on Galpha(s)-mediated signalling to engraft bone marrow. Nature 2009;459(7243):103–7.

[77] Frisch BJ, Porter RL, Gigliotti BJ, et al. In vivo prostaglandin E2 treatment alters the bone marrow microenvironment and preferentially expands short-term hematopoietic stem cells. Blood 2009;114(19):4054–63.

[78] Jee WS, Ueno K, Deng YP, et al. The effects of prostaglandin E2 in growing rats: increased metaphyseal hard tissue and cortico-endosteal bone formation. Calcif Tissue Int 1985;37(2):148–57.

[79] Weinreb M, Suponitzky I, Keila S. Systemic administration of an anabolic dose of PGE2 in young rats increases the osteogenic capacity of bone marrow. Bone 1997;20(6):521–6.

[80] Klein-Nulend J, Pilbeam CC, Harrison JR, et al. Mechanism of regulation of prostaglandin production by parathyroid hormone, interleukin-1, and cortisol in cultured mouse parietal bones. Endocrinology 1991;128(5):2503–10.

[81] Cherian PP, Cheng B, Gu S, et al. Effects of mechanical strain on the function of Gap junctions in osteocytes are mediated through the prostaglandin EP2 receptor. J Biol Chem 2003;278(44):43146–56.

[82] Tang Y, Wu X, Lei W, et al. TGF-beta1-induced migration of bone mesenchymal stem cells couples bone resorption with formation. Nat Med 2009;15(7):757–65.

[83] Yamazaki S, Iwama A, Takayanagi S, et al. TGF-beta as a candidate bone marrow niche signal to induce hematopoietic stem cell hibernation. Blood 2009;113(6):1250–6.

[84] Yamazaki S, Iwama A, Takayanagi S, et al. Cytokine signals modulated via lipid rafts mimic niche signals and induce hibernation in hematopoietic stem cells. EMBO J 2006;25(15):3515–23.

[85] Tothova Z, Kollipara R, Huntly BJ, et al. FoxOs are critical mediators of hematopoietic stem cell resistance to physiologic oxidative stress. Cell 2007;128(2):325–39.

[86] Tokoyoda K, Egawa T, Sugiyama T, et al. Cellular niches controlling B lymphocyte behavior within bone marrow during development. Immunity 2004;20(6):707–18.

[87] Jung Y, Wang J, Schneider A, et al. Regulation of SDF-1 (CXCL12) production by osteoblasts; a possible mechanism for stem cell homing. Bone 2006;38(4):497–508.

[88] Wu J, Purton LE, Rodda SJ, et al. Osteoblastic regulation of B lymphopoiesis is mediated by Gsalpha-dependent signalling pathways. Proc Natl Acad Sci U S A 2008;105(44):16976–81.

[89] Wu JY, Scadden DT, Kronenberg HM. Role of the osteoblast lineage in the bone marrow hematopoietic niches. J Bone Miner Res 2009;24(5):759–64.

[90] Kiel MJ, Morrison SJ. Uncertainty in the niches that maintain haematopoietic stem cells. Nat Rev Immunol 2008;8(4):290–301.

[91] Salter AB, Meadows SK, Muramoto GG, et al. Endothelial progenitor cell infusion induces hematopoietic stem cell reconstitution in vivo. Blood 2009;113(9):2104–7.

[92] Slayton WB, Li XM, Butler J, et al. The role of the donor in the repair of the marrow vascular niche following hematopoietic stem cell transplant. Stem Cells 2007;25(11):2945–55.

[93] Hooper AT, Butler JM, Nolan DJ, et al. Engraftment and reconstitution of hematopoiesis is dependent on VEGFR2-mediated regeneration of sinusoidal endothelial cells. Cell Stem Cell 2009;4(3):263–74.

[94] Dominici M, Rasini V, Bussolari R, et al. Restoration and reversible expansion of the osteoblastic hematopoietic stem cell niche after marrow radioablation. Blood 2009;114(11):2333–43.

[95] Schaniel C, Moore KA. Genetic models to study quiescent stem cells and their niches. Ann N Y Acad Sci 2009;1176:26–35.

[96] Sacchetti B, Funari A, Michienzi S, et al. Self-renewing osteoprogenitors in bone marrow sinusoids can organize a hematopoietic microenvironment. Cell 2007;131(2):324–36.

[97] Crisan M, Yap S, Casteilla L, et al. A perivascular origin for mesenchymal stem cells in multiple human organs. Cell Stem Cell 2008;3(3):301–13.

[98] Morikawa S, Mabuchi Y, Kubota Y, et al. Prospective identification, isolation, and systemic transplantation of multipotent mesenchymal stem cells in murine bone marrow. J Exp Med 2009;206(11):2483–96.

[99] Kollet O, Dar A, Shivtiel S, et al. Osteoclasts degrade endosteal components and promote mobilization of hematopoietic progenitor cells. Nat Med 2006;12(6):657–64.

[100] Naveiras O, Nardi V, Wenzel PL, et al. Bone-marrow adipocytes as negative regulators of the haematopoietic microenvironment. Nature 2009;460(7252):259–63.

[101] Walkley CR, Olsen GH, Dworkin S, et al. A microenvironment-induced myeloproliferative syndrome caused by retinoic acid receptor gamma deficiency. Cell 2007;129(6):1097–110.

[102] Lane SW, Scadden DT, Gilliland DG. The leukemic stem cell niche: current concepts and therapeutic opportunities. Blood 2009;114(6):1150–7.

[103] Ishikawa F, Yoshida S, Saito Y, et al. Chemotherapy-resistant human AML stem cells home to and engraft within the bone-marrow endosteal region. Nat Biotechnol 2007;25(11):1315–21.

[104] Wei J, Wunderlich M, Fox C, et al. Microenvironment determines lineage fate in a human model of MLL-AF9 leukemia. Cancer Cell 2008;13(6):483–95.

[105] Jin L, Hope KJ, Zhai Q, et al. Targeting of CD44 eradicates human acute myeloid leukemic stem cells. Nat Med 2006;12(10):1167–74.

[106] Krause DS, Lazarides K, von Andrian UH, et al. Requirement for CD44 in homing and engraftment of BCR-ABL-expressing leukemic stem cells. Nat Med 2006;12(10):1175–80.

[107] Kuhn R, Schwenk F, Aguet M, et al. Inducible gene targeting in mice. Science 1995;269(5229):1427–9.

# 5

# Osteoblasts and their Signaling Pathways: New Frontiers for Linkage to the Immune System

### Jane B. Lian[1,2], Ellen M. Gravallese[4], Gary S. Stein[1,3]

[1]DEPARTMENTS OF CELL BIOLOGY, [2]ORTHOPEDICS AND PHYSICAL REHABILITATION,
[3]CANCER CENTER, [4]DEPARTMENT OF MEDICINE, DIVISION OF RHEUMATOLOGY,
UNIVERSITY OF MASSACHUSETTS MEDICAL SCHOOL, WORCESTER, MA, USA

## CHAPTER OUTLINE

| | |
|---|---|
| Introduction | 102 |
| **Bone as a connective tissue, organ and supportive environment for other tissues** | 103 |
|     Bone structure | 103 |
|     Bone marrow niches | 104 |
|     Osteoblasts-lineage cells | 105 |
| **Functional activities of osteoblast lineage cells** | 108 |
|     Coupled bone resorption and formation | 108 |
|     Mechanotransduction and bone viability | 109 |
|     Hormonal regulation and responses | 110 |
| **Regulatory networks controlling the progression of osteoblast differentiation** | 113 |
|     Morphogens | 113 |
|     Mechanisms of transcriptional regulation | 115 |
|         *Transcription factors for control of osteoblast differentiation* | 116 |
|     Specificity for osteoblast lineage determination in nuclear microenvironments | 119 |
|         *MicroRNA regulation of bone formation and osteoblast differentiation* | 122 |
| **Osteoblast and immune cell crosstalk: A peek into the future** | 124 |
| **Immune responses and bone pathology: Effects of inflammation on bone in rheumatoid arthritis** | 125 |
|     Pathology and cytokine activities | 125 |
|     Repair mechanisms in RA | 126 |
|     Regulation of Wnt signaling in RA | 127 |
| **Acknowledgments** | 128 |
| **References** | 128 |

# Introduction

Bone is both a connective tissue and an organ whose activities are regulated in concert with other tissues. Maintaining bone tissue function and structure is a dynamic process. It occurs in the adult through a cyclic process of resorption by osteoclasts and formation by osteoblasts, mostly at sites on trabecular bone. This process of "bone turn over" plays a central role in mineral homeostasis and in the response of the skeleton to mechanical forces, providing signals to build new bone throughout the skeleton and to maintain growth and skeletal integrity. Bone is also involved in the tight regulation of serum calcium and phosphate ion concentrations by releasing stored mineral from hydroxyapatite into the circulation in response to resorptive stimuli such as parathyroid hormone or $1,25(OH)_2D_3$. In contrast calcitonin secretion occurs if serum calcium is elevated beyond the normal range. This hormone abruptly ceases the bone resorption process by inhibiting osteoclast activity and initiating the remodeling process of bone formation. The field of osteoimmunology has emerged from the initial discovery of the physical and functional interaction between bone-forming osteoblast lineage cells that express receptor activator of nuclear factor kappa B ligand (RANKL) on their surface and osteoclast mononuclear precursors expressing the RANK receptor. This intraction is essential for the fusion and differentiation of mononuclear pre-osteoclasts to multinucleated cells with bone-resorbing activity. The recognition that immune cells produce RANKL and secrete cytokines that stimulate osteoclastogenesis and bone resorption has led to important new concepts in the pathogenesis of inflammatory bone diseases. However, there is far less known about the connection between the immune system and osteoblasts. It is clear that osteoblasts support the survival of hematopoietic lineage cells in bone marrow niches, as well as early steps in the maturation of T and B cells. Parallel pathways are operative in osteoblasts and the hematopoietic cells that influence the differentiation and activities of subpopulations of bone-forming and immune lineage cells. While there is a paucity of direct evidence for the specific pathways that might regulate this process, understanding the responses of osteoblast lineage cells and their functional activities provides insights for future studies.

The goal of this chapter is to review the basic biology of osteoblasts and the regulation of their differentiation in the context of analogous pathways in immune cells. By identifying mechanisms regulating the differentiation of mesenchymal and hematopoietic cell lineages, we have discovered the essential regulatory factors and activation signals that commit cells to distinct phenotypes, as well as negative regulators that suppress differentiation to alternative phenotypes. It is reasonable to postulate that mechanisms exist to regulate crosstalk between osteoblast lineage and immune cells in the bone microenvironment that are mediated by the secretion of factors from these cells. In this chapter, we describe the current understanding of the regulatory roles that osteoblast lineage cells play in supporting activities of the hematopoietic stem cell niche from which immune cells originate. Finally we provide examples of bone diseases that are models for understanding osteoblast responses to immune cells.

# Bone as a connective tissue, organ and supportive environment for other tissues

## Bone structure

The structural and metabolic properties of bone tissue arise from the organization and interactions of functionally distinct cell populations. During development, bone tissue originates from the mesoderm and forms by two distinct processes. *Membranous bone formation*, as for example in the developing calvarium, occurs by condensation of mesenchymal stem cells (MSCs) that directly differentiate into osteoblasts. A marrow space will separate two plates of bone tissue which expand by appositional growth from the periosteum and the endosteal surface. The membranous process of formation and expansion of the cranial vault is regulated by the periosteum and suture tissue from which the osteoprogenitors arise. This process is under the regulation of FGF signaling and homeodomain proteins. Mutations in these pathways result in craniosynostosis and craniofacial abnormalities [1–3]. The formation and continued growth of all long bones proceeds by the *endochondral bone formation process*. Here, MSCs proliferate, condense and differentiate into a cartilage anlagen defining a template for bone. The anlagen develops a bone collar under the control of an intricate signaling cascade of parathyroid hormone-related peptide (PTHrP) which regulates the pace of chondrocytes proliferation and Indian hedgehog (Ihh) which control the rate of bone collar formation inducing the hypertrophic zone. The result is the formation of the proximal and distal growth plates which are separated by a medullary cavity that houses the marrow. The growth plate undergoes a well-defined sequence of events that includes chondrocyte proliferation, hypertrophy, and the production of a temporary calcified cartilage matrix. This is then replaced by bony spicules which arise from osteoprogenitor cells that are abundant on the marrow. Readers are referred to excellent reviews on the development of the growth plate [4, 5] for additional information about this process.

The overall structure of a mature endochondral bone includes many different tissues that contribute to the maintenance of the mechanical and organ functions of bone. These must be considered in order to fully understand the complex mechanisms that regulate neighboring immune cells in the bone marrow. The *periosteum* covering bone and the continuous perichondrium apposed to cartilage in growing bone are connective tissue membranes separating bone from muscle. This endosteal covering of bone consists of fibroblasts together with mesenchymal stem cells and more committed osteochondroprogenitor cells. Deeper towards the bone surface, pre-osteoblasts can be identified behind a row of surface osteoblasts at sites of active bone formation. Alternately, flattened lining cells cover quiescent bone surfaces on either periosteal or endosteal surfaces of bone. The latter is the surface that faces the marrow. Dispersed throughout the periosteum is a vascular network. The *solid bone* of the cortex is composed of *osteons* in which a central longitudinal Haversian canal houses the blood vessels and nerve cells. This canal is surrounded by layers of bone matrix with individual osteocytes residing in lacunae and forming a syncytium of interconnected cells through an extensive network of osteocytic cellular processes lying in canaliculi. Blood vessel channels also run in the horizontal direction (Volkman canals). The *trabecular bone* is comprised of the plates or spicules of bone surrounded by marrow in the metaphysis that provide inner structural strength to bone, but importantly also carry out metabolic functions. Trabecular bone exposed to the marrow vasculature and hematopoietic lineage cells can readily recruit monocyte precursors for fusion into multinucleated osteoclasts on the bone surface. The *medullary cavity* houses the bone marrow. This also contains a bedding of stroma in which there are mesenchymal cells that can differentiate into skeletal tissue and blood vessels. The bone marrow is rich in maturing hematopoietic-derived blood lineage cells, which eventually are released into the circulation. It is in the medullary cavity at the endosteum where osteoblasts, hematopoietic stems cells, their lineage-derived osteoclasts, and immune cells, all interact for survival and differentiation [6].

## Bone marrow niches

Marrow is highly vascularized and is believed to contain several "niches" of different cell types (reviewed in [7, 8]). A "niche" is defined as a "limited supportive microenvironment where specialized cells reside (quiescent), undergo self renewal and differentiate" [9]. Stem cell renewal is supported by BMP, Wnt, and FGF signaling. The hematopoietic stem cell (HSC) niche is a unique environment of low oxygen in which hypoxia is believed to be involved in maintenance of HSC survival and cell cycle control [10]. The normal HSC "niche" was first defined by the location of spindle-shaped, N-cadherin+, CD45– expressing osteoblasts, a subset of the osteoblast-lining cells located on the endosteal surfaces of bone [11–13]. Stimulation of osteoclastic bone resorption, and degradation of the endosteal surface, mobilizes HSCs, highlighting the critical role of both osteoclasts and osteoblasts in regulation of the HSC "niche" [14, 15]. A vascular "niche" composed of endothelial cells, supports the homing of HSCs to the endosteum and the release of lineage cells into the circulation [16]. The present paradigm is that the osteoblast "niche" maintains HSCs in a quiescent state, while the vascular "niche" promotes proliferation. There is a growing awareness of a metastatic cancer stem cell niche but knowledge of the defining markers of cancer stem cells or a tumor-initiating cell is currently rudimentary (reviewed in [17, 18]).

Hematopoietic stem cells give rise to all the blood lineages including the mononuclear precursors that differentiate into osteoclasts and the immune cell progenitors in the bone marrow that mature in the thymus to become T-lymphocytes. The close relationship between the requirement for osteoblasts and HSCs was identified by several in vivo studies showing that if osteoblasts were depleted in mice, so too were HSCs; and, if the unique subset of osteoblast-lining cells was increased, as in the BMPR1A knock-out mice or in PTH/PTHrP transgenic mice, the quiescent HSC population increased [11, 12]. The endosteal HSC niche can also be stimulated by sonic hedgehog, increasing both osteoblasts and HSC progenitors, while decreasing the more differentiated lymphocyte compartment [19].

Many studies have suggested specific interacting and/or complementary factors that influence the communication between osteoblasts and HSC. These influence survival or differentiation but are not established as essential requirements. For example, Notch signaling maintains progenitor pools for both osteoblasts [20, 21] and HSCs [22], and osteoblasts can promote HSC repopulation by increasing Notch signaling between HSCs and osteoblasts [23, 24]. The chemokine SDF-1 is expressed in stroma and osteoblast-lineage cells, and HSCs express its receptor CXCR4, which can support the homing of HSCs to bone [25]. Many proteins in the bone niche, including N-cadherin, β-catenin, VCAM, osteopontin with associated integrins and the calcium sensing receptor, appear to facilitate the adhesion of HSCs (reviewed in [7, 8]).

Bone marrow stromal cells produce two factors, CXCL12 and IL-7, which provide niches for lymphoid progenitors and immune memory cells [26]. The earliest B lymphocyte, pre-pro-B, appears to locate in the vicinity of CXCL12-expressing stromal cells, while the more mature pro-Bs are associated with IL-7-expressing osteoprogenitors [26]. PTH also stimulates B-cell development consistent with an increase in osteoprogenitor cells and IL-7 [12; 27]. The significance of the contribution of PTH to the functional activities of the bone cell–lymphoid cell niche was demonstrated in mice lacking the PTHR/G protein-coupled receptor subunit alpha. These animals were depleted of bone marrow β-lymphocyte precursors and circulating β lymphocytes [28]. Adhesion is an important component in the development of lymphoid cells. Lin–Sca1+Rag2– bone marrow cells in vitro require attachment to osteoblasts for maturation to lymphopoiesis by a process that is mediated by IL-7, SDF-1, and VCAM-1 [27]. Significantly, eliminating more mature cells of the osteoblast lineage in 2.3 col2.3Delta-TK transgenic mice, caused B-cell lineage cells in the bone to be depleted prior to the decline in HSCs [27]. In summary, there appear to be defined niches operating in the bone marrow microenvironment where osteoblasts-lineage cells appear to have somewhat selective functions for supporting the HSCs, and their development to other lineages, including memory T cells and B cells. While the HSC lineage has been well defined in terms of the identification of markers that indicate each stage of differentiation, the osteoblast lineage population is not characterized to the same extent, when many more subpopulations of osteogenic MSCs are described by markers in the future HSC–osteoblast interactions may be better defined.

## Osteoblasts-lineage cells

There are morphologically distinct bone-forming lineage cells which develop the various bone structures and mediate the specialized functions of the mammalian skeleton. *Osteoprogenitor cells* (OPCs) arise from mesenchymal stem cells (MSCs) that are recruited from the bone marrow, the circulation, or the periosteum. Marrow preparations take advantage of the adherent population for enrichment of mesenchymal stromal cells from the hematopoietic populations. The STRO-1+ fraction of rat and adult human bone marrow contains the osteogenic precursors but this lineage marker is not expressed in mouse marrow. For mouse, Sca1 and CD34 are useful markers, particularly for C57/BL6 mice [29]. Osteoprogenitors can be tri- or bi-potential and are directed to specific cell fates by transcriptional regulators that represent "master switches". Commitment of the MSC to the OPC is under the control of numerous development signaling factors, including the TGF/BMP superfamily and the Wnt family of secreted glycoproteins (detailed below), but ultimately determined by tissue-specific transcription factors (Figure 5-1). One mechanism for tissue specification is the ability of intracellular mediators of a developmental signal to interact with a cell-type-specific transcription factor. Such co-regulatory complexes have been demonstrated for the BMP/Smad intracellular receptors forming a complex with MyoD for muscle cell differentiation [30] and with Runx2 for osteoblastogenesis [31–33]. Potency of the master transcription factors is reflected by their ability to transdifferentiate cell phenotypes through forced expression of a transcriptional regulator in vitro (Figure 5-1). For example, by expressing either peroxisome proliferation-activated receptor γ2 (PPARγ2) in pre-osteoblasts or Runx2 in pre-adipocytes, their phenotype can be changed [34]. Hence, the fatty marrow that occurs with aging and is associated with osteoporosis may in part be a consequence of deregulated activity of the master transcription factor. Committed phenotypes from MSCs have a certain degree of "plasticity" and may dedifferentiate during proliferation and post-mitotically assume a different phenotype dependent on the local cellular

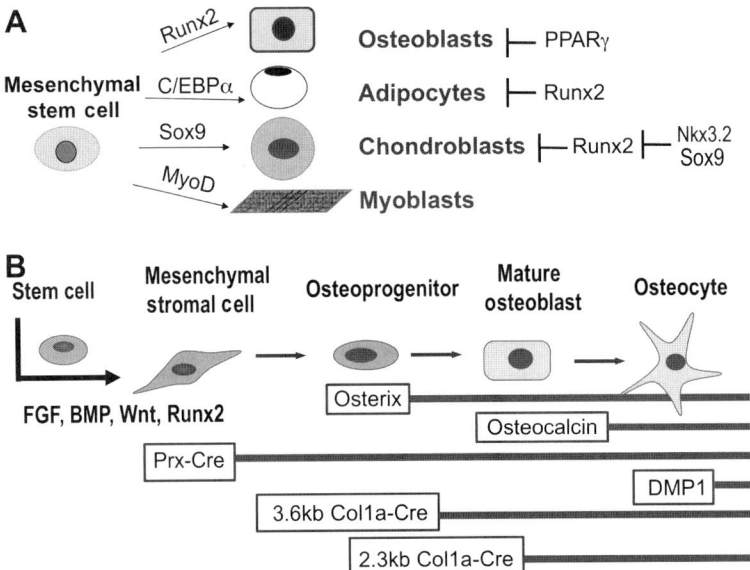

FIGURE 5-1 Lineage determination from mesenchymal stem cells. **(A)** Master transcription factors (TFs) essential for commitment to the indicated phenotypes are shown. In vitro overexpression of the PPARγ will block the osteoblast phenotype and Runx2 prevent adipocyte differentiation. Stromal cells isolated from marrow pluripotent mesenchymal cells (MSCs) are Runx2-positive and become downregulated by increases in TFs required for other lineages, e.g., Nkx3.2 and Sox9 in chondroblasts. **(B)** Illustrates stages in the osteogenic lineage in which genes can be excised in mice using Cre drivers with characterized promoters expressing at the indicated stage (reviewed in [218]). Please refer to color plate section.

environment. An example is the ability of pericytes and endothelial cells to enter the osteogenic lineage leading to pathologic vascular calcification [35]. Further along the osteoprogenitor pathway is the *preosteoblast*. This cell has the ability to divide, but in a non-dividing state becomes recognizable near the bone surface by its proximity to surface osteoblasts. The preosteoblasts have robust Runx2 expression, and serves as one of the earliest markers of the osteoblast phenotype, low levels of alkaline phosphatase enzyme activity.

*Osteoblasts* are the non-proliferating mature cuboidal cells on the active bone-forming surface that synthesize the bone matrix. Formation of the bone matrix includes the principal component Type I collagen, which constitutes 90% of the organic compartment, along with several minor collagens including type V, and type XII, which assist in organizing the fibrils. Alkaline phosphatase expression is further induced during this "matrix maturation" stage when specialized, non-collagenous, bone proteins are secreted to render the ECM capable of mineral deposition. Mineralization results in increased

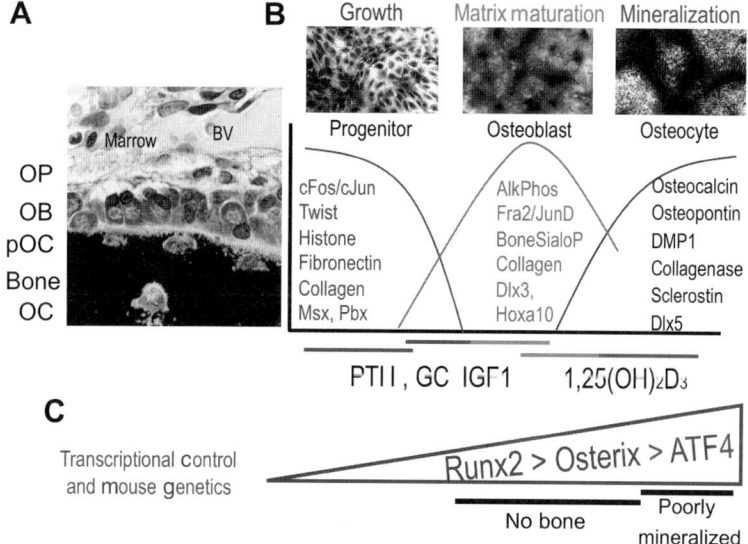

**FIGURE 5-2 Stages and markers of osteoblast growth and differentiation. (A)** Histologic section of the surface of bone showing organization of osteoprogenitor (OP) cells behind a row of mature osteoblasts (OB) on the bone surface secreting an osteoid non-mineralized matrix (pale blue color). As hydroxyapatite deposits in the osteoid, OBs become preosteocytes (pOC) until fully surrounded by mineral (bone, black, von Kossa stain) and differentiated to osteocytes (OC). **(B)** Isolation of osteoblasts from perinatal calvarial tissue can recapitulate the stages of osteoblast maturation from the proliferating progenitor to the postproliferative osteoblast and osteocyte. Growth stage (Toluidine-blue-stained cells) are characterized by proliferation markers (histone) and transcription factors that suppress genes expressed in OBs. As cells multilayer, they become alkaline-phosphatase-positive, designated as the matrix maturation stage, with maximal secretion, collagens and non-collagenous proteins such as bone sialoprotein, osteocalcin, osteopontin. This allows the matrix to mineralize, designated mineralization stage. In the osteocyte E11, MEPE, Dentin matrix protein 1 and Sclerostin, a Wnt pathway inhibitor, serve as identification markers for this population of bone cells. Hormones that have a major effect at each stage are indicated. **(C)** Transcription factors in hierarchical order are shown in the triangle as these have been proven in mouse models to be essential for initiation of bone formation and mineralization. Lines indicate the phenotype in the null mouse for each gene. Please refer to color plate section.

expression of osteocalcin, osteopontin, thrombospondin, bone sialoprotein, that can be considered markers of the mature osteoblast (Figure 5-2). These ECM proteins all contribute to bone remodeling [36–43]. Osteoblasts are joined on the bone surface by adhering junctions comprised of cadherins which are calcium-dependent transmembrane proteins that are important for cell–cell adhesion. Different cadherins in mesenchymal cells appear to define their differentiation pathways. The major cadherins expressed in osteoblasts include N-cadherin and cadherin-11, which are critical for embryonic limb development [44, 45].

The *osteocyte* is the terminal differentiation stage of the osteoblast, Osteocytes are encased in bone and support bone structure and viability. These cells appear negative for alkaline phosphatase in bone, but can produce large amounts of osteocalcin and the dentin matrix protein (DMP1) which serve as markers that can identify these cells. Sclerostin, an inhibitor of Wnt signaling [46], is also highly expressed in osteocytes, which is one mechanism by which these cells regulate the amount of bone that is formed. Osteocytes are metabolically and electrically coupled to each other and to osteoblasts on the bone surface through gap junction protein complexes which are principally comprised of Connexin 43 [46, 47]. Gap junction formation is essential for osteocyte maturation, activity, and survival [44, 48]. Osteocytes can reside for long periods, even decades, in healthy bone that is not turned over. In aging bone, empty lacunae are observed indicating osteocytes have undergone apoptosis by disruption of their intercellular communication systems [49]. In mouse models of accelerated aging, osteopenic bone induced by ovariectomy or glucocorticoid treatment has been identified. This finding confirmed that programmed cell death contributes to active bone loss [50]. The bisphosphonates, which inhibit bone resorption and are used to treat osteoporosis, have also been found to prevent osteoblast and osteocyte apoptosis [51].

# Functional activities of osteoblast lineage cells

The significance of the distinct lineage populations of the osteoblast is their specialization in to distinct functional activities, which includes interactions and responses with hematopoietic lineage cells. The bone-remodeling cycle is best defined in terms of osteoblast and immune cell interactions that stimulate bone resorption. In addition, the mechanosensory function and hormonal responses that ultimately connect to the vascular system are likely influenced by changes in myeloid and immune cell populations. However, these latter interactions are unexplored at the present time.

## Coupled bone resorption and formation

One key function of osteoblast lineage cells from the osteoprogenitor stage to the active osteoblast on bone surfaces is the regulation of osteoclastogenesis and bone turnover (detailed in other chapters of this book). The multinucleated *osteoclasts* are unique to bone. They contain specialized properties for resorbing bone, including the ruffled border membranes, which are generated when osteoclasts attach to the bone surface. Sealing zones which exclude the resorbing space from the extracellular environment, and the ability to create an acidic compartment in resorbing space for mineral dissolution. Osteoclasts derive from the HSC lineage and are found on bone surfaces in a resorbing lacunae flanked by other cell lineages. The requirement for stromal osteoprogenitors or osteoblasts in mediating osteoclast differentiation is linked to the role of osteotrophic hormones and cytokines in the development of preosteoclasts and the formation of the multinucleated osteoclasts. Stromal cells and more committed osteoprogenitor-cells express high levels of RANKL compared to the osteocyte and secrete M-CSF and OPG which also regulate hematopoietic lineage cell differentiation into osteoclasts, the precursors of the osteoclast express the receptor c-fms and RANK which bind M-CSF and RANKL respectively [14, 52]. The calcitrophic hormones $1,25(OH)_2D_3$ and parathyroid hormone (PTH) promote osteoclastogenesis at multiple stages of differentiation and activation through receptors on osteoblast lineage cells. In response to these hormones, osteoblast lineage cells produce RANKL bound to the cell surface which interacts with RANK on osteoclast lineage cells to promote osteoclast differentiation, activity, and bone resorption. It is now appreciated that numerous stimulators of bone resorption, including glucocorticoids, T4/T3, TNF$\alpha$, IL-1, IL-11, and PGE-2, regulate osteoblast expression of RANKL [14]. Osteoblasts are also the major source of the colony-stimulating factor 1 (CSF1) which is also regulated by $1,25(OH)_2D_3$ and PTH hormones. Osteoblasts secrete several cytokines, including IL-1, IL-6, and TNF$\alpha$ that have mitogenic effects on osteoclast progenitors. Prostaglandins produced by bone cells are a potent stimulator of bone formation, and resorption [53]. Toll-like receptors are also expressed on osteoblasts and activation of these receptors induces secretion of RANKL and TNF$\alpha$ [54]. Thus osteoblasts are coupled to osteoclastic resorption through many physiological signals.

In addition to this feed-forward mechanism for promoting bone resorption through the coupling of osteoblast and osteoclast activities, there are reversal and protective mechanisms for regulating bone turnover. Osteoblasts secrete osteoprotegerin, a soluble decoy receptor for RANKL and an inhibitor of osteoclastogenesis. To further ensure osteoclastogenesis, the calcitrophic hormones, as well as glucocorticoids, also inhibit OPG production in addition to stimulating RANKL. The ephrin signaling system is a bidirectional regulator of bone homeostasis that operates through coupling of ephrin B2 in osteoclasts with ephrin B4 in osteoblasts. Activation of this system results in inhibition of osteoclast differentiation and stimulation of bone formation [55]. To complete the remodeling cycle, bone formation at the resorption site (designated the bone-remodeling unit) is stimulated by the release of growth factors from bone during resorption which promotes both proliferation and differentiation of osteoprogenitors [56].

## Mechanotransduction and bone viability

Osteocytes, which constitute the majority of bone in the skeleton, are the mechano-transducting cells of bone within the mineralized matrix, these cells are in direct communication with each other and surface osteoblasts through their cellular processes [46, 57, 58]. The primary function of the osteoblast (or lining cell)–osteocyte continuum is considered to be the conversion of stress signals (stretching, bending) into biological activity. The flow of extracellular fluid in response to mechanical forces throughout the canalicular system induces a spectrum of immediate cellular responses in osteocytes, including release of prostaglandin $PGE_2$, and nitrous oxide [59]. An important anabolic response of the osteocyte to mechanical loading in bone is stimulation of the Wnt β-catenin pathway, NFATc1 and Runx2 [60–62]. The mechanosensing mechanisms of osteocytes which are transduced to the nucleus to modulate gene transcription are beginning to be characterized. Bone is a dynamic and adaptive tissue and the loading of forces on a skeleton does not require strenuous exercise to provide a stimulus for MSCs to enhance osteogenesis and reduce adipogenesis [63]. In vitro studies suggest that fluid flow, an exogenous mechanical signal over progenitor cells, can promote osteogenic differentiation [64]. While a specific mechanoreceptor has not been identified, much knowledge has been gained related to the intracellular connections and the intracellular signaling pathways that are activated (reviewed in [46, 57]). One of the cellular structures that is considered to mediate mechanosensory stimulus are the solitary cilium found in bone and cartilage [65]. How this cascade of events results in to an anabolic stimulus, shifting MSC differentiation into osteoblasts, is a complex process involving all osteoblast lineage cells.

## Hormonal regulation and responses

Steroid and polypeptide hormones contribute to the growth of osteoprogenitors and differentiation of osteoblasts. The steroid hormones primarily promote the progression to mature osteoblasts through direct gene regulation of bone-related proteins. Polypeptide hormones on the other hand mediate growth responses, (PTH), or relay responses to and from bone that affect bone mass, leptin. PTH stimulates growth of osteoprogenitor populations in addition to its response to maintaining mineral homeostasis and promoting bone resorption [66]. When used clinically, the anabolic and catabolic effects of PTH can be regulated by the frequency and duration of administration. Leptin has multiple functions that include the regulation of energy homeostasis, bone metabolism, and proinflammatory immune responses. It is an adipokine secreted largely by adipose tissue, but also by human osteoblasts at the mineralization stage. Leptin is recognized as an important regulator of bone cell functions and appears to control bone growth and osteoblast activities through several distinct mechanisms (reviewed in [67]). It was characterized as a systemic factor that could contribute to decreased bone mass in a null mouse model acting through the hypothalamus [68, 69]. Adrenomedullin is a member of the calcitonin gene-related peptide family and is secreted by many tissues. It increases cellular proliferation in cultured calvarial cells and has an anabolic effect on the skeleton in vivo (reviewed in [70]). Of interest, adrenomedullin production in prostate cancer cells was one of the factors suggested to promote osteoblastic metastatic disease.

The sex steroids at physiologic levels have anabolic effects on bone but through different mechanisms. Androgens directly stimulate osteoblast proliferation [71], while estrogen effects mediated by ERα protect the skeleton from osteoclastic activity (reviewed in [72]). Estrogens and their agonists induce osteoprotegerin (OPG) in osteoblasts, a factor which inhibits osteoclastic differentiation. Androgens which improve body composition, muscle function and stimulate periosteum and trabecular bone formation were shown in a genome-wide microarray screen to decrease Axin and Axin2, negative regulators of β-catenin and increase β-catenin protein [73, 74]. This finding indicates one mechanism for androgen's anabolic effects on bone [71].

Glucocorticoids (GCs) are necessary for embryonic development [75, 76]. Endogenous glucocorticoid signaling is required to maintain normal bone structure and normal suture craniofacial development [77] (Figure 5-3). Osteoblast-targeted disruption of GC signaling impairs differentiation and cortical bone mass [78]. GCs have complex effects on osteoblasts and the adult skeleton both directly on gene regulation in osteoblasts and indirectly by affecting multiple metabolic pathways [79]. Dexamethasone promotes the differentiation of MSCs to osteoblasts in vitro, but depletes the osteoprogenitor pool. One mechanism by which it reduces osteoprogenitors is upregulation of BMP and Wnt antagonists in osteoblasts, and by inhibition of bone formation [80–82]. Dexamethasone induces follistatin and the BMP2 Dan inhibitor which inhibit mineralization of osteoblast matrix [83]; although this effect can be rescued by exogenous BMP2 [84–86]. Dexamethasone inhibits the anabolic effects of Wnt signaling in bone (see next section for description of Wnt on effects on the skeleton) by increasing DKK1 and sFRP-1 inhibitors of the pathway. These actions ultimately diminish the transcriptional activity of LEF/TCF, which are downstream regulators of Wnt-dependent gene expression [87, 88]. GCs also inhibit the kinases that phosphorylate GSK3B and allow β-catenin to enter the nucleus. GSK3β, mediator of Wnt signaling appears to be the target GC-induced bone loss [89, 90]. However, GSK3β was found to phosphorylate GC receptor and change the profile of target genes which resulted in attenuated GC-dependent osteoblast death [92]. See Figure 5-3 (panel B) for multiple "hit" points in the Wnt pathway affected by GCs. When used therapeutically as an anti-inflammatory, dexamethasone can contribute to osteoporosis by inducing cell apoptosis of osteoblasts [93, 94]. Further, glucocorticoids attenuate the anabolic effects of PTH [95]. Another complication of GCs is their suppression of bone formation via the osteoclast. This occurs because bone resorption is the initial step for inducing bone remodeling and formation [96, 97]. Overall, these multiple targets of glucocorticoids bring to realization the profound negative effects of this steroid on the adult skeleton.

The active hormone of vitamin $D_3$, $1,25(OH)_2D_3$ ($VD_3$), is a potent regulator of gene transcription, increasing or decreasing expression of numerous phenotypic genes that are consistent with more differentiated osteoblasts, (e.g. enhanced osteocalcin synthesis). The mechanisms for $VD_3$ effects are multiple through interactions of the vitamin D receptor complex (VDR) with numerous co-activators and other transcriptional regulators, such including Runx2 [98–100]. In addition, in response to BMP2-induced

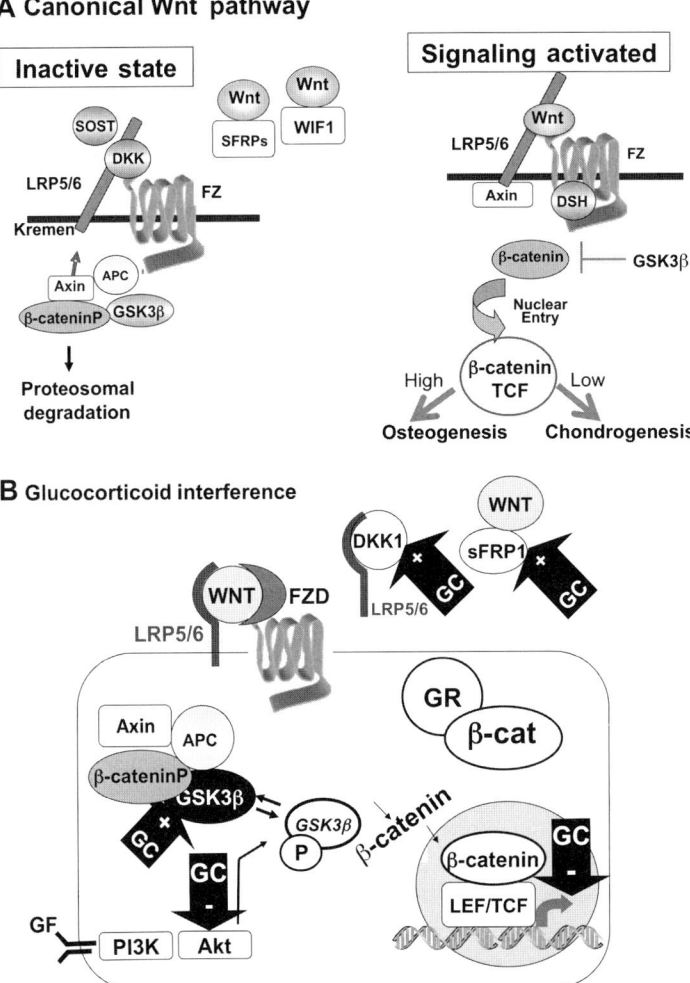

FIGURE 5-3 The canonical Wnt pathway (A) and disruption by glucocorticoids (B). (A) The inactive state is shown when the frizzled–LRP5/6 receptor complex is bound by multiple inhibitors, DKK , SOST, and Kremen. In addition Wnt signaling can be regulated by soluble inhibitors, WIF1 (Wnt inhibitory factor), and SFRPs (secreted frizzled related proteins) that bind Wnt ligands. In the absence of ligand binding, β-catenin, the intracellular mediator of Wnt signaling, is held in a complex with Axin, APC, and GSK3β, which phosphorylates β-catenin, resulting in its proteosomal degradation. With activation of the Wnt receptor, the intracellular β-catenin complex components are reorganized, releasing stabilized β-catenin (in the absence of phosphorylation) and allowing β-catenin nuclear translocation and complexing with the canonical transcription factor TCF/Lef. Several mouse genetic models have established that high cellular levels of β-catenin will drive an MSC to osteogenesis while lower levels can promote chondrogenesis (see text for references). (B) Multiple effects of glucocorticoids (GC) on Wnt signaling. Suppression of bone formation by dexamethasone is in part contributed by direct effects of glucocorticoids that inhibit Wnt signaling at multiple levels of the pathway as illustrated. GCs increases expression of Wnt inhibitors DKK1 and sFRP that target the receptor β-catenin degradation. The glucocorticoid receptor can bind to β-catenin reducing effectiveness of its activity for gene transcription. Together these multiple "hits" on canonical Wnt signaling decrease transcription of genes essential for bone formation. Please refer to color plate section.

osteoblastogenesis, 1,25(OH)$_2$D$_3$ increases expression of c-Myc and its association with promoters of VDR-dependent genes.

This brief survey of hormonal effects on the skeleton has been presented to call attention to immunologists of the spectrum of activities that converge within bone tissue, and has not discussed the complexity of regulation of these hormonal activities on bone metabolism. The growing knowledge of circulating factors from the hypothalamus–pituitary axis and the gut as an energy source that involves the brain–bone axis, together with the influence of fat metabolism on bone and the cannabinoid CB1 and CB2 receptors being expressed in bone cells and responsive to cannabinoids, are all complex regulatory circuits that may also alter any interactions of bone activities with the hematopoietic and immune cells (reviewed in [101–103]).

# Regulatory networks controlling the progression of osteoblast differentiation

The development of a tissue and its programs for cell phenotype differentiation are governed by a cascade of negative and positive regulatory factors that permit the temporal expression of genes. These factors and genes drive the maturation of osteogenic lineage cells to form bone, and include (but are not limited to) morphogenetic secreted signaling proteins, transcription factors, and the recently identified microRNAs that regulate embryonic bone formation and bone mass in the adult [104, 105]. These pathways are entangled in webs of intracellular signaling cascades and feed-forward and feed-reverse mechanisms that are all needed for the regulation of osteoblast differentiation. It is instructive to consider the major modulators of osteoblast maturation, and those factors that also represent markers for the stages of osteogenesis.

## Morphogens

The *BMP/TGFβ and Wnt signaling* factors are the prominent developmental signals for commitment to the osteoblast phenotype. There are over 30 BMPs and members of the TGF superfamily which affect the development of many tissues through five type I- and six type II-specific receptors on the cell surface. Both BMP and TGFβ ligands transduce their signals to intracellular receptor Smads (R-Smads), phosphorylating them for dimerization with Smad4, a common mediator having DNA-binding properties. Smads can also become phosphorylated through MAPK signaling independent of BMP/TGFβ receptors. The activated Smad complex enters the nucleus and forms multimeric complexes on gene promoters with other co-regulators and transcription factors. There are also a number of inhibitory Smads (I-Smads) regulating this pathway. External regulatory factors secreted from cells, (Chordin, Noggin and Dan), can neutralize BMP activity as well [106]. At another level, ubiquitin ligases degrade Smads; an example is the TGFβ1-induced Smurf1. BMP2/4/7 are the most osteogenic of the BMP family and are used clinically for bone reconstruction and to promote fracture healing [107]. Mouse models of deletion of different BMPs and their receptors have revealed those that are critical for fetal development, and those that have selective effects, such as in bone regeneration or repair [108–110].

The Wnt family of 19 signaling proteins are secreted glycoproteins that are involved in embryonic development, tissue induction, and axial polarity. Wnt proteins signal through several, different pathways. Two of these employ any number of available frizzled receptors; the canonical Wnt-β-catenin pathway and the non-canonical planer cell polarity (PCP) pathways. Canonical Wnt ligand/frizzled receptor interactions are regulated by the co-receptors LRP5, LRP6, a novel LRP4 receptor [111] and SOST signaling. β-catenin is found in a cytosolic complex with the tumor suppressor APC, the serine kinase CK1, Axin, and glycogen synthase kinase 3β (GSK-3β). The latter keeps β-catenin phosphorylated for proteosomal degradation until the receptor complex is activated. The binding of a Wnt ligand triggers a chain of events that recognizes the complexes (Figure 5-3A). Axin1 engages the LRP5 receptor and disheveled (DSH) forms of complex that block GSK-3β association with β-catenin leading to its dephosphorylation. This in turn that frees β-catenin to then enter the nucleus and complex with either of the two transcriptional regulators of canonical Wnt signaling, TCF or LEF. In contrast, the PCP pathway leads to the activation of small GTPases (Rho and Rac1), and stress kinases JNK and ROCK kinases to promote cytoskeletal remodeling and changes in cell adhesion and motility. Another non-canonical Wnt-calcium signaling pathway is mediated through G-protein receptors. This increases cytoplasmic free calcium leading to stimulation of PKC, a calcium calmodulin-mediated kinase and the calcineurin pathway for the activation of the AP1 and NFAT transcription factors on gene promoters. The combination of ligand and receptors that are differentially expressed in distinct cell populations mediates some degree of specificity of the response. However, these aspects of Wnt signaling are still being explored. Wnt ligands are associated with either canonical or non-canonical Wnt signaling and several Wnts can activate or influence multiple pathways. Limb morphogenesis and skeletal patterning are greatly influenced by specific Wnt ligands. Wnt3a and Wnt7a are expressed in the lumbar vertebrae. Wnt9a and Wnt14 are required for joint formation [112] and Wnt4 and Wnt5a promote chondrocyte maturation. A key Wnt required for bone formation, established in vivo, is Wnt10b which increases osteoblastogenesis and postnatal bone mass and inhibits adipogenesis [113, 114].

Canonical Wnt signaling is now recognized as one of the key events for normal bone development and the regulation of bone mass in the adult [115, 116]. The discovery came from studies in humans with genetic disorders that were characterized by: (1) a high bone mass phenotype which was due to an activating mutation in LRP5 that promoted increased bone formation [117, 118]; and (2) an inactivating mutation in LRP5 that caused the osteoporosis pseudoghioma syndrome [119]. These relationships were definitively established genetically in multiple mouse models that replicated the human phenotype. Various genetic mouse models of gain and loss of β-catenin function revealed that low/moderate cellular β-catenin levels promoted chondrogenesis, while high levels differentiated mesenchymal cells toward the osteoblast lineage [120, 121].

Activity of canonical Wnt signaling is exquisitely controlled by a number of inhibitors that function extra- and intracellularly. The association of DKK1 with LRP5 prevents Wnt ligands from binding to the frizzled receptor. A number of secreted frizzled related proteins (SFRPs 1, 2, 3, and 4) engage Wnt ligands which in turn prevents the interactions of the Wnt ligands with their receptor complex [122, 123]. Several inhibitors of Wnt signaling are being considered as anabolic therapies for osteoporosis, but caution must be exercised as high activity of Wnt signaling is associated with several cancers [124]. Of note is the fact that SFRP1 is silenced in many tumor types [125]. Inhibition of GSK-3B inhibitor is one approach for raising β-catenin nuclear levels. Another viable approach is targeting sclerostin, an inhibitor of Wnt signaling that is encoded by the SOST gene. Mutations of this gene were first identified as causing a sclerotic disorder (Van Buchem's disease) and later established as a product of a limited number of cell types, one of which is the osteocyte, the major cell of the adult skeleton [126, 127]. Sclerostin inhibits Wnt signaling by binding to the LRP5/6 receptors in mesenchymal stromal cells, thereby regulating the continued formation of bone. Administration of anti-sclerostin antibody in cynomolgus monkeys increased bone formation, bone mineral density, and bone strength, offering promise as a treatment for osteoporosis [128]. In fact, a component of the anabolic effects of PTH is in part the result of PTH decreasing SOST levels [129, 130]. The advantages of stimulating Wnt signaling and in this way enhancing fracture repair, have also been well documented [115, 131, 132]. Expression of different Wnts during chondrocyte and osteoblast differentiation and during fracture repair has been examined in a number of studies [133]. However, their functional roles during fracture healing need to be explored. There is still much to learn about Wnt signaling in the skeleton.

## Mechanisms of transcriptional regulation

Many classes of DNA-binding proteins have been shown to regulate the transcription of osteoblast-related genes, and genetic mouse models expressing or deleting a transcriptional regulator have elucidated their contribution to bone formation. Transcription factor binding to DNA motif occurs with simultaneous binding of co-regulatory factors in multimeric complexes on gene promoters. The co-regulatory factors in these complexes can be various chromatin remodeling factors for epigenetic control, mediators of signaling pathways, or other transcription factors. These complexes control gene expression by contributing to either attenuation or enhancement of transcription, thereby regulating expression of a gene at a specific stage of cell differentiation. Family members of DNA-binding proteins can dimerize with different partners in the same family; for example Hox, homeodomain proteins, or AP-1 factors or with other transcription factors. Also, transcriptional activity can be regulated by unique inhibitors recognizing a specific transcription factor resulting in another level of specificity for cellular activity. For example, the C/EBP family members are broadly distributed in many cell types and their activity is regulated by several inhibitors (e.g., CHOP) that are uniquely targeted to C/EBP factors and regulate C/EBP activity in bone [134].

Finally, there is growing awareness that the fidelity of gene expression necessitates coordination of transcription factor metabolism and organization of genes and regulatory proteins within the three-dimensional context of the nuclear architecture [135]. The regulatory machinery that governs genetic and epigenetic control of gene expression is compartmentalized in nuclear microenvironments [136]. Temporal and spatial parameters of regulatory complex organization and assembly are functionally linked to biological control and are compromised in disease states, (e.g., with the onset and progression of tumorigenesis [137]). All these components of the gene regulatory machinery contribute to precise levels of transcriptional control and specificity.

*Transcription factors for control of osteoblast differentiation*

The developmental expression levels of many transcription factors during osteoblast maturation reflect their roles as key determinants of osteoblast differentiation. The helix-loop-helix *HLH proteins* (Id, Twist, Dermo, Scleraxis) are expressed at peak levels in proliferating progenitor cells and are important for expansion of osteoprogenitors. HLH proteins function as repressors of genes representing the mature bone phenotype and are in effect negative regulators of osteoblast differentiation. These proteins must be downregulated for osteoblast differentiation to proceed [2]. The mechanism for Twist inhibition of osteogenesis has been shown to involve inhibition of Runx2-mediated transcription by interacting with the Runx2 DNA-binding domain [138].

The *Hox genes* and the subclass of *homeodomain proteins (HD)* are the major signaling proteins involved in segmentation of the vertebrate body which define the positions where bone structures will develop [139]. Human mutations in Hox genes and null mutations in mouse models have identified defects in rib and vertebrate transformations and limb shortening, while mutations in homeodomain proteins (Msx and Dlx) result in a spectrum of craniofacial abnormalities in development [140]. However, it is now appreciated that the functions of homeobox factors extend far beyond patterning the skeleton to regulating the differentiation of bone-forming cells in the adult skeleton during the turnover of bone tissue. Indeed, several cDNA-profiling studies revealed that commitment of pluripotent cell lines to the osteoblast phenotype or the differentiation of pre-osteoblast cells by the osteogenic bone morphogenetic protein BMP2, resulted in the induction of a temporal cascade of homeobox factors that included inhibitors and activators of differentiation. Negative regulators like Msx 1, Msx2, and Dlx2, were upregulated within an hour, followed by a sequential expression of Hoxa10, Dlx3, and Dlx5 factors coinciding with the appearance of Runx2 [141–143]. Msx-1 must be turned off in periosteal progenitors for progression of osteoblast differentiation in vivo [144]. The homeodomain proteins, Msx1/2 and Dlx2/3, decline during osteoblast differentiation, while Dlx5 reaches peak levels when alkaline phosphatase and osteocalcin are expressed. These factors are found to be temporally associated with the chromatin of genes in (ChIP assays). These results confirm that they occupy sites in the promoter at specific times in the developmental progression of osteoblast differentiation which facilitates their ability to up- and down-regulate transcription [141, 143]. These mechanistic studies have contributed to defining a regulatory network of Hox/HD proteins that are involved in osteoblastogenesis. Msx2 is well-characterized for suppressing genes in osteoprogenitor cells that are found in mature osteoblasts (e.g., osteocalcin). However, recent studies have shown Msx2 can promote mineralization of Vascular Tissue by acting through Wnt signaling [145].

*AP-1 factors* include the fos (c-fos, deltafosB, fra1, fra2) and jun (c-jun, jun-D, jun-B) oncogene-encoded transcription factors. These proteins regulate cell cycle and differentiation-related genes in cartilage and bone by forming homo- or hetero-dimeric complexes at AP-1 regulatory elements, particularly in response to growth factor signaling. By forming co-regulatory complexes with different partners, cellular levels of transcription can be finely modulated, or selectively repressed or enhanced, as occurs on the osteocalcin and MMP13 gene [146]. Genetic studies have established that several AP-1 family members are essential for normal bone development and osteoblast differentiation (reviewed in [147,148]).

In studies characterizing mice having gene deletion of *Runx2/Cbfa1/AML3*, the first transcription factor shown to be a master regulator gene essential for bone formation, was established. Mutations in Runx2 are the cause of the human disorder cleidocranial dysplasia (CCD). Runx2 is also abnormally expressed at high levels on metastatic cancer cells [149]. It is one member of a three-member gene family in mammalian species, each of which is being essential for commitment and cell differentiation. Runx1/AML1 is required for definitive hematopoieses and mutations in this factor due to chromosomal translocations are the basis of many leukemias. Runx1 and its CBFβ DNA-binding partners are critical determinants of HSC fate. They also function to silence and activate sub genes in populations of T cells [150–152]. Runx3 is required for gut development and deficiencies in this family member are linked to gastric cancers. Runx3 are also involved in immune cell differentiation [153]. Thus all three Runx factors function as tumor suppressors. Both Runx1 and Runx3 overlap in expression with Runx2 in subpopulations of skeletal cells [154]. While all three Runx proteins are present in MSCs and early osteoprogenitor cells, Runx1 in vivo is highly expressed in the periosteum and perichondrium. Runx2 and Runx3 are both strongly expressed in the hypertrophic zone. Runx2 is essential for hypertrophic chondrocyte maturation, long bone development, and membranous bone formation. The three Runx factors require an essential DNA-binding partner protein CBFβ, that interacts with the Runt homology DNA-binding domain to promote Runx interactions with ATGGCT/G motifs in gene promoters. Absence of functional CBFβ in hematopoietic cells or osteoblasts results in abnormalities as severe as gene deletion of the Runx factors in the mouse [155, 156]. Transcriptional control of Runx2 expression involves two promoters. The P1 promoter transcribes the major isoform in differentiated osteoblasts. Runx2 has more than seven Runx2 binding sites in its promoter; hence is autoregulated. Its promoter sequence also contains Multiple Hox and homeodomain sites, indicating mechanisms for early expression of Runx2 during development. [143, 157]. Runx2 still remains the first factor in the hierarchical chain of essential factors for bone formation. Runx2 target genes include those contributing not only to bone formation (osteocalcin, osteopontin, bone sialoprotein, galectin, TGFβ receptor I, and dentin sialophosphoprotein-1), but also to bone turnover (collagenase 3, osteoprotegerin, and RANKL). Studies in transgenic mice overexpressing Runx2 in osteoblasts or chondrocytes indicate that Runx2 levels must be tightly controlled in skeletal cells, as these mice exhibit osteoblast maturational defects and osteopenia [158].

*ATF4*, like Runx2, was discovered as an osteocalcin DNA-binding protein. Loss-of-function in the mouse revealed a delay in mineralization. ATF4 forms a complex with Runx2 and SATB2, an AT-rich homeobox protein, to regulate transcription and bone formation [159]. Thus, the three factors Runx2, Osterix (see below), and ATF may coordinate a temporal sequence gene of regulation essential for both the earliest stage of commitment of MSCs to osteoblasts (mostly requiring Runx2) and the final stages of osteoblast differentiation and mineralized tissue formation (more dependent on Osterix and ATF4) (see Figure 5-1).

*Zn finger transcription factors (TFs) and adapter proteins.* These proteins represent a spectrum of regulatory factors, several of which have a significance in regulating bone formation. *Osterix* is a zinc finger protein and a bone-specific novel family member of the SP (specificity proteins) class, designated SP-7. It was discovered as a BMP2-inducible protein. The null mouse, like the Runx2 null mouse, also dies at birth with an absence of mineralized tissue. Since Runx2 expression is unaltered in Osterix null mice, it is concluded that Runx2 is expressed earlier than Osterix [160]. Other SP factors (SP1 and SP3) are present in osteoblasts and cooperate with other transcription factors to regulate bone-related genes. SP6 is important for normal limb development and is expressed in the AER and regulated by canonical Wnt signaling, but it has not been studied in mature osteoblasts [161]. Several Kruppel-like factors (the second class of zinc finger TFs) are expressed in osteoblasts. Glis3 acts synergistically with SHH and BMP to stimulate osteoblast differentiation, in part by activating a positive regulator of bone formation, FGF18 [162]. Zfp521 has biphasic effects. It promotes in vivo bone formation when expressed in mature osteoblasts under the control of the osteocalcin promoter [163]. However, in vitro, this zinc finger factor binds to Runx2, decreasing its transcriptional activity. The zinc finger adapter proteins of the *Schnurri* class are potent regulators of immune cell development and bone homeostasis. Schnurri 3 (Shn3)-deficient mice have a normal skeleton at birth, but exhibit increased bone mass as they age. This is due to loss-of-function related to degradation of Runx2 through Shn3 recruitment of the E3 ubiquitin ligase, WWP1, which binds to Runx2 at the PPXY motif (reviewed in [164]). Schnurri-2 deficient mice on the other hand, exhibited reduced bone remodeling and osteopenia resulting from a decrease in both osteoblast and osteoclast functions [165]. This may be secondary to the fact that Shn2 cooperates with BMP Smads and C/EBP to promote adipogenesis, thereby depleting the pool of osteoprogenitors [166].

## Specificity for osteoblast lineage determination in nuclear microenvironments

The characterization of Runx2 as a bone-specific transcription factor is contributing to novel concepts for the requirement of "master" transcription factors to function as tissue-specific determinants of a lineage [167]. Runx2 has the ability to contribute to regulating chromatin organization for transcription of bone-specific genes activated at a stage of differentiation. This has been clearly documented for the repression and activation of the osteocalcin gene [168]. Runx2 binding initiates the positioning of a single nucleosome between the proximal promoter, containing the homeodomain site and basal transcription machinery, separated from the distal promoter containing the hormone-responsive vitamin D domain. Mutations in the three Runx sites in the osteocalcin genes prevent osteocalcin from responding to hormones and growth factors due to the absence of nuclease hypersensitive sites [169]. The interaction of Runx2 with a spectrum of histone-modifying enzymes that either promote acetylation (e.g., by p300) to increase transcription or inhibit target gene transcription through deacetylation of histone proteins by multiple histone deacetylase (HDAC) enzymes, allows Runx2 to temporally regulate genes. Runx2 thereby has epigenetic function in modifying chromatin organization of target genes to control accessibility of regulatory factors to target gene promoters [137].

Such modifications in chromatin structure involving Runx2 are critical for osteoblast differentiation and normal bone development as shown by null mutation and overexpression of several HDACs which affect bone development [170]. An additional novel concept of epigenetic control by Runx factors and other lineage-specific transcription factors was recently established by the finding that a master transcription factor remained bound to mitotic chromosomes in proliferating cells (Figure 5-4). This was shown for Runx1 in hematopoietic cells, Runx2 in osteoblasts, C/EBPα in adipocytes (but not PPARγ), and MyoD in muscle cells [171–173]. Although the factors are associated with genes on chromosomes these genes are not being transcribed during mitosis. These findings suggest a mechanism for assuring phenotype stability during mitosis. After cell division, phenotypic genes required for maintaining the phenotype will be expressed. One set of genes is the ribosomal genes which have as many as 50–60 Runx binding sites, (Figure 5-4). Indeed each one of the master transcription factors will bind to ribosomal DNA during mitosis and regulate ribosomal genes and protein synthesis in the nucleoli after mitosis is complete (Figure 5-4).

A unique feature of all Runx factors is a 31-amino-acid module located in the C-terminus and designated as the nuclear matrix targeting signal (NMTS) (Figure 5-5). The NMTS traffics Runx2 to subnuclear domains visualized as Runx foci, facilitating the formation of multimeric complexes of Runx2 with co-regulatory proteins on gene promoters. This provides a level of specificity for their major functional property which is to act as a platform protein that interacts with a spectrum of co-regulatory proteins that mediate chromatin remodeling, developmental signals, growth control, hormone

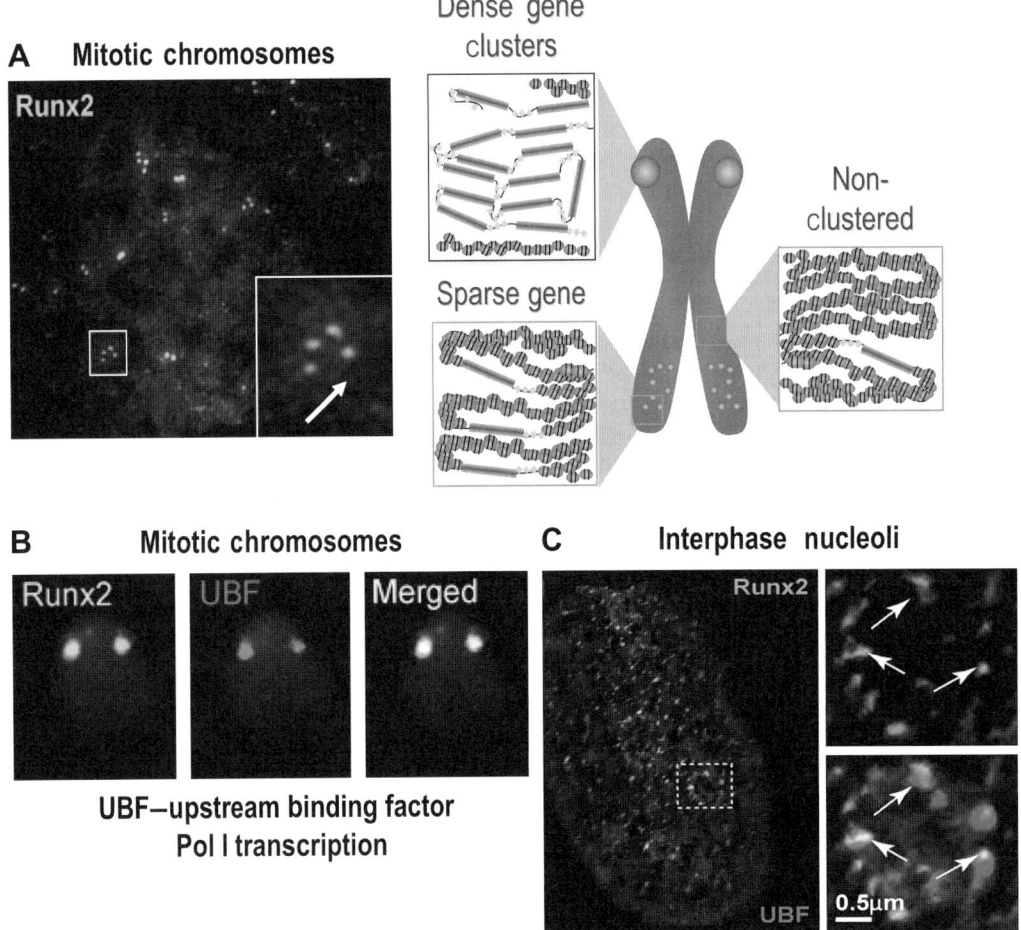

**FIGURE 5-4 Novel epigenetic functions of Runx2. (A)** Runx2 is associated with genes on mitotic chromosomes, but the genes are not transcribed, indicating a novel level of epigenetic control [172, 219]. The Runx2 foci appear as large spots (dense clusters) and small foci illustrated in the right panel. Analyses show that the associated genes include Smads, Runx2, VEGF, and ribosomal genes to name a few [172]. These findings suggest that these genes on mitotic chromosomes with Runx2 are being "book marked" for stability of the osteoblast phenotype when the cell division is completed.
**(B)** Pairs of large Runx2 foci on eccentric chromosomes in regions are associated with ribosomal genes (UBF is a marker). Runx2 is also found to downregulate protein synthesis by binding to ribosomal genes that contain from 40 to 60 Runx2 sites. **(C)** Runx is also found in interphase nucleus in the periphery of the nucleolus where protein synthesis occurs. Please refer to color plate section.

activities, and other physiologic functions. These co-regulators include CBFβ1, which increases DNA binding, histone deacetylases, the Wnt pathway regulator LEF-1, AP-1 factors in response to growth factors, hormone receptors (the VDR, GR and AR), and most significantly TGFβ/BMP-responsive SMAD proteins that bind to Runx2 in the NMTS at amino acid residues which are critical for binding to the nuclear matrix scaffold (Figure 5-5). The PPXY motif in Runx2 interacts with several WW domain proteins that result in

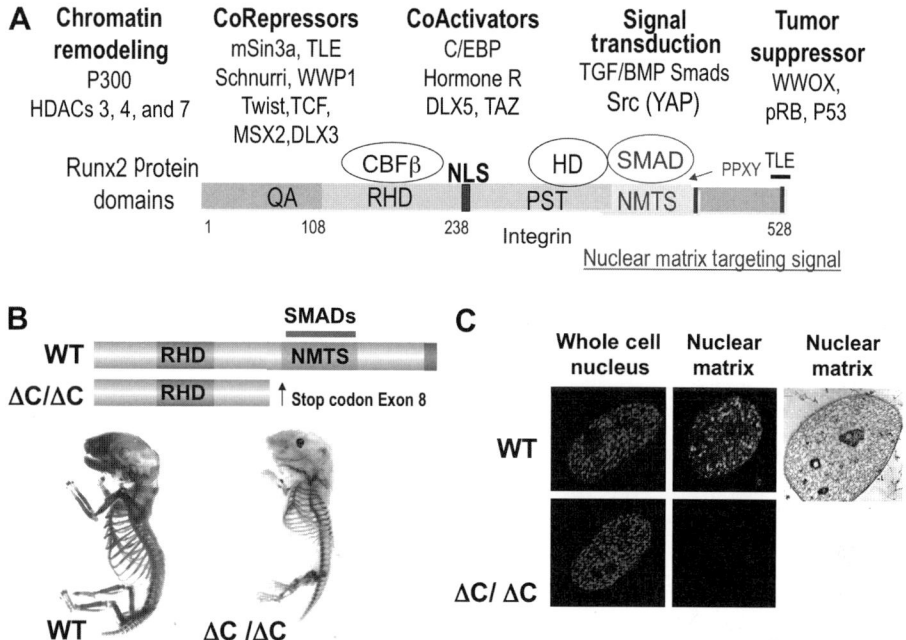

**FIGURE 5-5 Runx2 is a scaffolding protein with regulatory complexes organized in subnuclear domains. (A)** Classes of co-repressors and co-activators forming protein–protein interactions with Runx2 reflect mechanisms by which Runx2 can both activate and repress genes during development of the osteoblast phenotype. These co-regulatory proteins have been mapped to either the N- or C-terminus. The N-terminus contains a unique QA stretch found only in Runx2 and the DNA-binding module, the runt homology domain (RHD), that requires the CBFβ partner protein. In the C-terminus is located a region of phosphorylation sites (PST), activator (NMTS, nuclear matrix targeting signal), and repressor (PPXY, TLE) domains. Note the interaction site for SMAD lies within the NMTS and requires three amino acids (HTY). **(B)** The biological significance of the NMTS has been established by a knock-in mutation at exon 8. The loss of the segment of the C-terminus results in a phenotype analogous to the complete null mouse. Runx2 DC mutant protein enters the nucleus (whole cell nucleus stained for Runx2, IHC) but is not bound to the nuclear matrix (NM scaffold remains following high salt extraction of soluble chromatin). **(C)** Loss of Runx2 binding to the nuclear matrix scaffold in Runx2$^{\Delta c/\Delta c}$ mice. An electron micrograph of the filamentous nuclear is shown. These studies have established a critical role in Runx2 functional activity mediated by forming co-regulatory protein complexes on the nuclear matrix for specificity of bone formation. Please refer to color plate section.

either repression or enhancement of gene transcription, depending on the cell context. Among these WW domain proteins are the Src-responsive YAP protein, WWP1 which interacts with Schnurri3 to degrade Runx2, Smurf which functions 1 in response to TGFβ and also targets Runx2 to the proteosome, the tumor suppressor WWOX1 and TAZ, a co-activator. Negative regulators include STAT-1, which sequestes Runx2 in the cytoplasm. Groucho/TLE binds to a VWRPY motif and is a developmental signal that suppresses Runx2 activity, while a Groucho homolog Grg5 enhances Runx2 transcription and is particularly critical for growth plate development. Other domains of Runx2 interact with C/EBP, VDR, TCF/LEF homeodomain (HD) proteins and pRB tumor suppressor. These specialized properties of Runx2 provide a basis for its ability to support the integration of key signaling pathways in osteogenesis.

*MicroRNA regulation of bone formation and osteoblast differentiation*

MicroRNAs include a class of small non-coding RNAs (~22 nt) that regulate roughly more than 90% of the genome in higher organisms [175, 176]. They are processed from RNA by two enzymes, Drosha in the nucleus produces pre-MiR of about 70 nt in length. These pre-miRs are then further cleaved to the mature miRs in the cytoplasm by the enzyme Dicer. These miRs bind to complementary sequences in the 3'UTR of genes blocking translation of the protein (by stable binding) or causing degradation of the mRNA (by a deadenylation mechanism). This post-transcriptional control mechanism finely adjusts cellular levels of proteins. It is now well established that these small RNAs are involved in almost every biological process, including early development, lineage commitment, growth and differentiation, cell death, and metabolic control. The requirement for this architectural organization of Runx2 with other factors for bone formation was established by an in vivo. A knock-in mutation of Runx2 in the mouse, which deleted the Runx2 C-terminus containing the NMTS resulted in a lethal phenotype at birth. A complete absence of mineralized skeleton occured, analogous to the Runx2 null mouse phenotype [174].

The recent discovery that miRNAs regulate in vitro osteoblast differentiation and in vivo bone formation has provided insights into the potent activity of miRs in regulating osteogenesis. This appears to occur by a synchronized expression of multiple miRs that constitute a program of changing microRNAs which promote osteogenesis [104, 177–181]. Our group and others have shown that the osteogenic BMP2 downregulates a cohort of miRs that inhibit bone formation by targeting a set of osteogenic activators [104, 178]. One of these, miR-133, directly targets Runx2, an early BMP response gene, while another, miR-135, targets Smad5, a key transducer of the BMP2 osteogenic signal. Both miRNAs functionally inhibit differentiation of osteoprogenitors by attenuating Runx2 and Smad5 pathways that synergistically contribute to bone formation [32]. The key finding is that BMP2 controls bone cell determination by downregulating expression of multiple miR-NAs. This, in turn releases the osteogenic pathway components (BMP co-receptors, Wnt proteins, transcription factors) that are required for osteoblast lineage commitment from inhibition (Figure 5-6). Thus, our studies establish a novel mechanism for BMP morphogens in which there is selective induction of a tissue-specific phenotype and suppression of alternative lineages.

Following commitment, microRNAs also control the differentiation of osteoblasts to osteocytes. It was shown in a profiling study of miRs during the differentiation of murine pre-osteoblast MC3T3 cells, that there was a continuous upregulation of ~60 miRs from the growth period to the osteocyte/mineralization stage [178]. Among these, miR-29b was identified as one of the significantly enhanced microRNAs. Exogenous expression of miR-29b promoted osteoblastic differentiation, reflected by increased levels of osteogenic marker genes, including Runx2 and alkaline phosphatase. The mechanism for this positive osteogenic effect was established by identifying that miR-29 targeted at least five negative regulators of osteoblast differentiation, including HDAC4, TGFβ3, ACVR2A

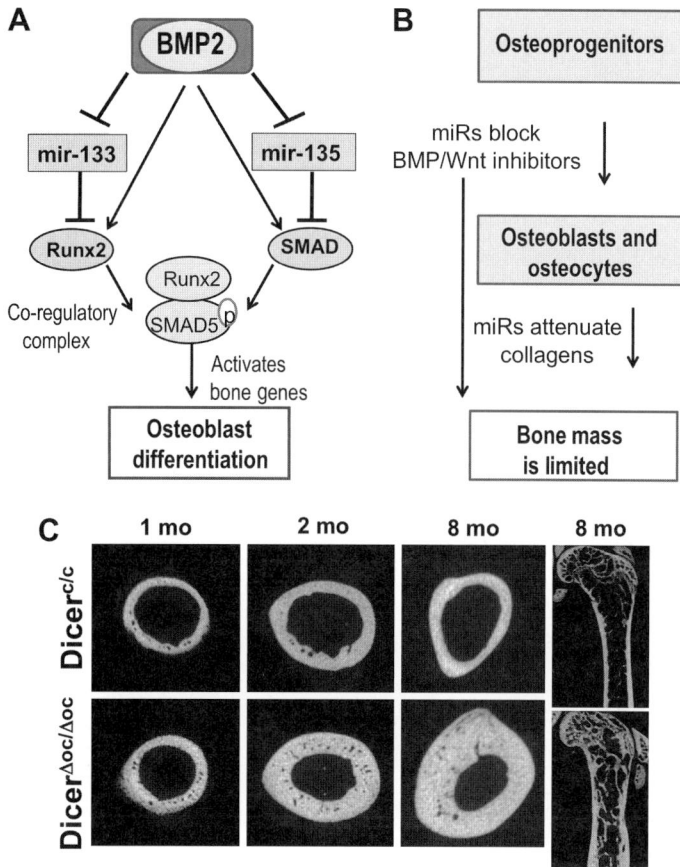

**FIGURE 5-6 MicroRNA regulation of osteogenesis. (A, D)** Schematic illustrations of characterized mature miRs. **(A)** Induction of C2C12 premyogenic cells into the osteoblast lineage by BMP2 results in downregulation of miRs that negatively regulate Runx2 (miR133) and BMP2 SMAD5 (miR135). BMP2 therefore relieves repression of these two factors essential for osteoblastogenesis. **(B)** During MC3T3 osteoblast differentiation (days 0–28 in differentiation medium), miRs are continuously upregulated and have selective functions as illustrated to either downregulate inhibitors of bone formation from osteoprogenitor proliferation stage to the osteoblast/osteocyte mineralization stage when the synthesis of bone matrix proteins is attenuated to prevent fibrosis. **(C)** Excision in osteoblast lineage cells by osteocalcin-Cre of the Dicer enzyme which processes precursor miRs to mature functional miRs, results in a high bone mass phenotype. Beginning from 1 month, cortical expansion and trabecular bone are increased at indicated ages. Single slice images from μCT three-dimensional data acquisition are shown [104]. Please refer to color plat section.

(an activin receptor), CTNNBIP1 (β-catenin inhibitory protein), and DUSP2 (a dual specific phosphatase). MiR-29b also targets Col1a1, Col5a3, and Col4a2, but this activity occurred at the late mineralization stage when collagen mRNA levels are decreased in response to the increasing levels of miR-29 [177]. We proposed this as a mechanism to attenuate collagen synthesis to allow for the orderly organization of accumulated collagen fibrils that would accommodate mineral deposition or prevent fibrosis or prevent excessive matrix deposition. MiR-29b was previously documented to be muscle-related and decreased during myocardial infarct leading to fibrosis as it targeted type 1 collagen [182]. MiR-29a was shown also to downregulate osteonectin (SPARC), a small non-collagenous protein that binds to collagens [183]. Thus, both inhibitors of osteogenesis and numerous components of the bone ECM are under the coordinate control of multiple microRNAs (illustrated in Figure 5-6B). This hypothesis along with the biological significance of upregulation of many miRs in mature osteoblasts, was indeed corroborated by conditional deletion of the Dicer enzyme, which generates the functionally mature miRs, in mature osteoblasts in mice by osteocalcin-Cre [104]. This excision in mature osteoblasts resulted in a two fold increase in bone volume by 2–3 months after birth (Figure 5-6C). Elevated collagen synthesis was demonstrated, providing evidence that bone mass accrual is regulated by miR control of matrix proteins and anabolic pathways. These findings provide an opportunity for therapeutic approaches to increase bone mass by antagonizing miRs expressed in bone.

## Osteoblast and immune cell crosstalk: A peek into the future

The connections between osteoblasts and the HSC/immune cell niches and inflammatory-cells and cytokines that stimulate osteoclastogenesis are provided by multiple lines of evidence and were recently reviewed [14]. In further exploring other modes of crosstalk, it is likely that common pathways used for differentiating distinct cell populations may have complementary effects on osteoblast and immune cells. The Wnt signaling pathway is a potential avenue for providing complementary activities that result in a biological outcome that is dependent on both the mesenchymal and hematopoietic lineages. TCF1 and LEF1, mediators of canonical Wnt signaling, are known to be critical for both osteoblastogenesis (see above) and the development of lymphoid cells [23, 184]. The thymocyte, for example, exhibits dynamic changes in expression between naïve T cells where TCF/LEF1 are highly expressed, effector T cells in which these transcription factors are downregulated and memory T cells where they are upregulated again. There are a number of Wnt ligands, produced by stromal and hematopoietic lineage cells in the bone marrow microenvironments that may have complementary effects on each other, reflecting potential crosstalk [185]. Canonical Wnt3a, present in stromal cells, promotes the growth of pro-B cells. Wnt5b, a non-canonical Wnt expressed by human bone marrow B-cell progenitors is a Wnt that promotes osteoblast differentiation [186]. Wnt10b is produced by osteoblasts and increases osteoblast number and postnatal bone mass [187]. Wnt10b is also a growth factor for hematopoietic stem cells in human CD34-expressing bone marrow cells and is produced by macrophages [188]. Intermittent PTH increases the production of Wnt10b by the bone marrow CD8+ T cells, and thereby may facilitate lymphocytes to activate canonical Wnt signaling in preosteoblasts [113, 189]. It is of interest to note that T-cell null mice display diminished Wnt signaling, as well as osteoblast proliferation and differentiation [189]. The synergistic environment of pre-osteoblasts and the hematopoietic niche is in part supported by a Wnt pathway that is mediated by crosstalk between these two cell populations. It is reasonable to postulate that a Wnt ligand produced by one population contributes to supporting the properties of others in the hematopoietic niche.

Another opportunity for crosstalk between osteoblast and hematopoietic lineage cells for consideration is through Runx signaling. As described above, Runx2 is essential for osteogenesis and Runx1 is required for definitive hematopoiesis. But with each cell phenotype, the Runx family of proteins, Runx1, 2, and 3 are operating to regulate progenitor cell differentiation into distinct subtypes. Runx1 and Runx3 regulate immune cell differentiation and activation and the selection of T-cell populations [152, 190]. Furthermore the cross-regulation of Runx1 by Runx3 influences the proliferation of B cells following infection with Epstein-Barr virus. The downregulation of Runx1 requires interaction of Runx3 at its VWRPY motif with the suppressor co-regulatory protein TLE [191]. Whether or not Runx1 and Runx3 in osteogenic lineage cells contribute to expression of secreted factors that also influence T- and B-cell properties in the bone marrow environment remains a question for future exploration.

# Immune responses and bone pathology: Effects of inflammation on bone in rheumatoid arthritis

## Pathology and cytokine activities

Rheumatoid arthritis (RA) is an inflammatory joint disease that is characterized by infiltration of the synovial tissues lining the joint by inflammatory cells. This inflammation is accompanied by proliferation and activation of cells present within the synovial lining. Cells within the inflamed synovium and associated pannus tissue produce pro-inflammatory cytokines and factors that lead to the destruction of cartilage and bone matrix. One important cytokine is tumor necrosis factor-alpha (TNF$\alpha$), synthesized by activated macrophages, synovial lining cells, and activated T cells within the RA synovium [192, 193]. Among its diverse pathologic effects, TNF$\alpha$ induces production of other pro-inflammatory cytokines in RA joints, including interleukin (IL)-1, IL-6, and IL-8 [194] upregulates adhesion molecule expression in the synovium that, in turn, attracts leukocytes into affected joints [195], increases metalloproteinase production by synovial lining cells [196], and inhibits proteoglycan production by chondrocytes within cartilage [197].

Bone erosions in RA occur at the interface of the pannus and bone. Pannus tissue invasion into cortical and trabecular bone is seen in histologic sections from patients with RA. Many of the cytokines expressed in RA synovial tissues are pro-osteoclastogenic, and it has been demonstrated that osteoclasts are required for the local destruction of bone in RA joints (erosions) [198]. Included among these RA-induced cytokines are RANKL, produced by activated T cells and synovial fibroblasts in RA [199], TNFα, IL-1 IL-6, and IL-17. As a consequence of the actions of these cytokines, bone-resorbing osteoclasts are produced, activated in the pannus tissue and subsequently invade bone, leading to erosions. TNFα is a cytokine that exhibits a number of effects on bone remodeling. It regulates the abundance of osteoclast precursors in bone marrow by the induction of c-Fms expression [200], aids in the activation of osteoclasts by enhancing RANK signaling [201] and induces expression of RANKL and M-CSF by bone marrow stromal cells [202].

It is important to recognize that several of the pro-inflammatory cytokines secreted by immune cells at the site of inflammation and bone resorption in the RA joint can also inhibit osteoblast activity, and thus may dampen the repair response of osteoblasts. For example, TNFα impacts osteoblasts in in vitro culture systems by inhibiting osteoblast differentiation [203]. However, the role of osteoblasts in focal articular bone loss in RA has received less attention than the role of osteoclasts in this process.

## Repair mechanisms in RA

New and highly effective therapies have been introduced over the past decade for the treatment of patients with RA. It is of interest that although RA patients receiving these potent therapies demonstrate slowing or inhibition of the progression of bone erosions on X-ray, healing of erosions has been considered to occur infrequently in observational studies. However, repair of bone erosions with formation of new bone has been demonstrated in RA patients treated with disease modifiers [204, 205]. Two studies conducted by the OMERACT Subcommittee on Healing of Erosions have confirmed that repair of erosions can occur in patients with RA [206]. Further studies demonstrated a strong association between clinical disease remission and evidence of repair of erosions by X-ray, suggesting that inflammation within the bone microenvironment may play a role in suppressing the ability of osteoblasts to produce bone and thereby repair or "heal" erosions.

In a recent study from our laboratory using an animal model of RA, we demonstrated that the formation of mineralized bone is compromised at sites of inflammation and bone erosion [207]. One would expect that bone formation rates would increase at sites of inflammation and bone erosion, as bone has been lost at these sites. However, no difference was observed in the rate of mineralized bone formation at bone surfaces in inflamed areas compared to similar areas in mice without arthritis. In addition, there was a paucity of osteoblast-lineage cells at the inflamed erosion sites expressing maturation markers, but there were abundant cells expressing Runx2, a marker of early osteoblast-lineage cells. Furthermore, expression of alkaline phosphatase, a marker of the mineralization phase of bone formation, was markedly limited at sites of bone erosion. These results support the hypothesis that inflammation in RA may inhibit the capacity of osteoblasts at the erosion site to repair eroded bone. Our findings are consistent with those of a prior study demonstrating a paucity of mature osteocalcin-expressing cells within arthritic navicular bone in the rat model of antigen-induced arthritis [208]. These studies support the concept that inflammation impairs osteoblast differentiation and function, resulting in fewer sites at which bone formation and mineralization occurs, thereby impairing bone repair at these sites.

## Regulation of Wnt signaling in RA

A key pathway regulating the induction of osteoblast differentiation is the Wnt signaling pathway (described above). This pathway has recently been implicated in the regulation of osteoblast differentiation and function within sites of inflammation-induced bone erosion. In mice with inflammatory arthritis (models of RA), increased expression of members of the DKK and sFRP families of WNT signaling antagonists is observed in arthritic synovial tissues when compared to non-arthritic tissues [207, 209]. These include DKK1 and sFRP1, and these factors demonstrated a different pattern of temporal expression over the course of arthritis in one murine model. In a previous study examining human RA synovial tissues, expression of the sFRP-1 and -4 genes was seen, predominantly in fibroblast-like cell-enriched populations, and expression of the sFRP-3 gene was noted predominantly in macrophage-rich populations [210]. In addition, serum DKK1 levels were shown to be elevated in arthritic hTNF.Tg mice and in RA patients compared to normals [209]. Finally, TNFα treatment induced DKK1 expression in both human and mouse arthritic synovial fibroblasts [209]. The exact cell types responsible for the expression of DKK and sFRP family members in RA and their individual roles in bone remodeling in this disease are currently under investigation in our laboratory.

A role for DKK1 in the regulation of bone remodeling in RA was recently elucidated in an important "proof-of-principle" study using a neutralizing antibody for DKK1 in the hTNF.Tg mouse [209]. Treatment with this blocking antibody from the onset of arthritis resulted in protection from articular bone erosion, and resulted in the onset of periosteal osteophyte formation, which is otherwise rarely observed in this murine model. Protection from erosion was due at least in part to enhanced OPG expression, supporting a role for active Wnt signaling in the inhibition of osteoclastogenesis [209, 211]. Importantly, in support of the inhibition of bone formation by DKK1, increased osteoid deposition was seen within tarsal bones of affected hind paws with DKK1 blockade, in addition to the formation of periosteal osteophytes. However, the direct effect of activation of Wnt signaling via DKK1 blockade on the differentiation of osteoblasts at sites of erosion, and the capacity of osteoblasts to form mineralized bone, was not assessed.

Other aspects of the microenvironment within the RA joint may also play a role in the inhibition of Wnt signaling and osteoblast function. Oxidative stress directly inhibits Wnt signaling in osteoblast-lineage cells by sequestering β-catenin. Thus preventing β-catenin from inducing expression of genes within the canonical Wnt signaling pathway, leads to a decrease in alkaline phosphatase activity [212]. In addition, hypoxia increases DKK1 expression in myeloma cells [213]. Hypoxia is thus likely to play a role in bone erosion in RA, as the metabolic activity of cells present within the arthritic joint contributes to a hypoxic and low oxygen tension. This is accompanied by decreases in pH and both low oxygen tension and low PH could directly impact osteoblasts as well as osteoclasts [214, 215].

It is important to note that some antagonists of Wnt signaling may actually help to maintain joint integrity. For example, deficiency in sFRP3 (or FRZB) in mice led to a worsening of cartilage damage in an antigen-induced arthritis model [216]. In agreement with this observation, sFRP3 has been shown to bind directly to matrix-metalloproteinases (MMP)-2 and -3, enzymes that are thought to contribute to cartilage destruction in degenerative joint disease. A direct interaction of sFRP1 with RANKL has also been reported [217]. This finding suggests that sFRP1 may actually downregulate RANKL activity, and thus impair osteoclastogenesis. Taken together, these numerous effects of Wnt signaling in RA should be considered in the context of designing new anabolic therapies that stimulate bone formation in skeletal repair and osteoporosis by modulating the Wnt pathway.

Prevention of articular bone erosion is an important goal in the treatment of patients with RA. In addition, the ability to repair existing erosions would provide restoration of a more normal architecture to the joint. Further studies will be needed to clearly define the function of each of the Wnt signaling inhibitors in healthy joints, as well as the exact function of these factors in rheumatic diseases that damage and destroy bone.

## Acknowledgments

The authors thank Charlene Baron for assistance with figure preparations and Judy Rask for manuscript preparation. The National Institutes of Health grants supporting the research program related to this chapter include P01AR048818 (GSS), P01CA082834 (GSS), R01AR039588 (GSS), R37DE012528 (JBL), R01AR055952 (EMG) and a Rheumatoid Arthritis Initiative Grant from the American College of Rheumatology (EMG). The contents of this chapter are solely the responsibility of the authors and do not necessarily represent the official views of the National Institutes of Health.

## References

[1] Marie PJ, Kaabeche K, Guenou H. Roles of FGFR2 and twist in human craniosynostosis: insights from genetic mutations in cranial osteoblasts. Front Oral Biol 2008;12:144–59.

[2] Marie PJ, Coffin JD, Hurley MM. FGF and FGFR signaling in chondrodysplasias and craniosynostosis. J Cell Biochem 2005 December 1;96(5):888–96.

[3] Alappat S, Zhang ZY, Chen YP. Msx homeobox gene family and craniofacial development. Cell Res 2003 December;13(6):429–42.

[4] Kronenberg HM. Developmental regulation of the growth plate. Nature 2003 May 15;423(6937): 332–6.

[5] Horton WA, Degnin CR. FGFs in endochondral skeletal development. Trends Endocrinol Metab 2009 September;20(7):341–8.

[6] Taichman RS. Blood and bone: two tissues whose fates are intertwined to create the hematopoietic stem-cell niche. Blood 2005 April 1;105(7):2631–9.

[7] Yin T, Li L. The stem cell niches in bone. J Clin Invest 2006 May;116(5):1195–201.

[8] Wu JY, Scadden DT, Kronenberg HM. Role of the osteoblast lineage in the bone marrow hematopoietic niches. J Bone Miner Res 2009 May;24(5):759–64.

[9] Schofield R. The relationship between the spleen colony-forming cell and the haemopoietic stem cell. Blood Cells 1978;4(1-2):7–25.

[10] Eliasson P, Jonsson JI. The hematopoietic stem cell niche: low in oxygen but a nice place to be. J Cell Physiol 2010 January;222(1):17–22.

[11] Zhang J, Niu C, Ye L, Huang H, He X, Tong WG, et al. Identification of the haematopoietic stem cell niche and control of the niche size. Nature 2003 October 23;425(6960):836–41.

[12] Calvi LM, Adams GB, Weibrecht KW, Weber JM, Olson DP, Knight MC, et al. Osteoblastic cells regulate the haematopoietic stem cell niche. Nature 2003 October 23;425(6960):841–6.

[13] Lo CC, Klein RJ, Scadden DT. Analysis of the hematopoietic stem cell niche. Curr Protoc Stem Cell Biol 2007 November. Chapter 2:Unit.

[14] Takayanagi H. Osteoimmunology and the effects of the immune system on bone. Nat Rev Rheumatol 2009 December;5(12):667–76.

[15] Porter RL, Calvi LM. Communications between bone cells and hematopoietic stem cells. Arch Biochem Biophys 2008 May 15;473(2):193–200.

[16] Kiel MJ, Yilmaz OH, Iwashita T, Yilmaz OH, Terhorst C, Morrison SJ. SLAM family receptors distinguish hematopoietic stem and progenitor cells and reveal endothelial niches for stem cells. Cell 2005 July 1;121(7):1109–21.

[17] Lane SW, Scadden DT, Gilliland DG. The leukemic stem cell niche: current concepts and therapeutic opportunities. Blood 2009 August 6;114(6):1150–7.

[18] Iwasaki H, Suda T. Cancer stem cells and their niche. Cancer Sci 2009 July;100(7):1166–72.

[19] Kiuru M, Hidaka C, Hubner RH, Solomon J, Krause A, Leopold PL, et al. Sonic hedgehog expands diaphyseal trabecular bone altering bone marrow niche and lymphocyte compartment. Mol Ther 2009 August;17(8):1442–52.

[20] Engin F, Lee B. NOTCHing the bone: insights into multi-functionality. Bone 2010 February;46(2).274–00.

[21] Zanotti S, Canalis E. Notch and the skeleton. Mol Cell Biol 2010 February;30(4):886–96.

[22] Varnum-Finney B, Xu L, Brashem-Stein C, Nourigat C, Flowers D, Bakkour S, et al. Pluripotent, cytokine-dependent, hematopoietic stem cells are immortalized by constitutive Notch1 signaling. Nat Med 2000 November;6(11):1278–81.

[23] Chitteti BR, Cheng YH, Poteat B, Rodriguez-Rodriguez S, Goebel WS, Carlesso N, et al. Impact of interactions of cellular components of the bone marrow microenvironment on hematopoietic stem and progenitor cell function. Blood 2010 February 12.

[24] Weber JM, Calvi LM. Notch signaling and the bone marrow hematopoietic stem cell niche. Bone 2010 February;46(2):281–5.

[25] Kortesidis A, Zannettino A, Isenmann S, Shi S, Lapidot T, Gronthos S. Stromal-derived factor-1 promotes the growth, survival, and development of human bone marrow stromal stem cells. Blood 2005 May 15;105(10):3793–801.

[26] Tokoyoda K, Zehentmeier S, Chang HD, Radbruch A. Organization and maintenance of immunological memory by stroma niches. Eur J Immunol 2009 August;39(8):2095–9.

[27] Zhu J, Garrett R, Jung Y, Zhang Y, Kim N, Wang J, et al. Osteoblasts support B-lymphocyte commitment and differentiation from hematopoietic stem cells. Blood 2007 May 1;109(9):3706–12.

[28] Wu JY, Purton LE, Rodda SJ, Chen M, Weinstein LS, McMahon AP, et al. Osteoblastic regulation of B lymphopoiesis is mediated by Gs{alpha}-dependent signaling pathways. Proc Natl Acad Sci U S A 2008 November 4;105(44):16976–81.

[29] Peister A, Mellad JA, Larson BL, Hall BM, Gibson LF, Prockop DJ. Adult stem cells from bone marrow (MSCs) isolated from different strains of inbred mice vary in surface epitopes, rates of proliferation, and differentiation potential. Blood 2004 March 1;103(5):1662–8.

[30] Kobayashi N, Goto K, Horiguchi K, Nagata M, Kawata M, Miyazawa K, et al. c-Ski activates MyoD in the nucleus of myoblastic cells through suppression of histone deacetylases. Genes Cells 2007 March;12(3):375–85.

[31] Kollias HD, Perry RL, Miyake T, Aziz A, McDermott JC. Smad7 promotes and enhances skeletal muscle differentiation. Mol Cell Biol 2006 August;26(16):6248–60.

[32] Javed A, Bae JS, Afzal F, Gutierrez S, Pratap J, Zaidi SK, et al. Structural coupling of Smad and Runx2 for execution of the BMP2 osteogenic signal. J Biol Chem 2008 January 18;283(13):8412–22.

[33] Zaidi SK, Sullivan AJ, van Wijnen AJ, Stein JL, Stein GS, Lian JB. Integration of Runx and Smad regulatory signals at transcriptionally active subnuclear sites. Proc Natl Acad Sci U S A 2002 June 11;99(12):8048–53.

[34] Lafage-Proust MH, Thomas T, Guignandon A, Malaval L, Rattner A, Vico L. Plasticity of osteoprogenitor cells. Joint Bone Spine 2007 December;74(6):536–9.

[35] Shao JS, Cheng SL, Sadhu J, Towler DA. Inflammation and the osteogenic regulation of vascular calcification: a review and perspective. Hypertension 2010 March;55(3):579–92.

[36] Delany AM, Hankenson KD. Thrombospondin-2 and SPARC/osteonectin are critical regulators of bone remodeling. J Cell Commun Signal 2009 October 28;3:227–38.

[37] Noda M, Denhardt DT. Regulation of osteopontin gene expression in osteoblasts. Ann N Y Acad Sci 1995 April 21;760:242–8.

[38] Thurner PJ, Chen CG, Ionova-Martin S, Sun L, Harman A, Porter A, et al. Osteopontin deficiency increases bone fragility but preserves bone mass. Bone 2010 February 18.

[39] Monfoulet L, Malaval L, Aubin JE, Rittling SR, Gadeau AP, Fricain JC, et al. Bone sialoprotein, but not osteopontin, deficiency impairs the mineralization of regenerating bone during cortical defect healing. Bone 2010 February;46(2):447–52.

[40] Kavukcuoglu NB, Patterson-Buckendahl P, Mann AB. Effect of osteocalcin deficiency on the nanomechanics and chemistry of mouse bones. J Mech Behav Biomed Mater 2009 August;2(4):348–54.

[41] McGuigan F, Kumar J, Ivaska KK, Obrant KJ, Gerdhem P, Akesson K. Osteocalcin gene polymorphisms influence concentration of serum osteocalcin and enhance fracture identification. J Bone Miner Res 2010 January 29 [Epub ahead of print].

[42] Ueland T, Fougner SL, Godang K, Lekva T, Schurgers LJ, Scholz H, et al. Associations between body composition, circulating interleukin-1 receptor antagonist, osteocalcin, and insulin metabolism in active acromegaly. J Clin Endocrinol Metab 2010 January;95(1):361–8.

[43] Hinoi E, Gao N, Jung DY, Yadav V, Yoshizawa T, Myers Jr MG, et al. The sympathetic tone mediates leptin's inhibition of insulin secretion by modulating osteocalcin bioactivity. J Cell Biol 2008 December 29;183(7):1235–42.

[44] Civitelli R. Cell-cell communication in the osteoblast/osteocyte lineage. Arch Biochem Biophys 2008 May 15;473(2):188–92.

[45] Hay E, Laplantine E, Geoffroy V, Frain M, Kohler T, Muller R, et al. N-cadherin interacts with axin and LRP5 to negatively regulate Wnt/beta-catenin signaling, osteoblast function, and bone formation. Mol Cell Biol 2009 February;29(4):953–64.

[46] Bonewald LF, Johnson ML. Osteocytes, mechanosensing and Wnt signaling. Bone 2008 April; 42(4):606–15.

[47] Turner CH, Warden SJ, Bellido T, Plotkin LI, Kumar N, Jasiuk I, Danzig J, Robling AG. Mechanobiology of the skeleton. Sci Signal 2009;2:pt3.

[48] Jiang JX, Siller-Jackson AJ, Burra S. Roles of gap junctions and hemichannels in bone cell functions and in signal transmission of mechanical stress. Front Biosci 2007;12:1450–62.

[49] Boyce BF, Xing L, Jilka RJ, Bellido T, Weinstein RS, Parfitt AM, et al. Apoptosis in bone cells. In: Bilezikian JP, Raisz LG, Rodan GA, editors. Principles of Bone Biology. 2nd ed. San Diego: Academic Press; 2002. p. 151–68.

[50] Bellido T, Huening M, Raval-Pandya M, Manolagas SC, Christakos S. Calbindin-$D_{28K}$ is expressed in osteoblastic cells and suppresses their apoptosis by inhibiting caspase-3 activity. J Biol Chem 2000 June 1;275:26328–32.

[51] Plotkin LI, Manolagas SC, Bellido T. Transduction of cell survival signals by connexin-43 hemichannels. J Biol Chem 2002 March 8;277(10):8648–57.

[52] Faccio R, Takeshita S, Zallone A, Ross FP, Teitelbaum SL. c-FMS and the alphavbeta3 integrin collaborate during osteoclast differentiation. J Clin Invest 2003;111:749–58.

[53] Graham S, Gamie Z, Polyzois I, Narvani AA, Tzafetta K, Tsiridis E, et al. Prostaglandin EP2 and EP4 receptor agonists in bone formation and bone healing: in vivo and in vitro evidence. Expert Opin Investig Drugs 2009 June;18(6):746–66.

[54] Bar-Shavit Z. Taking a toll on the bones: regulation of bone metabolism by innate immune regulators. Autoimmunity 2008 April;41(3):195–203.

[55] Edwards CM, Mundy GR. Eph receptors and ephrin signaling pathways: a role in bone homeostasis. Int J Med Sci 2008;5(5):263–72.

[56] Canalis E. Growth factor control of bone mass. J Cell Biochem 2009 November 1;108(4):769–77.

[57] Riddle RC, Donahue HJ. From streaming-potentials to shear stress: 25 years of bone cell mechanotransduction. J Orthop Res 2009 February;27(2):143–9.

[58] Knothe Tate ML, Falls TD, McBride SH, Atit R, Knothe UR. Mechanical modulation of osteochondroprogenitor cell fate. Int J Biochem Cell Biol 2008;40(12):2720–38.

[59] Rubin J, Rubin C, Jacobs CR. Molecular pathways mediating mechanical signaling in bone. Gene 2006 February 15;367:1–16.

[60] Robinson JA, Chatterjee-Kishore M, Yaworsky PJ, Cullen DM, Zhao W, Li C, et al. Wnt/beta-catenin signaling is a normal physiological response to mechanical loading in bone. J Biol Chem 2006 October 20;281(42):31720–8.

[61] Sen B, Styner M, Xie Z, Case N, Rubin CT, Rubin J. Mechanical loading regulates NFATc1 and beta-catenin signaling through a GSK3beta control node. J Biol Chem 2009 December 11;284(50):34607–17.

[62] Ziros PG, Basdra EK, Papavassiliou AG. Runx2: of bone and stretch. Int J Biochem Cell Biol 2008;40(9):1659–63.

[63] Luu YK, Capilla E, Rosen CJ, Gilsanz V, Pessin JE, Judex S, et al. Mechanical stimulation of mesenchymal stem cell proliferation and differentiation promotes osteogenesis while preventing dietary-induced obesity. J Bone Miner Res 2009 January;24(1):50–61.

[64] Arnsdorf EJ, Tummala P, Kwon RY, Jacobs CR. Mechanically induced osteogenic differentiation – the role of RhoA, ROCKII and cytoskeletal dynamics. J Cell Sci 2009 February 15;122(Pt 4):546–53.

[65] Whitfield JF. The solitary (primary) cilium – a mechanosensory toggle switch in bone and cartilage cells. Cell Signal 2008 June;20(6):1019–24.

[66] Kousteni S, Bilezikian JP. The cell biology of parathyroid hormone in osteoblasts. Curr Osteoporos Rep 2008 June;6(2):72–6.

[67] Kawai M, Devlin MJ, Rosen CJ. Fat targets for skeletal health. Nat Rev Rheumatol 2009 July;5(7): 365–72.

[68] Ducy P, Amling M, Takeda S, Priemel M, Schilling AF, Beil FT, et al. Leptin inhibits bone formation through a hypothalamic relay: a central control of bone mass. Cell 2000 January 21;100(2):197–207.

[69] Takeda S, Elefteriou F, Levasseur R, Liu X, Zhao L, Parker KL, et al. Leptin regulates bone formation via the sympathetic nervous system. Cell 2002 November 1;111(3):305–17.

[70] Naot D, Cornish J. The role of peptides and receptors of the calcitonin family in the regulation of bone metabolism. Bone 2008 November;43(5):813–18.

[71] Wiren KM, Semirale AA, Hashimoto JG, Zhang XW. Signaling pathways implicated in androgen regulation of endocortical bone. Bone 2010 March;46(3):710–23.

[72] Imai Y, Nakamura T, Matsumoto T, Takaoka K, Kato S. Molecular mechanisms underlying the effects of sex steroids on bone and mineral metabolism. J Bone Miner Metab 2009;27(2):127–30.

[73] Dao DY, Yang X, Chen D, Zuscik M, O'Keefe RJ. Axin1 and Axin2 are regulated by TGF-and mediate Cross-talk between TGF-and WnT signaling pathways. Ann NY Acad Sci 2007;1116:82–99.

[74] Haÿ E, Laplantine E, Coeoffr YV, Frain M, Kohler T, Müller R, Mario PJ. N-cadherin interacts with axin and LRP5 to negatively regulate Wnt/beta-catenin signaling osteoblast function, and bone formation. Mol Cell Biol 2009;29:953–64.

[75] Zhou H, Mak W, Kalak R, Street J, Fong-Yee C, Zheng Y, et al. Glucocorticoid-dependent Wnt signaling by mature osteoblasts is a key regulator of cranial skeletal development in mice. Development 2009 February;136(3):427–36.

[76] Zhou H, Mak W, Zheng Y, Dunstan CR, Seibel MJ. Osteoblasts directly control lineage commitment of mesenchymal progenitor cells through Wnt signaling. J Biol Chem 2008 January 25;283(4):1936–45.

[77] Gentile MA, Nantermet PV, Vogel RL, Phillips R, Holder D, Hodor P, et al. Androgen-mediated improvement of body composition and muscle function involves a novel early transcriptional program including IGF1, mechano growth factor, and induction of {beta}-catenin. J Mol Endocrinol 2010 January;44(1):55–73.

[78] Sher LB, Harrison JR, Adams DJ, Kream BE. Impaired cortical bone acquisition and osteoblast differentiation in mice with osteoblast-targeted disruption of glucocorticoid signaling. Calcif Tissue Int 2006 August;79(2):118–25.

[79] Canalis E. Mechanisms of glucocorticoid-induced osteoporosis. Curr Opin Rheumatol 2003 July;15(4):454–7.

[80] Kream BE, Lukert BP. Clinical and basic aspects of glucocorticoid action in bone. In: Bilezikian JP, Raisz LG, Rodan GA, editors. Principles of Bone Biology. 2nd ed. San Diego: Academic Press; 2002. p. 723–40.

[81] Weinstein RS, Jilka RL, Parfitt AM, Manolagas SC. Inhibition of osteoblastogenesis and promotion of apoptosis of osteoblasts and osteocytes by glucocorticoids. Potential mechanisms of their deleterious effects on bone. J Clin Invest 1998;102(2):274–82.

[82] Noble BS, Reeve J. Osteocyte function, osteocyte death and bone fracture resistance. Mol Cell Endocrinol 2000 January 25;159(1-2):7–13.

[83] Kanazawa I, Yamaguchi T, Yano S, Hayashi K, Yamauchi M, Sugimoto T. Inhibition of the mevalonate pathway rescues the dexamethasone-induced suppression of the mineralization in osteoblasts via enhancing bone morphogenetic protein-2 signal. Horm Metab Res 2009 August;41(8):612–16.

[84] Luppen CA, Chandler RL, Noh T, Mortlock DP, Frenkel B. BMP-2 vs. BMP-4 expression and activity in glucocorticoid-arrested MC3T3-E1 osteoblasts: Smad signaling, not alkaline phosphatase activity, predicts rescue of mineralization. Growth Factors 2008 August;26(4):226–37.

[85] Smith E, Coetzee GA, Frenkel B. Glucocorticoids inhibit cell cycle progression in differentiating osteoblasts via glycogen synthase kinase-3beta. J Biol Chem 2002 May 17;277(20):18191–7.

[86] Leclerc N, Noh T, Cogan J, Samarawickrama DB, Smith E, Frenkel B. Opposing effects of glucocorticoids and Wnt signaling on Krox20 and mineral deposition in osteoblast cultures. J Cell Biochem 2008 April 15;103(6):1938–51.

[87] Mak W, Shao X, Dunstan CR, Seibel MJ, Zhou H. Biphasic glucocorticoid-dependent regulation of Wnt expression and its inhibitors in mature osteoblastic cells. Calcif Tissue Int 2009 December; 85(6):538–45.

[88] Smith E, Frenkel B. Glucocorticoids inhibit the transcriptional activity of LEF/TCF in differentiating osteoblasts in a glycogen synthase kinase-3beta-dependent and -independent manner. J Biol Chem 2005 January 21;280(3):2388–94.

[89] Wang FS, Ko JY, Weng LH, Yeh DW, Ke HJ, Wu SL. Inhibition of glycogen synthase kinase-3beta attenuates glucocorticoid-induced bone loss. Life Sci 2009 November 4;85(19-20):685–92.

[90] Abraham SM, Lawrence T, Kleiman A, Warden P, Medghalchi M, Tuckermann J, et al. Antiinflammatory effects of dexamethasone are partly dependent on induction of dual specificity phosphatase 1. J Exp Med 2006 August 7;203(8):1883–9.

[91] Horsch K, de WH, Schuurmans MM, Allie-Reid F, Cato AC, Cunningham J, et al. Mitogen-activated protein kinase phosphatase 1/dual specificity phosphatase 1 mediates glucocorticoid inhibition of osteoblast proliferation. Mol Endocrinol 2007 December;21(12):2929–40.

[92] Galliher-Beckley AJ, Williams JG, Collins JB, Cidlowski JA. Glycogen synthase kinase 3beta-mediated serine phosphorylation of the human glucocorticoid receptor redirects gene expression profiles. Mol Cell Biol 2008 December;28(24):7309–22.

[93] Jilka RL, Weinstein RS, Parfitt AM, Manolagas SC. Quantifying osteoblast and osteocyte apoptosis: challenges and rewards. J Bone Miner Res 2007 October;22(10):1492–501.

[94] Hofbauer LC, Rauner M. Minireview: live and let die: molecular effects of glucocorticoids on bone cells. Mol Endocrinol 2009 October;23(10):1525–31.

[95] Oxlund H, Ortoft G, Thomsen JS, Danielsen CC, Ejersted C, Andreassen TT. The anabolic effect of PTH on bone is attenuated by simultaneous glucocorticoid treatment. Bone 2006 August;39(2):244–52.

[96] Kim HJ, Zhao H, Kitaura H, Bhattacharyya S, Brewer JA, Muglia LJ, et al. Dexamethaone suppresses bone formation via the osteoclast. Adv Exp Med Biol 2007;602:43–6.

[97] Hayashi K, Yamaguchi T, Yano S, Kanazawa I, Yamauchi M, Yamamoto M, et al. BMP/Wnt antagonists are upregulated by dexamethasone in osteoblasts and reversed by alendronate and PTH: potential therapeutic targets for glucocorticoid-induced osteoporosis. Biochem Biophys Res Commun 2009 February 6;379(2):261–6.

[98] Pike JW, Zella LA, Meyer MB, Fretz JA, Kim S. Molecular actions of 1,25-dihydroxyvitamin D3 on genes involved in calcium homeostasis. J Bone Miner Res 2007 December;22(Suppl. 2):V16–19.

[99] Paredes R, Arriagada G, Cruzat F, Villagra A, Olate J, Zaidi K, et al. The bone-specific transcription factor RUNX2 interacts with the 1α,25-dihydroxyvitamin D3 receptor to upregulate rat osteocalcin gene expression in osteoblastic cells. Mol Cell Biol 2004;24(20):8847–61.

[100] Marcellini S, Bruna C, Henriquez JP, Albistur M, Reyes AE, Barriga EH, et al. Evolution of the interaction between Runx2 and VDR, two transcription factors involved in osteoblastogenesis. BMC Evol Biol 2010;10:78.

[101] Rosen CJ, Ackert-Bicknell C, Rodriguez JP, Pino AM. Marrow fat and the bone microenvironment: developmental, functional, and pathological implications. Crit Rev Eukaryot Gene Expr 2009; 19(2):109–24.

[102] Bab I, Zimmer A, Melamed E. Cannabinoids and the skeleton: from marijuana to reversal of bone loss. Ann Med 2009;41(8):560–7.

[103] Idris AI, Sophocleous A, Landao-Bassonga E, Canals M, Milligan G, Baker D, et al. Cannabinoid receptor type 1 protects against age-related osteoporosis by regulating osteoblast and adipocyte differentiation in marrow stromal cells. Cell Metab 2009 August;10(2):139–47.

[104] Gaur T, Hussain S, Mudhasani R, Parulkar I, Colby JL, Frederick D, et al. Dicer inactivation in osteoprogenitor cells compromises fetal survival and bone formation, while excision in differentiated osteoblasts increases bone mass in the adult mouse. Dev Biol 2010 January 15;340(1):10–21.

[105] Harfe BD, McManus MT, Mansfield JH, Hornstein E, Tabin CJ. The RNaseIII enzyme Dicer is required for morphogenesis but not patterning of the vertebrate limb. Proc Natl Acad Sci U S A 2005 August 2;102(31):10898–903.

[106] Gazzerro E, Canalis E. Bone morphogenetic proteins and their antagonists. Rev Endocr Metab Disord 2006 June;7(1-2):51–65.

[107] Axelrad TW, Einhorn TA. Bone morphogenetic proteins in orthopaedic surgery. Cytokine Growth Factor Rev 2009 October;20(5-6):481–8.

[108] Kamiya N, Ye L, Kobayashi T, Lucas DJ, Mochida Y, Yamauchi M, et al. Disruption of BMP signaling in osteoblasts through type IA receptor (BMPRIA) increases bone mass. J Bone Miner Res 2008 December;23(12):2007–17.

[109] Tsuji K, Bandyopadhyay A, Harfe BD, Cox K, Kakar S, Gerstenfeld L, et al. BMP2 activity, although dispensable for bone formation, is required for the initiation of fracture healing. Nat Genet 2006 December;38(12):1424–9.

[110] Tsuji K, Cox K, Bandyopadhyay A, Harfe BD, Tabin CJ, Rosen V. BMP4 is dispensable for skeletogenesis and fracture-healing in the limb. J Bone Joint Surg Am 2008 February;90(Suppl. 1):14–18.

[111] Choi HY, Dieckmann M, Herz J, Niemeier A. Lrp4, a novel receptor for Dickkopf 1 and sclerostin, is expressed by osteoblasts and regulates bone growth and turnover in vivo. PLoS ONE 2009;4(11): e7930.

[112] Hartmann C. Skeletal development – Wnts are in control. Mol Cells 2007 October 31;24(2):177–84.

[113] Bennett CN, Ouyang H, Ma YL, Zeng Q, Gerin I, Sousa KM, et al. Wnt10b increases postnatal bone formation by enhancing osteoblast differentiation. J Bone Miner Res 2007 December;22(12): 1924–32.

[114] Wright WS, Longo KA, Dolinsky VW, Gerin I, Kang S, Bennett CN, et al. Wnt10b inhibits obesity in ob/ob and agouti mice. Diabetes 2007 February;56(2):295–303.

[115] Chen Y, Alman BA. Wnt pathway, an essential role in bone regeneration. J Cell Biochem 2009 February 15;106(3):353–62.

[116] Collette NM, Genetos D, Murugesh D, Harland RM, Loots GG. Genetic evidence that SOST inhibits WNT signaling in the limb. Dev Biol 2010 March 30.

[117] Boyden LM, Mao J, Belsky J, Mitzner L, Farhi A, Mitnick MA, et al. High bone density due to a mutation in LDL-receptor-related protein 5. N Engl J Med 2002 May 16;346(20):1513–21.

[118] Little RD, Carulli JP, Del Mastro RG, Dupuis J, Osborne M, Folz C, et al. A mutation in the LDL receptor-related protein 5 gene results in the autosomal dominant high-bone-mass trait. Am J Hum Genet 2002 January;70(1):11–19.

[119] Ai M, Heeger S, Bartels CF, Schelling DK. Clinical and molecular findings in osteoporosis-pseudoglioma syndrome. Am J Hum Genet 2005 November;77(5):741–53.

[120] Day TF, Guo X, Garrett-Beal L, Yang Y. Wnt/beta-catenin signaling in mesenchymal progenitors controls osteoblast and chondrocyte differentiation during vertebrate skeletogenesis. Dev Cell 2005 May;8(5):739-50.

[121] Hill TP, Spater D, Taketo MM, Birchmeier W, Hartmann C. Canonical Wnt/beta-catenin signaling prevents osteoblasts from differentiating into chondrocytes. Dev Cell 2005 May;8(5):727-38.

[122] Rey JP, Ellies DL. Wnt modulators in the biotech pipeline. Dev Dyn 2010 January;239(1):102-14.

[123] Hoeppner LH, Secreto FJ, Westendorf JJ. Wnt signaling as a therapeutic target for bone diseases. Expert Opin Ther Targets 2009 April;13(4):485-96.

[124] Krause U, Harris S, Green A, Ylostalo J, Zeitouni S, Lee N, et al. Pharmaceutical modulation of canonical Wnt signaling in multipotent stromal cells for improved osteoinductive therapy. Proc Natl Acad Sci U S A 2010 March 2;107(9):4147-52.

[125] Bovolenta P, Esteve P, Ruiz JM, Cisneros E, Lopez-Rios J. Beyond Wnt inhibition: new functions of secreted Frizzled-related proteins in development and disease. J Cell Sci 2008 March 15;121 (Pt 6):737-46.

[126] Van Bezooijen RL, ten DP, Papapoulos SE, Lowik CW. SOST/sclerostin, an osteocyte-derived negative regulator of bone formation. Cytokine Growth Factor Rev 2005 June;16(3):319-27.

[127] ten Dijke P, Krause C, de Gorter DJ, Lowik CW, Van Bezooijen RL. Osteocyte-derived sclerostin inhibits bone formation: its role in bone morphogenetic protein and Wnt signaling. J Bone Joint Surg Am 2008 February;90(Suppl. 1):31-5.

[128] Ominsky MS, Vlasseros F, Jolette J, Smith SY, Stouch B, Doellgast G, et al. Two doses of sclerostin antibody in cynomolgus monkeys increases bone formation, bone mineral density, and bone strength. J Bone Miner Res 2010 January 8.

[129] Keller H, Kneissel M. SOST is a target gene for PTH in bone. Bone 2005 August;37(2):148-58.

[130] Kramer I, Loots GG, Studer A, Keller H, Kneissel M. Parathyroid hormone (PTH) induced bone gain is blunted in SOST overexpressing and deficient mice. J Bone Miner Res 2009 July 13.

[131] Gaur T, Wixted JJ, Hussain S, O'Connell SL, Morgan EF, Ayers DC, et al. Secreted frizzled related protein 1 is a target to improve fracture healing. J Cell Physiol 2009;220(1):174-81.

[132] Secreto FJ, Hoeppner LH, Westendorf JJ. Wnt signaling during fracture repair. Curr Osteoporos Rep 2009 July;7(2):64-9.

[133] Bais M, McLean J, Sebastiani P, Young M, Wigner N, Smith T, et al. Transcriptional analysis of fracture healing and the induction of embryonic stem cell-related genes. PLoS ONE 2009;4(5):e5393.

[134] Pereira RC, Stadmeyer LE, Smith DL, Rydziel S, Canalis E. CCAAT/Enhancer-binding protein homologous protein (CHOP) decreases bone formation and causes osteopenia. Bone 2007 March;40(3):619-26.

[135] Zaidi SK, Young DW, Choi JY, Pratap J, Javed A, Montecino M, et al. The dynamic organization of gene-regulatory machinery in nuclear microenvironments. EMBO Rep 2005 February;6(2):128-33.

[136] Stein GS, Lian JB, van Wijnen AJ, Stein JL, Montecino M, Javed A, et al. Runx2 control of organization, assembly and activity of the regulatory machinery for skeletal gene expression. Oncogene 2004 May 24;23(24):4315-29.

[137] Zaidi SK, Young DW, Javed A, Pratap J, Montecino M, van WA, et al. Nuclear microenvironments in biological control and cancer. Nat Rev Cancer 2007 June;7(6):454-63.

[138] Isenmann S, Arthur A, Zannettino AC, Turner JL, Shi S, Glackin CA, et al. TWIST family of basic helix-loop-helix transcription factors mediate human mesenchymal stem cell growth and commitment. Stem Cells 2009 October;27(10):2457-68.

[139] Iimura T, Denans N, Pourquie O. Establishment of Hox vertebral identities in the embryonic spine precursors. Curr Top Dev Biol 2009;88:201-34.

[140] Depew MJ, Simpson CA, Morasso M, Rubenstein JL. Reassessing the Dlx code: the genetic regulation of branchial arch skeletal pattern and development. J Anat 2005 November;207(5):501–61.

[141] Hassan MQ, Javed A, Morasso MI, Karlin J, Montecino M, van Wijnen AJ, et al. Dlx3 transcriptional regulation of osteoblast differentiation: temporal recruitment of Msx2, Dlx3, and Dlx5 homeodomain proteins to chromatin of the osteocalcin gene. Mol Cell Biol 2004;24(20):9248–61.

[142] Hassan MQ, Tare RS, Lee S, Mandeville M, Morasso MI, Javed A, et al. BMP2 commitment to the osteogenic lineage involves activation of Runx2 by Dlx3 and a homeodomain transcriptional network. J Biol Chem 2006;281(52):40515–26.

[143] Hassan MQ, Tare R, Lee SH, Mandeville M, Weiner B, Montecino M, et al. HOXA10 controls osteoblastogenesis by directly activating bone regulatory and phenotypic genes. Mol Cell Biol 2007;27(9):3337–52.

[144] Zakany J, Kmita M, Duboule D. A dual role for Hox genes in limb anterior-posterior asymmetry. Science 2004 June 11;304(5677):1669–72.

[145] Shao JS, Cheng SL, Pingsterhaus JM, Charlton-Kachigian N, Loewy AP, Towler DA. Msx2 Promotes Cardiovascular Calcification by activating paracrine Wnt signals. J Clin Invest 2005;115:1210–20.

[146] Selvamurugan N, Jefcoat SC, Kwok S, Kowalewski R, Tamasi JA, Partridge NC. Overexpression of Runx2 directed by the matrix metalloproteinase-13 promoter containing the AP-1 and Runx/RD/Cbfa sites alters bone remodeling in vivo. J Cell Biochem 2006 October 1;99(2):545–57.

[147] Wagner EF. Bone development and inflammatory disease is regulated by AP-1 (Fos/Jun). Ann Rheum Dis 2010 January;69(Suppl. 1):i86–8.

[148] Rowe GC, Choi CS, Neff L, Horne WC, Shulman GI, Baron R. Increased energy expenditure and insulin sensitivity in the high bone mass DeltaFosB transgenic mice. Endocrinology 2009 January;150(1):135–43.

[149] Pratap J, Lian JB, Javed A, Barnes GL, van Wijnen AJ, Stein JL, et al. Regulatory roles of Runx2 in metastatic tumor and cancer cell interactions with bone. Cancer Metastasis Rev 2006 December;25(4):589–600.

[150] Cohen Jr MM. Perspectives on RUNX genes: an update. Am J Med Genet A 2009 December;149A(12):2629–46.

[151] Link KA, Chou FS, Mulloy JC. Core binding factor at the crossroads: determining the fate of the HSC. J Cell Physiol 2010 January;222(1):50–6.

[152] Collins A, Littman DR, Taniuchi I. RUNX proteins in transcription factor networks that regulate T-cell lineage choice. Nat Rev Immunol 2009 February;9(2):106–15.

[153] Inoue K, Shiga T, Ito Y. Runx transcription factors in neuronal development. Neural Develop 2008;3:20.

[154] Smith N, Dong Y, Lian JB, Pratap J, Kingsley PD, van Wijnen AJ, et al. Overlapping expression of Runx1(Cbfa2) and Runx2(Cbfa1) transcription factors supports cooperative induction of skeletal development. J Cell Physiol 2005;203:133–43.

[155] Kundu M, Javed A, Jeon JP, Horner A, Shum L, Eckhaus M, et al. Cbfbeta interacts with Runx2 and has a critical role in bone development. Nat Genet 2002 December;32(4):639–44.

[156] Miller J, Horner A, Stacy T, Lowrey C, Lian JB, Stein G, et al. The core-binding factor beta subunit is required for bone formation and hematopoietic maturation. Nat Genet 2002 December;32(4):645–9.

[157] Hassan MQ, Saini S, Gordon JA, van Wijnen AJ, Montecino M, Stein JL, et al. Molecular switches involving homeodomain proteins, HOXA10 and RUNX2 regulate osteoblastogenesis. Cells Tissues Organs 2009;189(1-4):122–5.

[158] Liu W, Toyosawa S, Furuichi T, Kanatani N, Yoshida C, Liu Y, et al. Overexpression of Cbfa1 in osteoblasts inhibits osteoblast maturation and causes osteopenia with multiple fractures. J Cell Biol 2001 October 1;155(1):157–66.

[159] Dobreva G, Chahrour M, Dautzenberg M, Chirivella L, Kanzler B, Farinas I, et al. SATB2 is a multifunctional determinant of craniofacial patterning and osteoblast differentiation. Cell 2006 June 2;125(5):971–86.

[160] Nakashima K, Zhou X, Kunkel G, Zhang Z, Deng JM, Behringer RR, et al. The novel zinc finger-containing transcription factor osterix is required for osteoblast differentiation and bone formation. Cell 2002 January 11;108(1):17–29.

[161] Talamillo A, Delgado I, Nakamura T, de-Vega S, Yoshitomi Y, Unda F, et al. Role of Epiprofin, a zinc-finger transcription factor, in limb development. Dev Biol 2010 January 15;337(2):363–74.

[162] Beak JY, Kang HS, Kim YS, Jetten AM. Kruppel-like zinc finger protein Glis3 promotes osteoblast differentiation by regulating FGF18 expression. J Bone Miner Res 2007 August;22(8):1234–44.

[163] Wu M, Hesse E, Morvan F, Zhang JP, Correa D, Rowe GC, et al. Zfp521 antagonizes Runx2, delays osteoblast differentiation in vitro, and promotes bone formation in vivo. Bone 2009 April;44(4): 528–36.

[164] Jones DC, Glimcher LH. Regulation of bone formation and immune cell development by schnurri proteins. Adv Exp Med Biol 2010;658:117–22.

[165] Saita Y, Takagi T, Kitahara K, Usui M, Miyazono K, Ezura Y, et al. Lack of Schnurri-2 expression associates with reduced bone remodeling and osteopenia. J Biol Chem 2007 April 27;282(17): 12907–15.

[166] Jin W, Takagi T, Kanesashi SN, Kurahashi T, Nomura T, Harada J, et al. Schnurri-2 controls BMP-dependent adipogenesis via interaction with Smad proteins. Dev Cell 2006 April;10(4):461–71.

[167] Stein GS, Zaidi SK, Stein JL, Lian JB, van Wijnen AJ, Montecino M, et al. Transcription-factor-mediated epigenetic control of cell fate and lineage commitment. Biochem Cell Biol 2009 February;87(1):1–6.

[168] Montecino M, Stein GS, Stein JL, Lian JB, van Wijnen AJ, Carvallo L, et al. Vitamin D control of gene expression: temporal and spatial parameters for organization of the regulatory machinery. Crit Rev Eukaryot Gene Expr 2008;18(2):163–72.

[169] Javed A, Gutierrez S, Montecino M, van Wijnen AJ, Stein JL, Stein GS, et al. Multiple Cbfa/AML sites in the rat osteocalcin promoter are required for basal and vitamin D responsive transcription and contribute to chromatin organization. Mol Cell Biol 1999;19:7491–500.

[170] Jensen ED, Nair AK, Westendorf JJ. Histone deacetylase co-repressor complex control of Runx2 and bone formation. Crit Rev Eukaryot Gene Expr 2007;17(3):187–96.

[171] Young DW, Pratap J, Javed A, Weiner B, Ohkawa Y, van Wijnen A, et al. SWI/SNF chromatin remodeling complex is obligatory for BMP2-induced, Runx2-dependent skeletal gene expression that controls osteoblast differentiation. J Cell Biochem 2005 March 1;94(4):720–30.

[172] Young DW, Hassan MQ, Yang X-Q, Galindo M, Javed A, Zaidi SK, et al. Mitotic retention of gene expression patterns by the cell fate determining transcription factor Runx2. Proc Natl Acad Sci U S A 2007;104(9):3189–94.

[173] Ali SA, Zaidi SK, Dacwag CS, Salma N, Young DW, Shakoori AR, et al. Phenotypic transcription factors epigenetically mediate cell growth control. Proc Natl Acad Sci U S A 2008 May 6;105 (18):6632–7.

[174] Choi J-Y, Pratap J, Javed A, Zaidi SK, Xing L, Balint E, et al. Subnuclear targeting of Runx/Cbfa/AML factors is essential for tissue-specific differentiation during embryonic development. Proc Natl Acad Sci U S A 2001;98(15):8650–5.

[175] Bartel DP. MicroRNAs: genomics, biogenesis, mechanism, and function. Cell 2004 January 23; 116(2):281–97.

[176] Pawlicki JM, Steitz JA. Nuclear networking fashions pre-messenger RNA and primary microRNA transcripts for function. Trends Cell Biol 2010 January;20(1):52–61.

[177] Li Z, Hassan MQ, Jafferji M, Garzon R, Croce CM, van Wijnen AJ, et al. Biological functions of miR-29b contribute to positive regulation of osteoblast differentiation. J Biol Chem 2009 April 2;284(23):15676–84.

[178] Li Z, Hassan MQ, Volinia S, van Wijnen AJ, Stein JL, Croce CM, et al. A microRNA signature for a BMP2-induced osteoblast lineage commitment program. Proc Natl Acad Sci U S A 2008;105(37):13906–11.

[179] Luzi E, Marini F, Sala SC, Tognarini I, Galli G, Brandi ML. Osteogenic differentiation of human adipose tissue-derived stem cells is modulated by the miR-26a targeting of the SMAD1 transcription factor. J Bone Miner Res 2008 February;23(2):287–95.

[180] Itoh T, Nozawa Y, Akao Y. MicroRNA-141 and -200a are involved in bone morphogenetic protein-2-induced mouse pre-osteoblast differentiation by targeting distal-less homeobox 5. J Biol Chem 2009 July 17;284(29):19272–9.

[181] Huang J, Zhao L, Xing L, Chen D. MicroRNA-204 regulates Runx2 protein expression and mesenchymal progenitor cell differentiation. Stem Cells 2010 February 1;28(2):357–64.

[182] van Rooij E, Sutherland LB, Thatcher JE, DiMaio JM, Naseem RH, Marshall WS, et al. Dysregulation of microRNAs after myocardial infarction reveals a role of miR-29 in cardiac fibrosis. Proc Natl Acad Sci U S A 2008 September 2;105(35):13027–32.

[183] Kapinas K, Kessler CB, Delany AM. miR-29 suppression of osteonectin in osteoblasts: regulation during differentiation and by canonical Wnt signaling. J Cell Biochem 2009 June 29;108(1):216–24.

[184] Smerdel-Ramoya A, Zanotti S, Deregowski V, Canalis E. Connective tissue growth factor enhances osteoblastogenesis in vitro. J Biol Chem 2008 August 15;283(33):22690–9.

[185] Fleming HE, Janzen V, Lo CC, Guo J, Leahy KM, Kronenberg HM, et al. Wnt signaling in the niche enforces hematopoietic stem cell quiescence and is necessary to preserve self-renewal in vivo. Cell Stem Cell 2008 March 6;2(3):274–83.

[186] Staal FJ, Luis TC, Tiemessen MM. WNT signalling in the immune system: WNT is spreading its wings. Nat Rev Immunol 2008 August;8(8):581–93.

[187] Bennett CN, Longo KA, Wright WS, Suva LJ, Lane TF, Hankenson KD, et al. Regulation of osteoblastogenesis and bone mass by Wnt10b. Proc Natl Acad Sci U S A 2005 March 1;102(9):3324–9.

[188] Van Den Berg DJ, Sharma AK, Bruno E, Hoffman R. Role of members of the Wnt gene family in human hematopoiesis. Blood 1998 November 1;92(9):3189–202.

[189] Terauchi M, Li JY, Bedi B, Baek KH, Tawfeek H, Galley S, et al. T lymphocytes amplify the anabolic activity of parathyroid hormone through Wnt10b signaling. Cell Metab 2009 September;10(3):229–40.

[190] Puig-Kroger A, Corbi A. RUNX3: a new player in myeloid gene expression and immune response. J Cell Biochem 2006 July 1;98(4):744–56.

[191] Brady G, Whiteman HJ, Spender LC, Farrell PJ. Downregulation of RUNX1 by RUNX3 requires the RUNX3 VWRPY sequence and is essential for Epstein-Barr virus-driven B-cell proliferation. J Virol 2009 July;83(13):6909–16.

[192] MacNaul KL, Hutchinson NI, Parsons JN, Bayne EK, Tocci MJ. Analysis of IL-1 and TNF-alpha gene expression in human rheumatoid synoviocytes and normal monocytes by in situ hybridization. J Immunol 1990 December 15;145(12):4154–66.

[193] Choy EH, Panayi GS. Cytokine pathways and joint inflammation in rheumatoid arthritis. N Engl J Med 2001 March 22;344(12):907–16.

[194] Butler DM, Maini RN, Feldmann M, Brennan FM. Modulation of proinflammatory cytokine release in rheumatoid synovial membrane cell cultures. Comparison of monoclonal anti TNF-alpha antibody with the interleukin-1 receptor antagonist. Eur Cytokine Netw 1995 July;6(4):225–30.

[195] Chin JE, Winterrowd GE, Krzesicki RF, Sanders ME. Role of cytokines in inflammatory synovitis. The coordinate regulation of intercellular adhesion molecule 1 and HLA class I and class II antigens in rheumatoid synovial fibroblasts. Arthritis Rheum 1990 December;33(12):1776–86.

[196] Burrage PS, Mix KS, Brinckerhoff CE. Matrix metalloproteinases: role in arthritis. Front Biosci 2006;11:529–43.

[197] Hardingham TE, Bayliss MT, Rayan V, Noble DP. Effects of growth factors and cytokines on proteoglycan turnover in articular cartilage. Br J Rheumatol 1992;31(Suppl. 1):1–6.

[198] Pettit AR, Ji H, von SD, Muller R, Goldring SR, Choi Y, et al. TRANCE/RANKL knockout mice are protected from bone erosion in a serum transfer model of arthritis. Am J Pathol 2001 November; 159(5):1689–99.

[199] Gravallese EM, Manning C, Tsay A, Naito A, Pan C, Amento E, et al. Synovial tissue in rheumatoid arthritis is a source of osteoclast differentiation factor. Arthritis Rheum 2000 February;43(2):250–8.

[200] Yao Z, Li P, Zhang Q, Schwarz EM, Keng P, Arbini A, et al. Tumor necrosis factor-alpha increases circulating osteoclast precursor numbers by promoting their proliferation and differentiation in the bone marrow through upregulation of c-Fms expression. J Biol Chem 2006 April 28;281(17):11846–55.

[201] Lam J, Takeshita S, Barker JE, Kanagawa O, Ross FP, Teitelbaum SL. TNF-alpha induces osteoclastogenesis by direct stimulation of macrophages exposed to permissive levels of RANK ligand. J Clin Invest 2000 December;106(12):1481–8.

[202] Wei S, Kitaura H, Zhou P, Ross FP, Teitelbaum SL. IL-1 mediates TNF-induced osteoclastogenesis. J Clin Invest 2005 February;115(2):282–90.

[203] Gilbert L, He X, Farmer P, Boden S, Kozlowski M, Rubin J, et al. Inhibition of osteoblast differentiation by tumor necrosis factor-alpha. Endocrinology 2000 November;141(11):3956–64.

[204] Rau R, Wassenberg S, Herborn G, Perschel WT, Freitag G. Identification of radiologic healing phenomena in patients with rheumatoid arthritis. J Rheumatol 2001 December;28(12):2608–15.

[205] Ideguchi H, Ohno S, Hattori H, Senuma A, Ishigatsubo Y. Bone erosions in rheumatoid arthritis can be repaired through reduction in disease activity with conventional disease-modifying antirheumatic drugs. Arthritis Res Ther 2006;8(3):R76.

[206] Sharp JT, Van Der Heijde D, Boers M, Boonen A, Bruynesteyn K, Emery P, et al. Repair of erosions in rheumatoid arthritis does occur. Results from 2 studies by the OMERACT Subcommittee on Healing of Erosions. J Rheumatol 2003 May;30(5):1102–7.

[207] Walsh NC, Reinwald S, Manning CA, Condon KW, Iwata K, Burr DB, et al. Osteoblast function is compromised at sites of focal bone erosion in inflammatory arthritis. J Bone Miner Res 2009 September;24(9):1572–85.

[208] Schett G, Middleton S, Bolon B, Stolina M, Brown H, Zhu L, et al. Additive bone-protective effects of anabolic treatment when used in conjunction with RANKL and tumor necrosis factor inhibition in two rat arthritis models. Arthritis Rheum 2005 May;52(5):1604–11.

[209] Diarra D, Stolina M, Polzer K, Zwerina J, Ominsky MS, Dwyer D, et al. Dickkopf-1 is a master regulator of joint remodeling. Nat Med 2007 February;13(2):156–63.

[210] Ijiri K, Nagayoshi R, Matsushita N, Tsuruga H, Taniguchi N, Gushi A, et al. Differential expression patterns of secreted frizzled related protein genes in synovial cells from patients with arthritis. J Rheumatol 2002 November;29(11):2266–70.

[211] Schett G. Joint remodelling in inflammatory disease. Ann Rheum Dis 2007 November;66(Suppl. 3):iii42–4.

[212] Almeida M, Han L, Martin-Millan M, O'Brien CA, Manolagas SC. Oxidative stress antagonizes Wnt signaling in osteoblast precursors by diverting beta-catenin from T cell factor- to forkhead box O-mediated transcription. J Biol Chem 2007 September 14;282(37):27298–305.

[213] Colla S, Zhan F, Xiong W, Wu X, Xu H, Stephens O, et al. The oxidative stress response regulates DKK1 expression through the JNK signaling cascade in multiple myeloma plasma cells. Blood 2007 May 15;109(10):4470–7.

[214] Arnett TR, Gibbons DC, Utting JC, Orriss IR, Hoebertz A, Rosendaal M, et al. Hypoxia is a major stimulator of osteoclast formation and bone resorption. J Cell Physiol 2003 July;196(1):2–8.

[215] Utting JC, Robins SP, Brandao-Burch A, Orriss IR, Behar J, Arnett TR. Hypoxia inhibits the growth, differentiation and bone-forming capacity of rat osteoblasts. Exp Cell Res 2006 June 10;312(10):1693–702.

[216] Lories RJ, Peeters J, Bakker A, Tylzanowski P, Derese I, Schrooten J, et al. Articular cartilage and biomechanical properties of the long bones in Frzb-knockout mice. Arthritis Rheum 2007 December;56(12):4095–103.

[217] Hausler KD, Horwood NJ, Chuman Y, Fisher JL, Ellis J, Martin TJ, et al. Secreted frizzled-related protein-1 inhibits RANKL-dependent osteoclast formation. J Bone Miner Res 2004 November;19(11):1873–81.

[218] Lyons K. Animal models: genetic manipulation. In: Rosen CJ, editor. Primer on the Metabolic Bone Diseases and Disorders of Mineral Metabolism. 7th ed. Washington, DC: American Society for Bone and Mineral Research; 2008. p. 45–51.

[219] Young DW, Hassan MQ, Pratap J, Galindo M, Zaidi SK, Lee SH, et al. Mitotic occupancy and lineage-specific transcriptional control of rRNA genes by Runx2. Nature 2007;445(7126):442–6.

# 6

# The Osteoclast: The Pioneer of Osteoimmunology

Roberta Faccio[1], Yongwon Choi[2], Steven L. Teitelbaum[3], Hiroshi Takayanagi[4]

[1]DEPARTMENT OF ORTHOPEDICS, WASHINGTON UNIVERSITY, ST. LOUIS, MO, USA, [2]DEPARTMENT OF PATHOLOGY AND LABORATORY MEDICINE, UNIVERSITY OF PENNSYLVANIA SCHOOL OF MEDICINE, PHILADELPHIA, PA, USA, [3]DEPARTMENT OF PATHOLOGY AND IMMUNOLOGY, WASHINGTON UNIVERSITY, ST. LOUIS, MO, USA, [4]DEPARTMENT OF CELL SIGNALING, GRADUATE SCHOOL, TOKYO MEDICAL AND DENTAL UNIVERSITY, YUSHIMA, BUNKYO-KU, TOKYO, JAPAN

## CHAPTER OUTLINE

Introduction .................................................................................................................. 142
RANKL and RANK, an osteoclastogenic cytokine and its receptor ........................... 143
TRAF6: the multifunctional signaling molecule activated by RANK ......................... 144
What happens downstream of TRAF6? ..................................................................... 146
The role of NF-κB in osteoclast differentiation .......................................................... 147
The critical role of AP-1 transcription factors ............................................................ 148
MAPKs activated by RANKL ....................................................................................... 149
NFATc1 is a master transcription factor of osteoclast differentiation ...................... 149
Autoamplification of NFATc1 and its epigenetic regulation ...................................... 151
Transcriptional control governed by NFATc1 ............................................................ 152
Co-stimulatory receptors for RANK signal: FcRγ and DAP12 .................................... 152
Importance of ITAM co-stimulatory signals in humans: Nasu-Hakola disease ........ 156
Additional co-stimulatory signals involved in osteoclastogenesis ............................ 157
Receptors signaling through DAP12 .......................................................................... 157
Src family kinases: activation of ITAM signaling ....................................................... 158
Syk kinase: downstream of DAP12/FcRγ? ................................................................. 159
PLCγ2: enzyme and adaptor molecule ....................................................................... 159
Tec kinases: integrating RANK and ITAM signaling .................................................. 160
Negative regulatory role of DAP12 ............................................................................. 161
M-CSF and c-Fms: a road to proliferation and survival ............................................ 162
M-CSF signaling .......................................................................................................... 162
Erk and PI3K signaling ................................................................................................ 163

The osteoclast's job: bone resorption .................................................................................. 164
Osteoclast cytoskeleton: the podosomes and the sealing zone ................................................. 165
Osteoclast cytoskeleton: the microtubules and the sealing zone ............................................... 165
Osteoclast functional structure: the ruffled border ...................................................................... 166
Osteoclast and bone matrix: role of αvβ3 integrin ....................................................................... 168
Integrin-associated proteins ........................................................................................................... 171
M-CSF and the osteoclast cytoskeleton ....................................................................................... 174
Conclusion ...................................................................................................................................... 175
References ..................................................................................................................................... 175

# Introduction

Bone is a dynamic tissue that continuously undergoes coupled resorption and formation, mediated by osteoclasts and osteoblasts, respectively. This process, called bone remodeling, is a prerequisite for normal skeletal homeostasis throughout life. Bone remodeling is regulated by a variety of factors, such as cytokines, chemokines, hormones, and biochemical stimuli. While the physiological necessity of bone remodeling is incompletely understood, its imbalance often prompts disorders such as post-menopausal osteoporosis and the peri-articular bone destruction of inflammatory arthritis [1–3]. Remodeling also probably maintains structural integrity by replacing effete bone with new as attenuated resorption suppresses formation. In consequence, weak bone accumulates, predisposing to fracture despite less compromised skeletal mass.

Osteoclasts are large, multinucleated cells with the unique capacity to degrade the organic and inorganic matrices of bone. Because they are fundamental to the pathogenesis of virtually all diseases associated with bone loss, the insights gained into the mechanisms by which they form and degrade bone are clinically significant. The study of osteoclasts has eventuated in antiresorptive therapy, the most common means of treating the pathologically diminished skeleton. Thus, therapeutic interventions to maintain skeletal health require optimizing physiological bone resorption, and hence, understanding the mechanisms by which osteoclasts resorb bone.

Osteoclasts are members of the monocyte/macrophage family and, as such, share many of the features of immune cells. Perhaps the most important discovery in our understanding of these cells is the finding that receptor activator of nuclear factor-κB ligand (RANKL) is the key molecule regulating their differentiation. This observation permits the generation of virtually pure populations of osteoclasts, in culture, thus facilitating dissection of the intracellular mechanisms that are involved in the cell's differentiation.

The capacity to generate osteoclasts, in virtual purity, has also yielded insights into the mechanisms by which they degrade bone, much of which involves cytoskeletal organization. Such organization facilitates the formation of an isolated microenvironment between the osteoclast and bone into which osteoclasts secrete matrix-degrading molecules that are additional candidate therapeutic targets [4].

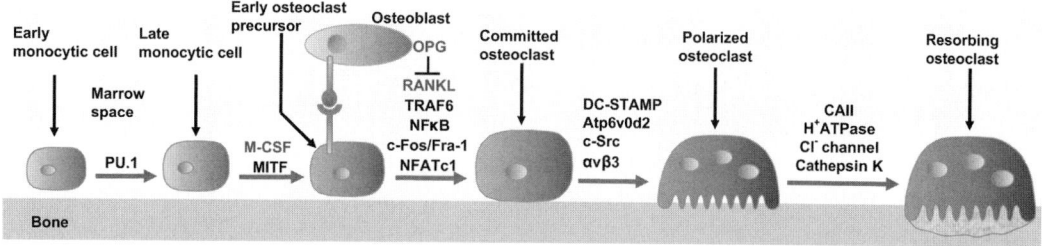

**FIGURE 6-1 Regulation of osteoclast formation and function.** The osteoclast is a member of the monocyte/macrophage family. Early non-specific differentiation along the osteoclast pathway is dependent on PU.1 and the MITF family of transcription factors, as well as the macrophage proliferation and survival cytokine M-CSF. Activation of RANK by osteoblast-expressed RANK ligand (RANKL) commits the cell to the osteoclast fate, which is mediated by signaling molecules such as AP-1 transcription factors, tumor necrosis factor receptor associated factor 6 (TRAF6), nuclear factor κB (NFκB), c-Fos, and Fra-1. RANKL-stimulated osteoclastogenesis is inhibited by the RANKL decoy receptor osteoprotegerin (OPG). Committed osteoclasts express the fusogenic genes DC-STAMP and Atp6v0d2, allowing formation of the multinucleated cell. The initial event in the development of the resorptive capacity of the mature osteoclast is its polarization, which requires c-Src and the $\alpha v \beta 3$ integrin. Once polarized, the osteoclast mobilizes the mineralized component of bone. Bone mobilization is achieved through the acidifying molecules, carbonic anhydrase II (CAII), an electrogenic H+ATPase and a charge-coupled $Cl^-$ channel. Cathepsin K mediates bone organic matrix degradation. Please refer to color plate section.

While the discovery of RANKL has been central to enhancing mechanistic insights into osteoclast formation and function, osteopetrotic patients and animals have provided confirmation of the skeletal importance of this cell. Osteopetrosis is a rare congenital disease wherein bone mass is present in substantial excess, due to the failure of osteoclast mobilization or function [1]. Development of this condition in the context of mutated genes, in association with in vitro molecular studies, has yielded a scheme wherein a host of proteins have been ascribed with specific roles in the progression of osteoclast differentiation and the capacity of the mature cell to resorb bone (Figure 6-1). Thus, in the past decade a combination of remarkable in vitro and in vivo tools have taken the osteoclast from an enigmatic, interesting-appearing cell, to one whose biology is increasingly clear.

# RANKL and RANK, an osteoclastogenic cytokine and its receptor

Osteoclast differentiation is supported by mesenchymal cells (bone-marrow stromal cells or osteoblasts) through intercellular contact. An in vitro osteoclast formation system in which bone marrow cells are co-cultured with osteoblasts was established in the late 1980s [5]. The fact that osteoclastogenesis in this system requires precursor contact with osteoblasts indicated the bone-forming cell expressed a membrane-residing molecule, which promotes osteoclast differentiation. This molecule, originally named osteoclast differentiation factor (ODF), proved to be RANKL. In addition, M-CSF, secreted by osteoblasts, provides the survival signal to osteoclast lineage cells [6, 7]. RANKL was cloned in 1998 [8, 9], one year after identification of osteoprotegerin (OPG), which competes with the receptor RANK, for RANKL, thus attenuating osteoclastogenesis [9–11].

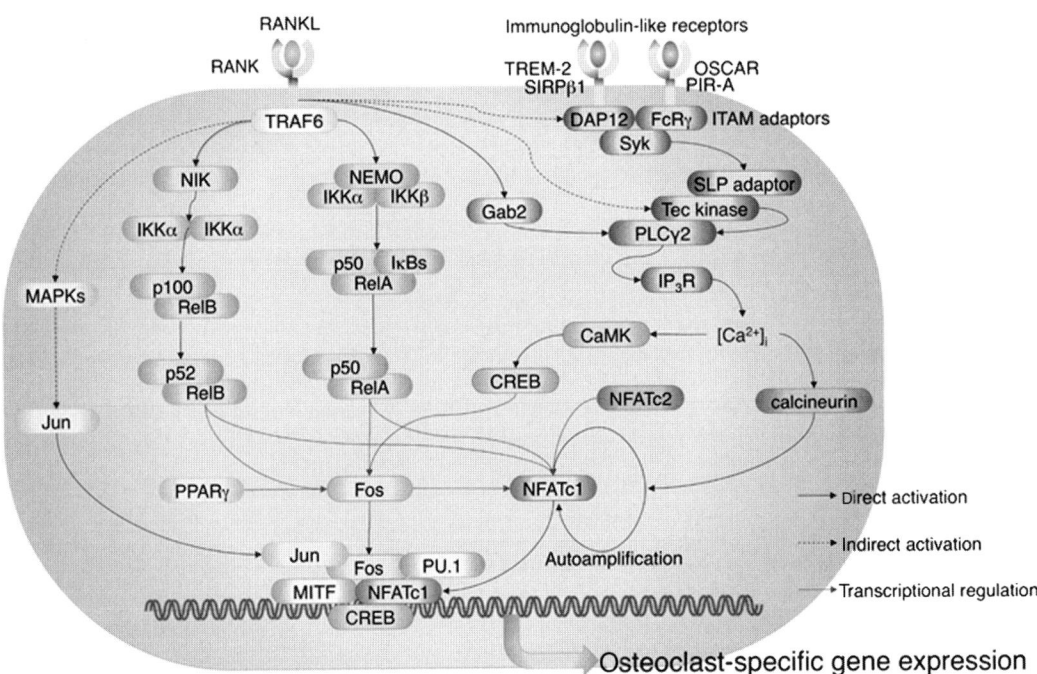

FIGURE 6-2 **RANK signaling in osteoclast differentiation.** RANKL stimulation leads to activation of several signaling pathways including MAPK and the canonical/noncanonical NF-κB pathways through TRAF6 as well as a Ca$^{2+}$ pathway through the ITAM adaptors for immunoglobulin-like receptors such as OSCAR, TREM-2, SIRPβ1, and PIR-A. NF-κB pathways contribute to induction of Fos and NFATc1, which are essential transcription factors for osteoclast differentiation. The Fos induction is also mediated by CREB and PPARγ. NFATc1 is activated by a Ca$^{2+}$ signal downstream of Ig-like receptors through the tyrosine phosphorylation of signaling molecules such as Syk, Tec kinases, SLP adaptors, and PLCγ2. Finally, NFATc1 orchestrates the transcription of osteoclast-specific genes together with AP-1, CREB, PU.1, and MITF. Please refer to color plate section.

RANKL, a type II membrane protein, belongs to the tumor necrosis factor (TNF) superfamily and contains C-terminal receptor-binding and transmembrane domains. RANKL and RANK are indispensable for osteoclastogenesis [12, 13], and human mutations in the *RANK* gene cause familial expansile osteolysis [14] and autosomal-recessive osteopetrosis [15]. While M-CSF promotes proliferation and survival of the osteoclast precursor cells [7], RANKL directly controls their differentiation. Molecular events during osteoclastogenesis have thus come to be understood in the context of RANKL–RANK signaling (Figure 6-2).

# TRAF6: the multifunctional signaling molecule activated by RANK

RANK, the signaling receptor for RANKL, was initially identified through a large-scale analysis of genes expressed in dendritic cells [16] and it is a type I membrane protein sharing high homology with CD40. RANK lacks intrinsic enzymatic activity in its intracellular domain, and the analyses of molecules associating with the cytoplasmic domain of RANK revealed that it transduces a signal by recruiting adaptor molecules such as the TRAF family of proteins [17–21]. The TRAF family contains seven members (TRAF1–7), which mainly mediate signals induced by TNF family cytokines and pathogen-associated molecular patterns (PAMPs) [22–24]. The cytoplasmic tail of RANK contains three TRAF6-binding sites and two sites for the binding of other TRAF family members including TRAF2, TRAF3, and TRAF5 [18–20, 25–28]. TRAF1 has been shown to bind to RANK in an in vitro binding assay and an overexpression study [18, 19]. Despite these findings, the phenotype of a variety of knockout mice identified TRAF6 as the major adaptor molecule linking RANK to osteoclastogenesis. Two groups independently showed that $Traf6^{-/-}$ mice developed severe osteopetrosis due to impaired bone resorption [29, 30]. However, it was controversial as to whether TRAF6 is essential for the differentiation of osteoclasts since the two groups attributed the osteopetrotic phenotype to different types of defect in the osteoclast lineage: the absence of osteoclasts [30] vs. the formation of dysfunctional osteoclasts [29]. A third mutant line was generated, and these mice also exhibited an osteopetrotic phenotype, due to a severe impairment of osteoclast differentiation [31]. Additional in vitro experiments further support the importance of TRAF6 in osteoclast differentiation [27, 28, 31]. In contrast to the essential role of TRAF6 in osteoclastogenesis, the contributions of TRAF2, TRAF3, and TRAF5 seem to be relatively limited. For instance, the binding sites for TRAF2, TRAF3, and TRAF5 in RANK are dispensable for osteoclast formation [32]. In addition, osteoclast formation from $Traf2$-deficient or $Traf5$-deficient osteoclast progenitor cells is only minimally reduced [33, 34].

Although it is necessary for osteoclast differentiation, TRAF6 is not the only molecule that associates with RANK to mediate osteoclastogenic signaling. It has been reported that Grb2 associated binder 2 (Gab2) interacts with RANK [35–37]. Gab proteins are scaffolding adapters for signaling molecules and are phosphorylated on tyrosine upon their stimulation by various cytokines, growth factors, and antigen receptors. In osteoclast precursor cells, Gab2 is phosphorylated in response to RANKL stimulation and associates with RANK through the highly conserved domain (HCD), which is distinct from TRAF6 binding sites. The essential role of Gab2 in osteoclast differentiation was revealed by the finding that $Gab2^{-/-}$ mice develop osteopetrosis due to impaired osteoclast differentiation [35]. Moreover, a peptide that targets the HCD has an inhibitory effect on osteoclast differentiation [38]. These studies establish the significant role of TRAF6-dependent and -independent pathways in osteoclast differentiation.

TRAF6 is composed of an N-terminal RING finger domain, a series of zinc fingers, a coiled-coil domain, and a C-terminal TRAF domain. TRAF6 activation leads to downstream activation of the mitogen-activated protein kinase (MAPK) and NF-κB pathways [39, 40]. In other cell types, it has been suggested that upon activation via TRAF6 homo-oligomerization, the RING finger ubiquitin E3 ligase domain complexes with a K63-specific E2 conjugating enzyme (Ubc13/Uev1a, or possibly UbcH7) to mediate attachment of non-degradative K63-linked ubiquitin chains to TRAF6 substrates, specifically TRAF6 itself [39]. These chains recruit factors, like the adaptors TAB2/3, which contain atypical zinc finger domains with a special affinity for binding K63-linked ubiquitin chains in certain cell types [41]. However, deletion analysis indicated that the RING finger domain of TRAF6 is dispensable for the formation of osteoclasts [42]. In addition, lysine-deficient TRAF6 could rescue RANKL-mediated NF-κB and MAPK activation, and osteoclast differentiation in *Traf6*-deficient precursor cells [43]. Interestingly, lysine-deficient TRAF6 cannot be ubiquitinated itself but it can still induce ubiquitination of IKK [43]. Thus, it is likely that TRAF6 autoubiquitination is dispensable for activation of downstream cascades, such as MAPK and IKK pathways.

## What happens downstream of TRAF6?

It has been shown in the context of osteoclastogenesis that TRAF6 forms a signaling complex containing RANK and TAK1-binding protein (TAB) 2, which results in TGF-β-activated kinase (TAK)1 activation [44]. Mammalian TAK1 was initially discovered through complementation screening of a murine cDNA library in a *Ste11p/MAPKKK*-deficient strain of yeast [45] and recent studies have confirmed the role of TAK1 in a number of different biological contexts. In RANKL signaling, dominant negative forms of TAK1 and TAB2 inhibit the NF-κB activation induced by RANKL, and endogenous TAK1 has been shown to be activated in response to RANKL stimulation in a mouse macrophagic cell line, RAW 264.7 [44]. Thus, in vitro experiments suggest that TAK1 is involved in the TRAF6-mediated activation of NF-κB and MAPKs, but the physiological relevance and precise molecular mechanism of this interaction remain to be elucidated. It is also unresolved how TRAF6 function is regulated by its interaction with other proteins. Atypical protein kinase C (aPKC)-interacting protein p62 is one of the candidate molecules involved in the regulation of TRAF6. It has been shown that p62 is upregulated and binds to TRAF6 during osteoclastogenesis [46]. A null mutation of the *p62* gene in osteoclast precursor cells causes severe impairment of osteoclast formation in a culture system. However, *p62*-deficient mice exhibited no defects in osteoclasts unless they were challenged by osteoclastogenic stimuli such as PTH-related peptide (PTHrP) [46]. On the other hand, *p62* mutations in humans are associated with the development of Paget's disease of bone (PDB), a disorder characterized by focal, disorganized, and increased bone turnover and the generation of very large osteoclasts [47]. A P392L amino acid substitution is the most commonly observed mutation, which has been linked to this condition. Expression of mutant p62 (P392L) specifically in the osteoclast lineage in transgenic mice produced a phenotype of low bone volume due to hypersensitivity to RANKL [48]. These results indicate that p62 is dispensable but involved in the regulation of osteoclast differentiation.

## The role of NF-κB in osteoclast differentiation

The NF-κB family of transcription factors consists of five members, p50 (processed from its precursor, p105), p52 (processed from its precursor, p100), RelA, RelB, and c-Rel, encoded by *Nfkb1*, *Nfkb2*, *Rela*, *Relb* and *Rel*, respectively, which share an N-terminal Rel homology domain (RHD) responsible for DNA binding and dimerization [49, 50]. The transcription activation domain (TAD) necessary for the positive regulation of gene expression is present only in RelA, RelB, and c-Rel, whereas p50 and p52 lack TADs. Whereas RelB preferentially associates with p100 as well as its processed form p52, RelA and c-Rel associate with p50.

This NF-κB activation depends on two pathways [49, 50]. One is the canonical and the other is the noncanonical NF-κB signaling pathway. The canonical pathway involves activation of the IκB kinase (IKK) complex, including IKKα, IKKβ, and NF-κB essential modulator (NEMO, or IKKγ), leading to phosphorylation and subsequent degradation of IκBs. This pathway targets p50:RelA and p50:c-Rel dimers. Among the constituents of the IKK complex, IKKβ is essential for the activation of the canonical pathway. The other is the noncanonical pathway that selectively activates p52:RelB dimers and this pathway is dependent on the activation of IKKα homodimers by the upstream kinase NF-κB-inducing kinase (NIK). IKKα phosphorylates p100 and induces the processing to p52, resulting in nuclear translocation of the p52:RelB heterodimer.

The in vivo role of NF-κB in osteoclast differentiation was first observed in p50/p52 double knockout mice. Although mice lacking either p50 or p52 have no obvious bone disorder, mice doubly deficient in p50 and p52 developed severe osteopetrosis due to a defect in osteoclast differentiation [51, 52]. However, a series of subsequent reports on NF-κB have suggested that more complex mechanisms may be involved. A cell-permeable peptide that blocks the association of NEMO with IKKs has been shown to efficiently inhibit osteoclast formation in vitro and inflammatory bone destruction in vivo [53]. Furthermore, targeted disruption of the IKKβ gene results in an impairment of osteoclastogenesis both in vitro and in vivo [54]. In addition, recently it was shown that $Rela^{-/-}$ cells do not differentiate into osteoclasts due to reduced survival [55]. These results suggest that the canonical pathway is indispensable for osteoclast differentiation in vivo.

Accumulating evidence also suggests the functional importance of the noncanonical pathway. Osteoclast differentiation of $NIK^{-/-}$ osteoclast precursor cells is completely blocked in vitro [56]. In addition, trabecular bone volume is slightly but significantly increased in $NIK^{-/-}$ mice [57] and alymphoplasia (*aly/aly*) mice, which carry a point mutation in the *NIK* gene [58, 59]. Interestingly, although NIK is thought to be important for activation of p52:RelB, loss of NIK inhibits the nuclear translocation of not only RelB but also RelA [56], indicating that NIK also regulates both canonical and noncanonical pathways in osteoclast differentiation. When RelA or RelB is overexpressed in NIK-deficient cells, only RelB can rescue the defect in osteoclast differentiation [57].

The details of the role of NF-κB in osteoclast differentiation are evolving. Further studies are needed to obtain a complete and accurate understanding of the complex regulatory mechanisms by which the NF-κB pathway is involved in osteoclast formation and function.

## The critical role of AP-1 transcription factors

RANK also activates the transcription factor complex AP-1, partly through an induction of its critical component, c-Fos [60]. The AP-1 transcription factor is a dimeric complex composed of the Fos (c-Fos, FosB, Fra-1, Fra-2), Jun (c-Jun, JunB, JunD), and ATF (ATFa, ATF2, ATF3, ATF4, B-ATF) proteins [61]. Mice lacking c-Fos develop severe osteopetrosis due to a complete block of osteoclast differentiation [62, 63]. The expression of c-Fos in response to RANKL is regulated by calcium ($Ca^{2+}$)/calmodulin-dependent protein kinases (CaMKs)-cyclic AMP-responsive element-binding protein (CREB) pathway [64], NF-κB [65], and peroxisome proliferator-activated receptor-γ (PPAR-γ), which is an activator of adipogenesis and a repressor of osteoblast differentiation [66].

Another member of the Fos family Fra-1, which is a transcriptional target of c-Fos during osteoclast differentiation, compensates for the loss of *c-Fos* both in vivo and in vitro [60], but *Fra-1*-deficient mice do not exhibit osteopetrosis [67]. This suggests that Fra-1 has an ability to compensate for the loss of *c-Fos* but is not the exclusive downstream signal molecule of the c-Fos pathway. FosB or Fra-2 also rescues the differentiation blockade of *c-Fos*-deficient osteoclast precursor cells in vitro, but their abilities to compensate are relatively weak [60]. Importantly, it is reported that transgenic mice expressing dominant negative c-Jun under the control of the *TRAP* promoter exhibit osteopetrosis [68]. Since dominant negative c-Jun inhibits AP-1 activity by binding to the Fos, Jun, and ATF families of proteins, this transgenic mouse confirms that AP-1 activity is critical for osteoclastogenesis. It should be noted, however, that the results from studies of dominant negative c-Jun provide no definitive information on the selective role of c-Jun itself, because it inhibits all AP-1 proteins.

In contrast to the critical role of c-Fos, the role of Jun family proteins, which are partners to Fos family members in the AP-1 complex, is redundant. Although mice lacking Jun family proteins such as c-Jun and JunB are embryonically lethal, conditional knockout mice are viable and a deficiency in JunB or c-Jun leads to a considerable decrease in osteoclast formation, but not to the complete blockade of this process. This result suggests that Jun members can substitute for each other during osteoclastogenesis [69, 70]. Although the precise composition of the AP-1 dimers has not been determined in the physiological context, a recent study showed that an AP-1 dimer composed of c-Fos and any Jun protein induces osteoclastogenesis but that Fra-1 has different partners [71]. Another Fos family protein, Fra-2, was shown to regulate osteoclast number and size in vivo [72]. Although *Fra-2*$^{-/-}$ mice die within a week after birth, newborn knockout mice have an increased size and number of osteoclasts. Mice lacking leukemia inhibitory factor (LIF), a transcriptional target of Fra-2 in osteoclasts, also show a similar phenotype. In addition, the expression level of hypoxia-inducible factor (HIF) 1α and Bcl-2 is increased through the suppression of a prolyl-hydroxylase, PHD2, leading to the prolonged survival of osteoclasts [72].

It is unclear how c-Fos plays such an exclusive role among the AP-1 proteins. One possibility is that, although AP-1 complexes bind to similar DNA sequences, an AP-1 complex containing c-Fos may have a selective affinity for some of the target genes. Alternatively, it is possible that c-Fos is crucial for AP-1 interaction with a specific transcriptional partner, which is required in osteoclastogenesis. Whatever the detailed mechanisms, c-Fos is clearly required for osteoclastogenesis. However, the next question that arises is what are the targets of transcription factors, like NF-κB and AP1, which are essential for osteoclast differentiation?

## MAPKs activated by RANKL

A series of in vitro experiments have suggested that MAPKs play an important role in osteoclastogenesis, but in vivo evidence has yet to be obtained confirming this hypothesis. MAPKs are involved in the activation of AP-1 components [73] and therefore may have a role in osteoclastogenesis by modulating AP-1 activity, but the molecular mechanisms by which this occurs are not well understood. Mammals express at least four distinctly regulated groups of MAPKs, p38-MAPKs (p38α/β/γ/δ), JNK1/2/3, ERK1/2, and ERK5 and many of these MAPKs have been shown to be activated downstream of RANK. Based on the effect of the specific inhibitor of p38α and β (SB203580) in RAW 264.7 cells, it is suggested that p38α and/or β are involved in osteoclast formation [74]. Recently, it was shown that p38α is dominantly expressed in the osteoclast lineages and the specific deletion of the p38α gene in osteoclast lineage cells results in a partial blockade of osteoclastogenesis in vivo [75]. These results indicate that p38α is important but other isoform(s) may also contribute to osteoclast differentiation. However, the functions of the other p38 isoforms (β, γ and δ) in osteoclasts remain to be elucidated. MEKs (ERK1/2 kinases) are also activated by RANKL. Curiously, inhibition of ERK activity by an MEK inhibitor does not suppress osteoclastogenesis [76] but rather potentiates it [77], which suggests that the ERK pathway negatively regulates osteoclastogenesis. Although mice with *JNK1/2/3* genes individually inactivated are viable and do not show obvious bone abnormalities, in vitro study has indicated that at least JNK1 is involved, albeit partially, in osteoclastogenesis [69]. Taken together, it appears that although RANKL activates a variety of MAPKs in vitro, their critical roles in osteoclast differentiation are largely unknown. Therefore, more detailed in vivo analyses are needed to obtain conclusive evidence of the roles that MAPKs have in osteoclastogenesis.

## NFATc1 is a master transcription factor of osteoclast differentiation

NF-κB and AP-1 are activated by RANKL in the early phase of osteoclastogenesis, but these transcription factors are also activated by other cytokines, which are not capable of inducing osteoclast differentiation. These observations suggested that RANKL had an as yet unknown target gene specifically linked to osteoclast differentiation. Based on a genome-wide search for genes expressed in the later phase, NFATc1 was shown to be a transcription factor that was most potently induced by RANKL [2].

The NFAT transcription factor, which was originally identified in the context of T-cell activation, is known to be involved in the function and development of diverse cells in other biological systems, such as the cardiovascular and muscular systems [78]. The NFAT family comprises five members including NFATc1-4 and NFAT5. Except for NFAT5, which is activated in response to osmotic stress, all of the NFAT family members are mainly regulated by the serine/threonine phosphatase calcineurin, which is activated by intracellular calcium ions. Dephosphorylation of the serine residues in NFATs by calcineurin leads to exposure of their nuclear-localization signal and translocation into the nucleus. Consistent with the critical role of NFATs in osteoclastogenesis, calcineurin inhibitors such as FK506 and cyclosporin A strongly inhibited osteoclastogenesis [2]. The necessary and sufficient role of *NFATc1* in osteoclastogenesis is suggested by the in vitro observation that $NFATc1^{-/-}$ embryonic stem cells do not differentiate into osteoclasts and the finding that ectopic expression of NFATc1 causes bone marrow-derived precursor cells to undergo osteoclast differentiation in the absence of RANKL [2]. Although *NFATc1*-deficient mice die at E13.5 due to cardiac valve defects, the in vivo significance of NFATc1 in osteoclast differentiation was revealed by fetal liver complementation experiments [79], by rescue of embryonic lethality through a targeted expression in the heart [80], and targeted disruption of the *NFATc1* gene in osteoclast lineages [81].

Since NFATc1 induction was shown to be impaired in $TRAF6^{-/-}$ cells [2], it has been suggested that *NFATc1* is one of the key target genes of NF-κB in the early phase of osteoclastogenesis. This idea is also supported by the observations that an NF-κB inhibitor suppressed RANKL-stimulated induction of NFATc1 [82] and that NFATc1 induction is also impaired in *p50/p52*-deficient cells [65]. Immediately after RANKL stimulation, NF-κB is recruited to the *NFATc1* promoter, which contains κB sites [79]. Another molecule, NFATc2, is also recruited to the *NFATc1* promoter within minutes of RANKL stimulation and activates the *NFATc1* promoter in cooperation with NF-κB [79]. This is an important step toward the robust induction of NFATc1. NFATc2 has the capacity to induce osteoclastogenesis when overexpressed. However, the physiological role of NFATc2 in osteoclastogenesis seems to be limited to the initial induction of *NFATc1* and can be substituted, possibly by other NFATs. This hypothesis stems from the finding that *NFATc2*-deficient mice have no obvious defect in osteoclast differentiation [79].

NFATc1 induction by RANKL is also completely abrogated in *c-Fos*-deficient cells [2]. c-Fos is recruited to the *NFATc1* promoter 24 h after RANKL stimulation. At this time, the main NFAT member recruited to the *NFATc1* promoter is NFATc1 itself, and an AP-1 complex containing c-Fos may cooperate with NFATc1 to enable the robust induction of *NFATc1* (i.e. autoamplification of NFATc1, see below). Thus, *NFATc1* is a common target gene of both of the essential transcription factors NF-κB and AP-1 during osteoclastogenesis.

# Autoamplification of NFATc1 and its epigenetic regulation

Since NFATc1 and NFATc2 play a redundant role in the immune system (e.g. cytokine production from lymphocytes is not affected unless both genes are disrupted [83]), the question arises as to how NFATc1 plays such an exclusive function in osteoclastogenesis. Interestingly, osteoclast formation in $NFATc1^{-/-}$ cells is recovered by forced expression of not only NFATc1 but also NFATc2 [79], indicating no functional difference between NFATc1 and NFATc2 during osteoclast differentiation and further suggesting a difference in the mechanisms of their transcriptional regulation. The mRNA of *NFATc1* is induced selectively and potently by RANKL, while *NFATc2* mRNA is expressed constitutively in precursor cells at a low level. Importantly, FK506, which suppresses the activity of NFAT through an inactivation of calcineurin, downregulates the induction of *NFATc1* but not *NFATc2*, strongly suggesting that *NFATc1* is selectively autoregulated by NFAT during osteoclastogenesis. ChIP experiments revealed that NFATc1 is recruited to the *NFATc1* but not the *NFATc2* promoter 24 h after RANKL stimulation, and the occupancy persists during the terminal differentiation of osteoclasts, indicating that the autoamplification mechanism by NFATc1 is specifically operative in the *NFATc1* promoter. Since NFAT binding sites are found in both the *NFATc1* and the *NFATc2* promoters, the promoter sequence thus cannot be the answer. It is generally accepted that cell-specific transcriptional regulatory proteins are not sufficient to initiate differentiation. Changes at the level of both higher-order chromatin structure and chromatin organization at individual genes are also essential. Histone acetylation is thought to be a marker of the transcriptionally active chromatin structure, and transcriptional co-activators such as CBP and PCAF have histone acetyltransferase activity. The rate of histone acetylation in the *NFATc1* promoter is increased gradually after RANKL stimulation, and methylation of histone H3 lysine 4, which is characteristic of a transcriptionally active locus, is also upregulated exclusively in the *NFATc1* promoter but not in the *NFATc2* promoter. Conversely, the *NFATc2* promoter is constantly associated with methylated DNA-binding proteins, such as methyl-CpG-binding protein 2 (MeCP2), suggesting that epigenetic modification of the *NFATc2* promoter is responsible for the muted pattern of gene expression [79]. Thus, contrasting epigenetic modification of the *NFATc1* and the *NFATc2* promoters might explain their unique spatiotemporal induction pattern during osteoclastogenesis. In conclusion, the essential role of the *NFATc1* gene is determined not only by the function of the encoded protein but also by an NFATc1-specific gene regulatory mechanism. However, MeCP2 is dispensable for osteoclast differentiation in vivo based on a study of $MeCP2^{-/-}$ mice [84]. Thus, it remains to be determined as to how such a specific epigenetic regulation functions specifically in osteoclasts.

## Transcriptional control governed by NFATc1

Accumulating evidence suggests that a number of osteoclast-specific genes are directly regulated by NFATc1. Based on promoter analyses, the *TRAP* [2, 85, 86], *calcitonin receptor* [2, 85–87], *cathepsin K* [76, 86], *β3 integrin* genes [88], *Atp6v0d2* [89], and *DC-STAMP* [89, 90] are modulated by NFATc1. The osteoclast-specific immunoreceptor osteoclast-associated receptor (OSCAR) is also controlled by NFATc1 [86, 91].

NFATc1 activity is regulated by other transcription factors. The AP-1 complex is known to be a transcriptional partner of NFAT in lymphocytes [92]. Likewise, an NFATc1: AP-1 complex is also important for the induction of the *TRAP* and *calcitonin receptor* genes as well as the autoamplification of NFATc1 during osteoclast differentiation [2]. It has also been shown that NFATc1 cooperates with PU.1 and MITF in the activation of the *cathepsin K* and the *OSCAR* promoters [76, 86]. It is noteworthy that both PU.1 and MITF, which are important for the survival of osteoclast precursor cells, also participate in osteoclast-specific gene induction at the terminal stage of differentiation. In addition, CREB, which is activated downstream of CaMK, regulates gene expressions in cooperation with NFATc1. Thus, NFATc1 forms an osteoclast-specific transcriptional complex, containing AP-1 (Fos/Jun), PU.1, MITF, and CREB for the efficient induction of osteoclast-specific genes. Taken together, the process of osteoclast differentiation can be divided into four stages in the context of transcriptional control: (1) induction of c-Fos mediated by NF-κB, CREB, and PPARγ; (2) initial induction of NFATc1 initiated by NFATc2 and NF-κB; (3) autoamplification of NFATc1; and (4) expression of a number of osteoclast-specific genes such as *cathepsin K*, *TRAP*, *calcitonin receptor*, and *OSCAR* induced by a transcriptional complex containing NFATc1 and cooperators such as AP-1, PU.1, CREB, and MITF (Figure 6-3).

In addition to positive regulation of NFATc1, recent reports also indicate that NFATc1 activity is negatively regulated by other transcription factors such as interferon regulatory factor-8 (IRF-8) [93] and v-maf musculoaponeurotic fibrosarcoma oncogene family, protein B (MafB) [94]. These factors directly associate with NFATc1 to suppress its transcriptional activity. Furthermore, the expression of IRF-8 and MafB is repressed by Blimp1, a transcriptional repressor, induced by RANKL stimulation [95].

## Co-stimulatory receptors for RANK signal: FcRγ and DAP12

Although the importance of the $Ca^{2+}$–NFATc1 pathway has been established, until recently it remained unclear how RANKL specifically activates $Ca^{2+}$ signals leading to the induction of NFATc1. As RANK belongs to the TNF receptor family, it is unlikely that RANK directly initiates $Ca^{2+}$ signaling. The screening of osteoclast-specific genes has shed light on a novel type of receptor. The immunoglobulin-like receptor OSCAR is reported to be involved in the cell–cell interaction between osteoblasts and osteoclasts [96]. OSCAR is expressed specifically in preosteoclasts and mature osteoclasts. Its

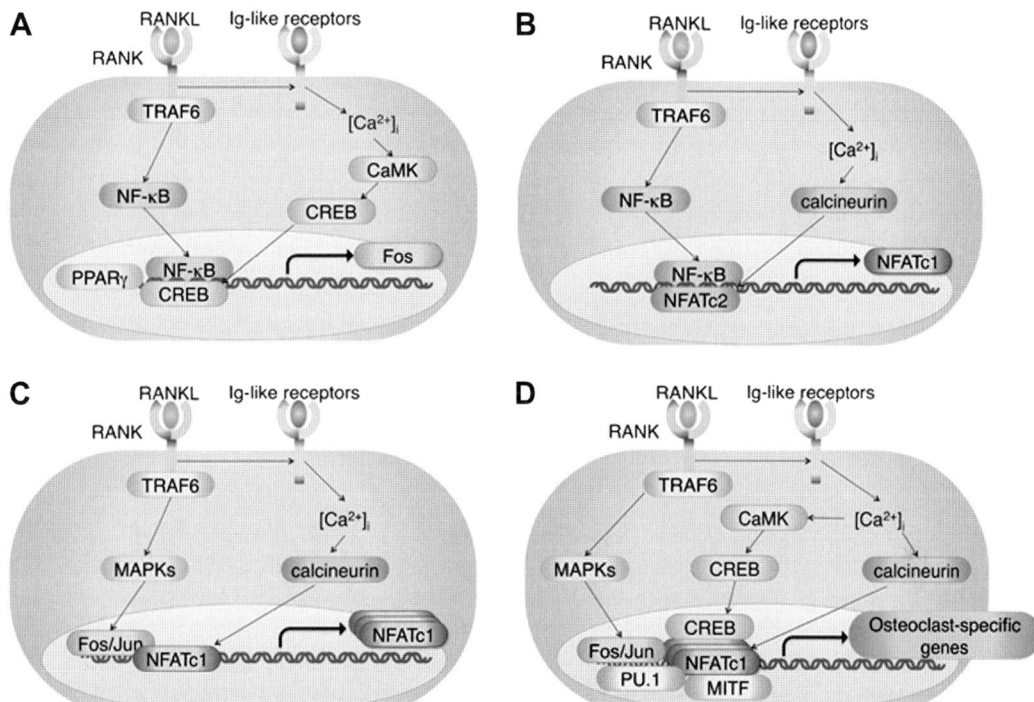

FIGURE 6-3 Temporal regulation of gene expression by transcription factors during osteoclast differentiation. **(A) Induction of Fos.** RANKL binding to RANK results in the activation of NF-κB. At the same time, RANK activation results in the phosphorylation of ITAM adaptors, DAP12 and FcRγ. Activation of NF-κB and CREB in the downstream of TRAF6 and ITAM adaptors, respectively, leads to induction of c-Fos at the early stage of osteoclast differentiation. The induction of c-Fos also requires PPARγ activity. **(B) Initial induction of NFATc1.** NFATc1 is initially induced by NF-κB and NFATc2. **(C) Autoamplification of NFATc1.** NFATc1 and AP-1 transcription factors are essential for the robust induction of NFATc1. The NFATc1 promoter is epigenetically activated through histone acetylation and NFATc1 binds to an NFAT-binding site in its own promoter. **(D) Terminal differentiation of osteoclasts.** NFATc1 works together with other transcription factors, such as AP1, PU.1, MITF, and CREB, to induce various osteoclast-specific genes, including TRAP, cathepsin K, and the calcitonin receptor. Please refer to color plate section.

putative-ligand (OSCAR-L) is expressed primarily in osteoblasts/stromal cells [96]. Following ligand binding, OSCAR must associate with the adaptor molecule Fc receptor common γ subunit (FcRγ), expressed on the same osteoclastic cell, to transmit intracellular signals. FcRγ consists of a small extracellular domain which lacks ligand-binding capacity, a transmembrane motif negatively charged, which is important for interaction with immune receptors (commonly aspartic acid in the adaptor and lysine or arginine in the receptor) and a cytoplasmic tail containing an immune tyrosine adaptor motif (ITAM), which consists of a highly conserved sequence $D/ExxYxxL/I(x_{6-8})YxxL/I$ [97]. FcRγ was shown to modulate calcium influx and upregulate the NFAT transcription factor in several immune cells [98]. Similarly, the OSCAR/FcRγ complex modulates RANKL-mediated activation of calcium signaling in osteoclasts, leading to the upregulation of NFATc1 [91]. However, $FcR\gamma^{-/-}$ mice did not display any obvious osteoclast defect in vivo or in vitro, suggesting that osteoclasts may express other adaptors controlling NFATc1 upregulation and the cell differentiation [99].

Like macrophages, neutrophils and NK cells, osteoclast precursor cells express another ITAM-harboring adaptor, DNAX-activating protein 12 (DAP12). Similarly to FcRγ, DAP12 associates with other surface receptors expressed on the osteoclast precursors themselves. Those include triggering receptor expressed in myeloid cells (TREM) -2 and -3, signaling regulatory protein β1 (SIRP β1) [99], and myeloid DAP12 associated lectin-1 (MDL-1) [100]. However, their ligands remain unknown. It has been reported that deletion of DAP12 affected in vitro osteoclast formation by modulating RANKL-induced calcium influx and NFATc1 activation [99]. However, other studies showed that few $DAP12^{-/-}$ osteoclasts form in vitro but have major spreading defects, suggesting that DAP12 also regulates osteoclast function [101, 102]. Indeed, a normal number of osteoclasts is present in the bones of $DAP12$-deficient mice and these animals exhibit mild osteopetrosis [101]. It is still not clear why osteoclasts are present in some circumstances but not in others. The apparent conflict between the in vitro and in vivo findings can be in part explained by the discovery that $DAP12$-deficient bone marrow precursors are able to generate multinucleated, albeit dysfunctional osteoclasts, when co-cultured with osteoblasts.

Since the OSCAR/OSCAR-L pair, via association with FcRγ, controls osteoclast and osteoblast interactions that are important for osteoclastogenesis, it was hypothesized that FcRγ might transmit compensatory signals in the absence of DAP12. Mice deficient in both FcRγ and DAP12 were generated by two distinct groups [99, 103]. Both found that $DAP12^{-/-}FcR\gamma^{-/-}$ mice exhibit severe osteopetrosis with very few osteoclasts in vivo. In vitro, no osteoclasts were found, either in co-culture with osteoblasts or in the presence of exogenous RANKL and M-CSF, due to defects in NFATc1 activation. The retroviral expression of DAP12 into $DAP12^{-/-}FcR\gamma^{-/-}$ cells efficiently rescued osteoclast differentiation and function. In contrast, the expression of DAP12Y65F mutant, which has a mutation in the ITAM motif, fails to generate fully spread, functional cells, indicating that the ITAMs play a critical role in the osteoclast [99, 103].

These findings led to a model for the differential contribution of DAP12 and FcRγ ITAM signaling during osteoclast development [99]. WT osteoclasts form either in the presence of exogenous RANKL/M-CSF or in co-culture with osteoblasts, where osteoclastogenic cytokines and co-stimulatory ligands expressed by the osteoblasts (such as OSCAR-L) drive the osteoclast differentiation process. In the RANKL/M-CSF system, the ITAM co-stimulatory signals are provided by DAP12 adaptors when they bind co-receptors that are expressed on osteoclast precursor cells and their as yet unidentified ligands. Under this circumstance, activation of the FcRγ pathway is not required for osteoclast formation (Figure 6-4A). When DAP12 is absent, ITAM co-stimulatory signals are not transmitted resulting in defective upregulation of osteoclastogenic genes. However, $DAP12^{-/-}$ osteoclasts can fully form in co-culture because osteoblasts provide OSCAR-L, activating the OSCAR/FcRγ pair and transmitting ITAM-dependent signals

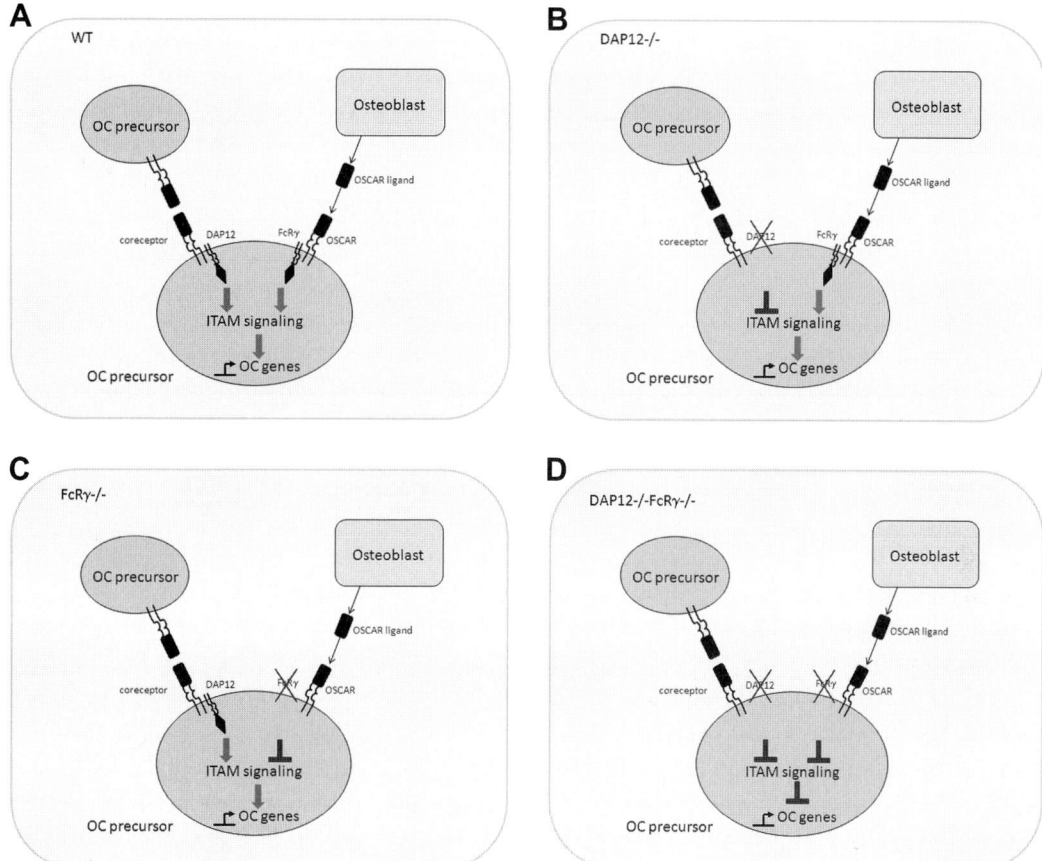

**FIGURE 6-4 Role of DAP12 and FcRγ in osteoclastogenesis. (A)** In the presence of exogenous RANKL, the ITAM co-stimulatory signals are provided by DAP12, its co-receptors and as yet unidentified ligands, expressed on osteoclast precursor cells. Under this circumstance, activation of FcRγ is not required for osteoclast formation. **(B)** In the absence of DAP12, ITAM co-stimulatory signals are not activated by exogenous RANKL, resulting in defective upregulation of osteoclastogenic genes. However, in the co-culture system, osteoblasts provide OSCAR-L and activate the OSCAR/FcRγ pair in $DAP12^{-/-}$ osteoclasts. $DAP12^{-/-}$ osteoclasts can differentiate but remain dysfunctional. **(C)** In the absence of FcRγ, DAP12, its associated receptors and their ligands can transmit ITAM-dependent co-stimulatory signals from adjacent osteoclast precursors, whether in co-culture or in the presence of exogenous RANKL, allowing their differentiation. **(D)** When both DAP12 and FcRγ are deleted, the two ITAM pathways are blocked, and osteoclastogenesis is not observed in vivo or under any in vitro condition since co-stimulatory signals cannot be delivered. Please refer to color plate section.

(Figure 6-4B). In the absence of FcRγ, DAP12, its associated receptors and their ligands, are still present and the $FcR\gamma^{-/-}$ precursor cells can receive ITAM-dependent co-stimulatory signals from adjacent osteoclast precursors, either in co-culture or in the presence of RANKL and M-CSF, allowing their differentiation (Figure 6-4C). When both DAP12 and FcRγ are deleted, the two ITAM pathways are blocked, and osteoclastogenesis is not observed in vivo or under any in vitro condition since co-stimulatory signals cannot be delivered (Figure 6-4D).

Additional in vitro evidence further documents the activating role of these immunoglobulin-like receptors in osteoclastogenesis. Triggering of FcRγ or DAP12 adaptors by antibody crosslinking can accelerate RANKL-induced osteoclast differentiation [104]. Similarly, anti-TREM2 mAb treatment with RANKL and M-CSF enhances the formation of multinuclear TRACP$^+$ osteoclasts compared to treatment with control mAb [104]. However, in the absence of RANKL, stimulation of ITAM-associated receptors alone is unable to induce osteoclast differentiation, suggesting that these receptor-mediated signals act cooperatively with the RANKL signal [99, 104]. Therefore, ITAM-mediated signals are thought to be co-stimulatory signals for RANK, in analogy to the co-stimulatory signals that are required for activation of immune cells.

## Importance of ITAM co-stimulatory signals in humans: Nasu-Hakola disease

The importance of the ITAM-harboring adaptors and their associated receptors in bone metabolism is further supported by the clinical manifestations of patients with polycystic lipomembranous osteodysplasia with sclerosing leukoencephalopathy (PLOSL), also known as Nasu-Hakola disease [105]. Typical manifestations include skeletal abnormalities and psychosis. Mutations in *DAP12* and *TREM-2* genes are responsible for these effects [106]. In vitro, peripheral blood mononuclear cells isolated from DAP12- and TREM-2-deficient PLOSL patients show inefficient and delayed osteoclast differentiation and a remarkably reduced bone resorption capability [107, 108]. However, patients with Nasu-Hakola disease develop distinct clinical manifestations such as focal bone cysts and osteoporotic fractures in all limb bones, especially in the bones of the wrists, hands, ankles, and feet, instead of the osteopetrotic phenotype, which is observed in the $DAP12^{-/-}$ mice or as one would expect from in vitro studies, using mutant human cells [106]. Current knowledge of DAP12/TREM-2 signaling does not provide adequate explanations for this paradox. It is known that the innate immune receptors that associate with *DAP12* and FcRγ are not identical between species and that some homologous receptors have differential cell expression between humans and mice. It is not clear whether the biological differences observed in the *DAP12*-deficient mice and humans are due to differential receptor expression, age, environmental or mechanical force differences, or differential interactions of these receptors with osteoblasts or the bony matrix.

Another plausible explanation would be that the locally increased bone loss seen in the Nasu-Hakola patients may be dependent on confined factors (endocrine, paracrine, etc.), affecting the differentiation or activation of osteoclasts at specific sites. This latter possibility is supported by the unexpected response of $Dap12^{-/-}/FcR\gamma^{-/-}$ mice to ovariectomy (OVX) [109]. Ovariectomized $Dap12^{-/-}/FcR\gamma^{-/-}$ mice showed resistance to lumbar vertebral body (LVB) trabecular bone loss. However, approximately 40% bone loss was detected in both femurs and tibias of these same mice. Although both $DAP12^{-/-}/FcR\gamma^{-/-}$ and WT mice lost an equal percentage of trabecular bone (BV/TV), because of the higher basal bone mass in the double deficient mice (53.6% BV/TV in $Dap12^{-/-}/FcR\gamma^{-/-}$ versus 5.7% BV/TV in WT), the total amount of bone loss in these mice was 10-fold higher than in WT controls [109]. The local increase in bone loss in long bones but not in the vertebrae following OVX is therefore reminiscent of the phenotype of Nasu-Hakola patients.

## Additional co-stimulatory signals involved in osteoclastogenesis

The presence of small numbers of osteoclasts in $DAP12^{-/-}/FcR\gamma^{-/-}$ mice under basal condition, and their activation leading to profound bone loss following OVX, would suggest the existence of other activating adaptors or pathways that can compensate for the lack of the ITAM signaling and induce the formation of mature, fully resorptive osteoclasts in vivo. A candidate molecule that might operate in the absence of FcRγ and DAP12 could be DAP10, a membrane-associated signal adaptor cognate of DAP12 in the immune system [110, 111]. DAP10 is expressed broadly on hematopoietic cells, including monocytes and macrophages [110, 111], NK cells, and $CD8^+$ T cells [111]. DAP10 specifically associates with NKG2D in NK cells, the only known DAP10 partner receptor that recognizes stress-induced MHC class I-like ligands [110]. The ligand for DAP10 in the bone marrow has not yet been discovered. In contrast to DAP12, DAP10 is unique in that it does not have an ITAM in its cytoplasmic domain but, instead, contains a different tyrosine-based motif, YINM. In NK cells, the YINM motif of DAP10 is involved in the recruitment of PI3K and Grb2 and in triggering $Ca^{2+}$ signaling and cytotoxicity. However, $DAP10^{-/-}$ mice do not exhibit an osteopetrotic phenotype; conversely these mice display a mild reduction of bone mass with age [100]. When bred together with $DAP12$-deficient animals, no additional effects due to lack of DAP10 were detected, suggesting that the function of DAP10 in the osteoclasts depends on DAP12 [100].

## Receptors signaling through DAP12

Because of their small extracellular domains, DAP12 and FcRγ cannot bind ligands directly but must associate with immunoglobulin superfamily receptors to transmit external stimuli inside the cell. For this reason, the transmembrane region of DAP12 and FcRγ plays a critical role in signal transduction, since this is the motif where interactions between the negatively charged aspartic acid in the adaptor and the positively charged lysine or arginine in the receptor take place. In immune cells, immune receptors as well as cytokine receptors, integrins, and other adhesion molecules utilize the ITAM-containing adaptors for signaling [112].

DAP12 is required for RANKL-induced osteoclast formation [99] and for integrin-mediated bone resorption [113]. DAP12, RANK, and several downstream signaling molecules colocalize in caveolin-rich membrane domains, forming an osteoclastogenic signaling complex. It remains to be established whether this complex consists of direct interaction of RANK and ITAM-related molecules or if it is simply dependent on the localization of constituent molecules to specific membrane domains.

Perhaps better described is the involvement of FcRγ and DAP12 in leukocyte β1, β2, and β3 integrin signaling [112]. Lack of DAP12 impairs integrin-mediated spreading, migration, and even cytokine production in macrophages and neutrophil phagocytosis. The linking of DAP12 to integrin signaling was based on the finding that the signaling pathways described as modulating leukocyte immunoreceptors were also required for activation of integrin downstream signals, such as Src family and Syk kinases, SLP-76 and PLCγ (reviewed in [112]). Interestingly, these same signaling pathways are also important in osteoclasts. Specifically, cooperation between αvβ3 integrin and DAP12 is essential to ensure proper adhesion of osteoclasts to bone and to produce resorption.

DAP12 also signals through the M-CSF receptor, c-Fms, in osteoclast precursors [113, 114]. In the absence of DAP12, monocytes display reduced proliferative responses to M-CSF as well as poor survival at limited doses of the cytokine [114]. Consequently, inefficient regeneration of myeloid cells is observed after transplantation of DAP12-deficient marrow progenitor cells. c-Fms activation by M-CSF stabilizes and promotes nuclear translocation of beta-catenin, which activates cell cycle genes. M-CSF also signals through DAP12 in mature osteoclasts via Syk kinase activation, and regulates cell spreading and bone resorption [113]. The mechanism by which DAP12 is coupled to integrins and c-Fms is unclear. It is possible that ITAM adaptors are coupled to these receptors via direct binding; although there are no biochemical data to validate this hypothesis. The fact that the transmembrane arginine residue of DAP12 is required for c-Fms and αvβ3 signaling in osteoclastic cells suggests an associated immunoreceptor mediates these interactions [113]. Alternatively, integrins, c-Fms, and ITAM adaptors could associate through lipid raft colocalization or intermediary proteins.

# Src family kinases: activation of ITAM signaling

Independent of which receptor activates DAP12 and FcRγ, phosphorylation of the two tyrosine residues in the ITAM motif is a critical step in derivative signaling [99]. Src-family kinases are responsible for phosphorylation of ITAM receptors in immune cells [112, 115]. Src inhibition dampens DAP12 phosphorylation and the activation of its effectors, thus arresting osteoclastogenesis [37]. However, the exact Src family kinase members involved in ITAM signaling in response to RANKL stimulation has yet to be determined. While c-Src deletion leads to an osteopetrotic phenotype, this Src family member is required for osteoclast function but not differentiation, since numerous multinucleated dysfunctional osteoclasts are present in the long bones of c-Src-deficient mice [116]. Thus, *c-Src* likely induces phosphorylation of DAP12 following integrin-mediated adhesion or M-CSF stimulation but not in response to RANKL.

In immune cells, the Src family kinase Lyn phosphorylates ITAM-containing receptors. However, this same enzyme suppresses ITIM signals, indicating the complex role of this Src family kinase member as both a positive and negative regulator of co-stimulatory molecules. Lyn deletion promotes formation and function of osteoclasts highly sensitive to RANKL [117]. Thus, in the context of osteoclastogenesis, Lyn is a negative regulator of RANKL-dependent ITAM phosphorylation, establishing this enzyme as the second Src family kinase to impact the bone resorbing polykaryon and the first to mediate its generation.

## Syk kinase: downstream of DAP12/FcRγ?

Once phosphorylated, the ITAM motifs of DAP12 and FcRγ provide high-affinity docking sites for the tandem SH2 domains of spleen tyrosine kinase (Syk), which is then recruited to the receptor complex. Next, Syk triggers activation of molecules including phospholipase Cγ (PLCγ) and TEC-family kinases [3, 112]. Syk-deficient osteoclasts form in vitro, but their spreading is impaired. Indeed, *Syk*-deficient bone-marrow chimeras, generated by transplant of $Syk^{-/-}$ fetal liver cells into lethally irradiated WT recipients, exhibited increased bone mass with normal osteoclast number [118]. These findings are reminiscent of the osteoclast phenotype of $c\text{-}Src^{-/-}$ and $DAP12^{-/-}$ mice, suggesting that Syk might not be recruited to the ITAM motifs of DAP12 and FcRγ in response to RANKL, but insteads transmits ITAM signals downstream of the αvβ3 integrin and M-CSF receptors.

## PLCγ2: enzyme and adaptor molecule

PLCγ is the downstream of Syk or ZAP70 tyrosine kinases in immune cells. RANKL promotes PLCγ phosphorylation in the context of DAP12 and FcRγ [37]. Once phosphorylated, PLCγ cleaves the membrane phospholipid, phosphatidylinositol-4,5-bisphosphate (PIP$_2$) to generate inositol-1,4,5-trisphosphate (IP$_3$) and diacylglycerol (DAG) [119]. IP$_3$ induces release of endoplasmic reticulum Ca$^{2+}$, which, via calcineurin, prompts the NFAT family of transcription factors to translocate into the nucleus and activate the osteoclastogenic machinery.

The PLCγ family consists of two isoforms, PLCγ1 and PLCγ2, and both proteins require phosphorylation of specific tyrosine residues to activate their catalytic activity [119]. PLCγ1 deficiency is lethal and deletion of PLCγ2 impairs B-cell maturation, NK cytotoxicity, neutrophil degranulation, and DC-mediated T-cell activation. $PLC\gamma2^{-/-}$ mice also exhibit defective osteoclast differentiation and function, inducing an osteopetrotic phenotype [37, 120]. Osteoclasts fail to form even when $PLC\gamma2^{-/-}$ marrow cells are cultured in vitro with WT stromal cells, suggesting that additional ITAM signals provided by activation of the OSCAR/FcRγ axis are not sufficient to rescue the differentiation defect.

PLCγ2 is a complex molecule with enzymatic activity and adaptor function. Due to its ability to generate IP$_3$, thus increasing intracellular calcium levels, re-expression of a catalytically inactive PLCγ2 mutant is unable to rescue the defect RANKL-mediated

NFATc1 upregulation and thus osteoclast differentiation [37]. Similarly, treatment with U73122, a non-specific PLCγ inhibitor, abrogates formation of the multinucleated cell [37]. These data support a model in which the catalytic activity of PLCγ2, downstream of FcRγ and DAP12, is required for NFATc1 upregulation (see Figure 6-2). PLCγ2 adaptor function, independent of the catalytic activity, is also required for osteoclastogenesis. PLCγ2 contains two tandem SH2 domains, which bind phosphorylated tyrosines in Syk, Vav3, or Gab2, and an SH3 motif, which recognizes the E3 ubiquitin ligase, c-Cbl. While an SH3 mutation does not impact osteoclast formation, an SH2 mutant affects NF-κB and AP1 activation in response to RANKL. In consequence, an intact SH2 domain is required to rescue in osteoclastogenesis that is seen in $PLC\gamma2^{-/-}$ cells [37, 120]. This effect of PLCγ2 on NF-κB and AP1 activation is independent of ITAMs, as DAP12/FcRγ deficient osteoclasts respond normally to RANKL stimulation [99].

The adaptor protein, Gab2, a crucial component of RANK signaling and osteoclastogenesis, is required to activate NF-κB and JNK, but not NFATc1 [35]. PLCγ2 associates with Gab2, in immune cells, suggesting the same complex regulates RANKL signaling in osteoclasts. PLCγ2 binds Gab2 in response to RANKL, mediates Gab2 recruitment to RANK, and is required for Gab2 phosphorylation, thereby activating AP1 and NF-κB [36, 37]. This dual ability of PLCγ2 to affect both NFATc1 upregulation and NF-κB and AP1 activation is likely to be responsible for the blockade of osteoclastogenesis induced by the absence of PLCγ2.

Despite the critical role of PLCγ2 in osteoclast formation and RANKL signaling in vitro, fewer osteoclasts are still present in vivo [37]. Furthermore, the osteopetrosis of $PLC\gamma2^{-/-}$ mice is not as severe as that of those lacking DAP12 and FcRγ. This conundrum may reflect the fact that only M-CSF and RANKL are used in vitro to induce osteoclastogenesis, while in vivo a milieu of cytokines probably triggers osteoclast differentiation. However, neither TNF, LPS, nor an abundance of RANKL and M-CSF rescues osteoclast formation in vitro in cells from $PLC\gamma2^{-/-}$ mice or increase their number in vivo [37]. Similarly, $PLC\gamma2^{-/-}$ mice are protected from inflammatory-induced bone loss [121, 122]. These findings suggest that mechanisms other than a favorable micro-environment might be responsible for the few osteoclasts still present in vivo. Alternatively, PLCγ1, which is highly homologous to PLCγ2, could be compensatory in vivo, but no experimental evidence supports this hypothesis. PLCγ1 expression is not affected by PLCγ2 deletion but both molecules can be phosphorylated by RANKL and M-CSF [37].

## Tec kinases: integrating RANK and ITAM signaling

Tec family kinases are predominantly expressed in hematopoietic cells, where they respond to extracellular stimuli mediated by ITAM-containing proteins. They include Bmx, Bruton's tyrosine kinase Btk, inducible T-cell kinase (Itk), resting lymphocyte kinase (Rlk), and Tec. These molecules are activated by both T- and B-cell receptors [123]. A broad range of mutations in Btk causes X-linked agammaglobulinemia [124]. X-linked immunodeficiency (Xid), a similar, although less severe syndrome in mice, is also the product of a *Btk* point mutation [125]. Interestingly, bone marrow cells derived from Xid mice undergo retarded osteoclastogenesis [126]. The fusion defect of the Xid mutant osteoclasts is caused by decreased expression of NFATc1 and osteoclast fusion-related molecules, such as the d2 isoform of vacuolar H(+)-ATPase V0 domain and the dendritic cell-specific transmembrane protein (DC-STAMP) [126]. This deficiency was completely rescued by the introduction of a constitutively active form of NFATc1 in bone marrow-derived macrophages [126]. However, Btk deficiency does not produce any obvious bone phenotype, in vivo, suggesting compensation by other family members. Indeed, double deletion of Btk and Tec leads to severe osteopetrosis due to cell-autonomous blockade of osteoclast differentiation in vitro and in vivo [127]. Mechanistically, RANKL-induced tyrosine phosphorylation of PLCγ2 is markedly suppressed, as is RANKL-induced $Ca^{2+}$ oscillation required for *NFATc1* induction. Thus, Btk and Tec activate PLCγ-$Ca^{2+}$-NFATc1 [127].

RANKL phosphorylates Btk and Tec wherein they complex with the SH2-containing leukocyte protein (SLP) family adaptor, B-cell linker protein (BLNK). This event, which occurs in caveolin-rich membrane domains, is dependent on DAP12 and FcRγ adaptors [127]. Thus, it may be the molecular switch integrating RANK and ITAM signals. Combined deficiency of BLNK and another member of the SLP adaptors, SLP-76, severely impairs osteoclast differentiation in vitro. Despite the crucial role of BLNK and SLP-76 in osteoclastogenesis in vitro, the bone phenotype was not markedly changed in $BLNK^{-/-}$ $SLP-76^{-/-}$ mice, implying that other adaptor molecules, such as cytokine-dependent hematopoietic cell linker (CLNK), may compensate for the loss of BLNK and SLP-76 in vivo. Hence, an osteoclastogenic signaling complex composed of Tec kinases, PLCγ2, and adaptor proteins has a critical role in integrating RANK and ITAM signals in efficient $Ca^{2+}$-NFATc1 activation [127].

## Negative regulatory role of DAP12

We have amply described the positive regulatory roles of DAP12 in osteoclast development and during basal bone homeostasis. However, the OVX studies indicate that a condition of estrogen deficiency induces an ITAM adaptor-independent bypass mechanism allowing for enhanced osteoclastogenesis and activation in specific bony microenvironments [109]. More importantly, the OVX data suggest that DAP12 may actually have a bone-protective effect under certain circumstances.

Positive and negative regulatory roles of DAP12 have been reported in other systems, for example studies in macrophages suggest that DAP12 signals downregulate the inflammatory response following Toll-like receptor (TLR) signaling. *DAP12*-deficient macrophages were found to produce a higher concentration of inflammatory cytokines (TNF-α) in response to pathogenic stimuli reactive with TLR receptors (LPS/TLR4, synthetic bacterial lipopeptide-TLR2/1, CpG DNA-TLR9, zymosan-TLR2, poly(I:C)-TLR3, and peptidoglycan-TLR2) and showed increased extracellular regulated kinase 1 and 2 (ERK1/2) phosphorylation [128]. TREM-2 completely accounts for the increased cytokine production observed in $DAP12^{-/-}$ macrophages [129]. Similarly, loss of both *DAP12* and *FcRγ* enhanced the pro-inflammatory cytokine production and maturation of dendritic cells after TLR stimulation [130]. These activating and inhibitory functions of DAP12 and its associated receptors might be explained by a differential ability of DAP12 to couple with various receptors. It has been hypothesized that when coupled to receptors that bind high-avidity ligands, robust and sustained activation of DAP12 will result in cellular activation. By contrast, when DAP12 is coupled to receptors that recognize low-avidity ligands, DAP12 will be only partially activated, resulting in inhibitory signaling. In this circumstance, weak phosphorylation of the DAP12 and/or FcRγ immunoreceptor tyrosine-based activation motifs will lead to primarily SHP-1 phosphatase recruitment. SHP-1 can then block downstream pathways from other DAP12-/FcRγ-associated receptors or from completely heterologous receptors, such as the TLRs [131]. It is possible that such interactions also occur in the osteoclasts and could explain why in PLOSL patients osteolytic lesions are seen at specific sites, despite the blockade of osteoclast differentiation in vitro. Further identification of DAP12-associated receptors and their ligands in osteoclasts will be necessary to clarify the effects in bone.

## M-CSF and c-Fms: a road to proliferation and survival

M-CSF, also called CSF-1, mediates survival and proliferation of precursors of the monocyte/macrophage lineage and their differentiation into mature phagocytes. As the osteoclast is a member of this family, absence of this cytokine arrests its development, at least transiently. Specifically, the *op/op* mice, which fail to express functional M-CSF because of a point mutation in the *Csf1* gene, are osteopetrotic [132]. Administration of soluble M-CSF rescues their osteopetrosis. Confirming this cytokine's role in osteoclastogenesis, deletion of its receptor also induces osteopetrosis [133].

While M-CSF is produced constitutively by a range of cells, regulated increases in the secretion of this cytokine have pathological consequences in the context of the osteoclast (reviewed in [134]). In the absence of estrogen, the cause of postmenopausal osteoporosis, enhanced bone resorption is caused, at least in part, by increased secretion of M-CSF from marrow stromal cells. c-Fms activation also participates in the bone loss attending inflammatory arthritis. For example, inflammation-enhanced monocytic interleukin-1 (IL-1) and TNF-α stimulate release of IL-7 from stromal cells, which in turn prompts activated T cells to produce M-CSF. Finally, raised serum levels of PTH are invariably associated with increased resorption, one consequence of which is release of M-CSF from osteoblasts and stromal cells within the bone environment.

## M-CSF signaling

c-Fms is the sole receptor for M-CSF. However, recent evidence indicates c-Fms also recognizes IL-34 raising the possibility that this cytokine may also influence the osteoclast [135]. The ability of IL-34 to activate c-Fms may explain the recovery of osteoclasts, experienced by *op/op* mice as they age.

Upon ligand binding seven tyrosine residues in the cytoplasmic tail of c-Fms undergo phosphorylation. Binding of M-CSF to c-Fms results in dimerization and hence activation of the receptor tyrosine kinase, leading to autophosphorylation of selected tyrosine residues. Each phosphorylated species acts as a binding site for SH2 or phosphotyrosine binding (PTB) domain-containing proteins, which amplify and transduce the original signal [136]. c-FmsTyr-559, the binding site for c-Src, is the dominant residue modulating macrophage proliferation, which also requires Tyr-697 and Tyr-807 [137]. Interestingly, these same residues also transmit signals to the cytoskeleton [138].

## Erk and PI3K signaling

Erk1/2 (p42/44) and PI3K/Akt mediate macrophage proliferation. Binding of M-CSF to c-Fms recruits the adaptor protein complex Grb2/Sos toY697 in the cytoplasmic tail of c-Fms and Sos, acting as a guanosine exchange factor for Ras, and stimulates the Ras/Raf/MEK/Erk pathway [134]. In contrast, M-CSF produces almost no activation of the MAP kinases, JNK and p38. M-CSF also robustly stimulates the PI3K/Akt pathway in macrophages. In general, PI3K/Akt regulates cell proliferation via GSK3β and the FOXO family of transcription factors. In brief, GSK3β phosphorylates cyclin D1, leading to its rapid proteosomal degradation, while FOXO inhibits transcription of the same cyclin and increases the cell cycle inhibitors, p27 and p130. By phosphorylating GSK3β and FOXO, Akt suppresses their capacity to inhibit entry into the cell cycle [134].

The SHIP1-deficient mouse provides evidence that PI3K/Akt stimulates pre-osteoclasts to divide. SHIP1, as a $5'$ lipid phosphatase, diminishes phosphatidylinositol 3, 4, 5 trisphosphate ($PIP_3$), and hence deactivates Akt. Animals lacking SHIP1 have large hyper-resorptive osteoclasts and, in consequence, are severely osteoporotic [139]. Moreover, these terminally differentiated polykaryons are protected from programmed cell death, reflecting prolonged and more robust activation of the PI3K/Akt axis. The finding of decreased apoptosis in SHIP1-null osteoclasts suggested macrophages from the same animals would be similarly longer lived, thus contributing to the massive increase in their numbers, in vivo. Surprisingly, $SHIP1^{-/-}$ macrophages are not distinguished from WT by decreased apoptosis, but by increased proliferation and accelerated entry into the cell cycle [140]. Mechanistically, PI3K mediates M-CSF-induced macrophage proliferation via a process involving suppression of p27 [140]. Following cytokine treatment, rapid induction (via Erks) and stabilization (via PI3K/Akt/GSK3β) of D-type cyclins typically forms a cyclin D/cdk4 complex, which hyperphosphorylates the pocket protein, Rb. Thus, exposure of primary macrophages to M-CSF enhances expression of all three D cyclins and simulates Rb phosphorylation.

Apart from its positive effect on proliferation, M-CSF is also critical for survival. For example, withdrawal of RANKL, and particularly M-CSF, results in rapid osteoclast death [141]. Similarly, macrophages, which are derived from the same myeloid precursor as osteoclasts, require M-CSF to survive. While the major signals blocking apoptosis in macrophages and osteoclasts involve the Erk and Akt pathways, respectively, information concerning downstream signaling in these two cell types is limited. Overexpression of a constitutively active form of MEK1 promotes osteoclast survival [142], suggesting a central role for M-CSF-activated Erks. Alternatively, M-CSF-dependent survival of the mature cells requires synthesis of rapidly metabolized proteins that prolong longevity and/or block apoptosis [143]. Furthermore, M-CSF stimulates ubiquitination and hence degradation of Bim, a BCL-2 family member whose genetic deletion increases osteoclast survival [144]. Finally, short interfering RNA-mediated "knock down" of Akt1 and/or Akt2 and their downstream target mTOR in murine osteoclast precursors demonstrated that mTOR mediates the capacity of M-CSF to suppress Bim [145]. These data suggest a novel model in which M-CSF-driven mTOR activation is mediated by PI3K in an Akt-independent manner.

## The osteoclast's job: bone resorption

The critical function of the osteoclast is to degrade the organic and inorganic matrices of bone. Accumulation of bone-degrading molecules on the resorption surface requires physical intimacy between the osteoclast and bone and the creation of a microenvironment, which is functionally isolated from the general extracellular space. Osteoclasts accomplish this task by restructuring their actin cytoskeleton to form "a gasket-like" sealing zone, which surrounds the resorptive milieu (Figure 6-5). The cell secretes HCl via an electrogenic $H^+$ATPase (proton pump) and charge-coupled $Cl^-$ channel, ClC-7, thus acidifying the resorptive microenvironment. In consequence, the bone's organic matrix is

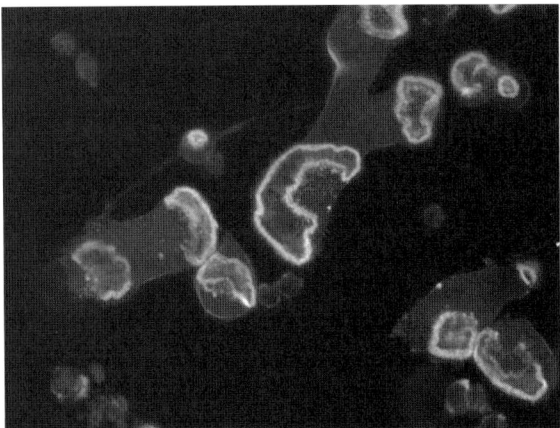

**FIGURE 6-5 Osteoclasts on bone form actin rings.** Bone marrow macrophages expressing GFP-actin were cultured on bone in the presence of M-CSF and RANKL to generate osteoclasts. After 5 days the cells were visualized by fluorescent microscopy. Note that individual cells contain multiple actin rings, indicating the generation of numerous resorptive microenvironments by a single osteoclast. (Courtesy of Dr. Wei Zou.)

demineralized and subsequently degraded by the lysosomal enzyme, cathepsin K. Abnormalities of components of the acidification pathway, such as the $H^+$/ATPase [146], cause osteopetrosis, a rare inherited disease in which the bones become very dense and fracture easily. Inactivating mutations of cathepsin K in humans cause pyknodysostosis, a disorder characterized by increased bone mass, dwarfism, and facial dysmorphism [147]. Thus, a combination of acidification and enzymatic digestion of inorganic and organic matrix components cooperate to fully remove bone within this well-defined area [148].

## Osteoclast cytoskeleton: the podosomes and the sealing zone

In most cells, such as fibroblasts, matrix attachment prompts formation of stable structures known as focal adhesions that contain both integrins and a host of signaling and cytoskeletal molecules, which mediate contact and formation of actin stress fibers. Osteoclasts, however, contain podosomes instead of focal adhesions [149, 150]. Podosomes consist of an F-actin core whose formation is regulated by CD44, Wiskott-Aldrich syndrome interacting protein (WIP), WASp, and the Arp2/3 complex [151, 152]. The F-actin-containing podosomal core is surrounded by a "cloud", consisting of filamentous actin and associated cytoskeleton-regulating proteins such as the αvβ3 integrin, vinculin, α-actinin, paxillin, and talin [153]. Both the actin core and the cloud participate in the osteoclast's adherence to substrate and thus the formation of its resorptive microenvironment [152, 154].

Virtually all studies of osteoclast podosomes, until recently, have been performed on cells cultured on glass or plastic substrate. In this circumstance podosomes are easily visualized, initially appearing as clusters of actin-containing, punctuate structures. With time, podosomes coalesce into a single actin belt localized entirely at the cell's periphery [151].

These observations suggested that the in vivo structure isolating the resorptive microenvironment, the sealing zone, also reflects podosomal organization. This conclusion, however, was challenged by striking differences in osteoclasts cultured on glass compared to mineralized substrate. Under non-stressed conditions, bone-residing osteoclasts do not form an actin belt and by fluorescent microscopy, podosomes are difficult to visualize [151]. While these distinctions challenge the biological significance of assessing the cytoskeleton of osteoclasts present on a non-mineralized substrate, it is now clear that the sealing zone of bone-residing osteoclasts also consists of podosome-containing structural units.

## Osteoclast cytoskeleton: the microtubules and the sealing zone

In other cells, microtubules and actin collaborate in cytoskeletal organization and the same appears true in the osteoclast. The capacity of microtubules to influence actin organization, in the bone-resorptive cell, depends upon tubulin acetylation, which, in turn, polymerizes and stabilizes these helical structures. Thus, there exist two pools of microtubules; those that are unstable and have a half-life of minutes and those that are polymerized and persist for hours. Polymerized microtubules are necessary for generating actin belts in glass-residing osteoclasts but not for podosomal clustering or actin-ring formation [154, 155]. The fact that sealing zone formation, in bone-resorbing polykaryons, is attended by microtubule acetylation underscores differences in formation of this structure and glass-induced actin rings [155].

The state of microtubule acetylation, and thus stability, is governed by the histone deacetylase, HDAC6, which depolymerizes tubulin. HDAC6's association with tubulin is regulated, in turn, by Cbl family proteins, which redundantly compete with the deacetylase for tubulin binding [156]. Alternatively, HDAC6 activity is believed to be induced by RhoA and thus organization of the cytoskeleton of glass-residing osteoclasts may be negatively regulated by this small GTPase. The physiological relevance of this observation is controversial as RhoA stabilizes the cytoskeleton of other cells and its inhibition diminishes the resorptive capacity and apical–basal polarity of osteoclasts cultured on bone [153]. Additionally, RhoA mediates formation of the actin ring and a constitutively active construct stimulates podosome formation, osteoclast motility, and bone resorption [153]. Dominant negative RhoA and C3 exoenzyme, which inactivate the GTPase, arrest these events [157]. Upon attachment to bone, RhoA binds GTP [158] and translocates to the cytoskeleton, indicating it is integrin-regulated, a hypothesis confirmed by the fact that matrix-induced RhoA activation is arrested in αvβ3 integrin-deficient osteoclasts [150]. Thus, the relationship to αvβ3-activation of RhoA and the GTPase's negative effect on microtubule stability remains enigmatic.

## Osteoclast functional structure: the ruffled border

While formation of the sealing zone is essential for optimal osteoclast function, its presence does not assure that the cell is capable of degrading bone. The ruffled border, in contrast, is a universal hallmark of an active osteoclast. This unique structure, encompassed by the sealing zone, contains the machinery by which the osteoclast secretes matrix-degrading molecules on to the bone surface. In fact, absence of the ruffled border completely blocks bone resorption and abnormalities of its structure eventuate in varying degrees of osteoclast dysfunction. Thus, the ruffled border is the cell's resorptive organelle and its arrested formation is a common component of many forms of osteoclast-autonomous osteopetrosis.

The ruffled border is a complex enfolding of the plasma membrane, which abuts and extends into the resorptive space (Figure 6-6). Although its genesis was long enigmatic, the ruffled border clearly is the most specific marker of the activated osteoclast as it appears only in the bone-degrading cell. Recent evidence establishes the lysosomal origin of this resorptive organelle [159]. Thus, upon contact with bone, matrix-derived signals, likely transduced through integrins, cathepsin K-bearing lysosomal structures are

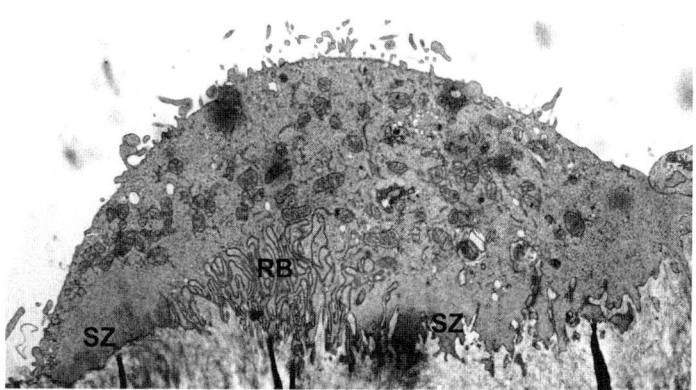

FIGURE 6-6 **Electron micrograph of osteoclast resident on bone.** The ruffled border (RB) is encompassed by sealing zone (SZ). (Courtesy of Dr. Haibo Zhao.)

induced to move towards, and insert into, the bone-apposed plasma membrane. The inserted vesicles thereby increase the complexity of the selected plasmalemma domain and discharge their cargo into the resorptive microenvironment. The ruffled border also contains the charge-coupled, HCl-producing, $H^+$ATPase and $Cl^-$ channel but whether they are delivered in the same vesicles as cathepsin K is unknown [159]. While there is evidence that the ruffled border-forming vesicles are transported to their destination via microtubles, this contention has yet to be proven.

Formation of the osteoclast ruffled border is reminiscent of vesicular exocytosis. In this process, eukaryotic cells release hydrophilic secretory products into the extracellular space or translocate specific functional proteins to the plasma membrane, perhaps mediated by GTPases such as Rabs 7, 9, and 3 [160–162]. The capacity of Rab GTPases to mediate cell polarization depends upon their prenylation, which permits membrane association. This lipid modification is precisely targeted by bisphosphonates, the osteoclast-inhibitors, which are presently the family of drugs most frequently administered for osteoporosis. Interestingly, osteoclasts of bisphosphonate-treated patients are often unable to polarize or form actin rings and ruffled borders, buttressing the concept that Rab proteins are involved in organizing the cell's resorptive machinery [163].

Fusion of secretory vesicles with the target plasma membrane is the common final step in exocytosis, which may be constitutive or regulated. In both circumstances the exocytic event is mediated by v- (vesicular) and t- (target) SNAREs (soluble N-ethylmaleimide-sensitive fusion protein (NSF) attachment protein (SNAP) receptors) [164]. Regulation of SNARE function requires synaptotagmins (Syt), a family of vesicular trafficking proteins that uniquely links the vesicle and target (in this case plasma membrane) [165].

Fifteen Syt isoforms have been identified in mammalian cells. Each Syt family member is distinctly distributed and exhibits distinct calcium- and phospholipid-binding affinities. While Syt I, II, III, V, and X are expressed predominantly in the nervous system and neuroendocrine cells, others are ubiquitous. Syt VII, which is broadly expressed, regulates calcium-dependent exocytosis of lysosomes in fibroblasts and macrophages, among

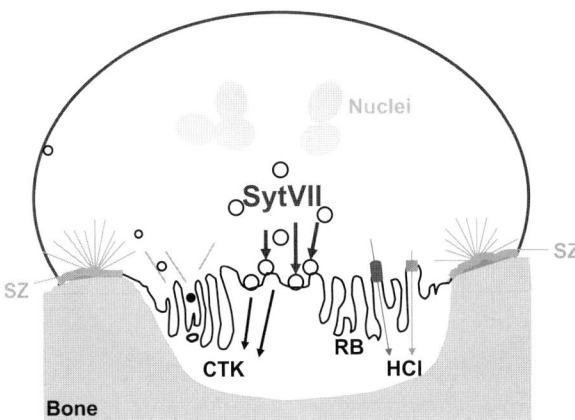

**FIGURE 6-7 Generation of the osteoclast ruffled border.** Lysosomal enzyme, vacuolar H+ATPase (proton pump) and chloride channel-containing vesicles are inserted into the bone-apposed plasma membrane, under the aegis of synaptotagmin VII (SytVII), permitting delivery of HCl and cathepsin K (CTK) into the resorptive microenvironment. (Courtesy of Dr. Haibo Zhao.) Ruffled border (RB); sealing zone (SZ). Please refer to color plate section.

others [166]. Given that macrophages derive from the same cell as the osteoclasts and osteoblasts are specialized fibroblasts, Syt VII is a candidate to regulate calcium-dependent secretory activities in both cell types. In fact, Syt VII is essential for formation of the ruffled border as it mediates fusion of secretory lysosomal vesicles to the bone-apposed plasma membrane [159] (Figure 6-7). In keeping with this observation, Syt VII-deficient polykaryons fail to resorb bone. The defective osteoclasts in $SyVII^{-/-}$ mice should predispose them to have increased bone mass. However, surprisingly these mice are osteoporotic. This counterintuitive observation reflects the fact that SytVII also regulates the secretion of bone matrix proteins by osteoblasts and hence osteogenesis. Thus, SytVII deficiency dampens both bone formation and resorption with the former deficiency predominating; thereby diminishing skeletal mass.

The ruffled border also participates in the removal of products of bone degradation from the resorptive microenvironment. This event predominantly involves transcytosis, as the efficacy of the sealing zone appears to substantially limit the size of molecules moving between it and the bone surface [167]. Hence, the majority of degraded matrix molecules are endocytosed at the central area of the ruffled border or "uptake zone" [167]. Endocytosed vesicles then move along microtubules to fuse with the apical domain of the cell and discharge their contents into the general extracellular space [167]. Thus, the ruffled border represents a dynamic structure continuously being remodeled by insertion of exocytic and endocytic vesicles.

## Osteoclast and bone matrix: role of αvβ3 integrin

Integrins are α/β heterodimers with long extracellular domains. In most instances, the cytoplasmic region is relatively short, consisting of 40–70 amino acids. Because they are the principal cell/matrix attachment molecules, integrins are likely candidates to mediate osteoclast/bone recognition. Members of the β1 family of these heterodimers, which recognize collagen, fibronectin, and laminin, are present on osteoclasts. However, a series of in vitro and in vivo experiments indicated that αvβ3 is the principal integrin mediating bone resorption. This heterodimer, like all members of the "v" integrin family, recognizes the amino acid motif Arg-Gly-Asp (RGD), which is present in a variety of bone-residing proteins such as osteopontin and bone sialoprotein. Thus, osteoclasts attach to and spread on these substrates in an RGD-dependent manner, as demonstrated by the fact that competitive ligands arrest bone resorption, in vivo [168–170].

Definitive proof of the pivotal role that αvβ3 plays in the resorptive process came with the generation of the β3 subunit knockout mouse, which has progressively increasing bone mass due to osteoclast dysfunction [171]. This mouse serves as the key tool for determining the role of the osteoclast's most abundant integrin in its capacity to resorb bone. Failure to express αvβ3 results in a dramatic osteoclast phenotype, particularly with regards to the actin cytoskeleton. The $β3^{-/-}$ osteoclast forms abnormal ruffled borders, in vivo, and, whether generated in vitro or directly isolated from bone, the mutant cell fails to spread when plated on immobilized RGD ligand or mineralized matrix with physiological amounts of RANKL and M-CSF. Confirming their attenuated resorptive activity, $β3^{-/-}$ osteoclasts produce fewer and shallower resorptive lacunae. In keeping with the in vitro finding of attenuated bone resorption, $β3^{-/-}$ mice are substantially hypocalcemic [171].

αvβ3 is expressed by endothelial cells in pathological conditions, such as inflammation and tumor angiogenesis. In physiological circumstances, however, the integrin's distribution is largely confined to osteoclasts and the placenta. Importantly, the β3 chain is abundant in platelets where it associates with αIIb. Because the αIIbβ3 heterodimer regulates hemostasis, inactivating mutations of the β3 integrin subunit prompt the bleeding dyscrasia, Glanzmann's thrombasthenia, a condition also present in the $β3^{-/-}$ mouse. Glanzmann's disease has been associated with osteopetrosis [172] but, unlike the mutant mouse, there is no evidence that most affected patients have increased bone mass. This unexpected finding may relate to physiological compensation by the increased expression of β1 integrins with β3 deletion [173]. A reciprocal relationship also exists between αvβ3 and another RGD-recognizing integrin, αvβ5. Whereas αvβ3 is abundant on mature osteoclasts, and is a marker of their phenotype, it is absent on marrow macrophages, which express αvβ5 [171]. With exposure to RANKL, the β5 subunit disappears to be replaced by β3 [174]. Suggesting that αvβ5 exercises an inhibitory effect on osteoclast differentiation, the β5 integrin-deleted mouse responds to estrogen deprivation with increased osteoclastogenesis and accelerated bone loss [175]. Therefore, while αvβ3 and αvβ5 are compositionally similar, they are differentially expressed in response to RANKL and exert opposite effects on osteoclast function.

Like integrins in other cells [176], αvβ3, in the osteoclast, is maintained in a default, low-affinity state, mediated by charge interactions between the α and β subunits [150, 177]. Resolution of the αvβ3 crystal structure documents that when inactive, the ecto-domains of the integrin are in a bent, closed conformation with α and β cytoplasmic domains in close proximity. Upon activation, α and β cytoplasmic tails separate and the ecto-domains extend into a high-affinity state [176].

Integrins are activated by outside-in or inside-out signaling (Figure 6-8). In the case of outside-in activation, the heterodimers are clustered by their multivalent ligands; thereby increasing ligand avidity and changing their conformation to that of high-affinity receptors. The high-affinity ecto-domains cause the heterodimer's cytoplasmic regions to separate and initiate intracellular signaling [176]. On the other hand, inside-out activation is an indirect event in which other receptors, typically those induced by growth factors, transmit signals to the integrin's cytoplasmic regions. These α and β intracellular domains then separate and re-orient their associated ecto-domains. In many circumstances, overlap exists between inside-out and outside-in signaling. For example, inside-out activation prompts release of integrins from cytoskeletal restraints, thus permitting their lateral movement in the membrane leading to ligand-induced clustering and outside-in signals. Clustering of integrins involves the small GTPase Rap 1, deletion of which arrests the function of the αIIbβ3 integrin in platelets [178]. As will be discussed, inside-out αvβ3 signaling is a means by which M-CSF regulates the osteoclast cytoskeleton.

The osteoclast functions in a cyclical fashion, migrating to a candidate bone-resorptive site to which it attaches. It degrades the underlying matrix, ultimately detaches and re-initiates the cycle. During matrix attachment, the integrin, in the context of the podosome, is predominantly in a low-affinity state and confined to the actin ring

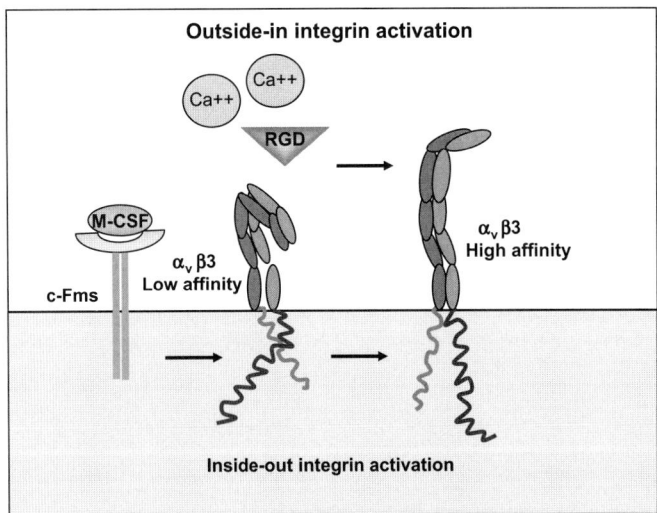

**FIGURE 6-8 Outside-in and Inside-out αvβ3 integrin activation.** αvβ3 integrin can exist in a low-affinity/low-binding conformation (inactive) or in a high-affinity/high-ligand-binding conformation (active state). Signals from outside the cells (e.g. integrin ligands and calcium cations) or inside the cells (e.g. signals from the M-CSF receptor, c-Fms) can both induce conformational changes allowing integrin activation. Please refer to color plate section.

[150, 177]. Localization of αvβ3 to the podosome requires intracellular signals mediated via the integrin's cytoplasmic domain [150]. Upon activation, αvβ3 leaves the podosome and transits to lamellipodia, which mediate osteoclast motility. During bone resorption, the heterodimer appears in the ruffled membrane [150, 177].

When first discovered, integrins were exclusively considered cell attachment molecules. It is now apparent that their capacity to transmit matrix-derived signals to the cell interior is at least as important. For example, precursors committed to the osteoclast phenotype, by exposure to M-CSF and RANKL, activate the MAP kinases Erk1/2 when placed on an αvβ3 substrate. Attenuation of the adhesion-induced signal in $β3^{-/-}$ cells confirms it is the product of the integrin [179]. Erk activation is often followed by accelerated cell division, suggesting that failure to stimulate the kinase would reduce the number of osteoclasts generated in vitro. While osteoclasts derived from $β3^{-/-}$ precursors are diminished in number, this does not reflect arrested division, but rather retarded differentiation [179]. In fact, the $β3$ integrin knockout mouse exhibits a 3.5-fold increase in osteoclast number, in vivo, indicating that influences other than integrin-mediated Erk activation are in play [171]. These mutant animals have an abundance of M-CSF in their marrow microenvironment, accounting for the profound osteoclasto-genesis [179]. While not yet proven, the increased M-CSF may result from targeting of osteoblasts and their precursors by excess parathyroid hormone [180], probably present in $β3^{-/-}$ mice, due to hypocalcemia. While surprising because of the hypothesized anti-apoptotic effect of matrix attachment, $β3^{-/-}$ osteoclasts are also long-lived because the unoccupied integrin transmits a caspase 8-mediated death signal, which does not occur in the absence of αvβ3 [181].

## Integrin-associated proteins

The capacity of integrins to transmit intracellular signals to the cytoskeleton heightened interest in the cytoplasmic molecules, mediating these events. As regards the osteoclast, the first pivotal observation regarding cytoskeletal organization came in 1991 with the generation of c-Src-deficient mice, whose phenotypic abnormalities are surprisingly predominantly skeletal [116]. Despite an increase in osteoclast number in $c\text{-}Src^{-/-}$ mice, a feature shared with animals lacking αvβ3, this mutant develops severe osteopetrosis due to dysfunctional bone resorption. The failure of $c\text{-}Src^{-/-}$ mice to resorb bone reflects abortive cytoskeletal organization, resulting in the inability to spread, in vitro, and an absence of normal ruffled membranes, in vivo [182]. The dramatic phenotype of the c-Src-deficient mouse established this proto-oncogene as a focus of osteoclast regulation. The tyrosine kinase serves as an adaptor protein, the function of which is independent of its capacity to induce tyrosine phosphorylation. This conclusion is based on the finding that reintroduction of a kinase-dead mutant partially rescues $c\text{-}Src^{-/-}$-induced osteo-petrosis [183]. On the other hand, complete reversal of these abnormalities also requires c-Src kinase activity [183].

The similar cytoskeletal phenotypes of c-Src- and αvβ3-deficient osteoclasts suggest a commonality of intracellular signaling. c-Src binds directly to the β3 subunit in the context of the platelet integrin, αIIbβ3. The same obtains in the osteoclast as regards αvβ3 [118]. Similar to its association with αIIbβ3, the proto-oncogene recognizes the β chain of αvβ3 constitutively. c-Src kinase activity, however, requires ligand occupancy of the integrin [118]. Both the capacity of c-Src to bind αvβ3 and subsequent activation of the kinase are mediated by PLCγ2, which, in turn, is stimulated by the integrin [120]. Activated c-Src reduces the strength of the association between the β3 cytoplasmic domain and the cytoskeleton and therefore probably mediates movement of αvβ3 to the migratory machinery of the cell in the form of lamellipodia. In fact, c-Src activation accelerates podosomal disassembly, most probably by phosphorylation of its substrate, cortactin. As a result, podosomes are more abundant in $c\text{-}Src^{-/-}$ osteoclasts and the cells develop focal adhesions and actin stress fibers, which are not encountered in highly motile, wild-type osteoclasts [184]. Consistent with this hypothesis, lentiviral-based suppression of cortactin expression completely blocks bone resorption [185]. Thus, the defect in $c\text{-}Src^{-/-}$ osteoclasts probably represents both failure to generate resorptive organelles, such as the ruffled border, and arrested migration on the bone surface.

A popular model holds that c-Cbl and the focal adhesion kinase family member *Pyk2*, mobilize c-Src to the integrin, under the aegis of PLCγ2 [184]. In this paradigm, αvβ3 occupancy induces phosphorylation of $Pyk2^{Y402}$, which then binds the SH2 domain of c-Src. $Pyk2^{-/-}$ mice have increased skeletal mass and the capacity of their osteoclasts to resorb bone is impaired [186]. Pyk2-deficient osteoclasts are unable to form a podosome belt at the cell periphery and more significantly do not generate normal sealing zones on bone.

Pyk2 contributes to the organization of the osteoclast cytoskeleton through its impact on microtubules. The kinase promotes tubulin acetylation, presumably by inhibiting RhoA. As in other circumstances, acetylated tubulin promotes microtubule stability and podosomal organization [186].

Integrin-induced $Pyk2^{Y402}$ activation typically represents $Ca^{2+}$-dependent autophosphorylation, but may also be mediated, in osteoclasts, by other tyrosine kinases [187]. Like c-Src, Pyk2 tyrosine phosphorylation requires PLCγ2 [120]. It has been postulated that PLCγ2 modulates Pyk2 phosphorylation via its catalytic activity, while PLCγ2 associates with c-Src, bringing the kinase to the β3 integrin cytoplasmic tail, through its adaptor SH2 motifs [120].

Regardless of the mechanism, the proposed association between phosphorylated $Pyk2^{Y402}$ and c-Src would prevent phospho-c-$Src^{Y527}$ from interacting with its own SH2 domain, thus relieving auto-inhibition of kinase function. c-Src recruits c-Cbl to its SH3 domain by serving as an adaptor protein. This event further increases c-Src auto kinase activity, thereby phosphorylating c-Cbl.

Once phosphorylated, c-Cbl converts from a c-Src kinase activator to an inhibitor as a result of two events. First, the phosphotyrosine-binding domain of phosphorylated c-Cbl occupies the c-Src kinase-essential residue, Y416. Second, c-Cbl is a ubiquitin E3 ligase and, as such, it has the potential to promote proteosomal degradation of the integrin-associated Pyk2/c-Src/c-Cbl complex [188]. This model proposes a temporal scenario in which c-Cbl initially enhances, and subsequently suppresses, c-Src activation, ultimately arresting αvβ3 function in osteoclasts.

Syk is a non-receptor tyrosine kinase, which, in the osteoclast, associates with the β3 cytoplasmic domain in a region close to that binding c-Src [118]. When in complex with the integrin, Syk is activated by c-Src, a key event in organizing the cell's cytoskeleton [118]. Syk's ability to prompt cytoskeletal organization is also regulated by c-Cbl. In this circumstance, Cbl interacts with Syk$^{Y317}$, ubiquitinates the kinase, and prevents its phosphorylation [189].

Syk also recognizes the ITAM-bearing adaptors, DAP12 and FcRγ [102, 103, 113, 118]. In contrast to its recognition of DAP12, however, Syk associates with the β3 integrin cytoplasmic domain in a non-phosphotyrosine-dependent manner [190]. This is of particular interest because the two tyrosines in the β3 cytoplasmic domain appear to function differently in the context of αvβ3 as compared to αIIbβ3. While the human Glanzmann's mutation, β3$^{S752P}$, also dramatically impacts the osteoclast, the platelet-inactivating alterations of the β3 tyrosines have no discernible impact on the bone-resorbing cell [191].

Syk kinase targets a number of cytoskeleton-regulating proteins, including the Vav family of guanine nucleotide exchange factors (GEFs). These proteins convert Rho GTPases from their inactive GDP to their active GTP conformation. Three Vav isoforms are identified. While Vav1 dominates in lymphocytes, Vav3 is uniquely expressed in abundance in osteoclasts [158]. In the osteoclast, Vav3 is phosphorylated by Syk but the process requires an association with Slp adaptor proteins, particularly Slp-76 [192].

Like *β3-, c-Src-, Slp* family-, or *Syk*-deficient osteoclasts, those derived from *Vav3*$^{-/-}$ mice fail to organize their cytoskeleton or optimally degrade mineralized matrix, in vitro. Importantly, the same mutant animals are osteopetrotic and do not respond to resorptive stimuli, in vivo [158]. Vav proteins are phosphorylated upon β3 integrin activation in the context of the platelet [193] and the same holds in the osteoclast. Activation of Vav3 is Syk-mediated and the GEF regulates the signaling capacity of the integrin in a reciprocal fashion; thus, adhesion-induced c-Src and Erk phosphorylation, which were attenuated in osteoclasts lacking Vav3. Alternatively, the β3 integrin/Syk/Vav complex appears to organize the cytoskeleton in a manner distinct from FAK and, by inference, Pyk2 [193].

Similar to RhoA, matrix-induced activation of Rac is blunted in the absence of αvβ3 [194]. While both small GTPases impact the actin cytoskeleton, Rac and Rho exert distinctive effects. Rac stimulation in osteoclast precursors prompts appearance of lamellipodia, thus forming the migratory front of the cell to which αvβ3 moves when activated. On the other hand, RhoA stimulates filamentous actin formation, which, in osteoclasts, represents organization of the sealing zone, a view conflicting with RhoA's disruption of the actin belt via microtubule destabilization. In any event, it is likely that RhoA principally affects cell adhesion, whereas Rac mediates the cytoskeleton's migratory machinery. Importantly, absence of Vav3 blunts Rac but not RhoA activity in the osteoclast [158].

Rac1 and 2 are mutually compensatory isoforms of the GTPase when expressed in osteoclasts. Effective deletion of both, however, produces severe osteopetrosis in which osteoclasts fail to organize their cytoskeleton. Absence of the related Rho family GTPase, Cdc42, also causes osteopetrosis but in this circumstance, the dominant mechanism is arrested osteoclast recruitment due to inhibited precursor proliferation and accelerated apoptosis of the mature polykaryon.

## M-CSF and the osteoclast cytoskeleton

M-CSF, which promotes osteoclastogenesis, coincidentally organizes the cytoskeleton of the mature bone-resorptive cell by two distinct mechanisms [179, 195]. One is αvβ3-independent, involving activated c-Fos. The biological significance of M-CSF's effects on the osteoclast cytoskeleton, in the absence of the integrin, is, however, unclear. For example, M-CSF-induction of $β3^{-/-}$ osteoclast spreading requires a concentration of cytokine one order of magnitude greater than that stimulating the same phenomenon via inside-out activation of the integrin [179]. Furthermore, even at high concentrations, M-CSF, in the absence of αvβ3, is incapable of activating Rho GTPases, which are key regulators of actin organization [179]. In keeping with this observation, distribution of cytoskeletal proteins in $β3^{-/-}$ osteoclasts remains abnormal even in the face of high-dose M-CSF [179]. Finally, M-CSF-treated $β3^{-/-}$ osteoclasts assume an atypical phenotype on bone, which they are incapable of resorbing [179]. Hence, the significance of M-CSF-induced cytoskeletal organization in the absence of αvβ3 is questionable.

On the other hand, M-CSF, targeting the β3 subunit cytoplasmic domain, alters the conformation of osteoclast αvβ3 to its high-affinity, ligand-binding state [150] (Figure 6-8). The consequences of this cytokine-induced, inside-out activation of αvβ3 are stimulation of the resorptive signaling pathway, comprising c-Src, Syk, DAP12, SLP-76, and Rho GTPases [113, 179, 195] (Figure 6-9). M-CSF activates these molecules in osteoclasts that are adherent to αvβ3 ligand-containing proteins or the cell's natural substrate, bone. These observations indicate that the cytoskeleton-organizing properties of M-CSF are initiated largely by inside-out αvβ3 activation, which in other cells requires talin.

Talin is a 270-kDa cytosolic protein, which binds the β3 cytoplasmic domain in osteoclasts and is one of a number of adaptors linking the integrin to the actin cytoskeleton. Association of talin with the β3 tail is also required to activate the integrin's ligand-binding motif [196]. In fact, in all cells studied, talin interacting with β cytoplasmic domains is a final common and necessary step in integrin activation. The high-affinity, talin-associated heterodimer avidly binds extracellular ligand, which stimulates transmission of intracellular signals, including those that organize the actin cytoskeleton.

Thus, the ability of the osteoclast to organize its cytoskeleton is central to its capacity to resorb bone. The αvβ3 integrin is a key regulator of this process exerting its effects by transmitting bone-matrix-derived signals to the cytoskeleton which are critical for the formation of the resorptive microenvironment.

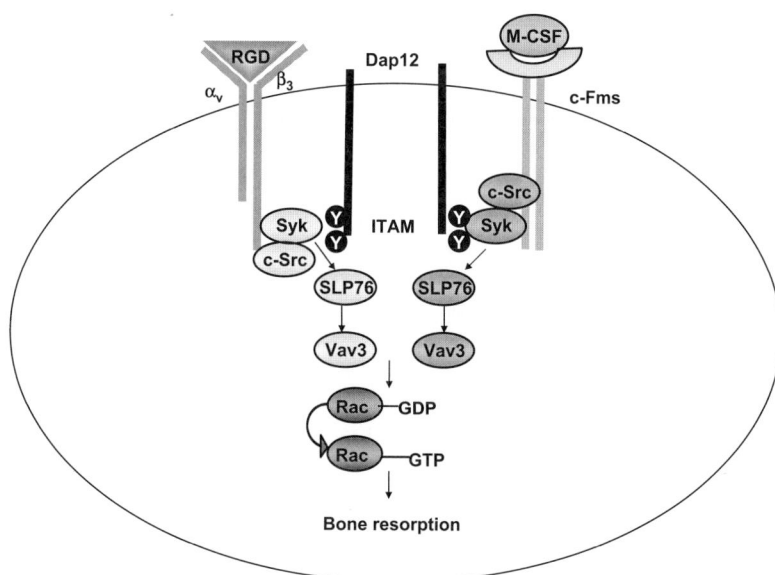

FIGURE 6-9 αvβ3 **integrin and c-Fms collaborate to organize the osteoclast cytoskeleton.** Activation of the αvβ3 integrin or the M-CSF receptor, c-Fms, organizes the osteoclast cytoskeleton by a mechanism involving stimulation of the tyrosine kinases, c-Src and Syk, phosphorylation of the ITAM-containing protein, DAP12, the adaptor, SLP-76, and the guanine nucleotide exchange factor, Vav3. This series of events activates the Rho GTPase family of transcription factors, particularly Rac, ultimately permitting the osteoclast to restructure its cytoskeleton and resorb bone. Whether this series of events represents independent activation of intracellular signals by αvβ3 and c-Fms or inside-out activation of the integrin by the cytokine receptor is unknown. Please refer to color plate section.

## Conclusion

The concept of osteoimmunology is based upon evidence that the classical immune system and skeletal cells share a commonality of regulation. Because it is derived from a precursor that also gives rise to the macrophage and the myeloid dendritic cell, this concept is particularly relevant to the osteoclast. Thus, the long predominant view that the skeleton is an organ unto itself in which its resident cells are regulated exclusively in an endocrine- or biomechanical-dependent manner has undergone modification. The realization that a host of immune cells speak to osteoclasts generates a new vision of skeletal biology and provides an additional framework for designing therapeutic strategies that address pathological bone loss.

## References

[1] Teitelbaum SL, Ross FP. Genetic regulation of osteoclast development and function. Nat Rev Genet 2003;4(8):638–49.

[2] Takayanagi H, Kim S, Kogal T, et al. Induction and activation of the transcription factor NFATc1 (NFAT2) integrate RANKL signaling in terminal differentiation of osteoclasts. Dev Cell 2002;3 (6):889–901.

[3] Takayanagi H. Osteoimmunology: shared mechanisms and crosstalk between the immune and bone systems. Nat Rev Immunol 2007;7(4):292–304.

[4] Boyle WJ, Simonet WS, Lacey DL. Osteoclast differentiation and activation. Nature 2003;423:337–42.

[5] Takahashi N, Akatsu T, Udagawa N, et al. Osteoblastic cells are involved in osteoclast formation. Endocrinology 1988;123:2600–2.

[6] Yoshida H, Hayashi S, Kunisada T, et al. The murine mutation osteopetrosis is in the coding region of the macrophage colony stimulating factor gene. Nature 1990;345:442–4.

[7] Lagasse E, Weissman IL. Enforced expression of Bcl-2 in monocytes rescues macrophages and partially reverses osteopetrosis in op/op mice. Cell 1997;89(7):1021–31.

[8] Lacey DL, Timms E, Tan HL, et al. Osteoprotegerin ligand is a cytokine that regulates osteoclast differentiation and activation. Cell 1998;93(2):165–76.

[9] Yasuda H, Shima N, Nakagawa N, et al. Osteoclast differentiation factor is a ligand for osteoprotegerin/osteoclastogenesis-inhibitory factor and is identical to TRANCE/RANKL. Proc Natl Acad Sci U S A 1998;95(7):3597–602.

[10] Simonet WS, Lacey DL, Dunstan CR, et al. Osteoprotegerin: a novel secreted protein involved in the regulation of bone density. Cell 1997;89(2):309–19.

[11] Tsuda E, Goto M, Mochizuki S, et al. Isolation of a novel cytokine from human fibroblasts that specifically inhibits osteoclastogenesis. Biochem Biophys Res Commun 1997;234(1):137–42.

[12] Dougall WC, Glaccum M, Charrier K, et al. RANK is essential for osteoclast and lymph node development. Genes Dev 1999;13(18):2412–24.

[13] Kong YY, Yoshida H, Sarosi I, et al. OPGL is a key regulator of osteoclastogenesis, lymphocyte development and lymph-node organogenesis. Nature 1999;397:315–23.

[14] Hughes AE, Ralston SH, Marken J, et al. Mutations in TNFRSF11A, affecting the signal peptide of RANK, cause familial expansile osteolysis. Nat Genet 2000;24(1):45–8.

[15] Sobacchi C, Frattini A, Guerrini MM, et al. Osteoclast-poor human osteopetrosis due to mutations in the gene encoding RANKL. Nat Genet 2007;39(8):960–2.

[16] Anderson DM, Maraskovsky E, Billingsley WL, et al. A homologue of the TNF receptor and its ligand enhance T-cell growth and dendritic-cell function. Nature 1997;390:175–9.

[17] Darnay BG, Haridas V, Ni J, Moore PA, Aggarwal BB. Characterization of the intracellular domain of receptor activator of NF-κB (RANK). Interaction with tumor necrosis factor receptor-associated factors and activation of NF-κb and c-Jun N-terminal kinase. J Biol Chem 1998; 273(32):20551–5.

[18] Galibert L, Tometsko ME, Anderson DM, Cosman D, Dougall WC. The involvement of multiple tumor necrosis factor receptor (TNFR)-associated factors in the signaling mechanisms of receptor activator of NF-κB, a member of the TNFR superfamily. J Biol Chem 1998;273 (51):34120–7.

[19] Wong BR, Josien R, Lee SY, et al. The TRAF family of signal transducers mediates NF-κB activation by the TRANCE receptor. J Biol Chem 1998;273(43):28355–9.

[20] Darnay BG, Ni J, Moore PA, Aggarwal BB. Activation of NF-κB by RANK requires tumor necrosis factor receptor-associated factor (TRAF) 6 and NF-κB-inducing kinase. Identification of a novel TRAF6 interaction motif. J Biol Chem 1999;274(12):7724–31.

[21] Wong BR, Bresser D, Kim N, et al. TRANCE, a TNF family member, activates Akt/PKB through a signaling complex involving TRAF6 and c-Src. Mol Cell 1999;4(6):1041–9.

[22] Inoue J, Ishida T, Tsukamoto M, et al. Tumor necrosis factor receptor-associated factor (TRAF) family: adapter proteins that mediate cytokine signaling. Exp Cell Res 2000;254(1):14–24.

[23] Kawai T, Akira S. Signaling to NF-κB by Toll-like receptors. Trends Mol Med 2007;13(11):460–9.

[24] Ha H, Han D, Choi Y. TRAF-mediated TNFR-family signaling. Curr Protoc Immunol Chapter 11, Unit 11 19D 2009.

[25] Hsu H, Lacey DL, Dunstan CR, et al. Tumor necrosis factor receptor family member RANK mediates osteoclast differentiation and activation induced by osteoprotegerin ligand. Proc Natl Acad Sci U S A 1999;96(7):3540–5.

[26] Ye H, Arron JR, Lamothe B, et al. Distinct molecular mechanism for initiating TRAF6 signalling. Nature 2002;418:443–7.

[27] Gohda J, Akiyama T, Koga T, et al. RANK-mediated amplification of TRAF6 signaling leads to NFATc1 induction during osteoclastogenesis. EMBO J 2005;24(4):790–9.

[28] Kadono Y, Okada F, Perchonock C, et al. Strength of TRAF6 signalling determines osteoclastogenesis. EMBO Rep 2005;6(2):171–6.

[29] Lomaga MA, Yeh WC, Sarosi I, et al. TRAF6 deficiency results in osteopetrosis and defective interleukin-1, CD40, and LPS signaling. Genes Dev 1999;13(8):1015–24.

[30] Naito A, Azuma S, Tanaka S, et al. Severe osteopetrosis, defective interleukin-1 signalling and lymph node organogenesis in TRAF6-deficient mice. Genes Cells 1999;4(6):353–62.

[31] Kobayashi T, Walsh PT, Walsh MC, et al. TRAF6 is a critical factor for dendritic cell maturation and development. Immunity 2003;19(3):353–63.

[32] Armstrong AP, Tometski ME, Glaccum M, et al. A RANK/TRAF6-dependent signal transduction pathway is essential for osteoclast cytoskeletal organization and resorptive function. J Biol Chem 2002;277(46):44347–56.

[33] Kanazawa K, Azuma Y, Nakano H, Kudo A. TRAF5 functions in both RANKL- and TNFα-induced osteoclastogenesis. J Bone Miner Res 2003;18(3):443–50.

[34] Kanazawa K, Kudo A. TRAF2 is essential for TNF-α-induced osteoclastogenesis. J Bone Miner Res 2005;20(5):840–7.

[35] Wada T, Nakashima T, Oliveira-dos-Santos AJ, et al. The molecular scaffold Gab2 is a crucial component of RANK signaling and osteoclastogenesis. Nat Med 2005;11(4):394–9.

[36] Taguchi Y, Gohda J, Koga T, Takayanagi H, Inoue J. A unique domain in RANK is required for Gab2 and PLCγ2 binding to establish osteoclastogenic signals. Genes Cells 2009;14(11):1331–45.

[37] Mao D, Epple H, Uthgenannt B, Novack DV, Faccio R. PLCγ2 regulates osteoclastogenesis via its interaction with ITAM proteins and GAB2. J Clin Invest 2006;116(11):2869–79.

[38] Kim H, Choi HK, Shin JH, et al. Selective inhibition of RANK blocks osteoclast maturation and function and prevents bone loss in mice. J Clin Invest 2009;119(4):813–25.

[39] Deng L, Wang C, Spencer E, et al. Activation of the IκB kinase complex by TRAF6 requires a dimeric ubiquitin-conjugating enzyme complex and a unique polyubiquitin chain. Cell 2000;103(2):351–61.

[40] Wang C, Deng L, Hong M, et al. TAK1 is a ubiquitin-dependent kinase of MKK and IKK. Nature 2001;412:346–51.

[41] Kanayama A, Seth RB, Sun L, et al. TAB2 and TAB3 activate the NF-κB pathway through binding to polyubiquitin chains. Mol Cell 2004;15(4):535–48.

[42] Kobayashi N, Kadono Y, Naito A, et al. Segregation of TRAF6-mediated signaling pathways clarifies its role in osteoclastogenesis. EMBO J 2001;20(6):1271–80.

[43] Walsh MC, Kim GK, Maurizio PL, Molnar EE, Choi Y. TRAF6 autoubiquitination-independent activation of the NF-κB and MAPK pathways in response to IL-1 and RANKL. PLoS One 2008; 3(12):e4064.

[44] Mizukami J, Takaesu G, Akatsuka H, et al. Receptor activator of NF-κB ligand (RANKL) activates TAK1 mitogen-activated protein kinase kinase kinase through a signaling complex containing RANK, TAB2, and TRAF6. Mol Cell Biol 2002;22(4):992–1000.

[45] Yamaguchi K, Shirakabe K, Shibuya K, et al. Identification of a member of the MAPKKK family as a potential mediator of TGF-β signal transduction. Science 1995;270(5244):2008–11.

[46] Duran A, Serrano M, Leitges M, et al. The atypical PKC-interacting protein p62 is an important mediator of RANK-activated osteoclastogenesis. Dev Cell 2004;6(2):303–9.

[47] Laurin N, Brown JP, Morissette J, Raymond V. Recurrent mutation of the gene encoding sequestosome 1 (SQSTM1/p62) in Paget disease of bone. Am J Hum Genet 2002;70(6):1582–8.

[48] Kurihara N, Hiruma Y, Shou H, et al. Mutation of the sequestosome 1 (p62) gene increases osteoclastogenesis but does not induce Paget disease. J Clin Invest 2007;117(1):133–42.

[49] Ghosh S, Karin M. Missing pieces in the NF-κB puzzle. Cell 2002;109:S81–96.

[50] Hayden MS, Ghosh S. Shared principles in NF-κB signaling. Cell 2008;132(3):344–62.

[51] Franzoso G, Carlson L, Xing L, et al. Requirement for NF-κB in osteoclast and B-cell development. Genes Dev 1997;11(24):3482–96.

[52] Iotsova V, Caamano J, Loy J, et al. Osteopetrosis in mice lacking NF-κB1 and NF-κB2. Nat Med 1997;3(11):1285–9.

[53] Jimi E, Aoki K, Saito H, et al. Selective inhibition of NF-κB blocks osteoclastogenesis and prevents inflammatory bone destruction in vivo. Nat Med 2004;10(6):617–24.

[54] Ruocco MG, Maeda S, Park JM, et al. IκB kinase (IKK)β, but not IKKα, is a critical mediator of osteoclast survival and is required for inflammation-induced bone loss. J Exp Med 2005;201(10):1677–87.

[55] Vaira S, Johnson T, Hirbe AC, et al. RelA/p65 promotes osteoclast differentiation by blocking a RANKL-induced apoptotic JNK pathway in mice. J Clin Invest 2008;118(6):2088–97.

[56] Novack DV, Yin L, Hagen-Stapleton A, et al. The IκB function of NF-κB2 p100 controls stimulated osteoclastogenesis. J Exp Med 2003;198(5):771–81.

[57] Vaira S, Alhawagri M, Anwisye I, et al. RelB is the NF-κB subunit downstream of NIK responsible for osteoclast differentiation. Proc Natl Acad Sci U S A 2008;105(10):3897–902.

[58] Maruyama T, Fukushima H, Nakao K, et al. Processing of the NF-κB2 precursor, p100, to p52 is critical for RANKL-induced osteoclast differentiation. J Bone Miner Res 2009.

[59] Soysa NS, Alles N, Weih D, et al. The pivotal role of the alternative NF-κB pathway in maintenance of basal bone homeostasis and osteoclastogenesis. J Bone Miner Res 2009.

[60] Matsuo K, Owens JM, Tonko M, et al. Fosl1 is a transcriptional target of c-Fos during osteoclast differentiation. Nat Genet 2000;24(2):184–7.

[61] Wagner EF, Eferl R. Fos/AP-1 proteins in bone and the immune system. Immunol Rev 2005;208(1):126–40.

[62] Johnson RS, Spiegelman BM, Papaioannou V. Pleiotropic effects of a null mutation in the c-fos proto-oncogene. Cell 1992;71(4):577–86.

[63] Wang ZQ, Ovitt C, Grigoriadis AE, et al. Bone and haematopoietic defects in mice lacking c-fos. Nature 1992;360:741–5.

[64] Sato K, Suematsu A, Nakashima T, et al. Regulation of osteoclast differentiation and function by the CaMK-CREB pathway. Nat Med 2006;12(12):1410–16.

[65] Yamashita T, Yao Z, Li F, et al. NF-κB p50 and p52 regulate receptor activator of NF-κB ligand (RANKL) and tumor necrosis factor-induced osteoclast precursor differentiation by activating c-Fos and NFATc1. J Biol Chem 2007;282(25):18245–53.

[66] Wan Y, Chong LW, Evans RM. PPAR-γ regulates osteoclastogenesis in mice. Nat Med 2007;13(12):1496–503.

[67] Eferl R, Hoebertz A, Schilling AF, et al. The Fos-related antigen Fra-1 is an activator of bone matrix formation. EMBO J 2004;23(14):2789–99.

[68] Ikeda F, Nishimura R, Matsubara T, et al. Critical roles of c-Jun signaling in regulation of NFAT family and RANKL-regulated osteoclast differentiation. J Clin Invest 2004;114(4):475–84.

[69] David JP, Sabapathy K, Hoffmann O, Idarraga MH, Wagner EF. JNK1 modulates osteoclastogenesis through both c-Jun phosphorylation-dependent and -independent mechanisms. J Cell Sci 2002;115(Pt 22):4317–25.

[70] Kenner L, Hoebertz A, Beil T, et al. Mice lacking JunB are osteopenic due to cell-autonomous osteoblast and osteoclast defects. J Cell Biol 2004;164(4):613–23.

[71] Bakiri L, Takada Y, Radolf M, et al. Role of heterodimerization of c-Fos and Fra1 proteins in osteoclast differentiation. Bone 2007;40(4):867–75.

[72] Bozec A, Bakiri L, Hoebertz A, et al. Osteoclast size is controlled by Fra-2 through LIF/LIF-receptor signalling and hypoxia. Nature 2008;454(7201):221–5.

[73] Chang L, Karin M. Mammalian MAP kinase signalling cascades. Nature 2001;410:37–40.

[74] Li X, Udagawa N, Itoh K, et al. p38 MAPK-mediated signals are required for inducing osteoclast differentiation but not for osteoclast function. Endocrinology 2002;143(8):3105–13.

[75] Bohm C, Hayer S, Kilian A, et al. The alpha-isoform of p38 MAPK specifically regulates arthritic bone loss. J Immunol 2009;183(9):5938–47.

[76] Matsumoto M, Kogawa M, Wada S, et al. Essential role of p38 mitogen-activated protein kinase in cathepsin K gene expression during osteoclastogenesis through association of NFATc1 and PU.1. J Biol Chem 2004;279(44):45969–79.

[77] Hotokezaka H, Sakai E, Kanaoka K, et al. U0126 and PD98059, specific inhibitors of MEK, accelerate differentiation of RAW264.7 cells into osteoclast-like cells. J Biol Chem 2002;277(49):47366–72.

[78] Crabtree GR, Olson EN. NFAT signaling: choreographing the social lives of cells. Cell 2002;109 (Suppl):S67–79.

[79] Asagiri M, Sato K, Usami T, et al. Autoamplification of NFATc1 expression determines its essential role in bone homeostasis. J Exp Med 2005;202(9):1261–9.

[80] Winslow MM, Pan M, Starbuck M, et al. Calcineurin/NFAT signaling in osteoblasts regulates bone mass. Dev Cell 2006;10(6):771–82.

[81] Aliprantis AO, Ueki Y, Sulyanto R, et al. NFATc1 in mice represses osteoprotegerin during osteoclastogenesis and dissociates systemic osteopenia from inflammation in cherubism. J Clin Invest 2008;118(11):3775–89.

[82] Takatsuna H, Asagiri M, Kubota T, et al. Inhibition of RANKL-induced osteoclastogenesis by (−)-DHMEQ, a novel NF-κB inhibitor, through downregulation of NFATc1. J Bone Miner Res 2005; 20(4):653–62.

[83] Peng SL, Gerth AJ, Ranger AM, Glimcher LH. NFATc1 and NFATc2 together control both T and B cell activation and differentiation. Immunity 2001;14(1):13–20.

[84] O'Connor RD, Zayzafoon M, Farach-Carson MC, Schanen NC. Mecp2 deficiency decreases bone formation and reduces bone volume in a rodent model of Rett syndrome. Bone 2009;45 (2):346–56.

[85] Matsuo K, Galson DL, Zhao C, et al. Nuclear factor of activated T-cells (NFAT) rescues osteoclastogenesis in precursors lacking c-Fos. J Biol Chem 2004;279(25):26475–80.

[86] Kim K, Kim JH, Lee J, et al. Nuclear factor of activated T cells c1 induces osteoclast-associated receptor gene expression during tumor necrosis factor-related activation-induced cytokine-mediated osteoclastogenesis. J Biol Chem 2005;280(42):35209–16.

[87] Anusaksathien O, Laplace C, Li X, et al. Tissue-specific and ubiquitous promoters direct the expression of alternatively spliced transcripts from the calcitonin receptor gene. J Biol Chem 2001;276(25):22663–74.

[88] Crotti TN, Flannery M, Walsh NC, et al. NFATc1 regulation of the human β3 integrin promoter in osteoclast differentiation. Gene 2006;372:92–102.

[89] Kim K, Lee SH, Ha Kim J, Choi Y, Kim N. NFATc1 induces osteoclast fusion via up-regulation of Atp6v0d2 and the dendritic cell-specific transmembrane protein (DC-STAMP). Mol Endocrinol 2008;22(1):176–85.

[90] Yagi M, Miyamoto T, Sawatani Y, et al. DC-STAMP is essential for cell-cell fusion in osteoclasts and foreign body giant cells. J Exp Med 2005;202(3):345–51.

[91] Kim Y, Sato K, Asagiri M, et al. Contribution of nuclear factor of activated T cells c1 to the transcriptional control of immunoreceptor osteoclast-associated receptor but not triggering receptor expressed by myeloid cells-2 during osteoclastogenesis. J Biol Chem 2005;280(38):32905–13.

[92] Chen L, Glover JN, Hogan PG, Rao A, Harrison SC. Structure of the DNA-binding domains from NFAT, Fos and Jun bound specifically to DNA. Nature 1998;392:42–8.

[93] Zhao B, Takami M, Yamada A, et al. Interferon regulatory factor-8 regulates bone metabolism by suppressing osteoclastogenesis. Nat Med 2009;15(9):1066–71.

[94] Kim K, Kim JH, Lee J, et al. MafB negatively regulates RANKL-mediated osteoclast differentiation. Blood 2007;109(8):3253–9.

[95] Nishikawa K, Nakashima T, Hayashi M, et al. Blimp1-mediated repression of negative regulators is required for osteoclast differentiation. Proc Natl Acad Sci U S A E Pub Ahead of Print 2010.

[96] Kim N, Takami M, Rho J, Josien R, Choi Y. A novel member of the leukocyte receptor complex regulates osteoclast differentiation. J Exp Med 2002;195(2):201–9.

[97] Reth M. Antigen receptor tail clue. Nature 1989;338(6214):383–4.

[98] Takai T. Roles of Fc receptors in autoimmunity. Nat Rev Immunol 2002;2(8):580–92.

[99] Koga T, Inui M, Inoue K, et al. Costimulatory signals mediated by the ITAM motif cooperate with RANKL for bone homeostasis. Nature 2004;428:758–63.

[100] Inui M, Kikuchi Y, Aoki N, et al. Signal adaptor DAP10 associates with MDL-1 and triggers osteoclastogenesis in cooperation with DAP12. Proc Natl Acad Sci USA 2009;106(12):4816–21.

[101] Kaifu T, Nakahara J, Inui M, et al. Osteopetrosis and thalamic hypomyelinosis with synaptic degeneration in DAP12-deficient mice. J Clin Invest 2003;111(3):323–32.

[102] Faccio R, Zou W, Colaianni G, Teitelbaum SL, Ross FP. High dose M-CSF partially rescues the Dap12−/− osteoclast phenotype. J Cell Biochem 2003;90(5):871–83.

[103] Mocsai A, Humphrey MB, Van Ziffle JA, et al. The immunomodulatory adapter proteins DAP12 and Fc receptor gamma-chain (FcRgamma) regulate development of functional osteoclasts through the Syk tyrosine kinase. Proc Natl Acad Sci U S A 2004;101(16):6158–63.

[104] Humphrey MB, Ogasawara K, Yao W, et al. The signaling adapter protein DAP12 regulates multinucleation during osteoclast development. J Bone Miner Res 2004;19(2):224–34.

[105] Hakola HP. Neuropsychiatric and genetic aspects of a new hereditary disease characterized by progressive dementia and lipomembranous polycystic osteodysplasia. Acta Psychiatr Scand 1972;232(Suppl):1–173.

[106] Paloneva J, Kestila M, Wu J, et al. Loss-of-function mutations in TYROBP (DAP12) result in a presenile dementia with bone cysts. Nat Genet 2000;25(3):357–61.

[107] Cella M, Buonsanti C, Strader C, et al. Impaired differentiation of osteoclasts in TREM-2-deficient individuals. J Exp Med 2003;198(4):645–51.

[108] Paloneva J, Mandelin J, Kiialainen A, et al. DAP12/TREM2 deficiency results in impaired osteoclast differentiation and osteoporotic Features. J Exp Med 2003;198(4):669–75.

[109] Wu Y, Torchia J, Yao W, et al. Bone microenvironment specific roles of ITAM adapter signaling during bone remodeling induced by acute estrogen-deficiency. PLoS ONE 2007;2(7):e586.

[110] Wu J, Song Y, Bakker AB, et al. An activating immunoreceptor complex formed by NKG2D and DAP10. Science 1999;285(5428):730–2.

[111] Chang C, Dietrich J, Harpur AG, et al. Cutting edge: KAP10, a novel transmembrane adapter protein genetically linked to DAP12 but with unique signaling properties. J Immunol 1999;163(9):4651–4.

[112] Hamerman AJ, Ni M, Killebrew RJ, Chu C-L, Lowell AC. The expanding roles of ITAM adapters FcRγ and DAP12 in myeloid cells. Immunol Rev 2009;232(1):42–58.

[113] Zou W, Reeve JL, Liu Y, Teitelbaum SL, Ross FP. DAP12 couples c-Fms activation to the osteoclast cytoskeleton by recruitment of Syk. Mol Cell 2008;31(3):422–31.

[114] Otero K, Turnbull IR, Poliani PL, et al. Macrophage colony-stimulating factor induces the proliferation and survival of macrophages via a pathway involving DAP12 and [beta]-catenin. Nat Immunol 2009;10(7):734–43.

[115] Jakus Z, Fodor S, Abram CL, Lowell CA, Mócsai A. Immunoreceptor-like signaling by [beta]2 and [beta]3 integrins. Trends Cell Biol 2007;17(10):493–501.

[116] Soriano P, Montgomery C, Geske R, Bradley A. Targeted disruption of the c-src proto-oncogene leads to osteopetrosis in mice. Cell 1991;64(4):693–702.

[117] Kim H-J, Zhang K, Zhang L, et al. The Src family kinase, Lyn, suppresses osteoclastogenesis in vitro and in vivo. Proc Natl Acad Sci U S A 2009;106(7):2325–30.

[118] Zou W, Kitaura H, Reeve J, et al. Syk, c-Src, the alphavbeta3 integrin, and ITAM immunoreceptors, in concert, regulate osteoclastic bone resorption. J Cell Biol 2007;176(6):877–88.

[119] Carpenter G, Ji Q-S. Phospholipase C-[gamma] as a signal-transducing element. Exp Cell Res 1999;253(1):15–24.

[120] Epple H, Cremasco V, Zhang K, et al. Phospholipase C{gamma}2 modulates integrin signaling in the osteoclast by affecting the localization and activation of Src kinase. Mol Cell Biol 2008;28(11):3610–22.

[121] Cremasco V, Graham BD, Novack VD, Swat W, Faccio R. Vav/Phospholipase Cgamma2-mediated control of a neutrophil-dependent murine model of rheumatoid arthritis. Arthritis Rheum 2008;58(9):2712–22.

[122] Cremasco V, Benasciutti E, Cella M, et al. Phospholipase C gamma 2 is critical for development of a murine model of inflammatory arthritis by affecting actin dynamics in dendritic cells. PLoS One 2010;5(1):e8909.

[123] Schmidt U, Boucheron N, Unger B, Ellmeier W. The Role of Tec family kinases in myeloid cells. Int Arch Allergy Immunol 2004;134(1):65–78.

[124] Tsukada S, Saffran DC, Rawlings DJ, et al. Deficient expression of a B cell cytoplasmic tyrosine kinase in human X-linked agammaglobulinemia. Cell 1993;72(2):279–90.

[125] Kerner JD, Appleby MW, Mohr RN, et al. Impaired expansion of mouse B cell progenitors lacking Btk. Immunity 1995;3(3):301–12.

[126] Lee SH, Kim T, Jeong D, Kim N, Choi Y. The Tec family tyrosine kinase Btk regulates RANKL-induced osteoclast maturation. J Biol Chem 2008;283(17):11526–34.

[127] Shinohara M, Koga T, Okamoto K, et al. Tyrosine kinases Btk and Tec regulate osteoclast differentiation by linking RANK and ITAM signals. Cell 2008;132(5):794–806.

[128] Hamerman JA, Tchao NK, Lowell CA, Lanier LL. Enhanced Toll-like receptor responses in the absence of signaling adaptor DAP12. Nat Immunol 2005;6(6):579–86.

[129] Turnbull IR, Gilfillan S, Cella M, et al. Cutting edge: TREM-2 attenuates macrophage activation. J Immunol 2006;177(6):3520–4.

[130] Chu CL, Yu YL, Shen KY, et al. Increased TLR responses in dendritic cells lacking the ITAM-containing adapters DAP12 and FcR&ggr. Eur J Immunol 2008;38(1):166–73.

[131] Kanamaru Y, Pfirsch S, Aloulou M, et al. Inhibitory ITAM signaling by Fc{alpha}RI-FcR{gamma} chain controls multiple activating responses and prevents renal inflammation. J Immunol 2008;180(4):2669–78.

[132] Wiktor-Jedrzejczak W, Bartocci A, Ferrante Jr AW, et al. Total absence of colony-stimulating factor 1 in the macrophage-deficient osteopetrotic (op/op) mouse. Proc Natl Acad Sci USA 1990;87(12):4828–32.

[133] Dai XM, Ryan GR, Hapel AJ, et al. Targeted disruption of the mouse colony-stimulating factor 1 receptor gene results in osteopetrosis, mononuclear phagocyte deficiency, increased primitive progenitor cell frequencies, and reproductive defects. Blood 2002;99(1):111–20.

[134] Ross FP, Teitelbaum SL. $\alpha_v\beta_3$ and macrophage colony-stimulating factor: partners in osteoclast biology. Immunol Rev 2005;208:88–105.

[135] Lin H, Lee E, Hestir K, et al. Discovery of a cytokine and its receptor by functional screening of the extracellular proteome. Science 2008;320(5877):807–11.

[136] Schlessinger J. Cell signaling by receptor tyrosine kinases. Cell 2000;103(2):211–25.

[137] Takeshita S, Faccio R, Chappel J, et al. c-Fms tyrosine 559 is a major mediator of M-CSF-induced proliferation of primary macrophages. J Biol Chem 2007;282(26):18980–90.

[138] Faccio R, Takeshita S, Colaianni G, et al. M-CSF regulates the cytoskeleton via recruitment of a multimeric signaling complex to c-Fms Tyr-559/697/721. J Biol Chem 2007;282(26):18991–9.

[139] Takeshita S, Namaba N, Zhao JJ, et al. SHIP-deficient mice are severely osteoporotic due to increased numbers of hyper-resorptive osteoclasts. Nat Med 2002;8(9):943–9.

[140] Zhou P, Kitaura H, Teitelbaum SL, et al. SHIP1 Negatively regulates proliferation of osteoclast precursors via Akt-dependent alterations in D-type cyclins and p27. J Immunol 2006;177(12):8777–84.

[141] Tanaka S, Nakamura I, Inoue J, Oda H, Nakamura K. Signal transduction pathways regulating osteoclast differentiation and function. J Bone Miner Metab 2003;21(3):123–33.

[142] Miyazaki T, Takayanagi H, Isshiki M, et al. In vitro and in vivo suppression of osteoclast function by adenovirus vector-induced csk gene. J Bone Miner Res 2000;15(1):41–51.

[143] Glantschnig H, Fisher JE, Wesolowski G, Rodan GA, Reszka AA. M-CSF, TNFalpha and RANK ligand promote osteoclast survival by signaling through mTOR/S6 kinase. Cell Death Differ 2003;10(10):1165–77.

[144] Akiyama T, Bouillet P, Miyazaki T, et al. Regulation of osteoclast apoptosis by ubiquitylation of proapoptotic BH3-only Bcl-2 family member Bim. EMBO J 2003;22(24):6653–64.

[145] Sugatani T, Hruska KA. Akt1/Akt2 and mammalian target of rapamycin/Bim play critical roles in osteoclast differentiation and survival, respectively, whereas Akt is dispensable for cell survival in isolated osteoclast precursors. J Biol Chem 2005;280(5):3583–9.

[146] Frattini A, Orchard PJ, Sobacchi C, et al. Defects in TCIRG1 subunit of the vacuolar proton pump are responsible for a subset of human autosomal recessive osteopetrosis. Nat Genet 2000;25(3):343–6.

[147] Hunt NP, Cunningham SJ, Adnan N, Harris M. The dental, craniofacial, and biochemical features of pyknodysostosis: a report of three new cases. J Oral Maxillofac Surg 1998;56(4):497–504.

[148] Novack DV, Teitelbaum SL. The osteoclast: friend or foe? Annu Rev Pathol 2008;3:457–84.

[149] Marchisio PC, Bergui L, Corbascio GC, et al. Vinculin, talin, and integrins are localized at specific adhesion sites of malignant B lymphocytes. Blood 1988;72(2):830–3.

[150] Faccio R, Novack DV, Zallone A, Ross FP, Teitelbaum SL. Dynamic changes in the osteoclast cytoskeleton in response to growth factors and cell attachment are controlled by beta3 integrin. J Cell Biol 2003;162(3):499–509.

[151] Saltel F, Chabadel A, Bonnelye E, Jurdic P. Actin cytoskeletal organisation in osteoclasts: a model to decipher transmigration and matrix degradation. Eur J Cell Biol 2008;87(8-9):459–68.

[152] Chabadel A, Banon-Rodriguez I, Cluet D, et al. CD44 and beta3 integrin organize two functionally distinct actin-based domains in osteoclasts. Mol Biol Cell 2007;18(12):4899–910.

[153] Saltel F, Destaing O, Bard F, Eichert D, Jurdic P. Apatite-mediated actin dynamics in resorbing osteoclasts. Mol Biol Cell 2004;15(12):5231–41.

[154] Destaing O, Sanjay A, Itzstein C, et al. The tyrosine kinase activity of c-Src regulates actin dynamics and organization of podosomes in osteoclasts. Mol Biol Cell 2008;19(1):394–404.

[155] Destaing O, Satel F, Gilquin B, et al. A novel Rho-mDia2-HDAC6 pathway controls podosome patterning through microtubule acetylation in osteoclasts. J Cell Sci 2005;118(Pt 13):2901–11.

[156] Purev E, Neff L, Horne WC, Baron R. c-Cbl and Cbl-b act redundantly to protect osteoclasts from apoptosis and to displace HDAC6 from beta-tubulin, stabilizing microtubules and podosomes. Mol Biol Cell 2009;20(18):4021–30.

[157] Chellaiah MA, Soga N, Swanson S, et al. Rho-A is critical for osteoclast podosome organization, motility, and bone resorption. J Biol Chem 2000;275(16):11993–2002.

[158] Faccio R, Teitelbaum SL, Fujikawa K, et al. Vav3 regulates osteoclast function and bone mass. Nat Med 2005;11(3):284–90.

[159] Zhao H, Ito Y, Chappel J, et al. Synaptotagmin VII regulates bone remodeling by modulating osteoclast and osteoblast secretion. Dev Cell 2008;14(6):914–25.

[160] Abu-Amer Y, Teitelbaum SL, Chappel JC, Schlesinger P, Ross FP. Expression and regulation of RAB3 proteins in osteoclasts and their precursors. J Bone Miner Res 1999;14(11):1855–60.

[161] Zhao H, Laitala-Leinonen T, Parikka V, Vaananen HK. Downregulation of small GTPase Rab7 impairs osteoclast polarization and bone resorption. J Biol Chem 2001;276(42):39295–302.

[162] Chieregatti E, Meldolesi J. Regulated exocytosis: new organelles for non-secretory purposes. Nat Rev Mol Cell Biol 2005;6(2):181–7.

[163] Sato M, Grasser W, Endo N, et al. Bisphosphonate action. Alendronate localization in rat bone and effects on osteoclast ultrastructure. J Clin Invest 1991;88(6):2095–105.

[164] Jahn R, Scheller RH. SNAREs – engines for membrane fusion. Nat Rev Mol Cell Biol 2006;7(9):631–43.

[165] Meldolesi J, Chieregatti E. Fusion has found its calcium sensor. Nat Cell Biol 2004;6(6):476–8.

[166] Andrews NW, Chakrabarti S. There's more to life than neurotransmission: the regulation of exocytosis by synaptotagmin VII. Trends Cell Biol 2005;15(11):626–31.

[167] Stenbeck G, Horton MA. A new specialized cell-matrix interaction in actively resorbing osteoclasts. J Cell Sci 2000;113(Pt 9):1577–87.

[168] Yamamoto M, Fisher JE, Gentile M, et al. The integrin ligand echistatin prevents bone loss in ovariectomized mice and rats. Endocrinology 1998;139(3):1411–19.

[169] Engleman VW, Nickols GA, Ross FP, et al. A peptidomimetic antagonist of the alpha(v)beta3 integrin inhibits bone resorption in vitro and prevents osteoporosis in vivo. J Clin Invest 1997;99(9):2284–92.

[170] Murphy MG, Cerchio K, Stoch SA, et al. Effect of L-000845704, an alphaVbeta3 integrin antagonist, on markers of bone turnover and bone mineral density in postmenopausal osteoporotic women. J Clin Endocrinol Metab 2005;90(4):2022–8.

[171] McHugh KP, Hodivala-Dilke K, Zheng MH, et al. Mice lacking beta3 integrins are osteosclerotic because of dysfunctional osteoclasts. J Clin Invest 2000;105(4):433–40.

[172] Yarali N, Fisgin T, Duru F, Kara A. Osteopetrosis and Glanzmann's thrombasthenia in a child. Ann Hematol 2003;82(4):254–6.

[173] Horton MA, Massey HM, Rosenbergy N, et al. Upregulation of osteoclast alpha2beta1 integrin compensates for lack of alphavbeta3 vitronectin receptor in Iraqi-Jewish-type Glanzmann thrombasthenia. Br J Haematol 2003;122(6):950–7.

[174] Inoue M, Namba N, Chappel J, Teitelbaum SL, Ross FP. Granulocyte macrophage-colony stimulating factor reciprocally regulates alphav-associated integrins on murine osteoclast precursors. Mol Endocrinol 1998;12(12):1955–62.

[175] Lane NE, Yao W, Nakamura MC, et al. Mice lacking the integrin beta5 subunit have accelerated osteoclast maturation and increased activity in the estrogen-deficient state. J Bone Miner Res 2005;20(1):58–66.

[176] Takagi J, Petre BM, Walz T, Springer TA. Global conformational rearrangements in integrin extracellular domains in outside-in and inside-out signaling. Cell 2002;110(5):511–99.

[177] Faccio R, Grano M, Colucci S, et al. Localization and possible role of two different alpha v beta 3 integrin conformations in resting and resorbing osteoclasts. J Cell Sci 2002;115(Pt 14):2919–29.

[178] de Bruyn KM, Zwartkruis FJ, de Rooij J, Akkerman JW, Bos JL. The small GTPase Rap1 is activated by turbulence and is involved in integrin [alpha]IIb[beta]3-mediated cell adhesion in human megakaryocytes. J Biol Chem 2003;278(25):22412–17.

[179] Faccio R, Takeshita S, Zallone A, Ross FP, Teitelbaum SL. c-Fms and the alphavbeta3 integrin collaborate during osteoclast differentiation. J Clin Invest 2003;111(5):749–58.

[180] Weir EC, Lowik CW, Paliwal I, Insogna KL. Colony stimulating factor-1 plays a role in osteoclast formation and function in bone resorption induced by parathyroid hormone and parathyroid hormone-related protein. J Bone Miner Res 1996;11(10):1474–81.

[181] Zhao H, Ross FP, Teitelbaum SL. Unoccupied alpha(v)beta3 integrin regulates osteoclast apoptosis by transmitting a positive death signal. Mol Endocrinol 2005;19(3):771–80.

[182] Boyce BF, Yoneda T, Lowe C, Soriano P, Mundy GR. Requirement of pp60c-src expression for osteoclasts to form ruffled borders and resorb bone in mice. J Clin Invest 1992;90(4):1622–7.

[183] Schwartzberg PL, Xing L, Hoffmann O, et al. Rescue of osteoclast function by transgenic expression of kinase-deficient Src in src−/− mutant mice. Genes Dev 1997;11(21):2835–44.

[184] Sanjay A, Houghton A, Neff L, et al. Cbl associates with Pyk2 and Src to regulate Src kinase activity, alpha(v)beta(3) integrin-mediated signaling, cell adhesion, and osteoclast motility. J Cell Biol 2001;152(1):181–95.

[185] Tehrani S, Faccio R, Chandrasekar I, Ross FP, Cooper JA. Cortactin has an essential and specific role in osteoclast actin assembly. Mol Biol Cell 2006;17(7):2882–95.

[186] Gil-Henn H, Hestaing O, Sims NA, et al. Defective microtubule-dependent podosome organization in osteoclasts leads to increased bone density in Pyk2(−/−) mice. J Cell Biol 2007;178(6):1053–64.

[187] Lakkakorpi PT, Bett AJ, Lipfert L, Rodan GA, Duong le T. PYK2 autophosphorylation, but not kinase activity, is necessary for adhesion-induced association with c-Src, osteoclast spreading, and bone resorption. J Biol Chem 2003;278(13):11502–12.

[188] Yokouchi M, Kondo T, Sanjay A, et al. Src-catalyzed phosphorylation of c-Cbl leads to the interdependent ubiquitination of both proteins. J Biol Chem 2001;276(37):35185–93.

[189] Zou W, Reeve JL, Zhao H, Ross FP, Teitelbaum SL. Syk tyrosine 317 negatively regulates osteoclast function via the ubiquitin-protein isopeptide ligase activity of Cbl. J Biol Chem 2009;284 (28):18833–9.

[190] Obergfell A, Eto K, et al. Coordinate interactions of Csk, Src, and Syk kinases with [alpha]IIb[beta]3 initiate integrin signaling to the cytoskeleton. J Cell Biol 2002;157(2):265–75.

[191] Feng X, Novack DV, Faccio R, et al. A Glanzmann's mutation in beta 3 integrin specifically impairs osteoclast function. J Clin Invest 2001;107(9):1137–44.

[192] Reeve JL, Zou W, Liu Y, et al. SLP-76 couples Syk to the osteoclast cytoskeleton. J Immunol 2009; 183(3):1804–12.

[193] Miranti CK, Leng L, Maschberger P, Brugge JS, Shattil SJ. Identification of a novel integrin signaling pathway involving the kinase Syk and the guanine nucleotide exchange factor Vav1. Curr Biol 1998;8(24):1289–99.

[194] Razzouk S, Lieberherr M, Cournot G. Rac-GTPase, osteoclast cytoskeleton and bone resorption. Eur J Cell Biol 1999;78(4):249–55.

[195] Nakamura I, Lipfert L, Rodan GA, Le TD. Convergence of alpha(v)beta(3) integrin- and macrophage colony stimulating factor-mediated signals on phospholipase Cgamma in prefusion osteoclasts. J Cell Biol 2001;152(2):361–73.

[196] Tadokoro S, Shattil SJ, Tai V, et al. Talin binding to integrin beta tails: a final common step in integrin activation. Science 2003;302(5642):103–6.

# 7

# The Effects of Immune Cell Products (Cytokines and Hematopoietic Cell Growth Factors) on Bone Cells

Joseph Lorenzo

UNIVERSITY OF CONNECTICUT, HEALTH CENTER, FARMINGTON, CT, USA

## CHAPTER OUTLINE

Receptor activator of nuclear factor-κB ligand (RANKL), receptor activator of nuclear factor-κB (RANK), and osteoprotegerin (OPG) .................................................. 188
Macrophage colony-stimulating factor ........................................................................ 189
Additional colony-stimulating factors ......................................................................... 190
Interleukin-1 .................................................................................................................. 191
Tumor necrosis factor ................................................................................................... 192
Additional TNF superfamily members ........................................................................ 193
    Fas-ligand ................................................................................................................ 193
    TNF-related apoptosis-inducing ligand (TRAIL) ..................................................... 194
    CD40 ligand ............................................................................................................. 194
Interleukin-6 .................................................................................................................. 194
Additional interleukin-6 family members ................................................................... 195
    Interleukin-11 .......................................................................................................... 195
    Leukemia inhibitory factor ..................................................................................... 196
    Oncostatin M .......................................................................................................... 196
Interleukin-7 .................................................................................................................. 196
Interleukin-8 and other chemokines ........................................................................... 197
    Interleukin-8 ............................................................................................................ 198
    CCL2 ......................................................................................................................... 198
    CCL3 ......................................................................................................................... 198
    CCL9 ......................................................................................................................... 198
    CXCL12 and CXCR4 ................................................................................................ 199
    CX3CR1 .................................................................................................................... 199
    CCR1 ......................................................................................................................... 199
    CCR2 ......................................................................................................................... 199

Osteoimmunology. DOI: 10.1016/B978-0-12-375670-1.10007-X
Copyright © 2011 by Elsevier Inc. All rights reserved.

Interleukin-10 .......................................................................................................................... 200
Interleukin-12 .......................................................................................................................... 200
Interleukin-15 .......................................................................................................................... 201
Interleukin-17, interleukin-23, and interleukin-27 ............................................................... 201
Interleukin-18 .......................................................................................................................... 202
Interferons .............................................................................................................................. 202
Additional cytokines ............................................................................................................... 203
References ............................................................................................................................... 204

# Receptor activator of nuclear factor-κB ligand (RANKL), receptor activator of nuclear factor-κB (RANK), and osteoprotegerin (OPG)

The characterization of the functions of RANKL and its receptors (RANK and OPG) has contributed significantly to the development of the field of osteoimmunology. RANKL is a TNF superfamily ligand (TNFSF11) and together with its receptors RANK and OPG (TNFRSF11A and TNFRSF11B, respectively) forms the critical paracrine system, which regulates osteoclast formation [1–3].

The discovery of RANKL, which has potent activity as a stimulator of both the formation of osteoclasts from precursor cells and the bone-resorbing activity of mature osteoclasts, clarified our understanding of how stromal and osteoblastic cells regulate bone resorption [4, 5]. RANKL directs the terminal differentiation of osteoclast precursor cells into mature osteoclasts. In addition, it stimulates and maintains the resorptive activity of mature cells. Importantly, this activity could be reproduced in vitro in the absence of bone marrow stromal cells [4–6].

RANKL-deficient mice have significant osteopetrosis and no osteoclasts, but a normal number of monocyte/macrophages [7]. They also fail to erupt teeth, which is a common finding in all causes of developmental osteopetrosis. Because they lack an organized marrow cavity, RANKL-deficient mice develop extramedullary hematopoiesis in their spleen and liver [7, 8]. Marrow stromal and osteoblastic cells produce RANKL, and regulation of its mRNA expression in murine marrow cell cultures correlates with activation of osteoclastogenesis [9]. Many well-known osteotropic factors, including cytokines and hormones appear to exert their primary osteoclastogenic activity by inducing RANKL expression in mesenchymal lineage cells [1, 10]. Conversely, the shedding of membrane-bound RANKL appears to be a mechanism for inhibiting osteoblast-mediated osteoclast formation by removing RANKL from the osteoblast surface. The process depends on the expression of matrix metalloproteinase 14 (MMP14) [11] since osteoclasts were increased in mice deficient in this enzyme.

OPG is a novel secreted inhibitor of osteoclast formation, which acts as a decoy receptor for RANKL [5, 6, 12]. It was initially identified as a soluble factor, which induced osteopetrosis when transgenically overexpressed in mice [12] and inhibited osteoclastogenesis in vitro [12, 13]. In marrow it is produced by a variety of cells, including stromal cells, B lymphocytes, and dendritic cells [14]. Besides RANKL, OPG also binds the TNF-like ligand TRAIL (TNF-related ligand) [15]. Mice that lack OPG have severe osteoporosis, an increased number of osteoclasts, and arterial calcification [16, 17]. The latter finding highlights a potential genetic link between osteoporosis and vascular calcification [10]. Overexpression of OPG in transgenic mice caused osteopetrosis, decreased osteoclast numbers, and extramedullary hematopoiesis [12].

The biologically active receptor for RANKL is RANK. It was first identified on dendritic cells [18], but it is also present on osteoclast precursors and mature osteoclasts [19]. RANK expression at the RNA level is detected in a variety of cell types and tissues [18]. RANK-deficient mice were demonstrated to phenocopy the defect in osteoclast development that was observed in the RANKL-knockout mouse, confirming the exclusive specificity of RANKL for osteoclast-expressed RANK [19]. In humans, gain-of-function mutations in RANK were found to be associated with familial expansile osteolysis and expansible skeletal hyperphosphatasia, which are diseases caused by excessive formation and activity of osteoclasts [20–29].

While osteoclast-like cells can form in vitro in the absence of RANK or TRAF6 signaling when exposed to a cocktail of cytokines and growth factors [30–32], the significance of this in vitro finding is questionable since osteoclasts are not detected in RANK-deficient animals [19, 33]. In most instances, cytokines and growth factors other than RANKL, which are produced at sites of inflammation or physiologically during bone turnover, act as co-factors that enhance or modulate the response of osteoclasts and their precursors to RANKL-RANK stimulation [34–36]. However, it was recently demonstrated that the p100 precursor of NF-κB inhibits the ability of tumor necrosis factor alpha (TNFα) to stimulate osteoclastogenesis [37]. It was found that in the absence of p100, TNFα stimulated osteoclastogenesis both in vitro and in vivo by a mechanism that was dependent on TRAF3.

## Macrophage colony-stimulating factor

In addition to RANKL, macrophage colony-stimulating factor (M-CSF) (also known as CSF-1) is important for normal osteoclast formation. This cytokine was originally identified by its ability to regulate macrophage formation [38]. However, it was subsequently shown that a spontaneous mouse mutant (*op/op*) with a phenotype of absent osteoclasts and defective macrophage/monocyte formation was deficient in M-CSF [39–41]. Injection of M-CSF into *op/op* mice corrected the defect in osteoclast formation and bone resorption [42], as did expression of the protein specifically in osteoblastic cells [43].

Stimulators of bone resorption can increase the production of M-CSF in bone [44–46], and multiple transcripts of M-CSF are produced by alternative splicing [47, 48]. Paradoxically, M-CSF was reported to inhibit the in vitro ability of osteoblasts to produce RANKL and, in this way, decrease osteoclastogenesis. This effect may be mediated by the expression of c-Fms on osteoblast lineage cells [49]. In vivo treatment with M-CSF increased osteoclast number and bone resorption as well as the rate of fracture repair [50]. The membrane-bound form of M-CSF is regulated by stimulators of resorption, and facilitates the differentiation of osteoclasts from precursor cells [45, 51]. This may be significant since in marrow cultures soluble M-CSF inhibited osteoclast-like cell formation that was stimulated by 1,25-dihydroxyvitamin $D_3$ [52, 53].

M-CSF has multiple effects on osteoclast precursors. It stimulates their replication and differentiation [54, 55] and it regulates their motility [56]. The receptor for M-CSF is the tyrosine kinase c-Fms (CSF-1R) [57, 58]. Signaling through c-Fms is also mediated by the immunoreceptor tyrosine-based activation motif (ITAM)-containing protein DAP12 and by β-catenin [54].

The role of M-CSF in regulating osteoclast apoptosis has also been examined. Addition of M-CSF to mature osteoclast cultures prolongs their survival [59, 60]. This response may be important for the development of the osteopetrotic phenotype in *op/op* mice, since transgenic expression in myeloid cells of Bcl-2, which blocks apoptosis, partially reversed the defects in osteoclast and macrophage development in these animals [61]. The effects of M-CSF on osteoclasts has been linked to activation of an Na/HCO [62] cotransporter [63]. M-CSF also is a potent stimulator of RANK expression in osteoclast precursor cells [54, 55] and is critical for expanding the osteoclast precursor pool size [64].

## Additional colony-stimulating factors

Like M-CSF, the colony-stimulating factors, granulocyte macrophage colony-stimulating factor (GM-CSF) and interleukin-3 (IL-3), affect osteoclast differentiation [53, 65, 66]. Both have complex actions that are dependent on the lineage stage of the myeloid precursor cells with which they interact. IL-3 has multiple effects on in vitro osteoclastogenesis [67–70]. Both stimulatory and inhibitory effects are reported, depending on the cells that are examined and the culture conditions. IL-3 is also reported to inhibit osteoblast differentiation in multiple myeloma [71].

In early multipotential myeloid precursors, GM-CSF inhibits RANKL-mediated osteoclastogenesis [72, 73] and enhances the number of osteoclast precursor cells [74, 75]. This cytokine drives a common myeloid precursor cell toward the dendritic cell lineage [62, 72]. One mechanism by which GM-CSF produces this effect is by increasing the shedding of c-Fms (M-CSF receptor) through upregulation of "a disintegrin and metalloproteinase 17" (ADAM 17), which is also known as tumor necrosis factor alpha converting enzyme or TACE [76].

GM-CSF also inhibits expression of monocyte chemotactic protein 1 (MCP-1) by osteoclast precursor cells [77]. MCP-1 is a chemokine involved in osteoclast motility. Both GM-CSF and IL-3 also inhibit expression of TNF receptors on myeloid precursor cells [78]. However, in prefusion osteoclast precursors, which are myeloid cells that have been stimulated with RANKL and M-CSF for 3 days to a point where they will shortly (within 6 hours) fuse into osteoclasts, treatment with GM-CSF + M-CSF enhanced osteoclastogenesis and mimicked the response to RANKL + M-CSF [79]. This increased osteoclastogenesis was mediated by upregulation of the cell fusion protein, dendritic cell specific antigen (DC-STAMP) in the prefusion osteoclasts [79]. It has also been shown that if multipotential myeloid precursor cells are cultured sequentially with GM-CSF and then with M-CSF + RANKL, they form osteoclasts and, in some instances, act as dendritic cells, which can present antigen to activate T-lymphocyte responses [80–82]. IL-3 and GM-CSF may also support osteoclast differentiation by stimulating M-CSF production [70].

At relatively high doses granulocyte-colony-stimulating factor (G-CSF) decreases bone mass in rodents when injected systemically [83, 84]. This response appeared to result from increased osteoclast formation and decreased osteoblast function. Similar effects are seen in humans [85]. G-CSF also mobilizes the migration of hematopoietic precursor cells from the bone marrow into the circulation [86] and it increases the number of circulating osteoclast precursor cells [87], which is probably related to its ability to increase osteoclast resorptive activity.

In mice, overexpression of G-CSF inhibited the ability of osteoblasts to respond to bone morphogenetic protein [88]. Short-term treatment of mice with G-CSF decreased endosteal and trabecular osteoblasts by increasing apoptosis and inhibiting osteoblast precursor cell differentiation [89]. Mice overexpressing G-CSF have increased bone resorption, which, in contrast to wild-type mice, was not increased with ovariectomy [90]. Targeted deletion of ADAM17 in mice using Cre/Lox technology and the Sox9 promoter to target Cre recombinase to chondroprogenitor cells, produced a phenotype of enhanced G-CSF production, osteoporosis, and extramedullary hematopoiesis [91]. These results suggest that production of ADAM17 on osteoblasts and/or chondroblasts regulates expression of G-CSF.

# Interleukin-1

There are two separate interleukin-1 (IL-1) genes, IL-1α and IL-1β, which have identical activities [92]. IL-1 is a potent peptide stimulator of in vitro bone resorption [93], which has potent in vivo actions [94]. Its effects on resorption appear to be both direct on osteoclasts [95] and indirect through its ability to stimulate RANKL production [96]. In addition, both RANKL- and 1,25-dihydroxyvitamin $D_3$-stimulated osteoclast formation in vitro are mediated, in part, by their effects on IL-1 production [35, 97, 98]. IL-1 also increases prostaglandin synthesis in bone [93, 99], which may enhance the resorptive activity of IL-1, since prostaglandins are potent resorption stimuli [100]. Direct stimulation of osteoclastogenesis by IL-1 in mixed murine stromal and hematopoietic cell cultures is dependent on RANKL but not tumor necrosis factor expression in the stromal/osteoblastic cells [101].

IL-1 is produced in bone [102], and its activity is present in bone marrow serum [103, 104]. One source of bone-cell-derived IL-1 is osteoclast precursor cells, which produce IL-1 when they interact with bone matrix [105]. There is also a natural inhibitor of IL-1, IL-1 receptor antagonist (IL-1ra), which is an analog of IL-1 that binds but does not activate the biologically important type I IL-1 receptors [106–108].

There are two known receptors for IL-1: type I and type II [109]. All known biologic responses to IL-1 appear to be mediated exclusively through the type I receptor [110]. IL-1 receptor type I requires interaction with a second protein, IL-1 receptor accessory protein (IRAcP), to generate post-receptor signals [111–113]. Signaling through type I receptors involves activation of specific TRAFs and NF-κB [114, 115]. IL-1 receptor type II is a decoy receptor that prevents activation of type I receptors [116]. One report found a decrease in the bone mass of mice that were deficient in the bioactive type I IL-1 receptor [117]. However, this has not been our experience [118].

Expression of myeloid differentiation factor 88 (MyD88) but not Toll/interleukin-1 receptor domain-containing adaptor inducing interferon-beta (TRIF) was necessary for IL-1 to stimulate RANKL production in osteoblasts and prolong the survival of osteoclasts [119]. Survival of osteoclasts by treatment with IL-1 appears to require PI3-kinase/AKT and ERK [120].

The effects of IL-1 on bone in inflammatory states, such as rheumatoid arthritis, are multiple and mediated by both direct and indirect mechanisms. IL-1 stimulates bone resorption and inhibits bone formation through, respectively, its effects on osteoclasts and osteoblasts [121–123]. In addition, it stimulates the production of a variety of secondary factors in the bone microenvironment including prostaglandins and GM-CSF, which have complex effects of their own on bone cells [124]. IL-1 also has been reported both to inhibit and to stimulate production of osteoprotegerin in various osteoblastic cell models in vitro [125, 126]. The stimulatory effect was dependent on p38 and ERK MAP kinases [126].

## Tumor necrosis factor

Like IL-1, tumor necrosis factor (TNF) represents a family of two related polypeptides (α and β) that are the products of separate genes [127–131]. TNFα and TNFβ have similar biologic activities and are both potent stimulators of bone resorption [93, 132, 133].

In vivo administration of TNFα was shown to increase the serum calcium of mice [133] and to stimulate new osteoclast formation and bone resorption [134]. Like IL-1, TNF also enhances the formation of osteoclast-like cells in bone marrow culture [133] through its ability to increase RANKL production [135]. However, RANKL is not the only cytokine that TNF stimulates in bone and many of these enhance the response to RANKL. For example, TNF stimulates osteoclast formation in mixed stromal cell/osteoclast precursor cell cultures by a mechanism that was partially dependent on the production of IL-1 [136–138]. In addition, TNF-induced osteolysis was found to be dependent on M-CSF production [139].

TNF can also directly stimulate osteoclast formation in vitro by a mechanism that is independent of RANK, since it occurred in cells from RANK-deficient mice [30, 31, 140]. However, the significance of this in vitro finding is questionable since in vivo administration of TNF to RANK-deficient mice caused only an occasional osteoclast to form [33]. It was recently demonstrated that TNF can stimulate osteoclastogenesis in vivo in mice that are deficient in the p100 precursor protein of NF-κB, which is a critical signaling molecule in RANKL-mediated stimulation of osteoclastogenesis and bone resorption [37]. These results demonstrate that there are crucial differences between RANKL and TNF in the cascade of signaling molecules each produces in osteoclast precursor cells.

Like IL-1, TNF binds to two cell surface receptors, TNF receptor 1 or p55, and TNF receptor 2 or p75 [141]. In contrast to IL-1, both receptors transmit biologic responses. However, the principal effects appear to be mediated through the type TNF receptor 1 [142, 143]. Mice deficient in TNF receptor 1 and TNF receptor 2 have been produced [144–146]. These animals appear healthy and are not reported to have an abnormal bone phenotype.

TNF can stimulate the expression of c-fms, the receptor for M-CSF, in osteoclast precursor cells [147] and, through this mechanism, increase the number of these cells [148]. It also enhances RANK signaling, which activates osteoclasts and their precursor cells [34] and it enhances expression of the co-stimulatory molecule, paired Ig-like receptor A (PIR-A), which enhances NFATc1 activation [149].

TNF is a potent inhibitor of osteoblast function and bone formation [150–153]. This effect appears direct and mediated by downregulation of the critical transcription factor genes RUNX2 and osterix [154, 155] as well as type 1 collagen [156] and osteocalcin [157, 158], which are essential for differentiated osteoblast function. It also stimulates osteoblast apoptosis [159, 160] and suppresses production of insulin-like growth factor-1 (IGF-1) in osteoblasts [161].

## Additional TNF superfamily members

### Fas-ligand

Fas-ligand (FasL), which binds its receptor Fas on responsive cells, regulates apoptosis and other cellular processes in multiple cell types [162]. In osteoblasts FasL inhibits differentiation through a caspase 8-mediated mechanism [163]. In osteoclasts addition of FasL to cultures of osteoclast precursor cells, which were also treated with M-CSF and RANKL, increased osteoclast formation. Osteoclast precursors and mature osteoclasts express Fas and FasL [164]. Expression of Fas was upregulated by RANKL treatment in the RAW 264.7 osteoclast precursor cell line and treatment of mature osteoclasts with Fas-induced apoptosis [165]. However, in contrast to their similar effects on osteoclastogenesis in cultures of precursor cells, there appear to be divergent roles of RANKL and FasL on mature osteoclast apoptosis. At high concentrations, RANKL inhibited the ability of FasL to induce this response [166]. The effect that FasL deficiency has on bone mass is controversial. One group has found that bone mass is decreased in FasL-deficient mice [165], while another found it to be increased [167]. However, the significance of studying bone mass in Fas or FasL-deficient mice is probably minimal since these models have a generalized lymphproliferative disorder, which activates a wide variety of immune responses that affect bone and makes it difficult to interpret the results of these studies.

Most recently there has been controversy about the role that FasL has on the effects of estrogen on bone. One group found in mice that stimulation of estrogen receptor α in osteoclasts enhanced FasL production by these cells, which, in turn, reduced rates of bone loss by increasing osteoclast apoptosis [168]. In contrast, a second group failed to detect expression of FasL in osteoclasts. Rather, they found that estrogen enhanced FasL production in osteoblasts. They speculated that estrogen-induced increases in FasL production in osteoblasts regulated osteoclast apoptosis through a paracrine mechanism [169].

## TNF-related apoptosis-inducing ligand (TRAIL)

TNF-related apoptosis-inducing ligand (TRAIL) is another TNF-superfamily member that has a wide variety of activities. Its effects on osteoclast function and bone are also controversial. Some groups have found that treatment of osteoclasts with TRAIL induced apoptosis [170] through effects that were mediated by the receptor TRAIL-R2 which is also known as "death receptor 5" (DR5) [171, 172]. In vivo, injection of TRAIL for 8 days in 4-week-old mice induced an increase in bone mass. In vitro this effect was associated with an increase in the cyclin-dependent kinase inhibitor (CDKI), p27$^{Kip1}$ through effects of TRAIL on the ubiquitin-proteosome pathway [173]. TRAIL may also be a factor in the effects that myeloma has on osteoblasts [174]. However, other groups have failed to find either in vitro or in vivo effects of recombinant TRAIL on osteoclasts or in vivo effects on bone mass [175].

## CD40 ligand

CD40 ligand (CD40L) is involved in the differentiation of naïve T-lymphocytes into $T_H1$ effector cells [176]. In humans, deficiency of CD40L causes X-linked hyper IgM (XHIM) syndrome. Bones of XHIM patients develop spontaneous fractures and are osteopenic [177]. Activated T-lymphocytes from XHIM patients have normal RANKL and deficient INF-γ production, which may contribute to the decreased bone mass, which is seen in these patients [177]. In addition, expression of CD40L in rheumatoid arthritis synovial cells induced RANKL expression in these cells and enhanced their ability to stimulate osteoclastogenesis, which suggests that this mechanism is involved in the effects of rheumatoid arthritis on bone [178]. The ability of parathyroid hormone (PTH) to stimulate osteoclastogenesis has been reported to involve induction of CD40L on T lymphocytes and the subsequent induction of responses in stromal cells, expressing the receptor CD40 [179].

# Interleukin-6

IL-6, like IL-1 and TNF, has a wide variety of activities on immune cell function and on the replication and differentiation of a number of cell types [180, 181]. Osteoblastic cells (both rodent and human) produce IL-6 and IL-6 receptors [182, 183]. Another source of IL-6 in the bone microenvironment is bone marrow stromal cells, which can produce IL-6 after they are stimulated with IL-1 and TNF [184]. The receptor for IL-6 is composed of two parts: a specific IL-6-binding protein (IL-6 receptor), which can be either membrane-bound or soluble, and gp130, an activator protein that is common to a number of cytokine receptors [185]. Soluble IL-6 receptor binds IL-6, and this complex can then activate cells that contain the gp130 signal peptide [185, 186]. The shedding of IL-6 receptor from osteoblasts is stimulated by IL-1 and TNFα [187].

The ability of IL-6 to affect bone resorption in vitro is variable and depends on the assay system that is used as both stimulatory and inhibitory effects have been observed [183, 188–192]. It appears that a major effect of IL-6 is to regulate osteoclast progenitor cell differentiation into mature osteoclasts [193, 194]. IL-6 also directly stimulates both RANKL and OPG mRNA production in bone [195] and it enhances production of prostaglandins [196]. In addition, one publication suggested that IL-6 can stimulate osteoclastogenesis in vitro by a RANKL-independent mechanism [197]. However, two publications have found IL-6 to directly inhibit RANKL signaling in osteoclast precursor cells and decrease osteoclast formation [191, 192]. IL-6 appears to mediate some of the increased resorption and bone pathology that is seen in the clinical syndromes of Paget's disease [185], hypercalcemia of malignancy [198], fibrous dysplasia [199], giant cell tumors of bone [200], inflammatory states mediated by TNF or RANKL [201] and Gorham–Stout disease [202]. There have been conflicting data about the role of IL-6 in parathyroid hormone (PTH)-mediated responses in bone as some investigators have found it critical [203] while others have not [204].

## Additional interleukin-6 family members

IL-6 is a member of a group of cytokines that share the gp130 activator protein in their receptor complex [205, 206]. Each family member utilizes unique ligand receptors to generate specific binding. Signal transduction through these receptors utilizes the JAK/STAT (Janus kinase/signal transduction and activators of transcription) pathway [185].

### Interleukin-11

IL-11 is produced by bone cells in response to a variety of resorptive stimuli [207]. It stimulates osteoclast formation in murine bone marrow cultures [208] and bone resorption in a variety of in vitro assays [209, 210]. Interestingly, it has no effect on isolated mature osteoclasts. In mice deficient in the specific IL-11 receptor, trabecular bone mass is increased. This effect appears to result from decreased bone turnover, which is associated with decreased in vitro osteoclast formation and resorption [211]. After ovariectomy-induced estrogen withdrawal, IL-11 receptor-deficient and wild-type mice lose bone mass at similar rates. This result demonstrates that IL-11 signaling is not involved in the effects of estrogen withdrawal on bone mass [211]. However, IL-11 does appear involved in the ability of mechanical stress to stimulate osteoblast activity in vivo through its effects on Wnt signaling [212, 213].

## Leukemia inhibitory factor

Leukemia inhibitory factor (LIF) is produced by bone cells in response to a number of resorption stimuli [214–216]. The effects of LIF on bone resorption are variable. In a number of in vitro model systems, LIF stimulated resorption by a prostaglandin-dependent mechanism [217]. However, in vitro it had inhibitory effects [218, 219]. In neonatal murine calvaria cultures, LIF stimulated production of both RANKL and OPG [195].

Local injections of LIF in vivo were shown to increase both resorption and formation parameters, as well as the thickness of the treated bones [220]. In mice that lacked the specific LIF receptor (LIF-R) and hence could not respond to LIF, bone volume was reduced and osteoclast number was increased six-fold [221]. Animals lacking LIF are characterized by giant osteoclasts, which are produced through mechanisms involving Fra-2, hypoxia, hypoxia-induced factor $1\alpha$ (HIF$1\alpha$), and Bcl-2 [222]. Expression of LIF appears downregulated during osteoblast differentiation [223] by a mechanism that is mediated by micro-RNAs [224].

## Oncostatin M

Oncostatin M was demonstrated to stimulate multinuclear cell formation in murine and human bone marrow cultures [186, 225]. However, these cells appeared to be macrophage polykaryons and not osteoclasts [225]. In contrast, oncostatin M inhibited osteoclast-like cell formation that was stimulated by 1,25-dihydroxyvitamin $D_3$ in human marrow cultures [225] and it decreased bone resorption rates in fetal mouse long bone cultures [226]. In vivo, overexpression of oncostatin M in transgenic mice induced a phenotype of osteopetrosis [227]. Hence, it appears that oncostatin M is predominantly an inhibitor of osteoclast formation and bone resorption [228]. However, oncostatin M can affect cellular responses in bone by binding to either the oncostatin M receptor (OSRM), which produces inhibitory effects on resorption, or the leukemia inhibitory factor receptor (LIFR), which promotes bone formation through inhibition of sclerostin expression [229].

Oncostatin M stimulates mesenchymal cells to differentiate towards osteoblasts and osteocytes and inhibits their differentiation toward adipocytes [230, 231].

The role of all IL-6 family members in osteoclast formation has to be examined in the light of data demonstrating that mice lacking the gp130 activator protein have an increased number of osteoclasts in their bones compared with normal animals [232]. Since gp130 is an activator of signal transduction for all members of the IL-6 family, this result argues that at least some IL-6 family members have a predominantly inhibitory effect on osteoclast formation and bone resorption.

# Interleukin-7

IL-7 is a cytokine that has diverse effects on the hematopoietic and immunologic systems [233] and is best known for its non-redundant role in supporting B- and T-lymphopoiesis. Studies have demonstrated that IL-7 also plays an important role in the regulation of bone homeostasis [234, 235]. However, the precise nature of how IL-7 affects osteoclasts and osteoblasts is controversial, since it has a variety of actions in different target cells. Systemic administration of IL-7 upregulated osteoclast formation in human peripheral blood cells by increasing osteoclastogenic cytokine production in T lymphocytes [236]. Furthermore, mice with global overexpression of IL-7 had a phenotype of decreased bone mass with increased osteoclasts and no change in osteoblasts [237]. Significantly, IL-7 did not induce bone resorption and bone loss in T-cell-deficient nude mice in vivo [238]. In addition, treatment of mice with a neutralizing anti-IL-7 antibody inhibited ovariectomy-induced proliferation of early T-cell precursors in the thymus, demonstrating that ovariectomy upregulates T-cell development through IL-7. This latter effect may be a mechanism by which IL-7 regulates ovariectomy-induced bone loss [239]. However, the interpretation of results from in vivo IL-7 treatment studies is complicated by secondary effects of IL-7, which result from the production of bone-resorbing cytokines by T cells in response to activation by this cytokine [236, 238, 240].

In contrast with previously reported studies [234, 236, 238], we found differential effects of IL-7 on osteoclastogenesis [241]. IL-7 inhibited osteoclast formation in murine bone marrow cells that were cultured for 5 days with M-CSF and RANKL [241]. IL-7-deficient mice had markedly increased osteoclast number and decreased trabecular bone mass compared to wild-type controls [242]. In addition, we found that trabecular bone loss after ovariectomy was similar in wild-type and IL-7-deficient mice [242]. Curiously, IL-7 mRNA levels in bone increase with ovariectomy and this effect may be linked to alterations in osteoblast function with estrogen withdrawal [235, 243]. Treatment of newborn murine calvaria cultures with IL-7-inhibited bone formation, as did injection of IL-7 above the calvaria of mice in vivo [235]. When IL-7 was overexpressed locally by osteoblasts, trabecular bone mass was increased compared with wild-type mice [244]. Furthermore, targeted overexpression of IL-7 in IL-7-deficient mice rescued the osteoporotic bone phenotype of the IL-7-deficient mice [245]. These studies indicated that the actions of IL-7 on bone cells are dependent on whether IL-7 is delivered systemically or locally. Production of IL-7 by osteoblasts appears critical for normal B-lymphopoiesis [246]. Induction of this cytokine in osteoblasts is mediated by $G_s\alpha$-dependent signaling.

## Interleukin-8 and other chemokines

Recruitment and homing of myeloid cells often occurs under the direction of chemokines and their receptors. This superfamily of relatively small proteins induce interactions through cognate G-protein-coupled receptors to initiate cytoskeletal rearrangement, adhesion, and directional migration [247, 248]. Chemokines can be divided into four branches, depending on the spacing and sequence motifs of their first cysteine (C) residues. These are CXC, CC, C and $CX_3C$, where X is any other amino acid [249, 250]. The majority of chemokine–receptor interactions occur through the CC and CXC chemokines, which are referred to as major, while C and $CX_3C$ chemokines are referred to as minor.

### Interleukin-8

Many cells produce chemokines that bind specific G-protein-coupled receptors. Interleukin-8 (IL-8), a CXC chemokine, is produced by osteoclasts [251] and stimulates osteoclastogenesis and bone resorption by a mechanism that is reported to be independent of the RANKL pathway [252–254]. IL-8 may also be produced by certain cancers and stimulate lytic bone lesions in metastatic disease [252–254]. Effects of IL-8 on bone may be partially mediated by upregulation of nitric oxide synthase expression in osteoclasts [255]. Production of IL-8 is stimulated by RANKL and it, in turn, enhances RANKL-induced osteoclast formation [256].

### CCL2

CCL2 (monocyte chemoattractant protein-1, MCP-1) is a potent chemokine for monocytes, and a variety of other immune cells. Its receptor is CCR2, which is expressed at high levels on monocytes [257]. CCL2 is produced by osteoblasts in response to parathyroid hormone and proinflammatory cytokines and may regulate the recruitment of osteoclast precursors to bone [258–260]. CCL2 may be involved in tooth eruption, since it is expressed by dental follicle cells [261–263]. Among the factors that stimulate CCL2 in the dental follicle are PTHrP [264], PDGF-BB, and FGF-2 [265]. However, CCL2 is not critical for tooth eruption since there were only minor changes in the temporal pattern of this process in CCL2-deficient mice [266]. CCL2 is induced by RANKL in mononuclear precursor cells [267] and it enhances osteoclast-like cell formation in these cells [77]. However, cells induced by treatment of cultures with CCL2 alone, while multinucleated and calcitonin receptor positive, did not resorb bone unless they were also exposed to RANKL [77].

### CCL3

CCL3 (macrophage inflammatory protein-1α, MIP-1α) is a direct stimulator of osteoclastogenesis that is expressed in bone and bone marrow cells [268–271] by a mechanism that is proposed to be independent of RANK activation [272]. In addition, it can enhance RANKL expression by stromal cells and osteoblasts in the bone microenvironment [273]. CCL3 mediates some of the osteolytic activity of multiple myeloma [274–276]. Activation of osteoclastogenesis by CCL3 involves the receptors CCR1 and CCR5 [277]. CCL3 and IL-8 also stimulate motility but suppress resorption in mature osteoclasts [278].

### CCL9

CCL9 (macrophage inflammatory peptide gamma, MIP-1γ), like CCL3, binds the receptor CCR1, and regulates osteoclast function [279]. Injection of M-CSF to induce osteoclastogenesis and bone resorption in osteopetrotic tl/tl rats, which lack M-CSF, caused a rapid (within 2 days) upregulation of CCL9 and CCR1 in the bones, and a rapid increase in osteoclastogenesis [280]. Furthermore, antibodies to CCL9 ameliorated the ability of

M-CSF injections to stimulate osteoclastogenesis in this model. RANKL appears to be an inducer of CCL9 and CCR1 in osteoclasts [281]. Induction of CCR1 by RANKL is dependent on NFATc1 expression [282]. CCL9 and other chemokines that bind CCR1 (CCL3, CCL5, and CCL7) are produced by osteoclasts, osteoblasts, and their precursors in bone. In addition, expression of these chemokines in differentiating osteoblasts is induced by proinflammatory cytokines like IL-1 and TNF [283].

## CXCL12 and CXCR4

CXCL12 (stromal cell derived factor-1, SDF-1) and its receptor CXCR4 are involved in a variety of cellular processes including hematopoietic cell homeostasis and immune responses [284]. Osteoclast precursor cells express CXCR4 [285] and expression of this receptor is downregulated by differentiation of these cells towards the osteoclast lineage [286, 287]. Treatment of the cell line RAW 264.7 with CXCL12 induced expression of matrix metalloproteinase 9 (MMP9), which may be a mechanism for the migration of precursor cells towards bone [285]. In human osteoclast precursor cells, CXCL12 stimulated migration and enhanced osteoclastogenesis in response to RANKL and M-CSF [285, 286]. Expression of CXCL12 is upregulated in osteoclasts when they differentiate on a calcium phosphate matrix [286]. Production of CXCL12 may also be involved in the recruitment of precursor cells, which form giant cell tumors of bone [288], and in the increased osteolysis that is seen in multiple myeloma [289].

## CX3CR1

CX3CR1 is a chemokine receptor that is present on most early myeloid lineage cells including osteoclast precursors [290, 291]. Its ligand is CX3CL1, which is produced by osteoblasts and appears to stimulate the migration of osteoclast precursors to the bone surface [292]. Antibody neutralization of CX3CR1 in both in vitro and in vivo models blocked osteoclast formation [292].

## CCR1

Inhibition of CCR1 expression with siRNA or by blocking NFATc1 activation with cyclosporin A, inhibited migration of RAW 264.7 cells (a model for osteoclast precursors) and murine bone marrow cells in Boyden chambers [282]. Furthermore, inhibition of CCR1 signaling with a mutated form of CCL5, which blocks the binding of CCR1 to its ligands, prevented osteoclast-like cell formation in murine bone marrow cultures [282]. In addition, antibody neutralization of the CCR1 ligand CCL9 inhibited RANKL-induced osteoclastogenesis by 60–70% in murine bone marrow cultures [281].

## CCR2

CCR2, which binds CCL2 and CCL7 (MCP-3), the CCR1 ligand appears to have major effects on osteoclasts. Mice deficient in CCR2 have increased bone mass, decreased osteoclast number, size, and resorptive activity and no defect in osteoblast function [293]. In addition, osteoclast formation in vitro was attenuated in CCR2-deficient mice and these animals lost less bone with ovariectomy than did wild-type mice.

Additional chemokine receptors that are produced on osteoclasts include CCR3 and CCR5 [277, 279].

## Interleukin-10

Interleukin-10 (IL-10) is produced by activated T and B lymphocytes [294]. It is a direct inhibitor of osteoclastogenesis [295, 296] and osteoblastogenesis [297]. This effect is associated with increased tyrosine phosphorylation of a variety of proteins in osteoclast precursor cells [298]. The direct effects of IL-10 on RANKL-stimulated osteoclastogenesis are associated with decreases in NFATc1 expression and reduced translocation of this transcription factor into the nucleus [299] as well as suppressed c-Fos and c-Jun expression [300]. Administration of IL-10 may have utility as a mechanism to control wear-induced osteolysis [301]. In dental follicle cells, which function to regulate tooth eruption, in vitro treatment with IL-10 inhibited RANKL production and enhanced OPG [302]. Hence, there appears also to be an indirect effect of IL-10 on osteoclastogenesis that is mediated by its ability to regulate RANKL and OPG production.

Treatment of bone marrow cell cultures with IL-10 suppressed the production of osteoblastic proteins and prevented the onset of mineralization [297]. IL-10 also inhibited the formation of OCL in bone marrow cultures without affecting macrophage formation or the resorptive activity of mature osteoclasts [303]. This effect appears to involve the production of novel phosphotyrosine proteins in osteoclast precursor cells [298]. IL-10 also stimulates a novel inducible nitric oxide synthase [255].

4-1BB is an inducible T-cell co-stimulatory molecule, which interacts with 4-1BB ligand. In vitro treatment of RANKL-stimulated osteoclast precursor cells with 4-1BB ligand enhanced IL-10 production. In addition, expression of IL-10 was greater in RANKL-stimulated wild-type osteoclast precursor cell cultures than in cultured cells from 41-BB-deficient mice [304]. These results imply that some effects of IL-10 on osteoclasts may be mediated through interactions of 4-1BB and 4-1BB ligand.

## Interleukin-12

Interleukin-12 (IL-12) is a cytokine that is produced by myeloid and other cell types. It induces $T_H1$ differentiation in T lymphocytes and the subsequent expression of interferon-$\gamma$ (INF-$\gamma$) [305]. IL-12 has an inhibitory effect on osteoclastogenesis. However, the mechanisms by which this effect occurs in vitro are controversial. Some authors have demonstrated direct inhibitory effects of IL-12 on RANKL-stimulated osteoclastogenesis in purified primary osteoclast precursors and RAW 264.7 cells [306]. This effect was associated with inhibition of NFATc1 expression in the osteoclast precursor cells. Interestingly, the inhibitory effects of IL-12 on osteoclastogenesis were absent in cells that were pretreated with RANKL [306]. In contrast, others have found that the inhibitory effects of IL-12 on osteoclastogenesis are indirect. One group demonstrated that the inhibitory effects of IL-12 are mediated by T lymphocytes and do not involve production of interferon-γ (INF-γ) [307]. A second group disputes this result and found inhibition of osteoclastogenesis by IL-12 in cells from T-lymphocyte-depleted cultures and cells from T-lymphocyte-deficient nude mice [308]. The latter authors also demonstrated that antibody neutralization of INF γ blocked some of the inhibitory effects of IL-12 on RANKL-stimulated osteoclast formation.

The effects of IL-12 on TNFα-induced osteoclastogenesis have been examined in vivo [309]. It was found that osteoclastogenesis, which was stimulated by injection of TNFα over the calvaria of mice, was decreased when the mice were also treated with IL-12. Furthermore, this effect was not altered by antibody neutralization of T lymphocytes in the mice. Induction of Fas by TNFα and FasL by IL-12 in bone was critical for this response [310].

## Interleukin-15

Interleukin-15 (IL-15), like IL-7, is a member of the interleukin-2 superfamily and shares many activities with IL-2 including the ability to stimulate lymphocytes. It has been shown to enhance osteoclast progenitor cell number in culture [311]. Production of IL-15 by T lymphocytes has been linked to the increased osteoclastogenesis and bone destruction seen in the bone lesions of rheumatoid arthritis [312]. Polymorphisms of the IL-15 gene have also been linked to variations in bone mineral density in women [313].

## Interleukin-17, interleukin-23, and interleukin-27

Interleukin-17 (IL-17) is a family of related cytokines, which are unique and contain at least six members (A–F) [314]. IL-17E is also called interleukin-25 [315]. These cytokines are central for the development of the adaptive immune response and the products of a subset of CD4 T lymphocytes with a unique cytokine expression profile, termed $T_H17$. This is in contrast to the more established T-lymphocyte cytokine-expressing subsets $T_H1$ and $T_H2$.

IL-17A was initially identified as a stimulator of osteoclastogenesis in mixed cultures of mouse hematopoietic cells and osteoblasts [316]. This enhanced resorptive activity was mediated through increased production of prostaglandin and RANKL. The direct effects of IL-17A on the differentiation of osteoclast precursor cells is controversial with some investigators finding stimulatory effects [317] and others finding it to be inhibitory [318]. Production of IL-17A in rheumatoid arthritis appears involved in the production of activated osteoclasts and bone destruction in involved joints [316, 319, 320]. Effects of IL-17 on osteoclastogenesis and bone resorption are enhanced by TNFα, which is also produced in the inflamed joints of patients with rheumatoid arthritis [321]. Inhibition of IL-17A in an antigen-induced arthritis model reduced the joint and bone destruction that is typically seen and decreased production of RANKL, IL-1β, and TNFα in the involved lesions [322].

Interleukin-23 (IL-23) is an IL-12-related cytokine composed of one subunit of p40, which it shares with IL-12, and one subunit of p19, which is unique [323]. It is critical for the differentiation of the $T_H17$ subset of T lymphocytes along with TGF-β and IL-6 [324]. IL-23 appears most important for expanding the population of $T_H17$ T lymphocytes. This subset of T lymphocytes, which produces RANKL, has a high osteoclastogenic potential that is mediated by their production of IL-17 [325]. In an LPS-induced model of inflammatory bone destruction, it was found that there was markedly less bone loss in mice that were deficient either in IL-17 or IL-23 [325]. Hence, production of both is involved in the bone loss in this model. IL-23 induces RANKL expression in CD4 T lymphocytes [326] and RANK expression in osteoclast precursor cells [327]. However, the actions of IL-23 on bone are complex. IL-23-deficient mice have decreased bone mass and in one report IL-23 inhibited osteoclastogenesis through actions that were mediated by CD4 T lymphocytes [328, 329].

A related cytokine, IL-27, was found to have inhibitory effects on osteoclastogenesis in murine bone marrow cultures that were mediated by T lymphocytes [329]. However, in another study the inhibitory effects of IL-27 on RANKL-stimulated osteoclastogenesis were direct and mediated by inhibition of c-Fos [330].

## Interleukin-18

Interleukin-18 (IL-18) is similar to IL-1 in its structure and a member of the IL-1 superfamily [331]. IL-18 synergizes with IL-12 to induce INF-γ production [332] and its levels are increased at sites of inflammation such as rheumatoid arthritis [333]. It is expressed by osteoblastic cells and is reported to inhibit osteoclast formation through a variety of mechanisms. These include its ability to stimulate GM-CSF [100], which is produced by T cells in response to IL-18 treatment [334]. It also stimulates INF-γ production in vivo in bone [335] and its inhibitory effects on osteoclastogenesis and bone resorption are enhanced by co-treatment with IL-12 [336]. IL-18 has been shown to indirectly stimulate osteoclastogenesis through its effects on T lymphocytes [337]. Finally, IL-18 has been shown to increase production of OPG [338]. In IL-18-overexpressing transgenic mice, osteoclasts were decreased; although, curiously, so was bone mass. These results indicate that there also may be effects of IL-18 on bone growth [335]. In confirmation of this hypothesis, it was demonstrated that PTH treatment of osteoblasts stimulated IL-18 production. In addition, the anabolic effects of intermittent PTH treatment on trabecular bone mass in IL-18-deficient mice were reduced [339]. IL-18 is also a mitogen for osteoblastic cells in vitro [340].

## Interferons

Interferon-γ (INF-γ) is a type II interferon with a wide variety of biologic activities. In vitro, INF-γ has inhibitory actions on bone resorption [341, 342]. These appear to be direct and are mediated by its effects on osteoclast progenitor cells. INF-γ inhibits the ability of 1,25-dihydroxyvitamin $D_3$, parathyroid hormone, and IL-1 to stimulate the formation of osteoclast-like cells in cultures of human bone marrow [343]. INF-γ also inhibits RANK signaling by accelerating the degradation of TRAF6 through activation of the ubiquitin/proteasome system [344]. However, it does not directly inhibit resorption in mature osteoclasts [345]. INF-γ is also reported to have stimulatory effects on resorption through its ability to stimulate RANKL and TNF-α production in T lymphocytes [346]. It is an inhibitor of osteoblast proliferation [340, 347, 348] and has variable effects on osteoblast differentiation [347, 349, 350].

The effects of INF-γ on bone in vivo are variable as both inhibitory and stimulatory effects have been reported. In mice with collagen-induced arthritis, loss of the INF-γ receptor (INF-γR) leads to increased bone destruction [351, 352]. Similarly, in mice that are injected over their calvaria with bacterial endotoxin, which activates Toll-like receptors (TLRs), loss of INF-γR resulted in an enhanced resorptive response [344]. This result is consistent with more recent findings demonstrating that the inhibitory effects of INF-γ on osteoclastogenesis are enhanced by activation of TLRs [353].

In contrast, intraperitoneal injection of INF-γ for 8 days in rats induced osteopenia [354]. In patients who have osteopetrosis, because they produce defective osteoclasts, administration of INF-γ stimulated bone resorption and appeared to partially reverse the disease. The latter effects are possibly due to the ability of INF-γ to stimulate osteoclast superoxide synthesis [355, 356], osteoclast formation in vivo [357], or a generalized immune response [358].

Type I interferons (INF-α and INF-β)) are typically produced in response to invading pathogens [359]. Mice deficient in the INF-α/β receptor component IFNAR1 have a reduction in trabecular bone mass and an increase in osteoclasts [360]. RANKL induces INF-β in osteoclasts, and INF-β, in turn, inhibits RANKL-mediated osteoclastogenesis by decreasing c-fos expression [360]. INF-α has also been shown to inhibit bone resorption in vitro although its mechanism of action is not as well studied as that of INF-γ and -β [361]. In vivo, INF-α had no effect on bone turnover [362].

## Additional cytokines

IL-4 and IL-13 are members of a group of locally acting factors that have been termed "inhibitory cytokines". The effects of IL-4 and IL-13 seem related and appear to affect both osteoblasts and osteoclasts. Transgenic mice that overexpress IL-4 had a phenotype of osteoporosis [363]. This effect may result from both an inhibition of osteoclast formation and activity [364, 36] and an inhibition of bone formation [366]. IL-13 and IL-4 inhibited IL-1-stimulated bone resorption by decreasing the production of prostaglandins and the activity of cyclooxygenase-2 [367]. They induce cell migration (chemotaxis) in osteoblastic cells [368] and they regulate the ability of osteoblasts and vascular endothelial cells to control OPG and RANKL production [369–371]. The direct inhibitory effects of IL-4 on osteoclast precursor cell maturation are more potent than that of IL-13 and involve effects on STAT6, NF-κB, peroxisome proliferator-activated receptor γ1, mitogen-activated protein kinase signaling, $Ca^{++}$ signaling, NFATc1, and c-Fos [370, 372–376]. IL-4 along with GM-CSF induces multipotential myeloid cell differentiation towards the dendritic cell lineage and away from the osteoclast lineage [76].

IL-32 is a cytokine that is involved in innate and adoptive immunity. It is produced by T lymphocytes, natural killer cells, and epithelial cells [377]. IL-32 stimulated the formation of multinuclear cells that were TRAP- and vitronectin receptor-positive but did not resorb. In addition, it inhibited resorption that was stimulated by RANKL [378].

Macrophage migration inhibitory factor (MIF) was initially identified as an activity in conditioned medium from activated T lymphocytes that inhibited macrophage migration in capillary tube assays [379]. Once purified and cloned [380], it became available for functional studies and was shown to have a variety of activities. In addition to T lymphocytes, it is produced by pituitary cells and activated macrophages. Mice that overexpress MIF globally were found to have high turnover osteoporosis [380]. In contrast, MIF-deficient mice failed to lose bone mass or increase osteoblast or osteoclast number in bone with ovariectomy [382]. Hence, MIF appears to be another mediator of the effects of estrogen withdrawal on bone. Estrogen downregulates MIF expression in activated macrophages [383]. A similar response may occur in bone or bone marrow and mediate some of the effects that ovariectomy has on bone mass. MIF is made by osteoblasts [384] and its production by these cells was upregulated by a variety of growth factors including TGF-β, FGF-2, IGF-II, and fetal calf serum [385]. In vitro, MIF increased MMP9 and MMP13 expression in osteoblasts [386] and inhibited RANKL-stimulated osteoclastogenesis by decreasing the fusion of precursors, possibly through its ability to inhibit the migration of these cells [387].

# References

[1] Suda T, Takahashi N, Udagawa N, Jimi E, Gillespie MT, Martin TJ. Modulation of osteoclast differentiation and function by the new members of the tumor necrosis factor receptor and ligand families. Endocr Rev 1999 Jun;20(3):345–57.

[2] Teitelbaum SL. Bone resorption by osteoclasts. Science 2000 Sep 1;289(5484):1504–8.

[3] Walsh MC, Kim N, Kadono Y, Rho J, Lee SY, Lorenzo J, et al. Osteoimmunology: interplay between the immune system and bone metabolism. Annu Rev Immunol 2006;24:33–63.

[4] Fuller K, Wong B, Fox S, Choi Y, Chambers TJ. TRANCE is necessary and sufficient for osteoblast-mediated activation of bone resorption in osteoclasts. J Exp Med 1998 Sep 7;188(5):997–1001.

[5] Lacey DL, Timms E, Tan HL, Kelley MJ, Dunstan CR, Burgess T, et al. Osteoprotegerin ligand is a cytokine that regulates osteoclast differentiation and activation. Cell 1998 Apr 17;93(2):165–76.

[6] Yasuda H, Shima N, Nakagawa N, Yamaguchi K, Kinosaki M, Mochizuki S, et al. Osteoclast differentiation factor is a ligand for osteoprotegerin osteoclastogenesis-inhibitory factor and is identical to TRANCE/RANKL. Proc Natl Acad Sci U S A 1998;95(7):3597–602.

[7] Kong YY, Yoshida H, Sarosi I, Tan HL, Timms E, Capparelli C, et al. OPGL is a key regulator of osteoclastogenesis, lymphocyte development and lymph-node organogenesis. Nature 1999 Jan 28;397(6717):315–23.

[8] Kim N, Odgren PR, Kim DK, Marks Jr SC, Choi Y. Diverse roles of the tumor necrosis factor family member TRANCE in skeletal physiology revealed by TRANCE deficiency and partial rescue by a lymphocyte-expressed TRANCE transgene. Proc Natl Acad Sci U S A 2000 Sep 26;97(20):10905–10.

[9] Lee SK, Lorenzo JA. Parathyroid hormone stimulates TRANCE and inhibits osteoprotegerin messenger ribonucleic acid expression in murine bone marrow cultures: correlation with osteoclast-like cell formation. Endocrinology 1999;140(8):3552–61.

[10] Boyle WJ, Simonet WS, Lacey DL. Osteoclast differentiation and activation. Nature 2003 May 15;423(6937):337–42.

[11] Hikita A, Yana I, Wakeyama H, Nakamura M, Kadono Y, Oshima Y, et al. Negative regulation of osteoclastogenesis by ectodomain shedding of receptor activator of NF-kappaB ligand. J Biol Chem 2006;281(48):36846–55.

[12] Simonet WS, Lacey DL, Dunstan CR, Kelley M, Chang MS, Luthy R, et al. Osteoprotegerin: a novel secreted protein involved in the regulation of bone density. Cell 1997 Apr 18;89(2):309–19.

[13] Tsuda E, Goto M, Mochizuki S, Yano K, Kobayashi F, Morinaga T, et al. Isolation of a novel cytokine from human fibroblasts that specifically inhibits osteoclastogenesis. Biochem Biophys Res Commun 1997 May 8;234(1):137–42.

[14] Yun TJ, Chaudhary PM, Shu GL, Frazer JK, Ewings MK, Schwartz SM, et al. OPG/FDCR-1, a TNF receptor family member, is expressed in lymphoid cells and is up-regulated by ligating CD40. J Immunol 1998 Dec 1;161(11):6113–21.

[15] Emery JG, McDonnell P, Burke MB, Deen KC, Lyn S, Silverman C, et al. Osteoprotegerin is a receptor for the cytotoxic ligand TRAIL [In Process Citation]. J Biol Chem 1998;273(23):14363–7.

[16] Bucay N, Sarosi I, Dunstan CR, Morony S, Tarpley J, Capparelli C, et al. Osteoprotegerin-deficient mice develop early onset osteoporosis and arterial calcification. Genes Dev 1998 May 1;12 (9):1260–8.

[17] Mizuno A, Amizuka N, Irie K, Murakami A, Fujise N, Kanno T, et al. Severe osteoporosis in mice lacking osteoclastogenesis inhibitory factor/osteoprotegerin [In Process Citation]. Biochem Biophy Res Comm 1998;247(3):610–15.

[18] Anderson DM, Maraskovsky E, Billingsley WL, Dougall WC, Tometsko ME, Roux ER, et al. A homologue of the TNF receptor and its ligand enhance T-cell growth and dendritic-cell function. Nature 1997 Nov 13;390(6656):175–9.

[19] Dougall WC, Glaccum M, Charrier K, Rohrbach K, Brasel K, De Smedt T, et al. RANK is essential for osteoclast and lymph node development. Genes Dev 1999 Sep 15;13(18):2412–24.

[20] Cecchini MG, Hofstetter W, Halasy J, Wetterwald A, Felix R. Role of CSF-1 in bone and bone marrow development. Mol Repr Devel 1997;46(1):75–83.

[21] Hayashi S, Miyamoto A, Yamane T, Kataoka H, Ogawa M, Sugawara S, et al. Osteoclast precursors in bone marrow and peritoneal cavity. J Cell Physiol 1997;170(3):241–7.

[22] Muguruma Y, Lee MY. Isolation and characterization of murine clonogenic osteoclast progenitors by cell surface phenotype analysis. Blood 1998;91(4):1272–9.

[23] Suda T, Takahashi N, Martin TJ. Modulation of osteoclast differentiation: Update 1995. Endcr Rev 1995;13(1):66–80.

[24] Takahashi N, Akatsu T, Udagawa N, Sasaki T, Yamaguchi A, Moseley JM, et al. Osteoblastic cells are involved in osteoclast formation. Endocrinology 1988;123:2600–2.

[25] Hughes AE, Ralston SH, Marken J, Bell C, MacPherson H, Wallace RG, et al. Mutations in TNFRSF11A, affecting the signal peptide of RANK, cause familial expansile osteolysis. Nat Genet 2000 Jan;24(1):45–8.

[26] Whyte MP, Hughes AE. Expansile skeletal hyperphosphatasia is caused by a 15-base pair tandem duplication in TNFRSF11A encoding RANK and is allelic to familial expansile osteolysis. J Bone Miner Res 2002 Jan;17(1):26–9.

[27] Whyte MP, Mumm S. Heritable disorders of the RANKL/OPG/RANK signaling pathway. J Musculoskelet Neuronal Interact 2004 Sep;4(3):254–67.

[28] Whyte MP, Obrecht SE, Finnegan PM, Jones JL, Podgornik MN, McAlister WH, et al. Osteoprotegerin deficiency and juvenile Paget's disease. N Engl J Med 2002 Jul 18;347(3):175–84.

[29] Whyte MP, Reinus WR, Podgornik MN, Mills BG. Familial expansile osteolysis (excessive RANK effect) in a 5-generation American kindred. Medicine (Baltimore) 2002 Mar;81(2):101–21.

[30] Kim N, Kadono Y, Takami M, Lee J, Lee SH, Okada F, et al. Osteoclast differentiation independent of the TRANCE-RANK-TRAF6 axis. J Exp Med 2005;202(5):589–95.

[31] Kobayashi K, Takahashi N, Jimi E, Udagawa N, Takami M, Kotake S, et al. Tumor necrosis factor alpha stimulates osteoclast differentiation by a mechanism independent of the ODF/RANKL-RANK interaction. J Exp Med 2000;191(2):275–86.

[32] Kudo O, Fujikawa Y, Itonaga I, Sabokbar A, Torisu T, Athanasou NA. Proinflammatory cytokine (TNFalpha/IL-1alpha) induction of human osteoclast formation. J Pathol 2002;198(2):220–7.

[33] Li J, Sarosi I, Yan XQ, Morony S, Capparelli C, Tan HL, et al. RANK is the intrinsic hematopoietic cell surface receptor that controls osteoclastogenesis and regulation of bone mass and calcium metabolism. Proc Natl Acad Sci U S A 2000;97(4):1566–71.

[34] Lam J, Takeshita S, Barker JE, Kanagawa O, Ross FP, Teitelbaum SL. TNF-alpha induces osteoclastogenesis by direct stimulation of macrophages exposed to permissive levels of RANK ligand. J Clin Invest 2000;106(12):1481–8.

[35] Lee SK, Gardner AE, Kalinowski JF, Jastrzebski SL, Lorenzo JA. RANKL-stimulated osteoclast-like cell formation in vitro is partially dependent on endogenous interleukin-1 production. Bone 2005;38(5):678–85.

[36] Takita C, Fujikawa Y, Itonaga I, Taira H, Kawashima M, Torisu T. Infliximab acts directly on human osteoclast precursors and enhances osteoclast formation induced by receptor activator of nuclear factor kappaB ligand in vitro. Mod Rheumatol 2005;15(2):97–103.

[37] Yao Z, Xing L, Boyce BF. NF-kappaB p100 limits TNF-induced bone resorption in mice by a TRAF3-dependent mechanism. J Clin Invest 2009 Sep 21;21(38716).

[38] Roth P, Stanley ER. The biology of CSF-1 and its receptor. Curr Top Microbiol Immunol 1992;181:141–67.

[39] Felix R, Cecchini MG, Hofstetter W, Elford PR, Stutzer A, Fleisch H. Impairment of macrophage colony-stimulating factor production and lack of resident bone marrow macrophages in the osteopetrotic op/op mouse. J Bone Min Res 1990;5:781–9.

[40] Wiktor Jedrzejczak W, Bartocci A, Ferrante Jr AW, Ahmed Ansari A, Sell KW, Pollard JW, et al. Total absence of colony-stimulating factor 1 in the macrophage-deficient osteopetrotic (op/op) mouse. Proc Natl Acad Sci U S A 1990;87:4828–32.

[41] Yoshida H, Hayashi S, Kunisada T, Ogawa M, Nishikawa S, Okamura H, et al. The murine mutation osteopetrosis is in the coding region of the macrophage colony stimulating factor gene. Nature 1990;345:442–4.

[42] Felix R, Cecchini MG, Fleisch H. Macrophage colony stimulating factor restores in vivo bone resorption in the op/op osteopetrotic mouse. Endocrinology 1990;127:2592–4.

[43] Abboud SL, Woodruff K, Liu C, Shen V, Ghosh-Choudhury N. Rescue of the osteopetrotic defect in op/op mice by osteoblast-specific targeting of soluble colony-stimulating factor-1. Endocrinology 2002;143(5):1942–9.

[44] Felix R, Fleisch H, Elford PR. Bone-resorbing cytokines enhance release of marophage colony-stimulating activity by the osteoblastic cell MC3T3- E1. Calcif Tissue Int 1989;44:356–60.

[45] Rubin J, Fan X, Thornton D, Bryant R, Biskobing D. Regulation of murine osteoblast macrophage colony-stimulating factor production by 1,25(OH)2D3. Calcif Tissue Int 1996;59(4):291–4.

[46] Weir EC, Horowitz MC, Baron R, Centrella M, Kacinski BM, Insogna KL. Macrophage colony-stimulating factor release and receptor expression in bone cells. J Bone Miner Res 1993;8(12):1507–18.

[47] Cerretti DP, Wignall J, Anderson D, Tushinski RJ, Gallis B, Cosman D. Membrane bound forms of human macrophage colony stimulating factor (M-CSF, CSF-1). Prog Clin Biol Res 1990;352:63–70.

[48] Cerretti DP, Wignall J, Anderson D, Tushinski RJ, Gallis BM, Stya M, et al. Human macrophage-colony stimulating factor: alternative RNA and protein processing from a single gene. Mol Immunol 1988;25(8):761–70.

[49] Wittrant Y, Gorin Y, Mohan S, Wagner B, Abboud-Werner SL. Colony-stimulating factor-1 (CSF-1) directly inhibits receptor activator of nuclear factor-{kappa}B ligand (RANKL) expression by osteoblasts. Endocrinology 2009 Oct 9;9:9.

[50] Sarahrudi K, Mousavi M, Grossschmidt K, Sela N, Konig F, Vecsei V, et al. The impact of colony-stimulating factor-1 on fracture healing: an experimental study. J Orthop Res 2009 Jan;27(1):36–41.

[51] Yao GQ, Sun BH, Hammond EE, Spencer EN, Horowitz MC, Insogna KL, et al. The cell-surface form of colony-stimulating factor-1 is regulated by osteotropic agents and supports formation of multinucleated osteoclast-like cells. J Biol Chem 1998;273(7):4119–28.

[52] Fan X, Biskobing DM, Fan D, Hofstetter W, Rubin J. Macrophage colony stimulating factor down-regulates MCSF-receptor expression and entry of progenitors into the osteoclast lineage. J Bone Min Res 1997;12(9):1387–95.

[53] Shinar DM, Sato M, Rodan GA. The effect of hemopoietic growth factors on the generation of osteoclast-like cells in mouse bone marrow cultures. Endocrinology 1990;126:1728–35.

[54] Otero K, Turnbull IR, Poliani PL, Vermi W, Cerutti E, Aoshi T, et al. Macrophage colony-stimulating factor induces the proliferation and survival of macrophages via a pathway involving DAP12 and beta-catenin. Nat Immunol 2009;10(7):734–43.

[55] Arai F, Miyamoto T, Ohneda O, Inada T, Sudo T, Brasel K, et al. Commitment and differentiation of osteoclast precursor cells by the sequential expression of c-Fms and receptor activator of nuclear factor kappaB (RANK) receptors. J Exp Med 1999;190(12):1741–54.

[56] Novack DV, Faccio R. Osteoclast motility: putting the brakes on bone resorption. Ageing Res Rev 2009 Sep 26;26:26.

[57] Sherr CJ, Ashmun RA, Downing JR, Ohtsuka M, Quan SG, Golde DW, et al. Inhibition of colony-stimulating factor-1 activity by monoclonal antibodies to the human CSF-1 receptor. Blood 1989;73(7):1786–93.

[58] Ashmun RA, Look AT, Roberts WM, Roussel MF, Seremetis S, Ohtsuka M, et al. Monoclonal antibodies to the human CSF-1 receptor (c-fms proto-oncogene product) detect epitopes on normal mononuclear phagocytes and on human myeloid leukemic blast cells. Blood 1989;73(3):827–37.

[59] Fuller K, Owens JM, Jagger CJ, Wilson A, Moss R, Chambers TJ. Macrophage colony-stimulating factor stimulates survival and chemotactic behavior in isolated osteoclasts. J Exp Med 1993;178:1733–44.

[60] Jimi E, Shuto T, Koga T. Macrophage colony-stimulating factor and interleukin-1α maintain the survival of osteoclast-like cells. Endocrinology 1995;136:808–11.

[61] Lagasse E, Weissman IL. Enforced expression of Bcl-2 in monocytes rescues macrophages and partially reverses osteopetrosis in op/op mice. Cell 1997;89(7):1021–31.

[62] Alnaeeli M, Penninger JM, Teng YT. Immune interactions with CD4+ T cells promote the development of functional osteoclasts from murine CD11c+ dendritic cells. J Immunol 2006;177(5):3314–26.

[63] Bouyer P, Sakai H, Itokawa T, Kawano T, Fulton CM, Boron WF, et al. Colony-stimulating factor-1 increases osteoclast intracellular pH and promotes survival via the electroneutral Na/HCO3 cotransporter NBCn1. Endocrinology 2007;148(2):831–40.

[64] Jacquin C, Gran DE, Lee SK, Lorenzo JA, Aguila HL. Identification of multiple osteoclast precursor populations in murine bone marrow. J Bone Min Res 2006;21(1):67–77.

[65] Lorenzo JA, Sousa SL, Fonseca JM, Hock JM, Medlock ES. Colony-stimulating factors regulate the development of multinucleated osteoclasts from recently replicated cells in vitro. J Clin Invest 1987 Jul;80(1):160–4.

[66] MacDonald BR, Mundy GR, Clark S, Wang EA, Kuehl TJ, Stanley ER, et al. Effects of human recombinant CSF-GM and highly purified CSF-1 on the formation of multinucleated cells with osteoclast characteristics in long-term bone marrow cultures. J Bone Miner Res 1986;1:227–33.

[67] Khapli SM, Mangashetti LS, Yogesha SD, Wani MR. IL-3 acts directly on osteoclast precursors and irreversibly inhibits receptor activator of NF-kappaB ligand-induced osteoclast differentiation by diverting the cells to macrophage lineage. J Immunol 2003;171(1):142–51.

[68] Yogesha SD, Khapli SM, Wani MR. Interleukin-3 and granulocyte-macrophage colony-stimulating factor inhibits tumor necrosis factor (TNF)-alpha-induced osteoclast differentiation by downregulation of expression of TNF receptors 1 and 2. J Biol Chem 2005;280(12):11759–69.

[69] Enelow RI, Sullivan GW, Carper HT, Mandell GL. Induction of multinucleated giant cell formation from in vitro culture of human monocytes with interleukin-3 and interferon-gamma: comparison with other stimulating factors. Am J Respir Cell Mol Biol 1992 Jan;6(1):57–62.

[70] Fujikawa Y, Sabokbar A, Neale SD, Itonaga I, Torisu T, Athanasou NA. The effect of macrophage-colony stimulating factor and other humoral factors (interleukin-1, -3, -6, and -11, tumor necrosis factor-alpha, and granulocyte macrophage-colony stimulating factor) on human osteoclast formation from circulating cells. Bone 2001 Mar;28(3):261–7.

[71] Ehrlich LA, Chung HY, Ghobrial I, Choi SJ, Morandi F, Colla S, et al. IL-3 is a potential inhibitor of osteoblast differentiation in multiple myeloma. Blood 2005;106(4):1407–14.

[72] Khapli SM, Mangashetti LS, Yogesha SD, Wani MR. IL-3 acts directly on osteoclast precursors and irreversibly inhibits receptor activator of NF-kappaB ligand-induced osteoclast differentiation by diverting the cells to macrophage lineage. J Immunol 2003;171(1):142–51.

[73] Udagawa N, Horwood NJ, Elliott J, Mackay A, Owens J, Okamura H, et al. Interleukin-18 (interferon-gamma-inducing factor) is produced by osteoblasts and acts via granulocyte/macrophage colony-stimulating factor and not via interferon-gamma to inhibit osteoclast formation. J Exp Med 1997;185(6):1005–12.

[74] Kurihara N, Chenu C, Miller M, Civin C, Roodman GD. Identification of committed mononuclear precursors for osteoclast-like cells formed in long term human marrow cultures. Endocrinology 1990;126(5):2733–41.

[75] Takahashi N, Udagawa N, Akatsu T, Tanaka H, Shionome M, Suda T. Role of colony-stimulating factors in osteoclast development. J Bone Miner Res 1991;6(9):977–85.

[76] Hiasa M, Abe M, Nakano A, Oda A, Amou H, Kido S, et al. GM-CSF and IL-4 induce dendritic cell differentiation and disrupt osteoclastogenesis through M-CSF receptor shedding by up-regulation of TNF-{alpha} converting enzyme (TACE). Blood 2009 Sep 17;17:17.

[77] Kim MS, Day CJ, Selinger CI, Magno CL, Stephens SJ, Morrison NA. MCP-1 induced human osteoclast-like cells are TRAP, NFATc1 and calcitonin receptor positive but require RANKL for bone resorption. J Biol Chem 2005.

[78] Yogesha SD, Khapli SM, Wani MR. Interleukin-3 and granulocyte-macrophage colony-stimulating factor inhibits tumor necrosis factor (TNF)-alpha-induced osteoclast differentiation by downregulation of expression of TNF receptors 1 and 2. J Biol Chem 2005;280(12):11759–69.

[79] Lee MS, Kim HS, Yeon JT, Choi SW, Chun CH, Kwak HB, et al. Granulocyte-macrophage colony-stimulating factor regulates fusion of mononuclear osteoclasts into bone-resorbing osteoclasts by activating the Ras/ERK pathway. J Immunol 2009.

[80] Lari R, Fleetwood AJ, Kitchener PD, Cook AD, Pavasovic D, Hertzog PJ, et al. Macrophage lineage phenotypes and osteoclastogenesis – complexity in the control by GM-CSF and TGF-beta. Bone 2007;40(2):323–36.

[81] Nomura K, Kuroda S, Yoshikawa H, Tomita T. Inflammatory osteoclastogenesis can be induced by GM-CSF and activated under TNF immunity. Biochem Biophys Res Commun 2008.

[82] Alnaeeli M, Teng YT. Dendritic cells: a new player in osteoimmunology. Curr Mol Med 2009 Sep;9(7):893–910.

[83] Soshi S, Takahashi HE, Tanizawa T, Endo N, Fujimoto R, Murota K. Effect of recombinant human granulocyte colony-stimulating factor (rh G-CSF) on rat bone: inhibition of bone formation at the endosteal surface of vertebra and tibia. Calcif Tissue Int 1996;58(5):337–40.

[84] Takamatsu Y, Simmons PJ, Moore RJ, Morris HA, To LB, Levesque JP. Osteoclast-mediated bone resorption is stimulated during short-term administration of granulocyte colony-stimulating factor but is not responsible for hematopoietic progenitor cell mobilization. Blood 1998;92(9):3465–73.

[85] Takamatsu Y, Simmons PJ, Moore RJ, Morris HA, To LB, Levesque JP. Osteoclast-mediated bone resorption is stimulated during short-term administration of granulocyte colony-stimulating factor but is not responsible for hematopoietic progenitor cell mobilization. Blood 1998;92(9):3465–73.

[86] Lapidot T, Petit I. Current understanding of stem cell mobilization: the roles of chemokines, proteolytic enzymes, adhesion molecules, cytokines, and stromal cells. Exp Hematol 2002;30(9):973–81.

[87] Purton LE, Lee MY, Torok-Storb B. Normal human peripheral blood mononuclear cells mobilized with granulocyte colony-stimulating factor have increased osteoclastogenic potential compared to nonmobilized blood. Blood 1996;87:1802–8.

[88] Kuwabara H, Wada T, Oda T, Yoshikawa H, Sawada N, Kokai Y, et al. Overexpression of the granulocyte colony-stimulating factor gene impairs bone morphogenetic protein responsiveness in mice. Lab Invest 2001;81(8):1133–41.

[89] Christopher MJ, Link DC. Granulocyte colony-stimulating factor induces osteoblast apoptosis and inhibits osteoblast differentiation. J Bone Miner Res 2008 Nov;23(11):1765–74.

[90] Oda T, Wada T, Kuwabara H, Sawada N, Yamashita T, Kokai Y. Ovariectomy fails to augment bone resorption and marrow B lymphopoiesis in granulocyte colony-stimulating factor transgenic mice. J Orthop Sci 2005;10(1):70–6.

[91] Horiuchi K, Kimura T, Miyamoto T, Miyamoto K, Akiyama H, Takaishi H, et al. Conditional inactivation of TACE by a Sox9 promoter leads to osteoporosis and increased granulopoiesis via dysregulation of IL-17 and G-CSF. J Immunol 2009;182(4):2093–101.

[92] Dinarello CA. Interleukin-1 and interleukin-1 antagonism. Blood 1991;77:1627–52.

[93] Lorenzo JA, Sousa S, Alander C, Raisz LG, Dinarello CA. Comparison of the bone-resorbing activity in the supernatants from phytohemagglutinin-stimulated human peripheral blood mononuclear cells with that of cytokines through the use of an antiserum to interleukin 1. Endocrinology 1987;121:1164–70.

[94] Sabatini M, Boyce B, Aufdemorte T, Bonewald L, Mundy GR. Infustions of recombinant human interleukins 1 alpha and 1 beta cause hypercalcemia in normal mice. Proc Natl Acad Sci U S A 1988;85:5235–9.

[95] Jimi E, Nakamura I, Duong LT, Ikebe T, Takahashi N, Rodan GA, et al. Interleukin 1 induces multinucleation and bone-resorbing activity of osteoclasts in the absence of osteoblasts/stromal cells Exp Cell Res 1999;247(1):84–93.

[96] Hofbauer LC, Lacey DL, Dunstan CR, Spelsberg TC, Riggs BL, Khosla S. Interleukin-1beta and tumor necrosis factor-alpha, but not interleukin-6, stimulate osteoprotegerin ligand gene expression in human osteoblastic cells. Bone 1999;25(3):255–9.

[97] Lee SK, Kalinowski J, Jastrzebski S, Lorenzo JA. 1,25 (OH)(2) vitamin D(3)-stimulated osteoclast formation in spleen-osteoblast cocultures is mediated in part by enhanced IL-1alpha and receptor activator of NF-kappaB ligand production in osteoblasts. J Immunol 2002;169(5):2374–80.

[98] Nakamura I, Jimi E. Regulation of osteoclast differentiation and function by interleukin-1. Vitam Horm 2006;74:357–70.

[99] Sato K, Fujii Y, Asano S, Ohtsuki T, Kawakami M, Kasono K, et al. Recombinant human interleukin 1 alpha and beta stimulate mouse osteoblast-like cells (MC3T3-E1) to produce macrophage-colony stimulating activity and prostaglandin E2. Biochem Biophy Res Comm 1986;141:285–91.

[100] Klein DC, Raisz LG. Prostaglandins: stimulation of bone resorption in tissue culture. Endocrinology 1970;86:1436–40.

[101] Ma T, Miyanishi K, Suen A, Epstein NJ, Tomita T, Smith RL, et al. Human interleukin-1-induced murine osteoclastogenesis is dependent on RANKL, but independent of TNF-alpha. Cytokine 2004;26(3):138–44.

[102] Lorenzo JA, Sousa SL. Van Den Brink-Webb SE, Korn JH. Production of both interleukin-1α and β by newborn mouse calvaria cultures. J Bone Min Res 1990;5:77–83.

[103] Kawaguchi H, Pilbeam CC, Vargas SJ, Morse EE, Lorenzo JA, Raisz LG. Ovariectomy enhances and estrogen replacement inhibits the activity of bone marrow factors that stimulate prostaglandin production in cultured mouse calvaria. J Clin Invest 1995;96:539–48.

[104] Miyaura C, Kusano K, Masuzawa T, Chaki O, Onoe Y, Aoyagi M, et al. Endogenous bone-resorbing factors in estrogen deficiency: cooperative effects of IL-1 and IL-6. J Bone Min Res 1995;10:1365–73.

[105] Yao Z, Xing L, Qin C, Schwarz EM, Boyce BF. Osteoclast precursor interaction with bone matrix induces osteoclast formation directly by an interleukin-1-mediated autocrine mechanism. J Biol Chem 2008 Apr 11;283(15):9917–24.

[106] Arend WP, Welgus HG, Thompson C, Eisenberg SP. Biological properties of recombinant human monocyte-derived interleukin 1 receptor antagonist. J Clin Invest 1990;85:1694–7.

[107] Eisenberg SP, Evans RJ, Arend WP, Verderber E, Brewer MT, Hannum CH, et al. Primary structure and functional expression from complementary DNA of a human inteìeukin-1 receptor antagonist. Nature 1990;343:341–6.

[108] Hannum CH, Wilcox CJ, Arend WP, Joslin FG, Dripps DJ, Heimdal PL, et al. Interleukin-1 receptor antagonist activity of a human interleukin-1 inhibitor. Nature 1990;343:336–40.

[109] Dinarello CA. Blocking interleukin-1 in disease. Blood Purif 1993;11:118–27.

[110] Sims JE, Gayle MA, Slack JL, Alderson MR, Bird TA, Giri JG, et al. Interleukin 1 signaling occurs exclusively via the type I receptor. Proc Natl Acad Sci U S A 1993;90:6155–9.

[111] Huang J, Gao X, Li S, Cao Z. Recruitment of IRAK to the interleukin 1 receptor complex requires interleukin 1 receptor accessory protein. Proc Natl Acad Sci U S A 1997;94(24):12829–32.

[112] Korherr C, Hofmeister R, Wesche H, Falk W. A critical role for interleukin-1 receptor accessory protein in interleukin-1 signaling. Eur J Immunol 1997;27(1):262–7.

[113] Wesche H, Korherr C, Kracht M, Falk W, Resch K, Martin MU. The interleukin-1 receptor accessory protein (IL-1RAcP) is essential for IL-1-induced activation of interleukin-1 receptor-associated kinase (IRAK) and stress-activated protein kinases (SAP kinases). J Biol Chem 1997;272(12):7727–31.

[114] Eder J. Tumour necrosis factor alpha and interleukin 1 signalling: do MAPKK kinases connect it all? Trends Pharmacol Sci 1997;18(9):319–22.

[115] Martin MU, Falk W. The interleukin-1 receptor complex and interleukin-1 signal transduction. Eur Cytokine Netw 1997;8(1):5–17.

[116] Colotta F, Re F, Muzio M, Bertini R, Polentarutti N, Sironi M, et al. Interleukin-1 type II receptor: a decoy target for IL-1 that is regulated by IL-4. Science 1993;261:472–5.

[117] Bajayo A, Goshen I, Feldman S, Csernus V, Iverfeldt K, Shohami E, et al. Central IL-1 receptor signaling regulates bone growth and mass. Proc Natl Acad Sci U S A 2005;102(36):12956–61.

[118] Vargas SJ, Naprta A, Glaccum M, Lee SK, Kalinowski J, Lorenzo JA. Interleukin-6 expression and histomorphometry of bones from mice deficient for receptors for interleukin-1 or tumor necrosis factor. J Bone Min Res 1996;11:1736–40.

[119] Sato N, Takahashi N, Suda K, Nakamura M, Yamaki M, Ninomiya T, et al. MyD88 but not TRIF is essential for osteoclastogenesis induced by lipopolysaccharide, diacyl lipopeptide, and IL-1alpha. J Exp Med 2004 Sep 6;200(5):601–11.

[120] Lee ZH, Lee SE, Kim CW, Lee SH, Kim SW, Kwack K, et al. IL-1alpha stimulation of osteoclast survival through the PI 3-kinase/Akt and ERK pathways. J Biochem 2002;131(1):161–6.

[121] Boyce BF, Aufdemorte TB, Garrett R, Yates AJP, Mundy GR. Effects of interleukin-1 on bone turnover in normal mice. Endocrinology 1989;125:1142.

[122] Canalis E. Interleukin-1 has independent effects on deoxyribonucleic acid and collagen synthesis in cultures of rat calvariae. Endocrinology 1986;118:74–81.

[123] Tsuboi M, Kawakami A, Nakashima T, Matsuoka N, Urayama S, Kawabe Y, et al. Tumor necrosis factor-alpha and interleukin-1beta increase the Fas-mediated apoptosis of human osteoblasts. J Lab ClinMed 1999;134(3):222–31.

[124] Niki Y, Takaishi H, Takito J, Miyamoto T, Kosaki N, Matsumoto H, et al. Administration of cyclooxygenase-2 inhibitor reduces joint inflammation but exacerbates osteopenia in IL-1 alpha transgenic mice due to GM-CSF overproduction. J Immunol 2007 Jul 1;179(1):639–46.

[125] Tanabe N, Maeno M, Suzuki N, Fujisaki K, Tanaka H, Ogiso B, et al. IL-1 alpha stimulates the formation of osteoclast-like cells by increasing M-CSF and PGE2 production and decreasing OPG production by osteoblasts. Life Sci 2005 Jun 24;77(6):615–26.

[126] Lambert C, Oury C, Dejardin E, Chariot A, Piette J, Malaise M, et al. Further insights in the mechanisms of interleukin-1beta stimulation of osteoprotegerin in osteoblast-like cells. J Bone Miner Res 2007 Sep;22(9):1350–61.

[127] Beutler B, Cerami A. Cachectin and tumour necrosis factor as two sides of the same biological coin. Nature 1986;320:584–8.

[128] Beutler B, Cerami A. Cachectin: more than a tumor necrosis factor. N Eng J Med 1987;316:379–85.

[129] Old LJ. Tumor necrosis factor (TNF). Science 1985;230:630–2.

[130] Oliff A. The role of tumor necrosis factor (cachectin) in cachexia. Cell 1988;54:141–2.

[131] Paul NL, Ruddle NH. Lymphotoxin. Ann Rev Immunol 1988;6:407–38.

[132] Bertolini DR, Nedwin GE, Bringman TS, Smith DD, Mundy GR. Stimulation of bone resorption and inhibition of bone formation in vitro by human tumour necrosis factors. Nature 1986;319:516–18.

[133] Tashjian Jr AH, Voelkel EF, Lazzaro M, Goad D, Bosma T, Levine L. Tumor necrosis factor-alpha (cachectin) stimulates bone resorption in mouse calvaria via a prostaglandin-mediated mechanism. Endocrinology 1987;120:2029–36.

[134] Stashenko P, Dewhirst FE, Peros WJ, Kent RL, Ago JM. Synergistic interactions between interleukin 1, tumor necrosis factor, and lymphotoxin in bone resorption. J Immunol 1987;138:1464–8.

[135] Hofbauer LC, Lacey DL, Dunstan CR, Spelsberg TC, Riggs BL, Khosla S. Interleukin-1beta and tumor necrosis factor-alpha, but not interleukin-6, stimulate osteoprotegerin ligand gene expression in human osteoblastic cells. Bone 1999;25(3):255–9.

[136] Wei S, Kitaura H, Zhou P, Ross FP, Teitelbaum SL. IL-1 mediates TNF-induced osteoclastogenesis. J Clin Invest 2005;115(2):282–90.

[137] Zwerina J, Redlich K, Polzer K, Joosten L, Kronke G, Distler J, et al. TNF-induced structural joint damage is mediated by IL-1. Proc Natl Acad Sci U S A 2007 Jul 10;104(28):11742–7.

[138] Polzer K, Joosten L, Gasser J, Distler JH, Ruiz G, Baum W, et al. IL-1 is essential for systemic inflammatory bone loss TNF-induced structural joint damage is mediated by IL-1. Ann Rheum Dis 2009 Feb 5, Jul 10;5(28):5.

[139] Kitaura H, Zhou P, Kim HJ, Novack DV, Ross FP, Teitelbaum SL. M-CSF mediates TNF-induced inflammatory osteolysis. J Clin Invest 2005;115(12):3418–27.

[140] Azuma Y, Kaji K, Katogi R, Takeshita S, Kudo A. Tumor necrosis factor-alpha induces differentiation of and bone resorption by osteoclasts. J Biol Chem 2000;275(7):4858–64.

[141] Fiers W, Sim E. Tumor Necrosis Factor. The Natural Immune System: Humoral Factors. Oxford: IRL Press at Oxford University Press; 1993. p. 65.

[142] Zhang YH, Heulsmann A, Tondravi MM, Mukherjee A, Abu-Amer Y. Tumor necrosis factor-alpha (TNF) stimulates RANKL-induced osteoclastogenesis via coupling of TNF type 1 receptor and RANK signaling pathways. J Biol Chem 2001;276(1):563–8.

[143] Abbas S, Zhang YH, Clohisy JC, Abu-Amer Y. Tumor necrosis factor-alpha inhibits pre-osteoblast differentiation through its type-1 receptor. Cytokine 2003;22(1-2):33–41.

[144] Erickson SL, de SFJ, Kikly K, Carver-Moore K, Pitts-Meek S, Gillett N, et al. Decreased sensitivity to tumour-necrosis factor but normal T-cell development in TNF receptor-2-deficient mice. Nature 1994;372(6506):560–3.

[145] Pfeffer K, Matsuyama T, Kundig TM, Wakeham A, Kishihara K, Shahinian A, et al. Mice deficient for the 55 kd tumor necrosis factor receptor are resistant to endotoxic shock, yet succumb to L. monocytogenes infection. Cell 1993;73:457–67.

[146] Rothe J, Lesslauer W, Lotscher H, Lang Y, Koebel P, Kontgen F, et al. Mice lacking the tumour necrosis factor receptor 1 are resistant to TNF-mediated toxicity but highly susceptible to infection by Listeria monocytogenes. Nature 1993;364:798–802.

[147] Yao Z, Li P, Zhang Q, Schwarz EM, Keng P, Arbini A, et al. Tnf increases circulating osteoclast precursor numbers by promoting their proliferation and differentiation in the bone marrow through up-regulation of c-fms expression. J Biol Chem 2006;281(17):11846–55.

[148] Yao Z, Li P, Zhang Q, Schwarz EM, Keng P, Arbini A, et al. Tumor necrosis factor-alpha increases circulating osteoclast precursor numbers by promoting their proliferation and differentiation in the bone marrow through up-regulation of c-Fms expression. J Biol Chem 2006 Apr 28;281(17):11846–55.

[149] Ochi S, Shinohara M, Sato K, Gober HJ, Koga T, Kodama T, et al. Pathological role of osteoclast costimulation in arthritis-induced bone loss. Proc Natl Acad Sci U S A 2007 Jul 3;104(27):11394–9.

[150] Bertolini DR, Nedwin GE, Bringman TS, Smith DD, Mundy GR. Stimulation of bone resorption and inhibition of bone formation in vitro by human tumour necrosis factors. Nature 1986;319:516–8.

[151] Canalis E. Effects of tumor necrosis factor on bone formation in vitro. Endocrinology 1987;121:1596–604.

[152] Stashenko P, Dewhirst FE, Peros WJ, Kent RL, Ago JM. Synergistic interactions between interleukin 1, tumor necrosis factor, and lymphotoxin in bone resorption. Journal of Immunology 1987;138:1464–8.

[153] Nanes MS, McKoy WM, Marx SJ. Inhibitory effects of tumor necrosis factor-alpha and interferon-gamma on deoxyribonucleic acid and collagen synthesis by rat osteosarcoma cells (ROS 17/2.8). Endocrinology 1989;124:339–45.

[154] Gilbert L, He X, Farmer P, Rubin J, Drissi H, Van Wijnen AJ, et al. Expression of the osteoblast differentiation factor RUNX2 (Cbfa1/AML3/Pebp2alphaA)is inhibited by tumor necrosis factor-alpha. J Biol Chem 2002;277.

[155] Gilbert LC, Rubin J, Nanes MS. The p55 TNF Receptor Mediates TNF Inhibition of Osteoblast Differentiation Independent of Apoptosis. Am J Physiol Endocrinol Metab 2004.

[156] Mori L, Iselin S, De Libero G, Lesslauer W. Attenuation of collagen-induced arthritis in 55-kDa TNF receptor type 1 (TNFR1)-IgG1-treated and TNFR1-deficient mice. J Immunol 1996;157: 3178–82.

[157] Gowen M, MacDonald BR, Russell GG. Actions of recombinant human gamma-interferon and tumor necrosis factor alpha on the proliferation and osteoblastic characteristics of human trabecular bone cells in vitro. Arth and Rheum 1988;31:1500–7.

[158] Nanes MS, Rubin J, Titus L, Hendy GN, Catherwood B. Tumor necrosis factor-α inhibits 1,25-dihydroxyvitamin D3-stimulated bone Gla protein synthesis in rat osteosarcoma cells (ROS 17/2.8) by a pretranslational mechanism. Endocrinology 1991;128:2577–82.

[159] Kitajima I, Soejima Y, Takasaki I, Beppu H, Tokioka T, Maruyama I. Ceramide-induced nuclear translocation of NF-kappaB is a potential mediator of the apoptotic response to TNF-α in murine clonal osteoblasts. Bone 1996;19:263–70.

[160] Hock JM, Krishnan V, Onyia JE, Bidwell JP, Milas J, Stanislaus D. Osteoblast apoptosis and bone turnover. J Bone Miner Res 2001;16(6):975–84.

[161] Scharla SH, Strong DD, Mohan S, Chevalley T, Linkhart TA. Effect of tumor necrosis factor-alpha on the expression of insulin-like growth factor I and insulin-like growth factor binding protein 4 in mouse osteoblasts. Eur J Endocrinol 1994 Sep;131(3):293–301.

[162] Wesche DE, Lomas-Neira JL, Perl M, Chung CS, Ayala A. Leukocyte apoptosis and its significance in sepsis and shock. J Leukoc Biol 2005;78(2):325–37.

[163] Kovacic N, Lukic IK, Grcevic D, Katavic V, Croucher P, Marusic A. The Fas/Fas ligand system inhibits differentiation of murine osteoblasts but has a limited role in osteoblast and osteoclast apoptosis. J Immunol 2007;178(6):3379–89.

[164] Park H, Jung YK, Park OJ, Lee YJ, Choi JY, Choi Y. Interaction of fas ligand and fas expressed on osteoclast precursors increases osteoclastogenesis. J Immunol 2005;175(11):7193–201.

[165] Wu X, McKenna MA, Feng X, Nagy TR, McDonald JM. Osteoclast apoptosis: the role of fas in vivo and in vitro. Endocrinology 2003;144(12):5545–55.

[166] Wu X, Pan G, McKenna MA, Zayzafoon M, Xiong WC, McDonald JM. RANKL regulates Fas expression and Fas-mediated apoptosis in osteoclasts. J Bone Min Res 2005;20(1):107–16.

[167] Katavic V, Lukic IK, Kovacic N, Grcevic D, Lorenzo JA, Marusic A. Increased bone mass is a part of the generalized lymphoproliferative disorder phenotype in the mouse. J Immunol 2003,170(3).1540–7.

[168] Nakamura T, Imai Y, Matsumoto T, Sato S, Takeuchi K, Igarashi K, et al. Estrogen prevents bone loss via estrogen receptor alpha and induction of Fas ligand in osteoclasts. Cell 2007 Sep 7;130 (5):811–23.

[169] Krum SA, Miranda-Carboni GA, Hauschka PV, Carroll JS, Lane TF, Freedman LP, et al. Estrogen protects bone by inducing Fas ligand in osteoblasts to regulate osteoclast survival. EMBO J 2008 Feb 6;27(3):535–45.

[170] Roux S, Lambert-Comeau P, Saint-Pierre C, Lepine M, Sawan B, Parent JL. Death receptors, Fas and TRAIL receptors, are involved in human osteoclast apoptosis. Biochem Biophys Res Commun 2005;333(1):42–50.

[171] Zauli G, Rimondi E, Celeghini C, Milani D, Secchiero P. Dexamethasone counteracts the anti-osteoclastic, but not the anti-leukemic, activity of TNF-related apoptosis inducing ligand (TRAIL). J Cell Physiol 2010 Feb;222(2):357–64.

[172] Colucci S, Brunetti G, Cantatore FP, Oranger A, Mori G, Pignataro P, et al. The death receptor DR5 is involved in TRAIL-mediated human osteoclast apoptosis. Apoptosis 2007;12(9):1623–32.

[173] Zauli G, Rimondi E, Stea S, Baruffaldi F, Stebel M, Zerbinati C, et al. TRAIL inhibits osteoclastic differentiation by counteracting RANKL-dependent p27(Kip1) accumulation in pre-osteoclast precursors. J Cell Physiol 2007.

[174] Tinhofer I, Biedermann R, Krismer M, Crazzolara R, Greil R. A role of TRAIL in killing osteoblasts by myeloma cells. FASEB J 2006;20(6):759–61.

[175] Labrinidis A, Liapis V, Thai le M, Atkins GJ, Vincent C, Hay S, et al. Does Apo2L/TRAIL play any physiologic role in osteoclastogenesis? Blood 2008 Jun 1;111(11):5411–2. author reply 3.

[176] Loskog A, Totterman TH. CD40L – a multipotent molecule for tumor therapy. Endocr Metab Immune Disord Drug Targets 2007;7(1):23–8.

[177] Lopez-Granados E, Temmerman ST, Wu L, Reynolds JC, Follmann D, Liu S, et al. Osteopenia in X-linked hyper-IgM syndrome reveals a regulatory role for CD40 ligand in osteoclastogenesis. Proc Natl Acad Sci U S A 2007;104(12):5056–61.

[178] Lee HY, Jeon HS, Song EK, Han MK, Park SI, Lee SI, et al. CD40 ligation of rheumatoid synovial fibroblasts regulates RANKL-medicated osteoclastogenesis: evidence of NF-kappaB-dependent, CD40-mediated bone destruction in rheumatoid arthritis. Arth and Rheum 2006;54(6):1747–58.

[179] Gao Y, Wu X, Terauchi M, Li JY, Grassi F, Galley S, et al. T cells potentiate PTH-induced cortical bone loss through CD40L signaling. Cell Metab 2008 Aug;8(2):132–45.

[180] Akira S, Hirano T, Taga T, Kishimoto T. Biology of multifunctional cytokines: IL-6 and related molecules (IL-1 and TNF). FASEB J 1990;4:2860–7.

[181] Hirano T, Akira S, Taga T, Kishimoto T. Biological and clinical aspects of interleukin 6. Immunol Today 1990;11:443–9.

[182] Feyen JH, Elford P, Di Padova FE, Trechsel U. Interleukin-6 is produced by bone and modulated by parathyroid hormone. J Bone Miner Res 1989;4:633–8.

[183] Lowik CW, Van der PG, Bloys H, Hoekman K, Bijvoet OL, Aarden LA, et al. Parathyroid hormone (PTH) and PTH-like protein (PLP) stimulate interleukin-6 production by osteogenic cells: a possible role of interleukin-6 in osteoclastogeneis. Biochem Biophys Res Commun 1989;162:1546–52.

[184] Girasole G, Jilka RL, Passeri G, Boswell S, Boder G, Williams DC, et al. 17β-Estradiol inhibits interleukin-6 production by bone marrow-derived stromal cells and osteoblasts in vitro: a potential mechanism for the antiosteoporotic effect of estrogens. J Clin Invest 1992;89:883–91.

[185] Kishimoto T, Taga T, Akira S. Cytokine signal transduction. Cell 1994;76:253–62.

[186] Tamura T, Udagawa N, Takahashi N, Miyaura C, Tanaka S, Yamada Y, et al. Soluble interleukin-6 receptor triggers osteoclast formation by interleukin 6. Proc Natl Acad Sci U S A 1993;90:11924–8.

[187] Franchimont N, Lambert C, Huynen P, Ribbens C, Relic B, Chariot A, et al. Interleukin-6 receptor shedding is enhanced by interleukin-1beta and tumor necrosis factor alpha and is partially mediated by tumor necrosis factor alpha-converting enzyme in osteoblast-like cells. Arthritis Rheum 2005;52(1):84–93.

[188] Al-Humidan A, Ralston SH, Hughes DE, Chapman K, Aarden L, Russell RGG, et al. Interleukin-6 does not stimulate bone resorption in neonatal mouse calvariae. J Bone Min Res 1991;6:3–7.

[189] Ishimi Y, Miyaura C, Jin CH, Akatsu T, Abe E, Nakamura Y, et al. IL-6 is produced by osteoblasts and induces bone resorption. J Immunol 1990;145:3297–303.

[190] Linkhart TA, Linkhart SG, MacCharles DC, Long DL, Strong DD. Interleukin-6 messenger RNA expression and interleukin-6 protein secretion in cells isolated from normal human bone: regulation by interleukin-1. J Bone Min Res 1991;6:1285–94.

[191] Yoshitake F, Itoh S, Narita H, Ishihara K, Ebisu S. IL-6 directly inhibits osteoclast differentiation by suppressing rank signaling pathways. J Biol Chem 2008;283:11535–40.

[192] Duplomb L, Baud'huin M, Charrier C, Berreur M, Trichet V, Blanchard F, et al. IL-6 inhibits RANKL-induced osteoclastogenesis by diverting cells into the macrophage lineage: key role of Serine727 phosphorylation of STAT3. Endocrinology 2008;149:3688–97.

[193] Manolagas SC, Jilka RL. Mechanisms of disease: bone marrow, cytokines, and bone remodeling – emerging insights into the pathophysiology of osteoporosis. N Eng J Med 1995;332:305–11.

[194] Roodman GD. Interleukin-6: an osteotropic factor? J Bone Miner Res 1992;7:475–8.

[195] Palmqvist P, Persson E, Conaway HH, Lerner UH. IL-6, leukemia inhibitory factor, and oncostatin M stimulate bone resorption and regulate the expression of receptor activator of NF-kappaB ligand, osteoprotegerin, and receptor activator of NF-kappaB in mouse calvariae. J Immunol 2002;169(6):3353–62.

[196] Liu XH, Kirschenbaum A, Yao S, Levine AC. Cross-talk between the interleukin-6 and prostaglandin E(2) signaling systems results in enhancement of osteoclastogenesis through effects on the osteoprotegerin/receptor activator of nuclear factor-{kappa}B (RANK) ligand/RANK system. Endocrinology 2005;146(4):1991–8.

[197] Kudo O, Sabokbar A, Pocock A, Itonaga I, Fujikawa Y, Athanasou NA. Interleukin-6 and interleukin-11 support human osteoclast formation by a RANKL-independent mechanism. Bone 2003;32(1):1–7.

[198] Guise TA, Mundy GR. Cancer and bone. Endocrine Reviews 1998;19(1):18–54.

[199] Yamamoto T, Ozono K, Kasayama S, Yoh K, Hiroshima K, Takagi M, et al. Increased IL-6-production by cells isolated from the fibrous bone dysplasia tissues in patients with McCune-Albright syndrome. J Clin Invest 1996;98(1):30–5.

[200] Reddy SV, Takahashi S, Dallas M, Williams RE, Neckers L, Roodman GD. Interleukin-6 antisense deoxyoligonucleotides inhibit bone resorption by giant cells from human giant cell tumors of bone. J Bone Miner Res 1994;9(5):753–7.

[201] Axmann R, Bohm C, Kronke G, Zwerina J, Smolen J, Schett G. Inhibition of interleukin-6 receptor directly blocks osteoclast formation in vitro and in vivo. Arthritis Rheum 2009 Aug 27;60(9): 2747–56.

[202] Devlin RD, Bone III HG, Roodman GD. Interleukin-6: a potential mediator of the massive osteolysis in patients with Gorham-Stout disease. J Clin Endocrinol Metab 1996;81:1893–7.

[203] Grey A, Mitnick MA, Masiukiewicz U, Sun BH, Rudikoff S, Jilka RL, et al. A role for interleukin-6 in parathyroid hormone-induced bone resorption in vivo. Endocrinology 1999;140(10):4683–90.

[204] O'Brien CA, Jilka RL, Fu Q, Stewart S, Weinstein RS, Manolagas SC. IL-6 is not required for parathyroid hormone stimulation of RANKL expression, osteoclast formation, and bone loss in mice. Am J Physiol Endocrinol Metab 2005;289(5):E784–93.

[205] Manolagas SC, Bellido T, Jilka RL. New insights into the cellular, biochemical, and molecular basis of postmenopausal and senile osteoporosis: roles of IL-6 and gp130. [Review]. Int J Immunopharmacol 1995;17(2):109–16.

[206] Romas E, Udagawa N, Zhou H, Tamura T, Saito M, Taga T, et al. The role of gp130-mediated signals in osteoclast development: regulation of interleukin 11 production by osteoblasts and distribution of its receptor in bone marrow cultures. J Exp Med 1996;183:2581–91.

[207] Elias JA, Tang W, Horowitz MC. Cytokine and hormonal stimulation of human osteosarcoma interleukin-11 production. Endocrinology 1995;136:489–98.

[208] Girasole G, Passeri G, Jilka RL, Manolagas SC. Interleukin-11: a new cytokine critical for osteoclast development. J Clin Invest 1994;93:1516–24.

[209] Hill PA, Tumber A, Papaioannou S, Meikle MC. The cellular actions of interleukin-11 on bone resorption in vitro. Endocrinology 1998;139(4):1564–72.

[210] Morinaga Y, Fujita N, Ohishi K, Zhang Y, Tsuruo T. Suppression of interleukin-11-mediated bone resorption by cyclooxygenases inhibitors. J Cell Physiol 1998;175(3):247–54.

[211] Sims NA, Jenkins BJ, Nakamura A, Quinn JM, Li R, Gillespie MT, et al. Interleukin-11 receptor signaling is required for normal bone remodeling. J Bone Min Res 2005;20(7):1093–102.

[212] Kido S, Kuriwaka-Kido R, Imamura T, Ito Y, Inoue D, Matsumoto T. Mechanical stress induces interleukin-11 expression to stimulate osteoblast differentiation. Bone 2009 Dec;45(6):1125–32.

[213] Koyama Y, Mitsui N, Suzuki N, Yanagisawa M, Sanuki R, Isokawa K, et al. Effect of compressive force on the expression of inflammatory cytokines and their receptors in osteoblastic Saos-2 cells. Arch Oral Biol 2008 May;53(5):488–96.

[214] Greenfield EM, Horowitz MC, Lavish SA. Stimulation by parathyroid hormone of interleukin-6 and leukemia inhibitory factor expression in osteoblasts is an immediate-early gene response induced by cAMP signal transduction. J Biol Chem 1996;271:10984–9.

[215] Marusic A, Kalinowski JF, Jastrzebski S, Lorenzo JA. Production of leukemia inhibitory factor mRNA and protein by malignant and immortalized bone cells. J Bone Min Res 1993;8:617–24.

[216] Shiina-Ishimi Y, Abe E, Tanaka H, Suda T. Synthesis of colony-stimulating factor (CSF) and differentiation inducing factor (D-factor) by osteoblastic cells, clone MC3T3-E1. Biochem Biophys Res Comm 1986;134:400–6.

[217] Reid IR, Lowe C, Cornish J, Skinner SJ, Hilton DJ, Willson TA, et al. Leukemia inhibitory factor: a novel bone-active cytokine. Endocinology 1990;126:1416–20.

[218] Lorenzo JA, Sousa SL, Leahy CL. Leukemia inhibitory factor (LIF) inhibits basal bone resorption in fetal rat long bone cultures. Cytokine 1990;2:266–71.

[219] Van Beek E, Van der Wee-Pals L, van de Ruit M, Nijweide P, Papapoulos S, Lowik C. Leukemia inhibitory factor inhibits osteoclastic resorption, growth, mineralization, and alkaline phosphatase activity in fetal mouse metacarpal bones in culture. J Bone Min Res 1993;8(2):191–8.

[220] Cornish J, Callon K, King A, Edgar S, Reid IR. The effect of leukemia inhibitory factor on bone in vivo. Endocrinology 1993;132:1359–66.

[221] Ware CB, Horowitz MC, Renshaw BR, Hunt JS, Liggitt D, Koblar SA, et al. Targeted disruption of the low-affinity leukemia inhibitory factor receptor gene causes placental, skeletal, neural and metabolic defects and results in perinatal death. Development 1995;121:1283–99.

[222] Bozec A, Bakiri L, Hoebertz A, Eferl R, Schilling AF, Komnenovic V, et al. Osteoclast size is controlled by Fra-2 through LIF/LIF-receptor signalling and hypoxia. Nature 2008 Jul 10;454(7201):221–5.

[223] Falconi D, Oizumi K, Aubin JE. Leukemia inhibitory factor influences the fate choice of mesenchymal progenitor cells. Stem Cells 2007 Feb;25(2):305–12.

[224] Oskowitz AZ, Lu J, Penfornis P, Ylostalo J, McBride J, Flemington EK, et al. Human multipotent stromal cells from bone marrow and microRNA: regulation of differentiation and leukemia inhibitory factor expression. Proc Natl Acad Sci U S A 2008 Nov 25;105(47):18372–7.

[225] Heymann D, Guicheux J, Gouin F, Cottrel M, Daculsi G. Oncostatin M stimulates macrophage-polykaryon formation in long-term human bone-marrow cultures. Cytokine 1998;10(2):98–109.

[226] Jay PR, Centrella M, Lorenzo J, Bruce AG, Horowitz MC. Oncostatin-M: A new bone active cytokine that activates osteoblasts and inhibits bone resorption. Endocrinology 1996;137:1151–8.

[227] Malik N, Haugen HS, Modrell B, Shoyab M, Clegg CH. Developmental abnormalities in mice transgenic for bovine oncostatin M. Mol Cell Biol 1995;15:2349–58.

[228] Mundy GR. An OAF by any other name [editorial; comment]. Endocrinology 1996;137(4):114950.

[229] Walker EC, McGregor NE, Poulton IJ, Solano M, Pompolo S, Fernandes TJ, et al. Oncostatin M promotes bone formation independently of resorption when signaling through leukemia inhibitory factor receptor in mice. J Clin Invest 2010 Jan 4;4(40568).

[230] Song HY, Jeon ES, Kim JI, Jung JS, Kim JH. Oncostatin M promotes osteogenesis and suppresses adipogenic differentiation of human adipose tissue-derived mesenchymal stem cells. J Cell Biochem 2007 Aug 1;101(5):1238–51.

[231] Brounais B, David E, Chipoy C, Trichet V, Ferre V, Charrier C, et al. Long term oncostatin M treatment induces an osteocyte-like differentiation on osteosarcoma and calvaria cells. Bone 2009 May;44(5):830–9.

[232] Kawasaki K, Gao YH, Yokose S, Kaji Y, Nakamura T, Suda T, et al. Osteoclasts are present in gp130-deficient mice. Endocrinology 1997;138(11):4959–65.

[233] Namen AE, Lupton S, Hjerrild K, Wignall J, Mochizuki DY, Schmierer A, et al. Stimulation of B-cell progenitors by cloned murine interleukin-7. Nature 1988;333:571–3.

[234] Miyaura C, Onoe Y, Inada M, Maki K, Ikuta K, Ito M, et al. Increased B-lymphopoiesis by interleukin 7 induces bone loss in mice with intact ovarian function: similarity to estrogen deficiency. Proc Natl Acad Sci U S A 1997;94(17):9360–5.

[235] Weitzmann MN, Roggia C, Toraldo G, Weitzmann L, Pacifici R. Increased production of IL-7 uncouples bone formation from bone resorption during estrogen deficiency. J Clin Invest 2002;110(11):1643–50.

[236] Weitzmann MN, Cenci S, Rifas L, Brown C, Pacifici R. Interleukin-7 stimulates osteoclast formation by up-regulating the T-cell production of soluble osteoclastogenic cytokines. Blood 2000;96(5):1873–8.

[237] Salopek D, Grcevic D, Katavic V, Kovacic N, Lukic IK, Marusic A. Increased bone resorption and osteopenia are a part of the lymphoproliferative phenotype of mice with systemic over-expression of interleukin-7 gene driven by MHC class II promoter. Immunol Lett 2008 Dec 22;121(2):134–9.

[238] Toraldo G, Roggia C, Qian WP, Pacifici R, Weitzmann MN. IL-7 induces bone loss in vivo by induction of receptor activator of nuclear factor kappa B ligand and tumor necrosis factor alpha from T cells. Proc Natl Acad Sci U S A 2003;100(1):125–30.

[239] Ryan MR, Shepherd R, Leavey JK, Gao YH, Grassi F, Schnell FJ, et al. An IL-7-dependent rebound in thymic T cell output contributes to the bone loss induced by estrogen deficiency. Proc Natl Acad Sci U S A 2005;102(46):16735–40.

[240] Gendron S, Boisvert M, Chetoui N, Aoudjit F. Alpha1beta1 integrin and interleukin-7 receptor up-regulate the expression of RANKL in human T cells and enhance their osteoclastogenic function. Immunology 2008 Nov;125(3):359–69.

[241] Lee SK, Kalinowski JF, Jastrzebski SL, Puddington L, Lorenzo JA. Interleukin-7 is a direct inhibitor of in vitro osteoclastogenesis. Endocrinology 2003;144(8):3524–31.

[242] Lee SK, Kalinowski JF, Jacquin C, Adams DJ, Gronowicz G, Lorenzo JA. Interleukin-7 influences osteoclast function in vivo but is not a critical factor in ovariectomy-induced bone loss. J Bone Miner Res 2006;21(5):695–702.

[243] Sato T, Watanabe K, Masuhara M, Hada N, Hakeda Y. Production of IL-7 is increased in ovariectomized mice, but not RANKL mRNA expression by osteoblasts/stromal cells in bone, and IL-7 enhances generation of osteoclast precursors in vitro. J Bone Min Res 2007;25(1):19–27.

[244] Lee S, Kalinowski JF, Adams DJ, Aguila HL, Lorenzo JA. Osteoblast specific overexpression of human interleukin-7 increases femoral trabecular bone mass in female mice and inhibits in vitro osteoclastogenesis. J Bone Min Res 2004;19:S410.

[245] Lee S, Kalinowski J, Adams DJ, Aguila HL, Lorenzo JA. Osteoblast specific overexpression of human interlukin-7 rescues the bone phenotype of interleukin-7 deficient female mice. J Bone Min Res 2005;20:S48.

[246] Wu JY, Purton LE, Rodda SJ, Chen M, Weinstein LS, McMahon AP, et al. Osteoblastic regulation of B lymphopoiesis is mediated by Gs{alpha}-dependent signaling pathways. Proc Natl Acad Sci U S A 2008;105(44):16976–81.

[247] Butcher EC, Williams M, Youngman K, Rott L, Briskin M. Lymphocyte trafficking and regional immunity. Adv Immunol 1999;72:209–53.

[248] Campbell JJ, Butcher EC. Chemokines in tissue-specific and microenvironment-specific lymphocyte homing. Curr Opin Immunol 2000;12(3):336–41.

[249] Baggiolini M. Chemokines in pathology and medicine. J Intern Med 2001;250(2):91–104.

[250] Horuk R. Chemokine receptors. Cytokine Growth Factor Rev 2001;12(4):313–35.

[251] Rothe L, Collin-Osdoby P, Chen Y, Sunyer T, Chaudhary L, Tsay A, et al. Human osteoclasts and osteoclast-like cells synthesize and release high basal and inflammatory stimulated levels of the potent chemokine interleukin-8. Endocrinology 1998;139(10): 4353–63.

[252] Bendre MS, Gaddy-Kurten D, Mon-Foote T, Akel NS, Skinner RA, Nicholas RW, et al. Expression of interleukin 8 and not parathyroid hormone-related protein by human breast cancer cells correlates with bone metastasis in vivo. Cancer Res 2002;62(19):5571–9.

[253] Bendre MS, Margulies AG, Walser B, Akel NS, Bhattacharrya S, Skinner RA, et al. Tumor-derived interleukin-8 stimulates osteolysis independent of the receptor activator of nuclear factor-kappaB ligand pathway. Cancer Res 2005;65(23):11001–9.

[254] Bendre MS, Montague DC, Peery T, Akel NS, Gaddy D, Suva LJ. Interleukin-8 stimulation of osteoclastogenesis and bone resorption is a mechanism for the increased osteolysis of metastatic bone disease. Bone 2003;33(1):28–37.

[255] Sunyer T, Rothe L, Jiang XS, Osdoby P, Collin-Osdoby P. Proinflammatory agents, IL-8 and IL-10, upregulate inducible nitric oxide synthase expression and nitric oxide production in avian osteoclast-like cells. J Cell Biochem 1996;60:469–83.

[256] Kim MS, Day CJ, Morrison NA. MCP-1 is induced by RANKL, promotes osteoclast fusion and rescues GM-CSF suppression of osteoclast formation. J Biol Chem 2005.

[257] Rollins BJ. Chemokines. Blood 1997;90(3):909–28.

[258] Li X, Qin L, Bergenstock M, Bevelock LM, Novack DV, Partridge NC. Parathyroid hormone stimulates osteoblastic expression of MCP-1 to recruit and increase the fusion of pre-osteoclasts. J Biol Chem 2007.

[259] Rahimi P, Wang C, Stashenko P, Lee SK, Lorenzo JA, Graves DT. Monocyte chemoattractant protein-1 expression and monocyte recruitment in osseous inflammation in the mouse. Endocrinology 1995;136:2752–9.

[260] Zhu JF, Valente AJ, Lorenzo JA, Carnes D, Graves DT. Expression of monocyte chemoattractant protein 1 in human osteoblastic cells stimulated by proinflammatory mediators. J Bone Min Res 1994;9:1123–30.

[261] Que BG, Wise GE. Tooth eruption molecules enhance MCP-1 gene expression in the dental follicle of the rat. Dev Dyn 1998;212(3):346–51.

[262] Wise GE, Huang H, Que BG. Gene expression of potential tooth eruption molecules in the dental follicle of the mouse. Eur J Oral Sci 1999;107(6):482–6.

[263] Wise GE, Que BG, Huang H. Synthesis and secretion of MCP-1 by dental follicle cells – implications for tooth eruption. J Dent Res 1999;78(11):1677–81.

[264] Wise GE, Que BG, Huang H, Lumpkin SJ. Enhancement of gene expression in rat dental follicle cells by parathyroid hormone-related protein. Arch Oral Biol 2000;45(10):903–9.

[265] Bsoul S, Terezhalmy G, Abboud H, Woodruff K, Abboud SL. PDGF BB and bFGF stimulate DNA synthesis and upregulate CSF-1 and MCP-1 gene expression in dental follicle cells. Arch Oral Biol 2003;48(6):459–65.

[266] Graves DT, Alsulaimani F, Ding Y, Marks Jr SC. Developmentally regulated monocyte recruitment and bone resorption are modulated by functional deletion of the monocytic chemoattractant protein-1 gene. Bone 2002;31(2):282–7.

[267] Kim MS, Day CJ, Morrison NA. MCP-1 is induced by RANKL, promotes osteoclast fusion and rescues GM-CSF suppression of osteoclast formation. J Biol Chem 2005;280(16):16163–9.

[268] Kukita T, Nakao J, Hamada F, Kukita A, Inai T, Kurisu K, et al. Recombinant LD78 protein, a member of the small cytokine family, enhances osteoclast differentiation in rat bone marrow culture system. Bone Miner 1992;19(3):215.

[269] Kukita T, Nomiyama H, Ohmoto Y, Kukita A, Shuto T, Hotokebuchi T, et al. Macrophage inflammatory protein-1 alpha (LD78) expressed in human bone marrow: its role in regulation of hematopoiesis and osteoclast recruitment. Lab Invest 1997;76(3):399–406.

[270] Scheven BA, Milne JS, Hunter I, Robins SP. Macrophage-inflammatory protein-1α regulates pre-osteoclast differentiation in vitro. Biochem Biophy Res Comm 1999;254(3):773–8.

[271] Watanabe T, Kukita T, Kukita A, Wada N, Toh K, Nagata K, et al. Direct stimulation of osteoclastogenesis by MIP-1alpha: evidence obtained from studies using RAW264 cell clone highly responsive to RANKL. J Endocrinol 2004;180(1):193–201.

[272] Han JH, Choi SJ, Kurihara N, Koide M, Oba Y, Roodman GD. Macrophage inflammatory protein-1alpha is an osteoclastogenic factor in myeloma that is independent of receptor activator of nuclear factor kappaB ligand. Blood 2001;97(11):3349–53.

[273] Tsubaki M, Kato C, Manno M, Ogaki M, Satou T, Itoh T, et al. Macrophage inflammatory protein-1alpha (MIP-1alpha) enhances a receptor activator of nuclear factor kappaB ligand (RANKL) expression in mouse bone marrow stromal cells and osteoblasts through MAPK and PI3K/Akt pathways. Mol Cell Biochem 2007.

[274] Abe M, Hiura K, Wilde J, Moriyama K, Hashimoto T, Ozaki S, et al. Role for macrophage inflammatory protein (MIP)-1alpha and MIP-1beta in the development of osteolytic lesions in multiple myeloma. Blood 2002;100(6):2195–202.

[275] Choi SJ, Cruz JC, Craig F, Chung H, Devlin RD, Roodman GD, et al. Macrophage inflammatory protein 1-alpha is a potential osteoclast stimulatory factor in multiple myeloma. Blood 2000;96(2):671–5.

[276] Choi SJ, Oba Y, Gazitt Y, Alsina M, Cruz J, Anderson J, et al. Antisense inhibition of macrophage inflammatory protein 1-alpha blocks bone destruction in a model of myeloma bone disease. J Clin Invest 2001;108(12):1833–41.

[277] Choi J, Oba Y, Jelinek D, Ehrlich L, Lee W, Roodman D. Blocking CCR1 or CCR5 inhibits both osteoclast formation and increased alpha1-integrin expression induced by MIP-1alpha. Eur J Haematol 2003;70(4):272–8.

[278] Fuller K, Owens JM, Chambers TJ. Macrophage inflammatory protein-1α and IL-8 stimulate the motility but suppress the resorption of isolated rat osteoclasts. J Immunol 1995;154:6065–72.

[279] Lean JM, Murphy C, Fuller K, Chambers TJ. CCL9/MIP-1gamma and its receptor CCR1 are the major chemokine ligand/receptor species expressed by osteoclasts. J Cell Biochem 2002;87(4):386–93.

[280] Yang M, Mailhot G, MacKay CA, Mason-Savas A, Aubin J, Odgren PR. Chemokine and chemokine receptor expression during colony stimulating factor-1-induced osteoclast differentiation in the toothless osteopetrotic rat: a key role for CCL9 (MIP-1gamma) in osteoclastogenesis in vivo and in vitro. Blood 2006;107(6):2262–70.

[281] Okamatsu Y, Kim D, Battaglino R, Sasaki H, Spate U, Stashenko P. MIP-1{gamma} promotes receptor activator of NF-{kappa}B ligand-induced osteoclast formation and survival. J Immunol 2004;173(3):2084–90.

[282] Ishida N, Hayashi K, Hattori A, Yogo K, Kimura T, Takeya T. CCR1 acts downstream of NFAT2 in osteoclastogenesis and enhances cell migration. J Bone Min Res 2006;21(1):48–57.

[283] Yu X, Huang Y, Collin-Osdoby P, Osdoby P. CCR1 chemokines promote the chemotactic recruitment, RANKL development, and motility of osteoclasts and are induced by inflammatory cytokines in osteoblasts. J Bone Min Res 2004;19(12):2065–77.

[284] Gronthos S, Zannettino AC. The role of the chemokine CXCL12 in osteoclastogenesis. Trends Endocrinol Metab 2007;18(3):108–13.

[285] Wright LM, Maloney W, Yu X, Kindle L, Collin-Osdoby P, Osdoby P. Stromal cell-derived factor-1 binding to its chemokine receptor CXCR4 on precursor cells promotes the chemotactic recruitment, development and survival of human osteoclasts. Bone 2005;36(5):840–53.

[286] Grassi F, Piacentini A, Cristino S, Toneguzzi S, Cavallo C, Facchini A, et al. Human osteoclasts express different CXC chemokines depending on cell culture substrate: molecular and immunocytochemical evidence of high levels of CXCL10 and CXCL12. Histochem Cell Biol 2003;120(5):391–400.

[287] Yu X, Huang Y, Collin-Osdoby P, Osdoby P. Stromal cell-derived factor-1 (SDF-1) recruits osteoclast precursors by inducing chemotaxis, matrix metalloproteinase-9 (MMP-9) activity, and collagen transmigration. J Bone Min Res 2003;18(8):1404–18.

[288] Liao TS, Yurgelun MB, Chang SS, Zhang HZ, Murakami K, Blaine TA, et al. Recruitment of osteoclast precursors by stromal cell derived factor-1 (SDF-1) in giant cell tumor of bone. J Orthop Res 2005;23(1):203–9.

[289] Zannettino AC, Farrugia AN, Kortesidis A, Manavis J, To LB, Martin SK, et al. Elevated serum levels of stromal-derived factor-1alpha are associated with increased osteoclast activity and osteolytic bone disease in multiple myeloma patients. Cancer Res 2005;65(5):1700–9.

[290] Ishii M, Egen JG, Klauschen F, Meier-Schellersheim M, Saeki Y, Vacher J, et al. Sphingosine-1-phosphate mobilizes osteoclast precursors and regulates bone homeostasis. Nature 2009.

[291] Fogg DK, Sibon C, Miled C, Jung S, Aucouturier P, Littman DR, et al. A clonogenic bone marrow progenitor specific for macrophages and dendritic cells. Science 2006;311(5757):83–7.

[292] Koizumi K, Saitoh Y, Minami T, Takeno N, Tsuneyama K, Miyahara T, et al. Role of CX3CL1/fractalkine in osteoclast differentiation and bone resorption. J Immunol 2009 Dec 15;183(12):7825–31.

[293] Binder NB, Niederreiter B, Hoffmann O, Stange R, Pap T, Stulnig TM, et al. Estrogen-dependent and C-C chemokine receptor-2-dependent pathways determine osteoclast behavior in osteoporosis. NatMed 2009.

[294] Moore KW, de Waal MR, Coffman RL, O'Garra A. Interleukin-10 and the interleukin-10 receptor. Annu Rev Immunol 2001;19:683–765.

[295] Owens JM, Gallagher AC, Chambers TJ. IL-10 modulates formation of osteoclasts in murine hemopoietic cultures. J Immunol 1996;157:936–40.

[296] Xu LX, Kukita T, Kukita A, Otsuka T, Niho Y, Iijima T. Interleukin-10 selectively inhibits osteoclastogenesis by inhibiting differentiation of osteoclast progenitors into preosteoclast-like cells in rat bone marrow culture system. J Cell Physiol 1995;165:624–9.

[297] Van Vlasselaer P, Borremans B, Van Der Heuvel R, Van Gorp U, De Waal Malefyt R. Interleukin-10 inhibits the osteogenic activity of mouse bone marrow. Blood 1993;82(8):2361–70.

[298] Hong MH, Williams H, Jin CH, Pike JW. The inhibitory effect of interleukin-10 on mouse osteoclast formation involves novel tyrosine-phosphorylated proteins. J Bone Miner Res 2000;15(5):911–18.

[299] Evans KE, Fox SW. Interleukin-10 inhibits osteoclastogenesis by reducing NFATc1 expression and preventing its translocation to the nucleus. BMC Cell Biol 2007;19:4.

[300] Mohamed SG, Sugiyama E, Shinoda K, Taki H, Hounoki H, bdel-Aziz HO, et al. Interleukin-10 inhibits RANKL-mediated expression of NFATc1 in part via suppression of c-Fos and c-Jun in RAW264.7 cells and mouse bone marrow cells. Bone 2007.

[301] Carmody EE, Schwarz EM, Puzas JE, Rosier RN, O'Keefe RJ. Viral interleukin-10 gene inhibition of inflammation, osteoclastogenesis, and bone resorption in response to titanium particles. Arthritis Rheum 2002;46(5):1298–308.

[302] Liu D, Yao S, Wise GE. Effect of interleukin-10 on gene expression of osteoclastogenic regulatory molecules in the rat dental follicle. Eur J Oral Sci 2006;114(1):42–9.

[303] Owens J, Chambers TJ. Differential regulation of osteoclast formation: interleukin 10 (cytokine synthesis inhibitory factor) suppresses formation of osteoclasts but not macrophages in murine bone marrow cultures. J Bone Min Res 1995;10(Suppl. 1):S220.

[304] Shin HH, Lee JE, Lee EA, Kwon BS, Choi HS. Enhanced osteoclastogenesis in 4-1BB-deficient mice caused by reduced interleukin-10. J Bone Min Res 2006;21(12):1907–12.

[305] Hsieh CS, Macatonia SE, Tripp CS, Wolf SF, O'Garra A, Murphy KM. Development of TH1 CD4+ T cells through IL-12 produced by Listeria-induced macrophages. Science 1993;260(5107):547–9.

[306] Amcheslavsky A, Bar-Shavit Z. Interleukin (IL)-12 mediates the anti-osteoclastogenic activity of CpG-oligodeoxynucleotides. J Cell Physiol 2006;207(1):244–50.

[307] Horwood NJ, Elliott J, Martin TJ, Gillespie MT. IL-12 Alone and in synergy with IL-18 inhibits osteoclast formation. Vitro. J Immunol 2001;166(8):4915–21.

[308] Nagata N, Kitaura H, Yoshida N, Nakayama K. Inhibition of RANKL-induced osteoclast formation in mouse bone marrow cells by IL-12: involvement of IFN-gamma possibly induced from non-T cell population. Bone 2003;33(4):721–32.

[309] Yoshimatsu M, Kitaura H, Fujimura Y, Eguchi T, Kohara H, Morita Y, et al. IL-12 inhibits TNF-alpha induced osteoclastogenesis via a T cell-independent mechanism in vivo. Bone 2009.

[310] Kitaura H, Nagata N, Fujimura Y, Hotokezaka H, Yoshida N, Nakayama K. Effect of IL-12 on TNF-alpha-mediated osteoclast formation in bone marrow cells: apoptosis mediated by Fas/Fas ligand interaction. J Immunol 2002 Nov 1;169(9):4732–8.

[311] Ogata Y, Kukita A, Kukita T, Komine M, Miyahara A, Miyazaki S, et al. A novel role of IL-15 in the development of osteoclasts: inability to replace its activity with IL-2. J Immunol 1999;162(5):2754–60.

[312] Miranda-Carus ME, ito-Miguel M, Balsa A, Cobo-Ibanez T, Perez De AC, Pascual-Salcedo D, et al. Peripheral blood T lymphocytes from patients with early rheumatoid arthritis express RANKL and interleukin-15 on the cell surface and promote osteoclastogenesis in autologous monocytes. Arthritis Rheum 2006;54(4):1151–64.

[313] Koh JM, Oh B, Ha MH, Cho KW, Lee JY, Park BL, et al. Association of IL-15 polymorphisms with bone mineral density in postmenopausal korean women. Calcif Tissue Int 2009 Sep 11;11:11.

[314] Weaver CT, Hatton RD, Mangan PR, Harrington LE. IL-17 family cytokines and the expanding diversity of effector T cell lineages. Annu Rev Immunol 2007;25:821–52.

[315] Fort MM, Cheung J, Yen D, Li J, Zurawski SM, Lo S, et al. IL-25 induces IL-4, IL-5, and IL-13 and Th2-associated pathologies in vivo. Immunity 2001;15(6):985–95.

[316] Kotake S, Udagawa N, Takahashi N, Matsuzaki K, Itoh K, Ishiyama S, et al. IL-17 in synovial fluids from patients with rheumatoid arthritis is a potent stimulator of osteoclastogenesis. J Clin Invest 1999;103(9):1345–52.

[317] Yago T, Nanke Y, Ichikawa N, Kobashigawa T, Mogi M, Kamatani N, et al. IL-17 induces osteoclastogenesis from human monocytes alone in the absence of osteoblasts, which is potently inhibited by anti-TNF-alpha antibody: a novel mechanism of osteoclastogenesis by IL-17. J Cell Biochem 2009 Sep 2;2:2.

[318] Kitami S, Tanaka H, Kawato T, Tanabe N, Katono-Tani T, Zhang F, et al. IL-17A suppresses the expression of bone resorption-related proteinases and osteoclast differentiation via IL-17RA or IL-17RC receptors in RAW264.7 cells. Biochimie 2010 Jan 2;2:2.

[319] Lubberts E, Joosten LAB, Chabaud M, Van den Bersselaar L, Oppers B, Coenen-de Roo CJJ, et al. IL-4 gene therapy for collagen arthritis suppresses synovial IL-17 and osteoprotegerin ligand and prevents bone erosion. J Clin Invest 2000;105(12):1697–710.

[320] Lubberts E, Van Den BL, Oppers-Walgreen B, Schwarzenberger P, Coenen-De Roo CJ, Kolls JK, et al. IL-17 promotes bone erosion in murine collagen-induced arthritis through loss of the receptor activator of NF-kappaB ligand/osteoprotegerin balance. J Immunol 2003;170(5):2655–62.

[321] Van Bezooijen RL, Farih-Sips HC, Papapoulos SE, Lowik CW. Interleukin-17: a new bone acting cytokine in vitro. J Bone Min Res 1999;14(9):1513–21.

[322] Koenders MI, Lubberts E, Oppers-Walgreen B, Van Den BL, Helsen MM, Di Padova FE, et al. Blocking of interleukin-17 during reactivation of experimental arthritis prevents joint inflammation and bone erosion by decreasing RANKL and interleukin-1. Am J Pathol 2005;167(1):141–9.

[323] Kastelein RA, Hunter CA, Cua DJ. Discovery and biology of IL-23 and IL-27: related but functionally distinct regulators of inflammation. Annu Rev Immunol 2007;25:221–42.

[324] Bettelli E, Carrier Y, Gao W, Korn T, Strom TB, Oukka M, et al. Reciprocal developmental pathways for the generation of pathogenic effector TH17 and regulatory T cells. Nature 2006;441(7090):235–8.

[325] Sato K, Suematsu A, Okamoto K, Yamaguchi A, Morishita Y, Kadono Y, et al. Th17 functions as an osteoclastogenic helper T cell subset that links T cell activation and bone destruction. J Exp Med 2006 Nov 27;203(12):2673–82.

[326] Ju JH, Cho ML, Moon YM, Oh HJ, Park JS, Jhun JY, et al. IL-23 induces receptor activator of NF-kappaB ligand expression on CD4+ T cells and promotes osteoclastogenesis in an autoimmune arthritis model. J Immunol 2008;181(2):1507–18.

[327] Chen L, Wei XQ, Evans B, Jiang W, Aeschlimann D. IL-23 promotes osteoclast formation by upregulation of receptor activator of NF-kappaB (RANK) expression in myeloid precursor cells. Eur J Immunol 2008 Oct;38(10):2845–54.

[328] Quinn JM, Sims NA, Saleh H, Mirosa D, Thompson K, Bouralexis S, et al. IL-23 inhibits osteoclastogenesis indirectly through lymphocytes and is required for the maintenance of bone mass in mice. J Immunol 2008 Oct 15;181(8):5720–9.

[329] Kamiya S, Nakamura C, Fukawa T, Ono K, Ohwaki T, Yoshimoto T, et al. Effects of IL-23 and IL-27 on osteoblasts and osteoclasts: inhibitory effects on osteoclast differentiation. J Bone Miner Metab 2007;25(5):277–85.

[330] Furukawa M, Takaishi H, Takito J, Yoda M, Sakai S, Hikata T, et al. IL-27 abrogates receptor activator of NF-{kappa}B ligand-mediated osteoclastogenesis of human granulocyte-macrophage colony-forming unit cells through STAT1-dependent inhibition of c-Fos. J Immunol 2009.

[331] Orozco A, Gemmell E, Bickel M, Seymour GJ. Interleukin 18 and periodontal disease. J Dent Res 2007;86(7):586–93.

[332] Okamura H, Tsutsui H, Komatsu T, Yutsudo M, Hakura A, Tanimoto T, et al. Cloning of a new cytokine that induces IFN-gamma production by T cells. Nature 1995;378:88–91.

[333] Yamamura M, Kawashima M, Taniai M, Yamauchi H, Tanimoto T, Kurimoto M, et al. Interferon-gamma-inducing activity of interleukin-18 in the joint with rheumatoid arthritis. Arthritis Rheum 2001;44(2):275–85.

[334] Horwood NJ, Udagawa N, Elliott J, Grail D, Okamura H, Kurimoto M, et al. Interleukin 18 inhibits osteoclast formation via T cell production of granulocyte macrophage colony-stimulating factor. J Clin Invest 1998;101(3):595–603.

[335] Kawase Y, Hoshino T, Yokota K, Kuzuhara A, Nakamura M, Maeda Y, et al. Bone malformations in interleukin-18 transgenic mice. J Bone Min Res 2003;18(6):975–83.

[336] Yamada N, Niwa S, Tsujimura T, Iwasaki T, Sugihara A, Futani H, et al. Interleukin-18 and interleukin-12 synergistically inhibit osteoclastic bone-resorbing activity. Bone 2002;30(6):901–8.

[337] Dai SM, Nishioka K, Yudoh K. Interleukin (IL) 18 stimulates osteoclast formation through synovial T cells in rheumatoid arthritis: comparison with IL1 beta and tumour necrosis factor alpha. Ann Rheum Dis 2004;63(11):1379–86.

[338] Makiishi-Shimobayashi C, Tsujimura T, Iwasaki T, Yamada N, Sugihara A, Okamura H, et al. Interleukin-18 up-regulates osteoprotegerin expression in stromal/osteoblastic cells. Biochem Biophy Res Comm 2001;281(2):361–6.

[339] Raggatt LJ, Qin L, Tamasi J, Jefcoat Jr SC, Shimizu E, Selvamurugan N, et al. Interleukin-18 is regulated by parathyroid hormone and is required for its bone anabolic actions. J Biol Chem 2008 Mar 14;283(11):6790–8.

[340] Cornish J, Gillespie MT, Callon KE, Horwood NJ, Moseley JM, Reid IR. Interleukin-18 is a novel mitogen of osteogenic and chondrogenic cells. Endocrinology 2003;144(4):1194–201.

[341] Gowen M, Mundy GR. Actions of recombinant interleukin 1, interleukin 2, and interferon-gamma on bone resorption in vitro. J Immunol 1986;136:2478–82.

[342] Peterlik M, Hoffmann O, Swetly P, Klaushofer K, Koller K. Recombinant gamma-interferon inhibits prostaglandin-mediated and parathyroid hormone-induced bone resorption in cultured neonatal mouse calvaria. FEBS Let 1985;185:287–90.

[343] Takahashi N, Mundy GR, Roodman GD. Recombinant human interferon-gamma inhibits formation of human osteoclast-like cells. J Immunol 1986;137:3544–9.

[344] Takayanagi H, Ogasawara K, Hida S, Chiba T, Murata S, Sato K, et al. T-cell-mediated regulation of osteoclastogenesis by signalling cross-talk between RANKL and IFN-gamma. Nature 2000 Nov 30;408(6812):600–5.

[345] Hattersley G, Dorey E, Horton MA, Chambers TJ. Human macrophage colony-stimulating factor inhibits bone resorption by osteoclasts disaggregated from rat bone. J Cell Physiol 1988;137:199–203.

[346] Gao Y, Grassi F, Ryan MR, Terauchi M, Page K, Yang X, et al. IFN-gamma stimulates osteoclast formation and bone loss in vivo via antigen-driven T cell activation. J Clin Invest 2007;117(1):122–32.

[347] Gowen M, MacDonald BR, Russell GG. Actions of recombinant human gamma-interferon and tumor necrosis factor alpha on the proliferation and osteoblastic characteristics of human trabecular bone cells in vitro. Arth and Rheum 1988;31:1500–7.

[348] Nanes MS, McKoy WM, Marx SJ. Inhibitory effects of tumor necrosis factor-alpha and interferon-gamma on deoxyribonucleic acid and collagen synthesis by rat osteosarcoma cells (ROS 17/2.8). Endocrinology 1989;124:339–45.

[349] Shen V, Kohler G, Jeffrey JJ, Peck WA. Bone-resorbing agents promote and interferon-gamma inhibits bone cell collagenase production. J Bone Min Res 1988;3(6):657–66.

[350] Smith DD, Gowen M, Mundy GR. Effects of interferon-gamma and other cytokines on collagen synthesis in fetal rat bone cultures. Endocrinology 1987;120:2494–9.

[351] Manoury-Schwartz B, Chiocchia G, Bessis N, Abehsira-Amar O, Batteux F, Muller S, et al. High susceptibility to collagen-induced arthritis in mice lacking IFN-gamma receptors. J Immunol 1997 Jun 1;158(11):5501–6.

[352] Vermeire K, Heremans H, Vandeputte M, Huang S, Billiau A, Matthys P. Accelerated collagen-induced arthritis in IFN-gamma receptor-deficient mice. J Immunol 1997 Jun 1;158(11):5507–13.

[353] Ji JD, Park-Min KH, Shen Z, Fajardo RJ, Goldring SR, McHugh KP, et al. Inhibition of RANK expression and osteoclastogenesis by TLRs and IFN-{gamma} in human osteoclast precursors. J Immunol 2009 Nov 4;4:4.

[354] Mann GN, Jacobs TW, Buchinsky FJ, Armstrong EC, Li M, Ke HZ, et al. Interferon-gamma causes loss of bone volume in vivo and fails to ameliorate cyclosporin A-induced osteopenia. Endocrinology 1994;135:1077–83.

[355] Key Jr LL, Ries WL, Rodriguiz RM, Hatcher HC. Recombinant human interferon gamma therapy for osteopetrosis. J Pediatr 1992;121:119–24.

[356] Key Jr LL, Rodriguiz RM, Willi SM, Wright NM, Hatcher HC, Eyre DR, et al. Long-term treatment of osteopetrosis with recombinant human interferon gamma. N Eng J Med 1995;332-9:1594.

[357] Vignery A, Niven-Fairchild T, Shepard MH. Recombinant murine interferon-gamma inhibits the fusion of mouse alveolar macrophages in vitro but stimulates the formation of osteoclast-like cells on implanted syngenic bone particles in mice in vivo. J Bone Min Res 1990;5:637–44.

[358] Schoenborn JR, Wilson CB. Regulation of interferon-gamma during innate and adaptive immune responses. Adv Immunol 2007;96:41–101.

[359] Takayanagi H. Inflammatory bone destruction and osteoimmunology. J Periodontal Res 2005;40(4):287–93.

[360] Takayanagi H, Kim S, Matsuo K, Suzuki H, Suzuki T, Sato K, et al. RANKL maintains bone homeostasis through c-Fos-dependent induction of interferon-beta. Nature 2002 Apr 18;416(6882):744–9.

[361] Avnet S, Cenni E, Perut F, Granchi D, Brandi ML, Giunti A, et al. Interferon-alpha inhibits in vitro osteoclast differentiation and renal cell carcinoma-induced angiogenesis. Int J Oncol 2007;30(2):469–76.

[362] Goodman GR, Dissanayake IR, Gorodetsky E, Zhou H, Ma YF, Jee WS, et al. Interferon-alpha, unlike interferon-gamma, does not cause bone loss in the rat. Bone 1999;25(4):459–63.

[363] Lewis DB, Liggitt HD, Effmann EL, Motley ST, Teitelbaum SL, Jepsen KJ, et al. Osteoporosis induced in mice by overproduction of interleukin 4. Proc Natl Acad Sci U S A 1993;90:11618–22.

[364] Nakano Y, Watanabe K, Morimoto I, Okada Y, Ura K, Sato K, et al. Interleukin-4 inhibits spontaneous and parathyroid hormone-related protein-stimulated osteoclast formation in mice. J Bone Min Res 1994;9:1533–9.

[365] Shioi A, Teitelbaum SL, Ross FP, Welgus HG, Suzuki H, Ohara J, et al. Interleukin 4 inhibits murine osteoclast formation in vitro. J Cell Biochem 1991;47:272–7.

[366] Okada Y, Morimoto I, Ura K, Nakano Y, Tanaka Y, Nishida S, et al. Short-term treatment of recombinant murine interleukin-4 rapidly inhibits bone formation in normal and ovariectomized mice. Bone 1998;22(4):361–5.

[367] Onoe Y, Miyaura C, Kaminakayashiki T, Nagai Y, Noguchi K, Chen QR, et al. IL-13 and IL-4 inhibit bone resorption by suppressing cyclooxygenase-2-dependent prostaglandin synthesis in osteoblasts. J Immunol 1996;156:758–64.

[368] Lind M, Deleuran B, Yssel H, Fink-Eriksen E, Thestrup-Pedersen K. IL-4 and IL-13, but not IL-10, are chemotactic factors for human osteoblasts. Cytokine 1995;7(1):78–82.

[369] Palmqvist P, Lundberg P, Persson E, Johansson A, Lundgren I, Lie A, et al. Inhibition of hormone and cytokine-stimulated osteoclastogenesis and bone resorption by interleukin-4 and interleukin-13 is associated with increased osteoprotegerin and decreased RANKL and RANK in a STAT6-dependent pathway. J Biol Chem 2006;281(5):2414–29.

[370] Yamada A, Takami M, Kawawa T, Yasuhara R, Zhao B, Mochizuki A, et al. Interleukin-4 inhibition of osteoclast differentiation is stronger than that of interleukin-13 and they are equivalent for induction of osteoprotegerin production from osteoblasts. Immunology 2007;120(4):573–9.

[371] Stein NC, Kreutzmann C, Zimmermann SP, Niebergall U, Hellmeyer L, Goettsch C, et al. Interleukin-4 and interleukin-13 stimulate the osteoclast inhibitor osteoprotegerin by human endothelial cells through the STAT6 pathway. J Bone Miner Res 2008 May;23(5):750–8.

[372] Bendixen AC, Shevde NK, Dienger KM, Willson TM, Funk CD, Pike JW. IL-4 inhibits osteoclast formation through a direct action on osteoclast precursors via peroxisome proliferator-activated receptor gamma 1. Proc Natl Acad Sci U S A 2001;98(5):2443–8.

[373] Kamel Mohamed SG, Sugiyama E, Shinoda K, Hounoki H, Taki H, Maruyama M, et al. Interleukin-4 inhibits RANKL-induced expression of NFATc1 and c-Fos: a possible mechanism for down-regulation of osteoclastogenesis. Biochem Biophy Res Comm 2005;329(3):839–45.

[374] Mangashetti LS, Khapli SM, Wani MR. IL-4 inhibits bone-resorbing activity of mature osteoclasts by affecting NF-kappa B and Ca2+ signaling. J Immunol 2005;175(2):917–25.

[375] Moreno JL, Kaczmareck M, Keegan AD, Tondravi M. IL-4 suppresses both osteoclast development and mature osteoclast function by a STAT6-dependent mechanism: irreversible inhibition of the differentiation program activated by RANKL. Blood 2003;102(3):1078–86.

[376] Wei S, Wang MW, Teitelbaum SL, Ross FP. Interleukin-4 reversibly inhibits osteoclastogenesis via inhibition of NF-{kappa}B and MAP kinase signaling. J Biol Chem 2001;277(8):6622–30.

[377] Felaco P, Castellani ML, De Lutiis MA, Felaco M, Pandolfi F, Salini V, et al. IL-32: a newly-discovered proinflammatory cytokine. J Biol Regul Homeost Agents 2009 Jul-Sep;23(3):141–7.

[378] Mabilleau G, Sabokbar A. Interleukin-32 promotes osteoclast differentiation but not osteoclast activation. PLoS One 2009;4(1):e4173.

[379] Baugh JA, Bucala R. Macrophage migration inhibitory factor. Crit Care Med 2002;30(Suppl. 1):S27–35.

[380] Weiser WY, Temple PA, Witek-Giannotti JS, Remold HG, Clark SC, David JR. Molecular cloning of a cDNA encoding a human macrophage migration inhibitory factor. Proc Natl Acad Sci U S A 1989;86(19):7522–6.

[381] Onodera S, Sasaki S, Ohshima S, Amizuka N, Li M, Udagawa N, et al. Transgenic mice over-expressing macrophage migration inhibitory factor (MIF) exhibit high-turnover osteoporosis. J Bone Min Res 2006;21(6):876–85.

[382] Oshima S, Onodera S, Amizuka N, Li M, Irie K, Watanabe S, et al. Macrophage migration inhibitory factor-deficient mice are resistant to ovariectomy-induced bone loss. FEBS Lett 2006;580(5):1251–6.

[383] Ashcroft GS, Mills SJ, Lei K, Gibbons L, Jeong MJ, Taniguchi M, et al. Estrogen modulates cutaneous wound healing by downregulating macrophage migration inhibitory factor. J Clin Invest 2003;111(9):1309–18.

[384] Onodera S, Suzuki K, Matsuno T, Kaneda K, Kuriyama T, Nishihira J. Identification of macrophage migration inhibitory factor in murine neonatal calvariae and osteoblasts. Immunology 1996;89(3):430–5.

[385] Onodera S, Suzuki K, Kaneda K, Fujinaga M, Nishihira J. Growth factor-induced expression of macrophage migration inhibitory factor in osteoblasts: relevance to the plasminogen activator system. Semin Thromb Hemost 1999;25(6):563–8.

[386] Onodera S, Nishihira J, Iwabuchi K, Koyama Y, Yoshida K, Tanaka S, et al. Macrophage migration inhibitory factor up-regulates matrix metalloproteinase-9 and -13 in rat osteoblasts. Relevance to intracellular signaling pathways. J Biol Chem 2002;277(10):7865–74.

[387] Jacquin C, Koczon-Jaremko B, Aguila HL, Leng L, Bucala R, Kuchel GA, et al. Macrophage migration inhibitory factor inhibits osteoclastogenesis. Bone 2009 Oct;45(4):640–9.

# 8

# Interactions Among Osteoblasts, Osteoclasts, and Other Cells in Bone

T. John Martin[1,2], Natalie A. Sims[1,2], Julian M. W. Quinn[3]

[1]ST VINCENT'S INSTITUTE OF MEDICAL RESEARCH,
[2]UNIVERSITY OF MELBOURNE DEPARTMENT OF MEDICINE, FITZROY, VICTORIA, AUSTRALIA,
[3]PRINCE HENRY'S INSTITUTE, MONASH MEDICAL CENTRE, CLAYTON, VICTORIA, AUSTRALIA

## CHAPTER OUTLINE

**Introduction** ... 228
**Osteoblasts, lining cells, and osteocytes** ... 228
  Osteoblast differentiation ... 229
  Osteoblast-derived growth factors and cytokines ... 231
    Cells of bone: osteoclasts ... 231
  The ontogenic relationship between osteoclasts and macrophages ... 232
  Control of osteoclast formation by the osteoblast lineage ... 233
  Regulation of the pro-osteoclastic signal from the osteoblast ... 235
**Bone modeling and remodeling** ... 238
**Cell communication within the bone multicellular unit** ... 240
  Initiation of remodeling ... 240
  Cessation of resorption in the BMU ... 240
  Reversal phase ... 241
  Coupling of formation to resorption ... 242
    Communication within the osteoblast lineage determines bone formation levels ... 243
    Can osteoclasts generate bone-forming activities? ... 244
    Remodeling in the anabolic action of PTH ... 245
    Local PTHrP or systemic PTH? ... 246
**Cell communication from beyond the bone multicellular unit** ... 248
  Influence of immune cells – T and B lymphocytes ... 248
    Activated T cells can stimulate bone resorption ... 249
    T-cell effects influenced by hormones ... 250
    B cells as a source of RANKL and OPG ... 250
    Lymphocyte-mediated inhibition of osteoclasts ... 251
  Osteomacs ... 253

Summary and conclusions – remodeling and the coupling of bone formation to resorption .. 254
Acknowledgments ........................................................................................................ 255
References.................................................................................................................... 255

# Introduction

In approaching the way in which cells function to generate this complicated organ, the skeleton, under the control of hormones, cytokines, the central and sympathetic nervous systems, and being itself recently realized to function as an endocrine organ, we should recall that study of cells isolated and cultured from bone only began at the end of the 1970s. What is known now is that the cells of bone, including, but not limited to, osteoblasts and osteoclasts, communicate with each other through many local signaling processes. Furthermore, each cell lineage is subject to influences from cells of the immune and nervous systems, in order to achieve the very tight control of bone modeling and remodeling that is necessary to preserve the structural integrity and form of the skeleton and allow it to repair injury. This chapter will describe the cells of bone, and how they communicate with each other and with neighboring cells in the bone microenvironment.

# Osteoblasts, lining cells, and osteocytes

The mature osteoblast is responsible for synthesis and secretion of the organic matrix of bone. Osteoblasts are derived from multipotential mesenchymal stem cells that become marrow stromal cells which can subsequently form multiple lineages including: osteoblasts, chondrocytes, myoblasts, and adipocytes. Bone formation results from a series of events: proliferation of primitive mesenchymal cells, differentiation into osteoblast precursor cells (osteoprogenitors, preosteoblasts), maturation of osteoblasts, formation of matrix, and finally mineralization of matrix. The mature osteoblast is readily recognized by its function, but the word "osteoblast" is often used loosely to describe many cells of the osteoblast lineage, including those that are carrying out functions other than the synthesis of bone. For example, with the discovery of receptor activator of NFκB ligand (RANKL) as the essential regulator of osteoclast formation, derived from the osteoblast lineage (see below), osteoclasts are often described as formed by "osteoblast-derived RANKL", but osteoclast formation does not take place adjacent to mature, synthesizing osteoblasts. Rather it is likely that cells earlier in the osteoblast lineage serve this function, a property readily shown in early-stage stromal cells co-cultured with hemopoietic cells [1]. Similarly the "hemopoietic niche" is very often referred to as being regulated by osteoblasts, but it is most likely that the cells responsible are earlier in the progression of the osteoblast lineage than the cells that form bone [2].

The life history of the osteoblast lineage, which also includes the osteocyte and the bone-lining cell, as well as mesenchymal precursors in the marrow space (or the

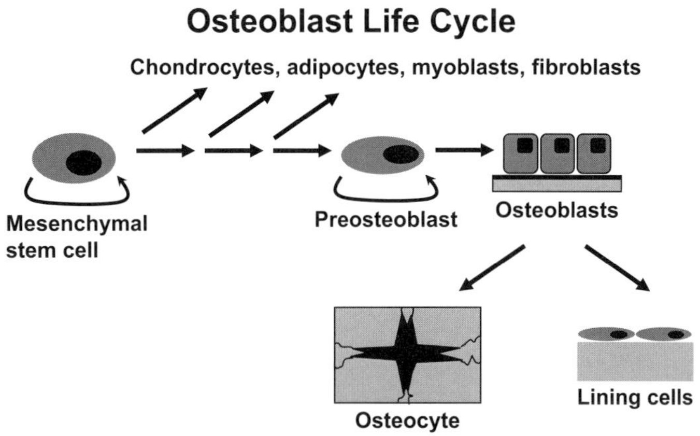

## Osteoblast Life Cycle

**FIGURE 8-1** The osteoblast develops from a mesenchymal precursor that is capable of differentiation also into cartilage, fat, or muscle cells.

periosteum), is summarized in Figure 8-1. It seems likely that at the end of the remodeling sequence, when matrix synthesis is no longer required, osteoblasts lose their synthetic capacity and either become bone-lining cells or become trapped behind the advancing calcification front, and are embedded in bone as osteocytes. The osteocytes within individual lacunae communicate with each other and with surface osteoblasts or lining cells by an extensive system of cell microprocesses within canaliculi throughout the bone matrix. Osteocytes comprise by far the most abundant cells of bone – greater than 90%. They have long been thought to function as sensors of pressure changes and micro-damage. However, only recently have new approaches allowed studies of osteocyte function, largely as a result of discoveries of mutations in genes associated with either osteoporosis or high bone mass syndromes. Indeed, ideas about the control of bone formation have been significantly revised following the discovery of sclerostin, the protein product of the *sost* gene. It is produced in bone exclusively by osteocytes and is a powerful inhibitor of bone formation, sparking renewed interest in the role of the osteocyte in bone modeling and remodeling.

## Osteoblast differentiation

Osteoblast differentiation is achieved through the co-ordinated transcription of a particular sequence of target genes. Many of the major advances in understanding transcriptional regulation of osteoblast differentiation are also relevant to osteoblast–osteoclast communication. Runx2 (cbfa1) is an essential transcriptional regulator of the transition from mesenchymal cell to osteoprogenitor. Mice rendered null for *Runx2* die at birth because of complete failure of osteoblast formation [3]. Runx2 also controls bone cell function by maintaining the differentiated phenotype of the mature osteoblast. Transgenic overexpression of a dominant negative form of Runx2 postnatally in mice led to decreased production of endogenous Runx2, as well as diminished expression of later genes associated with osteoblast differentiation, reduced bone formation rate despite normal osteoblast numbers, and osteopenia [4]. Runx2 is central to replenishment of osteoblasts after bone loss and a key requirement in restoring bone. It is also a connection between osteoblast and osteoclast regulation through its ability to activate the osteoprotegerin (OPG) promoter and enhance production of this inhibitor of osteoclast formation [5].

The search for transcription factors other than Runx2 led to identification of a novel zinc finger-containing transcription factor, osterix (Osx), which is specifically expressed in bone. Mice rendered null for the *osx* gene did not develop mineralized bone. Like *Runx2−/−* mice they had an entirely cartilaginous skeleton and died at birth [6]. In *osx−/−* mice, Runx2 mRNA was expressed at the same level as in the wild type, but the mutant mice had no Osx mRNA. This suggests that Osx is an important transcription factor in osteoblast differentiation that functions downstream from runx2, and perhaps acts co-operatively with it. Manipulation of Osx remains a route for exploration in the search for effective therapies for metabolic bone diseases, as do a number of other transcriptional regulators of osteoblast differentiation.

Many other transcription factors play a role in osteoblast differentiation, and their functions form a complex network of overlapping pathways that modify osteoblast differentiation in response to locally acting cytokines or systemic factors including parathyroid hormone (PTH). These may act by controlling Runx2 expression (upstream factors), such as Msx2 [7] and Twist1 [8, 9]. Others enhance Runx2 action, such as C/EBPβ and C/EBPδ [10] or antagonise Runx2, such as ZFP521 [11]. Work on these transcription factors has highlighted the close relationship between the osteoblast and adipocyte lineages, since a number of these transcription factors have reciprocal effects on these two cell lineages.

The importance of transcription regulation by β-catenin following activation of the Wnt pathway in osteoblast differentiation and bone formation has been evident since the discoveries that gain of function mutation in LRP5 leads to a high bone mass phenotype [12], and loss of function mutations lead to the osteoporosis-pseudoglioma syndrome [13, 14]. Activation of the canonical Wnt signaling pathway leads to stabilization of β-catenin in the cytoplasm through inhibition of glycogen synthase kinase (GSK)-3β-mediated phosphorylation, resulting in accumulation of cytoplasmic β-catenin, followed by its translocation to the nucleus and transcriptional activation of specific gene targets. Such activation in mesenchymal cells results in inhibition of chondrocyte differentiation and promotion of osteoblast activity [14]. Much evidence has emerged that canonical Wnt activation favors osteoblast differentiation and bone formation, with several approaches used to establish this role of Wnt signaling. These include development of genetically manipulated mouse models to recapitulate the human mutation syndromes [15–17], in vitro studies showing direct effects of Wnt signaling on osteoblast differentiation, co-operative effects with bone morphogenetic protein [14], and increases in bone formation in vivo with pharmacological inhibition of GSK-3β [18].

However, the effects of the Wnt signaling pathway on bone are not restricted to direct actions upon osteoblasts. Mice with an active mutant β-catenin, driven by the α1 (I) collagen promoter, have been constructed. These animals had constitutively activated Wnt signaling in osteoblasts with no significant alteration in osteoblast. Instead they showed a severe form of osteopetrosis, including failed tooth eruption [19]. This phenotype appeared to be due to failure of osteoclast formation, caused by increased production of OPG by osteoblast lineage cells in which active β-catenin was expressed. A similar osteopetrotic syndrome due to failed osteoclast development and increased OPG was observed in mice lacking the adenomatous polyposis coli protein (APC) in osteoblasts [20]. The main function of APC in the canonical Wnt signaling pathway is in a complex of proteins with GSK-3β where it contributes to maintaining the normal degradation of β-catenin. Its absence in osteoblast lineage cells leads to accumulation of β-catenin, resulting in cell-autonomous activation of Wnt signaling. β-catenin accumulation specifically in osteoblasts does not appear to be required for normal osteoblastic function, however, since no bone formation defect was observed in mice specifically deficient in LRP5 in osteoblasts [21].

## Osteoblast-derived growth factors and cytokines

Production and activation of bone growth factors is likely a vital step in stimulating bone formation in response to hormones, physical processes, and drugs. The osteoblast lineage produces growth factors that have profound stimulatory effects on bone formation as locally derived activities in bone. The most important and effective of these, IGF-1, TGFβ, fibroblast growth factors (FGFs), and the BMPs, are being investigated for possible application as local therapies in the healing of fractures and bone defects, but their effects in other organ systems preclude their use as systemic therapies. Osteoblasts also produce a range of cytokines, including interleukin-11, oncostatin-M, interleukin-6, and other locally acting factors such as PTHrP, which regulate osteoblast differentiation. These local factors will be discussed below in relation to the local processes involved in the coupling of bone formation to resorption.

### Cells of bone: osteoclasts

The only cell capable of resorbing bone efficiently is the osteoclast, a large multinucleated non-proliferating cell derived from myeloid progenitors of the monocyte–macrophage lineage. Thus, it is clearly distinct in its origin from the osteoblast, which is derived from a multipotential mesenchymal progenitor.

Osteoclasts were originally identified by Kölliker [22], who named them "ostoklasts", and suggested that they were responsible for bone resorption. Ultimately known as osteoclasts, they were recognized to possess unique ultrastructural characteristics, which both distinguished them from other cell types and enabled them to be motile and efficiently resorb bone [23, 24]. Apart from their multinuclearity, a striking feature of the osteoclast is the presence of a "ruffled border", which is a complex structure of deeply interfolded finger-like projections of the plasma and cytoplasmic membranes, adjacent to the bone surface and through which bone-resorbing acids and enzymes pass [25, 26].

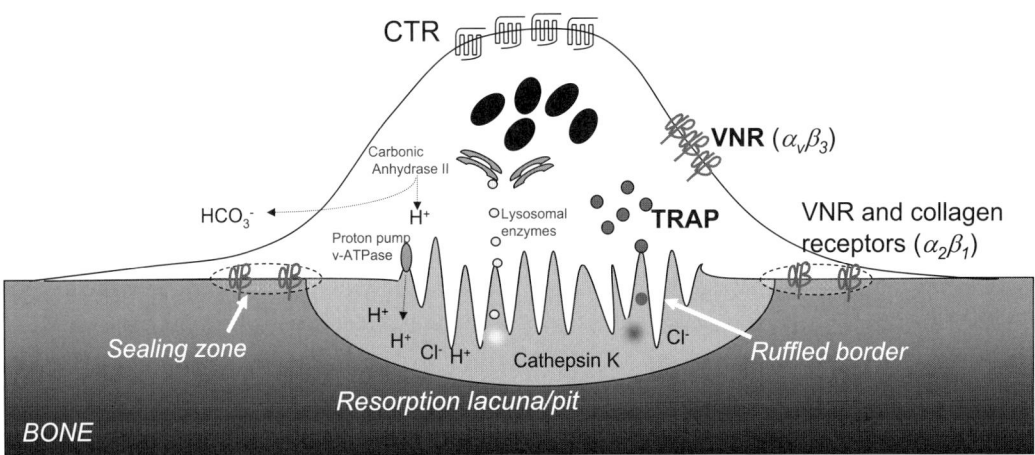

**FIGURE 8-2** The multinucleate osteoclast adheres to bone, seals off a resorption space that it acidifies (see text for details). CTR, calcitonin receptor; TRAP, tartrate-resistant acid phosphatase; VNR, vitronectin receptor.

Adjacent to and surrounding the ruffled border is the clear zone. This is an area of cytoplasm devoid of cellular organelles except for numerous cytoplasmic actin filaments. The clear zone is also known as the "sealing zone", since the plasma membrane in this region comes into very close apposition with the bone surface to ensure osteoclast attachment, and to separate the bone-resorbing area beneath the ruffled border from the unresorbed area, so as to maintain a favorable microenvironment for bone resorption [26]. Osteoclasts bring about dissolution of bone mineral by creating an acid microcompartment under the ruffled border, adjacent to the bone surface.

Osteoclasts are rich in tartrate-resistant acid phosphatase (TRAP), which is a commonly used histochemical marker for osteoclasts, although not exclusive to those cells. It is nevertheless a convenient marker for cells generated in vitro when combined with identification of calcitonin receptors [27] and the ability of the cells to form resorption pits on thin slices of cortical bone or dentine. Some other properties are indicated in Figure 8-2, including possession of vitronectin receptors, cathepsin K, vacuolar ATP-ase, and chloride-7 channels. This combination of properties provides the phenotype that equips osteoclasts uniquely to resorb bone efficiently.

## The ontogenic relationship between osteoclasts and macrophages

The origin of osteoclasts is considered in detail in Chapter 2 but it is relevant to this chapter to discuss certain aspects. Early views of osteoclast origin were that they were derived from cells of a connective tissue (later known as mesenchymal) origin. Thus Tonna [28] concluded that osteoclasts arise from fusion of osteoblasts and that osteoclasts can dissociate again into osteogenic precursor cells, and Young [29] believed that osteoclasts and osteoblasts originate from a common osteoprogenitor cell, returning at a later stage to the osteoprogenitor pool. These and similar views of Rasmussen and

## Chapter 8 • Interactions Among Osteoblasts, Osteoclasts, and Other Cells in Bone

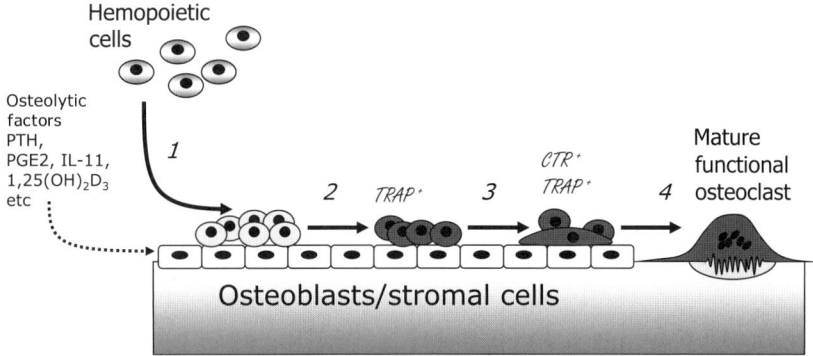

**FIGURE 8-3** The formation of osteoclasts from hemopoietic precursors requires contact with osteoblast-lineage cells, and the osteoclast formation process in these circumstances is stimulated by a number of hormones and cytokines (see text for details).

Bordier [30] were subsequently superseded in the face of compelling evidence for the derivation of osteoclasts from the monocyte/macrophage lineage [31], with many demonstrations that osteoclasts can be generated from proliferating progenitors that can be considered either as macrophages or as immature macrophage precursors [32–34] (Figure 8-3). There is also evidence that osteoclasts can derive from early hematopoietic progenitors, including certain types of early B-cell progenitors that display significant lineage plasticity and may form macrophages and osteoclasts [35, 36]. Specifically, mice lacking the *Pax5* gene display a greatly increased osteoclast formation and early-onset bone loss that is probably due to increased number of osteoclast progenitors [37]; *Pax5* encodes a transcription factor BSAP, essential for B-cell maturation [38], which suppresses expression of gene products not normally found in B cells.

Despite their highly specialized function as bone-resorbing cells, osteoclasts display many histochemical similarities and functional parallels with tissue and inflammatory macrophages [26]. Nevertheless, mature osteoclasts normally express only low levels of MHC class II molecules, CD14 and receptors for immunoglobulin Fc and complement [39–41]. In addition, they have not been described as antigen-presenting cells. A bone tissue-specific macrophage has been identified recently (osteomacs). This will be considered separately (see below).

### Control of osteoclast formation by the osteoblast lineage

Shortly after it became possible to culture bone cells in vitro, it was suggested that osteoblast-lineage cells might program the activity and generation of osteoclasts. The observations that isolated osteoblasts of various origins responded to bone-resorbing hormones and possessed receptors for these factors, in addition to the lack of evidence for receptors or direct responses to these hormones in osteoclasts, led to the concept that bone-resorbing factors must act first on osteoblast lineage cells, most likely bone-lining cells. This action was proposed to release factors that influence the formation and bone-resorbing activity of osteoclasts [42, 43]. Furthermore, since osteoclasts are derived from hemopoietic progenitors and not from a local bone cell, Chambers came to the same conclusion by arguing that since the osteoclast derives from a "wandering" cell, it made sense to have its activity programmed by an authentic bone cell, i.e. the osteoblast [44]. The first direct evidence in support of the importance of intercellular communication in bone came in the early 1980s, when it was shown that osteoclasts, which were isolated from newborn rat or mouse bone, required the presence of contaminating osteoblastic cells to be fully active and resorb bone [45, 46].

Work over the next few years established the concept of an osteoblast-derived stimulator of osteoclast-mediated resorption [45, 47, 48] and by the mid-1980s the field was at the stage of accepting that intercellular signaling may be an important mechanism in bone cell biology. Among the questions raised were whether a single stromal/osteoblastic cell factor existed that is responsible for osteoclast activation, and if so is it cell-associated or secreted? Would the formation of osteoclasts be at least as important as regulation of their activity? The next real advances came with the development of methods to study osteoclast formation in vitro.

Several in vitro systems provided strong evidence that accessory cells are indeed necessary for the generation of osteoclasts from hemopoietic precursors. Burger et al. [49] used a co-culture system in which hemopoietic cells from embryonic mouse liver were co-cultured with fetal long bone rudiments from which the periosteum had been stripped, to show that living bone cells are required for osteoclast development. However, it was the development of murine bone marrow cultures and reproducible assays of osteoclast formation that led to major advances [50, 51]. These were used first to show that treatment with bone-resorbing agents such as 1,25-dihydroxy-vitamin $D_3$ ($1,25(OH)_2D_3$) could promote osteoclast formation in a dose-dependent manner, with osteoclast quantitation carried out by counting TRAP-positive multinucleated cells that were also CT-receptor-positive by receptor autoradiography. In the course of these studies, Takahashi et al. [52] made an observation that turned out to be a crucial one. They noted consistently that more than 90% of the TRAP-positive mononucleated cell clusters and multinucleated cells formed in mouse marrow cultures in response to bone-resorbing stimuli were located near colonies of alkaline phosphatase-positive mononucleated cells (possibly osteoblasts). They regarded this as a strong indication that osteoblastic cells are involved in osteoclast formation, in addition to the evidence produced in the few earlier years of their influence on osteoclast activity. They set out to determine whether close contact between osteoclast progenitors and osteoblastic cells was necessary in order for osteoclast formation to occur.

They did so and established beyond doubt that osteoclast formation requires a contribution from cells of the osteoblast lineage (outlined in Figure 8-3). In doing this they provided the concepts and techniques that set the scene for the discovery of osteoclast control by RANKL, RANK, and OPG. Their first, relatively simple experiment was remarkably informative. They prepared osteoblast-rich cultures from newborn mouse calvariae and grew them in co-culture with mouse spleen cells, and on treatment with $1,25(OH)_2D_3$, osteoclasts were formed. Most importantly though, separation of osteoblastic cells from spleen cells by a 0.45-µm membrane filter in co-cultures prevented osteoclast-like cell formation [50], indicating that direct contact is required between the two cell types in order for osteoclast formation to occur (Figure 8-3). Similar results were obtained with the bone-marrow-derived stromal cell lines MC3T3-G2/PA6, ST2, and KS-4 [1, 53], any of which could be substituted for primary osteoblastic/stromal cells in co-cultures with spleen cells, to facilitate the formation of osteoclast-like cells in the presence of $1,25(OH)_2D_3$.

## Regulation of the pro-osteoclastic signal from the osteoblast

With increasing acceptance of the concept that cells of the osteoblast lineage control osteoclast formation and activity by a contact-dependent mechanism, it was important to understand how this process was regulated. Three distinct separate signaling mechanisms were all capable of promoting this function. Prostaglandin (PG)-induced osteoclast formation in mouse bone marrow cultures was found to be mediated by a mechanism involving cAMP [54]. The potencies of the PGs in this respect were greatest for $PGE_1$ and $PGE_2$, followed by $PGF_{2\alpha}$, which correlated closely with their relative potencies in increasing cAMP production in osteoblastic and bone marrow cells and in increasing bone resorption in organ culture [42]. Likewise, PTH and PTHrP, acting through their common receptor, promoted osteoclast formation in marrow cultures by a cAMP-dependent mechanism [55], and the effect of interleukin-1 (IL-1) resulted from the generation of PGE2 as an intermediate effector [56].

A second signaling mechanism for regulation was provided by the steroid hormone, $1,25(OH)_2D_3$, which had very similar effects on osteoclast formation in marrow cultures and in co-cultures of osteoblastic with hemopoietic cells [52]. $1,25(OH)_2D_3$ uses an entirely different signaling system, in which it combines with its receptor and translocates to the nucleus to influence transcription through a vitamin-D-responsive element in target genes including RANKL [57].

Finally, a membrane-bound receptor complex involving a 130-kDa glycoprotein (gp130) [58] provides for osteoclast formation under the influence of the group of cytokines that use this signaling mechanism. In mouse co-cultures, simultaneous treatment with IL-6 and its soluble receptor (sIL-6R) induced osteoclast formation, but when added separately they were ineffective. The other cytokines in this group, IL-11, leukemia inhibitory factor (LIF), and oncostatin M (OSM), all of which use gp130 as a common transducer, also stimulated osteoclast formation. In following up this observation, using cells from IL-6R-overexpressing transgenic mice in crossover co-cultures with cells from wild-type mice, expression of IL-6R by osteoblastic cells was shown to be indispensable for the induction of osteoclasts [59]. This clear demonstration that IL-6 stimulation of osteoclast formation required the cytokine to act upon the osteoblast, despite the fact that osteoclasts possessed its receptor [60], illustrated the power of using ex vivo experimentation with cells from genetically modified animals to study osteoclast formation, an approach that has proved itself repeatedly since that time.

Thus, the concept that stromal/osteoblastic cells regulated osteoclastogenesis, and, in turn, were regulated by a number of circulating and local factors was firmly established. Despite the fact that they fell into three main classes with respect to their initial signaling mechanisms (Figure 8-3), it seemed that a common pathway for these agents was a membrane-bound stromal factor [51]. It was assumed that these agents must converge in their actions at some stage before finally generating this crucial membrane factor [61].

Of the many multifunctional cytokines that had some role in osteoclast formation, none provided an explanation for the molecular regulation of osteoclast formation and activity that was evident from the foregoing studies. In a form of murine osteopetrosis resulting from a mutation in the coding region of the M-CSF gene in the *op/op* mouse [62, 63], M-CSF was found to play an essential role in both proliferation and differentiation of osteoclast progenitors [56]. On the other hand M-CSF inhibited the bone-resorbing activity of isolated osteoclasts. Bone resorption in organ culture was reduced by M-CSF, GM-CSF, and IL-3 [64], and all three cytokines inhibited the generation of osteoclasts in mouse bone-marrow cultures. The conclusion from these and other observations was that none of these hemopoietic growth factors fulfilled criteria including the specific membrane-associated activity required for osteoclast formation that was so strongly suggested by existing data.

The discovery of OPG, a soluble member of the TNF receptor superfamily, revealed it as a powerful inhibitor of osteoclast formation [65, 66]. This provided the means of identifying and cloning the elusive osteoclastogenic factor, which proved to be a TNF ligand family member that came to be called receptor activator of nuclear factor κB ligand (RANKL), the common factor mediating osteoclast formation in response to all known stimuli [67, 68]. Osteoblasts/stromal cells are also the source of M-CSF, which plays a crucial role in osteoclast formation by promoting the proliferation of precursors. As a membrane protein, RANKL fulfilled the predictions of earlier work, that osteoclast differentiation required contact-dependent activation of hemopoietic precursors. The communication with the hemopoietic lineage results from RANKL binding to its receptor, RANK, on osteoclast precursors, thereby initiating signaling essential for osteoclast differentiation. Figure 8-4 summarizes these control mechanisms, illustrating the regulated production of RANKL that is essential for osteoclast formation and maintenance of activity. The bone-resorbing cytokines and hormones, with disparate signaling mechanisms, converge in promoting RANKL production, with the decoy receptor, OPG, having an essential physiological role as a paracrine inhibitor of osteoclast formation, produced by the osteoblasts and binding RANKL to limit its activation of osteoclast formation through its receptor, RANK.

All of these discoveries have been validated by studies in genetically altered mice, establishing clearly the essential physiological role of these TNF ligand and receptor family members in controlling osteoclast formation and activity. First, transgenic

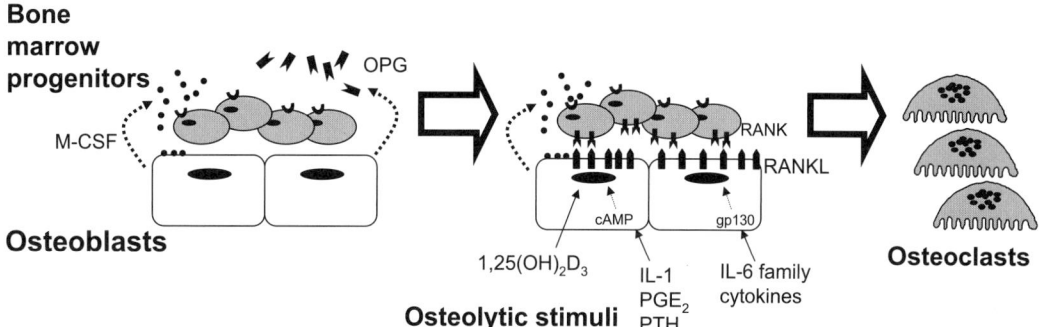

**FIGURE 8-4** Osteoblasts and related stromal cells produce soluble and membrane-bound M-CSF, which causes c-fms+ progenitors to proliferate and to upregulate expression of RANK. The osteoblasts also secrete high levels of OPG, the soluble decoy receptor of RANKL. Osteoblastic RANKL expression is enhanced, and OPG secretion decreased, by three types of osteolytic factors: 1,25(OH)$_2$D$_3$ which binds VDR, IL-6 family cytokines that act via gp130-dependent receptors, and factors that induce osteoblast cAMP responses.

overexpression of OPG in mice results in osteopetrosis because of failure to form osteoclasts [65]. Genetic ablation of OPG, on the other hand, leads to severe high-turnover osteoporosis [69]. Results of ex vivo studies of bone organ cultures and cells from the latter animals were consistent with the concept of OPG being constantly produced, and acting physiologically as a paracrine "brake" on RANKL production, with its production varied under the influence of local and hormonal factors [70]. Removal of the stimulatory pathway by genetic ablation of RANKL also resulted in osteopetrosis since RANKL is necessary for normal osteoclast formation [71]. Finally, genetic ablation of RANK also leads to osteopetrosis [72]. Because this signaling pathway also functions in immune cells, RANK null mice have severe abnormalities in that system, reflecting an intriguing link with the immune system, which was evident very early in the discovery process.

These crucial local factors link the immune system and bone metabolism. Two other groups identified and cloned RANKL, but these groups were interested in its role in the immune response [73, 74]. They identified and characterized a TNF-related activation-induced cytokine (TRANCE) during a search for apoptosis-regulatory genes in murine T-cell hybridomas and found it to be predominantly expressed on T cells and in lymphoid organs and controlled by the T-cell receptor through a calcineurin-regulated pathway. The putative receptor for TRANCE was detected on mature dendritic cells [73]. In studying the processing and presentation of antigens by dendritic cells to T cells, Anderson et al. [74] characterized receptor activator of NF-κB (RANK), a new member of the TNF receptor family derived from dendritic cells, and its ligand RANKL, which they recognized to be identical to TRANCE [73]. Furthermore, production of soluble RANKL by activated T cells directly stimulated osteoclast formation in vitro and in vivo [75], with important pathophysiological implications in inflammatory bone diseases.

Physiological control of bone resorption is thus dependent on the function of RANKL, but this is not restricted to its actions in promoting osteoclast formation. As predicted from the earlier demonstration of osteoclast activation through contact with osteoblastic cells [45], RANKL also modified osteoclast survival and activity [76]. The concepts resulting from these discoveries are summarized in Figure 8-4.

By treating with RANKL and M-CSF it became possible to prepare osteoclasts in relatively large numbers without the participation of stromal/osteoblastic precursors, including the preparation of human osteoclasts from peripheral blood [77, 78]. As predicted, important osteotropic hormones and cytokines, including $1,25(OH)_2D_3$, IL-11, PTH, and $PGE_2$, stimulated osteoblast lineage RANKL production, triggering the development of osteoclasts, and the same treatments reduced production of the RANKL decoy receptor, OPG [75, 79].

RANKL also has the ability to limit its own osteoclastogenic effect by promoting interferon-β (IFN-β) production by monocytic osteoclast precursors [80]. This inhibits osteoclast formation by preventing RANKL-induced expression of *c-fos*. The latter was known as an essential transcription factor in osteoclast differentiation, since *c-fos* null mice were osteopetrotic because of failed osteoclast formation [81].

The functions of RANKL extend to pathological states of increased bone resorption, where increased local RANKL production contributes to osteoclast-mediated bone destruction in disease states such as breast cancer metastases to bone [82] in rheumatoid arthritis [75], multiple myeloma [83], and osteoporosis [84].

# Bone modeling and remodeling

During fetal bone development and during bone repair, woven bone, characterized by a random (woven) organization of its collagen, is produced. It is then remodeled to form lamellar bone, with arrays of collagen fibers in parallel, which is the form that constitutes most of the mature skeleton. In compact (cortical) bone, lamellar bone is formed as a solid mass enclosing the marrow cavity and constituting approximately 80% of the total bone mass. In cancellous (trabecular) bone the mass is a spongy one, traversing the marrow cavity. It is mainly found in the vertebrae, the flat bones and in the juxta-articular epiphyses of the long bones, forming an important part of the stromal microenvironment of the bone marrow.

In the adult human skeleton, approximately 5–10% of the existing bone is replaced every year by bone remodeling. The remodeling process, which continues throughout adult life, is an integral part of the calcium homeostatic system and provides the mechanism for resorptive removal of old bone, adaptation to mechanical stress, repair of micro-damage and replacement of that bone by bone formation. In order to achieve this, bone is continuously resorbed and reformed at about 1–2 million microscopic remodeling foci per adult skeleton. Within each of these "basic multicellular units" (BMUs), focal resorption is carried out by hematopoietically derived osteoclasts and takes about 3 weeks per site, whereas the refilling of lost bone by osteoblasts, derived from bone marrow stromal cells and circulating precursors, takes about 3–4 months [85, 86].

The several purposes of remodeling, and the distribution of BMUs asynchronously throughout the skeleton, imply that there might be many varied signals used to initiate remodeling cycles. Thus, for example, the age of bone has been shown to play an important role in controlling osteoclast-mediated resorption, with significantly higher levels of osteoclast differentiation, resorption, and survival when osteoclast formation and activity were assessed on aged bones in comparison to young bones [87]. As a different example, the remodeling sequence is initiated when mechanical deformation or micro-cracks in old bone provoke signaling that leads to osteoclast formation and bone resorption (Figure 8-5). Although the initiating mechanisms might differ at various sites, the sequence of events is the same at BMUs that are both geographically and chronologically separated from each other throughout the skeleton. Both bone resorption and bone formation occur at the same site in these BMUs, so that there is no change in the shape of the bone [85, 86]. The cell supply to BMUs will be discussed later. In the case of trabecular bone, a given volume is more rapidly turned over than the same volume of cortical bone because it is fashioned with more surface upon which remodeling occurs. Provided bone remodeling remains balanced, with the same volumes of bone removed and formed within each BMU, no change in cortical thickness, trabecular number, thickness, or connectivity occurs. Circulating hormones contribute to this tight control, but the key influences are locally generated cytokines that are the signals mediating information transfer among osteoblasts, osteoclasts, immune cells, and constituents of the bone matrix.

In addition to remodeling, bone modeling on its periosteal surface is characterized by bone formation without prior bone resorption. This process, so vigorous during growth, establishes the adult size and shape of bone. Modeling and remodeling during growth achieves peak bone strength and continued remodeling during adulthood maintains the mechanical integrity of the skeleton.

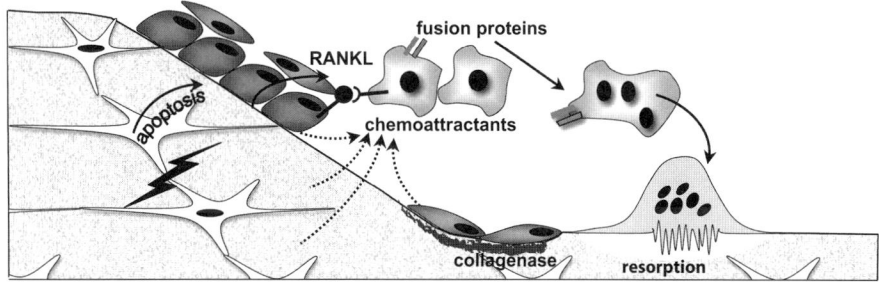

**FIGURE 8-5** Remodeling sequence can be initiated by damage, with signaling from the osteocyte to surface cells to promote osteoclast precursor recruitment and differentiation, under the influence of RANKL produced by the osteoblastic-lineage cells. The bone surface is prepared for attachment of the mature osteoclasts through collagenase digestion of surface protein by osteoblastic cells.

# Cell communication within the bone multicellular unit

## Initiation of remodeling

The very nature of the remodeling process, occurring as it does in different parts of the skeleton at different times, highlights the importance of locally generated and regulated factors in the process. Initiation of the cycle, say by mechanical strain, damage, or signals from cytokines or systemic factors, would generate local signals, leading to digestion by collagenase of the thin layer of non-mineralized matrix under the lining cells to expose the mineralized matrix which osteoclasts can resorb [88–90]. As well as RANKL production by osteoblast-lineage cells in response to cytokines and systemic factors, osteocytes are likely an important source of signaling to surface cells of the need either for increased bone formation or the initiation of resorption [91–93]. According to this model of initiation of resorption, osteoclasts or their late precursors near to final maturation and activation must be available close to the sites where they are required. Only recently, evidence was obtained for this possibility, with the discovery of the "osteoclast niche" – cell-cycle-arrested quiescent osteoclast precursors (QuOPs) that are RANK and c-fms positive and can be detected along bone surfaces near to osteoblasts [94]. QuOPs are capable of rapid differentiation into osteoclasts, and in addition to providing a potential source of osteoclasts for the rapid initiation of the remodeling sequence, their existence might explain why injection of PTH in vivo can result in the induction of active osteoclasts in less than 30 minutes [24].

## Cessation of resorption in the BMU

Once a BMU site is activated, resorption within that site must be limited. If it were to continue unchecked, resorption would be excessive. An important unanswered question about osteoclast behavior and the control of resorption in the remodeling cycle is: how does each osteoclast know when to stop resorbing? The process is likely to finish with osteoclast death, for which a number of contributory mechanisms have been investigated, but its regulation in vivo remains obscure. Osteoclasts phagocytose osteocytes, which might provide a mechanism to remove the signal for resorption [95, 96]. TGFβ, which can promote osteoclast apoptosis, is available when it is resorbed from matrix and activated by acid pH [97]. A direct effect of estrogen to enhance osteoclast apoptosis through Fas ligand mediation has been identified in mouse genetic experiments [98]. The B-cell lymphoma 2 family members, Bcl-xL and Bcl2, are both anti-apoptotic for osteoclasts [99, 100], although their actions are not restricted to apoptosis in bone. Osteoclast-specific deletion of $Bcl_{XL}$ surprisingly led to osteopenia due to enhanced osteoclast activity, despite increased apoptosis. Global deletion of Bcl2, however, leads to increased bone mass and reduced osteoclast numbers, presumably due to enhanced apoptosis [100]. Some further insights into control of apoptosis arise from genetic and pharmacological studies showing that inhibition of acidification of the resorption space by blockade of either ClC-7 or the V-type H+ATPase of the osteoclast results in prolonged osteoclast survival [101–103]. This might suggest a role for acidification in determining osteoclast lifespan, perhaps through TGFβ activation. Although candidates have been identified that influence osteoclast apoptosis, no studies yet exist that reveal apoptotic mechanisms operating specifically in the BMU.

A novel aspect of signaling in osteoclast development that could be relevant to the limitation of osteoclast generation and activity within the BMU is the induction of IFN-β by RANKL derived from osteoblast-lineage cells [80]. This provides an appealing mechanism that could contribute to inhibition of osteoclast formation within the BMU (see above). A further mechanism of osteoclast restraint within the BMU might come from reverse signaling through osteoblast-derived EphB4 acting on ephrinB2 in the osteoclast to restrict osteoclast formation [104]. If this process does indeed operate in vivo, it would require direct contact between osteoclasts and the osteoblast-lineage source of RANKL. Although such proximity is not readily seen in histological sections, the topography of cells in a BMU might support this. It was noted in a study of the reversal phase of bone remodeling that virtually all osteoclasts attached to bone in a BMU were in close contact with osteoblast-lineage bone-lining cells, and cytoplasmic extensions of these cells contained small vesicles [105].

## Reversal phase

Toward the end of resorption, mononuclear cells are seen at the bottom of resorption pits, where they remove demineralized collagen and prepare the pits for the engagement of osteoblasts to restore bone [106]. The mononuclear cells ascribed this function have sometimes been fibroblasts, but more commonly macrophages. Recent work, applying light and electron microscopy to a number of in vitro models, describes how bone-lining cells, through activation of matrix metalloproteinases, carry out the task of cleaning resorption pits after osteoclasts have completed their task [105]. To do this they must be able to respond to cytokines and growth factors that most likely govern these steps, and must use specific properties that they acquire and retain as part of the osteoblast lineage, even though they have not retained the ability of mature osteoblasts to make copious amounts of bone matrix.

In the reversal phase these cells nevertheless can lay down a thin layer of collagen along the Howship's lacuna, closely associated with a cement line containing an abundance of osteopontin, which is produced by both osteoclasts and osteoblasts. This is an RGD-containing extracellular matrix protein, which interacts with integrin receptors $\alpha_v\beta_3$ in osteoclasts and primarily $\alpha_v\beta_5$ in osteoblasts. These receptors were shown not only to mediate cell attachment of both cell types to the bone matrix, but also to act as signal-transducing receptors [107]. The influence of osteopontin on osteoclast or osteoblast activity is not fully established, but it appears to be required for a normal response of bone to mechanical unloading, matrix mineralization [108], and PTH-induced osteoclast formation [109]. The presence of osteopontin on the reversal line raises the possibility that it may be one of the signals for cessation of osteoclast activity or initiation of osteoblastic bone formation or possibly both.

## Coupling of formation to resorption

The maintenance of trabecular and cortical bone mass and architecture in accordance with loading requirements demands that bone resorption and formation be coordinated, such that a high or low level of focal resorption is usually associated with a similar but not necessarily identical change in the level of focal bone formation in the BMU. The theory that resorption is followed by an equal amount of formation came to be known as "coupling". Although it has long been appreciated that bone formation is tightly coupled to bone resorption in normal adult bone turnover [85, 110, 111], this coupling can be dissociated in some circumstances, for example during skeletal growth, in postmenopausal osteoporosis, and in some, but not all, osteopetrotic mutations [102].

A favored hypothesis for some years is that coupling is regulated by bone-formation factors released from the bone matrix during bone resorption [112, 113]. Indeed, a large number of substances which are mitogenic to osteoblasts or which stimulate bone formation in vivo can be extracted from bone matrix [114]. These include insulin-like growth factor (IGF) I and II, transforming growth factor $\beta$ (TGF$\beta$) 1 and 2, bone morphogenetic proteins (BMPs) 2, 3, 4, 6, and 7, and platelet-derived growth factor (PDGF). The amount of bone resorbed by the osteoclast would determine the concentration of factors released, thereby modifying bone formation in a manner proportional to resorption. Several questions should be considered regarding the role of TGF$\beta$ and other growth factors in the coupling of bone formation to bone resorption: (1) which cells produce them and under what circumstances; (2) do they stimulate bone formation in vivo; (3) can they be released from the matrix in an active form and in controlled amounts during bone resorption; (4) is there evidence for an increase in the abundance of these substances at sites of bone remodeling; and (5) are there regulated mechanisms by which they are activated? The latter is most important, relating to the time course and the distance between the resorption and formation processes, and whether the amount of growth factors, and therefore the extent of activation of the osteoblast lineage, can be controlled with sufficient precision in this way.

One specific function ascribed to TGF$\beta$-1 is that upon its release from bone by resorption, it can induce bone marrow stromal cell migration to resorbed surfaces, making them available for osteoblast differentiation under the influence of other growth factors and cytokines that become available in that environment [115]. Such a mechanism might explain how osteoblast precursors, which are many microns away from sites of osteoclastic resorption, are regulated. It seems unlikely, though, that this would be an exclusive mechanism of providing osteoblast precursors to a BMU. Of the several potential sources of osteoblast precursors, one is that lining cells, the single layer of flattened cells that have ceased their bone-forming function, can revert to that activity. Other likely sources are adjacent marrow stromal cell precursors and blood-borne osteoblast precursors that could be presented to the BMU from capillaries within the sinus structure provided by the BRC underneath its canopy [116–118]. Osteoblast progenitors are associated with vascular structures in the marrow and several studies suggest there may also be common progenitors giving rise to cells forming the blood vessel and pluripotent perivascular cells [119–123].

*Communication within the osteoblast lineage determines bone formation levels*

The coupling of osteoblast-mediated bone formation to bone resorption requires understanding of how osteoblast-lineage cells, forming bone within the BMU during balanced bone remodeling, replace virtually precisely the amount of bone that has been lost. Some evidence indicates that once the formation process is initiated, the participating cells themselves sense spatial limits, most likely through chemical communication within that population of cells [124]. Growth factors and cytokines produced by the osteoblasts are candidate mediators, as is gap junction communication between the osteoblasts themselves [125]. The ability of cells to respond to topography is not unique to osteoblasts, and it is not clear whether osteoblasts preferentially form matrix on disrupted bone because they "sense" the change in physical dimensions of the bone or because they detect a change in the composition of the surface, and both scenarios are possible. Osteoblast precursors have been shown to respond to changes in surface topography, whether the change is larger or smaller than the cell itself [126]. Altered nanotopography induces the formation of osteoblast filipodia, important for topographical sensing, followed by cytoskeletal changes involving cell adhesion and differentiation as well as altered expression of osteocalcin and osteopontin. Regulation from outside the BMU population could also be provided by osteocytes, most notably by the production of sclerostin, a powerful inhibitor of bone formation that is suppressed by several stimuli that increase bone mass.

Another possible local control mechanism within the osteoblast lineage comes with the finding that PTH and PTHrP promote production by osteoblasts of ephrinB2 which acts through its receptor, EphB4, to promote osteoblast differentiation and bone formation within the BMU [127]. Ephrin/Eph family members are recognized as local mediators of cell function in a diverse range of cells through contact-dependent processes in development and in maturity [128, 129]. A particular feature is their capacity for bi-directional signaling, in that when an ephrin acts upon its corresponding Eph receptor tyrosine kinase, the latter can signal in the reverse direction (Figure 8-2) by promoting rapid phosphorylation on highly conserved tyrosine residues within the PDZ domain of the ephrin ligand [130, 131]. The main ligand for EphB4 is ephrinB2, with ephrinB1 interacting with much less affinity [131]. Some evidence suggests that osteoclast-derived ephrinB2 acts through a contact-dependent mechanism on EphB4 in osteoblasts, to promote osteoblast differentiation and bone formation, and that through reverse signaling, osteoblast-derived EphB4 acts upon ephrinB2 in osteoclasts to suppress osteoclast formation [104]. The latter mechanism would require osteoblast–osteoclast contact, with evidence for the proximity of EphB4-positive osteoblast precursors to osteoclasts being provided by De Freitas et al. [132]. In the same work ephrinB2-positive osteoblasts were close to bone-formation sites, with the overall conclusion that the ephrin/Eph signaling pathway could function both within the osteoblast lineage to favor formation of bone [127], and between osteoblast and osteoclast to limit resorption [104]. These mechanisms warrant further study since pharmacological manipulation of ephrinB2/EphB4 signaling might be an attractive means of regulating the volumes of bone formed in the BMU.

Secondly, among the newly recognized functions of the osteocyte, its role in limiting bone formation is having a major impact on thinking in bone biology. Mutations in the *sost* gene were found responsible for the rare sclerosing bone dysplasias, sclerosteosis and Van Buchem disease. Each is characterized by a greatly increased amount of bone. The *sost* gene product, sclerostin, is produced exclusively in bone by osteocytes, and is a negative regulator of bone formation, inhibiting Wnt signaling through binding to the LRP5/6, thus preventing its participation in the receptor complex that activates the Wnt pathway and bone formation [133]. Production of sclerostin by osteocytes is rapidly decreased by treatment with PTH or PTHrP [134, 135], as well as by mechanical loading [92] and actions of cytokines including OSM, CT-1, and LIF [136]. For each of these stimuli, removal of sclerostin as a constitutive inhibitor of bone formation could at least partly explain the accompanying increased bone formation. Physiologically, rapid changes in osteocyte production of sclerostin could signal to surface cells to limit the filling of remodeling spaces by osteoblasts, in addition to keeping lining cells in a quiescent state on non-remodeling bone surfaces.

A further controlling influence upon osteoblast differentiation from within the osteoblast lineage comes from a study that identified a mechanism in which mature osteoblasts directed the differentiation of early mesenchymal osteoblast precursors through activation of canonical Wnt signaling [137]. Such a mechanism has the added advantage that its regulation would be susceptible to inhibition by osteoblast-derived inhibitors of Wnt signaling, such as a dickopf (Dkk) protein or secreted frizzled related protein (sFRP). Zhou et al. [137] provided evidence for inhibition of the process by sFRP1.

*Can osteoclasts generate bone-forming activities?*

Observations made in genetically manipulated mice and human genetics suggest that the osteoclast itself could also be the source of an activity that contributes to the fine control of osteoblast function in bone remodeling. In individuals with the osteopetrotic syndrome, ADOII, due to inactivating mutations in the chloride-7 channel (ClC-7), bone resorption is deficient because of failure of the osteoclast acidification process. Bone formation in these patients is nevertheless normal, rather than diminished as might be expected because of the greatly impaired resorption [138]. Furthermore, in mice deficient in c-*src* [139, 140], cathepsin K [141], or tyrosine phosphatase epsilon [142], bone resorption is inhibited without inhibition of formation, and v-ATPase V0 subunit D2-deficient mice exhibit failure of fusion of osteoclast precursors accompanied by increased bone formation [143]. In these knockout mouse lines osteoclast resorption is greatly reduced by the mutation, although osteoclast numbers are not reduced. Indeed in some, osteoclast numbers are actually increased because of reduced osteoclast apoptosis. A possibility is that these osteoclasts, although unable to resorb bone, are nevertheless capable of generating a factor (or factors) contributing to bone formation. This might even apply to the TRAP-expressing mononuclear osteoclast precursors, which accumulate in the v-ATPase-deficient mice [143]. On the other hand, mice lacking *c-fos,* which are unable to generate osteoclasts, have reduced bone formation as well as resorption [81]. Studies with other mutant mice, with specific inactivation of each of the two alternative signaling pathways through gp130, led to the conclusion that resorption alone was insufficient to promote the coupled bone formation, but active osteoclasts are the likely source [144, 145].

In a search for osteoclast products that might contribute to the coupling process, evidence was obtained in mice, using genetic and pharmacological approaches, that osteoclast-derived ephrinB2 acts through a contact-dependent mechanism on EphB4, its receptor in the osteoblast, to promote osteoblast differentiation and bone formation (see above) [104]. EphB4-positive osteoblast precursors have been noted to gather close to osteoclasts in mouse bone [132]. There are likely to be several such factors, and another possibility is cardiotropin-1 (CT-1), a member of the family of cytokines that signal through the gp130 transducer and is expressed in differentiated osteoclasts but not in the osteoblast lineage [146]. As well as indirectly stimulating osteoclast differentiation through stimulation of RANKL production, CT-1, like LIF and IL-11, powerfully stimulates bone formation using a mechanism that begins with rapidly increased expression of the C/EBPδ transcription factor. Intriguingly, CT-1, like oncostatin M, LIF and PTH, also profoundly decreases sclerostin mRNA expression by osteocytes [136], thus introducing the concept that osteoclast products might communicate with the osteoblast lineage by signaling directly to the osteocyte. CT-1 may be a coupling factor that signals from the osteoclast to the osteoblast in more than one way to promote bone formation and contribute to the regulated process of remodeling in the BMU.

## Remodeling in the anabolic action of PTH

Some observations that led to interest in the possibility of an osteoclast influence on bone formation were those related to the anabolic effect of PTH, which has been developed as a highly effective anabolic therapy for the skeleton, despite its best-known action as a resorptive hormone. The anabolic effect requires that PTH be given in an intermittent, rather than a continuous, mode [147]. This has been obtained by daily injections, which rapidly achieve a transient peak level in blood [148]. There are two general mechanisms proposed for the PTH anabolic effect, which require its direct action upon the osteoblast lineage. One is the promotion of the differentiation of committed osteoblast precursors [149], the other being the inhibition of osteoblast apoptosis [150]. Since PTH affects both bone formation and resorption, and since the activities of osteoclasts and osteoblasts are linked through the normal process of bone remodeling, it is likely that the anabolic effect of PTH also relates either directly or indirectly to bone remodeling. This could mean either that PTH treatment activates new BMUs, or that the balance of formation against resorption is improved within BMUs, or a combination of the two. Consistent with this view are the observations that the anabolic effect of PTH is greater on trabecular and endocortical bone [147, 151]. The PTH effect is particularly marked on the endocortical surface, which is very actively remodeling in old age [152].

The anabolic effect of PTH was significantly reduced in sheep when an osteoclast-inhibiting bisphosphonate (tiludronate) was co-administered [153]. Treatment of patients with osteoporosis concomitantly with PTH and a bisphosphonate resulted in significant early blunting of the anabolic response to PTH [154, 155]. After cessation of bisphosphonate treatment, full responsiveness to PTH can eventually return. Some, but not all, studies of the PTH anabolic effect in rats treated concomitantly with bisphosphonates have also shown impaired anabolic responses. The conclusion that osteoclasts are required for the anabolic response to PTH came from studies in mice rendered null for the *c-fos* gene, which are osteopetrotic because they cannot develop osteoclasts, and furthermore fail to show an anabolic response to PTH [156]. A further study in *c-fos* null mice showed that administration of PTH to these mice led to proliferation of osteoblast precursors but no increase in bone formation [132].

If any osteoclast function is required for the anabolic effect of PTH, how is that connection made since osteoclasts do not express a functional PTH receptor? Treatment of rats with a single subcutaneous injection of PTH resulted in a transient increase in mRNA for RANKL and a decrease in that for OPG, with maximum effect at 1 h, and returning to control within 2 h [157, 158]. This led to the suggestion that a subtle or transient increase in osteoclast formation or activation might be needed to prepare the bone surface for new matrix deposition. Finally and most importantly, Holtrop et al. [24] showed that intravenous injection of PTH in young rats resulted in transient activation of osteoclasts in vivo, evident within 30 minutes, and followed only some hours later at high PTH doses by increased osteoclast number. As indicated earlier, this rapid response might be explained by the existence of QuOPs in the proposed osteoclast niche [94].

The foregoing data lead to the suggestion that what is needed for full expression of the anabolic action of PTH, in addition to its direct effects on committed preosteoblasts, is a transient effect on the osteoclast, achieved by promoting *activation,* rather than *formation,* of osteoclasts [159]. This is an important distinction, and the precise way in which the osteoclast is involved in the anabolic process needs to be clearly understood because of the implications for sequential or combined use of therapeutic resorption inhibitors and anabolic agents, and for the development of new anabolic agents. A further indirect action of PTH that has been discovered in work with genetically manipulated mice as a contributor to the anabolic effect is its action upon T cells to increase production of Wnt 10b, which in turn promotes osteoblast differentiation [160]. This adds a new dimension to the complexity of PTH action.

*Local PTHrP or systemic PTH?*

With the focus of this chapter on communication mechanisms between osteoclasts and osteoblasts and how these can be influenced by other cell types, it would be an omission if we were not to put PTHrP into perspective, as the local ligand for the PTH/PTHrP receptor (PTHR1). The discovery of PTHrP production in bone raised the possibility that this molecule has important local actions in bone, perhaps even being the primary ligand for the receptor it shares with PTH. Targeted disruption of the genes for PTHrP or PTHR1 in mice resulted in death in the perinatal period with gross skeletal abnormalities, consistent with chondrodysplasia [161]. Further investigation of this phenomenon demonstrated that PTHrP plays a central role in endochondral bone formation.

It seemed likely that PTHrP is also involved in intramembranous bone formation. In the rabbit, cells of the osteoblast lineage express PTHrP mRNA and protein throughout the entire sequence of intramembranous bone formation, with prominent production by cuboidal, actively synthesizing osteoblasts and newly embedded osteocytes, and weaker expression in lining cells on the mineralized trabeculae [162]. Together with the finding of PTHrP mRNA in rat and mouse cells of the osteoblast lineage [163], this supports a role for PTHrP in the differentiation of mesenchymal precursors to mature cells of the osteogenic lineage.

Further investigations of PTHrP mutant mice provided evidence to suggest that PTHrP is equally important for the orderly commitment of precursor cells to the osteogenic lineage and subsequent maturation and/or function. Although PTHrP null mice die around birth, PTHrP heterozygote mice are phenotypically normal at birth, but by 3 months of age exhibit osteopenia characterized by a marked decrease in trabecular thickness and connectivity [164]. This bone deficit, associated with PTHrP haploinsufficiency, pointed to the likelihood of PTHrP playing a significant part in the maintenance of normal bone. Proof came when specific ablation of PTHrP in osteoblasts resulted in mice with impaired bone formation, both in vitro and in vivo [165]; thus, identifying a central physiological role for PTHrP in the regulation of bone formation. The results of the mouse genetic experiments indicate that there is a critical role for PTHrP in bone remodeling.

What are the ways in which PTHrP as a paracrine/autocrine factor in bone can contribute? In order for PTHrP to stimulate bone formation by enhancing osteoblast differentiation and reducing osteoblast apoptosis, control mechanisms must exist to ensure that only short-lived, high levels of PTHrP are available to local targets, which favor bone formation. This is because persistently increased local PTHrP levels would favor increased osteoclast formation, through stimulation of RANKL production. PTHrP release needs to be exquisitely regulated in terms of concentration, location, and time, so that it is presented only briefly to these target cells. On the other hand, the spatiotemporal controls might be such that excessive osteoclast formation is much less likely under these conditions in comparison to when PTH is presented systemically to the whole skeleton. Possible mechanisms could involve cytokine-mediated and/or neural control of PTHrP gene transcription, mRNA stability, or proteolytic protein processing.

# Cell communication from beyond the bone multicellular unit

## Influence of immune cells – T and B lymphocytes

The formation and function of both osteoclasts and osteoblasts are regulated by cellular signals and cytokines that also play significant roles in the immune system, and there are numerous examples of T-cell influence on bone cells that have provided important insights into the regulation of bone metabolism. While studies of mice lacking lymphocytes suggest that the immune system does not play an essential role in normal bone homeostasis, many immune disorders, particularly inflammatory disorders, are associated with significant localized or systemic loss of bone mass. The very many cytokines and factors produced by lymphocytes that modify osteoclast and osteoblast generation and activity are summarized in Table 8-1.

Table 8-1 Summary of actions of some of the immune-cell-derived cytokines discussed in the text (this list is not exhaustive). The degree of cytokine production is in most cases context dependent, as is the magnitude of the effects.

| Cytokine | Immune cell source | Effect: OC diffn/activity | Effect: OB diffn/activity | Refs |
|---|---|---|---|---|
| RANKL | T(act), B, MM | ↑ | | 188 |
| | | | | 83 |
| | | | | 73 |
| | | | | 68 |
| OPG | T, B | ↓ | – | 229 |
| | | | | 230 |
| IFN-γ | T | ↓ | ↓ | 231 |
| | | | | 232 |
| GM-CSF | T | ↓ | – | 201 |
| IL-3 | T | ↓ | – | 233 |
| IL-4 | T | ↓ | ↓ or ↑ | 231 |
| | | | | 234 |
| | | | | 235 |
| IL-10 | T | ↓ | ↑ | 236 |
| | | | | 237 |
| IL-13 | T | ↓ | ↓ | 238 |
| | | | | 235 |
| sFRPs | T | ↓ | ↓ | 239 |
| | | | | 240 |
| Wnt family | T | – | ↑ | 14 |
| | | | ↑ | 160 |
| DKK1 | MM | – | ↓ | 241 |
| | | | | 242 |
| IL-17 | T | ↑ | – | 175 |

Act, activated; Diffn, differentiation; T, T cell; B, B cell; MM, Multiple myeloma cell.

## Activated T cells can stimulate bone resorption

Activated T cells produce significant amounts of soluble RANKL, enough to support in vitro osteoclast formation [74, 166–168]. This has been shown to occur also in vivo [74, 167–170] and T-cell-derived RANKL is implicated in osteolysis associated with inflammatory disorders (Figure 8-6). It should be noted, however, that such disorders also show elevated levels of pro-inflammatory cytokines that greatly stimulate osteoblast lineage RANKL expression, such that the direct and unique contributions of T cells are hard to determine. Since T-cell-derived RANKL depends on activation state, it has also been proposed that systemic levels of antigen presentation affect bone mass [171]. However, RANKL levels provoked by inflammatory lesions are subject to neutralization by circulating OPG. The role of T-cell-derived RANKL in normal physiological bone metabolism is uncertain but is clearly not essential, as mice lacking T cells do not lack osteoclasts.

Th2 cytokines IL-4 and IL-13 reduce RANKL expression by T cells and enhance their production of OPG [170, 172]. While it was proposed that Th1 cells are the major T-cell source of RANKL [173], later work demonstrated that Th17 cells generated in vitro

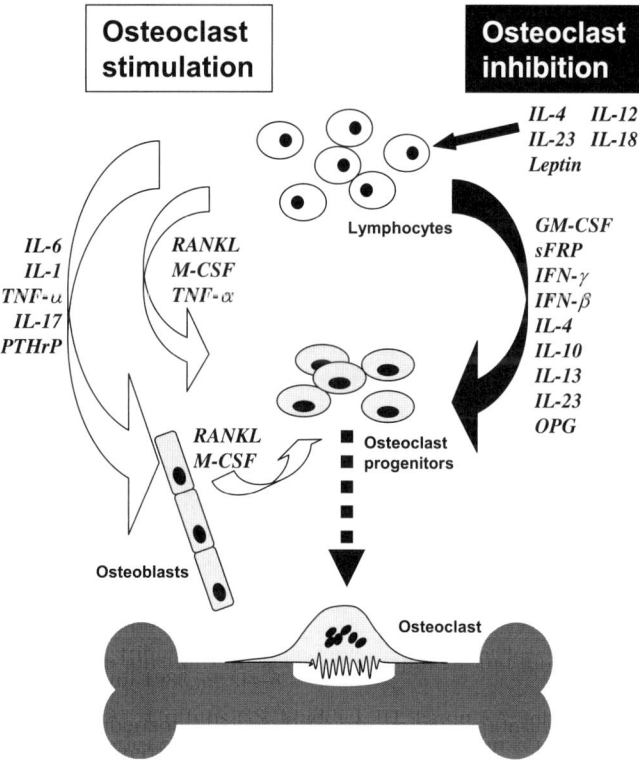

**FIGURE 8-6** Lymphocytes produce RANKL and OPG, and a number of cytokines that increase or decrease osteoclast formation, either by direct action upon hemopoietic precursors or by influencing osteoblast production of RANKL or M-CSF.

produced RANKL upon activation, while Th1 and Th2 cells instead produced osteoclast inhibitors [168]. Th17 cells are IL-17-secreting T helper (CD4$^+$) cells [174]. IL-23 acts through receptors in these cells to indirectly stimulate maturation and to stimulate cytokine secretion (notably IL-17) [168, 174].

Th17 cells, IL-17, and IL-23 have been identified as critical agents for maintaining chronic inflammation in a number of autoimmune disease models [174], and are probably of wider pathophysiological importance. The association of bone loss with chronic inflammation raises the possibility that activated Th17 cells may stimulate bone resorption as well as inflammation through production of soluble RANKL and IL-17. This is supported by evidence that IL-17 is a strong stimulator of RANKL production by osteoblast lineage cells, is abundant in inflamed joints and has been implicated in periarticular bone destruction in rheumatoid arthritis [175]. While this points to activated Th17 cell involvement in osteoclast differentiation, the participation of other RANKL-producing cells such as CD8$^+$ T cells, and other IL-17 sources, such as $\gamma\delta$ T cells [176, 177] requires further investigation.

*T-cell effects influenced by hormones*

Many hormones and cytokines that act directly on bone cells, including PTH, 1,25(OH)$_2$-D$_3$ [178] and estrogen [179], also influence cells of the immune system, and may influence bone metabolism in part through immunoregulatory and anti-inflammatory actions. T cells are themselves also a source of many factors (including PTHrP) that induce RANKL production in osteoblasts [180]. In particular, T cells may participate in the catabolic action of PTH [181]. Furthermore, the anabolic effects of PTH also appear to require direct action on CD8+ T cells to modify Wnt10b production, thereby promoting osteoblast differentiation and bone formation [160].

Estrogen withdrawal in females results in loss of bone mass, with mediating roles proposed for circulating pro-inflammatory cytokines IL-1, IL-6, and M-CSF, which are all elevated by ovariectomy [182, 183]. Lymphocytes express estrogen receptors, and estrogen loss in humans is reported to cause a decline in T-cell subsets [179, 184]. Conversely, ovariectomy in mice increases T-cell activation and TNF secretion [185, 186]. Although lymphocyte-deficient nude mice have been reported to fail to lose bone upon ovariectomy, a response restored by adoptive transfer of wild-type T cells [171, 185, 186], others have reported that nude mice and two strains of mice with severe T-cell deficiencies responded to ovariectomy with bone loss [187]. Thus, lymphocyte involvement in estrogen-deficiency-related bone loss remains to be clarified.

*B cells as a source of RANKL and OPG*

B cells also produce RANKL and OPG [188]. They are the source of a large proportion of the OPG in bone marrow, production of which increases with T-cell stimulation [189]. Ovariectomy, which causes significant bone loss, also increases B220$^+$ pre-B cell numbers [190]. The link between ovariectomy-induced bone loss and the change in

B-cell numbers remains to be clarified. Elevated levels of IL-7, a major B-cell growth factor produced by stromal cells, also cause bone loss [191, 192], although this may occur via inflammatory cytokine induction [193]. In sum, there is little direct evidence that B-cell-derived RANKL significantly affects bone resorption, but given the large B-cell numbers in bone marrow there may arise circumstances where it does. For example, multiple myeloma, a B-cell-derived malignancy, causes much bone loss and multiple myeloma cells are a rich source of RANKL [83]. Since RANKL also influences pro-B-cell development, ovariectomy-stimulated RANKL production may also affect B-cell development [72, 167, 194, 195], suggesting that regulation may occur in both directions.

*Lymphocyte-mediated inhibition of osteoclasts*

T cells, including naïve T cells [196], are a significant source of factors that inhibit osteoclasts and their formation, including IFN-γ, GM-CSF, IL-3, osteoprotegerin, IL-4, IL-10, IL-13, osteoclast inhibitory lectin (OCIL), and secreted frizzled-related proteins (sFRPs) [197]; these are summarized in Figure 8-6. Some of these factors, e.g. GM-CSF, IL-3, and sFRPs, block osteoclast formation when RANKL levels are high but do not negatively influence osteoclast-lineage cells when RANKL is absent; they may thus restrain damage to bone that is caused by T-cell-derived RANKL.

IFNγ is a particularly potent osteoclast differentiation inhibitor and macrophage activator, produced by T cells, which causes proteasome-dependent degradation of the RANK adaptor protein TRAF6 [198]. Its actions may overlap with IFNβ, a *c-fos*-dependent autocrine inhibitor of osteoclasts [80]. However, the pluripotency of IFNβ and IFNγ means their in vivo effects are not simple. IFNγ causes augmented antigen presentation by stimulating MHC class II expression, which, in turn, raises T-cell activation levels [171] with possible osteolytic results. The effects of IFNs on osteoclasts may thus depend on the degree to which they are localized to the bone microenvironment.

GM-CSF secretion by T cells can be induced by several Th1 cytokines (see above), and in hematopoietic cells stimulated by RANKL and M-CSF in vitro, GM-CSF potently inhibits osteoclast formation. In human cells at least, this action of GM-CSF is attributable to suppression of the chemokine MCP-1/CCL-2 [199], a critical autocrine factor, and a stimulator of osteoclast fusion [189]. GM-CSF may also favor dendritic cell formation at the expense of osteoclasts, since they share a common precursor. However, it is interesting to note in this respect that dendritic cells themselves may serve as a source of osteoclast progenitors [200].

IL-18 was the first osteoclast inhibitory cytokine to be documented to act through naïve T cells [201]. IL-18 participates in the development of inflammation by strongly stimulating secretion of IFNγ by Th1 cells, B cells, NK cells, and macrophages [202, 203]. The inhibitory action of IL-18 on osteoclast formation depends wholly on GM-CSF induction in $CD4^+$ and $CD8^+$ T cells [201, 204]. Osteoblasts also produce IL-18 in response to PTH [205], display receptors for IL-18, and respond to it with increased proliferation [206] and OPG production [207]. These results are consistent with IL-18's osteoclast-inhibitory effects. It should be noted, however, that IL-18 can also enhance production of RANKL by activated T cells [208], consistent with the notion that T-cell activation status strongly influences the indirect effects of cytokines on osteoclasts.

Like IL-18, the heterodimeric Th1-stimulating cytokine IL-12 inhibits osteoclast formation in the presence of T cells [209] and mediates antiosteolytic effects of TLR9 ligand, CpG-oligodeoxynucleotide [210]. IFNγ and GM-CSF do not mediate these actions of IL-12 in osteoclastogenic cultures and its mechanism of action is unknown. Strikingly, however, IL-12 acts on osteoclast formation synergistically with IL-18 [209], resulting in extremely potent actions when low concentrations of IL-12 and IL-18 are combined. Such synergy between IL-18 and IL-12 is also noted in their immune actions [203, 211–213].

The dimeric IL-12-related Th17 cytokine IL-23 has roles in chronic inflammation, outlined above, but also inhibits osteoclast formation in the presence of $CD4^+$ (but not $CD8^+$) T cells [214], consistent with action via Th17 cells. This action is entirely GM-CSF-dependent [214, 215] and synergy between IL-23 and IL-18 also occurs. However, IL-23 also displays similar actions mediated by γδ T cells [177]. Mice lacking IL-23 display low trabecular bone mass, possibly by increased resorption at or near the growth plate [214].

In addition to the above Th1- and Th17-acting cytokines, the Th2 cytokine IL-4 is known as an alternative activator of macrophages and powerful blocker of osteoclast differentiation [172, 216]. In addition, IL-4 has a T-cell-dependent osteoclast inhibitory action evident only when precursors are in contact with hematopoietic progenitors unresponsive to IL-4 treatment [216]. It remains to be seen whether other cytokines have lymphocyte-dependent actions on osteoclasts that are similarly accompanied by direct inhibitory actions on osteoclast formation.

Regulatory T cells (Treg cells) suppress immune responses to maintain immune system homeostasis and tolerance to self-antigens, and are stimulated by IL-35. Treg cells reduce joint destruction in an arthritis model by elevating production of osteoclast-inhibiting cytokines [217]. Other powerful inhibitory influences of Treg cells on osteoclasts have also been identified [218, 219]. These include including a direct inhibitory action on osteoclast formation through the transmembrane protein CTLA-4 [220], and the inhibitory ligand of B7 co-stimulatory molecules CD80 and CD86. Clarification of CTLA-4 action may explain the direct Treg cell influences on osteoclasts. However, the powerful effects of Treg cells on immune responses suggest that they have other indirect influences on bone physiology.

In summary, T cells produce a number of mediators that influence osteoclast differentiation. Activated T cells produce factors, including RANKL, that strongly stimulate osteolysis, consistent with the association between inflammation and bone loss. However, T cells (particularly naïve T cells) also mediate antiosteoclastogenic actions. The importance of these influences is not well defined in vivo and is likely to be highly context-dependent. B-cell influence on bone resorption is also controversial but their abundance in bone marrow makes it likely that their production of RANKL and OPG has some influence on bone mass.

## Osteomacs

Recent work points to the presence of a myeloid population that has specific functions in the bone environment. These cells have been called "osteomacs" [221, 222]. There are many examples in other tissues of resident myeloid-macrophage cells that influence tissue function and repair in those sites, for example Kupffer cells in the liver, microglial cells in brain. Some possible roles in bone have been explored in the recent work, but the clues to this have been evident for a long time. Just as monocytes are rapidly recruited to sites of tissue infection, injury, or inflammation, so too do macrophages accumulate early and are prominently present during fracture repair [223, 224]. This and other evidence [221] argues for the existence and many functions of bone-specific macrophages.

Osteomacs were identified as tissue myeloid cells by staining for the antigen F4/80, and defined by their presence near the bone surface, where they intercalate with bone-forming osteoblasts in both mouse and human bone [222]. These cells were consistently co-isolated with osteoblasts in murine calvarial osteoblast cultures and organ explants, to the extent that myeloid cells were estimated to constitute as much as 15% of cells, isolated from mouse calvariae [222]. Notable from the functional point of view was that removal of these cells by antibody selection resulted in calvarial cultures that had greatly reduced mineralization, and production of osteocalcin mRNA. These properties were restored by co-cultivating the depleted calvarial cells with extracted macrophages, separated by a permeable membrane, suggesting generation from macrophages of a soluble factor enhancing osteoblast differentiation. In addition to these in vitro observations, osteomacs were identified as canopies covering sites where modeling was taking place. These cells have distinctive structures, covering large areas of mature osteoblasts on growing surfaces in young mice [222]. In a genetically engineered mouse in which apoptosis was induced in myeloid cells by fas-ligand treatment (MAFIA mouse), complete loss of bone-forming surfaces occurred, raising the possibility that the canopy of osteal myeloid cells might play a role in vivo that favored bone formation at modeling sites. When an induced defect in bone was produced in the same engineered mice, healing was greatly impaired when macrophage apoptosis was induced [225]. Hence, there might be a role for osteomacs at remodeling sites as well, but this has yet to be confirmed [226].

Identification of osteomacs provides a new pathway of intercellular communication in bone, with implications for both osteoblast and osteoclast biology. The evidence so far is focused upon signals from the osteomacs to osteoblasts, which might be directly relevant to the mechanisms by which cells communicate in bone remodeling to achieve coupling of bone formation to resorption. An important question though is whether osteomacs are precursors of osteoclasts in vivo. The plasticity of the myeloid lineage ensures that myeloid cells of all types can be differentiated into osteoclasts, given the right conditions in vitro, and osteomacs are no exception [226–228]. Thus for example, the finding that as much as 15% of mouse calvarial "osteoblasts" are "macrophages" could explain why some groups have successfully used crude calvarial preparations to generate osteoclasts in vitro [227, 228]. If calvarial osteomacs are adjacent to osteoblasts producing RANKL, this could also tell us why resorbing agents such as parathyroid hormone (PTH) and active vitamin D can rapidly induce formation of osteoclasts in calvarial organ cultures. None of this argues the case for osteomacs in vivo providing a significant source of osteoclast precursors in bone, although it is one of the interesting questions to be addressed. Just as macrophages and monocytes in other tissues do not generate osteoclasts because they are not in contact with cell sources of RANKL, so also would co-location adjacent to RANKL-producing cells be required in bone. It might be noted that in work identifying the "osteoclast niche" [94], F4/80+ cells in the metaphyseal microenvironment did not express RANKL. There might be situations in which appropriate co-location is found to occur, but these are not apparent so far in the best-studied aspect of osteomac function, its possible role in modeling.

Certainly the advent of this new player in the dynamics of intercellular communication is one of great interest that requires substantial further investigation.

## Summary and conclusions – remodeling and the coupling of bone formation to resorption

The important regulatory role of coupling activity, produced in the process of resorption, is to drive the osteoblast lineage to replace the necessary amount of bone in each BMU. Characterization of the process and its relationship to the actions of hormones and cytokines upon bone, constitute major outstanding problems in bone cell biology. For many years the concept of a coupling factor to ensure that bone formation closely matched resorption in the BMU appeared to focus on the idea of one coupling factor, or at least one major mechanism – somewhat analogous to what was discovered about control of osteoclast formation in the late 1990s. As more has been learned, however, it has become clear that the coupling is a complex process, made up of many stages with different contributors at all stages. Figure 8-7 attempts to summarize the pathways that are currently receiving attention. Precursors of osteoclasts arrive from the blood or marrow and are programmed to become bone-resorbing osteoclasts. Resorption of growth factors from bone can contribute to coupling, so too can changes in the releases of formation inhibitors – e.g. sclerostin – from osteocytes, and the osteoclast itself, with its precursors, probably also contribute activities that influence formation in the remodeling process.

Perhaps we should not be surprised that the breakdown of bone requires quite a simple control mechanism but its replacement is much more complex. From the evolutionary point of view, bone remodeling is a survival mechanism that responds to the acute emergencies of skeletal damage or pressure changes, or the less urgent but nevertheless crucial requirement to remove old bone. Once bone has been removed though, its replacement must be achieved with precision, with the supply of precursor cells assured, so as to carefully regulate differentiation to a precise endpoint.

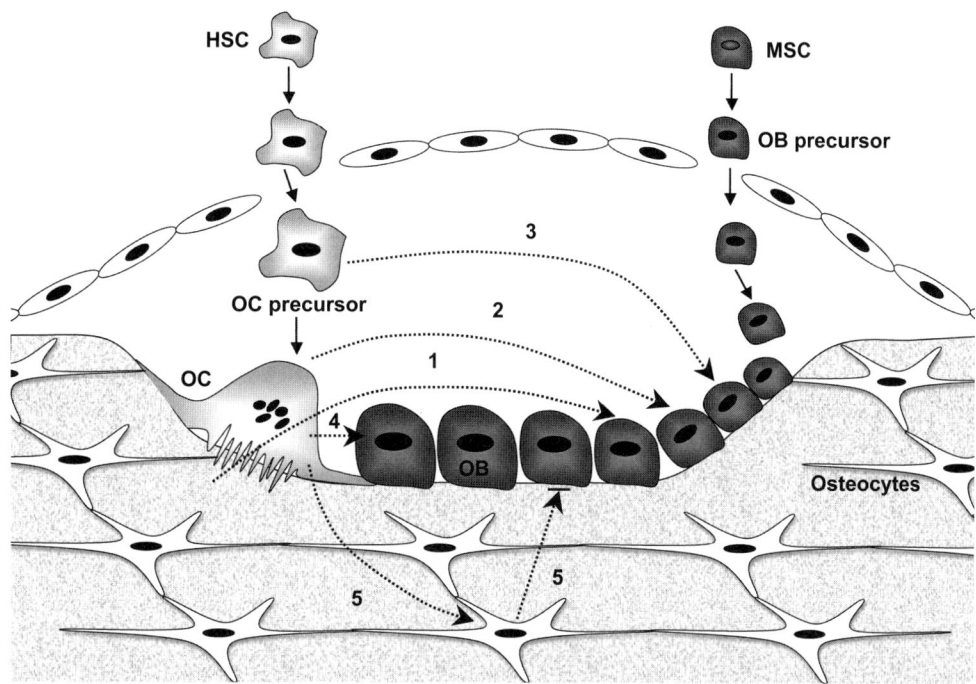

**FIGURE 8-7** The several pathways involved in the coupling of bone formation to resorption in the BMU. 1: growth factors from matrix; 2: osteoclast products; 3: osteoclast precursor products; 4: osteoclast–osteoblast contact; 5: signals from osteoclast to osteocyte and osteocyte to surface osteoblasts. See text for details and references.

## Acknowledgments

Work from the authors' laboratories was supported by a Program Grant (No. 345401) from the National Health and Medical Research Council of Australia.

## References

[1] Udagawa N, Takahashi N, Akatsu T, et al. The bone marrow-derived stromal cell lines MC3T3-G2/PA6 and ST2 support osteoclast-like cell differentiation in cocultures with mouse spleen cells. Endocrinology 1989;125(4):1805–13.

[2] Askmyr M, Sims NA, Martin TJ, Purton LE. What is the true nature of the osteoblastic hematopoietic stem cell niche? Trends Endocrinol Metab 2009;20(6):303–9.

[3] Karsenty G. Minireview: transcriptional control of osteoblast differentiation. Endocrinology 2001;142(7):2731–3.

[4] Ducy P, Starbuck M, Priemel M, et al. A Cbfa1-dependent genetic pathway controls bone formation beyond embryonic development. Genes Dev 1999;13(8):1025–36.

[5] Thirunavukkarasu K, Miles RR, Halladay DL, et al. Stimulation of osteoprotegerin (OPG) gene expression by transforming growth factor-beta (TGF-beta). Mapping of the OPG promoter region that mediates TGF-beta effects. J Biol Chem 2001;276(39):36241–50.

[6] Nakashima K, Zhou X, Kunkel G, et al. The novel zinc finger-containing transcription factor osterix is required for osteoblast differentiation and bone formation. Cell 2002;108(1):17–29.

[7] Shirakabe K, Terasawa K, Miyama K, Shibuya H, Nishida E. Regulation of the activity of the transcription factor Runx2 by two homeobox proteins, Msx2 and Dlx5. Genes Cells 2001;6(10): 851–6.

[8] el Ghouzzi V, Le Merrer M, Perrin-Schmitt F, et al. Mutations of the TWIST gene in the Saethre-Chotzen syndrome. Nat Genet 1997;15(1):42–6.

[9] Howard TD, Paznekas WA, Green ED, et al. Mutations in TWIST, a basic helix-loop-helix transcription factor, in Saethre-Chotzen syndrome. Nat Genet 1997;15(1):36–41.

[10] Gutierrez S, Javed A, Tennant DK, et al. CCAAT/enhancer-binding proteins (C/EBP) beta and delta activate osteocalcin gene transcription and synergize with Runx2 at the C/EBP element to regulate bone-specific expression. J Biol Chem 2002;277(2):1316–23.

[11] Wu M, Hesse E, Morvan F, et al. Zfp521 antagonizes Runx2, delays osteoblast differentiation in vitro, and promotes bone formation in vivo. Bone 2009;44(4):528–36.

[12] Boyden LM, Mao J, Belsky J, et al. High bone density due to a mutation in LDL-receptor-related protein 5. N Engl J Med 2002;346(20):1513–21.

[13] Gong Y, Slee RB, Fukai N, et al. LDL receptor-related protein 5 (LRP5) affects bone accrual and eye development. Cell 2001;107(4):513–23.

[14] Rawadi G, Vayssiere B, Dunn F, Baron R, Roman-Roman S. BMP-2 controls alkaline phosphatase expression and osteoblast mineralization by a Wnt autocrine loop. J Bone Miner Res 2003;18(10):1842–53.

[15] Babij P, Zhao W, Small C, et al. High bone mass in mice expressing a mutant LRP5 gene. J Bone Miner Res 2003;18(6):960–74.

[16] Kato M, Patel MS, Levasseur R, et al. Cbfa1-independent decrease in osteoblast proliferation, osteopenia, and persistent embryonic eye vascularization in mice deficient in Lrp5, a Wnt coreceptor. J Cell Biol 2002;157(2):303–14.

[17] Morvan F, Boulukos K, Clement-Lacroix P, et al. Deletion of a single allele of the Dkk1 gene leads to an increase in bone formation and bone mass. J Bone Miner Res 2006;21(6):934–45.

[18] Kulkarni NH, Onyia JE, Zeng Q, et al. Orally bioavailable GSK-3alpha/beta dual inhibitor increases markers of cellular differentiation in vitro and bone mass in vivo. J Bone Miner Res 2006;21(6):910–20.

[19] Glass 2nd DA, Bialek P, Ahn JD, et al. Canonical Wnt signaling in differentiated osteoblasts controls osteoclast differentiation. Dev Cell 2005;8(5):751–64.

[20] Holmen SL, Zylstra CR, Mukherjee A, et al. Essential role of beta-catenin in postnatal bone acquisition. J Biol Chem 2005;280(22):21162–8.

[21] Yadav VK, Ryu JH, Suda N, et al. Lrp5 controls bone formation by inhibiting serotonin synthesis in the duodenum. Cell 2008;135(5):825–37.

[22] Kolliker A. Die Normal Resorption des Knochengewebes und ihre Bedeutung die Entstehung der Typischen Knochenformen. Liepzig: Vogel FCW 1873.

[23] Holtrop ME, King GJ. The ultrastructure of the osteoclast and its functional implications. Clin Orthop Relat Res 1977;123:177–96.

[24] Holtrop ME, King GJ, Cox KA, Reit B. Time-related changes in the ultrastructure of osteoclasts after injection of parathyroid hormone in young rats. Calcif Tissue Int 1979;27(2):129–35.

[25] Vaes G. Cellular biology and biochemical mechanism of bone resorption. A review of recent developments on the formation, activation, and mode of action of osteoclasts. Clin Orthop Relat Res 1988;231:239–71.

[26] Baron R. Molecular mechanisms of bone resorption: therapeutic implications. Rev Rhum Engl Ed 1996;63(10):633–8.

[27] Nicholson GC, Moseley JM, Sexton PM, Mendelsohn FA, Martin TJ. Abundant calcitonin receptors in isolated rat osteoclasts. Biochemical and autoradiographic characterization. J Clin Invest 1986;78(2):355–60.

[28] Tonna EA. Osteoclasts and the aging skeleton: a cytological, cytochemical and autoradiographic study. Anat Rec 1960;137:251–69.

[29] Young RW. Cell proliferation and specialization during endochondral osteogenesis in young rats. J Cell Biol 1962;14:357–70.

[30] Rasmussen H, Bordier P. The Physiological Basis of Metabolic Bone Disease. Baltimore: Williams and Wilkins, Waverley Press; 1974.

[31] Walker DG. Bone resorption restored in osteopetrotic mice by transplants of normal bone marrow and spleen cells. Science 1975;190(4216):784–5.

[32] Udagawa N, Takahashi N, Akatsu T, et al. Origin of osteoclasts: mature monocytes and macrophages are capable of differentiating into osteoclasts under a suitable microenvironment prepared by bone marrow-derived stromal cells. Proc Natl Acad Sci U S A 1990;87(18):7260–4.

[33] Hattersley G, Kerby JA, Chambers TJ. Identification of osteoclast precursors in multilineage hemopoietic colonies. Endocrinology 1991;128(1):259–62.

[34] Quinn JM, Whitty GA, Byrne RJ, Gillespie MT, Hamilton JA. The generation of highly enriched osteoclast-lineage cell populations. Bone 2002;30(1):164–70.

[35] Nutt SL, Heavey B, Rolink AG, Busslinger M. Commitment to the B-lymphoid lineage depends on the transcription factor Pax5. Nature 1999;401(6753):556–62.

[36] Borrello MA, Palis J, Phipps RP. The relationship of CD5+ B lymphocytes to macrophages: insights from normal biphenotypic B/macrophage cells. Int Rev Immunol 2001;20(1):137–55.

[37] Horowitz MC, Xi Y, Pflugh DL, et al. Pax5-deficient mice exhibit early onset osteopenia with increased osteoclast progenitors. J Immunol 2004;173(11):6583–91.

[38] Delogu A, Schebesta A, Sun Q, Aschenbrenner K, Perlot T, Busslinger M. Gene repression by Pax5 in B cells is essential for blood cell homeostasis and is reversed in plasma cells. Immunity 2006;24(3):269–81.

[39] Hogg N, Shapiro IM, Jones SJ, Slusarenko M, Boyde A. Lack of Fc receptors on osteoclasts. Cell Tissue Res 1980;212(3):509–16.

[40] Athanasou NA, Quinn J. Immunophenotypic differences between osteoclasts and macrophage polykaryons: immunohistological distinction and implications for osteoclast ontogeny and function. J Clin Pathol 1990;43(12):997–1003.

[41] Sato T, Abe E, Jin CH, et al. The biological roles of the third component of complement in osteoclast formation. Endocrinology 1993;133(1):397–404.

[42] Martin TJ, Partridge NC, Greaves M, Atkins D, Ibbotson KJ. Prostaglandin effects on bone and role in cancer hypercalcaemia. In: MacIntyre I, Szelke M, editors. Molecular Endocrinology. Amsterdam: Elsevier; 1979. p. 251–64.

[43] Rodan GA, Martin TJ. Role of osteoblasts in hormonal control of bone resorption – a hypothesis. Calcif Tissue Int 1981;33(4):349–51.

[44] Chambers TJ. The cellular basis of bone resorption. Clin Orthop Relat Res 1980;151:283–93.

[45] Chambers TJ. Osteoblasts release osteoclasts from calcitonin-induced quiescence. J Cell Sci 1982;57:247–60.

[46] Chambers TJ. The pathobiology of the osteoclast. J Clin Pathol 1985;38(3):241–52.

[47] McSheehy PM, Chambers TJ. Osteoblast-like cells in the presence of parathyroid hormone release soluble factor that stimulates osteoclastic bone resorption. Endocrinology 1986;119(4):1654–9.

[48] McSheehy PM, Chambers TJ. Osteoblastic cells mediate osteoclastic responsiveness to parathyroid hormone. Endocrinology 1986;118(2):824–8.

[49] Burger EH, van der Meer JW, Nijweide PJ. Osteoclast formation from mononuclear phagocytes: role of bone-forming cells. J Cell Biol 1984;99(6):1901–6.

[50] Takahashi N, Akatsu T, Udagawa N, et al. Osteoblastic cells are involved in osteoclast formation. Endocrinology 1988;123(5):2600–2.

[51] Suda T, Takahashi N, Martin TJ. Modulation of osteoclast differentiation. Endocr Rev 1992;13(1):66–80.

[52] Takahashi N, Yamana H, Yoshiki S, et al. Osteoclast-like cell formation and its regulation by osteotropic hormones in mouse bone marrow cultures. Endocrinology 1988;122(4):1373–82.

[53] Yamashita T, Asano K, Takahashi N, et al. Cloning of an osteoblastic cell line involved in the formation of osteoclast-like cells. J Cell Physiol 1990;145(3):587–95.

[54] Akatsu T, Takahashi N, Debari K, et al. Prostaglandins promote osteoclastlike cell formation by a mechanism involving cyclic adenosine 3',5'-monophosphate in mouse bone marrow cell cultures. J Bone Miner Res 1989;4(1):29–35.

[55] Akatsu T, Takahashi N, Udagawa N, et al. Parathyroid hormone (PTH)-related protein is a potent stimulator of osteoclast-like multinucleated cell formation to the same extent as PTH in mouse marrow cultures. Endocrinology 1989;125(1):20–7.

[56] Akatsu T, Takahashi N, Udagawa N, et al. Role of prostaglandins in interleukin-1-induced bone resorption in mice in vitro. J Bone Miner Res 1991;6(2):183–9.

[57] Kitazawa R, Kitazawa S. Vitamin D(3) augments osteoclastogenesis via vitamin D-responsive element of mouse RANKL gene promoter. Biochem Biophys Res Commun 2002;290(2):650–5.

[58] Tamura T, Udagawa N, Takahashi N, et al. Soluble interleukin-6 receptor triggers osteoclast formation by interleukin 6. Proc Natl Acad Sci U S A 1993;90(24):11924–8.

[59] Udagawa N, Takahashi N, Katagiri T, et al. Interleukin (IL)-6 induction of osteoclast differentiation depends on IL-6 receptors expressed on osteoblastic cells but not on osteoclast progenitors. J Exp Med 1995;182(5):1461–8.

[60] Gao Y, Morita I, Maruo N, Kubota T, Murota S, Aso T. Expression of IL-6 receptor and GP130 in mouse bone marrow cells during osteoclast differentiation. Bone 1998;22(5):487–93.

[61] Martin TJ, Ng KW. Mechanisms by which cells of the osteoblast lineage control osteoclast formation and activity. J Cell Biochem 1994;56(3):357–66.

[62] Felix R, Cecchini MG, Fleisch H. Macrophage colony stimulating factor restores in vivo bone resorption in the op/op osteopetrotic mouse. Endocrinology 1990;127(5):2592–4.

[63] Yoshida H, Hayashi S, Kunisada T, et al. The murine mutation osteopetrosis is in the coding region of the macrophage colony stimulating factor gene. Nature 1990;345(6274):442–4.

[64] Lorenzo JA, Sousa SL, Fonseca JM, Hock JM, Medlock ES. Colony-stimulating factors regulate the development of multinucleated osteoclasts from recently replicated cells in vitro. J Clin Invest 1987;80(1):160–4.

[65] Simonet WS, Lacey DL, Dunstan CR, et al. Osteoprotegerin: a novel secreted protein involved in the regulation of bone density. Cell 1997;89(2):309–19.

[66] Tsuda E, Goto M, Mochizuki S, et al. Isolation of a novel cytokine from human fibroblasts that specifically inhibits osteoclastogenesis. Biochem Biophys Res Commun 1997;234(1):137–42.

[67] Lacey DL, Timms E, Tan HL, et al. Osteoprotegerin ligand is a cytokine that regulates osteoclast differentiation and activation. Cell 1998;93(2):165–76.

[68] Yasuda H, Shima N, Nakagawa N, et al. Osteoclast differentiation factor is a ligand for osteoprotegerin/osteoclastogenesis-inhibitory factor and is identical to TRANCE/RANKL. Proc Natl Acad Sci U S A 1998;95(7):3597–602.

[69] Bucay N, Sarosi I, Dunstan CR, et al. Osteoprotegerin-deficient mice develop early onset osteoporosis and arterial calcification. Genes Dev 1998;12(9):1260–8.

[70] Udagawa N, Takahashi N, Yasuda H, et al. Osteoprotegerin produced by osteoblasts is an important regulator in osteoclast development and function. Endocrinology 2000;141(9):3478–84.

[71] Kong YY, Yoshida H, Sarosi I, et al. OPGL is a key regulator of osteoclastogenesis, lymphocyte development and lymph-node organogenesis. Nature 1999;397(6717):315–23.

[72] Dougall WC, Glaccum M, Charrier K, et al. RANK is essential for osteoclast and lymph node development. Genes Dev 1999;13(18):2412–24.

[73] Wong BR, Josien R, Lee SY, et al. TRANCE (tumor necrosis factor [TNF]-related activation-induced cytokine), a new TNF family member predominantly expressed in T cells, is a dendritic cell-specific survival factor. J Exp Med 1997;186(12):2075–80.

[74] Anderson DM, Maraskovsky E, Billingsley WL, et al. A homologue of the TNF receptor and its ligand enhance T-cell growth and dendritic-cell function. Nature 1997;390(6656):175–9.

[75] Horwood NJ, Elliott J, Martin TJ, Gillespie MT. Osteotropic agents regulate the expression of osteoclast differentiation factor and osteoprotegerin in osteoblastic stromal cells. Endocrinology 1998;139(11):4743–6.

[76] Jimi E, Akiyama S, Tsurukai T, et al. Osteoclast differentiation factor acts as a multifunctional regulator in murine osteoclast differentiation and function. J Immunol 1999;163(1):434–42.

[77] Quinn JM, Neale S, Fujikawa Y, McGee JO, Athanasou NA. Human osteoclast formation from blood monocytes, peritoneal macrophages, and bone marrow cells. Calcif Tissue Int 1998;62(6):527–31.

[78] Matsuzaki K, Udagawa N, Takahashi N, et al. Osteoclast differentiation factor (ODF) induces osteoclast-like cell formation in human peripheral blood mononuclear cell cultures. Biochem Biophys Res Commun 1998;246(1):199–204.

[79] Lee SK, Lorenzo JA. Parathyroid hormone stimulates TRANCE and inhibits osteoprotegerin messenger ribonucleic acid expression in murine bone marrow cultures: correlation with osteoclast-like cell formation. Endocrinology 1999;140(8):3552–61.

[80] Takayanagi H, Kim S, Matsuo K, et al. RANKL maintains bone homeostasis through c-Fos-dependent induction of interferon-beta. Nature 2002;416(6882):744–9.

[81] Grigoriadis AE, Wang ZQ, Cecchini MG, et al. c-Fos: a key regulator of osteoclast-macrophage lineage determination and bone remodeling. Science 1994;266(5184):443–8.

[82] Thomas RJ, Guise TA, Yin JJ, et al. Breast cancer cells interact with osteoblasts to support osteoclast formation. Endocrinology 1999;140(10):4451–8.

[83] Lai FP, Cole-Sinclair M, Cheng WJ, et al. Myeloma cells can directly contribute to the pool of RANKL in bone bypassing the classic stromal and osteoblast pathway of osteoclast stimulation. Br J Haematol 2004;126(2):192–201.

[84] Eghbali-Fatourechi G, Khosla S, Sanyal A, Boyle WJ, Lacey DL, Riggs BL. Role of RANK ligand in mediating increased bone resorption in early postmenopausal women. J Clin Invest 2003;111(8):1221–30.

[85] Parfitt AM. Skeletal heterogeneity and the purposes of bone remodelling: implications for the understanding of osteoporosis. In: Marcus R, Feldman D, Kelsey J, editors. Osteoporosis. San Diego, CA: Academic Press; 1996. p. 315–39.

[86] Martin TJ, Seeman E. New mechanisms and targets in the treatment of bone fragility. Clin Sci (Lond) 2007;112(2):77–91.

[87] Henriksen K, Leeming DJ, Byrjalsen I, et al. Osteoclasts prefer aged bone. Osteoporos Int 2007;18(6):751–9.

[88] Chambers TJ, Darby JA, Fuller K. Mammalian collagenase predisposes bone surfaces to osteoclastic resorption. Cell Tissue Res 1985;241(3):671–5.

[89] Delaisse JM, Eeckhout Y, Vaes G. Bone-resorbing agents affect the production and distribution of procollagenase as well as the activity of collagenase in bone tissue. Endocrinology 1988;123(1):264–76.

[90] Henriksen K, Sorensen MG, Nielsen RH, et al. Degradation of the organic phase of bone by osteoclasts: a secondary role for lysosomal acidification. J Bone Miner Res 2006;21(1):58–66.

[91] Bonewald LF, Johnson ML. Osteocytes, mechanosensing and Wnt signaling. Bone 2008;42(4):606–15.

[92] Robling AG, Bellido T, Turner CH. Mechanical stimulation in vivo reduces osteocyte expression of sclerostin. J Musculoskelet Neuronal Interact 2006;6(4):354.

[93] Kogianni G, Mann V, Noble BS. Apoptotic bodies convey activity capable of initiating osteoclastogenesis and localized bone destruction. J Bone Miner Res 2008;23(6):915–27.

[94] Mizoguchi T, Muto A, Udagawa N, et al. Identification of cell cycle-arrested quiescent osteoclast precursors in vivo. J Cell Biol 2009;184(4):541–54.

[95] Elmardi AS, Katchburian MV, Katchburian E. Electron microscopy of developing calvaria reveals images that suggest that osteoclasts engulf and destroy osteocytes during bone resorption. Calcif Tissue Int 1990;46(4):239–45.

[96] Suzuki R, Domon T, Wakita M. Some osteocytes released from their lacunae are embedded again in the bone and not engulfed by osteoclasts during bone remodeling. Anat Embryol (Berl) 2000;202(2):119–28.

[97] Hughes DE, Boyce BF. Apoptosis in bone physiology and disease. Mol Pathol 1997;50(3):132–7.

[98] Nakamura T, Imai Y, Matsumoto T, et al. Estrogen prevents bone loss via estrogen receptor alpha and induction of Fas ligand in osteoclasts. Cell 2007;130(5):811–23.

[99] Iwasawa M, Miyazaki T, Nagase Y, et al. The antiapoptotic protein Bcl-xL negatively regulates the bone-resorbing activity of osteoclasts in mice. J Clin Invest 2009;119(10):3149–59.

[100] Nagase Y, Iwasawa M, Akiyama T, et al. The anti-apoptotic molecule Bcl-2 regulates the differentiation, activation and survival of both osteoblasts and osteoclasts. J Biol Chem 2009.

[101] Karsdal MA, Henriksen K, Sorensen MG, et al. Acidification of the osteoclastic resorption compartment provides insight into the coupling of bone formation to bone resorption. Am J Pathol 2005;166(2):467–76.

[102] Karsdal MA, Martin TJ, Bollerslev J, Christiansen C, Henriksen K. Are nonresorbing osteoclasts sources of bone anabolic activity? J Bone Miner Res 2007;22(4):487–94.

[103] Henriksen K, Gram J, Schaller S, et al. Characterization of osteoclasts from patients harboring a G215R mutation in ClC-7 causing autosomal dominant osteopetrosis type II. Am J Pathol 2004;164(5):1537–45.

[104] Zhao C, Irie N, Takada Y, et al. Bidirectional ephrinB2-EphB4 signaling controls bone homeostasis. Cell Metab 2006;4(2):111–21.

[105] Everts V, Delaisse JM, Korper W, et al. The bone lining cell: its role in cleaning Howship's lacunae and initiating bone formation. J Bone Miner Res 2002;17(1):77–90.

[106] Villanueva AR, Sypitkowski C, Parfitt AM. A new method for identification of cement lines in undecalcified, plastic embedded sections of bone. Stain Technol 1986;61(2):83–8.

[107] Hynes RO. Integrins: versatility, modulation, and signaling in cell adhesion. Cell 1992;69(1):11–25.

[108] Ishijima M, Rittling SR, Yamashita T, et al. Enhancement of osteoclastic bone resorption and suppression of osteoblastic bone formation in response to reduced mechanical stress do not occur in the absence of osteopontin. J Exp Med 2001;193(3):399–404.

[109] Ihara H, Denhardt DT, Furuya K, et al. Parathyroid hormone-induced bone resorption does not occur in the absence of osteopontin. J Biol Chem 2001;276(16):13065–71.

[110] Parfitt AM. The coupling of bone formation to bone resorption: a critical analysis of the concept and of its relevance to the pathogenesis of osteoporosis. Metab Bone Dis Relat Res 1982;4(1):1–6.

[111] Martin T, Gooi JH, Sims NA. Molecular mechanisms in coupling of bone formation to resorption. Crit Rev Eukaryot Gene Expr 2009;19(1):73–88.

[112] Howard GA, Bottemiller BL, Turner RT, Rader JI, Baylink DJ. Parathyroid hormone stimulates bone formation and resorption in organ culture: evidence for a coupling mechanism. Proc Natl Acad Sci U S A 1981;78(5):3204–8.

[113] Mohan S, Baylink DJ. Bone growth factors. Clin Orthop Relat Res 1991;263:30–48.

[114] Baylink DJ, Finkelman RD, Mohan S. Growth factors to stimulate bone formation. J Bone Miner Res 1993;8(Suppl. 2):S565–72.

[115] Tang Y, Wu X, Lei W, et al. TGF-beta1-induced migration of bone mesenchymal stem cells couples bone resorption with formation. Nat Med 2009;15(7):757–65.

[116] Hauge EM, Qvesel D, Eriksen EF, Mosekilde L, Melsen F. Cancellous bone remodeling occurs in specialized compartments lined by cells expressing osteoblastic markers. J Bone Miner Res 2001;16(9):1575–82.

[117] Eghbali-Fatourechi GZ, Modder UI, Charatcharoenwitthaya N, et al. Characterization of circulating osteoblast lineage cells in humans. Bone 2007;40(5):1370–7.

[118] Andersen TL, Sondergaard TE, Skorzynska KE, et al. A physical mechanism for coupling bone resorption and formation in adult human bone. Am J Pathol 2009;174(1):239–47.

[119] Doherty MJ, Ashton BA, Walsh S, Beresford JN, Grant ME, Canfield AE. Vascular pericytes express osteogenic potential in vitro and in vivo. J Bone Miner Res 1998;13(5):828–38.

[120] Howson KM, Aplin AC, Gelati M, Alessandri G, Parati EA, Nicosia RF. The postnatal rat aorta contains pericyte progenitor cells that form spheroidal colonies in suspension culture. Am J Physiol Cell Physiol 2005;289(6):C1396–407.

[121] Matsumoto T, Kawamoto A, Kuroda R, et al. Therapeutic potential of vasculogenesis and osteogenesis promoted by peripheral blood CD34-positive cells for functional bone healing. Am J Pathol 2006;169(4):1440–57.

[122] Modder UI, Khosla S. Skeletal stem/osteoprogenitor cells: current concepts, alternate hypotheses, and relationship to the bone remodeling compartment. J Cell Biochem 2008;103(2):393–400.

[123] Otsuru S, Tamai K, Yamazaki T, Yoshikawa H, Kaneda Y. Circulating bone marrow-derived osteoblast progenitor cells are recruited to the bone-forming site by the CXCR4/stromal cell-derived factor-1 pathway. Stem Cells 2008;26(1):223–34.

[124] Gray C, Boyde A, Jones SJ. Topographically induced bone formation in vitro: implications for bone implants and bone grafts. Bone 1996;18(2):115–23.

[125] Stains JP, Civitelli R. Gap junctions in skeletal development and function. Biochim Biophys Acta 2005;1719(1-2):69–81.

[126] Dalby MJ, McCloy D, Robertson M, et al. Osteoprogenitor response to semi-ordered and random nanotopographies. Biomaterials 2006;27(15):2980–7.

[127] Allan EH, Hausler KD, Wei T, et al. EphrinB2 regulation by PTH and PTHrP revealed by molecular profiling in differentiating osteoblasts. J Bone Miner Res 2008;23(8):1170–81.

[128] Gale NW, Holland SJ, Valenzuela DM, et al. Eph receptors and ligands comprise two major specificity subclasses and are reciprocally compartmentalized during embryogenesis. Neuron 1996;17(1):9–19.

[129] Pasquale EB. Eph receptor signalling casts a wide net on cell behaviour. Nat Rev Mol Cell Biol 2005;6(6):462–75.

[130] Lu Q, Sun EE, Klein RS, Flanagan JG. Ephrin-B reverse signaling is mediated by a novel PDZ-RGS protein and selectively inhibits G protein-coupled chemoattraction. Cell 2001;105(1):69–79.

[131] Murai KK, Pasquale EB. 'Eph'ective signaling: forward, reverse and crosstalk. J Cell Sci 2003;116 (Pt 14):2823–32.

[132] Luiz de Freitas PH, Li M, Ninomiya T, et al. Intermittent PTH administration stimulates pre-osteoblastic proliferation without leading to enhanced bone formation in osteoclast-less c-fos (−/−) mice. J Bone Miner Res 2009;24(9):1586–97.

[133] van Bezooijen RL, Roelen BA, Visser A, et al. Sclerostin is an osteocyte-expressed negative regulator of bone formation, but not a classical BMP antagonist. J Exp Med 2004;199(6):805–14.

[134] Keller H, Kneissel M. SOST is a target gene for PTH in bone. Bone 2005;37(2):148–58.

[135] Bellido T, Ali AA, Gubrij I, et al. Chronic elevation of parathyroid hormone in mice reduces expression of sclerostin by osteocytes: a novel mechanism for hormonal control of osteoblastogenesis. Endocrinology 2005;146(11):4577–83.

[136] Walker EC, McGregor NE, Poulton IJ, et al. Oncostatin M promotes bone formation independently of resorption through the leukemia inhibitory factor receptor. J Clin Invest 2010;120(2):582–92.

[137] Zhou H, Mak W, Zheng Y, Dunstan CR, Seibel MJ. Osteoblasts directly control lineage commitment of mesenchymal progenitor cells through Wnt signaling. J Biol Chem 2008;283(4):1936–45.

[138] Del Fattore A, Peruzzi B, Rucci N, et al. Clinical, genetic, and cellular analysis of 49 osteopetrotic patients: implications for diagnosis and treatment. J Med Genet 2006;43(4):315–25.

[139] Soriano P, Montgomery C, Geske R, Bradley A. Targeted disruption of the c-src proto-oncogene leads to osteopetrosis in mice. Cell 1991;64(4):693–702.

[140] Kornak U, Kasper D, Bosl MR, et al. Loss of the ClC-7 chloride channel leads to osteopetrosis in mice and man. Cell 2001;104(2):205–15.

[141] Pennypacker B, Shea M, Liu Q, et al. Bone density, strength, and formation in adult cathepsin K (−/−) mice. Bone 2009;44(2):199–207.

[142] Chiusaroli R, Knobler H, Luxenburg C, et al. Tyrosine phosphatase epsilon is a positive regulator of osteoclast function in vitro and in vivo. Mol Biol Cell 2004;15(1):234–44.

[143] Lee SH, Rho J, Jeong D, et al. v-ATPase V0 subunit d2-deficient mice exhibit impaired osteoclast fusion and increased bone formation. Nat Med 2006;12(12):1403–9.

[144] Sims NA, Jenkins BJ, Quinn JM, et al. Glycoprotein 130 regulates bone turnover and bone size by distinct downstream signaling pathways. J Clin Invest 2004;113(3):379–89.

[145] Martin TJ, Sims NA. Osteoclast-derived activity in the coupling of bone formation to resorption. Trends Mol Med 2005;11(2):76–81.

[146] Walker E, McGregor N, Poulton I, et al. Cardiotrophin-1 is an osteoclast-derived stimulus of bone formation required for normal bone remodeling. J Bone Miner Res 2008;23:2025–32.

[147] Neer RM, Arnaud CD, Zanchetta JR, et al. Effect of parathyroid hormone (1-34) on fractures and bone mineral density in postmenopausal women with osteoporosis. N Engl J Med 2001;344(19):1434–41.

[148] Frolik CA, Black EC, Cain RL, et al. Anabolic and catabolic bone effects of human parathyroid hormone (1-34) are predicted by duration of hormone exposure. Bone 2003;33(3):372–9.

[149] Dobnig H, Turner RT. Evidence that intermittent treatment with parathyroid hormone increases bone formation in adult rats by activation of bone lining cells. Endocrinology 1995; 136(8):3632–8.

[150] Jilka RL. Molecular and cellular mechanisms of the anabolic effect of intermittent PTH. Bone 2007;40(6):1434–46.

[151] Jiang Y, Zhao JJ, Mitlak BH, Wang O, Genant HK, Eriksen EF. Recombinant human parathyroid hormone (1-34) [teriparatide] improves both cortical and cancellous bone structure. J Bone Miner Res 2003;18(11):1932–41.

[152] Foldes J, Parfitt AM, Shih MS, Rao DS, Kleerekoper M. Structural and geometric changes in iliac bone: relationship to normal aging and osteoporosis. J Bone Miner Res 1991;6(7):759–66.

[153] Delmas PD, Vergnaud P, Arlot ME, Pastoureau P, Meunier PJ, Nilssen MH. The anabolic effect of human PTH (1-34) on bone formation is blunted when bone resorption is inhibited by the bisphosphonate tiludronate – Is activated resorption a prerequisite for the in vivo effect of PTH on formation in a remodeling system? Bone 1995;16(6):603–10.

[154] Black DM, Greenspan SL, Ensrud KE, et al. The effects of parathyroid hormone and alendronate alone or in combination in postmenopausal osteoporosis. N Engl J Med 2003;349(13):1207–15.

[155] Finkelstein JS, Hayes A, Hunzelman JL, Wyland JJ, Lee H, Neer RM. The effects of parathyroid hormone, alendronate, or both in men with osteoporosis. N Engl J Med 2003;349(13):1216–26.

[156] Demiralp B, Chen HL, Koh AJ, Keller ET, McCauley LK. Anabolic actions of parathyroid hormone during bone growth are dependent on c-fos. Endocrinology 2002;143(10):4038–47.

[157] Ma YL, Cain RL, Halladay DL, et al. Catabolic effects of continuous human PTH (1–38) in vivo is associated with sustained stimulation of RANKL and inhibition of osteoprotegerin and gene-associated bone formation. Endocrinology 2001;142(9):4047–54.

[158] Onyia JE, Bidwell J, Herring J, Hulman J, Hock JM. In vivo, human parathyroid hormone fragment (hPTH 1-34) transiently stimulates immediate early response gene expression, but not proliferation, in trabecular bone cells of young rats. Bone 1995;17(5):479–84.

[159] Martin TJ. Does bone resorption inhibition affect the anabolic response to parathyroid hormone? Trends Endocrinol Metab 2004;15(2):49–50.

[160] Terauchi M, Li JY, Bedi B, et al. T lymphocytes amplify the anabolic activity of parathyroid hormone through Wnt10b signaling. Cell Metab 2009,10(3).229–40.

[161] Amizuka N, Warshawsky H, Henderson JE, Goltzman D, Karaplis AC. Parathyroid hormone-related peptide-depleted mice show abnormal epiphyseal cartilage development and altered endochondral bone formation. J Cell Biol 1994;126(6):1611–23.

[162] Kartsogiannis V, Moseley J, McKelvie B, et al. Temporal expression of PTHrP during endochondral bone formation in mouse and intramembranous bone formation in an in vivo rabbit model. Bone 1997;21(5):385–92.

[163] Suda N, Gillespie MT, Traianedes K, et al. Expression of parathyroid hormone-related protein in cells of osteoblast lineage. J Cell Physiol 1996;166(1):94–104.

[164] Amizuka N, Karaplis AC, Henderson JE, et al. Haploinsufficiency of parathyroid hormone-related peptide (PTHrP) results in abnormal postnatal bone development. Dev Biol 1996;175 (1):166–76.

[165] Miao D, He B, Jiang Y, et al. Osteoblast-derived PTHrP is a potent endogenous bone anabolic agent that modifies the therapeutic efficacy of administered PTH 1-34. J Clin Invest 2005;115 (9):2402–11.

[166] Gendron S, Boisvert M, Chetoui N, Aoudjit F. Alpha1beta1 integrin and interleukin-7 receptor up-regulate the expression of RANKL in human T cells and enhance their osteoclastogenic function. Immunology 2008.

[167] Kong YY, Feige U, Sarosi I, et al. Activated T cells regulate bone loss and joint destruction in adjuvant arthritis through osteoprotegerin ligand. Nature 1999;402(6759):304–9.

[168] Sato K, Suematsu A, Okamoto K, et al. Th17 functions as an osteoclastogenic helper T cell subset that links T cell activation and bone destruction. J Exp Med 2006;203(12):2673–82.

[169] Horwood NJ, Kartsogiannis V, Quinn JM, Romas E, Martin TJ, Gillespie MT. Activated T lymphocytes support osteoclast formation in vitro. Biochem Biophys Res Commun 1999;265(1):144–50.

[170] Josien R, Wong BR, Li HL, Steinman RM, Choi Y. TRANCE, a TNF family member, is differentially expressed on T cell subsets and induces cytokine production in dendritic cells. J Immunol 1999;162(5):2562–8.

[171] Gao Y, Grassi F, Ryan MR, et al. IFN-gamma stimulates osteoclast formation and bone loss in vivo via antigen-driven T cell activation. J Clin Invest 2007;117(1):122–32.

[172] Stein NC, Kreutzmann C, Zimmermann SP, et al. Interleukin-4 and interleukin-13 stimulate the osteoclast inhibitor osteoprotegerin by human endothelial cells through the STAT6 pathway. J Bone Miner Res 2008;23(5):750–8.

[173] Kotake S, Nanke Y, Mogi M, et al. IFN-gamma-producing human T cells directly induce osteoclastogenesis from human monocytes via the expression of RANKL. Eur J Immunol 2005;35(11):3353–63.

[174] Iwakura Y, Ishigame H. The IL-23/IL-17 axis in inflammation. J Clin Invest 2006;116(5):1218–22.

[175] Kotake S, Udagawa N, Takahashi N, et al. IL-17 in synovial fluids from patients with rheumatoid arthritis is a potent stimulator of osteoclastogenesis. J Clin Invest 1999;103(9):1345–52.

[176] Kawai T, Matsuyama T, Hosokawa Y, et al. B and T lymphocytes are the primary sources of RANKL in the bone resorptive lesion of periodontal disease. Am J Pathol 2006;169(3):987–98.

[177] Lockhart E, Green AM, Flynn JL. IL-17 production is dominated by gammadelta T cells rather than CD4 T cells during Mycobacterium tuberculosis infection. J Immunol 2006;177(7):4662–9.

[178] Adorini L, Penna G. Control of autoimmune diseases by the vitamin D endocrine system. Nat Clin Pract Rheumatol 2008.

[179] Scariano JK, Emery-Cohen AJ, Pickett GG, Morgan M, Simons PC, Alba F. Estrogen receptors alpha (ESR1) and beta (ESR2) are expressed in circulating human lymphocytes. J Recept Signal Transduct Res 2008;28(3):285–93.

[180] Quinn JM, Saleh H. Modulation of osteoclast function in bone by the immune system. Mol Cell Endocrinol 2009;310(1-2):40–51.

[181] Gao Y, Wu X, Terauchi M, et al. T cells potentiate PTH-induced cortical bone loss through CD40L signaling. Cell Metab 2008;8(2):132–45.

[182] Jilka RL, Hangoc G, Girasole G, et al. Increased osteoclast development after estrogen loss: mediation by interleukin-6. Science 1992;257(5066):88–91.

[183] Kimble RB, Vannice JL, Bloedow DC, et al. Interleukin-1 receptor antagonist decreases bone loss and bone resorption in ovariectomized rats. J Clin Invest 1994;93(5):1959–67.

[184] Linton PJ, Dorshkind K. Age-related changes in lymphocyte development and function. Nat Immunol 2004;5(2):133–9.

[185] Cenci S, Weitzmann MN, Roggia C, et al. Estrogen deficiency induces bone loss by enhancing T-cell production of TNF-alpha. J Clin Invest 2000;106(10):1229–37.

[186] Roggia C, Gao Y, Cenci S, et al. Up-regulation of TNF-producing T cells in the bone marrow: a key mechanism by which estrogen deficiency induces bone loss in vivo. Proc Natl Acad Sci U S A 2001;98(24):13960–5.

[187] Lee SK, Kadono Y, Okada F, et al. T lymphocyte-deficient mice lose trabecular bone mass with ovariectomy. J Bone Miner Res 2006;21(11):1704–12.

[188] Manabe N, Kawaguchi H, Chikuda H, et al. Connection between B lymphocyte and osteoclast differentiation pathways. J Immunol 2001;167(5):2625–31.

[189] Li Y, Toraldo G, Li A, et al. B cells and T cells are critical for the preservation of bone homeostasis and attainment of peak bone mass in vivo. Blood 2007;109(9):3839–48.

[190] Onoe Y, Miyaura C, Ito M, Ohta H, Nozawa S, Suda T. Comparative effects of estrogen and raloxifene on B lymphopoiesis and bone loss induced by sex steroid deficiency in mice. J Bone Miner Res 2000;15(3):541–9.

[191] Miyaura C, Onoe Y, Inada M, et al. Increased B-lymphopoiesis by interleukin 7 induces bone loss in mice with intact ovarian function: similarity to estrogen deficiency. Proc Natl Acad Sci U S A 1997;94(17):9360–5.

[192] Valenzona HO, Pointer R, Ceredig R, Osmond DG. Prelymphomatous B cell hyperplasia in the bone marrow of interleukin-7 transgenic mice: precursor B cell dynamics, microenvironmental organization and osteolysis. Exp Hematol 1996;24(13):1521–9.

[193] Hartgring SA, Bijlsma JW, Lafeber FP, van Roon JA. Interleukin-7 induced immunopathology in arthritis. Ann Rheum Dis 2006;65(Suppl. 3):iii69–74.

[194] Yun TJ, Tallquist MD, Aicher A, et al. Osteoprotegerin, a crucial regulator of bone metabolism, also regulates B cell development and function. J Immunol 2001;166(3):1482–91.

[195] Kim D, Mebius RE, MacMicking JD, et al. Regulation of peripheral lymph node genesis by the tumor necrosis factor family member TRANCE. J Exp Med 2000;192(10):1467–78.

[196] Shinoda K, Sugiyama E, Taki H, et al. Resting T cells negatively regulate osteoclast generation from peripheral blood monocytes. Bone 2003;33(4):711–20.

[197] Gillespie MT. Impact of cytokines and T lymphocytes upon osteoclast differentiation and function. Arthritis Res Ther 2007;9(2):103.

[198] Takayanagi H, Ogasawara K, Hida S, et al. T-cell-mediated regulation of osteoclastogenesis by signalling cross-talk between RANKL and IFN-gamma. Nature 2000;408(6812):600–5.

[199] Kim MS, Day CJ, Morrison NA. MCP-1 is induced by receptor activator of nuclear factor-{kappa}B ligand, promotes human osteoclast fusion, and rescues granulocyte macrophage colony-stimulating factor suppression of osteoclast formation. J Biol Chem 2005;280(16):16163–9.

[200] Speziani C, Rivollier A, Gallois A, et al. Murine dendritic cell transdifferentiation into osteoclasts is differentially regulated by innate and adaptive cytokines. Eur J Immunol 2007;37(3):747–57.

[201] Horwood NJ, Udagawa N, Elliott J, et al. Interleukin 18 inhibits osteoclast formation via T cell production of granulocyte macrophage colony-stimulating factor. J Clin Invest 1998;101(3):595–603.

[202] McInnes IB, Liew FY, Gracie JA. Interleukin-18: a therapeutic target in rheumatoid arthritis? Arthritis Res Ther 2005;7(1):38–41.

[203] Sareneva T, Julkunen I, Matikainen S. IFN-alpha and IL-12 induce IL-18 receptor gene expression in human NK and T cells. J Immunol 2000;165(4):1933–8.

[204] Udagawa N, Horwood NJ, Elliott J, et al. Interleukin-18 (interferon-gamma-inducing factor) is produced by osteoblasts and acts via granulocyte/macrophage colony-stimulating factor and not via interferon-gamma to inhibit osteoclast formation. J Exp Med 1997;185(6):1005–12.

[205] Raggatt LJ, Qin L, Tamasi J, et al. Interleukin-18 is regulated by parathyroid hormone and is required for its bone anabolic actions. J Biol Chem 2008;283(11):6790–8.

[206] Cornish J, Gillespie MT, Callon KE, Horwood NJ, Moseley JM, Reid IR. Interleukin-18 is a novel mitogen of osteogenic and chondrogenic cells. Endocrinology 2003;144(4):1194–201.

[207] Makiishi-Shimobayashi C, Tsujimura T, Iwasaki T, et al. Interleukin-18 up-regulates osteoprotegerin expression in stromal/osteoblastic cells. Biochem Biophys Res Commun 2001;281(2):361–6.

[208] Dai SM, Nishioka K, Yudoh K. Interleukin (IL) 18 stimulates osteoclast formation through synovial T cells in rheumatoid arthritis: comparison with IL1 beta and tumour necrosis factor alpha. Ann Rheum Dis 2004;63(11):1379–86.

[209] Horwood NJ, Elliott J, Martin TJ, Gillespie MT. IL-12 alone and in synergy with IL-18 inhibits osteoclast formation in vitro. J Immunol 2001;166(8):4915–21.

[210] Amcheslavsky A, Bar-Shavit Z. Interleukin (IL)-12 mediates the anti-osteoclastogenic activity of CpG-oligodeoxynucleotides. J Cell Physiol 2006;207(1):244–50.

[211] Watford WT, Moriguchi M, Morinobu A, O'Shea JJ. The biology of IL-12: coordinating innate and adaptive immune responses. Cytokine Growth Factor Rev 2003;14(5):361–8.

[212] Chang JT, Segal BM, Nakanishi K, Okamura H, Shevach EM. The costimulatory effect of IL-18 on the induction of antigen-specific IFN-gamma production by resting T cells is IL-12 dependent and is mediated by up-regulation of the IL-12 receptor beta2 subunit. Eur J Immunol 2000;30(4):1113–9.

[213] Yoshimoto T, Takeda K, Tanaka T, et al. IL-12 up-regulates IL-18 receptor expression on T cells, Th1 cells, and B cells: synergism with IL-18 for IFN-gamma production. J Immunol 1998;161(7):3400–7.

[214] Quinn JM, Sims NA, Saleh H, et al. IL-23 inhibits osteoclastogenesis indirectly through lymphocytes and is required for the maintenance of bone mass in mice. J Immunol 2008;181(8):5720–9.

[215] Aggarwal S, Ghilardi N, Xie MH, de Sauvage FJ, Gurney AL. Interleukin-23 promotes a distinct CD4 T cell activation state characterized by the production of interleukin-17. J Biol Chem 2003;278(3):1910–1.

[216] Mirosavljevic D, Quinn JM, Elliott J, Horwood NJ, Martin TJ, Gillespie MT. T-cells mediate an inhibitory effect of interleukin-4 on osteoclastogenesis. J Bone Miner Res 2003;18(6):984–93.

[217] Kelchtermans H, Geboes L, Mitera T, Huskens D, Leclercq G, Matthys P. Activated CD4+CD25+ regulatory T cells inhibit osteoclastogenesis and collagen-induced arthritis. Ann Rheum Dis 2008.

[218] Zaiss MM, Axmann R, Zwerina J, et al. Treg cells suppress osteoclast formation: a new link between the immune system and bone. Arthritis Rheum 2007;56(12):4104–12.

[219] Kim YG, Lee CK, Nah SS, Mun SH, Yoo B, Moon HB. Human CD4+CD25+ regulatory T cells inhibit the differentiation of osteoclasts from peripheral blood mononuclear cells. Biochem Biophys Res Commun 2007;357(4):1046–52.

[220] Axmann R, Herman S, Zaiss M, et al. CTLA-4 directly inhibits osteoclast formation. Ann Rheum Dis 2008.

[221] Pettit AR, Chang MK, Hume DA, Raggatt LJ. Osteal macrophages: a new twist on coupling during bone dynamics. Bone 2008;43(6):976–82.

[222] Chang MK, Raggatt LJ, Alexander KA, et al. Osteal tissue macrophages are intercalated throughout human and mouse bone lining tissues and regulate osteoblast function in vitro and in vivo. J Immunol 2008;181(2):1232–44.

[223] Ushiku C, Jiang X, Wang L, Adams DJ, Rowe D. Cellular events that precede the formation of the fracture callus. J Bone Miner Res 2009;(Suppl. 1).

[224] Hankemeier S, Grassel S, Plenz G, Spiegel HU, Bruckner P, Probst A. Alteration of fracture stability influences chondrogenesis, osteogenesis and immigration of macrophages. J Orthop Res 2001;19(4):531–8.

[225] Alexander KS, Raggatt LJ, Chang MK, et al. Osteomacs are critical for optimal intramembranous bone formation in a tibial defect model of bone healing. J Bone Miner Res 2009.

[226] Raggatt LJ, Chang MK, Alexander K, et al. Osteomacs: osteoclast precursors during inflammatory. J Bone Miner Res 2009;24(Suppl. 1).

[227] Kanatani M, Sugimoto T, Kano J, Kanzawa M, Chihara K. Effect of high phosphate concentration on osteoclast differentiation as well as bone-resorbing activity. J Cell Physiol 2003;196(1):180–9.

[228] Kanatani M, Sugimoto T, Kaji H, et al. Stimulatory effect of bone morphogenetic protein-2 on osteoclast-like cell formation and bone-resorbing activity. J Bone Miner Res 1995;10(11): 1681–90.

[229] Li X, Warmington KS, Niu Q, et al. Treatment with an anti-sclerostin antibody increased bone mass by stimulating bone formation in aged male rats. Journal of Bone and Mineral Research 2007;22(Suppl. 1):S36.

[230] Yun TJ, Chaudhary PM, Shu GL, et al. OPG/FDCR-1, a TNF receptor family member, is expressed in lymphoid cells and is up-regulated by ligating CD40. J Immunol 1998;161(11):6113–21.

[231] Lacey DL, Erdmann JM, Teitelbaum SL, Tan HL, Ohara J, Shioi A. Interleukin 4, interferon-gamma, and prostaglandin E impact the osteoclastic cell-forming potential of murine bone marrow macrophages. Endocrinology 1995;136(6):2367–76.

[232] Smith DD, Gowen M, Mundy GR. Effects of interferon-gamma and other cytokines on collagen synthesis in fetal rat bone cultures. Endocrinology 1987;120(6):2494–9.

[233] Khapli SM, Mangashetti LS, Yogesha SD, Wani MR. IL-3 acts directly on osteoclast precursors and irreversibly inhibits receptor activator of NF-kappa B ligand-induced osteoclast differentiation by diverting the cells to macrophage lineage. J Immunol 2003;171(1):142–51.

[234] Ueno K, Katayama T, Miyamoto T, Koshihara Y. Interleukin-4 enhances in vitro mineralization in human osteoblast-like cells. Biochem Biophys Res Commun 1992;189(3):1521–6.

[235] Ura K, Morimoto I, Watanabe K, Saito K, Yanagihara N, Eto S. Interleukin (IL)-4 and IL-13 inhibit the differentiation of murine osteoblastic MC3T3-E1 cells. Endocr J 2000;47(3):293–302.

[236] Xu LX, Kukita T, Kukita A, Otsuka T, Niho Y, Iijima T. Interleukin-10 selectively inhibits osteoclastogenesis by inhibiting differentiation of osteoclast progenitors into preosteoclast-like cells in rat bone marrow culture system. J Cell Physiol 1995;165(3):624–9.

[237] Dresner-Pollak R, Gelb N, Rachmilewitz D, Karmeli F, Weinreb M. Interleukin 10-deficient mice develop osteopenia, decreased bone formation, and mechanical fragility of long bones. Gastroenterology 2004;127(3):792–801.

[238] Palmqvist P, Lundberg P, Persson E, et al. Inhibition of hormone and cytokine-stimulated osteoclastogenesis and bone resorption by interleukin-4 and interleukin-13 is associated with increased osteoprotegerin and decreased RANKL and RANK in a STAT6-dependent pathway. J Biol Chem 2006;281(5):2414–29.

[239] Hausler KD, Horwood NJ, Chuman Y, et al. Secreted frizzled-related protein-1 inhibits RANKL-dependent osteoclast formation. J Bone Miner Res 2004;19(11):1873–81.

[240] Yao W, Cheng Z, Shahnazari M, Dai W, Johnson ML, Lane NE. Overexpression of secreted frizzled-related protein 1 inhibits bone formation and attenuates PTH bone anabolic effects. J Bone Miner Res 2009.

[241] Tian E, Zhan F, Walker R, et al. The role of the Wnt-signaling antagonist DKK1 in the development of osteolytic lesions in multiple myeloma. N Engl J Med 2003;349(26):2483–94.

[242] Qiang YW, Chen Y, Stephens O, et al. Myeloma-derived Dickkopf-1 disrupts Wnt-regulated osteoprotegerin and RANKL production by osteoblasts: a potential mechanism underlying osteolytic bone lesions in multiple myeloma. Blood 2008.

# 9

# The Role of the Immune System in the Development of Osteoporosis

Ulrike I. Mödder, B. Lawrence Riggs, Sundeep Khosla

ENDOCRINE RESEARCH UNIT, COLLEGE OF MEDICINE, MAYO CLINIC, ROCHESTER, MN, USA

## CHAPTER OUTLINE

Introduction .................................................................................................................. 270
Patterns of age-related bone loss and fractures ............................................................ 270
    Age-related bone loss .............................................................................................. 270
    Incidence of osteoporotic fractures with aging ......................................................... 273
Relationship between sex steroid production and bone loss ......................................... 274
    Role in women ........................................................................................................ 274
    Role in men ............................................................................................................. 274
Sex steroid regulation of bone remodeling ................................................................... 276
    Overview ................................................................................................................. 276
    Effects of sex steroids on bone resorption and calcium homeostasis ....................... 277
    Estrogen effects on bone formation ........................................................................ 278
    Estrogen effects on osteocyte function ................................................................... 279
Estrogen and the immune system ................................................................................. 279
    Overview ................................................................................................................. 279
    Effects on immune cytokines .................................................................................. 279
        *The RANKL/OPG/RANK system* ........................................................................ 279
        *Proinflammatory cytokines* ............................................................................... 282
    Effects on immune effector cells ............................................................................. 284
        *T lymphocytes* ................................................................................................... 284
        *B lymphocytes* ................................................................................................... 286
    Summary of estrogen effects on regulating bone resorption .................................... 286
Potential effects of estrogen deficiency on general immunity ....................................... 286
Effects of aging on bone loss independent of changes in sex steroid production ......... 288
Secondary causes of osteoporosis mediated by the immune system ............................ 289
Prospects for treatment of osteoporosis by regulation of immune cytokines ................. 290
Summary and conclusions ............................................................................................ 291
Acknowledgments ......................................................................................................... 291
References .................................................................................................................... 292

# Introduction

The past decade has witnessed tremendous advances in our understanding of the pathogenesis of osteoporosis. Newer imaging approaches, including central and peripheral quantitative computed tomography (QCT) and high-resolution peripheral QCT (HRpQCT), have helped to more clearly define the overall pattern of bone mass acquisition and bone loss in both sexes. In addition, fundamental advances in bone biology, such as the discovery and characterization of the receptor-activator of nuclear factor kappa B ligand (RANKL)/RANK/osteoprotegerin (OPG) system, have provided new insights into the mediators of excessive bone resorption following sex steroid deficiency. Concurrent with this greater understanding of bone loss at the clinical and basic levels has been the increasing recognition of the close interactions between bone cells and immune cells, mediated by cytokines produced by both types of cells. In this chapter, we review the current data on patterns of bone loss with aging and gender that lead to osteoporosis in humans and the mechanisms by which these changes lead to osteoporosis, with particular emphasis on the potential role of immune cytokines in mediating these changes. We also focus on the key role of estrogen deficiency in bone loss in both women and men, acting both by altering the regulatory cytokines and by other mechanisms. These effects are superimposed on age effects on bone cell function that are independent of both estrogen deficiency and alterations in the production and function of immune cytokines.

# Patterns of age-related bone loss and fractures

## Age-related bone loss

Aging is associated with significant bone loss in both women and men (reviewed in [1]). Figure 9-1, which draws on a number of cross-sectional and longitudinal studies using areal bone mineral density (aBMD) by dual-energy X-ray absorptiometry (DXA), depicts the overall pattern of bone loss in both genders. As shown, the menopause in women is associated with a rapid loss of trabecular bone, in the vertebrae, pelvis, and ultra-distal forearm. There is a less dramatic loss of cortical bone (present in the long bones of the body and as a thin rim around the vertebrae and other sites of trabecular bone) following the menopause. Approximately 8–10 years following the menopause, a slow, age-related phase of bone loss in both trabecular and cortical bone becomes apparent and continues throughout life. Since men lack the equivalent of a menopause, they generally do not exhibit this rapid phase of bone loss. However, men have a very similar pattern of slow, age-related bone loss.

While DXA has provided important insights into the patterns of age-related bone loss, its utility is limited by the fact that it cannot clearly separate trabecular from cortical bone or provide information on possible changes in bone size or geometry with age. In recent studies, Riggs et al. [2] utilized central and peripheral quantitative computed

Chapter 9 • The Role of the Immune System in the Development of Osteoporosis   271

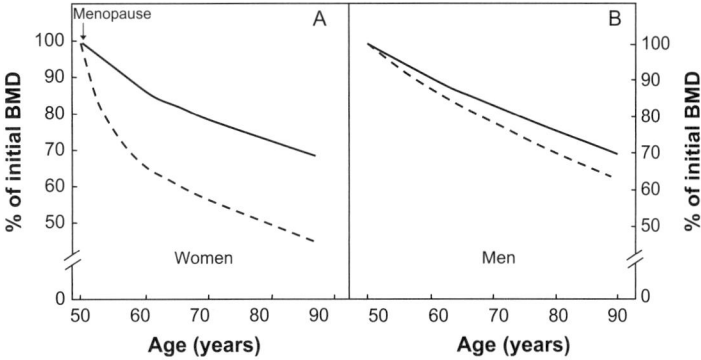

**FIGURE 9-1** Patterns of age-related bone loss in women and in men. Dashed lines represent trabecular bone and solid lines, cortical bone. The figure is a composite, based on multiple cross-sectional and longitudinal studies using DXA. See text for discussion. Reproduced from [134], with permission.

tomography (QCT) along with new image analysis software [3] to better define age-associated changes in bone volumetric density, geometry, and structure at different skeletal sites. As shown in Figure 9-2A, there were large decreases in volumetric BMD (vBMD) at the spine over life (predominantly trabecular bone), which seemed to begin even before middle life. These decreases were greater in women (~55%) than in men (~45%, $P < 0.001$). Even in this cross-sectional study, there was an apparent small midlife acceleration in the slope of the decrease in women that accounted for much of their significantly greater decrease in vertebral vBMD over life compared with men. In contrast to this pattern of changes in trabecular vBMD at the spine, cortical vBMD at the

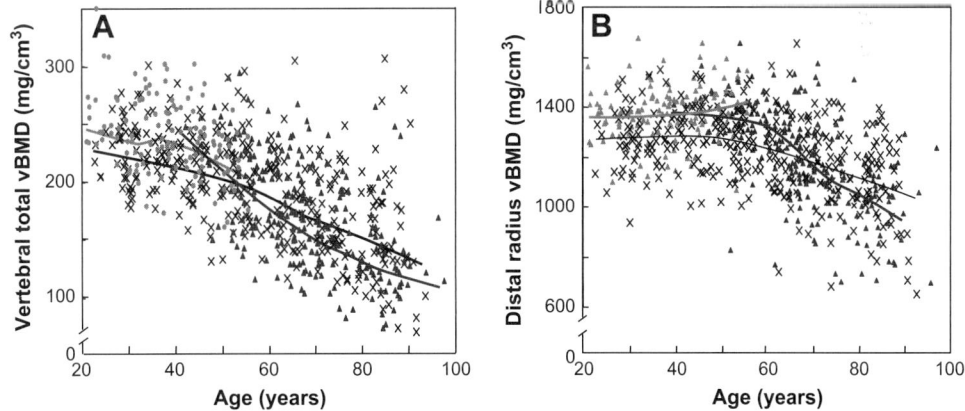

**FIGURE 9-2** **(A)** Values for vBMD (mg/cm$^3$) of the total vertebral body in a population sample of Rochester, Minnesota, women and men between the ages of 20 and 97 years. Individual values and smoother lines are given for premenopausal women in red, for postemenopausal women in blue, and for men in black. **(B)** Values for cortical vBMD at the distal radius in the same cohort. Color code is as in Panel A. All changes with age were significant ($P < 0.05$). Reproduced from Riggs et al. [2], with permission. Please refer to color plate section.

radius showed little change until midlife in either women or men (Figure 9-2B). Thereafter, there were linear decreases in both sexes, but the decreases were greater in women (28%) than in men (18%, $P < 0.001$). Aging was also associated with increases in bone cross-sectional area at various sites due to continued periosteal apposition throughout life. However, bone marrow space increased even more due to ongoing bone resorption: thus, because endocortical resorption increased even more than periosteal apposition, there was a net decrease in cortical area and thickness [2]. However, this process also resulted in outward displacement of the cortex, which increased the strength of bone to bending stresses and partially offset the decrease in bone strength resulting from decreased cortical area. These cross-sectional changes were subsequently confirmed by longitudinal data, which provided essentially identical results, including the documentation of substantial trabecular bone loss at multiple sites beginning in the third decade, well before the menopause in women or the onset of significant sex steroid deficiency in men [4]. Collectively, these findings indicate that age-related changes in bone are complex. Some are beneficial to bone strength, such as periosteal apposition with outward cortical displacement. However, others are deleterious, such as increased endocortical resorption, increased cortical porosity, and large decreases in trabecular and cortical vBMD.

The recent application of HRpQCT at the distal radius and tibia has also provided important new information on changes in trabecular and cortical microstructure with aging. This technology uses a voxel size of 89 μm to essentially obtain an "in vivo" bone biopsy at these sites, although a limitation of this approach is that central sites such as the spine and hip cannot be scanned due to the radiation doses needed for such high resolutions. Importantly, in cadaveric specimens, the bone microstructure variables assessed using this approach correlate extremely well ($R > 0.95$) with even higher-resolution μCT, which is generally considered the "gold standard" technique [5]. Using HRpQCT at the wrist in a population-based cross-sectional study involving 324 women and 278 men age 21–97 years [6], we found that relative to young women (age 20–29 years), young men had greater trabecular bone volume/tissue volume (BV/TV; by 26%) and trabecular thickness (TbTh, by 28%) but similar values for trabecular number (TbN) and trabecular separation (TbSp). Between ages 20 and 90 years, cross-sectional decreases in BV/TV were similar in women (−27%) and in men (−26%), but whereas women had significant decreases in TbN (−13%) and increases in TbSp (+24%), these parameters had little net change over life in men. However, TbTh decreased to a greater extent in men (−24%) than in women (−18%). These population-based structural data thus demonstrated that while decreases in trabecular BV/TV with age are similar in men and women, the structural basis for this decrease is quite different between the sexes. Over life, women undergo loss of trabeculae with an increase in TbSp, whereas men begin young adult life with thicker trabeculae and primarily sustain trabecular thinning, with no net change in TbN or TbSp. This has important biomechanical consequences, since decreases in TbN have been shown to have a much greater impact on bone strength compared to decreases in TbTh. These findings may help explain the lower life-long risk of fractures in men, and specifically their virtual immunity to age-related increases in distal forearm fractures.

## Incidence of osteoporotic fractures with aging

The age-associated changes in bone mass and structure described above lead, in turn, to a marked increase in the incidence of osteoporotic fractures in both sexes. As shown in Figure 9-3A, distal forearm (Colles') fractures increase sharply in women soon after the menopause and then plateau after 10–15 years postmenopausally. The increase in the incidence of vertebral fractures after menopause is more gradual but, in contrast to Colles' fractures, vertebral fractures continue to increase throughout life. The rise in hip fractures follows that in vertebral fractures, and hip fractures increase markedly late in life. Interestingly, men do not appear to have a measurable increase in Colles' fractures with age (Figure 9-3B), which may, in part, be due to their larger bones. However, with increasing age, there is a clear increase in the incidence of both vertebral and hip fractures in men, although the onset of these fractures is delayed by about 10 years as compared to women, likely due to the absence of the menopause and the associated accelerated bone loss present in women.

Based on these types of data, it has been estimated that 4 out of every 10 white women aged 50 years or older in the United States will experience a hip, spine, or wrist fracture sometime during the remainder of their lives; 13% of white men in this country will also suffer one of these fractures [7]. While the risk of these fractures is lower in non-white women and men, it is still substantial. Collectively, osteoporotic fractures result in a significant financial burden on society, with estimated direct care expenditures of $17 billion in 2005 [8].

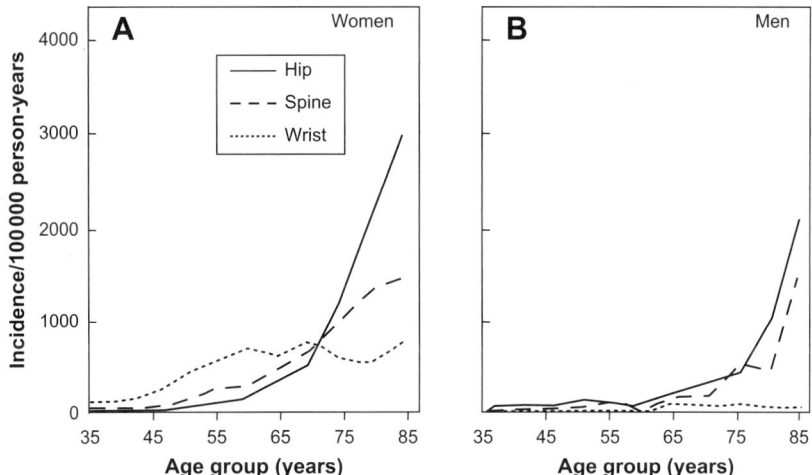

**FIGURE 9-3** Age-specific incidence rates for proximal femur (hip), vertebral (spine), and distal forearm (wrist) fractures in Rochester, Minnesota, women (A) and men (B). Adapted from Cooper and Melton [135], with permission.

# Relationship between sex steroid production and bone loss

## Role in women

As noted earlier, the menopause triggers a rapid phase of bone loss in women that can be prevented by estrogen replacement [9, 10] and clearly results from loss of ovarian function. During the menopausal transition, serum estradiol levels fall to 10–15% of their premenopausal level, although levels of serum estrone, a 4-fold weaker estrogen, fall to about 25–35% of the premenopausal level [11]. Serum testosterone levels also decrease following the menopause [12], although to a lesser extent, since testosterone continues to be produced by the adrenal cortex and by the ovarian interstitium. Bone resorption, as assessed by biochemical markers, increases by 90% at the menopause, whereas bone formation markers increase by only 45% [13]. This imbalance between bone resorption and bone formation leads to accelerated bone loss. The rapid bone loss in this phase produces an increased outflow of calcium from bone into the extracellular pool. However, hypercalcemia is prevented by compensatory increases in urinary calcium excretion [14], decreases in intestinal calcium absorption [15], and a partial suppression of parathyroid hormone (PTH) secretion [16].

The rapid phase of bone loss following the menopause generally subsides after 4–8 years. This is followed by a slow, continuous phase, which lasts indefinitely. Although both menopause and age effects play important roles in bone loss in aging women, the preponderance of evidence indicates that the menopause is more important, at least up to age 70 years. In a unique case-control study, Richelson et al. [17] studied a group of 14 women who had undergone oophorectomy in young adulthood, 22 years earlier, a group of 14 normal perimenopausal women, and a group of 14 women who were 22 years postmenopausal. They found that bone density was similar in the two groups who differed in age but not in years after menopause. These results suggest that estrogen deficiency, and not aging, is the predominant cause of bone loss during the first two decades after menopause. Moreover, rates of bone loss in elderly women correlate inversely with serum estrogen levels and their increases in bone turnover respond favorably to estrogen therapy [1, 16].

## Role in men

As noted previously, while osteoporosis is more common in women, men lose half as much bone with aging and have one-third as many fragility fractures as women [16]. Since most men do not develop overt hypogonadism with aging, the prevailing opinion has been that sex steroid deficiency is not a major cause of age-related bone loss in men. However, it is now clear that the failure of earlier studies to find major decreases in serum levels of total sex steroids with aging in men was because these earlier studies did not account for the confounding effect of a greater than two-fold age-related rise in levels of serum sex-hormone-binding globulin (SHBG) [11]. It is now generally believed that circulating sex steroids that are bound to SHBG have restricted access to target tissues, whereas the 1–3% fraction in serum that is free and the 35–55% fraction that is loosely bound to albumin are readily accessible. While there are a variety of methods to assess the "bioavailable" or non-SHBG-bound sex steroids, several groups have reported substantial decreases in serum levels of free or bioavailable sex steroid levels with aging [11, 18]. Data from a population of 346 men from Rochester, Minnesota, are shown in Table 9-1. The precise cause of the age-related increase in serum SHBG levels and the failure of the hypothalamic–pituitary–testicular axis to compensate for this and maintain free or bioavailable sex steroids at young normal levels is presently unclear and the focus of ongoing studies.

As shown in Table 9-1, aging in men is associated with substantial decreases in bioavailable testosterone and estrogen levels. The traditional notion had been that since testosterone is the major sex steroid in men, it was the decrease in bioavailable testosterone levels that would perhaps most closely be associated with bone loss in men. The initial attempts to address this issue came from cross-sectional observational studies in which sex steroid levels were related to BMD at various sites in cohorts of adult men. Slemenda and colleagues [19] found, somewhat surprisingly, that BMD at various sites in 93 healthy men over age 55 years correlated with serum estradiol levels (correlation coefficients, depending on the site, of +0.21 to +0.35, $P$ values 0.01 to 0.05) and inversely with serum testosterone levels (correlation coefficients of –0.20 to –0.28, $P$ values 0.03 to 0.10). Subsequent to this report, other similar cross-sectional studies have demonstrated significant positive associations between BMD and estrogen levels in men [11, 18, 20–24], particularly circulating bioavailable estradiol levels. These cross-sectional findings have also been confirmed by longitudinal data. Thus, we [25] studied young (22–39 years) and older (60–90 years) men in whom rates of change in BMD at various sites over 4 years were related to sex steroid levels. These two different age groups permitted a separate comparison of the possible effects of estrogen on the final stages of skeletal maturation versus age-related bone loss. Forearm sites (distal radius and ulna) provided the clearest data, perhaps due to the greater precision of peripheral site measurements as compared to central sites such as the spine or hip. In the younger men, BMD at the forearm sites

**Table 9-1** Changes in serum sex steroids and gonadotropins over life in a random sample of 346 Rochester, Minnesota, men aged 23–90 years. Adapted from Khosla et al. [11]

| Hormone | Percent change |
| --- | --- |
| Bioavailable estrogen | −47** |
| Bioavailable testosterone | −64** |
| SHBG | +124** |
| Luteinizing hormone | +285** |
| Follicle-stimulating hormone | +505** |

**$P < 0.005$.

increased by 0.42–0.43%/year, whereas in the older men, BMD at these sites declined by 0.49–0.66%/year. Both the increase in BMD in the younger men as well as the decrease in BMD in the older men were more closely associated with serum bioavailable estradiol levels than with testosterone levels.

While these studies helped to establish that estrogen levels are associated with skeletal maintenance in males, they could not definitively establish causal relationships. In order to address this issue, Falahati-Nini et al. [26] performed a direct interventional study to distinguish between the relative contributions of estrogen versus testosterone in regulating bone resorption and formation in normal elderly men. Endogenous estrogen and testosterone production were suppressed in 59 elderly men using a combination of a long-acting GnRH agonist and an aromatase inhibitor. Physiologic estrogen and testosterone levels were maintained by simultaneously placing the men on estrogen and testosterone patches, delivering doses of sex steroids that mimicked circulating estradiol and testosterone levels in this age group. Using selective withdrawal of either estrogen or testosterone, or both, these investigators demonstrated that even in men, the dominant sex-steroid-regulating bone resorption hormone was estrogen, and not testosterone. Using a somewhat different study design, Leder et al. [27] subsequently confirmed an independent effect of testosterone on bone resorption; although the data in the aggregate clearly favor a more prominent effect of estrogen on the control of bone resorption in men.

It appears, therefore, that similar to women, declining bioavailable estrogen levels in men play a significant role in mediating age-related bone loss. However, declining bioavailable testosterone levels may also contribute, since as demonstrated above, testosterone does have some antiresorptive effects and appears to be important for the maintenance of bone formation [26]. Moreover, it provides the substrate for aromatization to estradiol. In addition, at least in rodents, testosterone has been shown to enhance periosteal apposition [28]. Since larger bones are more resistant to fracture, effects of testosterone, which increase bone size in men, may also provide important protection against fracture risk.

# Sex steroid regulation of bone remodeling

## Overview

Given the key role described above for decreases in sex steroid concentrations and, in particular, for estrogen deficiency in mediating age-related bone loss in both sexes, there has been considerable interest over the years in better defining the possible mechanisms for estrogen regulation of bone metabolism. The evidence from a large number of cellular, animal, and human studies indicates that estrogen has three fundamental effects on bone metabolism: (1) it inhibits the activation of bone remodeling and the initiation of new basic multicellular units (BMUs); (2) it inhibits differentiation and promotes apoptosis of osteoclasts, thereby reducing bone resorption; and (3) while estrogen suppresses self-renewal of early mesenchymal progenitors, it promotes the commitment and differentiation and prevents apoptosis of osteoblastic cells, thereby maintaining bone formation at the cellular level. Each of these actions of estrogen is reviewed below. In further support of these effects, all three cell types have been shown to possess functioning estrogen receptors (for review, see [1]).

## Effects of sex steroids on bone resorption and calcium homeostasis

Estrogen conserves bone mass mainly by preventing increases in bone turnover and maintaining a balance between bone resorption and bone formation. When estrogen is deficient, bone turnover increases but bone resorption increases more than bone formation, leading to bone loss and negative calcium balance. These changes are particularly evident when serum estrogen levels fall relatively rapidly, as occurs following menopause or oophorectomy in women, events that lead to the initiation of a rapid phase of bone loss, as described earlier. In a population-based study of 653 women, Garnero et al. [13] found that menopause induced a net increase in biochemical markers for bone formation of about 45%, whereas those for bone resorption increased by about 90%. Furthermore, these changes were reversed by estrogen replacement [1, 16]. During this initial period of rapid bone loss in the first few years after menopause, there were significant decreases in serum PTH [29]. However, as the early rapid phase of bone loss transitions into the slow phase of sustained bone loss, the decreased levels of serum PTH increase above levels found in premenopausal women. Moreover, this increase is progressive and continues over life, reaching serum levels in advanced old age of about 50% higher than mean premenopausal levels. Similar, but smaller, increases were found in aging men [11]. We have suggested [1, 16] that bone loss during the rapid bone loss phase during the early years after menopause results from loss of the direct restraining effect of estrogen on bone turnover, acting through estrogen receptors on bone cells. This leads to a rapid outpouring of calcium from bone and to an initial partial suppression of PTH secretion. Both the early, rapid phase of bone loss and the subsequent, sustained phase are related, at least in part, to estrogen deficiency, since both respond to estrogen therapy [1, 16]. Since it is unclear whether estrogen can act directly on the parathyroid glands, this suggests that estrogen is acting indirectly in both phases and by separate mechanisms.

We have also suggested [1, 16] that this later phase is driven predominantly by loss of estrogen action on extraskeletal calcium homeostasis. In addition to the well-established effect of estrogen on bone cells, considerable evidence shows that estrogen improves calcium balance by enhancing both intestinal calcium absorption and renal tubular reabsorption of calcium. Both tissues contain functioning estrogen receptors [1, 16]. The negative calcium balance induced by loss of these effects during estrogen deficiency may be the major mechanism for the progressive secondary hyperparathyroidism that occurs with aging in both sexes (see [1, 16] for a more comprehensive discussion of these extraskeletal effects of estrogen).

## Estrogen effects on bone formation

In addition to its profound effects on bone resorption, estrogen is also clearly important for the maintenance of bone formation. Perhaps the most direct evidence for this comes from human studies showing that acute (3 weeks) estrogen deficiency either in women [30] or in men [26] is associated with a fall in bone formation markers. However, due to the subsequent "coupling" of bone formation with resorption, bone formation increases over time, so that when chronically estrogen-deficient women are studied, both bone resorption and bone formation markers are increased [13]. Nonetheless, while the increase in bone remodeling and in bone resorption in the setting of sex steroid deficiency is accompanied by a coupled increase in bone formation at the tissue level, at each BMU, there remains a gap between bone resorption and formation, with formation unable to keep up with resorption, resulting in a net loss of bone. By inference, therefore, sex steroid deficiency is associated with a defect in bone formation. Of interest, recent work by Chang and colleagues [31] has demonstrated that estrogen deficiency in mice is associated with a marked increase in NF-κB activity in osteoblastic cells. Moreover, these investigators found that inhibiting NF-κB action attenuated bone loss following ovariectomy by reducing the gap between bone resorption and bone formation. Thus, NF-κB may be a key mediator of the relative deficit in bone formation that is present in vivo following estrogen deficiency.

It also appears that the effects of estrogen on progenitor and osteoblastic cells may be stage-specific. Thus, consistent with the overall effects of estrogen on reducing bone remodeling, estrogen does reduce the self-renewal of early mesenchymal progenitors [32]. Perhaps the most consistent effects of estrogen are on inducing commitment of precursor cells to the osteoblast at the expense of the adipocyte lineage [33, 34] and on preventing apoptosis of osteoblastic cells [35]. Estrogen has also been shown to enhance osteoblast differentiation [34], although data on effects of estrogen on osteoblastic differentiation are more variable and seem to depend on the model system used. Estrogen increases production of IGF-I [36], TGF-β [37], and procollagen synthesis by osteoblastic cells in vitro [36] and increases osteoblast life-span by decreasing osteoblast apoptosis [38, 39]. Direct evidence that estrogen can stimulate bone formation after cessation of skeletal growth was provided by Khastgir et al. [40] who obtained iliac biopsies for histomorphometry in 22 elderly women (mean age, 65 years) before and 6 years after percutaneous administration of supraphysiologic dosages of estrogen. They found a 61% increase in cancellous bone volume and a 12% increase in the wall thickness of trabecular packets. Tobias and Compston [41] have reported similar results. It is unclear whether these results represent only pharmacologic effects or are an augmentation of physiologic effects of estrogen that are ordinarily not large enough to detect. Nonetheless, accumulating data do implicate estrogen deficiency as a contributing cause of decreased bone formation that is seen with aging.

### Estrogen effects on osteocyte function

At the tissue level, estrogen clearly reduces bone turnover and brings bone resorption and bone formation into closer balance, as assessed both by bone histology and by bone turnover markers [42]. Given the increasing evidence that osteocytes may regulate the activation of bone remodeling via connections with bone-lining cells [43], it is likely that much of the antiremodeling effects of estrogen are mediated via the osteocyte. Indeed, withdrawal of estrogen is associated with increased apoptosis of osteocytes both in rodents [35] and in humans [44]; although, the specific molecular mechanisms by which this leads to increased remodeling on the bone surface remain unclear.

In addition, much evidence indicates that osteocytes can sense mechanical strain. Frost [45] previously suggested that estrogen deficiency alters the sensing of mechanical loading by the skeleton, perhaps through effects on osteocytes in bone. Thus, for a given level of mechanical loading, bone mass may be perceived by these cells as being excessive in the setting of estrogen deficiency, leading to bone loss. Once sufficient bone is lost, however, increased mechanical loading on the remaining bone may then serve to limit additional bone loss, accounting for the cessation of the rapid phase of bone loss following estrogen deficiency. In support of this, the Lanyon group [46] has shown that estrogen receptors in osteocytes are able to modulate the transmission of signals from "mechanoreceptors" in osteocytes via their extended processes in canaliculae to lining cells on bone surfaces, and that this process can be blocked using estrogen receptor antagonists.

## Estrogen and the immune system

### Overview

Estrogen acts to regulate bone turnover, at least in part, via bone cells through high-affinity estrogen receptors. However, it is now apparent that much of this effect, particularly as it relates to the regulation of bone resorption, is through regulation of the production of immune cytokines produced by bone cells themselves, as well as by cells of the immune system, such as T and B lymphocytes. These cytokines then feed back on bone cells, especially on osteoclast progenitor cells and on mature osteoclasts, to regulate their function. It is of considerable interest that the major cytokines that regulate bone cell function include most of the cytokines that regulate immune function. This suggests a close homology between these two critical biological systems.

### Effects on immune cytokines

#### The RANKL/OPG/RANK system

In addition to inhibiting the activation of bone remodeling, estrogen also directly and indirectly suppresses bone resorption. It is this action of estrogen that is most closely linked to the immune system and cytokines traditionally associated with regulation of immune cells. Since the final effector system regulating the differentiation and lifespan of osteoclasts is the RANKL/OPG/RANK system (reviewed in detail in [47]), there has been considerable interest in defining possible direct or indirect effects of estrogen on the components of this system. To assess this in humans, our group [48] isolated bone marrow mononuclear cells expressing RANKL on their surfaces by two-color flow cytometry using FITC-conjugated OPG-Rc as a probe. The cells were characterized as preosteoblastic marrow stromal cells (MSCs), B lymphocytes, or T lymphocytes by using antibodies against alkaline phosphatase, CD20 and CD3, respectively, in 12 premenopausal women (Group A), 12 early postmenopausal women (Group B), and 12 age-matched, estrogen-treated postmenopausal women (Group C). As shown in Figure 9-4, fluorescence intensity of OPG-Fc-FITC, an index of the surface concentration of RANKL per cell, was increased in Group B over Groups A and C by two- to three-fold for MSCs, B cells, T cells, and total RANKL-expressing cells. Moreover, in the merged groups, RANKL expression per cell correlated directly with the bone resorption markers, serum CTX and urine NTX, in all three cell types and inversely with serum estradiol levels for total RANKL-expressing cells. Thus, these data demonstrated that upregulation of RANKL on bone marrow cells is an important determinant of increased bone resorption induced by estrogen deficiency. In a subsequent study, Taxel and colleagues [49] used a specific antibody to RANKL and flow cytometry to demonstrate that, compared to premenopausal women, postmenopausal women had an over three-fold greater percentage of bone marrow cells expressing RANKL. Moreover, treatment of the postmenopausal women with estrogen reduced the percentage of RANKL-expressing cells to those present in premenopausal women. Further support for an effect of estrogen deficiency in upregulating RANKL expression in vivo was provided by Yoneda et al. [50] who found that in a murine model of rheumatoid arthritis, ovariectomy was associated with increased RANKL mRNA expression in synovial cells. Collectively, these studies provide strong evidence that estrogen deficiency is associated with increased RANKL production and/or

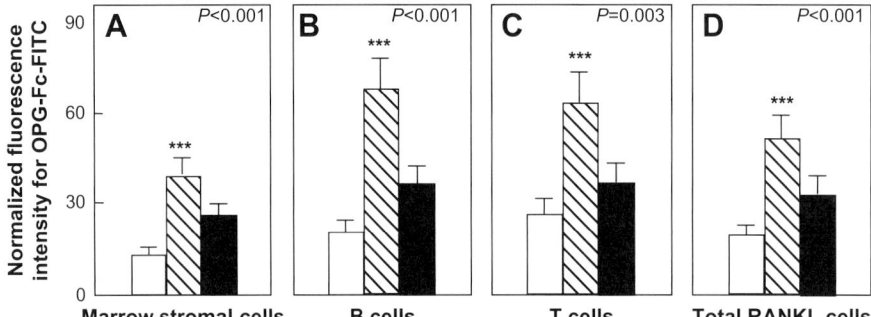

FIGURE 9-4 Changes in osteoprotegerin (OPG)-Fc-FITC fluorescence as an index of mean RANKL surface concentration per cell are shown for premenopausal women (A: white bars), estrogen-deficient postmenopausal women (B: stippled bars), and estrogen-treated postmenopausal women (C: black bars). P-values by ANOVA are as indicated. ***, $P < 0.001$ versus the premenopausal women. Adapted from Eghbali-Fatourechi et al. [48], with permission.

an expansion in RANKL-expressing cells in the bone microenvironment. These data also point to RANKL as a logical therapeutic target to inhibit bone resorption in post-menopausal women and indeed recent studies using denosumab, a fully humanized monoclonal antibody to RANKL, have demonstrated the efficacy of blocking RANKL action both in preventing bone loss [51, 52] and in reducing fracture risk [52] in post-menopausal women.

Androgens also appear to have suppressive effects on RANKL production. Thus, Kawano et al. [53] found that male androgen receptor knockout mice had high bone turnover and reduced trabecular and cortical bone mass, and these changes were associated with an upregulation of the RANKL mRNA in osteoblasts. Consistent with these data, Proell et al. [54] showed that orchiectomy in rats was associated with increased bone turnover and with a three-fold increase in free soluble RANKL concentrations in bone marrow supernatants, harvested from the proximal femur, and these changes could be prevented by testosterone treatment. Similar findings have been reported by Li et al. [55], who found that RANKL concentrations in bone marrow plasma and in bone marrow cell extracts were increased (by ~100%) following orchiectomy in rats. Moreover, these investigators also found a significant inverse correlation between testosterone and RANKL levels measured in marrow cell extracts, while marrow plasma RANKL correlated positivity with marrow plasma TRACP5b, an osteoclast marker. While it is possible (and perhaps even likely) that the observed effects of testosterone on RANKL production are mediated via aromatization to estrogens, these findings nonetheless provided a strong rationale for blocking RANKL activity to prevent hypogonadal bone loss in men. Thus, Smith et al. [56] used denosumab in men with prostate cancer undergoing androgen deprivation therapy for their cancer and showed that blocking RANKL activity with this agent prevented bone loss and significantly reduced the incidence of new vertebral fractures in these men.

In addition to suppressing RANKL production, estrogen also stimulates OPG mRNA and protein levels in osteoblastic cells. Hofbauer et al. [57] found that 17β-estradiol dose-dependently increased OPG mRNA and protein levels (by 370% and 320%, respectively) in conditionally immortalized human osteoblastic cells, transfected with ERα. 17β-estradiol also dose-dependently increased OPG mRNA and protein levels in normal human osteoblasts (by 60% and 73%, respectively). Similar findings were reported by Saika et al. [58], who found that 17β-estradiol stimulated OPG mRNA and protein levels in the mouse stromal cell line, ST-2. Interestingly, in this system, testosterone had no significant effect on OPG production. Hofbauer et al. [59] subsequently found that in human osteoblastic cells, transfected with the androgen receptor, 5α-dihydrotestosterone dose-dependently inhibited OPG mRNA and protein concentrations. Consistent with these in vitro findings, we [60] showed, using the subjects from the study described earlier involving selective withdrawal of estrogen or testosterone [26], that estrogen and testosterone had opposite effects on circulating OPG levels, with estrogen stimulating and testosterone (in the absence of aromatization to estrogen) inhibiting circulating OPG levels. These findings may explain, at least in part, why estrogen has more potent effects on inhibiting bone resorption compared to testosterone.

Estrogen also appears to have effects on RANK signaling in osteoclast precursors; although, there is disagreement among studies regarding whether estrogen affects the concentration of RANK on the cell surface. Thus, in studies in postmenopausal women, Clowes et al. [61] found that while estrogen treatment decreased the proportion of bone marrow mononuclear cells, expressing the late osteoclast phenotype marker, calcitonin receptor, there were no significant effects of estrogen on cell surface expression of RANK, TNF receptor 1, or receptors for the osteoclast co-stimulatory molecules, TREM2 or OSCAR. Similarly, Shevde et al. [62] found that while estrogen suppressed RANKL-induced osteoclast differentiation of the murine monocytic cell line, RAW264.7, it had no effect on RANK mRNA or protein levels. In contrast to these findings, Binder et al. [63] have reported that ovariectomy in mice is associated with increased RANK expression on bone marrow cells; although a potential problem with these findings is that the cells were not characterized as being in the monocytic/osteoclast versus other bone marrow lineages.

In contrast to the largely negative data with regards to regulation of RANK expression, there is evidence that estrogen has direct actions on monocytic/osteoclastic cells to inhibit RANKL-induced osteoclast differentiation by modulating various aspects of RANK signaling. Thus, several studies have found that, in RAW264.7 cells, estrogen blocked RANKL/M-CSF-induced AP-1-dependent transcription by suppressing c-Jun expression and its phosphorylation by c-Jun N-terminal kinase [62, 64]. Similar effects of androgens on inhibiting RANK signaling have also been reported [65]. In addition, estrogen inhibits RANKL-induced IκB degradation and NF-κB nuclear localization in osteoclastic cells [66], likely through a non-genomic mechanism involving sequestration of the RANKL-signaling intermediate, TRAF6, by ERα complexed with the scaffolding protein, BCAR1 [67].

There is considerable in vitro and in vivo evidence, then, that estrogen regulates bone resorption by effects on the RANKL/OPG/RANK system. Estrogen suppresses RANKL and stimulates OPG production in osteoblastic cells, while at the same time suppressing RANK signaling in osteoclastic cells. These actions, combined with direct effects of estrogen on inducing apoptosis of osteoclasts [68, 69], likely explain its potent anti-resorptive effects.

*Proinflammatory cytokines*

Despite the evidence noted above for effects of estrogen on the RANKL/OPG/RANK system, it remains unclear what proportion of these effects are direct versus indirect effects of estrogen. For example, RANKL production is upregulated in osteoblastic cells by TNF-α, IL-1β, IL-11, and PGE2 [70, 71], and is downregulated by TGF-β [72]. Thus, estrogen regulation of RANKL and, in turn, bone resorption could be indirectly mediated via suppression of one or more of these cytokines. In addition, a number of these cytokines act upstream of RANKL in expanding the pool of preosteoclastic cells [73]. Perhaps the most definitive evidence supporting a role for TNF-α and IL-1β in mediating estrogen deficiency bone loss in vivo in humans comes from the work of Charatcharoenwitthaya and colleagues [30]. In this study, transdermal estradiol was administered to 42 early postmenopausal women for 60 days to suppress bone resorption. Estrogen treatment was then discontinued, and the subjects were randomly assigned to intervention groups receiving 3 weeks of injections with saline, the IL-1 receptor 1 blocker, anakinra, or the soluble p75 TNF receptor, etanercept (which binds and thereby inhibits TNF action). As shown in Figure 9-5, either IL-1 or TNF-α blockade reduced the estrogen-deficiency-induced increase in bone resorption by ~50%, although TNF-α blockade appeared to be more effective (due to potential toxicity, both blockers could not be administered simultaneously).

These data in humans are consistent with considerable work in rodents demonstrating important roles for these cytokines in mediating bone loss following estrogen deficiency. Thus, Lorenzo and colleagues [74] assessed the effects of ovariectomy on bone mass in mice that lacked the type I IL-1 receptor. At 3 weeks after ovariectomy, wild-type C57BL/6 mice lost 48% of trabecular bone volume in the proximal humerus. However, there was no significant bone loss at this site in the IL-1 receptor 1 knockout mice. Similarly, Roggia et al. [75] showed that TNF-deficient mice did not lose trabecular vBMD at the tibia following ovariectomy. However, cytokines in addition to IL-1 and TNF-α are likely also involved in mediating estrogen deficiency bone loss, since Poli et al. [76] found that female mice with homozygous deletion of the IL-6 gene have a normal amount of trabecular bone, but higher rates of bone turnover than control littermates. Moreover, estrogen deficiency induced by ovariectomy failed to induce any change either in bone

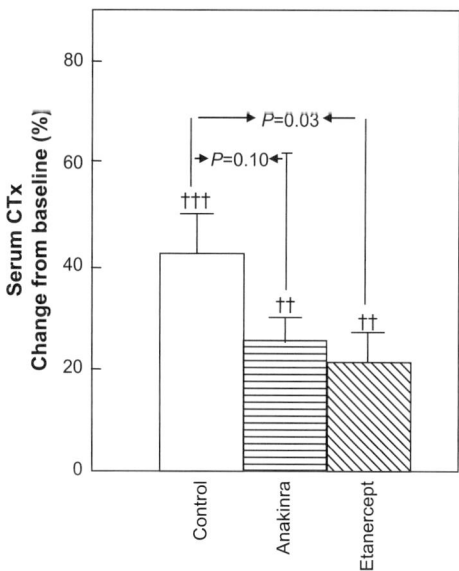

**FIGURE 9-5** Proportional change (%) in serum C-telopeptide of type I collagen (CTX) in postmenopausal women treated for 60 days with transdermal estradiol, made acutely estrogen-deficient, and treated with saline (control), an IL-1 blocker (anakinra), or a TNF blocker (etanercept). From Charatcharoenwitthaya et al. [30], with permission.

mass or bone remodeling rates in IL-6-deficient mice. Weitzmann and colleagues [77] observed that ovariectomy enhanced the production of IL-7 by murine bone marrow cells and that treatment of mice with a neutralizing antibody to IL-7 prevented ovariectomy-induced bone loss, at least as assessed by DXA measurements of the femur heads. In contrast to these findings, however, Lee et al. [78] found that IL-7 knockout mice clearly lost trabecular bone at the vertebrae and femurs following ovariectomy, although they failed to lose cortical bone.

In addition to these in vivo observations, in vitro studies in monocytes or preosteoclastic cells have shown that estrogen suppresses the production of IL-1 [79] and the signaling IL-1 receptor [80], but increases production of the decoy, soluble IL-1 receptor II, thereby reducing both the IL-1 signal and the responsiveness of osteoclast precursor cells to IL-1. Estrogen has also been shown to suppress the production of TNF-α [81] and M-CSF [82] by hematopoietic mononuclear cells as well as that of IL-6 and IL-6 receptor in bone marrow stromal cells [83, 84]. Similar suppressive effects of androgens on IL-6 production by bone marrow stromal cells have also been reported [85]. In addition, estrogen has also been shown to increase production of TGF-β by osteoblastic cells [37]; TGF-β, in turn, increases osteoblastic OPG production [72] and can directly induce apoptosis of osteoclasts [86].

## Effects on immune effector cells

### T lymphocytes

An expansion of T cells leading to increased TNF-α levels in the bone microenvironment has been proposed by Pacifici and colleagues to mediate the rapid bone loss following loss of estrogen [75, 87]. In support of this, Cenci et al. [87] reported that nude mice, which lack T lymphocytes, fail to lose bone following ovariectomy. However, this hypothesis is controversial, since studies using nude rats [88] and nude RAG2- or TCR-α-deficient mice (all of which lack functional T lymphocytes) have found that ovariectomy-induced trabecular bone loss in these models was similar to that seen in wild-type mice [89]. This is shown in Figure 9-6 where, using μCT, Lee et al. [89] found loss of trabecular bone following ovariectomy in the T-cell-deficient nude as well as RAG2 knockout mice that was similar to that observed in the wild-type mice (Figure 9-6A). Interestingly, while the RAG2 knockout mice, similar to the wild-type mice, lost cortical bone following ovariectomy, the nude mice were protected against ovariectomy-induced cortical bone loss (Figure 9-6B). These findings thus indicate that effects of T cells in mediating bone loss following ovariectomy may be bone-compartment-specific and may also vary between different T-cell-deficient models.

In addition to TNF-α, Gao and colleagues [90] have recently implicated IFN-γ in mediating bone loss following ovariectomy. IFN-γ is a major product of activated T cells that can function as a pro- or antiresorptive cytokine. Gao et al. [90] demonstrated that IFN-γ blunted osteoclast formation through direct targeting of osteoclast precursors, but indirectly stimulated osteoclast formation and promoted bone resorption by stimulating antigen-dependent T-cell activation and T-cell secretion of RANKL and TNF-α.

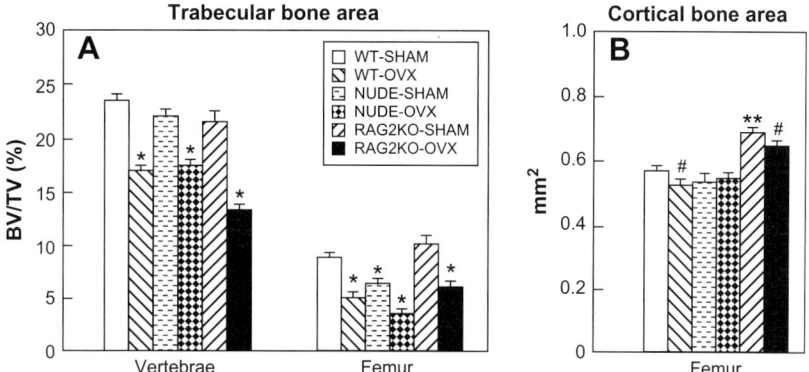

**FIGURE 9-6** Trabecular bone volume (BV/TV) of the vertebrae ($L_1$) and femur (A) and cortical volume of the femur (B) measured by μCT in wild-type (WT), nude mice, and RAG2 knockout mice that were either sham-operated (SHAM) or ovariectomized (OVX). *, significant effect of ovariectomy, $P < 0.05$. #, significant effect of ovariectomy, $P = 0.05$. +, significant difference between nude-SHAM and WT-SHAM mice, $P < 0.05$. **, significant difference between RAG2 knockout-SHAM and WT-SHAM mice, $P < 0.05$. From Lee et al. [89], with permission.

Studies in humans have provided some evidence in support of a role for T cells in the pathogenesis of osteoporosis. Fujita and colleagues [91] initially demonstrated over 20 years ago that there were abnormalities in T-lymphocyte subsets in patients with osteoporosis and spinal compression fractures compared to age-matched controls. These investigators found that the CD4/CD8 T-cell ratio was significantly higher in the osteoporotic patients, whereas total lymphocyte and T-cell counts were unchanged as compared to controls. Young normal control subjects had CD4/CD8 ratios similar to elderly non-osteoporotic subjects. Interestingly, administration of 0.5 μg/day of 1α-hydroxyvitamin $D_3$ for 2 months resulted in a significant fall in the CD4/CD8 ratio in the osteoporotic subjects. These findings were subsequently confirmed by Rosen et al. [92], who found a similar increase in the CD4/CD8 ratio in early postmenopausal women with osteoporosis, as well as a negative correlation between the CD4/CD8 ratio and spinal BMD, measured by dual-photon absorptiometry. In human male subjects made hypogonadal and selectively replaced with estrogen or testosterone, there was a trend ($P = 0.091$) for estrogen to suppress T-cell percentages in bone marrow, but no effect of either testosterone or estrogen on TNF-α mRNA levels in bone marrow CD3+ T cells [93]. These findings are not inconsistent with the work of the Pacifici group [75, 87], since in their studies in rodents, TNF-α production per cell was not increased by estrogen deficiency; rather, there was an overall expansion in the number of TNF-α-producing T cells in the bone marrow. Moreover, as noted earlier and shown in Figure 9-4, untreated postmenopausal women do have an increase in RANKL expression on T cells as compared to premenopausal women or postmenopausal women treated with estrogen, consistent with estrogen regulation (directly or indirectly) of RANKL expression by T cells [48]. Recent studies by D'Amelio and colleagues [94] have also found that peripheral blood T cells (identified on the basis of expression of CD3) from women with osteoporosis had increased mRNA levels for RANKL as compared to age-matched controls.

Interestingly, these investigators also found that TNF-α production by CD3+ T cells from the osteoporotic patients was increased relative to postmenopausal controls, but the latter did not differ with regards to T-cell TNF-α production as compared to premenopausal women.

*B lymphocytes*

As shown in Figure 9-4, estrogen deficiency is associated with increased RANKL expression not only by osteoblastic and T cells, but also by B cells [48]. These findings are consistent with the work of Kanematsu and colleagues [95], who also found that ovariectomy in mice was associated with increased RANKL expression by bone marrow B220+ B cells. However, studies in mice have found that estrogen deficiency is also associated with an increase in the number of bone marrow B220+ B cells [95, 96], which does not appear to be the case in humans [48]. In fact, in human males made hypogonadal and selectively replaced with either estrogen or testosterone, estrogen was associated with an increase (rather than a decrease) in the percentage of bone marrow CD19+ B cells [93]. Thus, while the human and mouse studies do indicate that RANKL expression by B cells is increased following estrogen deficiency, data on the effects of estrogen on B-cell numbers are conflicting, and these effects may differ between rodents and humans.

Interestingly, Li and colleagues [97] have recently reported that B cells are a major source of OPG in bone marrow and that mice deficient in B cells (μMT/μMT mice) had reduced bone mass and were deficient in bone marrow OPG levels. Despite these data, however, several investigators have found that trabecular bone loss following ovariectomy is similar in wild-type mice and mice that are deficient in the majority of their B cells [78, 89, 98]; thus, B cells do not appear to play a significant role in mediating estrogen-deficiency-induced bone loss.

## Summary of estrogen effects on regulating bone resorption

As is evident, estrogen regulates bone resorption through multiple mechanisms, likely involving several different cell types. Figure 9-7 provides an overall summary of these effects. Directly or indirectly, estrogen suppresses RANKL and increases OPG production by several different cell populations in the bone microenvironment. Estrogen also impairs RANK signaling in osteoclast-lineage cells. Increased production of TGF-β further increases OPG production and induces osteoclast apoptosis. Estrogen suppresses the production of a number of proinflammatory cytokines (TNF-α, IL-1, M-CSF, IL-6, PGE$_2$, GM-CSF) that not only regulate RANKL and OPG production, but also the osteoclast precursor pool.

# Potential effects of estrogen deficiency on general immunity

There is a large body of clinical and experimental evidence, which indicates that estrogen plays an important role in the regulation of general immunity. As with osteoclast differentiation and function, the key intermediate and effector cells of the immune

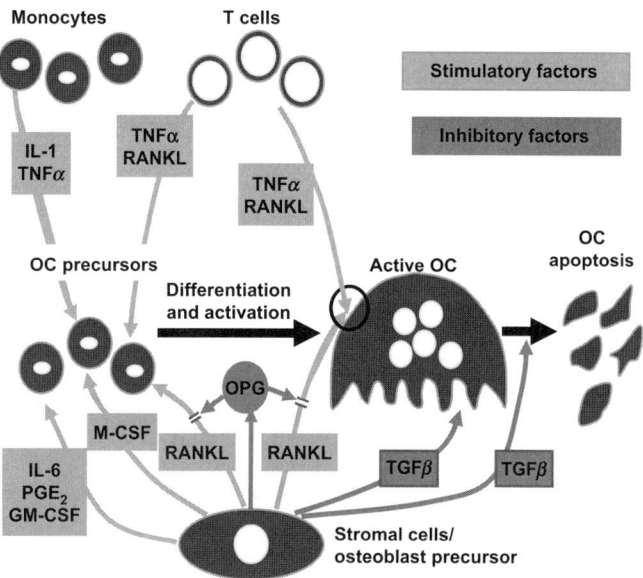

FIGURE 9-7 Model for effects of estrogen on osteoclast (OC) formation and function by cytokines in the bone marrow microenvironment. The positive (E+) and negative (E−) effects of estrogen on stimulatory and inhibitory factors are shown. Modified from Riggs et al. [136], with permission. Please refer to color plate section.

system contain estrogen receptors and their function and the elaboration of cytokines by them are, in part, estrogen-regulated. Also, there is substantial clinical evidence that gender affects the expression of immunity [99]. For example, the incidence of certain autoimmune diseases is higher in women than in men; the incidence is two- to three-fold higher for rheumatoid arthritis and multiple sclerosis whereas it is nine-fold higher for systemic lupus erythematosus. Moreover, women have a strengthened immune response to bacterial infection as compared with men. In contrast, other epidemiologic studies suggest that estrogen may in fact be protective against autoimmune diseases. Thus, pregnancy may result in remission of autoimmune diseases, and an upswing in the incidence of rheumatoid arthritis has been reported in women in the early years after menopause, whereas in men the incidence gradually increases over life.

Consistent with these divergent observational findings, experimental studies have also shown biphasic effects of estrogen in both stimulating and in suppressing immune function, depending on the conditions of the study. Several publications have reported that estrogen stimulates the differentiation and function of the major antigen-presenting cells, dendritic cells, and macrophages, from their precursors in the bone marrow [100]. However, other studies have shown that estrogen decreases activation of the class II protein transactivator protein (CIITP), the main facilitator of IFNγ-inducible class II major histocompatibility complex expression in brain cells [101]. Also, ovariectomy upregulates CIITP in mice, which increases antigen presentation by macrophages, resulting in enhanced activation and prolonged lifespan of T cells [102]. Finally, estrogen induces remission in a mouse model for multiple sclerosis acting through a suppression of dendritic function [103].

With respect to the pathogenesis of osteoporosis, most investigators have focused on the key effects of estrogen deficiency on stimulating a balance of immune cytokines that favors bone resorption [1, 47]. However, Weitzman and Pacifici [104] have suggested that osteoporosis may be caused by an alteration in general immunity. They hypothesize that estrogen deficiency leads to enhanced antigen presentation by dendritic cells and macrophages and stimulates the proliferation and activation of T cells. This increases osteoclast formation, which is further enhanced by the downregulation of antioxidant pathways in the estrogen-deficiency state, leading to an upswing in reactive oxygen species compounds. In support of their hypothesis, they have found that DO11.1 mice, a genetically engineered strain in which T cells recognize only a peptide epitope of ovalbumin not normally expressed in mammals, but not other antigens, do not lose bone after ovariectomy unless they are pretreated with ovalbumin [102].

The hypothesis that osteoporosis is a consequence of hypersensitized general immunity induced by estrogen deficiency has not as yet been generally accepted. In part, this is because the nature of the antigens leading to the postulated hyperstimulation of the immune system have not been clearly defined. However, Weitzman and Pacifici [104] point out that there is a pool of self and foreign antigens that is physiologically present in healthy animals and humans, one or more of which could maintain the antigenic response during estrogen deficiency in susceptible individuals. Clearly, more studies are needed to investigate this controversial but potentially important concept.

## Effects of aging on bone loss independent of changes in sex steroid production

In addition to the key effects of sex steroid deficiency in women and in men described above, it is clear that there are intrinsic changes in the skeletal metabolism that contribute to age-related bone loss. Some of these, discussed below, impact on a major signaling pathway, Wnt, involved in bone formation and resorption. Initial observations regarding the role of this pathway in bone metabolism came from loss- or gain-of-function mutations of the low-density lipoprotein receptor-related protein 5 or 6 (LRP-5 or LRP-6), which are co-receptors for secreted Wnt proteins [105]. In the canonical pathway Wnt signaling activates beta-catenin, which then regulates the commitment of multipotent mesenchymal progenitors to the osteoblast lineage, prevents osteoblast apoptosis, and increases OPG production, leading to both stimulation of bone formation and inhibition of bone resorption [105]. In a series of studies, Almeida and colleagues [35] have shown that in mice, aging is associated with an increase in markers of oxidative stress in osteoblastic cells, which results in an increase in Forkhead box O (FoxO) transcription factors. While FoxO induction is critical in the defense against oxidative stress, these investigators also found that by competing for cellular beta-catenin, the increase in

FoxO expression leads to a reduction in Wnt signaling in bone. Given the key role for Wnts in bone metabolism [105], these findings suggest that oxidative stress, which increases with aging but is accentuated by sex steroid deficiency [35], may be an important factor leading to impaired bone formation with aging.

There has also been considerable interest recently in the role of the nutrient-sensing NAD-dependent protein deacetylases, sirtuins, in aging phenotypes in a number of tissues [106]. Thus, it is of interest that Edwards et al. [107] found that mice with global deletion of sirtuin 1 (SIRT-1) have a decrease in bone mass associated with decreased bone formation and increased bone resorption. In further studies, these investigators showed that osteoblast-specific deletion of SIRT-1 results in low bone formation, whereas deletion of SIRT-1 in osteoclast precursor cells leads to an increase in bone resorption [108]. These findings therefore demonstrate that age-related changes in SIRT-1 activity may also contribute to age-related bone loss, at least in rodent models. Additional support for this hypothesis comes from data showing that the SIRT-1 agonist, resveratrol, preserves BMD in aging mice [109] and can prevent ovariectomy-induced bone loss [110].

In terms of the relationship of changes in the immune system with aging to alterations in bone metabolism, recent interest has focused on the role of senescent T cells in mediating age-related changes in a number of tissues, including bone. For example, it has been suggested that since elderly individuals have been exposed to multiple pathogens over life, this chronic immunological stimulation leads to the generation of expanded populations of memory CD8 T cells that can no longer replicate and undergo changes associated with senescence, including the production of a number of pro-inflammatory cytokines [111]. These senescent T cells can be identified based on lack of expression of CD28, and CD8+/CD28− T cells have been shown to accumulate with age [112–114]. Of interest, Pietschmann et al. [115] found that elderly patients with osteoporotic fractures had a significant increase in CD8+ T cells bearing markers suggestive of cell senescence. Thus, while considerable additional work in this area is needed, it does appear that the changes associated with immunosenescence related to aging may impact bone metabolism.

## Secondary causes of osteoporosis mediated by the immune system

There is considerable epidemiological evidence that many chronic inflammatory diseases (i.e. rheumatoid arthritis, celiac disease, inflammatory bowel disease) are associated not only with focal, but also systemic, bone loss. Thus, as a group, patients with rheumatoid arthritis are at greater risk of fractures overall, with most studies demonstrating increases in hip [116, 117], and some but not all studies finding increases in vertebral fractures [118]. Proposed mechanisms that are independent of the use of glucocorticoids and anti-immune compound include disease-related inflammation characterized by increased levels of pro-inflammatory cytokines [119]. Other possible causes include decreases in physical activity, associated with joint damage [120], and "accelerated aging" related to premature immunosenescence. Interestingly, the latter finding was also associated with the loss of CD28 expression on T cells, and occurs with normal aging [121]. In addition, relative sex steroid deficiency may predispose to rheumatoid arthritis in both women and men [122].

Celiac disease has also been associated with increased bone loss and fracture risk [123]. Bone resorption markers are increased in the serum of untreated celiac patients, and these markers are associated with an increase in IL-6 and in the RANKL/OPG ratio [124]. Peripheral blood mononuclear cells cultured in the presence of serum from patients with celiac disease formed increased numbers of osteoclasts, and expression of OPG was decreased in osteoblastic cultures from these patients [124]. Of interest, Riches and colleagues [125] have recently described a patient with celiac disease who presented with severe osteoporosis and high bone turnover related to autoantibodies against OPG. These investigators further detected lower levels of similar anti-OPG antibodies in an additional three of 15 patients with celiac disease, suggesting that a subset of patients with this disorder, and perhaps other autoimmune diseases, may have antibodies to OPG leading to increased bone resorption and subsequent bone loss.

Systemic bone loss has also been noted in patients with inflammatory bowel diseases, such as Crohn's disease and ulcerative colitis [126]. Serum OPG [127] as well as soluble RANKL [128] levels have been found to be elevated in these patients, with OPG levels correlating inversely with severity of bone loss [127]. Interestingly, OPG treatment of mice with IL-2-deficiency-induced ulcerative colitis prevented bone loss, and also improved the colitis activity due to reduced colonic dendritic cell survival [129].

# Prospects for treatment of osteoporosis by regulation of immune cytokines

As reviewed earlier, there is strong evidence that alterations of immune cytokines account for much of bone loss in postmenopausal women and in elderly men. Estrogen is no longer recommended for the treatment of osteoporosis because of adverse effects reported by the Women's Health Initiative and other studies [130]. Because of the key role of the RANKL/OPG/RANK system as the final regulator of the cascade of changes in proinflammatory cytokines, this would seem to serve as a reasonable initial step for treatment of osteoporosis using immune regulation. Thus, as previously noted, Cummings et al. [52] enrolled 7868 postmenopausal women with osteoporosis and randomly assigned them to receive 60 mg of denosumab (a monoclonal antibody to RANKL) or placebo subcutaneously every 6 months for 3 years. Compared to placebo, denosumab reduced the 3-year incidence of new vertebral fractures from 7.2% to 2.3% (a 68% decrease); of hip fractures from 1.2% to 0.7% (a 40% decrease); and of all non-vertebral fractures from 8.0% to 6.5% (a 20% decrease). In a smaller study of elderly men undergoing androgen deprivation for prostate cancer, Smith et al. [56] randomized 1468 men to a similar regimen of denosumab or placebo. After 36 months, patients in the denosumab group had a 1.5% incidence of new vertebral fractures compared to 3.9% in the placebo group (a 62% decrease). Concerns remain, however, regarding potential immune side-effects of neutralizing RANKL activity in humans [131], since RANKL is expressed not just on osteoblastic but also on immune cells [132]. While neither study found an increased rate of serious infections related to denosumab, as compared to placebo, Cummings et al. [52] did report statistically significant increases in the rates of eczema and hospitalizations for cellulitus. Previously, McClung and colleagues reported that 6/314 and 3/314 patients treated with denosumab developed neoplasms and serious infections, respectively, compared to 0/46 patients for either event in the placebo group [51]. These results, though not statistically significant, support ongoing surveillance of patients receiving denosumab, particularly when used in the community setting in patients with comorbidities, who might have been excluded from participating in clinical trials.

Neutralizing monoclonal antibodies directed against TNF and IL-1 have been shown to be effective in treating many autoimmune diseases [133]. Moreover, as described earlier, our group has demonstrated that neutralizing TNF-α and IL-1 were effective in improving biochemical markers of bone turnover in early postmenopausal women [30]. Whether these agents will produce favorable results on bone loss and fracture prevention with acceptable side-effects remains to be demonstrated.

## Summary and conclusions

It is evident that there is tremendous cross-talk between bone and the immune system and that immune cells and cytokines are likely involved in the pathogenesis of postmenopausal and age-related bone loss in both sexes. Recent years have seen tremendous advances in our understanding of estrogen regulation of the RANKL/RANK/OPG system and of pro-inflammatory cytokines impacting both this system as well as osteoclast precursor cells. T cells likely play an important role in mediating bone loss following estrogen deficiency, although whether their effects are localized to specific bone compartments (i.e., trabecular versus cortical bone) and whether their role in humans is as important as it appears to be in some mouse models remains to be defined. There is also growing interest in the role of age-related immunosenescence in mediating changes in multiple tissues, including bone. Finally, a number of immunological disorders result not only in focal but also systemic bone loss and osteoporosis through immune pathways.

## Acknowledgments

The authors would like to thank James Peterson for help with the figures and Amanda Oelkers for secretarial assistance. This work was supported by NIH Grants AG004875, AG028936, and AR027065.

# References

[1] Riggs BL, Khosla S, Melton LJ. Sex steroids and the construction and conservation of the adult skeleton. Endocr Rev 2002 Jun;23(3):279–302.

[2] Riggs BL, Melton III LJ, Robb RA, Camp JJ, Atkinson EJ, Peterson JM, et al. Population-based study of age and sex differences in bone volumetric density, size, geometry, and structure at different skeletal sites. J Bone Miner Res 2004;19:1945–54.

[3] Camp JJ, Karwoski RA, Stacy MC, Atkinson EJ, Khosla S, Melton LJ, et al. A system for the analysis of whole-bone strength from helical CT images. Proceedings of SPIE 2004;5369:74–88.

[4] Riggs BL, Melton LJI, Robb RA, Camp JJ, Atkinson EJ, McDaniel L, et al. A population-based assessment of rates of bone loss at multiple skeletal sites: evidence for substantial trabecular bone loss in young adult women and men. J Bone Miner Res 2008;23:205–14.

[5] Laib A, Ruegsegger P. Calibration of trabecular bone structure measurements of in vivo three-dimensional peripheral quantitative computed tomography with 28-microm-resolution micro-computed tomography. Bone 1999;24(1):35–9.

[6] Khosla S, Riggs BL, Atkinson EJ, Oberg AL, McDaniel LJ, Holets M, et al. Effects of sex and age on bone microstructure at the ultradistal radius: a population-based noninvasive in vivo assessment. J Bone Miner Res 2006;21(1):124–31.

[7] Cummings SR, Melton LJ. Epidemiology and outcomes of osteoporotic fractures. Lancet 2002;359(9319):1761–7.

[8] Burge R, Dawson-Hughes B, Solomon DH, Wong JB, King A, Tosteson A. Incidence and economic burden of osteoporosis-related fractures in the United States, 2005-2025. J Bone Miner Res 2007;22(3):465–75.

[9] Lindsay R, Aitkin JM, Anderson JB, Hart DM, MacDonald EB, Clarke AC. Long-term prevention of postmenopausal osteoporosis by oestrogen. Lancet 1976;i:1038–40.

[10] Genant HK, Cann CE, Ettinger B, Gordan GS. Quantitative computed tomography of vertebral spongiosa: a sensitive method for detecting early bone loss after oophorectomy. Ann Intern Med 1982;97:699–705.

[11] Khosla S, Melton III LJ, Atkinson EJ, O'Fallon WM, Klee GG, Riggs BL. Relationship of serum sex steroid levels and bone turnover markers with bone mineral density in men and women: a key role for bioavailable estrogen. J Clin Endocrinol Metab 1998 Jul;83(7):2266–74.

[12] Horton R, Romanoff E, Walker J. Androstenedione and testosterone in ovarian venous and peripheral plasma during ovariectomy for breast cancer. J Clin Endocrinol Metab 1966;26:1267–9.

[13] Garnero P, Sornay-Rendu E, Chapuy M, Delmas PD. Increased bone turnover in late post-menopausal women is a major determinant of osteoporosis. J Bone Miner Res 1996;11:337–49.

[14] Young MM, Nordin BEC. Effects of natural and artificial menopause on plasma and urinary calcium and phosphorus. Lancet 1967;2:118–20.

[15] Gennari C, Agnusdei D, Nardi P, Civitelli R. Estrogen preserves a normal intestinal responsiveness to 1,25-dihydroxyvitamin D3 in oophorectomized women. J Clin Endocrinol Metab 1990;71:1288–93.

[16] Riggs BL, Khosla S, Melton III LJ. A unitary model for involutional osteoporosis: estrogen deficiency causes both type I and type II osteoporosis in postmenopausal women and contributes to bone loss in aging men. J Bone Miner Res 1998;13:763–73.

[17] Richelson LS, Wahner HW, Melton LJI, Riggs BL. Relative contributions of aging and estrogen deficiency to postmenopausal bone loss. N Engl J Med 1984;311(20):1273–5.

[18] Greendale GA, Edelstein S, Barrett-Connor E. Endogenous sex steroids and bone mineral density in older women and men: the Rancho Bernardo study. J Bone Miner Res 1997;12:1833–43.

[19] Slemenda CW, Longcope C, Zhou L, Hui SL, Peacock M, Johnston C. Sex steroids and bone mass in older men: positive associations with serum estrogens and negative associations with androgens. J Clin Invest 1997;100:1755–9.

[20] Center JR, Nguyen TV, Sambrook PN, Eisman JA. Hormonal and biochemica parameters in the determination of osteoporosis in elderly men. J Clin Endocrinol Metab 1999;84:3626–35.

[21] Ongphiphadhanakul B, Rajatanavin R, Chanprasertyothin S, Piaseau N, Chailurkit L. Serum oestradiol and oestrogen-receptor gene polymorphism are associated with bone mineral density independently of serum testosterone in normal males. Clin Endocrinol 1998;49:803–9.

[22] van den Beld AW, de Jong FH, Grobbee DE, Pols HAP, Lamberts SWJ. Measures of bioavailable serum testosterone and estradiol and their relationships with muscle strength, bone density, and body composition in elderly men. J Clin Endocrinol Metab 2000;85:3276–82.

[23] Amin S, Zhang Y, Sawin CT, Evans SR, Hannan MT, Kiel DP, et al. Association of hypogonadism and estradiol levels with bone mineral density in elderly men from the Framingham study. [Comment]. Annals of Internal Medicine 2000;133(12):951–63.

[24] Szulc P, Munoz F, Claustrat B, Garnero P, Marchand F. Bioavailable estradiol may be an important determinant of osteoporosis in men: the MINOS study. J Clin Endocrinol Metab 2001;86:192–9.

[25] Khosla S, Melton LJ, Atkinson EJ, O'Fallon WM. Relationship of serum sex steroid levels to longitudinal changes in bone density in young versus elderly men. J Clin Endocrinol Metab 2001;86:3555–61.

[26] Falahati-Nini A, Riggs BL, Atkinson EJ, O'Fallon WM, Eastell R, Khosla S. Relative contributions of testosterone and estrogen in regulating bone resorption and formation in normal elderly men. J Clin Invest 2000;106(12):1553–60.

[27] Leder BZ, LeBlanc KM, Schoenfeld DA, Eastell R, Finkelstein JS. Differential effects of androgens and estrogens on bone turnover in normal men. J Clin Endocrinol Metab 2003;88:204–10.

[28] Turner RT, Wakley GK, Hannon KS. Differential effects of androgens on cortical bone histomorphometry in gonadectomized male and female rats. J Orthop Res 1990;8:612–17.

[29] Khosla SK, Atkinson EJ, Melton III LJ, Riggs BL. Effects of age and estrogen status on serum parathyroid hormone levels and biochemical markers of bone turnover in women: a population-based study. J Clin Endocrinol Metab 1997;82:1522–7.

[30] Charatcharoenwitthaya N, Khosla S, Atkinson EJ, McCready LK, Riggs BL. Effect of blockade of TNF-a and interleukin-1 action on bone resorption in early postmenopausal women. J Bone Miner Res 2007;22(5):724–9.

[31] Chang J, Wang Z, Tang E, Fan Z, McCauley L, Franceschi R, et al. Inhibition of osteoblastic bone formation by nuclear factor-kappaB. Nat Med 2009;15(6):682–9.

[32] Gregorio GB, Yamamoto M, Ali AA, Abe E, Roberson P, Manolagas SC, et al. Attenuation of the self-renewal of transit-amplifying osteoblast progenitors in the murine bone marrow by 17beta-estradiol. J Clin Invest 2001;107:803–12.

[33] Okazaki R, Inoue D, Shibata M, Saika M, Kido S, Ooka H, et al. Estrogen promotes early osteoblast differentiation and inhibits adipocyte differentiation in mouse bone marrow stromal cell lines that express estrogen receptor (ER) alpha or beta. Endocrinology 2002;143:2349–56.

[34] Dang ZC, Van Bezooijen RL, Karperien M, Papapoulos SE, Lowik CWGM. Exposure of KS483 cells to estrogen enhances osteogenesis and inhibits adipogenesis. J Bone Miner Res 2002;17:394–405.

[35] Almeida M, Martin-Millan M, Plotkin LI, Stewart SA, Roberson PK, Kousteni S, et al. Skeletal involution by age-associated oxidative stress and its acceleration by loss of sex steroids. J Biol Chem 2007;282:27285–97.

[36] Ernst M, Heath JK, Rodan GA. Estradiol effects on proliferation, messenger ribonucleic acid for collagen and insulin-like growth factor-I, and parathyroid hormone-stimulated adenylate cyclase activity in osteoblastic cells from calvariae and long bones. Endocrinology 1989;125:825–33.

[37] Oursler MJ, Cortese C, Keeting PE, Anderson MA, Bonde SK, Riggs BL, et al. Modulation of transforming growth factor-beta production in normal human osteoblast-like cells by 17beta-estradiol and parathyroid hormone. Endocrinology 1991;129:3313–20.

[38] Manolagas SC. Birth and death of bone cells: basic regulatory mechanisms and implications for the pathogenesis and treatment of osteoporosis. Endocr Rev 2000;21:115–37.

[39] Gohel A, McCarthy MB, Gronowicz G. Estrogen prevents glucocorticoid-induced apoptosis in osteoblasts in vivo and in vitro. Endocrinology 1999;140:5339–47.

[40] Khastgir G, Studd J, Holland N, Alaghband-Zadeh J, Fox S, Chow J. Anabolic effect of estrogen replacement on bone in postmenopausal women with osteoporosis: histomorphometric evidence in a longitudinal study. J Clin Endocrinol Metab 2001;86:289–95.

[41] Tobias JH, Compston JE. Does estrogen stimulate osteoblast function in postmenopausal women? Bone 1999;24:121–4.

[42] Lufkin EG, Wahner HW, O'Fallon WM, Hodgson SF, Kotowicz MA, Lane AW, et al. Treatment of postmenopausal osteoporosis with transdermal estrogen. Ann Intern Med 1992;117:1–9.

[43] Bonewald LF. Osteocyte messages from a bony tomb. Cell Metab 2007;5(6):410–11.

[44] Tomkinson A, Reeve J, Shaw RW, Noble BS. The death of osteocytes via apoptosis accompanies estrogen withdrawal in human bone. J Clin Endocrinol Metab 1997;82(9):3128–35.

[45] Frost HM. The Biomechanical "Face" of Osteoporosis: Emerging Views with Insights from the Utah Paradigm. Tokyo: Springer-Verlag; 1999.

[46] Lee K, Jessop HL, Suswillo RF, Zaman G, Lanyon L. Bone adaptation requires oestrogen receptor-alpha. Nature 2003;424:389.

[47] Kearns AE, Khosla S, Kostenuik PJ. Receptor activator of nuclear factor kappaB ligand and osteoprotegerin regulation of bone remodeling in health and disease. Endocr Rev 2008;29(2):155–92.

[48] Eghbali-Fatourechi G, Khosla S, Sanyal A, Boyle WJ, Lacey DL, Riggs BL. Role of RANK ligand in mediating increased bone resorption in early postmenopausal women. J Clin Invest 2003;111(8):1221–30.

[49] Taxel P, Kaneko H, Lee S-K, Aguila HL, Raisz LG, Lorenzo JA. Estradiol rapidly inhibits osteoclastogenesis and RANKL expression in bone marrow cultures in postmenopausal women: a pilot study. Osteoporos Int 2008;19:193–9.

[50] Yoneda T, Ishimaru N, Arakaki R, Kobayashi M, Izawa T, Moriyama K, et al. Estrogen deficiency accelerates murine autoimmune arthritis associated with receptor activator of nuclear factor-kappaB ligand-mediated osteoclastogenesis. Endocrinology 2004;145:2384–91.

[51] McClung MR, Lewiecki EM, Cohen SB, Bolognese MA, Woodson GC, Moffett AH, et al. Denosumab in postmenopausal women with low bone mineral density. N Engl J Med 2006;354:821–31.

[52] Cummings SR, San Martin J, McClung MR, Siris ES, Eastell R, Reid IR, et al. Denosumab for prevention of fractures in postmenopausal women with osteoporosis. N Engl J Med 2009;361(8):756–65.

[53] Kawano H, Sato T, Yamada T, Matsumoto T, Sekine K, Watanabe T, et al. Suppressive function of androgen receptor in bone resorption. Proc Natl Acad Sci U S A 200;100(16):9416–21.

[54] Proell V, Xu H, Schuler C, Weber K, Hofbauer LC, Erben RG. Orchiectomy upregulates free soluble RANKL in bone marrow of aged rats. Bone 2009;45(4):677–81.

[55] Li X, Ominsky MS, Stolina M, Warmington KS, Geng Z, Niu Q-T, et al. Increased RANK ligand in bone marrow of orchiectomized rats and prevention of their bone loss by the RANK ligand inhibitor osteoprotegerin. Bone 2009;45(4):669–76.

[56] Smith MR, Egerdie B, Toriz NH, Feldman R, Tammela TLJ, Saad F, et al. Denosumab in men receiving androgen-deprivation therapy for prostate cancer. N Engl J Med 2009;361:745–55.

[57] Hofbauer LC, Khosla S, Dunstan CR, Lacey DL, Spelsberg TC, Riggs BL. Estrogen stimulates gene expression and protein production of osteoprotegerin in human osteoblastic cells. Endocrinology 1999;140:4367–70.

[58] Saika M, Inoue D, Kido S, Matsumoto T. 17Beta-estradiol stimulates expression of osteoprotegerin by a mouse stromal cell line, ST-2, via estrogen receptor-alpha. Endocrinology 2001;142:2205–12.

[59] Hofbauer LC, Hicok KC, Chen D, Khosla S. Regulation of osteoprotegerin production by androgens and anti-androgens in human osteoblastic lineage cells. Eur J Endrcrinol 2002;147:269–73.

[60] Khosla S, Atkinson EJ, Dunstan CR, O'Fallon WM. Effect of estrogen versus testosterone on circulating osteoprotegerin and other cytokine levels in normal elderly men. J Clin Endocrinol Metab 2002;87:1550–4.

[61] Clowes JA, Eghbali-Fatourechi GZ, McCready L, Oursler MJ, Khosla S, Riggs BL. Estrogen action on bone marrow osteoclast lineage cells of postmenopausal women in vivo. Osteoporos Int 2009;20(5):761–9.

[62] Shevde NK, Bendixen AC, Dienger KM, Pike JW. Estrogens suppress RANK ligand-induced osteoclast differentiation via a stromal cell independent mechanism involving c-Jun repression. Proc Natl Acad Sci U S A 2000;97:7829–34.

[63] Binder NB, Niederreiter B, Hoffmann O, Stange R, Pap T, Stuinig TM, et al. Estrogen-dependent and C-C chemokine receptor-2-dependent pathways determine osteoclast behavior in osteoporosis. Nat Med 2009;15(4):417–24.

[64] Srivastava S, Toraldo G, Weitzmann MN, Cenci S, Ross FP, Pacifici R. Estrogen decreases osteoclast formation by down-regulating receptor activator of NF-kB ligand (RANKL)-induced JNK activation. J Biol Chem 2001;276:8836–40.

[65] Huber DM, Bendixen AC, Pathrose P, Srivastava S, Dienger KM, Shevde NK, et al. Androgens suppress osteoclast formation induced by RANKL and macrophage-colony stimulating factor. Endocrinology 2001;142:3800–8.

[66] Palacios VG, Robinson LJ, Borysenko CW, Lehmann T, Kalla SE, Blair HC. Negative regulation of RANKL-induced osteoclastic differentiation in RAW264.7 cells by estrogen and phytoestrogens. J Biol Chem 2005;280:13720–7.

[67] Robinson LJ, Yaroslavskity BB, Griswold RD, Zadorozny EV, Guo L, Tourkova IL, et al. Estrogen inhibits RANKL-stimulated osteoclastic differentiation of human monocytes through estrogen and RANKL-regulated interaction of estrogen receptor alpha with BCAR1 and Traf6. Exp Cell Res 2009;315:1287–301.

[68] Chen JR, Plotkin LI, Aguirre JI, Han L, Jilka RL, Kousteni S, et al. Transient versus sustained phosphorylation and nuclear accumulation of ERKs underlie anti-versus pro-apoptotic effects of estrogens. J Biol Chem 2005;280(6):4632–8.

[69] Nakamura T, Imai Y, Matsumoto T, Sato S, Takeuchi K, Igarashi K, et al. Estrogen prevents bone loss via estrogen receptor alpha and induction of fas ligand in osteoclasts. Cell 2007;130(5):811–23.

[70] Yasuda H, Shima N, Nakagawa N, Mochizuki SI, Yano K, Fujise N, et al. Identity of osteoclastogenesis inhibitory factor (OCIF) and osteoprotegerin (OPG): a mechanism by which OPG/OCIF inhibits osteoclastogenesis in vitro. Endocrinology 1998;39:1329–37.

[71] Hofbauer LC, Lacey DL, Dunstan CR, Spelsberg TC, Riggs BL, Khosla S. Interleukin-1beta and tumor necrosis factor-alpha, but not interleukin-6, stimulate osteoprotegerin ligand gene expression in human osteoblastic cells. Bone 1999;25(3):255–9.

[72] Takai H, Kanematsu M, Yano K, Tsuda E, Higashio K, Ikeda K, et al. Transforming growth factor-B stimulates the production of osteoprotegerin/osteoclastogenesis inhibitory factor by bone marrow stromal cells. J Biol Chem 1998;273:27091–6.

[73] Pacifici R. Estrogen, cytokines, and pathogenesis of postmenopausal osteoporosis. J Bone Miner Res 1996;11:1043–51.

[74] Lorenzo JA, Naprta A, Rao Y, Alander C, Glaccum M, Widmer M, et al. Mice lacking the type I interleukin-1 receptor do not lose bone mass after ovariectomy. Endocrinology 1998;139:3022–5.

[75] Roggia C, Gao Y, Cenci S, Weitzmann MN, Toraldo G, Isaia G, et al. Up-regulation of TNF-producing T cells in the bone marrow: a key mechanism by which estrogen deficiency induces bone loss in vivo. Proc Natl Acad Sci U S A 2001;98(24):13960–5.

[76] Poli V, Balena R, Fattori E, Markatos A, Yamamoto M, Tanaka H, et al. Interleukin-6 deficient mice are protected from bone loss caused by estrogen depletion. EMBO J 1994;13:1189–96.

[77] Weitzmann MN, Roggia C, Toraldo G, Weitzmann L, Pacifici R. Increased production of IL-7 uncouples bone formation from bone resorption during estrogen deficiency. J Clin Invest 2002;110:1643–50.

[78] Lee SK, Kalinowski JF, Jacquin C, Adams DJ, Gronowicz G, Lorenzo JA. Interleukin-7 influences osteoclast function in vivo but is not a critical factor in ovariectomy-induced bone loss. J Bone Miner Res 2006;21:695–702.

[79] Pacifici R, Rifas L, Teitelbaum S, Slatopolsky E, McCracken R, Bergfeld M, et al. Spontaneous release of interleukin 1 from human blood monocytes reflects bone formation in idiopathic osteoporosis. Proc Natl Acad Sci U S A 1987;84:4616–20.

[80] Sunyer T, Lewis J, Collin-Osdoby P, Osdoby P. Estrogen's bone-protective effects may involve differential IL-1 receptor regulation in human osteoclast-like cells. J Clin Invest 1999;103:1409–18.

[81] Pacifici R, Brown C, Puscheck E, Friedrich E, Slatopolsky E, Maggio D, et al. Effect of surgical menopause and estrogen replacement on cytokine release from human blood mononuclear cells. Proc Natl Acad Sci U S A 1991;88:5134–8.

[82] Srivastava S, Weitzmann MN, Kimble RB, Rizzo M, Zahner M, Milbrandt J, et al. Estrogen blocks M-CSF gene expression and osteoclast formation by regulating phosphorylation of Egr-1 and its interaction with Sp-1. J Clin Invest 1998;102:1850–9.

[83] Jilka RL, Hangoc G, Girasole G, Passeri G, Williams DC, Abrams JS, et al. Increased osteoclast development after estrogen loss: mediation by interleukin-6. Science 1992;257:88–91.

[84] Lin SC, Yamate T, Taguchi Y, Borba VZC, Girasole G, O'Brien CA, et al. Regulation of the gp130 subunits of the IL-6 receptor by sex steroids in the murine bone marrow. J Clin Invest 1997;100:1980–90.

[85] Bellido T, Girasole G, Jilka RL, Crabb D, Manolagas SC. Demonstration of androgen receptors in bone marrow stromal cells and their role in the regulation of transcription from the human interleukin-6 (IL-6) gene promoter. J Bone Miner Res 1993;8:S131.

[86] Hughes DE, Dai A, Tiffee JC, Li HH, Mundy GR, Boyce BF. Estrogen promotes apoptosis of murine osteoclasts mediated by TGF-beta. Nat Med 1996;2:1132–6.

[87] Cenci S, Weitzmann MN, Roggia C, Namba N, Novack D, Woodring J, et al. Estrogen deficiency induces bone loss by enhancing T-cell production of TNF-a. J Clin Invest 2000;106:1229–327.

[88] Sass DA, Liss T, Bowman AR, Rucinski B, Popoff SN, Pan Z, et al. The role of the T-lymphocyte in estrogen deficiency osteopenia. J Bone Miner Res 1997;12:479–86.

[89] Lee SK, Kadono Y, Okada F, Jacquin C, Koczon-Jareml B, Adams DJ, et al. T lymphocyte-deficient mice lose trabecular bone mass with ovariectomy. J Bone Miner Res 2006;21:1704–12.

[90] Gao Y, Grassi F, Ryan MR, Terauchi M, Page K, Yang X, et al. IFN-gamma stimulates osteoclast formation and bone loss in vivo via antigen-driven T cell activation. J Clin Invest 2007;117:122–32.

[91] Fujita T, Matsui T, Nakao Y, Watanabe S. T lymphocyte subsets in osteoporosis. Effect of 1-alpha hydroxyvitamin D3. Miner Electrolyte Metab 1984;10:375–8.

[92] Rosen CJ, Usiskin K, Owens M, Barlascini CO, Belsky M, Adler RA. T lymphocyte surface antigen markers in osteoporosis. J Bone Miner Res 1990;5:851–5.

[93] Sanyal A, Hoey KA, Modder UI, Lamsam JL, McCready LK, Peterson JM, et al. Regulation of bone turnover by sex steroids in men. J Bone Miner Res 2008;23(5):705–14.

[94] D'Amelio P, Grimaldi A, Di Bella S, Brianza SZM, Cristofaro MA, Tamone C, et al. Estrogen deficiency increases osteoclastogenesis up-regulating T cells activity: a key mechanism in osteoporosis. Bone 2008;43:92–100.

[95] Kanematsu M, Sato T, Takai H, Watanabe K, Ikeda K, Yamada Y. Prostaglandin E2 induces expression of receptor activator of nuclear factor-kB ligand/osteoprotegrin ligand on pre-B cells: implications for accelerated osteoclastogenesis in estrogen deficiency. J Bone Miner Res 2000;15:1321–9.

[96] Masuzawa T, Miyaura C, Onoe Y, Kusano K, Ohta H, Nozawa S, et al. Estrogen deficiency stimulates B lymphopoiesis in mouse bone marrow. J Clin Invest 1994;94:1090–7.

[97] Li Y, Toraldo G, Li A, Yang X, Zhang H, Qian WP, et al. B cells and T cells are critical for the preservation of bone homeostasis and attainment of peak bone mass in vivo. Blood 2007;109: 3839–48.

[98] Li Y, Li A, Yang X, Weitzmann MN. Ovariectomy-induced bone loss occurs independently of B cells. J Cell Biochem 2007;100:1370–5.

[99] Whitacre CC, Reingold SC, O'Looney PA. A gender gap in autoimmunity. Science 1999;283 (5406):1277–8.

[100] Kovats S, Carreras E. Regulation of dendritic cell differentiation and function by estrogen receptor ligands. Cell Immunol 2008;252(1-2):81–90.

[101] Adamski J, Ma Z, Nozell S, Benveniste EN. 17Beta-estradiol inhibits class II major histocompatibility complex (MHC) expression: influence on histone modifications and cbp recruitment to the class II MHC promoter. Mol Endocrinol 2004;18(8):1963–74.

[102] Cenci S, Toraldo G, Weitzmann MN, Roggia C, Gao Y, Qian WP, et al. Estrogen deficiency induces bone loss by increasing T cell proliferation and lifespan through IFN-gamma-induced class II transactivator. Proc Natl Acad Sci U S A 2003;100(18):10405–10.

[103] Liu HY, Buenafe AC, Matejuk A, Ito A, Zamora A, Dwyer J, et al. Estrogen inhibition of EAE involves effects on dendritic cell function. J Neurosci Res 2002;70(2):238–48.

[104] Weitzmann MN, Pacifici R. Estrogen deficiency and bone loss: an inflammatory tale. J Clin Invest 2006;116:1186–94.

[105] Khosla S, Westendorf JJ, Oursler MJ. Building bone to reverse osteoporosis and repair fractures. J Clin Invest 2008;118:421–8.

[106] Finkel T, Deng CX, Mostoslavsky R. Recent progress in the biology and physiology of sirtuins. Nature 2009;460(7255):587–91.

[107] Edwards JR, Zainabadi K, Elefteriou E, Connelly L, Yull F, Blackwell TS, et al. The aging associated gene SIRT-1 regulates osteoclast formation and bone mass in vivo. J Bone Miner Res 2007;22 (Suppl):S29.

[108] Edwards JR, Zainabadi K, Lwin ST, Elefteriou E, Munoz S, Moore MM, et al. The longevity genen SIRT-1 independently controls both osteoblast and osteoclast function. J Bone Miner Res 2008;23 (Suppl):S28.

[109] Baur JA, Pearson KJ, Price NL, Jamieson HA, Lerin C, Kalra A, et al. Resveratrol improves health and survival of mice on a high-calorie diet. Nature 2006;444(7117):337–42.

[110] Su JL, Yang CY, Zhao M, Kuo ML, Yen ML. Forkhead proteins are critical for bone morphogenetic protein-2 regulation and anti-tumor activity of resveratrol. J Biol Chem 2007;282(27): 19385–98.

[111] Effros RB. Replicative senescence of CD8 T cells: effect on human ageing. Exp Gerontol 2004;39 (4):517–24.

[112] Effros RB. Costimulatory mechanisms in the elderly. Vaccine 2000;18(16):1661–5.

[113] Boucher N, Defeu-Duchesne T, Vicaut E, Farge D, Effros RB, Schächter F. CD28 expression in T cell aging and human longevity. Exp Gerontol 1998;33(3):267–82.

[114] Looney RJ, Falsey A, Campbell D, Torres A, Kolassa J, Brower C, et al. Role of cytomegalovirus in the T cell changes seen in elderly individuals. Clin Immunol 1999;90(2):213–19.

[115] Pietschmann P, Grisar J, Thien R, Willheim M, Kerschan-Schindl K, Preisinger E, et al. Immune phenotype and intracellular cytokine production of peripheral blood mononuclear cell from postmenopausal patients with osteoporotic fractures. Exp Gerontol 2001;36(10):1749–59.

[116] Hooyman JR, Melton LJ, Nelson AM, O'Fallon WM, Riggs BL. Fractures after rheumatoid arthritis: a population-based study. Arthritis Rheum 1984;27:1353–61.

[117] Kanis JA, Oden A, Johnell O, Johansson H, De Laet C, Brown J, et al. The use of clinical risk factors enhances the performance of BMD in the prediction of hip and osteoporotic fractures in men and women. Osteoporos Int 2007;18:1033–46.

[118] Spector TD, Hall GM, McCloskey EV, Kanis JA. Risk of vertebral fracture in women with rheumatoid arthritis. BMJ 1993;306(6877):558.

[119] Haynes DR. Inflammatory cells and bone loss in rheumatoid arthritis. Arthritis Res Ther 2007;9(3):104.

[120] Haugeberg G, Ørstavik RE, Kvien TK. Effects of rheumatoid arthritis on bone. Curr Opin Rheumatol 2003;15(4):469–75.

[121] Weyand CM, Fulbright JW, Goronzy JJ. Immunosenescence, autoimmunity, and rheumatoid arthritis. Exp Gerontol 2003;38(8):833–41.

[122] Masi AT. Sex hormones and rheumatoid arthritis: cause or effect relationships in a complex pathophysiology. Clin Exp Rheumatol 1995;13(2):227–40.

[123] Vasquez H, Mazure R, Gonzalez D, Flores D, Pedreira S, Niveloni S, et al. Risk of fractures in celiac disease patients: a cross-sectional, case-control study. Am J Gastroenterol 2000;95(1):183–9.

[124] Taranta A, Fortunati D, Longo M, Rucci N, Iacomino E, Aliberti F, et al. Imbalance of osteoclastogenesis-regulating factors in patients with celiac disease. J Bone Miner Res 2004;19(7):1112–21.

[125] Riches PL, McRorie E, Fraser WD, Determann C, van't Hof R, Ralston SH. Osteoporosis associated with neutralizing autoantibodies against osteoprotegerin. N Engl J Med 2009;361(15):1459–65.

[126] Lopez I, Buchman AL. Metabolic bone disease in IBD. Curr Gastroenterol Rep 2000;2(4):317–22.

[127] Moschen AR, Kaser A, Enrich B, Ludwiczek O, Gabriel M, Obrist P, et al. The RANKL/OPG system is activiated in inflammatory bowel disease and relates to the state of bone loss. Gut 2005;54(4):479–87.

[128] Franchimont N, Reenaers C, Lambert C, Belaiche J, Bours V, Malaise M, et al. Increased expression of receptor activator of NF-kappaB ligand (RANKL), its receptor RANK and its decoy receptor osteoprotegerin in the colon of Crohn's disease patients. Clin Exp Immunol 2004;138(3):491–8.

[129] Ashcroft AJ, Cruickshank SM, Croucher PI, Perry MJ, Rollinson S, Lippitt JM, et al. Colonic dendritic cells, intestinal inflammation, and T cell-mediated bone destruction are modulated by recombinant osteoprotegerin. Immunity 2003;19(6):849–61.

[130] Rossouw JE, Anderson GL, Prentice RL, LaCroix AZ, Kooperberg C, Stefanick ML, et al. Risks and benefits of estrogen plus progestin in healthy postmenopausal women: principal results from the Women's Health Initiative randomized controlled trial. JAMA 2002;288(3):321–33.

[131] Khosla S. Increasing options for the treatment of osteoporosis. N Engl J Med 2009;361(8):818–20.

[132] Lacey DL, Timms E, Tan HL, Dunstan CR, Burgess T, Elliott R, et al. Osteoprotegerin ligand is a cytokine that regulates osteoclast differentiation and activation. Cell 1998;93:165–76.

[133] Venkateshan SP, Sidhu S, Malhotra S, Pandhi P. Efficacy of biologicals in the treatment of rheumatoid arthritis. a meta-analysis. Pharmacology 2009;83(1):1–9.

[134] Khosla S, Riggs BL. Pathophysiology of age-related bone loss and osteoporosis. Endocrinol Metab Clin N Am 2005;34:1015–30.

[135] Cooper C, Melton LJ. Epidemiology of osteoporosis. Trends Endocrinol Metab 1992;12/17/1992 (3):224–9.

[136] Riggs BL. The mechanisms of estrogen regulation of bone resorption. J Clin Invest 2000;106: 1203–4.

# 10

# The Role of the Immune System in the Bone Loss of Inflammatory Arthritis

Steven R. Goldring[1], Georg Schett[2]

[1]DEPARTMENTS OF ORTHOPEDICS AND RHEUMATOLOGY, HOSPITAL FOR SPECIAL SURGERY, WEILL MEDICAL COLLEGE OF CORNELL UNIVERSITY, NEW YORK, USA,
[2]DEPARTMENT OF INTERNAL MEDICINE III AND INSTITUTE FOR CLINICAL IMMUNOLOGY, UNIVERSITY OF ERLANGEN-NUREMBERG, ERLANGEN, GERMANY

## CHAPTER OUTLINE

Introduction .................................................................................................. 301
Bone disease associated with RA ................................................................ 302
Joint margin and subchondral bone loss in RA ........................................... 303
Peri-articular bone loss ............................................................................... 311
Generalized bone loss ................................................................................. 312
Seronegative spondyloarthropathies .......................................................... 315
Acknowledgment ......................................................................................... 319
References ................................................................................................... 319

## Introduction

The rheumatic diseases include a broad spectrum of disorders that share a propensity to affect the anatomic components of the joints and the adjacent peri-articular tissues. The diarthrodial joints are the most commonly affected sites. These joints join two opposing bone surfaces, which are covered by hyaline cartilage. The purpose of the hyaline cartilage is to provide an articulating surface that facilitates joint motion [1]. Diarthrodial joints are encased in a highly specialized lining membrane termed the synovium, which separates the joint cavity from the underlying connective tissues. This membrane consists of a mixed population of cells expressing phenotypic features of fibroblasts and macrophages. It provides a barrier that permits trafficking of nutrients from the underlying vasculature to the cells of the articular cartilage, and is also the source of products that interact with the cartilage and contribute to the unique "low friction" properties of the cartilage surfaces [1]. In patients with rheumatoid arthritis (RA), a prototypical form of inflammatory arthritis, the synovium is the site of the initial inflammatory process. This is characterized by infiltration of the synovial lining by cells indicative of an intense inflammatory immune-mediated process, including lymphocytes, plasma cells, endothelial cells, and activated macrophages [2, 3]. This process is accompanied by proliferation of the synovial-lining cells, the progressive extension of the expanding synovial tissue over the articular surface and the formation of a so-called pannus, which is derived from the Greek word meaning "mantle" or "covering". At the interface between the RA synovium and articular cartilage, there is evidence of destruction of the extracellular cartilage matrix. Where there is contact between the bone and synovium, as discussed in the following section, there is typically focal "erosion" of the cortical bone surfaces with invasion of the inflammatory tissue into the adjacent marrow spaces. In contrast to the diarthrodial joints, amphiarthroses lack a true joint cavity and synovial lining. Rather, these are characterized by a fibrocartilaginous union, which joins the two opposing bone surfaces [1]. The intervertebral discs are representative of amphiarthroses. The amphiarthroses are characteristically involved in a distinct subset of inflammatory arthropathies that are classified as seronegative spondyloarthropathies. These joint disorders include ankylosing spondylitis, reactive arthritis, and arthritis associated with psoriasis, inflammatory bowel disease, and juvenile-onset spondyloarthropathy. In these forms of inflammatory arthritis the spine, peripheral joints, periarticular structures, or all three sites are involved [4]. The joint inflammation that accompanies these disorders exhibits many of the same histopathological features of the joint pathology in RA, including synovial hyperplasia, immune cell infiltration, and pannus formation. As will be discussed in the following section, in contrast to RA, where the synovial lining is the initial site of the inflammatory process, in the spondyloarthropathies the inflammatory process initially localizes to the entheses, which are sites of tendon and ligament insertion [4]. This may in part account for the observed differences in the pattern of articular and periarticular skeletal remodeling observed in RA and the spondyloarthropathies.

It is important to note that in the rheumatic diseases, which are associated with inflammatory arthritis, the underlying disturbance in immune regulation that is responsible for the localized joint pathology may target extra-articular organs and tissues and these features may dominate the clinical picture. This chapter will not address the non-articular manifestation of these disorders, but it is relevant to note that involvement of organs, such as the kidneys, may profoundly affect skeletal tissue remodeling.

## Bone disease associated with RA

There are three principal forms of altered skeletal remodeling in patients with RA [5–10]. The first affects the immediate peri-articular cortical bone at the joint margins. This is the site where the inflamed synovial tissue is in direct contact with the bone surface, and the loss of bone at this site produces focal "bone erosions", which are characteristic of RA. A second pattern of altered bone remodeling affects the peri-articular trabecular and cortical bone and is characterized by the presence of periarticular bone loss. The third form of altered bone remodeling is characterized by generalized osteoporosis, involving the axial and appendicular skeleton at sites that are not involved directly in the inflammatory arthritis [11–15].

## Joint margin and subchondral bone loss in RA

Bromley and Woolley utilized histopathological techniques to characterize the cells involved in the pathogenesis of marginal joint erosions in retrieved joint tissues from patients with RA [16, 17]. They noted the presence of multinucleated cells with phenotypic features of osteoclasts at the interface where the advancing RA pannus came into contact with the bone surfaces and calcified cartilage. Similar cells were noted in the osteolytic lacunae at the junction of the subchondral bone and adjacent marrow space. Gravallese et al. utilized immunostaining and in situ hybridization techniques to more rigorously characterize the multinucleated cells at the bone pannus interface and noted that these cells express abundant tartrate-resistant acid phosphatase and the calcitonin receptor, phenotypic markers that distinguish the osteoclast from foreign body giant cells and other myeloid-lineage cells [18]. A hand radiograph from a patient with advanced RA is provided in Figure 10-1. There is evidence of severe articular bone destruction with multiple marginal joint erosions. A histopathologic section from the RA bone–pannus interface demonstrates the presence of multinucleated TRAP-positive osteoclasts in resorption lacunae on the bone surface. These descriptive observations do not preclude

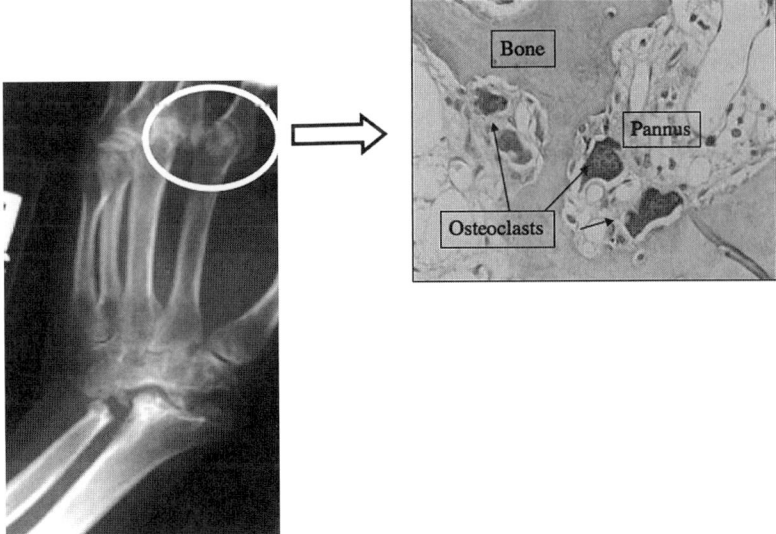

**FIGURE 10-1** Radiographic and histopathologic features of RA joint erosions. Hand radiograph from a patient with advanced RA demonstrating extensive articular bone destruction and accompanying joint erosion (highlighted by the circle). A representative histopathologic section from the bone pannus interface with osteoclasts in resorption lacunae on the bone surface. Please refer to color plate section.

the participation of other cell types in the bone-resorptive process, since there is evidence that macrophages and synovial fibroblasts have the ability to resorb bone, although their resorptive capacity is much less than that of osteoclasts [19, 20].

Important insights into the role of osteoclasts in the pathogenesis of marginal joint erosions was provided by the studies of Kong et al. [21] in which it was shown that blockade of the essential osteoclast-inducing factor, receptor activator of NF-κB ligand (RANKL), with osteoprotegerin (OPG) (the decoy receptor for RANKL) markedly reduced the number and extent of focal articular bone erosions in the rat model of adjuvant arthritis. Subsequently, several other laboratories confirmed the capacity of RANKL blockade with OPG to markedly attenuate bone erosions in different models of inflammatory arthritis [22, 23]. As will be discussed below, these studies also provided the first evidence indicating a critical role for RANKL in the pathogenesis of focal bone erosions in inflammatory arthritis.

More definitive evidence indicating that osteoclasts are required for bone resorption in inflammatory arthritis has been provided by studies employing genetic approaches utilizing animal models of inflammatory arthritis. The initial studies were conducted by Pettit et al. who generated inflammatory arthritis in mice lacking the gene for RANKL [24] by utilizing the serum transfer model of inflammatory arthritis that had been developed by Mathis and Benoist [25]. These animals lack the capacity to generate osteoclasts and exhibit a profound form of osteopetrosis. The authors noted that wild-type and RANKL-deficient mice developed comparable levels of synovitis and joint inflammation; however, the RANKL knockout mice, which lacked the capacity to form osteoclasts, failed to develop significant bone erosions. These findings provide definitive evidence that in this model of inflammatory arthritis, osteoclasts were essential for the generation of focal joint erosions. Subsequently, Redlich and co-workers provided further evidence indicating the requirement for osteoclasts in the pathogenesis of focal bone erosions in a murine model of RA [26]. They backcrossed TNF-α transgenic mice, which spontaneously develop a form of inflammatory arthritis resembling RA with mice lacking the *c-fos* gene. *C-fos* is a transcription factor that is essential for osteoclastogenesis, and the *c-fos* null mice lack the capacity to generate osteoclasts [27]. The *c-fos* null mice backcrossed with the TNF-α transgenic mice developed inflammatory arthritis that was comparable to their wild-type littermates, but failed to develop bone erosions. More recently, Li and co-workers backcrossed the TNF-α transgenic mice with mice lacking RANK, the receptor for RANKL [28]. These mice are similar to the *c-fos* knockout mice in that they lack the ability to form osteoclasts. As in the RANKL and *c-fos* knockout mice, the inability to form osteoclasts resulted in protection from the development of joint erosions, providing additional evidence that osteoclasts are required for the development of focal bone erosions in these murine models of inflammatory arthritis.

Evidence supporting an essential role for osteoclasts in the pathogenesis of focal joint erosions in patients with RA is provided by recent studies in which patients with early RA were treated with denosumab, a monoclonal antibody that blocks RANKL activity [29]. Utilizing MRI and radiographic imaging, the authors observed a significant reduction in articular bone erosions in the denosumab-treated individuals. Of interest, multiple studies have tested the efficacy of bisphosphonates to prevent articular and systemic bone loss in patients with RA. Although there is evidence of protection from systemic bone loss, in general these antiresorptive agents have not been effective in reducing focal joint destruction, with the exception of a recent publication involving the sequential administration of zoledronic acid [30]. The limitations and results of trials with bisphosphonates in RA subjects were recently reviewed [31].

The RA synovium is the source of a broad spectrum of cytokines and growth factors that modulate osteoclast differentiation and activation, including factors that are stimulatory as well as inhibitory. The progressive loss of bone at the joint margins and at sites of subchondral bone erosion provides evidence that osteoclastogenic and bone-resorptive influences predominate. Separation of RA synovial cells and in vitro cell culture models have been used to identify the cellular source of these regulatory products and the mechanisms by which they modulate bone remodeling. Co-culture studies have been employed to demonstrate that both synovial fibroblasts and specific T-cell subsets present in RA synovial tissues have the capacity to induce osteoclast precursors to differentiate into functional osteoclasts [21, 32–37]. Many of these pro-osteoclastogenic mediators are the same cytokines and factors that regulate immune cell function and activation that contributes to the initiation and perpetuation of the synovial inflammation. The factors with osteoclastogenic activity include interleukin-1 (IL-1), interleukin-6 (IL-6), interleukin-7 (IL-7), interleukin-11 (IL-11), interleukin-15 (IL-15), interleukin-17 (IL-17), monocyte/macrophage-colony stimulating factor (M-CSF), tumor necrosis factor-α (TNF-α), oncostatin M (OSM), leukemia inhibitory factor (LIF), prostaglandins of the E series, parathyroid hormone related peptide (PTHrP) (the factor associated with humoral hypercalcemia of malignancy), and importantly RANKL [7–10, 15]. Many of these osteoclast-inducing factors stimulate osteoclast differentiation and activation by acting on other cell types within the bone microenvironment or in the inflamed synovial tissues to stimulate these non-osteoclast-lineage cells to produce osteoclast-inducing factors. Cytokines including IL-7, IL-11, IL-17, and the humoral hypercalcemic factor PTHrP are representative of a subset of mediators that exert their effects to enhance osteoclastogenesis via this indirect mechanism. A schematic representation of the cellular interactions and synovial cell products that contribute to osteoclast formation at sites of RA joint erosions is shown in Figure 10-2.

Of the osteoclastogenic factors present within the RA synovium, particular attention has focused on RANKL, which was originally identified as a T-cell product that regulated dendritic cell function [38–40]. Demonstration that it is a product of both synovial fibroblasts and T cells strongly implicated this ligand as an essential mediator of the osteoclastic bone resorption in RA, and as described above, blockade of its activity in animal models of inflammatory arthritis and in patients with RA provided proof of concept for its pivotal role in the articular bone disease of RA.

The presence of T cells in the inflamed synovium represents a hallmark of the rheumatoid synovial lesion. However, the role of T cells in the pathogenesis and perpetuation

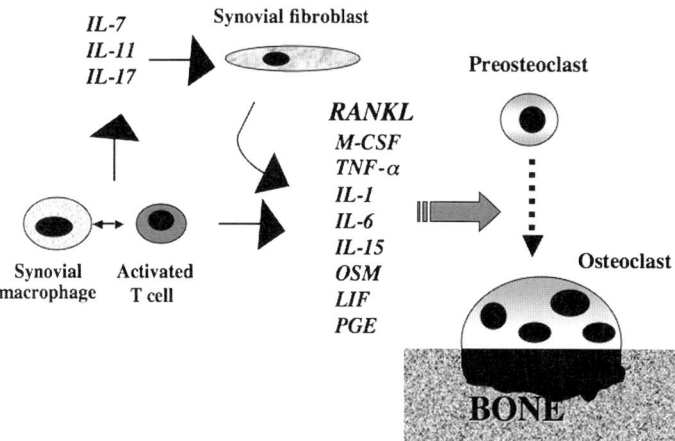

FIGURE 10-2 Role of RA synovial products in the pathogenesis of osteoclast-mediated joint erosions. Cytokines, growth factors, and other mediators produced by synovial fibroblasts, T cells, and macrophages act on osteoclast precursors in the RA synovium to induce their differentiation into osteoclasts. Some products of T cells and/or synovial macrophages enhance osteoclast formation indirectly by acting on fibroblasts or other mesenchymal cells in the synovium to release osteoclast-inducing factors.

of the RA synovial lesion is complex, in part because of the heterogeneity of the T-cell subsets that populate the synovial tissue. Among the T-cell subsets, there is good evidence implicating both the Th1 and Th17 subsets in the osteoclast-mediated bone resorption associated with the RA synovial lesion [9, 41–43]. Both T-cell subsets have the capacity to produce RANKL as well as several other cytokines and mediators with osteoclastogenic activity. Of interest, although the Th1 subset of $CD4^+$ T cells has the capacity to produce RANKL, in co-culture models with osteoclast precursors, they actually inhibit osteoclast formation. As will be discussed in the subsequent section, this inhibitory effect may be attributed to the concomitant production of interferon-γ (IFN-γ), which is a potent inhibitor of osteoclastogenesis [35].

Recent attention has focused on the role of the Th17 T-cell subset in RA pathogenesis and joint destruction. These cells produce multiple cytokines and inflammatory mediators including RANKL and TNF-α. They also are the primary, but not exclusive, source of IL-17, a cytokine with pleiotropic activities, including the capacity to induce RANKL and TNF-α in mesenchymal cells but also to enhance osteoclast formation directly [41, 42]. Because of its potent pro-inflammatory properties, IL-17 has emerged as a target for therapeutic intervention in RA and related inflammatory conditions and in part the efficacy of this intervention, at least in animal models of inflammatory arthritis, may be related to the associated inhibition of osteoclast-mediated bone destruction [42].

Among the osteoclastogenic factors produced by the RA synovium, TNF-α, IL-1, and IL-6 have emerged as three of the principal molecules that are targeted for treatment of the synovial inflammatory process and prevention of bone destruction [44]. TNF-α exerts its effect on osteoclastogenesis by acting directly on osteoclast precursors, as well as indirectly, by up-regulating the production of M-CSF and RANKL on mesenchymal cells [10, 45, 46]. Although there remains debate concerning whether TNF-α can induce osteoclastogenesis independent of RANKL, the results of studies in animal models of inflammatory arthritis in which RANK signaling is abrogated, either by RANKL gene deletion or knockout of RANK, the RANKL receptor, indicate that RANKL is required for TNF-induced osteoclast formation. It is likely that TNF plays a critical role in synergistically enhancing osteoclast-mediated bone resorption by interacting with RANKL. It has been suggested that TNF-α also contributes to osteoclast formation in the inflamed RA joint by up-regulating the expression of osteoclast-associated receptor (OSCAR) on osteoclasts and their precursors [47]. This cell-surface receptor interacts with the Fc receptor gamma chain to enhance osteoclast activity through a co-stimulatory pathway [48].

IL-1 shares with TNF-α the capacity to markedly up-regulate RANKL, as well as other osteoclast-inducing factors, and there is evidence in murine models of osteoclast formation that the IL-1 is essential for the osteoclastogenic effect of TNF-α [49]. Zwerina and co-workers demonstrated that mice lacking IL-1 were protected from bone loss in the TNF-transgenic model of inflammatory arthritis indicating that, at least in this animal model of inflammatory arthritis, the effects of TNF on osteoclast-mediated bone resorption are IL-1 dependent [50]. Similar to IL-1, IL-6 can also up-regulate RANKL and thus indirectly support osteoclast formation via interaction with mesenchymal cells. The direct effects of IL-6 on osteoclasts, however, are inhibitory rather than stimulatory. Cytokines engaging gp130-containing receptors have either inhibitory (IL-6) or stimulatory (IL-11, OSM, LIF) function on osteoclasts [51]. Interference with the binding of pro-osteoclastogenic cytokines to gp130 may at least in part explain the observation that neutralization IL-6R effectively blocks osteoclast formation and retards structural damage in RA patients. Of interest, there is evidence from clinical trials that inhibition of TNF-α in patients with RA may result in a dissociation of the beneficial effects of TNF-α blockade on the signs and symptoms of inflammation without a commensurate inhibition of focal articular bone erosions [52]. In part this may be related to effects of other pro-inflammatory products (such as IL-1 or IL-6) that have direct or indirect actions on osteoclast formation and activity and whose regulation and production are at least partially independent of TNF-α.

As discussed in the preceding section, besides producing pro-osteoclastogenic factors, cells within the RA synovium make inhibitors of osteoclast formation. These include IFN-γ, IFN-α/β, interleukin-4 (IL-4), interleukin-10 (IL-10), granulocyte monocyte-colony stimulating factor (GM-CSF), interleukin-12 (IL-12), interleukin-18 (IL-18), and interleukin-23 (IL-23) [10, 15, 45, 48]. The presence of progressive bone loss at sites of synovial inflammation indicates that the activities of these inhibitory molecules are not sufficient to counteract the pro-osteoclastogenic effects of other cytokines and factors. Many of these inhibitory factors are products of T-cell subsets that are present within the RA synovium, and some of these cells share the capacity to produce both stimulatory and inhibitory molecules. For example, activated Th1 cells produce IFN-γ, and the inhibitory effect of Th1 cells on osteoclast formation is abrogated in IFN-γ-deficient osteoclast precursors [43, 45, 48]. These cells also produce TNF-α, and in the presence of RANKL, the inhibitory effects of the interferon are overcome and osteoclast formation is enhanced. In contrast to the Th1 CD4$^+$ T cell, Th17 cells produce significantly less IFN-γ and when activated these cells up-regulate TNF and RANKL production, shifting the balance towards enhanced osteoclast differentiation and activation [7, 45, 48]. Osteoclast precursors also share the capacity to produce interferons and both type I and type II interferons are up-regulated during RANKL-induced osteoclast differentiation [45, 48]. These regulatory molecules and pathways provide a complex autocrine feedback system, which controls osteoclast-mediated bone resorption. Clearly this regulatory system is overridden in the presence of RA synovial inflammation, in part because of the production of potent osteoclastogenic cytokines such as TNF-α and IL-1 within the synovial lesion.

Regulatory T cells (Treg) represent an additional population of T cells, which are present within the RA synovium and likely contribute to the regulation of osteoclastogenesis [53–55]. Studies by Zaiss and co-workers indicated that the inhibitory effect of these cells on osteoclast formation required direct cell–cell contact and was mediated via interactions of CTLA-4 with CD80/CD86 on the surface of osteoclast precursors [55]. Treg-cell-derived transforming growth factor-β (TGF-β), IL-4, and IL-10 contributed to, but were not required for, the inhibitory effect. In contrast, Kim et al. observed that the inhibitory effects of Treg cells on osteoclastogenesis was independent of cell contact, but dependent on IL-4 and TGF-β [54]. Based on these in vitro cell culture models the essential regulatory role of Treg cells in RA has not yet been firmly established.

Studies by Ochi and co-workers [56] have provided further insights into the complex mechanisms by which T cells in the RA synovium contribute to enhanced osteoclast-mediated bone resorption. Using an in vitro osteoclast differentiation model, they showed that TNF-α induced expression of the paired Ig-like receptor-A (PIR-A) and to a lesser extent its intracellular signaling partner Fc receptor-γ (FcRγ) on osteoclast precursors. PIR-A upon ligation interacts with FcRγ to provide a co-stimulatory signal that acts in concert with RANK to enhance osteoclast formation. The authors demonstrated that osteoclast precursors, derived from mice lacking FcRγ were defective in TNF-induced osteoclastogenesis. To define the role of signaling via FcRγ in a model of inflammatory arthritis, the authors crossed the FcRγ knockout mice with hTNF transgenic (*hTNFtg*) mice. The onset and clinical course of the arthritis in the *hTNFtg/FcRγ$^{-/-}$* mice was comparable to the *hTNFtg* mice; however, osteoclast-mediated bone resorption was markedly attenuated in the *hTNFtg/FcRγ$^{-/-}$* mice. These results are consistent with the effects of this signaling pathway that were observed in the in vitro experiments. They also crossed the *hTNFtg* mice to mice lacking the β$_2$microglobulin gene ($β_2M^{-/-}$). MHC class I molecules function as ligands for PIR-A, and $β_2M$ like PIR-A is up-regulated by TNF-α. These mice, like the *hTNFtg/FcRγ$^{-/-}$* mice, exhibited impaired osteoclast formation, providing further evidence that this co-stimulatory pathway plays a role in the pathologic bone resorption that occurs in this model of inflammatory arthritis.

The availability of a source of myeloid lineage osteoclast precursors is essential to the development of osteoclast-mediated bone erosion at the bone–pannus interface. Analysis of retrieved synovial tissues from patients with RA reveals that the inflammatory tissue contains abundant osteoclast precursors [19, 35, 57, 58]. These myeloid-lineage cells are recruited to the site of inflammation in response to local inflammatory mediators. In the presence of appropriate stimuli, these cells have the capacity to differentiate into authentic osteoclasts, which in turn mediate the local bone-resorptive process. Numerous cytokines and chemokines are produced by the inflamed synovium. These include macrophage inflammatory protein-1α (MIP-1α), stromal cell-derived factor-1 (SDF-1), chemokine (C-C motif) ligand20 (CCL20), CCL13, and MIP-1β and all may participate in the recruitment of osteoclast precursors [59–62].

Imaging studies utilizing magnetic resonance imaging (MRI) have been particularly useful in assessing the anatomic and bone architectural changes associated with the marginal joint erosions and subchondral bone resorption associated with the RA synovial lesion. MRI evaluation permits the assessment of the extent and local distribution of the synovial and tenosynovial inflammation. Importantly, it also permits evaluation of the subchondral marrow compartment that is an active site of trabecular and subchondral inflammation and bone resorption. Alterations in the marrow compartment by the inflammatory process generates a signal that produces a characteristic bone marrow lesion, which is interpreted as "bone marrow edema" or "osteitis" (Figure 10-3). Examination of retrieved joint tissues from these anatomic sites reveals the presence of invading inflammatory cells with fibrous marrow replacement and the presence of lymphoid aggregates that are associated with a marked increase in vascular elements [63]. Importantly, the presence of bone marrow lesions predicts progression of bone erosions and peri-articular bone destruction [64, 65].

A striking feature of the marginal joint erosion in patients with RA is the virtual absence of bone repair [66]. In general, in physiologic remodeling, the activities of bone resorption and formation are exquisitely coupled. In a balanced state bone mass is neither gained nor lost during the remodeling cycle, although the shape and architecture of the bone may undergo modification. This balance in the activity of osteoclasts and osteoblasts is clearly disturbed within the rheumatoid joint. Insights into the pathogenesis of this remodeling disequilibrium were recently provided in a seminal paper by Diarra and co-workers. These authors examined the cellular and molecular events involved in the pathogenesis of bone loss in animal models of inflammatory arthritis and demonstrated that similar processes were occurring in patients with RA [67]. Analysis of retrieved joint tissues from three different animal models of inflammatory arthritis revealed the presence of the Wingless (Wnt) pathway inhibitor Dickkopf-1 (DKK-1) in the inflamed synovial tissue. They localized this potent inhibitor of bone formation to the synovial fibroblasts and endothelial cells and demonstrated, using in vitro cell cultures, that TNF-α markedly increased the production of DKK-1 in cultured synovial fibroblasts. A striking finding in their studies was the observation that animals with inflammatory arthritis that were treated with an antibody to DKK-1 experienced an increase in bone

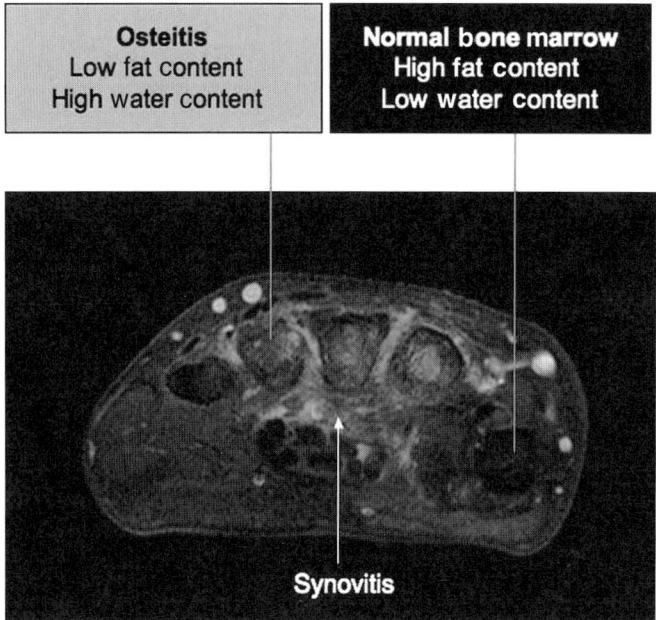

FIGURE 10-3 Magnetic resonance imaging (MRI) of the hand of a patient with rheumatoid arthritis. The image shows a STIR sequence through the base of the metacarpal bones. Normal bone marrow of the 1st and 5th metacarpal bone appear dark containing fat and only little water, whereas bone marrow of the 2nd, 3rd and 4th metacarpal appears bright and rich of water resembling osteitis. Synovitis is found around the metacarpal bases and also appears as water-rich tissue.

repair and also had a marked inhibition of focal articular bone resorption. This effect on bone resorption was attributed to an increase in the production of OPG, the potent inhibitor of RANKL. They speculated that the increase in OPG was attributable to enhanced Wnt signaling, resulting from DKK-1 inhibition, and the effects of Wnt-dependent up-regulation of β-catenin, which transcriptionally enhances OPG gene expression. A schematic representation of the interaction between osteoblast- and osteoclast-lineage cells and the role of synovial fibroblast-derived DKK-1 in inhibiting osteoblast differentiation is provided in Figure 10-4. Inhibition of DKK-1 restores osteoblast differentiation and bone formation and also results in enhanced OPG production, leading to inhibition of RANKL-induced osteoclastogenesis.

To confirm a role for DKK-1 in the inflamed joints of patients with RA, the authors analyzed retrieved joint tissues from RA patients undergoing joint replacement surgery. They found that DKK-1 was widely expressed in synovial fibroblasts, endothelial cells and in chondrocytes located in cartilage immediately adjacent to sites of pannus invasion. They also investigated the serum levels of DKK-1 in RA patients and found that DKK-1 levels were elevated compared to healthy controls and that the elevations correlated with measures of clinical disease activity. Of interest, they noted that in patients treated with anti-TNF therapy, the levels of DKK-1 were lower. In contrast, in patients with ankylosing

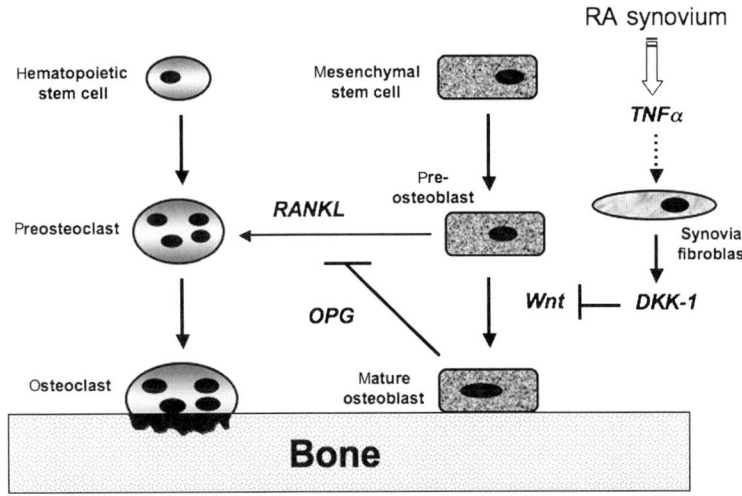

**FIGURE 10-4** Effect of DKK-1 inhibition on articular bone erosions in RA. TNF-α stimulates release of DKK-1 from RA synovial fibroblasts, resulting in inhibition of Wnt signaling and arrested osteoblast differentiation and impaired bone formation. Pre-osteoblasts produce RANKL, contributing to osteoclast formation and increased bone resorption. Blockade of DKK-1 restores Wnt pathway signaling and restores osteoblast differentiation and bone formation. Production of OPG by osteoblasts inhibits RANKL-dependent osteoclastogenesis.

spondylitis in whom there was evidence of increased peri-articular bone formation, the DKK-1 levels were below levels in a healthy control population.

More recently, Walsh et al. extended the observations of Diarra and co-workers to further characterize the role of inhibitors of bone formation using the *K/BXN* model of serum transfer arthritis [68]. Histomorphometric analyses were used to establish that there was uncoupling of bone resorption and formation at sites of bone erosions with arrest of osteoblast differentiation and defective bone accretion. They utilized quantitative polymerase chain reaction (qPCR) to evaluate the expression levels of a spectrum of Wnt pathway inhibitors in retrieved synovial specimens from arthritic and wild-type animals. These included several members of the DKK-1 and soluble frizzled related protein (sFRP) families, and immunostaining was utilized to confirm the local expression of DKK-1 and sFRP1 in synovial tissue at sites of erosions. These findings provide further insights into the cellular and molecular mechanisms involved in the pathogenesis of focal articular bone resorption in inflammatory arthritis. Importantly, they also provide new pathway targets for therapeutic intervention to restore joint repair and reduce bone loss in patients with RA and related forms of inflammatory arthritis.

## Peri-articular bone loss

In addition to the focal marginal joint erosions that are the radiologic hallmark of RA, there also is evidence of peri-articular bone loss at sites that are not in direct contact with the RA pannus or subchondral bone marrow lesions. This loss typically occurs early in the course of RA and frequently, but not always, precedes the development of the focal marginal bone erosions [69]. In recent studies by Hoff et al. [70], the authors employed digital X-ray radiogrammetry (DXR) to evaluate peri-articular hand bone mineral density (BMD) in a population of patients with early RA in order to determine whether the changes in the peri-articular bone density were predictive of the subsequent development of marginal joint erosions. They found that, as with previous studies using related techniques [71, 72], patients with peri-articular bone loss at 1 year developed significantly greater joint damage compared to patients without peri-articular BMD loss. Importantly, they found that these changes were an independent predictor of subsequent joint damage, which had a predictive power comparable to established disease biomarkers such as antibodies to citrullinated peptides and C-reactive protein.

Peri-articular bone loss occurs at sites that are remote from the synovial inflammatory process and this finding suggests that the pathogenic mechanisms leading to the bone loss involve mechanisms that differ from those associated with the RA pannus. Histomorphometric analyses of bone samples obtained from patients undergoing joint arthroplasty reveal the presence of increased remodeling of the peri-articular bone at sites that are remote from the RA pannus and marrow inflammation [73]. In this study the authors observed an increase in both bone formation and resorption indices. The progressive loss of bone, however, is evidence that bone resorption exceeds bone formation. Since joint inflammation is often accompanied by reduced joint motion, it has been suggested that decreased loading and disuse contributes to the increased bone resorption. Bone exhibits a unique capacity to rapidly adapt its structural organization to mechanical forces through the cellular processes of modeling and remodeling, and these changes are consistent with the immobilization and disuse that may accompany joint inflammation [74]. Alternately, increased bone resorption could be related to the release of cytokines from the synovium at sights of bone marrow inflammation, which could act locally to increase osteoclast-mediated bone resorption.

Guler-Yuksel and co-workers [75] examined the relationship of peri-articular BMD, marginal joint erosions, and alterations in generalized skeletal BMD in the hip and spine, measured by dual-energy X-ray absorptiometry (DEXA). They also evaluated the effects of treatments, including anti-TNF and glucocorticoid therapy, on bone at these sites. They found that across all treatment groups there was greater BMD loss in the hands than the hip or spine, and the loss of BMD in the hands preceded the changes in the hip and spine. These findings confirm the observations of other investigators [11, 76], who showed a similar temporal pattern of bone loss at these different sites.

## Generalized bone loss

RA and many of the related forms of inflammatory arthritis are often accompanied by systemic osteoporosis [13, 77–84]. Importantly, this decrease in bone mass is associated with an increase in the risk of fracture. For example, van Staa [84] examined the long-term risk of fracture in over 30, 000 RA patients utilizing the British General Practice Database.

Fracture rates were adjusted for smoking, body mass index, and several additional clinical risk factors. He found that patients with RA were at increased risk for fracture, and that the increased risk was attributable to a combination of disease activity and the use of oral glucocorticoids.

Several studies have demonstrated that the systemic bone loss in patients with RA correlates with disease activity, although as discussed above, the temporal sequence of bone loss at the joint margins, peri-articular bone, and systemically may be disassociated [69]. A study by Lodder et al. [79] was specifically designed to establish the relationship between disease activity, joint erosions, and systemic bone mass using BMD in a cohort of patients with low to moderately active RA. These authors found that there was an association between the extent of joint damage and the presence of low BMD at the hip. Solomon and co-workers performed similar analyses in a cohort of post-menopausal women with RA [85]. They also found that the hip BMD was associated with focal joint erosions, but that the association disappeared after multivariate adjustment. They concluded that, although bone mass and erosions may be correlated, the relationship was complex and influenced by multiple disease and treatment factors.

Multiple confounding patient factors have made it difficult to define the specific role of disease activity in systemic bone loss. These include the influence of sex, age, mobility, disease activity, and duration, and the concomitant use of glucocorticoids, non-steroidal anti-inflammatory drugs, and disease-modifying therapies, all of which have independent effects on bone metabolism. An additional factor is that the mechanisms involved in the pathogenesis of the systemic bone loss may be influenced not only by the disease activity but also by disease duration. For example, a longitudinal prospective study by Gough and co-workers [86] demonstrated that systemic loss of bone occurred early in the course of RA and that the magnitude of bone loss was associated with the level of disease activity. These findings were confirmed in the study by Als et al. [87].

The introduction of methotrexate and biologic therapies that target TNF-$\alpha$ have convincingly demonstrated the capacity of these interventions to retard or even prevent the development of local articular bone destruction. Several studies have addressed the efficacy of these therapies on systemic bone loss and provided evidence that amelioration of joint inflammation and arrest of articular destruction are accompanied by beneficial effects on systemic bone loss [71, 88–91]. For example, Haugeberg et al. recently completed the analysis of a cohort of patients with early RA treated with infliximab (the TNF blocker) and methotrexate and compared outcomes to those of patients treated with methotrexate alone [88]. They found that bone loss was significantly reduced in the anti-TNF plus methotrexate-treated group compared to the group treated only with methotrexate. Measures of disease activity and joint damage were independently associated with the effects of the therapies on systemic bone mass. Based on these results, the authors concluded that there was a causal link between joint inflammation and systemic bone loss. Interestingly, treatment with infliximab plus methotrexate arrested bone loss at the hip but not at the spine or hand as measured by BMD or dual energy X-ray absorptiometry.

Assessment of urine and serum biomarkers and histomorphometric analysis of bone specimens have been used to gain insights into the mechanisms involved in the systemic bone loss and remodeling disequilibrium in patients with RA. Analysis of bone biopsies from patients with RA indicate that the major cause of the decrease in bone mass is attributable to a decrease in bone formation rather than to an increase in bone resorption [92–94]. Results obtained from analysis of biochemical markers of bone turnover have yielded differing results, depending on the duration of the disease and the disease activity [86, 95–98]. Gough and co-workers [99] evaluated urinary markers of bone resorption and correlated these changes with the change in BMD at multiple sites in a group of patients with RA of less than 2 years' duration. They found that the markers of bone resorption were increased in subjects with RA compared to age-matched controls. Other investigators have made similar observations, indicating the presence of increased systemic bone resorption in early RA, and there is evidence that the systemic bone loss correlates with disease activity [86, 95, 97, 100]. Evaluation of serum and urine biomarkers of bone remodeling by Garnero and co-workers provides evidence that the increase in bone resorption observed in patients with early RA is accompanied by decreased bone formation, which is consistent with an uncoupling of systemic bone resorption and formation [101].

Evaluation of serum cytokine levels have provided additional insights into the mechanisms involved in the pathogenesis of systemic bone loss. Vis and co-workers [90] evaluated BMD, bone biomarkers, and serum OPG and RANKL levels in RA patients treated with infliximab for one year in an open label study. Urinary carboxyterminal telopeptide of type I collagen (CTx) and serum RANKL levels decreased and these changes correlated with the decrease in disease activity. There were minimal changes in serum OPG levels, although the RANKL/OPG ratio progressively decreased. The markers of bone formation increased initially but were not changed at the later time points. Geusens and co-workers [102] assessed RANKL and OPG levels and markers of disease activity in a group of patients with early RA who were treated with sulfasalazine alone or the combination of sulfasalazine, methrotrexate, and an initial course of high-dose prednisolone. They found that the initial pretreatment erythrocyte sedimentation rate and OPG/RANKL ratio independently predicted the 5-year radiographic progression of joint damage. They also found that the radiographic progression was greatest in patients with the highest ESR and lowest OPG/RANKL ratio.

In addition to RANKL and OPG, numerous studies have assessed serum levels of proinflammatory cytokines to gain insights into the mechanisms responsible for the systemic bone loss in RA. Rooney and co-workers recently reported a comprehensive review of synovial tissue and serum biomarkers of disease activity, therapeutic clinical response, and radiographic progression [103]. Patients with active RA were randomized to treatment with anakinra (an IL-1 receptor antagonist) alone or in combination with pegsunercept (a pegylated soluble TNF type I receptor that functionally acts as a TNF blocker). Arthroscopic synovoial biopsies and serum mediator levels were assessed at several time points. They found that T-cell infiltration, vascularity, and TGF-β expression correlated with the clinical course. In the serum, IL-6, matrix metalloproteinase-1 (MMP-1), MMP-3, and tissue inhibitor of metalloproteinase-1 (TIMP-1) levels exhibited predictive value with respect to the clinical disease activity. Of the serum markers, TIMP-1 at the initiation of therapy was predictive of a later therapeutic response. The extent of T-cell infiltration and the expression in the synovium of vascular endothelial growth factor (VEGF) and TGF-β pretreatment and serum levels of IL-6, IL-8, MMP-1, TIMP-1, soluble(s) TNF receptor I, sTNF receptor II, and IL-18 were correlated with radiographic progression. Of note, the levels of biomarker expression in the serum were independent of the levels measured in the synovium. Although these reports provide insights into the pathologic cellular and molecular events that are involved in deregulated bone remodeling in RA, more information and further study are needed to correlate these changes with specific biologic processes and to identify biomarkers that definitively predict clinical outcomes or permit rational approaches to therapeutic interventions.

## Seronegative spondyloarthropathies

The seronegative spondyloarthopathies include ankylosing spondylitis, reactive arthritis, and arthritis associated with psoriasis or inflammatory bowel disease, and juvenile-onset spondyloarthropathy. These forms of inflammatory arthritis exhibit extra-articular manifestations that differ from RA, and in contrast to RA, the joint inflammation affects the axial skeleton in addition to the diarthrodial joints. Importantly, the pattern and pathophysiology of peri-articular bone remodeling in the seronegative spondyloarthropathies differ from RA. Characteristically, the joint inflammation is asymmetrical, and the proximal as well as distal joints, which are not typically affected in RA, may also be involved. McGonagle and co-workers [104] utilized MRI and anatomic and histopathologic analysis to establish that the entheses (areas of tendinous or ligamentous attachment to bone) are the initial sites of inflammation in the spondyloarthropathies. The inflammatory infiltrate in the enthesis then appears to extend to the synovial lining, which exhibits many of the same histopathological features that characterize the joint pathology in RA, including synovial hyperplasia, lymphoid infiltration, and pannus formation. The earliest peri-articular bone resorption affects the insertion sites of juxta-articular tendon or ligamentous insertions. Subsequently, the extension of the inflammatory process to the joint margins and the development of synovial pannus may be accompanied by the development of marginal joint erosions, but in contrast to the marginal joint erosion in RA, this local inflammatory process may lead to calcification and ossification of the enthesis and adjacent periosteal bone (Figure 10-5).

As described above, Diarra and co-workers [67] analyzed sera from patients with ankylosing spondylitis and noted that in contrast to patients with RA, the levels of the Wnt pathway and bone-formation-inhibitor sclerostin were not elevated. More recently, this group of investigators [105] provided further evidence that sclerostin serum levels are reduced in patients with ankylosing spondylitis compared to healthy controls. Using serial serum samples they were able to provide longitudinal correlation of serum

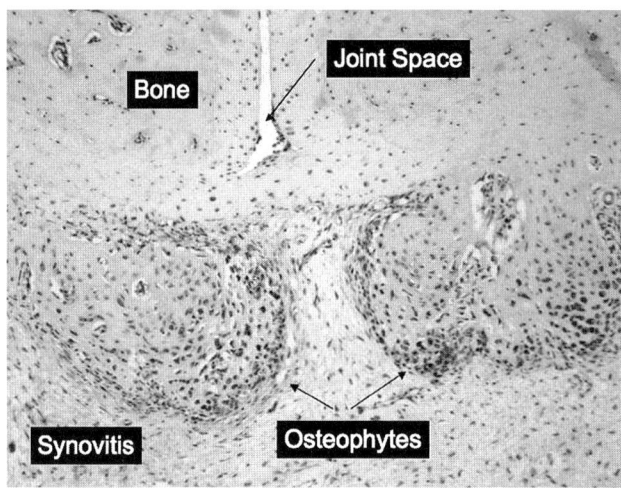

FIGURE 10-5 Osteophytes emerging from the peri-articular periosteum of a rat induced for adjuvant arthritis. The two lesions are surrounded by inflammatory tissue (synovitis). Brown-stained cells show expression of the proliferation marker Ki-67. Please refer to color plate section.

sclerostin levels with the subsequent formation of syndesmophytes and noted that low serum levels were significantly associated with the formation of new syndesmophytes. Syndesmophytes represent sites of ossification at the margins of vertebral bodies that develop in association with entheseal inflammation, and are a radiographic hallmark of the spondyloarthropathies. They also analyzed bone tissue from the joints of patients with ankylosing spondylitis and noted the virtual absence of sclerostin expression in osteocytes, providing a potential explanation for the enhanced bone formation.

To gain insights into the mechanisms involved in the excessive bone formation associated with ankylosing spondylitis, Braun and co-workers [106] studied the inflamed sacroiliac joints of patients with spondylarthritis, using immunohistological and in situ hybridization techniques. Analysis of synovial tissue revealed dense infiltrates of CD14-positive macrophages and $CD4^+$ and $CD8T^+$ lymphocytes. They also noted localized nodules containing active foci of endochondral ossification within the pannus. In situ hybridization revealed multiple inflammatory cells that expressed abundant message for TNF-$\alpha$ (but not IL-1$\beta$). Message for TGF-$\beta_2$ was noted in close proximity to the regions of new bone formation. Lories et al. [107] examined synovial biopsies from patients with spondylarthritis and RA for the expression of cytokines and bone morphogenic proteins (BMPs). They detected high levels of BMP-2 and BMP-6 expression in synovial fibroblasts and macrophages in the synovial tissue of patients with spondylarthritis and RA but not the controls, without inflammatory arthritis. The detection of elevated levels of these bone growth factors in the RA synovium was somewhat surprising in light of the absence of new bone formation at sites of synovial inflammation. These authors speculated that the new bone formation, associated with spondylarthritis such as ankylosing spondylitis, could be related to the anatomic site of the inflammatory process, which in spondylarthritis also involves the enthesis and adjacent periosteum. They also performed in vitro studies and showed that treatment of synovial fibroblasts with TNF-α or IL-1 increased BMP-2 and BMP-6 production.

More recently Schett and co-workers [108] examined the relationship between synovial inflammation and peri-articular bone formation in the murine models of collagen-induced and adjuvant arthritis. Although these models of inflammatory arthritis recapitulate many of the features of RA, unlike RA, there is evidence of extensive periosteal peri-articular new bone formation adjacent to sites of joint inflammation. In this respect, the new bone formation resembles the features of the spondyloarthropathies. They compared untreated mice to mice with either form of inflammatory arthritis or to mice treated with anti-TNF therapy or OPG. Parameters that they evaluated included: the sequential effects on joint inflammation, articular erosions, and peri-articular periosteal new bone formation. They found that periosteal new bone formation occurred within several days after the onset of the synovial inflammation and was preceded by the appearance of increased numbers of osteoclasts on the bone surfaces, which were associated with the development of small articular bone erosions. The authors speculated that the new bone formation was related to the activity of mesenchymal cells that appeared to be derived from the periosteum. The new bone formation was associated with the presence of hypertrophic chondrocytes that were forming new bone via endochondral ossification. Of note, they found that inhibition of TNF or RANKL did not prevent the growth of the new bone. This led the authors to conclude that the initial inflammatory process was sufficient to initiate the development of bone formation and this process was not dependent on the persistence of synovial inflammation or osteoclast-mediated bone resorption. This conclusion is supported by the observations of Lories and co-workers [109] in the DBA/1 model of inflammatory arthritis. These mice develop a spontaneous form of arthritis with features suggestive of psoriatic spondylarthritis, which is accompanied by extensive periosteal and peri-articular bone formation. As in the studies by Schett and co-workers, treatment of these mice with anti-TNF therapy, which reduces inflammation, did not affect the periosteal bone formation.

Studies by Maksymowych and co-workers [110] suggest that there is a relationship between inflammation and new bone formation in patients with spondylarthritis at least in the axial skeleton. These authors performed MRI and radiographs of the spine in a placebo-controlled trial of anti-TNF therapy in patients with ankylosing spondylarthritis. Their goal was to assess the effect that inhibition of marginal vertebral body inflammation had on the development of de novo syndesmophyte formation. They found that new syndesmophytes developed more frequently at the vertebral body margins with inflammation than at sites without inflammation and, in addition, that syndesmophytes developed more frequently at sites of previous inflammation, despite resolution of the inflammation after anti-TNF therapy. They concluded that inflammation played an essential role in the development of the syndesmophyte, but importantly, that resolution of inflammation did not prevent the subsequent development of new bone formation.

These conclusions are supported by results reported by van der Heijde and co-workers [111, 112], who showed that anti-TNF therapy, although producing significant benefit with respect to axial skeletal inflammation, did not prevent the development of syndesmophytes.

Vandooren and co-workers [113] have examined the relationship between expression of RANKL and OPG in patients with spondylarthritis to determine the effects of TNF blockade. The authors obtained synovial biopsies from patients with psoriatic and non-psoriatic spondylarthritis or RA and examined the tissue for expression of RANKL, OPG, RANK, and TRAP-positive osteoclasts, using semiquantitative scoring and digital image analysis. They also obtained paired biopsies from a subset of patients with spondylarthritis before and after anti-TNF therapy. They found RANKL expression was associated primarily with the synovial fibroblasts, as well as a minor subset of CD3 T cells, and demonstrated that the levels of RANKL, RANK, and OPG did not differ in the synovial tissue from patients with spondylarthritis or RA. Of note, however, TNF blockade did decrease the levels of synovial fibroblast expression of RANKL in a subset of patients with the best clinical response, providing a potential explanation for the beneficial effects of anti-TNF therapy on joint erosions in this subset of patients.

Despite the tendency of patients with spondylarthritis to exhibit increased local bone formation at sites of joint and vertebral body inflammation, in many patients there is evidence of axial skeletal osteopenia. The decrease in bone mass is most frequently observed in patients with spinal ankylosis in whom the bone loss has been attributed to immobilization [114]. However, studies by Will et al. [115] suggest that additional factors may be involved. They evaluated 25 patients with early ankylosing spondylitis, using dual-photon absorptiometry, and compared them to age- and sex-matched controls. The patients with spondylarthritis demonstrated a significant reduction in bone mineral density in the lumbosacral spine and hip even in the absence of bony ankylosis, and the authors concluded that generalized bone loss can occur early in disease before the development of spinal immobility related to the adverse effects of systemic inflammation on bone remodeling. This conclusion is supported by the observations of Frediani et al. [116], who compared BMD in a control population and a cohort of patients with psoriatic arthritis without axial skeletal involvement. Even in the absence of spondylarthritis, a majority of the patients with inflammatory peripheral psoriatic arthritis exhibited decreased BMD in the spine and hip. They noted an association with the functional status, patient age, and the number of years since menopause, although the decrease in BMD did not correlate with indices of inflammation or disease duration.

Several studies have demonstrated that the axial skeletal osteoporosis associated with the spondyloarthropathies is associated with an increased risk of fracture [117, 118]. Ralston et al. [118] prospectively evaluated a cohort of patients with spondylarthritis and observed that 15 of 111 patients developed radiographic evidence of vertebral compression fractures. Of note, systemic bone density in these patients was not decreased, suggesting that the osteoporosis was primarily localized to the axial spine.

## Acknowledgment

This work was supported in part by National Institute of Health Grant NIAMS R01 AR45472 and ACR-REF Within Our Reach: Finding a Cure for Rheumatoid Arthritis.

## References

[1] Goldring M, Goldring S. Biology of the normal joint. In: Firestein GS, Budd RG, Harris ED, McInnes IB, Ruddy S, Sergent JS, editors. Kelly's Textbook of Medicine. 8th ed. Philadelphia: Saunders Elsevier; 2009. p. 37–69.

[2] Krane SM, Conca W, Stephenson ML, Amento EP, Goldring MB. Mechanisms of matrix degradation in rheumatoid arthritis. Ann NY Acad Sci 1990;580:340–54.

[3] Gravallese EM, Goldring SR. Cellular mechanisms and the role of cytokines in bone erosions in rheumatoid arthritis. Arthritis Theum 2000;43:2143–51.

[4] McGonagle D, Benjamin M, Marzo-Ortega H, Emery P. Advances in the understanding of entheseal inflammation. Curr Rheumatol Rep 2002;4(6):500–6.

[5] Goldring SR. Pathogenesis of bone erosions in rheumatoid arthritis. Curr Opin Rheumatol 2002;14:406–10.

[6] Goldring SR, Gravallese EM. Pathogenesis of bone erosions in rheumatoid arthritis. Curr Opin Rheumatol 2000;12(3):195–9.

[7] Sato K, Takayanagi H. Osteoclasts, rheumatoid arthritis, and osteoimmunology. Curr Opin Rheumatol 2006;18(4):419–26.

[8] Schett G. Joint remodelling in inflammatory disease. Ann Rheum Dis 2007;66(Suppl. 3):iii42–4.

[9] Schett G. Osteoimmunology in rheumatic diseases. Arthritis Res Ther 2009;11(1):210.

[10] Schett G, Teitelbaum SL. Osteoclasts and arthritis. J Bone Miner Res 2009;24(7):1142–6.

[11] Deodhar AA, Woolf AD. Bone mass measurement and bone metabolism in rheumatoid arthritis: a review. Br J Rheumatol 1996;35(4):309–22.

[12] Goldring SR. Bone loss in chronic inflammatory conditions J Musculoskelet Neuronal Interact 2003;3(4):287–9, discussion 292–4.

[13] Goldring SR, Gravallese EM. Mechanisms of bone loss in inflammatory arthritis: diagnosis and therapeutic implications. Arthritis Research 2000;2:33–7.

[14] Schett G. Erosive arthritis. Arthritis Res Ther 2007;9(Suppl 1):S2.

[15] Walsh NC, Crotti TN, Goldring SR, Gravallese EM. Rheumatic diseases: the effects of inflammation on bone. Immunol Rev 2005;208:228–51.

[16] Bromley M, Woolley DE. Histopathology of the rheumatoid lesion; identification of cell types at sites of cartilage erosion. Arthritis Rheum 1984;27:857–63.

[17] Bromley M, Woolley DE. Chondroclasts and osteoclasts at subchondral sites of erosion in the rheumatoid joint. Arthritis Rheum 1984;27(9):968–75.

[18] Gravallese EM, Harada Y, Wang JT, Gorn AH, Thornhill TS, Goldring SR. Identification of cell types responsible for bone resorption in rheumatoid arthritis and juvenile rheumatoid arthritis. Am J Pathol 1998;152(4):943–51.

[19] Chang JS, Quinn JM, Demaziere A, et al. Bone resorption by cells isolated from rheumatoid synovium. Ann Rheum Dis 1992;51(11):1223–9.

[20] Pap T, Claus A, Ohtsu S, et al. Osteoclast-independent bone resorption by fibroblast-like cells. Arthritis Res Ther 2003;5(3):R163–73.

[21] Kong YY, Feige U, Sarosi I, et al. Activated T cells regulate bone loss and joint destruction in adjuvant arthritis through osteoprotegerin ligand. Nature 1999;402(6759):304–9.

[22] Redlich K, Hayer S, Maier A, et al. Tumor necrosis factor alpha-mediated joint destruction is inhibited by targeting osteoclasts with osteoprotegerin. Arthritis Rheum 2002;46:785–92.

[23] Romas E, Sims NA, Hards DK, et al. Osteoprotegerin reduces osteoclast numbers and prevents bone erosion in collagen-induced arthritis. Am J Pathol 2002;161(4):1419–27.

[24] Pettit AR, Ji H, von Stechow D, et al. TRANCE/RANKL knockout mice are protected from bone erosion in a serum transfer model of arthritis. Am J Pathol 2001;159(5):1689–99.

[25] Matsumoto I, Staub A, Benoist C, Mathis D. Arthritis provoked by linked T and B cell recognition of a glycolytic enzyme. Science 1999;286(5445):1732–5.

[26] Redlich K, Hayer S, Ricci R, et al. Osteoclasts are essential for TNF-alpha-mediated joint destruction. JCI 2002;110:1419–27.

[27] Grigoriadis AE, Wang ZQ, Cecchini MG, et al. c-Fos: a key regulator of osteoclast-macrophage lineage determination and bone remodeling. Science 1994;266(5184):443–8.

[28] Li P, Schwarz EM, O'Keefe RJ, Ma L, Boyce BF, Xing L. RANK signaling is not required for TNFalpha-mediated increase in CD11(hi) osteoclast precursors but is essential for mature osteoclast formation in TNFalpha-mediated inflammatory arthritis. J Bone Miner Res 2004;19(2):207–13.

[29] Cohen SB, Dore RK, Lane NE, et al. Denosumab treatment effects on structural damage, bone mineral density, and bone turnover in rheumatoid arthritis: a twelve-month, multicenter, randomized, double-blind, placebo-controlled, phase II clinical trial. Arthritis Rheum 2008;58(5):1299–309.

[30] Jarrett SJ, Conaghan PG, Sloan VS, et al. Preliminary evidence for a structural benefit of the new bisphosphonate zoledronic acid in early rheumatoid arthritis. Arthritis Rheum 2006;54(5):1410–14.

[31] Goldring SR, Gravallese EM. Bisphosphonates: environmental protection for the joint? Arthritis Rheum 2004;50(7):2044–7.

[32] Gravallese EM, Manning C, Tsay A, et al. Synovial tissue in rheumatoid arthritis is a source of osteoclast differentiation factor. Arthritis Rheum 2000;43(2):250–8.

[33] Horwood NJ, Kartsogiannis V, Quinn JM, Romas E, Martin TJ, Gillespie MT. Activated T lymphocytes support osteoclast formation in vitro. Biochem Biophys Res Commun 1999;265(1):144–50.

[34] Kotake S, Udagawa N, Hakoda M, et al. Activated human T cells directly induce osteoclastogenesis from human monocytes: possible role of T cells in bone destruction in rheumatoid arthritis patients. Arthritis Rheum 2001;44(5):1003–12.

[35] Takayanagi H, Ogasawara K, Hida S, et al. T-cell-mediated regulation of osteoclastogenesis by signalling cross-talk between RANKL and IFN-gamma. Nature 2000;408(6812):600–5.

[36] Takayanagi H, Iizuka H, Juji T, et al. Involvement of receptor activator of nuclear factor kappa-B ligand/osteoclast differentiation factor in osteoclastogenesis from synoviocytes in rheumatoid arthritis. Arthritis Rheum 2000;43:259–69.

[37] Weitzmann MN, Cenci S, Rifas L, Haug J, Dipersio J, Pacifici R. T cell activation induces human osteoclast formation via receptor activator of nuclear factor kappaB ligand-dependent and -independent mechanisms. J Bone Miner Res 2001;16(2):328–37.

[38] Anderson DM, Maraskovsky E, Billingsley WL, et al. A homologue of the TNF receptor and its ligand enhance T-cell growth and dendritic-cell function. Nature 1997;390(6656):175–9.

[39] Josien R, Wong BR, Li HL, Steinman RM, Choi Y. TRANCE, a TNF family member, is differentially expressed on T cell subsets and induces cytokine production in dendritic cells. J Immunol 1999;162(5):2562–8.

[40] Wong BR, Josien R, Lee SY, et al. TRANCE (tumor necrosis factor [TNF]-related activation-induced cytokine), a new TNF family member predominantly expressed in T cells, is a dendritic cell-specific survival factor. J Exp Med 1997;186(12):2075–80.

[41] Adamopoulos IE, Bowman EP. Immune regulation of bone loss by Th17 cells. Arthritis Res Ther 2008;10(5):225.

[42] Miossec P, Korn T, Kuchroo VK. Interleukin-17 and type 17 helper T cells. N Engl J Med 2009;361 (9):888–98.

[43] Sato K, Suematsu A, Okamoto K, et al. Th17 functions as an osteoclastogenic helper T cell subset that links T cell activation and bone destruction. J Exp Med 2006;203(12):2673–82.

[44] Klareskog L, Catrina AI, Paget S. Rheumatoid arthritis. Lancet 2009;373(9664):659–72.

[45] Takayanagi H. Osteoimmunology and the effects of the immune system on bone. Nat Rev Rheumatol 2009.

[46] Teitelbaum SL. Osteoclasts: what do they do and how do they do it? Am J Pathol 2007;170(2): 427–35.

[47] Herman S, Muller RB, Kronke G, et al. Induction of osteoclast-associated receptor, a key osteoclast costimulation molecule, in rheumatoid arthritis. Arthritis Rheum 2008;58(10):3041–50.

[48] Takayanagi H. Osteoimmunology: shared mechanisms and crosstalk between the immune and bone systems. Nat Rev Immunol 2007;7(4):292–304.

[49] Wei S, Kitaura H, Zhou P, Ross FP, Teitelbaum SL. IL-1 mediates TNF-induced osteoclastogenesis. J Clin Invest 2005;115(2):282–90.

[50] Zwerina J, Redlich K, Polzer K, et al. TNF-induced structural joint damage is mediated by IL-1. Proc Natl Acad Sci U S A 2007;104(28):11742–7.

[51] Axmann R, Bohm C, Kronke G, Zwerina J, Smolen J, Schett G. Inhibition of interleukin-6 receptor directly blocks osteoclast formation in vitro and in vivo. Arthritis Rheum 2009;60(9):2747–56.

[52] Smolen JS, van der Heijde DM, Aletaha D, et al. Progression of radiographic joint damage in rheumatoid arthritis: independence of erosions and joint space narrowing. Ann Rheum Dis 2009;68(10):1535–40.

[53] Esensten JH, Wofsy D, Bluestone JA. Regulatory T cells as therapeutic targets in rheumatoid arthritis. Nat Rev Rheumatol 2009;5(10):560–5.

[54] Kim YG, Lee CK, Nah SS, Mun SH, Yoo B, Moon HB. Human CD4+CD25+ regulatory T cells inhibit the differentiation of osteoclasts from peripheral blood mononuclear cells. Biochem Biophys Res Commun 2007;357(4):1046–52.

[55] Zaiss MM, Axmann R, Zwerina J, et al. Treg cells suppress osteoclast formation: a new link between the immune system and bone. Arthritis Rheum 2007;56(12):4104–12.

[56] Ochi S, Shinohara M, Sato K, et al. Pathological role of osteoclast costimulation in arthritis-induced bone loss. Proc Natl Acad Sci U S A 2007;104(27):11394–9.

[57] Itonaga I, Fujikawa Y, Sabokbar A, Murray DW, Athanasou NA. Rheumatoid arthritis synovial macrophage-osteoclast differentiation is osteoprotegerin ligand-dependent. J Pathol 2000; 192(1):97–104.

[58] Suzuki Y, Tsutsumi Y, Nakagawa M, et al. Osteoclast-like cells in an in vitro model of bone destruction by rheumatoid synovium. Rheumatology (Oxford) 2001;40(6):673–82.

[59] Burman A, Haworth O, Bradfield P, et al. The role of leukocyte-stromal interactions in chronic inflammatory joint disease. Joint Bone Spine 2005;72(1):10–16.

[60] Hintzen C, Quaiser S, Pap T, Heinrich PC, Hermanns HM. Induction of CCL13 expression in synovial fibroblasts highlights a significant role of oncostatin M in rheumatoid arthritis. Arthritis Rheum 2009;60(7):1932–43.

[61] Page G, Lebecque S, Miossec P. Anatomic localization of immature and mature dendritic cells in an ectopic lymphoid organ: correlation with selective chemokine expression in rheumatoid synovium. J Immunol 2002;168(10):5333–41.

[62] Zheng W, Li R, Pan H, et al. Role of osteopontin in induction of monocyte chemoattractant protein 1 and macrophage inflammatory protein 1beta through the NF-kappaB and MAPK pathways in rheumatoid arthritis. Arthritis Rheum 2009;60(7):1957–65.

[63] Jimenez-Boj E, Nobauer-Huhmann I, Hanslik-Schnabel B, et al. Bone erosions and bone marrow edema as defined by magnetic resonance imaging reflect true bone marrow inflammation in rheumatoid arthritis. Arthritis Rheum 2007;56(4):1118–24.

[64] Haavardsholm EA, Boyesen P, Ostergaard M, Schildvold A, Kvien TK. Magnetic resonance imaging findings in 84 patients with early rheumatoid arthritis: bone marrow oedema predicts erosive progression. Ann Rheum Dis 2008;67(6):794–800.

[65] Hetland ML, Ejbjerg B, Horslev-Petersen K, et al. MRI bone oedema is the strongest predictor of subsequent radiographic progression in early rheumatoid arthritis. Results from a 2-year randomised controlled trial (CIMESTRA). Ann Rheum Dis 2009;68(3):384–90.

[66] Goldring SR, Goldring MB. Eating bone or adding it: the Wnt pathway decides. Nat Med 2007;13(2):133–4.

[67] Diarra D, Stolina M, Polzer K, et al. Dickkopf-1 is a master regulator of joint remodeling. Nat Med 2007;13(2):156–63.

[68] Walsh NC, Reinwald S, Manning CA, et al. Osteoblast function is compromised at sites of focal bone erosion in inflammatory arthritis. J Bone Miner Res 2009;24(9):1572–85.

[69] Goldring SR. Periarticular bone changes in rheumatoid arthritis: pathophysiological implications and clinical utility. Ann Rheum Dis 2009;68(3):297–9.

[70] Hoff M, Haugeberg G, Odegard S, et al. Cortical hand bone loss after 1 year in early rheumatoid arthritis predicts radiographic hand joint damage at 5-year and 10-year follow-up. Ann Rheum Dis 2009;68(3):324–9.

[71] Haugeberg G, Green MJ, Conaghan PG, et al. Hand bone densitometry: a more sensitive standard for the assessment of early bone damage in rheumatoid arthritis. Ann Rheum Dis 2007;66(11):1513–17.

[72] Stewart A, Mackenzie LM, Black AJ, Reid DM. Predicting erosive disease in rheumatoid arthritis. A longitudinal study of changes in bone density using digital X-ray radiogrammetry: a pilot study. Rheumatology (Oxford) 2004;43(12):1561–4.

[73] Shimizu S, Shiozawa S, Shiozawa K, Imura S, Fujita T. Quantitative histologic studies on the pathogenesis of periarticular osteoporosis in rheumatoid arthritis. Arthritis Rheum 1985;28(1):25–31.

[74] Frost HM. Bone's mechanostat: a 2003 update. Anat Rec A Discov Mol Cell Evol Biol 2003;275(2):1081–101.

[75] Guler-Yuksel M, Allaart CF, Goekoop-Ruiterman YPM, et al. Changes in hand and generalized bone mineral density in patients with recent-onset rheumatoid arthritis. Ann Rheum Dis 2008.

[76] Devlin J, Lilley J, Gough A, et al. Clinical associations of dual-energy X-ray absorptiometry measurement of hand bone mass in rheumatoid arthritis. Br J Rheumatol 1996;35(12):1256–62.

[77] Forslind K, Keller C, Svensson B, Hafstrom I. Reduced bone mineral density in early rheumatoid arthritis is associated with radiological joint damage at baseline and after 2 years in women. J Rheumatol 2003;30(12):2590–6.

[78] Haugeberg G, Orstavik RE, Kvien TK. Effects of rheumatoid arthritis on bone. Curr Opin Rheumatol 2003;15(4):469–75.

[79] Lodder MC, de Jong Z, Kostense PJ, et al. Bone mineral density in patients with rheumatoid arthritis: relation between disease severity and low bone mineral density. Ann Rheum Dis 2004;63(12):1576–80.

[80] Orstavik RE, Haugeberg G, Mowinckel P, et al. Vertebral deformities in rheumatoid arthritis: a comparison with population-based controls. Arch Intern Med 2004;164(4):420–5.

[81] Orstavik RE, Haugeberg G, Uhlig T, et al. Incidence of vertebral deformities in 255 female rheumatoid arthritis patients measured by morphometric X-ray absorptiometry. Osteoporos Int 2005;16(1):35–42.

[82] Woolf AD. Osteoporosis in rheumatoid arthritis – the clinical viewpoint. Br J Rheum 1991;30:82–4.

[83] Sambrook PN, Spector TD, Seeman E, et al. Osteoporosis in rheumatoid arthritis. A monozygotic co-twin control study. Arthritis Rheum 1995;38(6):806–9.

[84] van Staa TP, Geusens P, Bijlsma JW, Leufkens HG, Cooper C. Clinical assessment of the long-term risk of fracture in patients with rheumatoid arthritis. Arthritis Rheum 2006;54(10):3104–12.

[85] Solomon DH, Finkelstein JS, Shadick N, et al. The relationship between focal erosions and generalized osteoporosis in postmenopausal women with rheumatoid arthritis. Arthritis Rheum 2009;60(6):1624–31.

[86] Gough AK, Peel NF, Eastell R, Holder RL, Lilley J, Emery P. Excretion of pyridinium crosslinks correlates with disease activity and appendicular bone loss in early rheumatoid arthritis. Ann Rheum Dis 1994;53:14–17.

[87] Als OS, Gotfredsen A, Riis BJ, Christiansen C. Are disease duration and degree of functional impairment determinanats of bone loss in rheumatoid arthritis? Ann Rhem Dis 1985;406–11.

[88] Haugeberg G, Conaghan PG, Quinn M, Emery P. Bone loss in patients with active early rheumatoid arthritis: infliximab and methotrexate compared with methotrexate treatment alone. Explorative analysis from a 12-month randomised, double-blind, placebo-controlled study. Ann Rheum Dis 2009;68(12):1898–901.

[89] Hoff M, Kvien TK, Kalvesten J, Elden A, Haugeberg G. Adalimumab therapy reduces hand bone loss in early rheumatoid arthritis: explorative analyses from the PREMIER study. Ann Rheum Dis 2009;68(7):1171–6.

[90] Vis M, Havaardsholm EA, Haugeberg G, et al. Evaluation of bone mineral density, bone metabolism, osteoprotegerin and receptor activator of the NFkappaB ligand serum levels during treatment with infliximab in patients with rheumatoid arthritis. Ann Rheum Dis 2006;65(11):1495–9.

[91] Guler-Yuksel M, Allaart CF, Goekoop-Ruiterman YP, et al. Changes in hand and generalised bone mineral density in patients with recent-onset rheumatoid arthritis. Ann Rheum Dis 2009;68(3):330–6.

[92] Compston JE, Vedi S, Croucher PI, Garrahan NJ, O'Sullivan MM. Bone turnover in non-steroid treated rheumatoid arthritis. Ann Rheum Dis 1994;53:163–6.

[93] Kroger H, Arnala I, Alhava EM. Bone remodeling in osteoporosis associated with rheumatoid arthritis. Calcif Tiss Int 1991;49:S90.

[94] Mellish RWE, O'Sullivan MM, Garrahan NJ, Compston JE. Iliac crest trabecular bone mass and structure in patients with non-steroid treated rheumatoid arthritis. Ann Rheum Dis 1987;46:830–6.

[95] Garnero P, Landewe R, Boers M, et al. Association of baseline levels of markers of bone and cartilage degradation with long-term progression of joint damage in patients with early rheumatoid arthritis: the COBRA study. Arthritis Rheum 2002;46(11):2847–56.

[96] Hall GM, Spector TD, Delmas PD. Markers of bone metabolism in postmenopausal women with rheumatoid arthritis. Effects of corticosteroids and hormone replacement therapy. Arthritis Rheum 1995;38:902–6.

[97] Iwamoto J, Takeda T, Ichimura S. Urinary cross-linked N-telopeptides of type I collagen levels in patients with rheumatoid arthritis. Calcif Tissue Int 2003.

[98] Wislowska M, Jakubicz D, Stepien K, Cicha M. Serum concentrations of formation (PINP) and resorption (Ctx) bone turnover markers in rheumatoid arthritis. Rheumatol Int 2009;29(12):1403–9.

[99] Gough A, Sambrook P, Devlin J, et al. Osteoclastic activation is the principal mechanism leading to secondary osteoporosis in rheumatoid arthritis. J Rheumatol 1998;25(7):1282–9.

[100] Sinigaglia L, Varenna M, Binelli L, et al. Urinary and synovial pyridinium crosslink concentrations in patients with rheumatoid arthritis and osteoarthritis. Ann Rheum Dis 1995;54(2):144–7.

[101] Garnero P, Jouvenne P, Buchs N, Delmas PD, Miossec P. Uncoupling of bone metabolism in rheumatoid arthritis patients with or without joint destruction: assessment with serum type I collagen breakdown products. Bone 1999;24(4):381–5.

[102] Geusens PP, Landewe RB, Garnero P, et al. The ratio of circulating osteoprotegerin to RANKL in early rheumatoid arthritis predicts later joint destruction. Arthritis Rheum 2006;54(6):1772–7.

[103] Rooney T, Roux-Lombard P, Veale DJ, Fitzgerald O, Dayer JM, Bresnihan B. Synovial tissue and serum biomarkers of disease activity, therapeutic response and radiographic progression. Analysis of a proof-of-concept randomized clinical trial of cytokine blockade. Ann Rheum Dis 2009.

[104] McGonagle D, Tan AL, Moller Dohn U, Ostergaard M, Benjamin M. Microanatomic studies to define predictive factors for the topography of periarticular erosion formation in inflammatory arthritis. Arthritis Rheum 2009;60(4):1042–51.

[105] Appel H, Ruiz-Heiland G, Listing J, et al. Altered skeletal expression of sclerostin and its link to radiographic progression in ankylosing spondylitis. Arthritis Rheum 2009;60(11):3257–62.

[106] Braun J, Bollow M, Neure L, et al. Use of immunohistologic and in situ hybridization techniques in the examination of sacroiliac joint biopsy specimens from patients with ankylosing spondylitis. Arthritis Rheum 1995;4:499–505.

[107] Lories RJ, Derese I, Ceuppens JL, Luyten FP. Bone morphogenetic proteins 2 and 6, expressed in arthritic synovium, are regulated by proinflammatory cytokines and differentially modulate fibroblast-like synoviocyte apoptosis. Arthritis Rheum 2003;48(10):2807–18.

[108] Schett G, Stolina M, Dwyer D, et al. Tumor necrosis factor alpha and RANKL blockade cannot halt bony spur formation in experimental inflammatory arthritis. Arthritis Rheum 2009;60(9):2644–54.

[109] Lories RJ, Matthys P, de Vlam K, Derese I, Luyten FP. Ankylosing enthesitis, dactylitis, and onychoperiostitis in male DBA/1 mice: a model of psoriatic arthritis. Ann Rheum Dis 2004;63(5):595–8.

[110] Maksymowych WP, Chiowchanwisawakit P, Clare T, Pedersen SJ, Ostergaard M, Lambert RG. Inflammatory lesions of the spine on magnetic resonance imaging predict the development of new syndesmophytes in ankylosing spondylitis: evidence of a relationship between inflammation and new bone formation. Arthritis Rheum 2009;60(1):93–102.

[111] van der Heijde D, Landewe R, Baraliakos X, et al. Radiographic findings following two years of infliximab therapy in patients with ankylosing spondylitis. Arthritis Rheum 2008;58(10):3063–70.

[112] van der Heijde D, Landewe R, Einstein S, et al. Radiographic progression of ankylosing spondylitis after up to two years of treatment with etanercept. Arthritis Rheum 2008;58(5):1324–31.

[113] Vandooren B, Cantaert T, Noordenbos T, Tak PP, Baeten D. The abundant synovial expression of the RANK/RANKL/Osteoprotegerin system in peripheral spondylarthritis is partially disconnected from inflammation. Arthritis Rheum 2008;58(3):718–29.

[114] Spencer D, Park W, Dick H, et al. Radiologic manifestations in 200 patients with ankylosing spondylitis; correlations with clinical features and HLA-B27. J Rheumatol 1979;6:305.

[115] Will R, Bhalla A, Palmer R, Ring F, Calin A. Osteoporosis in early ankylosing spondylitis; a primary pathological event? Lancet 1989;23:1483–5.

[116] Frediani B, Allegri A, Falsetti P, et al. Bone mineral density in patients with psoriatic arthritis. J Rheumatol 2001;28(1):138–43.

[117] Geusens P, Vosse D, van der Linden S. Osteoporosis and vertebral fractures in ankylosing spondylitis. Curr Opin Rheumatol 2007;19(4):335–9.

[118] Ralston SH, Urquhart GD, Brzeski M, Sturrock RD. Prevalence of vertebral compression fractures due to osteoporosis in ankylosing spondylitis. BMJ 1990;300:563–5.

# 11

# Inflammatory Bowel Disease and Bone

Francisco A. Sylvester[1], Anthony T. Vella[2]

[1]UNIVERSITY OF CONNECTICUT SCHOOL OF MEDICINE, HARTFORD, CT, USA,
[2]UNIVERSITY OF CONNECTICUT SCHOOL OF MEDICINE, FARMINGTON, CT, USA

## CHAPTER OUTLINE

| | |
|---|---|
| Introduction | 325 |
| The intestinal immune system | 326 |
| Innate immune functions and IBD pathogenesis | 327 |
| The adaptive immune system and IBD | 328 |
| Immunological differences between Crohn's disease and ulcerative colitis | 329 |
| IBD and osteoimmunology | 330 |
| Effects of IBD on bone mass | 330 |
| Mechanisms by which IBD affects bone mass | 331 |
|     General disease-related and treatment factors | 331 |
|     Disease-related immune factors | 332 |
|     RANKL and OPG | 334 |
| Conclusion | 335 |
| References | 335 |

## Introduction

Inflammatory bowel disease encompasses two main entities, Crohn's disease and ulcerative colitis. IBD affects approximately 1.2 million Americans, and affects children and adults. Both ulcerative colitis and Crohn's disease are associated with significant bone mass deficits. Although both cause chronic intestinal inflammation, they have important clinical and pathogenic differences that are relevant to how they affect the skeleton. Crohn's disease is characterized by patchy inflammation of the full thickness of the bowel wall, and can affect any part of the gastrointestinal tract. Most commonly, Crohn's disease localizes to the distal ileum and colon. Inflammation can perforate the intestinal wall and produce abscesses and fistulization into neighboring organs or the skin. Ulcerative colitis on the other hand typically produces continuous inflammation that begins in the rectum, which classically is limited to the lamina propria of the colon.

Although both Crohn's disease and ulcerative colitis can have extra-intestinal manifestations, Crohn's disease tends to affect bone mass more frequently and severely than ulcerative colitis, especially in growing children [1].

## The intestinal immune system

To understand the relationships between the intestinal immune system and bone metabolism in IBD, we will first describe its functions in health and disease. Mouse models of IBD have been invaluable in deriving much of this knowledge. In general, these models are characterized by inflammation of the colon, which may be acute and self-resolving or chronic and non-remitting, depending on the model [2]. Many of these models depend on bacterial colonization of the intestine, in the absence of which inflammation does not develop. This underscores the importance of bowel flora in the pathogenesis of IBD, as detailed below.

The intestine hosts the largest mass of immune cells in the body. Its luminal surface is exposed to the biggest load of foreign antigens. These come from multiple sources, including food, bacteria, and xenobiotics. These antigens are normally confined to the luminal surface of the intestine by a single layer of cells. This layer includes epithelial cells, goblet cells, and Paneth cells, each with specialized functions. These cells form a barrier that is both mechanical and functional. Epithelial cells adhere to each other with tight junctions. Mucus secreted by goblet cells forms a protective layer. Paneth cells secrete antibiotic peptides (e.g., α-defensins) into the mucus that control the concentration and adherence of bacteria to the surface epithelium. The mucus also contains secretory IgA, glycolytic and proteolytic enzymes that also help regulate the intestinal bacterial ecosystem and prevent bacterial penetration into the lamina propria [3].

Entry of bacteria and soluble antigens into the mucosa is normally very selective and carefully controlled. Specialized areas of the mucosa contain gut-associated lymphoid tissue (GALT) overlaid by modified intestinal epithelial cells called microfold (M) cells. The GALT holds the largest concentration of lymphocytes in the body. M cells capture, transport macromolecules and transfer them over to GALT dendritic cells. These cells then migrate to mesenteric lymph nodes that drain lymph from the intestine, and process and present antigen to naïve T cells. The complex luminal environment of the intestine is also sampled directly by intraepithelial projections of lamina propria dendritic cells. These cells interact with intestinal immune cells, which eliminate threats or become tolerant to these antigens. This cross-talk involves the generation of T regulatory cells (Treg), which dampen the inflammatory response. Resident macrophages act like scavengers to remove bacteria and debris, remarkably without activating an immune response. B cells undergo an isotype switch and secrete IgA into the lumen. In summary, physical confinement of the intestinal microbiota to the intestinal lumen, limitation of activation of innate immune cells, induction of adaptive immune tolerance to commensal microbes, and direct inhibition of effector T-cell differentiation and function ensure a mutually beneficial relationship between the host and the gut microbiota under normal or steady-state conditions.

For reasons that are not well understood, this delicate balance is perturbed in inflammatory bowel diseases (IBD). The primary events that lead to IBD involve a breakdown in the epithelial cell barrier, loss of tolerance to commensal intestinal bacteria, dysfunction of the innate immune system, and hyperactivation of the adaptive immune system. Patients with Crohn's disease and their healthy first-degree relatives appear to have a primary increase in mucosal permeability [4, 5], which may increase exposure to luminal antigens in the lamina propria. Recently, it has been proposed that defective macrophage function may be a basic pathogenic mechanism for IBD. In this model, large numbers of microorganisms that gain access to the intestinal mucosa are not effectively cleared, triggering secondary activation of T cells, tissue damage, and clinical symptoms [6]. This hypothesis is plausible, since primary defects in phagocytes such as chronic granulomatous disease, Hermansky-Pudlak syndrome, and glycogen storage disease-Ib can present with IBD-like features [7–9]. Secondary T-cell activation is largely responsible for IBD chronicity, as discussed in the next section.

## Innate immune functions and IBD pathogenesis

As mentioned, there is mounting evidence that defects in the innate immune system are critical to the development of IBD. The innate system is constituted by multiple cell types (dendritic cells, macrophages, neutrophils, eosinophils, natural killer T cells, and $\gamma\delta$ T cells). These cells have receptors called pattern recognition receptors (PRRs) that detect molecules common to multiple microorganisms, termed pathogen-associated molecular patterns (PAMPs) [10]. These invariable and predetermined receptors are encoded in the germ-line. PAMPs are recognized by PRRs such as Toll-like receptors (TLRs). Unlike the state of "controlled inflammation" described above, where macrophages clear microbes without activating inflammation, in the context of IBD activated macrophages express TLRs [11, 12]. Their recognition and binding to PAMPs stimulates the production of cytokines that amplifies the immune response leading to the activation of the adaptive immune system. Nucleotide-binding oligomerization domain (NOD)-2, which recognizes bacterial muramyl dipeptide inside of the cells [13], is defective in 20% of Caucasians with Crohn's disease. Loss-of-function mutations associated with three NOD2 polymorphisms are associated with stricturing and penetrating Crohn's disease of the distal ileum [14, 15].

Intestinal dendritic cells directly sample the luminal environment of the intestine, or receive antigen from M cells. Under normal circumstances, they process and present these antigens via MHC to T cells in the draining mesenteric lymph nodes. These intestinal dendritic cells express low levels of co-stimulators and cytokines, typically resulting in T-cell tolerance. In response, T cells produce cytokines (IL-4, IL-10, TGF-$\beta$) that promote the differentiation of Tregs, which dampen inflammation. However, in IBD dendritic cells, which are activated by PAMPs that migrate to the GALT, express high levels of cytokines and co-stimulatory molecules. These, in turn, induce effector T-cell differentiation [16]. Importantly, they stimulate expression of $\alpha4\beta7$, a gut-homing integrin in T cells in a process that involves retinoic acid [17]. As a result, activated T cells can reach the circulation and then return to the gut. A portion of these cells can migrate and stay in the bone marrow, and become activated by circulating dendritic cells, loaded with bacterial antigens [18]. This local immune response may have important implications for the function of bone cells in patients with IBD.

Although intestinal epithelial cells are not considered part of the classic innate immune system, they are also immunocompetent. They express TLR2, 4, 9 [19] and NOD1 and 2 [20, 21], so they can directly sense intestinal bacteria. These interactions strengthen the epithelial cell barrier and protect against the onset of colitis.

## The adaptive immune system and IBD

T cells represent the adaptive immune system in the intestine. They reside in the lamina propria and in the intraepithelial compartment. T cells that express the αβ T-cell receptor (TCR) reside in the lamina propria, whereas γδ TCR cells can be found in the intra-epithelial space. CD4+ αβ T cells are critical to the pathogenesis of IBD. In mesenteric lymph nodes αβ TCRs recognize cognate antigens presented by dendritic cells that have migrated from the intestinal mucosa after bacterial invasion or sampling of the luminal content. This constitutes signal 1 for T-cell activation, but a second signal is needed to fully activate the cell. Signal 2 is also present in the membrane of the antigen-presenting cell in the form of CD80 (B7.1) or CD86 (B7.2), which bind to CD28 on the surface of the T cell. However, binding of CD80/CD86 to CTLA-4 on the T-cell membrane inhibits T-cell activation. In addition, antigen binding to the TCR without co-stimulation typically renders the cell incapable of reacting to the antigen upon future re-exposure (a state called anergy) [2].

Activation of T cells in a defined cytokine context in the mesenteric lymph node directs them to a specific effector T helper (Th) phenotype. In the presence of IL-12 and IFN (interferon)-γ, T cells sequentially activate STAT (Signal Transducers and Activators of Transcription)-1 and STAT-4, acquire the ability to secrete IFN-γ, and are referred to as Th1 cells, which control intracellular pathogens through activation of innate cells like macrophages. IL-27 may also play a role in the differentiation of Th1 cells, especially in the intestine. The signature transcription factor for Th1 cells is T-bet (T-box transcription factor expressed in T cells). Th1 cells characteristically express CCR1, CCR5, and CXCR3 on their surface. On the other hand, Th2 cells are generated in the presence of IL-4, which activates STAT-6 and expression of the transcription factor GATA3. Th2 cells can contribute to allergic responses and help combat parasites through the synthesis of IL-4, IL-5, and IL-13. They can be recognized by their expression of the surface receptors CCR3 and CCR4. IL-6 and TGF-β direct cells toward a Th17 phenotype, which fight off fungi and extracellular bacteria. Th17 cells secrete IL-17, IL-21, IL-22, recruit neutrophils to the site of inflammation, and enhance epithelial barrier function. They are characterized by expression of the transcription factor ROR (retinoic acid receptor-related orphan nuclear receptor)-γt and the surface markers CCR4, CCR6, and IL-23R. IL-23 is a key cytokine in mediating IBD, which sustains a pathogenic pool of Th17 cells [22, 23]. Polymorphisms of the IL23R confer various levels of susceptibility to IBD [24]. IL-21 is also regarded as an important factor in human IBD perhaps by deregulating tolerance to microbial gut flora [25]. A fourth Th subpopulation called T follicular helper cells – T(FH) – secrete IL-21, express CXCR5 and Bcl-6 [26], and migrate into B-cell follicles where they provide support for B-cell function. They are also detected in gut-associated Peyer's patches in a CD155-dependent manner [27]. Although Th1 and Th17 cells are present in small numbers in the non-inflamed lamina propria, CD4+ T cells that produce IL-10 and Treg are abundant [28]. These cells suppress the intestinal immune response despite permanent exposure to an enormous load of foreign antigens.

These T-cell phenotypes are not fixed, and trans-differentiation may occur in response to changes in luminal environmental conditions [29, 30]. For example, Tregs have been shown to differentiate into T(FH) cells in Peyer's patches [27]. In addition, "dual expressors" of IFN-γ and Th17 have been reported [31].

Once activated, these effector T cells leave the mesenteric lymph nodes, enter the circulation, and can home to sites containing their cognate antigens. T cells activated by intestinal dendritic cells are imprinted to return to their original inductive site. They acquire mucosal addressins, including integrins and chemokine receptors, that enable them to home back to the intestine. For example, the integrin α4β7 binds to MAdCAM (Mucosal Addressin Cellular Adhesion Molecule)-1 and α4 to VCAM (Vascular Cell Adhesion Molecule)-1 in the vascular endothelium of specialized lamina propria post-capillary venules. The chemokine receptor CCR9 attaches to small intestinal epithelium CCL25 [32], CCR10 to colonic epithelial CCL28 [33], and CCR6 to CCL20 in small intestinal Peyer's patches [34]. These and other receptors are upregulated during intestinal inflammation, promoting cellular infiltration of specific mucosal sites. Activated T cells then secrete cytokines that increase intestinal and vascular permeability, blood flow, the influx of inflammatory cells, and the production of oxygen radicals, matrix metalloproteinases, and other enzymes, resulting in tissue damage and symptoms of IBD.

T cells express TLRs and can directly recognize PAMPs. There is experimental evidence that this can result in T-cell proliferation and production of pro-inflammatory factors in the intestinal lamina propria without the need for antigen presentation by dendritic cells or macrophages. This may be another mechanism by which microorganisms can regulate inflammation [35].

## Immunological differences between Crohn's disease and ulcerative colitis

Until recently, Crohn's disease was considered a Th1-mediated chronic inflammatory disease. This was based on the presence of CD4+ T cells expressing T-bet and IFN-γ in the inflamed intestine of mice and humans [36], the ability of an IFN-γ antibody to treat early IBD in mice, and the lack of colitis development in immunodeficient mice that received T-bet-deficient T cells. However, this paradigm was recently challenged by the observations that key molecules associated with the development, function, and maintenance of Th17 cells are abundant in the intestines of patients with Crohn's disease. Th17 cells are more numerous in the inflamed colon of IL-10-null mice and mice adoptively transferred with CD45RB$^{Hi}$ cells [22, 37]. Elson et al. have observed that Th17 cells, reactive to intestinal bacteria, produce colitis at much lower doses than Th1 cells in adoptive transfer models of colitis [23]. IL-23 is a critical cytokine in the perpetuation of the pathogenic Th17 cells pool, and plays other important Th17-independent roles in the pathogenesis of IBD. Genes involved in the IL-23/Th17 system such as IL-23R modulate susceptibility to IBD [38]. It is of interest that blockade of the common IL-12/IL-23 p40 subunit [39] and IL-23p19 ameliorate intestinal inflammation [23]. These hypotheses may not be mutually exclusive. It is possible that Th1 cells initiate the aberrant immune response, and that Th17 cells perpetuate it.

Compared to Crohn's disease, there is a paucity of data about the pathogenesis of ulcerative colitis, due to a relative lack of adequate experimental models. The inflammation in ulcerative colitis is considered a "modified Th2 response", given the abundant presence of IL-13 and other Th2-type cytokines, like IL-5 [40]. NKT cells, which mature in the thymus in response to lipid antigen presentation by the Cd1d complex, are the main source of IL-13 in ulcerative colitis [41]. In the oxazolone mouse colitis model, which is Th2-dependent, IL-4 is important in the early phase of inflammation, whereas IL-13 drives chronic disease [41]. The IL-23/Th17 system appears to be important in the development of ulcerative colitis as well [38].

## IBD and osteoimmunology

As mentioned above, IBD is characterized by a vigorous, chronic immune response to luminal microbial antigens. This response is produced by activated T and B cells, and results in the production of a variety of pro-inflammatory factors that damage the intestinal lining. It is well documented that both children and adults with IBD frequently have bone mass deficits, even before treatment is started. In this section we will describe the impact of IBD on bone health and possible inflammatory mechanisms by which IBD affects bone mass.

## Effects of IBD on bone mass

IBD can be diagnosed at any age, but in children it is most commonly diagnosed during adolescence, a time normally characterized by rapid linear growth and bone mineral accrual. Crohn's disease in particular is frequently associated with growth stunting and pubertal delay, which can affect the acquisition of bone mass. Laboratory, clinical, and bone biopsy evidence suggests that bone formation is markedly decreased in children with Crohn's disease at diagnosis, before anti-inflammatory treatment is instituted, while bone resorption is also reduced [1, 42, 43]. The second peak in incidence of IBD is in young adults in the fourth decade of life. In these patients there is bone biomarker evidence of decreased bone formation and increased bone resorption [44–46]. Therefore, intestinal inflammation appears to primarily suppress bone formation, with uncoupling of bone remodeling in adults.

The prevalence of bone mass deficits varies widely in IBD, depending on the population studied and how bone mass was measured [47]. The frequency of severe bone demineralization is ~10–20% in adults and children. Disease-related factors that are independent of treatment probably play an important role, since bone mass deficits are present from the time of diagnosis [1, 48]. In children with newly diagnosed Crohn's disease, the tibia exhibits decreased trabecular bone mineral density, expanded endocortical surface and reduced periosteal circumference, resulting in a mechanically weaker bone [48]. However, it is not clear whether these biomechanical alterations result in an increased risk of fractures in patients with IBD. Retrospective population-based cohort studies of fractures requiring hospital care in adults with IBD range from no increase to moderate increase compared to the general population [49–54]. However, these studies did not account for clinically silent vertebral fractures. Case series suggest that these may be present in up to 1:5 adults with IBD [55, 56], even in patients with normal BMD [57]. Vertebral fractures have also been reported in children with Crohn's disease [58]. More studies are needed to define the true risk of these fractures in adults and children.

## Mechanisms by which IBD affects bone mass

### General disease-related and treatment factors

The pathogenesis of bone loss in IBD is complex and likely involves disease and treatment factors. IBD, especially Crohn's disease, is commonly associated with decreased calorie-protein and micronutrient intake, which can adversely affect bone accrual and maintenance [59, 60]. Vitamin D deficiency is prevalent in patients with IBD, especially in northern latitudes [61], which may impair the intestinal absorption of calcium, and in the rodent model of colitis exacerbate inflammation [62]. Vitamin K, a co-factor in the γ-carboxylation of the glutamic acid of osteocalcin, may be deficient in IBD as well, affecting bone mineralization [63, 64]. Therefore, malnutrition may impair skeletal health in IBD.

Sex steroids and the growth hormone/insulin-like growth factor-1 (IGF-1) axis are frequently disturbed in active IBD. Delayed puberty is common in adolescent children with Crohn's disease and may affect the acquisition of peak bone mass [65]. Hypogonadism can be seen in adults and may impair bone remodeling [66]. Malnutrition and a decrease in growth hormone signaling in the liver and peripheral tissues act synergistically to decrease serum and tissue IGF-1 in children with active intestinal inflammation [67]. Low IGF-1 may impair longitudinal growth by chondrocytes, expansion of the outer cortical layer by periosteal osteoblasts, and recruitment of undifferentiated stromal cells into the osteoblast lineage [68, 69].

Large mechanical forces from skeletal muscles are important for the normal development and maintenance of bone mass [70]. Significant deficits in lean body mass, a surrogate for muscle mass, are prevalent in newly diagnosed children with Crohn's disease and persist despite clinical improvement [71, 72]. Sarcopenia may be a consequence of inflammation, malnutrition, glucocorticoid use, and decreased physical activity.

Medications to treat IBD can also impact bone cell function. Glucocorticoids (GC) are commonly used to induce remission in IBD. GC can inhibit bone formation, secondarily increase bone resorption, decrease muscle mass, and can increase fracture risk even at small doses [54, 73]. However, by decreasing inflammation, increasing appetite and weight gain and general well-being, GC may have positive effects on bone health as well. For this reason the role of GC in inducing bone deficits in IBD remains controversial. In addition, it is difficult to separate the effects of the underlying IBD from those of GC in the individual patient [74]. Infliximab, a TNF-α-specific antibody that is used to treat refractory IBD, improves biomarkers of bone formation in adults [75] and children [76], suggesting that adequate control of inflammation improves bone formation. Calcineurin inhibitors are used primarily in hospitalized patients with ulcerative colitis unresponsive to intravenous GC. Calcineurin is a calcium-binding phosphatase that regulates osteoblast and osteoclast function [77]. A substrate of calcineurin, nuclear factor of activated T cells (NFAT), is critically important in osteoblast and osteoclast differentiation [78]. Therefore, calcineurin inhibitors may modulate the development of osteoblasts and osteoclasts in patients with IBD.

## Disease-related immune factors

As described above, the intestine in IBD harbors activated epithelial cells, dendritic cells, macrophages, and T and B cells that secrete a vast array of inflammatory factors. These can spill over into the circulation and affect bone cell function (Figure 11-1). For example, it is well documented that serum IL-6 is elevated in active IBD [79, 80]. Serum from newly diagnosed children with Crohn's disease decreases bone formation in bone explants [81]. Antibody neutralization of serum IL-6 reverses the effects of serum in this model [43], suggesting that IL-6 decreases bone remodeling in IBD. TNF-α is a central cytokine in the pathogenesis of IBD. Mice injected with TNF-α and mice with TNBS (trinitrobenzene sulfonic acid)-induced colitis, have decreased expression of osteoblast genes involved in mineralization. Systemic administration of anti-TNF-α antibodies rescues this expression [82]. In humans, TNF-α-specific antibodies produce a significant increase in serum biomarkers of bone formation, especially in children with severe IBD [76]. Other proinflammatory factors produced by the inflamed intestine, such as IL-23 and IFN-γ, may also modulate bone cell function.

Importantly, it is not clear whether the beneficial effects of TNF-α neutralization on bone are secondary to a non-specific decrease in intestinal inflammation, to direct effects in the bone micro-environment, or both. Experimental evidence suggests that bone marrow cells become activated in the presence of intestinal inflammation and secrete TNF-α [83]. Although the mechanisms by which this happens are not yet understood,

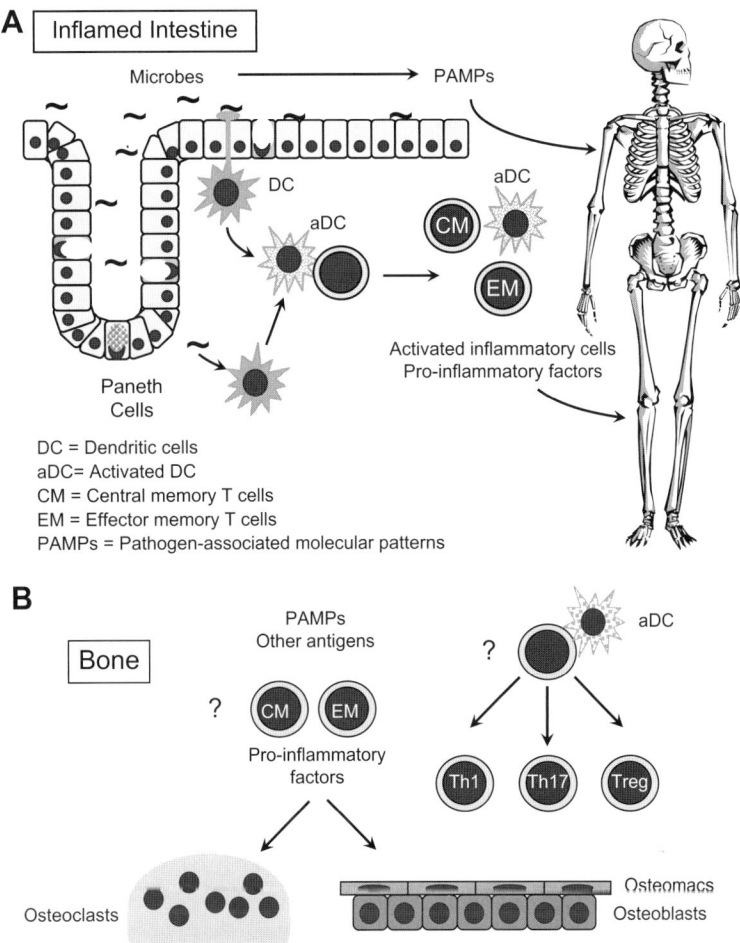

**FIGURE 11-1 (A)** The intestine is colonized with over 100 trillion microorganisms, which are contained in its lumen by a single layer of epithelial cells. In the steady state, the host tolerates this enormous antigenic load due to special adaptations of dendritic cells (DC), macrophages, and the presence of regulatory T cells. In IBD, these immune-suppressive mechanisms are lost, and dendritic cells, macrophages, and T cells become activated. T cells acquire immunological memory for their cognate antigen (central memory and effector memory cells), and intestinal mucosal addressins. T memory cells and dendritic cells leave mesenteric lymph nodes, Peyer's patches and intestinal lymphoid follicles, enter the circulation and home back to the intestine or establish residence in the bone marrow. In addition, pro-inflammatory factors, endotoxin, flagellin, and other bacterial products may enter the circulation and gain access to the bone microenvironment. **(B)** In bone, dendritic cells that carry antigen may activate the differentiation of T cells into effector cell lineages. The commitment to Th1, Th17, or Th2 cells will depend on the cytokine milieu. In addition, microbial antigens may be taken up by resident bone cells like osteomacs and activate immune responses that affect bone cell function. IBD is associated with innate immune defects that may facilitate the survival of gut-derived microorganisms, and generate vigorous adaptive immune response. Please refer to color plate section.

there are several possibilities. The bone marrow contains central memory CD8+ T cells that are capable of responding to antigen [84]. Circulating dendritic cells originating from the inflamed gut and loaded with antigen may encounter these cells and activate them [18, 85, 86]. The bone marrow can be the site where T cells first encounter antigen [87], so it is possible that PAMPs and other microbial antigens leaked by the intestine into the systemic circulation reach the bone marrow and after processing turn these cells on. Another subset of cells in the bone marrow may also be capable of cytokine secretion after antigen stimulation. This is a pool of effector memory CD4+ cells that originates in the inflamed gut and is maintained in the bone marrow by IL-7-producing cells [88]. These cells are capable of "transmitting" colitis when injected into an immunodeficient mouse recipient. Ex vivo activation results in high levels of TNF-α and IFN-γ. Therefore, several possible mechanisms exist that may facilitate the local production of cytokines in the bone microenvironment, affecting bone cell function.

Another possibility is that resident dendritic cells differentiate into functional, resorbing osteoclasts in bone under inflammatory conditions. The lack of skeletal defects in dendritic-cell-deficient mice suggests that they probably do not play a role in bone remodeling under homeostatic conditions [89]. However, in the bone environment immune interactions between immature dendritic cells, CD4+ T cells and microbial products, which may occur in IBD can stimulate their development into functional osteoclasts in mice [90, 91].

Autophagy is a process by which cells degrade defective organelles, proteins, and bacteria. Defects in the autophagy-related genes *ATG16L1* and *IRGM* are associated with susceptibility to Crohn's disease, but not ulcerative colitis [92–94]. It is unclear whether alterations in autophagy play any role in bone physiology in Crohn's disease. However, it is theoretically possible that osteoclasts, which are related to phagocytic cells with normally active autophagy, may be affected; but this remains to be tested.

Recently discovered macrophage-like cells (osteomacs), which overlay osteoblasts in endosteal and periosteal surfaces may also be a target of inflammatory responses. These cells may be activated by circulating microbial antigens or by neighboring immune cells and influence bone formative activity [95].

Finally, bone cells themselves may act as immunocompetent partners. Both osteoblasts and osteoclasts are capable of secreting pro-inflammatory factors when stimulated by bacterial products and cytokines [96, 97]. They express Toll-like and NOD-like receptors upon exposure to PAMPs, and therefore can directly sense microbial-derived molecules [98–100]. In consequence, translocation of microbial products into the circulation of patients with IBD may affect the function of these cells directly.

## RANKL and OPG

RANKL (receptor activator of nuclear factor-κB-ligand) is essential for normal osteoclast formation. RANKL binds to its receptor RANK on osteoclast precursors and stimulates their differentiation and activity. Osteoprotegerin (OPG) is a soluble decoy receptor for

RANKL [101]. Additionally, RANKL has well-defined immunologic effects, described in detail elsewhere in this textbook [102]. The discovery of RANKL and the fact that it is produced by activated T cells [103] sparked a search for a role in the pathogenesis of bone loss in IBD. Ashcroft et al. reported in the IL-2 null mouse model of inflammation that RANKL increased the survival of intestinal dendritic cells. Furthermore, exogenous OPG decreased the number of these cells, ameliorated the severity of colitis, and improved bone density [85]. Although OPG also improved bone mass in an adoptive transfer model of colitis, it had no effect on intestinal inflammation [83]. Clinical studies performed in adult patients reveal that the expression of OPG and RANKL is upregulated in the inflamed intestine [104], and that the expression of OPG predicts response to anti-TNF-$\alpha$ therapy [105]. OPG is constitutively expressed by intestinal epithelial cells, and TNF-$\alpha$ increases its production [106]. OPG is also expressed by macrophages and dendritic cells in patients with IBD [107, 108], suggesting that it plays an immunomodulatory role. This function may be a consequence of OPG binding to the pro-apoptic molecule TRAIL (tumor necrosis factor-related apoptosis-inducing ligand) [109]. IBD is characterized by decreased apoptosis of activated immune cells, and it is possible that OPG prolongs the survival of these cells by neutralizing TRAIL [110]. OPG is modestly elevated in serum from patients with IBD [111]. It has been suggested that this represents either a physiologic response to increased bone resorption or is a marker of active inflammation [112, 113].

## Conclusion

IBD is a complex clinical disorder, with various clinical phenotypes. It is likely that its effects on bone are mediated at least in part by inflammation. The skeleton may not be just a passive recipient of gut-derived inflammatory factors and cells, but an active participant. Thus immune reaction in bone may involve activated cells that originate in the intestine, and/or resident cells. In addition, primary defects in innate immunity and autophagy may affect the function of osteoblasts and osteoclasts. Much remains to be understood about these mechanisms, and the clinical consequences of bone mass deficits, including fractures.

## References

[1] Sylvester FA, Wyzga N, Hyams JS, Davis PM, Lerer T, Vance K, Hawker G, Griffiths AM. Natural history of bone metabolism and bone mineral density in children with inflammatory bowel disease. Inflamm Bowel Dis 2007;13:42–50.

[2] Maynard CL, Weaver CT. Intestinal effector T cells in health and disease. Immunity 2009;31: 389–400.

[3] Nieuwenhuis EE, Blumberg RS. The role of the epithelial barrier in inflammatory bowel disease. Adv Exp Med Biol 2006;579:108–16.

[4] Benjamin J, Makharia GK, Joshi YK. Association between intestinal permeability and anti-Saccharomyces cerevisiae antibodies in patients with Crohn's disease. Inflamm Bowel Dis 2008;14:1610–11.

[5] Hollander D, Vadheim CM, Brettholz E, Petersen GM, Delahunty T, Rotter JI. Increased intestinal permeability in patients with Crohn's disease and their relatives. A possible etiologic factor. Ann Intern Med 1986;105:883–5.

[6] Smith AM, Rahman FZ, Hayee B, Graham SJ, Marks DJ, Sewell GW, Palmer CD, Wilde J, Foxwell BM, Gloger IS, Sweeting T, Marsh M, Walker AP, Bloom SL, Segal AW. Disordered macrophage cytokine secretion underlies impaired acute inflammation and bacterial clearance in Crohn's disease. J Exp Med 2009;206:1883–97.

[7] Mahadeo R, Markowitz J, Fisher S, Daum F. Hermansky-Pudlak syndrome with granulomatous colitis in children. J Pediatr 1991;118:904–6.

[8] Marks DJ, Miyagi K, Rahman FZ, Novelli M, Bloom SL, Segal AW. Inflammatory bowel disease in CGD reproduces the clinicopathological features of Crohn's disease. Am J Gastroenterol 2009;104:117–24.

[9] Visser G, Rake JP, Fernandes J, Labrune P, Leonard JV, Moses S, Ullrich K, Smit GP. Neutropenia, neutrophil dysfunction, and inflammatory bowel disease in glycogen storage disease type Ib: results of the European Study on Glycogen Storage Disease type I. J Pediatr 2000;137:187–91.

[10] Medzhitov R, Janeway Jr CA. Innate immunity: the virtues of a nonclonal system of recognition. Cell 1997;91:295–8.

[11] Kamada N, Hisamatsu T, Honda H, Kobayashi T, Chinen H, Kitazume MT, Takayama T, Okamoto S, Koganei K, Sugita A, Kanai T, Hibi T. Human CD14+ macrophages in intestinal lamina propria exhibit potent antigen-presenting ability. J Immunol 2009;183:1724–31.

[12] Kamada N, Hisamatsu T, Okamoto S, Chinen H, Kobayashi T, Sato T, Sakuraba A, Kitazume MT, Sugita A, Koganei K, Akagawa KS, Hibi T. Unique CD14 intestinal macrophages contribute to the pathogenesis of Crohn disease via IL-23/IFN-gamma axis. J Clin Invest 2008;118:2269–80.

[13] Girardin SE, Boneca IG, Viala J, Chamaillard M, Labigne A, Thomas G, Philpott DJ, Sansonetti PJ. Nod2 is a general sensor of peptidoglycan through muramyl dipeptide (MDP) detection. J Biol Chem 2003;278:8869–72.

[14] Ogura Y, Bonen DK, Inohara N, Nicolae DL, Chen FF, Ramos R, Britton H, Moran T, Karaliuskas R, Duerr RH, Achkar JP, Brant SR, Bayless TM, Kirschner BS, Hanauer SB, Nunez G, Cho JH. A frameshift mutation in NOD2 associated with susceptibility to Crohn's disease. Nature 2001;411:603–6.

[15] Hugot JP, Chamaillard M, Zouali H, Lesage S, Cezard JP, Belaiche J, Almer S, Tysk C, O'Morain CA, Gassull M, Binder V, Finkel Y, Cortot A, Modigliani R, Laurent-Puig P, Gower-Rousseau C, Macry J, Colombel JF, Sahbatou M, Thomas G. Association of NOD2 leucine-rich repeat variants with susceptibility to Crohn's disease. Nature 2001;411:599–603.

[16] Hart AL, Al-Hassi HO, Rigby RJ, Bell SJ, Emmanuel AV, Knight SC, Kamm MA, Stagg AJ. Characteristics of intestinal dendritic cells in inflammatory bowel diseases. Gastroenterology 2005;129:50–65.

[17] Iwata M, Hirakiyama A, Eshima Y, Kagechika H, Kato C, Song S-Y. Retinoic acid imprints gut-homing specificity on T cells. Immunity 2004;21:527–38.

[18] Cavanagh LL, Bonasio R, Mazo IB, Halin C, Cheng G, van der Velden AW, Cariappa A, Chase C, Russell P, Starnbach MN, Koni PA, Pillai S, Weninger W, von Andrian UH. Activation of bone marrow-resident memory T cells by circulating, antigen-bearing dendritic cells. Nat Immunol 2005;6:1029–37.

[19] Ghadimi D, Vrese MD, Heller KJ, Schrezenmeir J. Effect of natural commensal-origin DNA on toll-like receptor 9 (TLR9) signaling cascade, chemokine IL-8 expression, and barrier integritiy of polarized intestinal epithelial cells. Inflamm Bowel Dis 2009.

[20] Barnich N, Aguirre JE, Reinecker HC, Xavier R, Podolsky DK. Membrane recruitment of NOD2 in intestinal epithelial cells is essential for nuclear factor-{kappa}B activation in muramyl dipeptide recognition. J Cell Biol 2005;170:21–6.

[21] Zilbauer M, Dorrell N, Elmi A, Lindley KJ, Schuller S, Jones HE, Klein NJ, Nunez G, Wren BW, Bajaj-Elliott M. A major role for intestinal epithelial nucleotide oligomerization domain 1 (NOD1) in eliciting host bactericidal immune responses to Campylobacter jejuni. Cell Microbiol 2007;9:2404–16.

[22] Yen D, Cheung J, Scheerens H, Poulet F, McClanahan T, McKenzie B, Kleinschek MA, Owyang A, Mattson J, Blumenschein W, Murphy E, Sathe M, Cua DJ, Kastelein RA, Rennick D. IL-23 is essential for T cell-mediated colitis and promotes inflammation via IL-17 and IL-6. J Clin Invest 2006;116:1310–16.

[23] Elson CO, Cong Y, Weaver CT, Schoeb TR, McClanahan TK, Fick RB, Kastelein RA. Monoclonal anti-interleukin 23 reverses active colitis in a T cell-mediated model in mice. Gastroenterology 2007;132:2359–70.

[24] Duerr RH, Taylor KD, Brant SR, Rioux JD, Silverberg MS, Daly MJ, Steinhart AH, Abraham C, Regueiro M, Griffiths A, Dassopoulos T, Bitton A, Yang H, Targan S, Datta LW, Kistner EO, Schumm LP, Lee AT, Gregersen PK, Barmada MM, Rotter JI, Nicolae DL, Cho JH. A genome-wide association study identifies IL23R as an inflammatory bowel disease gene. Science 2006;314:1461–3.

[25] Fantini MC, Monteleone G, MacDonald TT. IL-21 comes of age as a regulator of effector T cells in the gut. Mucosal Immunol 2008;1:110–15.

[26] Johnston RJ, Poholek AC, DiToro D, Yusuf I, Eto D, Barnett B, Dent AL, Craft J, Crotty S. Bcl6 and Blimp-1 are reciprocal and antagonistic regulators of T follicular helper cell differentiation. Science 2009;325:1006–10.

[27] Seth S, Ravens I, Kremmer E, Maier MK, Hadis U, Hardtke S, Forster R, Bernhardt G. Abundance of follicular helper T cells in Peyer's patches is modulated by CD155. Eur J Immunol 2009;39:3160–70.

[28] Maynard CL, Harrington LE, Janowski KM, Oliver JR, Zindl CL, Rudensky AY, Weaver CT. Regulatory T cells expressing interleukin 10 develop from Foxp3+ and Foxp3− precursor cells in the absence of interleukin 10. Nat Immunol 2007;8:931–41.

[29] Annunziato F, Cosmi L, Santarlasci V, Maggi L, Liotta F, Mazzinghi B, Parente E, Fili L, Ferri S, Frosali F, Giudici F, Romagnani P, Parronchi P, Tonelli F, Maggi E, Romagnani S. Phenotypic and functional features of human Th17 cells. J Exp Med 2007;204:1849–61.

[30] Jenner RG, Townsend MJ, Jackson I, Sun K, Bouwman RD, Young RA, Glimcher LH, Lord GM. The transcription factors T-bet and GATA-3 control alternative pathways of T-cell differentiation through a shared set of target genes. Proc Natl Acad Sci U S A 2009;106:17876–81.

[31] Lee YK, Mukasa R, Hatton RD, Weaver CT. Developmental plasticity of Th17 and Treg cells. Curr Opin Immunol 2009;21:274–80.

[32] Apostolaki M, Manoloukos M, Roulis M, Wurbel MA, Muller W, Papadakis KA, Kontoyiannis DL, Malissen B, Kollias G. Role of beta7 integrin and the chemokine/chemokine receptor pair CCL25/CCR9 in modeled TNF-dependent Crohn's disease. Gastroenterology 2008;134:2025–35.

[33] Ogawa H, Iimura M, Eckmann L, Kagnoff MF. Regulated production of the chemokine CCL28 in human colon epithelium. Am J Physiol Gastrointest Liver Physiol 2004;287:G1062–9.

[34] Wang C, Kang SG, Lee J, Sun Z, Kim CH. The roles of CCR6 in migration of Th17 cells and regulation of effector T-cell balance in the gut. Mucosal Immunol 2009;2:173–83.

[35] Maynard CL, Hatton RD, Helms WS, Oliver JR, Stephensen CB, Weaver CT. Contrasting roles for all-trans retinoic acid in TGF-β-mediated induction of Foxp3 and Il10 genes in developing regulatory T cells. J Exp Med 2009;206:343–57.

[36] Matsuoka K, Inoue N, Sato T, Okamoto S, Hisamatsu T, Kishi Y, Sakuraba A, Hitotsumatsu O, Ogata H, Koganei K, Fukushima T, Kanai T, Watanabe M, Ishii H, Hibi T. T-bet upregulation and subsequent interleukin 12 stimulation are essential for induction of Th1 mediated immunopathology in Crohn's disease. Gut 2004;53:1303–8.

[37] Izcue A, Hue S, Buonocore S, Arancibia-Carcamo CV, Ahern PP, Iwakura Y, Maloy KJ, Powrie F. Interleukin-23 restrains regulatory T cell activity to drive T cell-dependent colitis. Immunity 2008;28:559–70.

[38] Abraham C, Cho J. Interleukin-23/Th17 pathways and inflammatory bowel disease. Inflamm Bowel Dis 2009.

[39] Neurath MF, Fuss I, Kelsall BL, Stuber E, Strober W. Antibodies to interleukin 12 abrogate established experimental colitis in mice. J Exp Med 1995;182:1281–90.

[40] Fuss I, Neurath M, Boirivant M, Klein J, de la Motte C, Strong S, Fiocchi C, Strober W. Disparate CD4+ lamina propria (LP) lymphokine secretion profiles in inflammatory bowel disease. Crohn's disease LP cells manifest increased secretion of IFN-gamma, whereas ulcerative colitis LP cells manifest increased secretion of IL-5. J Immunol 1996;157:1261–70.

[41] Fuss IJ, Heller F, Boirivant M, Leon F, Yoshida M, Fichtner-Feigl S, Yang Z, Exley M, Kitani A, Blumberg RS, Mannon P, Strober W. Nonclassical CD1d-restricted NK T cells that produce IL-13 characterize an atypical Th2 response in ulcerative colitis. J Clin Invest 2004;113:1490–7.

[42] Ward LM, Rauch F, Matzinger MA, Benchimol EI, Boland M, Mack DR. Iliac bone histomorphometry in children with newly diagnosed inflammatory bowel disease. Osteoporos Int 2009.

[43] Sylvester FA, Wyzga N, Hyams JS, Gronowicz GA. Effect of Crohn's disease on bone metabolism in vitro: a role for interleukin-6. J Bone Miner Res 2002;17:695–702.

[44] Schoon EJ, Geerling BG, Van Dooren IM, Schurgers LJ, Vermeer C, Brummer RJ, Stockbrugger RW. Abnormal bone turnover in long-standing Crohn's disease in remission. Aliment Pharmacol Ther 2001;15:783–92.

[45] Robinson RJ, Iqbal SJ, Abrams K, Al-Azzawi F, Mayberry JF. Increased bone resorption in patients with Crohn's disease. Aliment Pharmacol Ther 1998;12:699–705.

[46] Schulte C, Dignass AU, Mann K, Goebell H. Reduced bone mineral density and unbalanced bone metabolism in patients with inflammatory bowel disease. Inflamm Bowel Dis 1998;4:268–75.

[47] Bernstein CN, Leslie WD, Leboff MS. AGA technical review on osteoporosis in gastrointestinal diseases. Gastroenterology 2003;124:795–841.

[48] Dubner SE, Shults J, Baldassano RN, Zemel BS, Thayu M, Burnham JM, Herskovitz RM, Howard KM, Leonard MB. Longitudinal assessment of bone density and structure in an incident cohort of children with Crohn's disease. Gastroenterology 2009;136:123–30.

[49] Bernstein CN, Blanchard JF, Leslie W, Wajda A, Yu BN. The incidence of fracture among patients with inflammatory bowel disease. A population-based cohort study. Ann Intern Med 2000;133:795–9.

[50] Vestergaard P, Krogh K, Rejnmark L, Laurberg S, Mosekilde L. Fracture risk is increased in Crohn's disease, but not in ulcerative colitis. Gut 2000;46:176–81.

[51] Vestergaard P, Mosekilde L. Fracture risk in patients with celiac disease, Crohn's disease, and ulcerative colitis: a nationwide follow-up study of 16,416 patients in Denmark. Am J Epidemiol 2002;156:1–10.

[52] Loftus Jr EV, Crowson CS, Sandborn WJ, Tremaine WJ, O'Fallon WM, Melton 3rd LJ. Long-term fracture risk in patients with Crohn's disease: a population-based study in Olmsted County, Minnesota. Gastroenterology 2002;123:468–75.

[53] Loftus Jr EV, Achenbach SJ, Sandborn WJ, Tremaine WJ, Oberg AL, Melton 3rd LJ. Risk of fracture in ulcerative colitis: a population-based study from Olmsted County, Minnesota. Clin Gastroenterol Hepatol 2003;1:465–73.

[54] van Staa TP, Cooper C, Leufkens HG, Bishop N. Children and the risk of fractures caused by oral corticosteroids. J Bone Miner Res 2003;18:913–18.

[55] Klaus J, Armbrecht G, Steinkamp M, Bruckel J, Rieber A, Adler G, Reinshagen M, Felsenberg D, von Tirpitz C. High prevalence of osteoporotic vertebral fractures in patients with Crohn's disease. Gut 2002;51:654–8.

[56] Heijckmann AC, Huijberts MS, Schoon EJ, Geusens P, de Vries J, Menheere PP, van der Veer E, Wolffenbuttel BH, Stockbrugger RW, Dumitrescu B, Nieuwenhuijzen Kruseman AC. High prevalence of morphometric vertebral deformities in patients with inflammatory bowel disease. Eur J Gastroenterol Hepatol 2008;20:740–7.

[57] Siffledeen JS, Siminoski K, Jen H, Fedorak RN. Vertebral fractures and role of low bone mineral density in Crohn's disease. Clin Gastroenterol Hepatol 2007;5:721–8.

[58] Semeao EJ, Stallings VA, Peck SN, Piccoli DA. Vertebral compression fractures in pediatric patients with Crohn's disease. Gastroenterology 1997;112:1710–13.

[59] Abrams SA, Silber TJ, Esteban NV, Vieira NE, Stuff JE, Meyers R, Majd M, Yergey AL. Mineral balance and bone turnover in adolescents with anorexia nervosa. J Pediatr 1993;123:326–31.

[60] Molgaard C, Thomsen BL, Michaelsen KF. Influence of weight, age and puberty on bone size and bone mineral content in healthy children and adolescents. Acta Paediatr 1998;87:494–9.

[61] Pappa HM, Gordon CM, Saslowsky TM, Zholudev A, Horr B, Shih MC, Grand RJ. Vitamin D status in children and young adults with inflammatory bowel disease. Pediatrics 2006;118:1950–61.

[62] Froicu M, Zhu Y, Cantorna MT. Vitamin D receptor is required to control gastrointestinal immunity in IL-10 knockout mice. Immunology 2006;117:310–18.

[63] Duggan P, O'Brien M, Kiely M, McCarthy J, Shanahan F, Cashman KD. Vitamin K status in patients with Crohn's disease and relationship to bone turnover. Am J Gastroenterol 2004;99:2178–85.

[64] Kuwabara A, Tanaka K, Tsugawa N, Nakase H, Tsuji H, Shide K, Kamao M, Chiba T, Inagaki N, Okano T, Kido S. High prevalence of vitamin K and D deficiency and decreased BMD in inflammatory bowel disease. Osteoporos Int 2009;20:935–42.

[65] Finkelstein JS, Neer RM, Biller BM, Crawford JD, Klibanski A. Osteopenia in men with a history of delayed puberty. N Engl J Med 1992;326:600–4.

[66] Clements D, Compston JE, Evans WD, Rhodes J. Hormone replacement therapy prevents bone loss in patients with inflammatory bowel disease. Gut 1993;34:1543–6.

[67] Difedele LM, He J, Bonkowski EL, Han X, Held MA, Bohan A, Menon RK, Denson LA. Tumor necrosis factor-α blockade restores growth hormone signaling in murine colitis. Gastroenterology 2005;128:1278–91.

[68] Yakar S, Rosen CJ, Beamer WG, Ackert-Bicknell CL, Wu Y, Liu JL, Ooi GT, Setser J, Frystyk J, Boisclair YR, LeRoith D. Circulating levels of IGF-1 directly regulate bone growth and density. J Clin Invest 2002;110:771–81.

[69] Rosen CJ. Insulin-like growth factor I and calcium balance: evolving concepts of an evolutionary process. Endocrinology 2003;144:4679–81.

[70] Frost HM. Bone "mass" and the "mechanostat": a proposal. Anat Rec 1987;219:1–9.

[71] Bechtold S, Alberer M, Arenz T, Putzker S, Filipiak-Pittroff B, Schwarz HP, Koletzko S. Reduced muscle mass and bone size in pediatric patients with inflammatory bowel disease. Inflamm Bowel Dis 2009.

[72] Sylvester FA, Leopold S, Lincoln M, Hyams JS, Griffiths AM, Lerer T. A two-year longitudinal study of persistent lean tissue deficits in children with Crohn's disease. Clin Gastroenterol Hepatol 2009;7:452–5.

[73] van Staa TP, Leufkens HG, Abenhaim L, Zhang B, Cooper C. Oral corticosteroids and fracture risk: relationship to daily and cumulative doses. Rheumatology (Oxford) 2000;39:1383–9.

[74] Leonard MB. Glucocorticoid-induced osteoporosis in children: impact of the underlying disease. Pediatrics 2007;119(Suppl 2):S166–74.

[75] Abreu MT, Geller JL, Vasiliauskas EA, Kam LY, Vora P, Martyak LA, Yang H, Hu B, Lin YC, Keenan G, Price J, Landers CJ, Adams JS, Targan SR. Treatment with infliximab is associated with increased markers of bone formation in patients with Crohn's disease. J Clin Gastroenterol 2006;40:55–63.

[76] Thayu M, Leonard MB, Hyams JS, Crandall WV, Kugathasan S, Otley AR, Olson A, Johanns J, Marano CW, Heuschkel RB, Veereman-Wauters G, Griffiths AM, Baldassano RN. Improvement in biomarkers of bone formation during infliximab therapy in pediatric Crohn's disease: results of the REACH study. Clin Gastroenterol Hepatol 2008;6:1378–84.

[77] Okamura H, Amorim BR, Wang J, Yoshida K, Haneji T. Calcineurin regulates phosphorylation status of transcription factor osterix. Biochem Biophys Res Commun 2009;379:440–4.

[78] Takayanagi H. The role of NFAT in osteoclast formation. Ann N Y Acad Sci 2007;1116:227–37.

[79] Stevens C, Walz G, Singaram C, Lipman ML, Zanker B, Muggia A, Antonioli D, Peppercorn MA, Strom TB. Tumor necrosis factor-alpha, interleukin-1 beta, and interleukin-6 expression in inflammatory bowel disease. Dig Dis Sci 1992;37:818–26.

[80] Kusugami K, Fukatsu A, Tanimoto M, Shinoda M, Haruta J, Kuroiwa A, Ina K, Kanayama K, Ando T, Matsuura T, et al. Elevation of interleukin-6 in inflammatory bowel disease is macrophage- and epithelial cell-dependent. Dig Dis Sci 1995;40:949–59.

[81] Hyams JS, Wyzga N, Kreutzer DL, Justinich CJ, Gronowicz GA. Alterations in bone metabolism in children with inflammatory bowel disease: an in vitro study. J Pediatr Gastroenterol Nutr 1997;24:289–95.

[82] Uno JK, Kolek OI, Hines ER, Xu H, Timmermann BN, Kiela PR. F.K.G. The role of tumor necrosis factor-$\alpha$ in down-regulation of osteoblast phex gene expression in experimental murine colitis. Gastroenterology 2006;131:497–509.

[83] Byrne FR, Morony S, Warmington K, Geng Z, Brown HL, Flores SA, Fiorino M, Yin SL, Hill D, Porkess V, Duryea D, Pretorius JK, Adamu S, Manuokian R, Danilenko DM, Sarosi I, Lacey DL, Kostenuik PJ, Senaldi G. CD4+CD45RBHi T cell transfer induced colitis in mice is accompanied by osteopenia which is treatable with recombinant human osteoprotegerin. Gut 2005;54:78–86.

[84] Mazo IB, Honczarenko M, Leung H, Cavanagh LL, Bonasio R, Weninger W, Engelke K, Xia L, McEver RP, Koni PA, Silberstein LE, von Andrian UH. Bone marrow is a major reservoir and site of recruitment for central memory CD8+ T cells. Immunity 2005;22:259–70.

[85] Ashcroft AJ, Cruickshank SM, Croucher PI, Perry MJ, Rollinson S, Lippitt JM, Child JA, Dunstan C, Felsburg PJ, Morgan GJ, Carding SR. Colonic dendritic cells, intestinal inflammation, and T cell-mediated bone destruction are modulated by recombinant osteoprotegerin. Immunity 2003;19:849–61.

[86] Alnaeeli M, Park J, Mahamed D, Penninger JM, Teng Y-TA. Dendritic cells at the osteo-immune interface: implications for inflammation-induced bone loss. J Bone Min Res 2007;22:775–80.

[87] Feuerer M, Beckhove P, Garbi N, Mahnke Y, Limmer A, Hommel M, Hammerling GJ, Kyewski B, Hamann A, Umansky V, Schirrmacher V. Bone marrow as a priming site for T-cell responses to blood-borne antigen. Nat Med 2003;9:1151–7.

[88] Nemoto Y, Kanai T, Makita S, Okamoto R, Totsuka T, Takeda K, Watanabe M. Bone marrow retaining colitogenic CD4+ T cells may be a pathogenic reservoir for chronic colitis. Gastroenterology 2007;132:176–89.

[89] McKenna HJ, Stocking KL, Miller RE, Brasel K, De Smedt T, Maraskovsky E, Maliszewski CR, Lynch DH, Smith J, Pulendran B, Roux ER, Teepe M, Lyman SD, Peschon JJ. Mice lacking flt3 ligand have deficient hematopoiesis affecting hematopoietic progenitor cells, dendritic cells, and natural killer cells. Blood 2000;95:3489–97.

[90] Alnaeeli M, Penninger JM, Teng YT. Immune interactions with CD4+ T cells promote the development of functional osteoclasts from murine CD11c+ dendritic cells. J Immunol 2006;177:3314–26.

[91] Wakkach A, Mansour A, Dacquin R, Coste E, Jurdic P, Carle GF, Blin-Wakkach C. Bone marrow microenvironment controls the in vivo differentiation of murine dendritic cells into osteoclasts. Blood 2008;112:5074–83.

[92] Hampe J, Franke A, Rosenstiel P, Till A, Teuber M, Huse K, Albrecht M, Mayr G, De La Vega FM, Briggs J, Gunther S, Prescott NJ, Onnie CM, Hasler R, Sipos B, Folsch UR, Lengauer T, Platzer M, Mathew CG, Krawczak M, Schreiber S. A genome-wide association scan of nonsynonymous SNPs identifies a susceptibility variant for Crohn disease in ATG16L1. Nat Genet 2007;39:207–11.

[93] Parkes M, Barrett JC, Prescott NJ, Tremelling M, Anderson CA, Fisher SA, Roberts RG, Nimmo ER, Cummings FR, Soars D, Drummond H, Lees CW, Khawaja SA, Bagnall R, Burke DA, Todhunter CE, Ahmad T, Onnie CM, McArdle W, Strachan D, Bethel G, Bryan C, Lewis CM, Deloukas P, Forbes A, Sanderson J, Jewell DP, Satsangi J, Mansfield JC, Cardon L, Mathew CG. Sequence variants in the autophagy gene IRGM and multiple other replicating loci contribute to Crohn's disease susceptibility. Nat Genet 2007;39:830–2.

[94] Fisher SA, Tremelling M, Anderson CA, Gwilliam R, Bumpstead S, Prescott NJ, Nimmo ER, Massey D, Berzuini C, Johnson C, Barrett JC, Cummings FR, Drummond H, Lees CW, Onnie CM, Hanson CE, Blaszczyk K, Inouye M, Ewels P, Ravindrarajah R, Keniry A, Hunt S, Carter M, Watkins N, Ouwehand W, Lewis CM, Cardon L, Lobo A, Forbes A, Sanderson J, Jewell DP, Mansfield JC, Deloukas P, Mathew CG, Parkes M, Satsangi J. Genetic determinants of ulcerative colitis include the ECM1 locus and five loci implicated in Crohn's disease. Nat Genet 2008;40:710–12.

[95] Chang MK, Raggatt LJ, Alexander KA, Kuliwaba JS, Fazzalari NL, Schroder K, Maylin ER, Ripoll VM, Hume DA, Pettit AR. Osteal tissue macrophages are intercalated throughout human and mouse bone lining tissues and regulate osteoblast function in vitro and in vivo. J Immunol 2008;181:1232–44.

[96] Shen F, Ruddy MJ, Plamondon P, Gaffen SL. Cytokines link osteoblasts and inflammation: microarray analysis of interleukin-17- and TNF-α-induced genes in bone cells. J Leukoc Biol 2005;77:388–99.

[97] Maruyama K, Sano G, Matsuo K. Murine osteoblasts respond to LPS and IFN-gamma similarly to macrophages. J Bone Miner Metab 2006;24:454–60.

[98] Kikuchi T, Matsuguchi T, Tsuboi N, Mitani A, Tanaka S, Matsuoka M, Yamamoto G, Hishikawa T, Noguchi T, Yoshikai Y. Gene expression of osteoclast differentiation factor is induced by lipopolysaccharide in mouse osteoblasts via Toll-like receptors. J Immunol 2001;166:3574–9.

[99] Madrazo DR, Tranguch SL, Marriott I. Signaling via Toll-like receptor 5 can initiate inflammatory mediator production by murine osteoblasts. Infect Immun 2003;71:5418–21.

[100] Marriott I, Rati DM, McCall SH, Tranguch SL. Induction of Nod1 and Nod2 intracellular pattern recognition receptors in murine osteoblasts following bacterial challenge. Infect Immun 2005;73:2967–73.

[101] Simonet WS, Lacey DL, Dunstan CR, Kelley M, Chang MS, Luthy R, Nguyen HQ, Wooden S, Bennett L, Boone T, Shimamoto G, DeRose M, Elliott R, Colombero A, Tan HL, Trail G, Sullivan J, Davy E, Bucay N, Renshaw-Gegg L, Hughes TM, Hill D, Pattison W, Campbell P, Boyle WJ, et al. Osteoprotegerin: a novel secreted protein involved in the regulation of bone density. Cell 1997;89:309–19.

[102] Anderson DM, Maraskovsky E, Billingsley WL, Dougall WC, Tometsko ME, Roux ER, Teepe MC, DuBose RF, Cosman D, Galibert L. A homologue of the TNF receptor and its ligand enhance T-cell growth and dendritic-cell function. Nature 1997;390:175–9.

[103] Kong YY, Yoshida H, Sarosi I, Tan HL, Timms E, Capparelli C, Morony S, Oliveira-dos-Santos AJ, Van G, Itie A, Khoo W, Wakeham A, Dunstan CR, Lacey DL, Mak TW, Boyle WJ, Penninger JM. OPGL is a key regulator of osteoclastogenesis, lymphocyte development and lymph-node organogenesis. Nature 1999;397:315–23.

[104] Franchimont N, Reenaers C, Lambert C, Belaiche J, Bours V, Malaise M, Delvenne P, Louis E. Increased expression of receptor activator of NF-kappa B ligand (RANKL), its receptor RANK and its decoy receptor osteoprotegerin in the colon of Crohn's disease patients. Clin Exp Immunol 2004;138:491–8.

[105] Arijs I, Li K, Toedter G, Quintens R, Van Lommel L, Van Steen K, Leemans P, De Hertogh G, Lemaire K, Ferrante M, Schnitzler F, Thorrez L, Ma K, Song XY, Marano C, Van Assche G, Vermeire S, Geboes K, Schuit F, Baribaud F, Rutgeerts P. Mucosal gene signatures to predict response to infliximab in patients with ulcerative colitis. Gut 2009.

[106] Vidal K, Serrant P, Schlosser B, van den Broek P, Lorget F, Donnet-Hughes A. Osteoprotegerin production by human intestinal epithelial cells: a potential regulator of mucosal immune responses. Am J Physiol Gastrointest Liver Physiol 2004;287:G836–44.

[107] Moschen AR, Kaser A, Enrich B, Ludwiczek O, Gabriel M, Obrist P, Wolf AM, Tilg H. The RANKL/OPG system is activated in inflammatory bowel disease and relates to the state of bone loss. Gut 2005;54:479–87.

[108] Schoppet M, Henser S, Ruppert V, Stubig T, Al-Fakhri N, Maisch B, Hofbauer LC. Osteoprotegerin expression in dendritic cells increases with maturation and is NF-kappaB-dependent. J Cell Biochem 2007;100:1430–9.

[109] Vitovski S, Phillips JS, Sayers J, Croucher PI. Investigating the interaction between osteoprotegerin and receptor activator of NF-kappaB or tumor necrosis factor-related apoptosis-inducing ligand: evidence for a pivotal role for osteoprotegerin in regulating two distinct pathways. J Biol Chem 2007;282:31601–9.

[110] Chino T, Draves KE, Clark EA. Regulation of dendritic cell survival and cytokine production by osteoprotegerin. J Leukoc Biol 2009.

[111] Bernstein CN, Sargent M, Leslie WD. Serum osteoprotegerin is increased in Crohn's disease: a population-based case control study. Inflamm Bowel Dis 2005;11:325–30.

[112] Sylvester FA, Davis PM, Wyzga N, Hyams JS, Lerer T. Are activated T cells regulators of bone metabolism in children with Crohn disease? J Pediatr 2006;148:461–6.

[113] Miheller P, Muzes G, Racz K, Blazovits A, Lakatos P, Herszenyi L, Tulassay Z. Changes of OPG and RANKL concentrations in Crohn's disease after infliximab therapy. Inflamm Bowel Dis 2007;13:1379–84.

# 12

# The Role of the Immune System in Fracture Healing

Brandon M. Steen, Louis C. Gerstenfeld, Thomas A. Einhorn

*DEPARTMENT OF ORTHOPAEDIC SURGERY, BOSTON UNIVERSITY SCHOOL OF MEDICINE, BOSTON, MA, USA*

## CHAPTER OUTLINE

Bone repair as a postnatal regenerative process ... 344
Fracture healing cascade ... 346
Role of mesenchymal stem cells in the modulation of immune function ... 350
Cytokines involved in fracture healing ... 351
    Tumor necrosis factor alpha ... 352
    RANK, RANKL, and OPG ... 353
    Interleukin-1 (IL-1) ... 353
    Interleukin-6 (IL-6) ... 354
Phase-specific roles of cytokines in fracture healing ... 354
    TNF-α ... 354
        *Inflammatory phase modulation* ... 354
        *TNF-α effects on chondrocytes and osteoblasts* ... 354
        *TNF-α effects on osteoclasts and remodeling* ... 355
    RANK ligand and osteoprotegerin ... 356
        *Role in early fracture repair* ... 356
        *Initiation of remodeling* ... 356
    IL-1 ... 357
        *Inflammatory phase regulation* ... 357
        *Regulator of callus remodeling* ... 357
    IL-6 ... 358
        *Inflammatory phase regulation* ... 358
        *Callus formation and maturation* ... 358
Role of non-steroidal anti-inflammatory drugs in fracture healing ... 358
    Biological effects of COX-2 inhibition ... 359
    Clinical effects of COX inhibitors on fracture healing ... 360
References ... 361

Osteoimmunology. DOI: 10.1016/B978-0-12-375670-1.10012-3
Copyright © 2011 by Elsevier Inc. All rights reserved.

# Bone repair as a postnatal regenerative process

While developmental processes usually terminate when animals reach maturity, morphogenetic processes may be reinitiated in specific tissues as a consequence of injury. Fracture healing and bone repair are among the unique processes of post natal tissue regeneration that are believed to mirror the ontological events that take place during embryological development of the skeleton (reviewed by [1–5]). Indeed many of the developmental processes and genes that are preferentially expressed in embryonic stem cells during the initiation and activation of the morphogenetic pathways of skeletal development are also expressed in fracture callus tissues [6]. It is generally believed that it is the recapitulation of these ontological processes during fracture healing that allows bone to be repaired without the development of scar tissue and ultimately leads to the regeneration of damaged tissue to its pre-injury structure. This is in contrast to the repair of soft tissue, which typically heals with some degree of fibroblastic scar formation. The interplay of a number of different tissues (vascular, hematopoietic, skeletal, and neural) is essential for the unimpeded regeneration of bone. Appropriate passage through this regenerative cascade is dependent on the proper orchestration of paracrine, autocrine, and systemic signaling pathways with the appropriate complement of stem cells needed to regenerate a skeletal organ [7, 8].

The cascade of events that is commonly described for fracture healing involves blood clot formation at the site of injury, an inflammatory phase, callus generation, primary bone formation, and secondary bone remodeling. While these processes take place in a consecutive temporal manner, they overlap significantly and represent a continuum of changing cell populations and signaling processes within the regenerating tissue. Following fracture, there is a disruption of the normal bone microenvironment which leads to the interactions of cell populations from the medullary space, periosteum, and enveloping muscular tissues. The signaling and cellular contributions from these different tissues and their microenvironments are unique and contribute to the heterogeneous nature of tissue formation at the fracture site [9]. Overviews of the histological and biological events of fracture healing and the known immune cells that are prevalent at each stage of fracture healing and the stages at which specific immune-produced or modifying cytokines are acting are presented in Figure 12-1.

It is of interest to note that the involvement of the immune system in fracture healing is not a new concept although much of the initial focus has been placed on the regulation of the events of fracture healing by immune-cell-associated cytokine signaling [10–14]. Unlike the processes associated with embryologic development and post natal growth that are regulated by ontogenetic and endocrinological mechanisms, fracture healing after trauma is initiated locally in response to regulatory mechanisms associated with inflammation and the innate immune response [7–9, 13, 14].

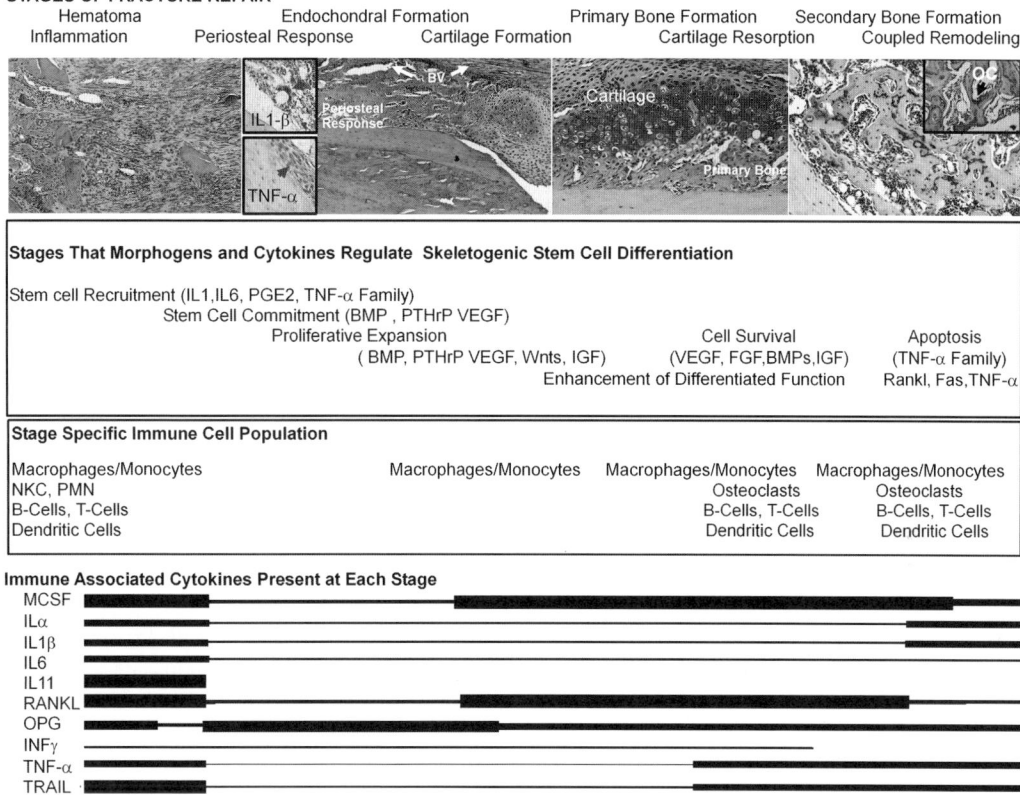

FIGURE 12-1 Summary of the multiple stages of fracture healing. Histological sections are presented for each stage and a summary of the various processes that are associated with each stage is presented. All histological specimens are from sagittal sections of mouse tibia transverse fractures and were stained with H&E or safranin O and fast green, micrographic images are at 200 × magnification. Section for the initial injury was taken from the fracture site 24 hours post injury (far left). Two small inserts depict immunohistological identification of the predominant cell types in the hematoma expressing IL1 (macrophages and monocytes) and TNF-α (periosteal-lining cells). Sections depicting the initial periosteal response and endochondral formation are from 7 days post injury (left middle). Arrows denote blood vessels (BV) of the vascular in-growth from the peripheral areas of the periosteum. Sections depicting the period of primary bone formation and cartilage tissue resorption are from 14 days post injury (middle right). Sections depicting the period of secondary bone formation are from 21 days post injury (far right). Callus sites. Insert depicts 400 × images of an osteoclast (∗chondroclast) resorbing an area of calcified cartilage. Summary of the stages of fracture repair, the major immune cells associated with each stage and their associated molecular processes and regulators are summarized in the three boxes below the histological sections. Bottom box depicts the relative levels of expression of various mRNAs that have been examined in our laboratories and are denoted by three line widths. The levels of expression are by percent over baseline for each and are not comparable between individual mRNAs. Data for expression levels for the pro-inflammatory cytokines and the ECM mRNAs were from Kon et al. [20], Lehmann et al. [77], and Gerstenfeld et al. [4]. Please refer to color plate section.

## Fracture healing cascade

Immediately after fracture, clot formation takes place adjacent to the injured bone. The hematoma is confined by the surrounding soft tissue envelope of the bone in cases of closed injury. Histological examination of fracture hematoma has shown that there is a large influx of immune cells including neutrophils, lymphocytes, and macrophages into the injured area, which become integrated into the initial hematoma that forms at the injury site. Over the subsequent 24–48 hours there is a predominance of T helper cells and macrophages, disproportionate to what is found in the systemic circulation [15], suggesting that they may be one of the primary sources of many of the initiating cytokines or morphogenetic factors that promote healing. Data suggesting that immune cells and their cytokine production within the hematoma are important in the initiation of fracture repair come from studies showing delayed healing or the development of non-union if the hematoma is removed after injury [16]. Disruption of hematoma formation has also been shown to lead to diminished mesenchymal stem cell recruitment from the periosteum [17]. More recent studies have focused on the production of angiogenic factors produced by macrophages and polymorphonuclear leukocytes (PMNs) in the hematoma [18]. It is also noteworthy that hematomas, which form in conjunction with fracture, have a much higher ratio of T helper to cytotoxic T cells than is seen in hematomas that form in response to muscle injury. In this context it is worth mentioning that the percentage of B cells is increased at the fracture site, but not in the muscle hematoma, over the first 4 hours after injury [19].

Consistent with the observation of the presence of T cells and B cells in the fracture hematoma, and in the absence of any observable exogenous pathogens, analysis of the transcriptome of fracture callus showed that although an adaptive immune response does not take place in a closed fracture, groups of genes associated with B-cell receptor signaling, natural killer cell-mediated cytotoxicity, and T-cell receptor signaling pathways were all transiently activated [6]. Interestingly, a second group of genes in these same pathways are downregulated. A summary of statistically significant pathways and associated genes in these pathways, which are related to adaptive immune functions and differentially upregulated in their expression, are presented in Table 12-1, while the selective subset of these genes that are downregulated in their expression are presented in Table 12-2. Such results suggest that different sets of genes in these pathways are either up- or downregulated in the context of fracture healing and are representative of the dual role that these cells might play in either initiating an adaptive response or the initiation of tissue repair. In this context these data suggest that only CD4-positive cells predominantly associated with MHCII pathways are activated. In contrast a much smaller number of molecules associated with MHCI pathways are upregulated and these can be related to the activation of NK cells. Interestingly, hematopoietic associated functions show predominant numbers of downregulated genes including genes associated with B-cell maturation, erythropoiesis, formation of magakaryocytes, and the formation of neutrophils.

**Table 12-1** Differentially upregulated genes of the transcriptome of fracture healing associated with immune function*

| Antigen presentation | Natural killer cell |
|---|---|
| CD4 T-cell receptor-activated pathway | ras-related c3 botulinum substrate 1 |
| MHC II pathway | zeta-chain (tcr) associated protein kinase |
| cd4 antigen | perforin 1 (pore forming protein) |
| cd74 antigen | protein phosphatase 3, catalytic subunit, alpha isoform |
| cathepsin s | histocompatibility 2, d region locus 1 |
| histocompatibility complex, class ii antigen-associated | src homology 2 domain-containing transforming protein c1 |
| histocompatibility 2, class ii, locus dma | natural cytotoxicity triggering receptor 1 |
| histocompatibility 2, class ii, locus mb1 | nuclear factor of activated t-cells, cytoplasmic, C2 |
| histocompatibility 2, t region locus 10 | sh2 domain protein 1a |
| histocompatibility 2, class ii antigen e, beta | vav2 oncogene |
| histocompatibility 2, t region locus 22 | TNF (ligand) superfamily, member 10 |
| histocompatibility 2, o region alpha locus | mitogen activated protein kinase 2 |
| histocompatibility 2, class ii antigen a, alpha | lymphocyte protein tyrosine kinase |
| class ii transactivaton | fc receptor, igg, low affinity iii |
| natural killer cell receptor-activated pathway | killer cell lectin-like receptor, subfamily a, member 7 |
| MHC1 pathway | killer cell lectin-like receptor, subfamily d, member 1 |
| transporter 1, atp-binding cassette, sub-family b (mdr/tap) | killer cell lectin-like receptor, subfamily a, member 4 |
| killer cell lectin-like receptor, subfamily d, member 1 | phosphoinositide-3-kinase, catalytic, gamma polypeptide |
| histocompatibility 2, m region locus | |

| Cytokine–cytokine receptor interactions | B-cell pathway |
|---|---|
| chemokine (c-c motif) ligand 21b | ras-related c3 botulinum substrate 1 |
| chemokine (c-c motif) receptor 5 | vav2 oncogene |
| chemokine (c-x3-c) receptor 1 | nuclear factor of κ light polypeptide gene enhancer in b-cells inhibitor, ε |
| chemokine (c-x3-c motif) ligand 1 | |
| chemokine (c motif) ligand 1 | protein phosphatase 3, catalytic subunit, α isoform |
| chemokine (c-x-c motif) ligand 2 | complement receptor 2 |
| chemokine (c-x-c motif) ligand 5 | interferon induced transmembrane protein 1 |
| burkitt lymphoma receptor 1 | nuclear factor of activated t-cells, cytoplasmic, C2 |
| chemokine (c-c motif) ligand 6 | phosphoinositide-3-kinase, catalytic, γ polypeptide |
| chemokine (c-c motif) ligand 9 | |
| cd40 antigen | |
| kit ligand | |
| leptin receptor | |
| colony stimulating factor 2 receptor, beta 1, | |
| tumor necrosis factor receptor superfamily, member 1b | |
| tumor necrosis factor receptor superfamily, member 25 | |
| tumor necrosis factor receptor superfamily, member 8 | |
| tumor necrosis factor receptor superfamily, member 18 | |
| tumor necrosis factor (ligand) superfamily, member 10 | |
| interleukin 2 receptor, gamma chain | |
| interleukin 4 receptor, alpha | |

*Data taken from Table 2 in Bais et al. [6]. Specific pathways and expressed genes that were associated with these pathways determined by analysis of the transcriptome of fracture healing. Pathways are as defined by Kyoto Encyclopedia of Genes and Genomes (KEGG) (http://www.genome.jp/kegg/pathway.html).

**Table 12-2** Differentially downregulated genes of the transcriptome of fracture healing associated with immune function*

| Hematopoietic cell lineage | B-cell receptor signaling pathway |
|---|---|
| cd2 antigen | cd22 antigen |
| cd7 antigen | cd19 antigen |
| cd19 antigen | ras-related c3 botulinum substrate 2 |
| cd22 antigen | protein phosphatase 3, catalytic subunit, β isoform |
| cd24a antigen | cd72 antigen |
| cd33 antigen | nuclear factor of activated t-cells C 1 |
| cd37 antigen | nuclear factor of activated t-cells, C4 |
| interleukin 1 receptor, type ii | nuclear factor of activated t-cells 5 |
| interleukin 2 receptor, alpha chain | phospholipase c, γ 2 |
| interleukin 1 alpha | phosphatidylinositol 3-kinase catalytic d polypeptide |
| fc receptor, ige, low affinity ii, alpha polypeptide | inositol polyphosphate-5-phosphatase d |
| colony stimulating factor 3 receptor | complement receptor 2 |
| complement receptor 2 | protein tyrosine phosphatase, non-receptor type 6 |
| alanyl (membrane) aminopeptidase | vav 1 oncogene |
| integrin alpha m | harvey rat sarcoma virus oncogene 1 |
| integrin alpha 2b | bruton agammaglobulinemia tyrosine kinase |
| integrin beta 3 | leukocyte immunoglobulin-like receptor, subfamily b |
| integrin alpha 4 | nuclear factor of κ light chain gene enhancer in b-cells 1, p105 |
| transferrin receptor | phosphoinositide-3-kinase, catalytic, γ polypeptide |
| glycoprotein ib, beta polypeptide | |
| glycoprotein 5 (platelet) | |

| Fc epsilon RI signaling | Natural killer cell-mediated cytotoxicity |
|---|---|
| phospholipase a2, group xiia | nuclear factor of activated t-cells, C1 |
| phospholipase a2, group iie | nuclear factor of activated t-cells C 4 |
| phospholipase c, gamma 2 | nuclear factor of activated t-cells 5 |
| phospholipase a2, group xiib | linker for activation of t cells |
| phospholipase a2, group iif | lymphocyte cytosolic protein 2 |
| phospholipase a2, group iid | protein phosphatase 3, catalytic subunit, β isoform |
| phosphoinositide-3-kinase, catalytic, γ polypeptide | histocompatibility 2, t region locus 23 |
| phosphatidylinositol 3-kinase catalytic δ | histocompatibility 2, d region locus 1 |
| polypeptiinositol polyphosphate-5-phosphatase dde | fc receptor, ige, high affinity i, γ polypeptide |
| fc receptor, ige, high affinity i, γ polypeptide | killer cell lectin-like receptor subfamily b member 1c |
| vav 1 oncogene | killer cell lectin-like receptor subfamily c, member 1 |
| lymphocyte cytosolic protein 2 | killer cell lectin-like receptor subfamily c, member 2 |
| ras-related c3 botulinum substrate 2 | phospholipase c, γ 2 |
| harvey rat sarcoma virus oncogene 1 | ras-related c3 botulinum substrate 2 |
| bruton agammaglobulinemia tyrosine kinase | phosphatidylinositol 3-kinase catalytic δ polypeptide |
| linker for activation of t cells | phosphoinositide-3-kinase, catalytic, γ |
| mitogen activated protein kinase 1 | mitogen activated protein kinase 1 |
| mitogen activated protein kinase kinase 3 | harvey rat sarcoma virus oncogene 1 |
| | vav 1 oncogene |
| | cd244 natural killer cell receptor 2b4 |
| | bh3 interacting domain death agonist |
| | integrin β 2-like |
| | integrin β 2 |
| | polypeptideprotein tyrosine phosphatase, non-receptor type 6 |
| | interferon (α and β) receptor 2 |

*Data taken from Table 2 in Bais et al. [6]. Specific pathways and expressed genes that were associated with these pathways determined by analysis of the transcriptome of fracture healing. Pathways are as defined by Kyoto Encyclopedia of Genes and Genomes (KEGG) (http://www.genome.jp/kegg/pathway.html).

During the inflammatory phase of fracture healing, multiple events occur in concert. Activated neutrophils and macrophages participate in removal of necrotic debris. Monocyte precursors begin to organize in preparation for mesenchymal and leukocyte cellular recruitment and differentiation. At this point in time, there is also an early spike in the number of inflammatory cytokines, interleukin (IL)-1, IL-6, and tumor necrosis factor (TNF)-α [9, 13, 20]. These cytokines have been found to have significantly different roles dependent on their temporal expression. During the inflammatory phase, these cytokines predominantly act to recruit the necessary cellular lineages that are needed for tissue regeneration to the area of injury. The functional role of these cytokines in initiating the expansion of these cell lineages and the activation of specific functions of these cells will be discussed in detail below. Within several days of this stage, histological analysis of the fracture site reveals the initiation of callus formation. Callus forms by two distinct processes that occur simultaneously. Adjacent to the periosteum in continuity with the fractured ends of the bones, osteoblastic cell expansion occurs and woven bone, made up of predominantly type I collagen, is deposited [2]. These cells are thought to arise via mesenchymal stem cell (MSC) differentiation from cells recruited to the site of injury. Studies examining the homing signals of SDF-1 and their affinity for CXCR-4 show increased uptake subperiosteally at the onset of callus formation, supporting the idea that stem cells are recruited from either surrounding tissues or systemically to this specific site [21]. This process of intramembranous ossification occurs peripherally at proximal and distal regions adjacent to the fracture and is characterized by the enlargement of the periosteum.

Within the first 2 weeks after fracture, the majority of newly synthesized tissue is made up of proliferating chondrocytes. Spatially these cells arise directly at the site of fracture and initially bridge the gap between the fracture ends, and seal off the medullary cavity. These cells proliferate, radially increasing the diameter of the callus. The cells secrete predominantly type II collagen into the surrounding matrix. While areas of cartilage proliferation are devoid of immune cells [15], histologic analysis has shown that there are Th2-positive cells in the developing fracture callus in areas of granulation and fibroblastic tissue formation adjacent to the cartilage. In particular, macrophages tend to be clustered at the periphery of developing cartilage islands, adjacent to areas of intramembranous ossification and where neovascularization is initiated. Interestingly, Th cells have also been found to be the predominant cell type at the sites of fracture non-unions [22]. The absence of Th cells in areas of bone regeneration, and the presence of these cells in areas of fibroblastic proliferation suggests that Th cells may play a role in the development of delayed bone healing.

Evidence for direct T-cell involvement in the control of fracture healing comes from studies over 20 years ago that showed negative effects of lymphocyte depletion on fracture healing [22–25]. More recent studies have investigated the specific role of γδ T cells. These cells are of interest since they play unique functions in the recognition of products of stressed or damaged cells through receptors that do not need the presence of an antigen-presenting cell. Although they produce interferon γ, a cytokine predominantly associated with Th1 cells, they can also produce Th2 cytokines and more specifically IL-17 through induction of Th17 cells. γδ T cells also have been shown to assist in the regulation of Th1 or Th2 phenotype of the CD4+ αβ T-cell responses [26, 27] and to carry out specialized functions in response to tissue injury and wound healing [28, 29]. Studies of fracture healing in mice deficient in γδ T cells displayed shorter time to union and earlier development of mature fracture callus [30] suggesting that these cells may control osteogenic stem cell recruitment, proliferation, or development.

While expansion of both soft and hard callus continues for the next several weeks after fracture to provide stiffness and support, areas of the cartilaginous callus begin to undergo resorption. Cartilage resorption cannot be initiated; however, until chondrocytes undergo hypertrophic maturation, the cartilage matrix calcifies, and vascular invasion is initiated [31]. At the end of the endochondral phase, chondrocytes embedded in the mineralized cartilage matrix undergo apoptosis, the matrix is resorbed by specific proteases and the infiltration of osteoclasts and angiogenesis is observed in the callus [32]. Hypertrophic chondrocyte apoptosis is mediated by TNF-α-regulated activation of TNFr1 and upregulation of Fas by TNF-α [20, 33]. The synergistic role of Fas and Fas ligand mediates chondrocyte apoptosis is demonstrated in mice that exhibit an autoimmune lupus-like syndrome in which Fas receptors are mutated (lrp/lrp). Even though these animals show elevated osteoclastic activity and osteopenia they have delayed removal of mineralized cartilage due to delayed apoptosis of the chondrocytes (unpublished data).

The programmed apoptosis of chondrocytes leaves a calcified shell of matrix into which blood vessel invasion proceeds. Osteoclastic resorption and simultaneous osteoblastic production of new type I collagen, and calcified matrix, will then establish an initial matrix of woven bone. During this period, new bone is laid down on the remnants of mineralized cartilage followed by a very prolonged period of skeletal tissue remodeling when the marrow space and hematopoietic compartment is re-established and the bone returns to its original structure. This period of coupled remodeling is mediated by a close association and re-establishment of the marrow hematopoietic niche. A subsequent prolonged period of remodeling then takes place that ultimately regenerates the original bone structure. Interestingly, complete inhibition of osteoclast formation by antibody blockade of the RANK receptor does not impede either chondrocyte apoptosis or vascular invasion, nor does it affect the continued deposition of new bone in the callus [34]. The coupled remodeling of the callus continues unabated for many months and ultimately forms a mature lamellar bone structure.

# Role of mesenchymal stem cells in the modulation of immune function

Fracture healing and bone regeneration that are stimulated by various types of orthopedic surgical procedures both involve the recruitment of large numbers of stem cells.

In general the stem cell that gives rise to the various skeletal tissues has been defined as the "mesenchymal stem cell" (MSC). This cell has been empirically defined as a multipotent stem cell that can give rise under various in vitro and in vivo experimental conditions to one of several different mesenchymal tissues (muscle, adipose, cartilage bone tendon, and ligaments) [35]. MSCs themselves are derived as adherent cells from any number of tissues including the bone marrow, periosteum, muscle, and fat [36–40].

It is of considerable interest, and of functional importance, that MSCs possess immunomodulatory activities with some studies showing that these cells are immunoprivileged [41, 42] or able to promote immunosuppressive activities [43]. One may speculate that these functions are important in both shifting the response in an injured tissue from an immune to a regenerative response as well as allowing these stem cells to grow across normally restricted tissues boundaries. The immunomodulatory activities of MSCs may also be important in the re-establishment of the hematopoietic niche within the marrow space during bone regeneration after injury. In the context of their immunomodulatory activities, MSCs have been shown to interact with various types of immune cells at various stages during bone healing.

Initially, after injury, MSCs will interact with natural killer cells that participate in the innate immune response. It is interesting to note that MSCs appear to suppress the proliferation and diminish the cytoxic activities of natural killer cells thereby enhancing MSC survival [44]. MSCs have also been shown to inhibit monocyte and/or hematopoietic stem cell differentiation into dendritic cells [43, 45, 46] while stimulating differentiated dendritic cells to switch from a pro- to an anti-inflammatory cytokine production mode. These coordinated activities are synergistic since both natural killer cells and dendritic cells function to enhance antigen presentation and adaptive response in the context of septic conditions after injury. MSCs also have been shown to directly affect both T-cell and B-cell growth and/or activities [47–49].

In regard to the role of MSCs and their interactions with T cells and B cells, in vitro studies have shown that MSCs can suppress the proliferation and activation of both types of cells [48, 50–54]. Concerning the inhibition of MSCs by T cells, it is of note that many factors that are initially produced during the period of injury promote MSC replication while being inhibitory to immune cells. However, the inhibitory effects of MSCs on B-cell proliferation are dependent on the cell numbers of MSCs, thus low numbers of MSCs are inhibitory, but higher numbers appear to stimulate B-cell proliferation [55]. Recent data have shown that B lymphopoiesis is mediated by $G_s\alpha$-dependent signaling pathways in osteoprogenitor cells [56], which would be consistent with the observed early effects on less differentiated MSCs that are initially recruited to the repair site.

## Cytokines involved in fracture healing

One of the most studied areas of immune involvement in fracture healing is the role of various cytokines and growth factors secreted by immune and mesenchymal cells in the local fracture environment. This focus has largely been driven by the potential for modulation of the activities of these cytokines for therapeutic purposes. The hematopoietic, inflammatory, and death receptor mediators are considered together since their activities are coordinated during repair. The collaboration of these factors' activities occurs during three successive waves of tissue resorption and remodeling over the time course of fracture healing. Immediately after injury, there is a massive influx of macrophages, neutrophils, and other cells associated with an innate immune response. Subsequent to the formation of the hematoma, the necrotic tissues that result from the injury must be cleared, the inflammatory response diminished, and the repair process initiated. Numerous inflammatory and immunomodulatory factors including prostaglandins (PGs), IL-1 IL-6, IL-11, TNF-α, TNF-β and lymphotoxin (LT)-α, macrophage colony stimulating factor (M-CSF), receptor activator of nuclear factor kappa B ligand (RANKL), osteoprotegerin (OPG), and interferon (INF)-γ [20] are induced within the first 3 days after fracture and factors such as PGs and TNF-α will carry out functions associated with both immune cells as well as those MSCs involved with the initiation of bone repair [9, 57, 58]. More recently, the discovery of similar signaling pathways in lymphoid tissue and bone, specifically the RANK/RANKL/OPG axis, has led to further inquiries about similarities and interactions between these two biological systems [59]. A summary of the major immunomodulatory factors that are expressed during fracture healing, their relative period of activity during the healing cascade, and their know target tissues, is presented at the bottom of Figure 12-1.

## Tumor necrosis factor alpha

After the initial discovery of tumor necrosis factor almost 30 years ago, a large superfamily of related peptides has been identified. Tumor necrosis factor alpha (TNF-α) is one of more than 20 members of the tumor necrosis factor family of cytokines and is the most studied in respect to its effects on bone tissue regeneration. Although the extracellular component of the TNF family of ligands tends to be fairly well conserved with regard to cysteine repeats, cytoplasmic domains of the different ligands are quite different, and can initiate a variety of different cellular responses. TNF-α acts as an arbitrator of the local tissue environment following injury. Both TNF-α and Fas L are known to initiate apoptosis through activation of independent intracellular pathways [33, 60]. TNFr1 is expressed by almost all cell types. Activation of TNFr1 on immune cells during the inflammatory phases of fracture healing induces apoptosis of multiple cell lines including macrophages, T cells, and B cells [61–63]. Activation of Fas by Fas L also initiates programmed cell death through an independent mechanism; however, Fas expression has been shown to be dependent on activation by TNF-α, implying that there is a clear connection between these two pathways. In the presence of mesenchymal cell populations, TNF-α plays a dichotomous role as both an initiator of cell death and as an arbitrator of cell survival and proliferation. In fracture tissue, TNF-α has the ability to bind to different receptors, TNFr1 and TNFr2. Preferential activation of one of these receptors is the basis for tissue regeneration and remodeling, via activation or inhibition of osteoclastogenesis [64, 65]. While not as well studied at present, TNF-α may also control bone regeneration via its effects on osteoblasts through activation of the Fra/AP1 pathway [66].

## RANK, RANKL, and OPG

Receptor activator of nuclear factor kappa B (RANK) and its ligand (RANKL) are members of the TNF superfamily of proteins. RANKL is expressed by macrophages and osteoblasts, and functions as an activator of RANK. RANK, expressed on osteoclast precursors, is a key regulator of osteoclastogenesis. RANK is also expressed by dendritic cells and is thought to play a role in T-cell–dendritic-cell interactions at extraosseous sites of lymphoid tissue development [67, 68]. Functionally, RANK has no intracellular enzymatic activity. Its actions on the cell are mediated through binding and activation of TRAF proteins, which in turn activate NF-κB and cFOS pathways [69–71]. Together, these pathways ultimately lead to activation of the intracellular promoter NFATc1, which has been shown to be essential for differentiation of mononuclear cells into the osteoclast progenitors [72, 73]. NFATc1 upregulates the production of TRAP, cathepsin K, and other proteins unique to the activated osteoclast [67]. Osteoprotegrin is a member of the TNF receptor family. It is a soluble protein that binds to RANKL, functioning as a competitive inhibitor of RANK activation. OPG has been found to be crucial in maintaining skeletal tissue homeostasis, as experiments involving constitutive overproduction have led to an osteopetrotic phenotype in mice while absence of OPG leads to early-onset severe osteoporosis [74, 75]. The ebb and flow of interaction between these three molecules ultimately regulates the delicate balance of bone formation and resorption, and at no time is this more crucial than during the hypermetabolic period of fracture healing.

## Interleukin-1 (IL-1)

IL-1 is a pro-inflammatory cytokine that exists in two subtypes, IL-1α and IL-1β. IL-1 is secreted by inflammatory cells that either migrate, or are activated at the site of injury. In the context of fracture healing, both cytokines are maximally expressed in the first 24 hours following a fracture. Macrophages have been identified as the predominant cell type secreting IL-1 during the early phases of fracture healing, although it has also been shown to be secreted by mesenchymal cells at low levels throughout the process of fracture repair. IL-1α subtype expression is inducible by addition of LPS, suggesting its role in upregulating inflammation in the presence of contamination. IL-1β is also maximally induced within 24 hours of injury, but it continues to be expressed at low levels throughout the repair process. IL-1β has been shown to be a potent stimulator of osteoblast proliferation [76]. IL-1 shares a similar temporal expression pattern to that of TNF-α and likely acts in conjunction with TNF-α to regulate the inflammation following injury [20]. IL-1β, along with TNF-α, also stimulates the breakdown of chondrocyte matrix via upregulation of metaloprotease expression [77]. IL-1 also serves as an activator of other pro-inflammatory cytokines, such as IL-6.

## Interleukin-6 (IL-6)

IL-6 is a multifunctional cytokine found to affect immunologic cellular development, the central nervous system, and bone metabolism [78]. It is part of a family of related cytokines that activate the gp-130 subunit in cells, upregulating intracellular STAT pathway factors [79]. Initially, it was thought that IL-6 was an indispensable factor for the initiation of osteoclastogenesis; however, further studies in IL-6 knockout mice have disproven this hypothesis [80, 81]. IL-6 has been shown to have modest effects on osteoblastic differentiation. The IL6 receptor, IL-6r, has the potential to be cleaved and solubilized. Binding of IL-6 to the soluble IL-6r was found to be necessary to exert its effects on osteoblastic cells. Activated IL-6 increases gp-130 phosphorylation in osteoblasts increasing local alkaline phosphatase activity. It is yet to be determined whether IL-6 primarily functions to indirectly promote osteoblast differentiation through upregulation of osteotropic factors, such as TGFβ and BMP-6, or by directly stimulating differentiation of osteoblast precursors. Studies in IL-6 knockout mice have shown no overt abnormality in bone homeostasis or remodeling; however, IL-6 knockout mice are protected against inflammatory joint destruction [80, 82, 83]. These findings suggest that IL-6 may play a role in stem cell commitment and differentiation, but is not crucial for normal development and repair to take place.

# Phase-specific roles of cytokines in fracture healing

## TNF-α

### Inflammatory phase modulation

TNF-α has been shown to have substantially different effects on fracture callus depending on the time at which it is expressed and the tissue it is acting upon. This dichotomous nature of TNF-α is likely related to its ability to bind different TNF receptors [9, 64]. Studies of murine fracture callus have shown the TNF-α expression peaks within 24 hours of injury and then returns during the transition from callus formation to primary remodeling. This cytokine likely plays an important role early post-injury by upregulating the inflammatory cascade. It functions as both a stimulator of proliferative, inflammatory cytokines and as a modulator of cell survival through selective activation of the apoptotic cascade [84–86]. TNF-α is a potent stimulator of IL-1 secretion which is important in the recruitment and proliferation of monocyte and mesenchymal cells. It is also capable of inducing selective apoptosis by multiple mechanisms including direct activation of the downstream apoptotic cascade via TNFr1 binding, and modulation of Fas receptor expression [33, 87].

### TNF-α effects on chondrocytes and osteoblasts

TNF-α has also been shown to be a potent stimulator of RANKL expression. With respect to bone, RANKL has previously been shown to be imperative for formation of osteoclasts and thought to be a marker for the initiation of coupled remodeling. RANKL functions as an internal barometer, with either inhibition or activation tipping the metabolic balance in favor of osteoclastic or osteoblastic activation. Support for this theory stems from research demonstrating that high levels of OPG and RANKL are present during periods of osteoblast proliferation and callus formation early in the course of fracture healing. This is also consistent with assays showing high levels of TNF-α and RANKL within the first few days following fracture. Osteoblastic proliferation is observed early in the course of fracture healing during periods when high levels of TNF-α and IL-1 are expressed. Interestingly, osteoblasts did not undergo apoptosis in vitro in the presence of TNF-α [9]. In fact, addition of exogenous TNF-α and IL-1 was shown to induce increased differentiated function by osteoblastic cell lines. Some controversy still exists regarding the effects of TNF-α on osteoblasts during early fracture repair. This may be an important area for future study in defining the complete role of cytokine cellular interactions at the initiation of fracture healing.

TNF-α expression increases a second time, approximately 1 week following fracture, during the onset of callus maturation and the initiation of remodeling. Again, high levels of OPG are present as chondrocytes proliferate and mature. OPG likely exerts a protective effect on cartilage resorption at this stage and competitively inhibits RANKL. Interestingly, TNF-α expression has been found in the chondrocytes within the callus, suggesting an autocrine function, initiating the onset of apoptosis. In vitro studies on proliferating and hypertrophic chondrocytes show induction of apoptosis with addition of Fas ligand or TNF-α. Additionally FAS receptor was found to be strongly upregulated following addition of TNF-α, along with other markers of apoptosis including caspase 2 and IL-1β converting enzyme (ICE/caspase 1) [33]. Other research utilizing TNFr knockout mice has shown that chondrocytes experience a profound delay in apoptosis. Numerous studies have demonstrated that animals with absent TNF-α function have hypertrophic callus with a delay in the onset of remodeling [9, 60, 77]. This important observation provides in vivo data affirming that alterations in TNF-α signaling lead to disruption of the normal bone repair process.

### TNF-α effects on osteoclasts and remodeling

The onset of osteoclastogenesis and remodeling follows closely behind the second peak of TNF-α expression, implicating TNF-α as the predominant stimulator of RANKL upregulation [20]. The activity of TNF-α is mediated through binding to its membrane-bound receptor TNFr1 found on osteoclast precursor cells. Interestingly, a second receptor, TNFr2, is also found. Knockout studies in mice have shown that absence of the TNFr1 receptor produces mice without osteoclasts at this stage of repair, while TNFr2 knockouts have a significant increase in the number of osteoclasts [88]. This observation provides a potential explanation for the dual role of TNF-α as both a stimulator and inhibitor of osteoclastogenesis. This concept was again demonstrated in an in vitro model in which osteoclasts were found to be stimulated and inhibited by mechanisms that were dependent on the concentration of TNF-α and the presence of osteoblasts [88]. Both membrane receptors act via downstream regulation of NF-κb. TNFr1 is an activator of NF-κb through activation of TRAF signaling molecules. TNFr2 has been shown to have a direct inhibitory effect on NF-κB through binding of TRIP. Further examination of the role that TNF receptors play in osteoclastogenesis has shown that TNFr2 can be solubilized and preferentially binds to the membrane-bound TNF-α, acting as a competitive osteoclastogenesis inhibitor. Binding of the soluble TNFr2 to membrane-bound TNF-α results in direct inhibition of NF-κB via binding to TRIP, resulting in inhibition of osteoclastogenesis [64]. The multiplicity of functions of TNF-α, as well as its dichotomous effects on RANKL activation, demonstrate the complex interactions that take place during bone repair. Taken in context, all of these different functions serve one ultimate purpose; to ensure proper temporal progression through the stages of fracture repair.

## RANK ligand and osteoprotegerin

### Role in early fracture repair

RANKL is found to be highly upregulated within the first 24 hours of fracture healing. This finding seems to contradict the primary role of RANKL as an inducer of osteoclastogenesis and a marker of bone remodeling. One theory suggests that early RANKL expression plays an important role in regulation of immunologic response, specifically maturation and proliferation of T helper cells, although little data exist to define its specific function. Experiments in RANKL knockout animals have shown that RANK/RANKL interaction within the immunologic niche is essential for activation of Th cells and maintenance of the local dendritic cell population [68]. In regards to RANKL's early effect on the mesenchymal cell population, OPG is also found to be highly expressed during the same time period, likely negating RANKL's effect on osteoclast precursors, via competitive inhibition of RANK [89]. Again, further study is necessary to define the role of RANK expression in early fracture healing.

### Initiation of remodeling

The role of RANKL in initiating osteoclastogenesis has been extensively studied. It is clearly the primary signaling pathway to initiate differentiation of osteoclast precursors into active, mature, multinucleated giant cells, capable of enzymatic bone resorption. The coupled catabolic effects of osteoclasts and anabolic effects of osteoblasts are the driving principle behind secondary remodeling of new bone tissue. Until recently, the mechanism by which initial remodeling occurred was not well understood. During maturation of the soft fracture callus, the chondrocytes actively calcify their surrounding matrix, providing further mechanical stability to the fractured bone. RNA evaluation of the fracture callus has defined the temporal expression of RANKL and OPG at the period of cartilage remodeling [20]. As the repair process progresses, RANKL expression begins to peak at a second time point approximately 1 week post fracture. Simultaneously OPG levels begin to fall, and shortly thereafter absorptive remodeling of the cartilaginous callus takes place. Along with an increase in the RANKL/OPG ratio, an increase in the presence of osteoclasts/chondroclasts is seen. Studies in TNF-α-deficient mice show that absence of TNF-α signaling correlates with a significant, 4-day delay in the onset of callus resorption. OPG levels in these animals remained significantly higher throughout the period of chondrocyte maturation. Eventually, osteoclastic migration into the callus took place, but was delayed almost a week in comparison to wild-type animals [89]. One noticeable difference in comparing the mechanism of bone remodeling and chondrocyte callus remodeling is the induction of RANKL expression. Cultured chondroblasts do not produce RANKL in the presence of TNF-α, suggesting that osteoclast activation is not achieved by the chondrocyte [90]. Concurrent vascularization of the callus may promote the influx of cells capable of stimulating osteoclastogenesis. This observation is in contradistinction to the osteoclast induction by surrounding osteoblasts observed in primary bone remodeling [89].

## IL-1

### Inflammatory phase regulation

Maximal induction of IL-1 in a closed fracture is seen within the first 24 hours following injury. Its pattern of expression mirrors that of TNF-α. Early IL-1 expression stimulates a robust upregulation of IL-6, a multifunctional cytokine that has multiple effects on the fracture healing cascade. IL-1 is also highly sensitive to induction by the presence of LPS, again suggesting that it plays an important role in regulating the initial inflammatory response, especially in the face of open contamination [20]. IL-1β has also been shown to promote the proliferation of osteoblasts in vitro, and upregulate expression of PGE2 [76, 91]. Its effects on osteoblast proliferation seem to be independent of the prostaglandin synthesis, as addition of indometacin does not inhibit the osteoblastic effects of IL-1β. In evaluating mRNA from these cells, it appears that osteoblastic marker protein such as parathormone receptor and osteocalcin are low, suggesting that while proliferation is increased, IL-1β may also function in differentiation. There appears to be interplay between prostaglandins and pro-inflammatory cytokines in regulating osteoblast function. While there is some evidence that IL-1β has effects independent of PGE2, there is also evidence that IL-1β upregulates expression of PGE2 and the EP4 receptor on osteoblasts. This likely represents an independent, yet related mechanism for induction of osteoblast proliferation [91]. Considerable interplay between different biochemical pathways has been observed repeatedly, again confirming that the regulation of fracture repair is an extremely complex interplay between hematopoietic and mesenchymal lineage cells and their trophic and signaling factors.

### Regulator of callus remodeling

IL-1β has a second peak of expression during the initiation of callus remodeling. Again, its expression mirrors that of TNF-α, and it is thought to be a stimulator of chondrogenic matrix degradation. In vitro IL-1β has been shown to increase production of matrix metalloproteases, implicating its role in both normal endochondral resorption and pathologic cartilage destruction as seen in the inflammatory arthropathies [20, 60].

## IL-6

### Inflammatory phase regulation

IL-6 is robustly induced within 24 hours of fracture occurrence [13]. Moderate increases in IL-1 have been shown to greatly increase IL-6 expression. IL-6, like other acute phase pro-inflammatory cytokines, is likely involved in initial trafficking and expansion of hematopoietic and mesenchymal cell lines. Its exact mechanism during the inflammatory phase of healing has yet to be elucidated; however, its temporal expression in fracture and osteotomy distraction models has provided insight to its function later in the bone repair process.

### Callus formation and maturation

IL-6 expression increases a second time post injury approximately 1 week after fracture [4]. In osteotomy distraction models its increased expression is triggered by the placement of tension across the osteotomy site. Immunoreactivity was found in cells of both hematopoietic and mesenchymal lineage. Induction of IL-6 did not lead to increased proliferation of mesenchymal cells at the site of osteotomy, but blocking of IL-6 activity did lead to significant decreased proliferation of mesenchymal cells [92]. In vivo studies of fractures in IL-6 knockout mice have yielded related observations. Mice lacking IL-6 activation showed increase in early callus volume with decreased mineralization, most evident 2–4 weeks after injury. The stiffness of the fracture calluses was decreased as well. Absence of IL-6 seemed to prolong the chondrogenic phase of healing, delaying osteoblastic proliferation, activation and maturation of the fracture callus. mRNA analysis of the calluses showed a decrease in chondroid matrix proteolytic factors including MMP-3, colagenase-3, and aggrecanase. Ultimately, fractures healed in IL-6 knockout mice with fairly normal architecture; however, alterations in callus integrity and a delay in calcification support the hypothesis that IL-6 plays an important, albeit non-essential role in regulating normal progression of fracture repair [93].

# Role of non-steroidal anti-inflammatory drugs in fracture healing

Modulation of the inflammatory cascade and its effects on immunological function after injury plays a pivotal role in initiation of bone repair. Non-steroidal anti-inflammatory drugs (NSAIDs) are widely used following injury to the musculoskeletal system. They exert their effects as both an analgesic and anti-inflammatory through inhibition of the cyclooxygenase (COX) family of enzymes [94]. Prostaglandin synthesis is mediated through enzymatic processing of arachidonic acid. Non-selective NSAIDs exert their effects on both COX-1 and COX-2 while the more selective Coxibs specifically target COX-2. Study of these different isotypes has revealed that COX-1 is widely expressed in a variety of cell types, and that inhibition of prostaglandin synthesis via COX-1 blockade leads to an unfavorable side-effect profile in extraskeletal tissues, on the function of the gastric mucosa and platelets [94, 95]. COX-2 expression is generally the specific isoenzyme that is primarily expressed in response to tissue injury locally, but its inhibition has little effect on platelet aggregation or gastric mucosal tissue [96]. These observations led to the development of a class of COX-2-selective inhibitors, in hopes of retaining anti-inflammatory properties while limiting systemic side effects.

## Biological effects of COX-2 inhibition

Numerous studies have demonstrated the biomechanical and histologic effects on COX inhibition on fracture healing [97–101]. Following these observations, other researchers sought to define the specific biologic effects that these drugs produce. COX-2 inhibition has been shown to decrease both endochondral and intramembranous bone formation [58]. In vitro nodule assays in bone marrow stromal cells, harvested from COX-2-null mice, show a dramatic decrease in mineralized nodule formation in osteoinductive media. Furthermore, mRNA analysis of the cultures showed significant reductions in the expression of osterix and RUNX2, which are genes that are both necessary for bone formation. In addition, administration of prostaglandin E2 (PGE-2) to these cultures completely rescued nodule formation [57]. Further studies on fracture calluses in COX-2 knockout mice showed that col2 and colX were present within newly synthesized callus of both control and COX-2-null animals. Histomorphologic analysis of the callus showed large areas of uncalcified cartilage, devoid of osteoblastic differentiation within the COX-2-null animals. These findings further suggest that expression of COX-2 is important for regulating proliferation and differentiation of osteoblasts. Chondrocyte differentiation does not appear to be affected significantly by lack of COX-2 expression [58]. Similarly, examination of bone repair tissue following traumatic drilling in rat femurs in animals treated with diclofenac showed an almost 50% reduction in osteoblasts at the site of injury 10 days later [101]. Tissue analysis also showed a significant decrease in type I collagen formation. Similar to in vitro studies by Kaspar et al. [102], the authors concluded that COX inhibition plays a pivotal role in early migration and differentiation of osteoblasts following bone injury. Other studies have demonstrated that the effects of prostaglandins, specifically PGE-2, are mediated through binding to the EP4 receptor on osteoblast precursors. In vitro osteoblast cultures treated with prostaglandins have demonstrated downregulation of pro-apoptotic caspases, suggesting that inhibition of apoptosis promotes osteoblastic proliferation at the site of injury [103]. Furthermore, blockade of the EP4 receptor in cultured osteoblasts inhibited upregulation of alkaline phosphatase in nodule formation seen with exogenous prostaglandin stimulation [104]. Activation of prostaglandin synthesis pathways has also been suggested to be modulated by interactions with pro-inflammatory cytokines. IL-1β has been shown to be a potent stimulator of prostaglandin synthesis and likely plays a role in modulating the effects of PGE-2 in the developing callus [91]. While the exact mechanisms contributing to this observation have not been fully defined, there is substantial evidence that COX inhibition detrimentally affects the proliferation of osteoblast progenitors following bony injury.

## Clinical effects of COX inhibitors on fracture healing

Because of the excellent anti-inflammatory and analgesic properties of COX inhibitors, they have been widely prescribed for decades for the treatment of post-traumatic musculoskeletal pain. As early as the 1970s, experiments were performed to evaluate the effects of these medications on acute bone repair. A substantial body of evidence has emerged in animal models suggesting that non-selective COX inhibitors have significant effects on healing bone [105–108]. Numerous studies in rats have demonstrated a delay in union after treatment with physiologic doses of indomethacin for periods of time ranging from 1 to 3 weeks [105, 106, 109]. In all cases there was a substantial delay in union, with decreased bending and torsional stiffness of the fracture, as well as decreased volume of fracture callus produced in comparison to animals treated with placebo.

Further animal studies have evaluated the effects of COX inhibition by examining fracture healing in COX-null animals [57, 58]. COX-2 knockout was associated with disruption of prostaglandin synthesis. Lack of prostaglandin production was shown to markedly decrease mesenchymal cell differentiation at the fracture site. Mice had failure of osteoblastogenesis, which resulted in an increased rate of non-union at the fracture site. Exogenous administration of prostaglandin E rescued the animals and they obtained near normal fracture healing.

With the development of COX-2-selective inhibitors, numerous studies have sought to investigate whether the effects of these medications on fracture repair were different from those of the earlier, non-selective COX inhibitors. Studies comparing non-selective COX inhibitors to COX-2-selective inhibitors in rats have shown that both classes of medications produce inhibition of fracture healing. There is some variation in the degree of inhibition seen between different COX-2 inhibitors studied in vivo. One study compared celecoxib, rofecoxib, COX-2 genetic knockout, and indomethacin following femur fracture. Both COX-2 inhibitors and the COX-2 knockout animals developed non-unions, while the animals in the non-selective COX inhibitor group achieved delayed union in comparison to controls. These results led the authors to conclude that prostaglandin synthesis is essential for fracture healing, and the effects of coxibs on fracture healing are most certainly related to COX inhibition and not alternate effects of the medication [57]. A second study comparing a different selective COX-2 inhibitor to a non-selective inhibitor again showed that both classes of medications delay healing, and again COX-2 inhibition was associated with a higher percentage of fracture non-unions. Negative effects on fracture callus stability and morphology at 3 and 5 weeks post-fracture were noted in both groups. They also defined the temporal expression of localized COX isoenzymes in animals at various stages following fracture through analysis of fracture callus mRNA.

COX-2 was found to be elevated 10-fold above baseline at 3–14 days post fracture. COX-1 expression was not elevated during this time interval. Both isoenzymes returned to baseline levels by 21 days [98]. This observation suggests that COX-2 expression is locally inducible, and plays a significant role in fracture repair relatively early in the course of healing. A related study from the same lab showed that prostaglandin inhibition was reversible with cessation of short-term NSAID therapy. Animals that received short-term therapy for 1 week post injury were found to have a rebound in prostaglandin synthesis in the callus after withdrawal of the medication. They also had regained callus strength and integrity by day 21 [110].

There is mounting clinical and biological data suggesting that COX inhibition most certainly has a negative impact on bone tissue repair. To date, no significant randomized controlled trials have been undertaken in human subjects. There are a handful of retrospective studies examining osseous union in patients treated with NSAIDs post injury [111–114]. These studies provide conflicting results on the effects on NSAID use. Two of the studies provided data that were highly suggestive of negative clinical impact of COX inhibition. Burd et al. [113], in a retrospective review of patients receiving 6 weeks of indometacin for heterotrophic ossification prophylaxis following acetabular fractures, noted that patients with concomitant long bone fractures had a four-fold higher incidence of non-union than those patients who did not receive indometacin. This study is unique in that the study population was on long-term NSAID therapy, whereas numerous studies have only looked at patients who have used NSAIDs intermittently for analgesia. The work of Burd et al. [113] seems to correlate significantly with observations in animal studies in the conclusion that COX inhibition effects fracture repair in a time- and dose-dependent manner [113]. COX inhibition has also been studied in a retrospective manner in patients undergoing spinal fusion [111, 115]. Glassman et al. [111] also showed that the non-union rate in patients on long-term NSAIDs was almost four-fold higher than in patients not taking NSAIDs. This observation was also seen by Deguchi et al. [115] in a similar retrospective review. Fusion rates in their study were 44% in patients taking NSAIDs versus 87% for patients not taking NSAIDs. While the paucity of well-designed trials makes it difficult to truly extrapolate the risk of using COX inhibitors in orthopedic trauma patients, there is mounting evidence that COX inhibition plays a negative role in fracture healing. Because of the widespread use and value of COX inhibitors for analgesia, it is important that future clinical research into this topic focuses on dosing, length of treatment, and patient stratification to better define the future role of COX inhibition in the treatment of pain from skeletal injury.

# References

[1] Bolander ME. Regulation of fracture repair by growth factors. Proc Soc Exp Biol Med 1992;200(2): 165–70.

[2] Einhorn TA. The cell and molecular biology of fracture healing. Clin Orthop Relat Res 1998; 355(Suppl):S7–21.

[3] Ferguson C, Alpern E, Miclau T, et al. Does adult fracture repair recapitulate embryonic skeletal formation? Mech Dev 1999;87:57–66.

[4] Gerstenfeld LC, Cullinane DM, Barnes GL, et al. Fracture healing as a post-natal developmental process: molecular, spatial, and temporal aspects of its regulation. J Cell Biochem 2003b;88(5): 873–84.

[5] Vortkamp A, Pathi S, Peretti GM, et al. Recapitulation of signals regulating embryonic bone formation during postnatal growth and in fracture repair. Mech Dev 1998;71:65–76.

[6] Bais M, McLean J, Sebastiani P, et al. Transcriptional analysis of fracture healing and the induction of embryonic stem cell-related genes. PLoS One 2009;**4**(5):e5393.

[7] Phillips AM. Overview of the fracture healing cascade. Injury 2005;36S:S5–7.

[8] Buckwalter JA, Einhorn TA, Marsh JL. Bone and joint healing. In: Bucholz RW, Heckman JD, editors. Rockwood and Green's Fractures in Adults. Philadelphia, Lippincott: Williams, and Wilkins; 2001. p. 245–71.

[9] Gerstenfeld LC, Cho TJ, Kon T, et al. Impaired fracture healing in the absence of TNF-alpha signaling: the role of TNF-alpha in endochondral cartilage resorption. J Bone Miner Res 2003a; 18(9):1584–92.

[10] Gowen M, Wood DD, Ihrie EJ, et al. Stimulation by human interleukin 1 of cartilage breakdown and production of collagenase and proteoglycanase by human chondrocytes but not by human osteoblasts in vitro. Biochim Biophys Acta 1984;797(2):186–93.

[11] Gowen M, Wood DD, Russell RG. Stimulation of the proliferation of human bone cells in vitro by human monocyte products with interleukin-1 activity. J Clin Invest 1985;75(4):1223–9.

[12] Turck CW, Dohlman JG, Goetzl EJ. Immunological mediators of wound-healing and fibrosis. J Cell Physiol Suppl 1987;5:89–93.

[13] Einhorn TA, Majeska RJ, Rush EB, et al. The expression of cytokine activity by fracture callus. J Bone Miner Res 1995;10(8):1272–81.

[14] Barnes GL, Kostenuik PJ, Gerstenfeld LC, et al. Growth factor regulation of fracture repair. J Bone Miner Res 1999;14(11):1805–15. Review. No abstract available.

[15] Andrew JG, Andrew SM, Freemont AJ, et al. Inflammatory cells in normal human fracture healing. Acta Orthop Scand 1994;65(4):462–6.

[16] Park SH, Silva M, Bahk WJ, et al. Effect of repeated irrigation and debridement on fracture healing in an animal model. J Orthop Res 2002;20(6):1197–204.

[17] Ozaki A, Tsunoda M, Kinoshita S, et al. Role of fracture hematoma and periosteum during fracture healing in rats: interaction of fracture hematoma and the periosteum in the initial step of the healing process. J Orthop Sci 2005;5(1):64–70.

[18] Timlin M, Toomey D, Condron C, et al. Fracture hematoma is a potent proinflammatory mediator of neutrophil function. J Trauma 2005;58(6):1223–9.

[19] Schmidt-Bleek K, Schell H, Kolar P, et al. Cellular composition of the initial fracture hematoma compared to a muscle hematoma: a study in sheep. J Orthop Res 2009;27(9):1147–51.

[20] Kon T, Cho TJ, Aizawa T, et al. Expression of osteoprotegerin, receptor activator of NF-kappaB ligand (osteoprotegerin ligand) and related proinflammatory cytokines during fracture healing. J Bone Miner Res 2001;16(6):1004–14.

[21] Kitaori T, Ito H, Schwarz EM, et al. Stromal cell-derived factor 1/CXCR4 signaling is critical for the recruitment of mesenchymal stem cells to the fracture site during skeletal repair in a mouse model. Arthritis Rheum 2009;60(3):813–23.

[22] Santavirta S, Konttinen YT, Nordstrom D, et al. Immunologic studies of nonunited fractures. Acta Orthop Scand 1992;63(6):579–86.

[23] Askalonov AA. Changes in some indices of cellular immunity in patients with uncomplicated and complicated healing of bone fractures. J Hyg Epidemiol Microbiol Immunol 1981;25:307–10.

[24] Askalonov AA, Gordienko SM, Avdyunicheva OE, et al. The role of T system immunity in reparatory regeneration of the bone tissue in animals. J Hyg Epidemiol Microbiol Immunol 1987;31:219–24.

[25] Hauser CJ, Zhou X, Joshi P, et al. The immune microenvironment of human fracture/soft-tissue hematomas and its relationship to systemic immunity. J Trauma 1997;42:895–903. Discussion 903–4.

[26] Ferrick DA, Schrenzel MD, Mulvania T, et al. Differential production of interferon- and interleukin-4 in response to Th1- and Th2-stimulating pathogens by T cells in vivo. Nature 1995; 373:255–7.

[27] Holoshitz J, Koning F, Coligan JE, et al. Isolation of CD4-CD8-mycobacteria-reactive T lymphocyte clones from rheumatoid arthritis synovial fluid. Nature 1989;339:226–9.

[28] Havran WL, Chien YH, Allison JP. Recognition of self antigens by skin-derived T cells with invariant antigen receptors. Science 1991;252:1430–2.

[29] Jameson J, Ugarte K, Chen N, et al. A role for gammadelta skin T cells in wound repair. Science 2002;296:747–9.

[30] Colburn NT, Zaal KJ, Wang F, et al. A role for gamma/delta T cells in a mouse model of fracture healing. Arthritis Rheum 2009;60(6):1694–703.

[31] Vu TH, Shipley JM, Bergers G, et al. MMP-9/gelatinase B is a key regulator of growth plate angiogenesis and apoptosis of hypertrophic chondrocytes. Cell 1998;93:411–22.

[32] Colnot C, Thompson Z, T Miclau, et al. Altered fracture repair in the absence of MMP9. Development 2003;130(17):4123–33.

[33] Aizawa T, Kon T, Einhorn TA, et al. Induction of apoptosis in chondrocytes by tumor necrosis factor-alpha. J Orthop Res 2001;19(5):785–96.

[34] Gerstenfeld LC, Sacks DJ, Pelis M, et al. Comparison of effects of the bisphosphonate alendronate versus the RANKL inhibitor denosumab on murine fracture healing. J Bone Miner Res 2009; 24(2):196–208.

[35] Dominici M, LeBlanc K, Mueller I, et al. Minimal criteria for defining multipotent mesenchymal stromal cells. The International Society for Cellular Therapy position statement. Cytotherapy 2006; 8(4):315–17.

[36] Friedenstein AJ, Petrakova KV, Kurolesova AI, et al. Heterotopic of bone marrow. Analysis of precursor cells for osteogenic and hematopoietic tissues. Transplantation 1968;6(2):230–47.

[37] Caplan AI. Mesenchymal stem cells. J Orthop Res 1991;9(5):641–50.

[38] Corsi KA, Pollett JB, Phillippi JA, et al. Osteogenic potential of postnatal skeletal muscle-derived stem cells is influenced by donor sex. J Bone Miner Res 2007;22(10):1592–602.

[39] Schäffler AC, Büchler C. Concise review: adipose tissue-derived stromal cells – basic and clinical implications for novel cell-based therapies. Stem Cells 2007;25(4):818–27. Review.

[40] Zhang X, Awad HA, O'Keefe RJ, et al. A perspective: engineering periosteum for structural bone graft healing. Clin Orthop Relat Res 2008;466(8):1777–87. Review.

[41] Stagg J, Galipeau J. Immune plasticity of bone marrow-derived mesenchymal stromal cells. Handb Exp Pharmacol 2007;180:45–66.

[42] Noel D, Djouad F, Bouffi C, et al. Multipotent mesenchymal stromal cells and immune tolerance. Leuk Lymphoma 2007;48(7):1283–9.

[43] Nauta AJ, Fibbe WE. Immunomodulatory properties of mesenchymal stromal cells. Blood 2007;110 (10):3499–506.

[44] Spaggiari GM, Capobianco A, Abdelrazik H, et al. Mesenchymal stem cells inhibit natural killer-cell proliferation, cytotoxicity, and cytokine production: role of indoleamine 2,3-dioxygenase and prostaglandin E2. Blood 2008;111(3):1327–33.

[45] Ramasamy R, Fazekasova H, Lam EW, et al. Mesenchymal stem cells inhibit dendritic cell differentiation and function by preventing entry into the cell cycle. Transplantation 2007;83(1):71–6.

[46] Jiang XX, Zhang Y, Liu B, et al. Human mesenchymal stem cells inhibit differentiation and function of monocyte-derived dendritic cells. Blood 2005;105(10):4120–6.

[47] Tse WT, Pendleton JD, Beyer WM, et al. Suppression of allogeneic T-cell proliferation by human marrow stromal cells: implications in transplantation. Transplantation 2003;15(75(3)):389–97.

[48] Di Nicola M, Carlo-Stella C, Magni M, et al. Human bone marrow stromal cells suppress T-lymphocyte proliferation induced by cellular or nonspecific mitogenic stimuli. Blood 2002;99(10):3838–43.

[49] Zappia E, Casazza S, Pedemonte E, et al. Mesenchymal stem cells ameliorate experimental autoimmune encephalomyelitis inducing T-cell anergy. Blood 2005;106(5):1755–61.

[50] Le Blanc K, Tammik C, Rosendahl K, et al. HLA expression and immunologic properties of differentiated and undifferentiated mesenchymal stem cells. Exp Hematol 2003a;31(10):890–6.

[51] Le Blanc K, Tammik L, Sundberg B, et al. Mesenchymal stem cells inhibit and stimulate mixed lymphocyte cultures and mitogenic responses independently of the major histocompatibility complex. Scand J Immunol 2003b;57(1):11–20.

[52] Bartholomew A, Sturgeon C, Siatskas M, et al. Mesenchymal stem cells suppress lymphocyte proliferation in vitro and prolong skin graft survival in vivo. Exp Hematol 2002;30(1):42–8.

[53] Krampera M, Glennie S, et al. Bone marrow mesenchymal stem cells inhibit the response of naive and memory antigen-specific T cells to their cognate peptide. Blood 2003;101(9):3722–9.

[54] Augello A, Tasso R, Negrini SM, et al. Bone marrow mesenchymal progenitor cells inhibit lymphocyte proliferation by activation of the programmed death 1 pathway. Eur J Immunol 2005;35(5):1482–90.

[55] Corcione A, Benvenuto F, Ferretti E, et al. Human mesenchymal stem cells modulate B-cell functions. Blood 2006;107(1):367–72.

[56] Wu JY, Purton LE, Rodda SJ, et al. Osteoblastic regulation of B lymphopoiesis is mediated by Gs {alpha}-dependent signaling pathways. Proc Natl Acad Sci U S A 2008;105(44):16976–81.

[57] Simon AM, Manigrasso MB, O'Connor JP. Cyclo-oxygenase 2 function is essential for bone fracture healing. J Bone Miner Res 2002;17(6):963–76.

[58] Zhang X, Schwarz EM, Young DA, et al. Cyclooxygenase-2 regulates mesenchymal cell differentiation into the osteoblast lineage and is critically involved in bone repair. J Clin Invest 2002;109(11):1405–15.

[59] Walsh MC, Kim N, Kadono Y, et al. Osteoimmunology: interplay between the immune system and bone metabolism. Annu Rev Immunol 2006;24:33–63.

[60] Cho TJ, Lehmann W, Edgar C, et al. Tumor necrosis factor alpha activation of the apoptotic cascade in murine articular chondrocytes is associated with the induction of metalloproteinases and specific pro-resorptive factors. Arthritis Rheum 2003;48(10):2845–54.

[61] Ju ST, Panka DJ, Cui H, et al. Fas(CD95)/FasL interactions required for programmed cell death after T-cell activation. Nature 1995;373(6513):444–8.

[62] Kiener PA, Davis PM, Starling GC, et al. Differential induction of apoptosis by Fas-Fas ligand interactions in human monocytes and macrophages. J Exp Med 1997;185(8):1511–16.

[63] Schultz DR, Harrington Jr WJ. Apoptosis: programmed cell death at a molecular level. Semin Arthritis Rheum 2003;32(6):345–69.

[64] Abu-Amer Y, Erdmann J, Alexopoulou L, et al. Tumor necrosis factor receptors types 1 and 2 differentially regulate osteoclastogenesis. J Biol Chem 2000;275(35):27307–10.

[65] Zhang YH, Heulsmann A, Tondravi MM, et al. Tumor necrosis factor-alpha (TNF) stimulates RANKL-induced osteoclastogenesis via coupling of TNF type 1 receptor and RANK signaling pathways. J Biol Chem 2001;276(1):563–8.

[66] Yamaguchi T, Takada Y, Maruyama K, et al. Fra-1/AP-1 Impairs inflammatory responses and chondrogenesis in fracture healing. J Bone Miner Res 2009;24(12):2056–65.

[67] Dougall WC, Glaccum M, Charrier K, et al. RANK is essential for osteoclast and lymph node development. Genes Dev 1999;13(18):2412–24.

[68] Kong YY, Yoshida H, Sarosi I, et al. OPGL is a key regulator of osteoclastogenesis, lymphocyte development and lymph-node organogenesis. Nature 1999;397(6717):315–23.

[69] Wong BR, Josien R, Lee SY, et al. The TRAF family of signal transducers mediates NF-kappaB activation by the TRANCE receptor. J Biol Chem 1998;273(43):28355–9.

[70] Kobayashi N, Kadono Y, Naito A, et al. Segregation of TRAF6-mediated signaling pathways clarifies its role in osteoclastogenesis. EMBO J 2001;20(6):1271–80.

[71] Gohda J, Akiyama T, Koga T, et al. RANK-mediated amplification of TRAF6 signaling leads to NFATc1 induction during osteoclastogenesis. EMBO J 2005;24(4):790–9.

[72] Crabtree GR, Olson EN. NFAT signaling: choreographing the social lives of cells. Cell 2002;109 (Suppl):S67–79.

[73] Takayanagi H, Kim S, Koga T, et al. Induction and activation of the transcription factor NFATc1 (NFAT2) integrate RANKL signaling in terminal differentiation of osteoclasts. Dev Cell 2002;3 (6):889–901.

[74] Simonet WS, Lacey DL, Dunstan CR, et al. Osteoprotegerin: a novel secreted protein involved in the regulation of bone density. Cell 1997;89(2):309–19.

[75] Bucay N, Sarosi I, Dunstan CR, et al. Osteoprotegerin-deficient mice develop early onset osteoporosis and arterial calcification. Genes Dev 1998;12(9):1260–8.

[76] Modrowski D, Godet D, Marie PJ. Involvement of interleukin 1 and tumour necrosis factor alpha as endogenous growth factors in human osteoblastic cells. Cytokine 1995;7(7):720–6.

[77] Lehmann W, Edgar CM, Wang K, et al. Tumor necrosis factor alpha (TNF-alpha) coordinately regulates the expression of specific matrix metalloproteinases (MMPS) and angiogenic factors during fracture healing. Bone 2005;36(2):300–10.

[78] Kamimura D, Ishihara K, Hirano T. IL-6 signal transduction and its physiological roles: the signal orchestration model. Rev Physiol Biochem Pharmacol 2003;149:1–38.

[79] Sims NA, Jenkins BJ, Quinn JM, et al. Glycoprotein 130 regulates bone turnover and bone size by distinct downstream signaling pathways. J Clin Invest 2004;113(3):379–89.

[80] Roodman GD. Interleukin-6: an osteotropic factor? J Bone Miner Res 1992;7(5):475–8.

[81] Palmqvist P, Persson E, Conaway HH, et al. IL-6, leukemia inhibitory factor, and oncostatin M stimulate bone resorption and regulate the expression of receptor activator of NF-kappa B ligand, osteoprotegerin, and receptor activator of NF-kappa B in mouse calvariae. J Immunol 2002; 169(6):3353–62.

[82] Alonzi T, Fattori E, Lazzaro D, et al. Interleukin 6 is required for the development of collagen-induced arthritis. J Exp Med 1998;187(4):461–8.

[83] Wong PK, Campbell IK, Egan PJ, et al. The role of the interleukin-6 family of cytokines in inflammatory arthritis and bone turnover. Arthritis Rheum 2003;48(5):1177–89.

[84] Bertolini DR, Nedwin GE, Bringman TS, et al. Stimulation of bone resorption and inhibition of bone formation in vitro by human tumour necrosis factors. Nature 1986;319(6053):516–18.

[85] Kimble RB, Bain S, Pacifici R. The functional block of TNF but not of IL-6 prevents bone loss in ovariectomized mice. J Bone Miner Res 1997;12(6):935–41.

[86] Nagata S. Apoptosis by death factor. Cell 1997;88(3):355–65.

[87] Lee FY, Choi YW, Behrens FF, et al. Programmed removal of chondrocytes during endochondral fracture healing. J Orthop Res 1998;16(1):144–50.

[88] Balga R, Wetterwald A, Portenier J, et al. Tumor necrosis factor-alpha: alternative role as an inhibitor of osteoclast formation in vitro. Bone 2006;39(2):325–35.

[89] Gerstenfeld LC, Cho TJ, Kon T, et al. Impaired intramembranous bone formation during bone repair in the absence of tumor necrosis factor-alpha signaling. Cells Tissues Organs 2001;169(3):285–94.

[90] Kim N, Odgren PR, Kim DK, et al. Diverse roles of the tumor necrosis factor family member TRANCE in skeletal physiology revealed by TRANCE deficiency and partial rescue by a lymphocyte-expressed TRANCE transgene. Proc Natl Acad Sci U S A 2000;97(20):10905–10.

[91] Watanabe Y, Namba A, Honda K, et al. IL-1beta stimulates the expression of prostaglandin receptor EP4 in human chondrocytes by increasing production of prostaglandin E2. Connect Tissue Res 2009;50(3):186–93.

[92] Cho TJ, Kim JA, Chung CY, et al. Expression and role of interleukin-6 in distraction osteogenesis. Calcif Tissue Int 2007;80(3):192–200.

[93] Yang X, Ricciardi BF, Hernandez-Soria A, et al. Callus mineralization and maturation are delayed during fracture healing in interleukin-6 knockout mice. Bone 2007;41(6):928–36.

[94] Vane JR, Botting RM. Mechanism of action of aspirin-like drugs. Semin Arthritis Rheum 1997;26(6 Suppl. 1):2–10.

[95] Davies P, Bailey PJ, Goldenberg MM, et al. The role of arachidonic acid oxygenation products in pain and inflammation. Annu Rev Immunol 1984;2:335–57.

[96] Fitzgerald GA, Patrono C. The coxibs, selective inhibitors of cyclooxygenase-2. N Engl J Med 2001;345(6):433–42.

[97] Beck A, Krischak G, Sorg T, et al. Influence of diclofenac (group of nonsteroidal anti-inflammatory drugs) on fracture healing. Arch Orthop Trauma Surg 2003;123(7):327–32.

[98] Gerstenfeld LC, Thiede M, Seibert K, et al. Differential inhibition of fracture healing by non-selective and cyclooxygenase-2 selective non-steroidal anti-inflammatory drugs. J Orthop Res 2003c;21(4):670–5.

[99] Brown KM, Saunders MM, Kirsch T, et al. Effect of COX-2-specific inhibition on fracture-healing in the rat femur. J Bone Joint Surg Am 2004;86-A(1):116–23.

[100] Krischak GD, Augat P, Blakytny R, et al. The non-steroidal anti-inflammatory drug diclofenac reduces appearance of osteoblasts in bone defect healing in rats. Arch Orthop Trauma Surg 2007a;127(6):453–8.

[101] Krischak GD, Augat P, Sorg T, et al. Effects of diclofenac on periosteal callus maturation in osteotomy healing in an animal model. Arch Orthop Trauma Surg 2007b;127(1):3–9.

[102] Kaspar D, Hedrich CM, Schmidt C, et al. Diclofenac inhibits proliferation and matrix formation of osteoblast cells. Unfallchirurg 2005;108(1):18, 20–4.

[103] Weinreb M, Shamir D, Machwate M, et al. Prostaglandin E2 (PGE2) increases the number of rat bone marrow osteogenic stromal cells (BMSC) via binding the EP4 receptor, activating sphingosine kinase and inhibiting caspase activity. Prostaglandins Leukot Essent Fatty Acids 2006;75(2):81–90.

[104] Shamir D, Keila S, Weinreb M. A selective EP4 receptor antagonist abrogates the stimulation of osteoblast recruitment from bone marrow stromal cells by prostaglandin E2 in vivo and in vitro. Bone 2004;34(1):157–62.

[105] Sudmann E, Dregelid E, Bessesen A, et al. Inhibition of fracture healing by indomethacin in rats. Eur J Clin Invest 1979;9(5):333–9.

[106] Allen HL, Wase A, Bear WT. Indomethacin and aspirin: effect of nonsteroidal anti-inflammatory agents on the rate of fracture repair in the rat. Acta Orthop Scand 1980;51(4):595–600.

[107] Engesaeter LB, Sudmann B, Sudmann E. Fracture healing in rats inhibited by locally administered indomethacin. Acta Orthop Scand 1992;63(3):330–3.

[108] Altman RD, Latta LL, Keer R, et al. Effect of nonsteroidal antiinflammatory drugs on fracture healing: a laboratory study in rats. J Orthop Trauma 1995;9(5):392–400.

[109] Bo J, Sudmann E, Martin PF. Effect of indomethacin on fracture healing in rats. Acta Orthop Scand 1976;47(6):588–99.

[110] Gerstenfeld LC, Al-Ghawas M, Alkhiary YM, et al. Selective and nonselective cyclooxygenase-2 inhibitors and experimental fracture-healing. Reversibility of effects after short-term treatment. J Bone Joint Surg Am 2007;89(1):114–25.

[111] Glassman SD, Rose SM, Dimar JR, et al. The effect of postoperative nonsteroidal anti-inflammatory drug administration on spinal fusion. Spine (Phila Pa 1976) 1998;23(7):834–8.

[112] Giannoudis PV, MacDonald DA, Matthews SJ, et al. Nonunion of the femoral diaphysis. The influence of reaming and non-steroidal anti-inflammatory drugs. J Bone Joint Surg Br 2000;82(5): 655–8.

[113] Burd TA, Hughes MS, Anglen JO. Heterotopic ossification prophylaxis with indomethacin increases the risk of long-bone nonunion. J Bone Joint Surg Br 2003;85(5):700–5.

[114] Bhattacharyya T, Levin R, Vrahas MS, et al. Nonsteroidal antiinflammatory drugs and nonunion of humeral shaft fractures. Arthritis Rheum 2005;53(3):364–7.

[115] Deguchi M, Rapoff AJ, Zdeblick TA, et al. Posterolateral fusion for isthmic spondylolisthesis in adults: analysis of fusion rate and clinical results. J Spinal Disord 1998;11(6):459–64.

# 13

# The Role of the Immune System and Bone Cells in Acute and Chronic Osteomyelitis

Brendan F. Boyce[1,2], Lianping Xing[1,2], Edward M. Schwarz[1,3]

[1]CENTER FOR MUSCULOSKELETAL RESEARCH, UNIVERSITY OF ROCHESTER MEDICAL CENTER, ROCHESTER, NY, USA, [2]DEPARTMENT OF PATHOLOGY AND LABORATORY MEDICINE, UNIVERSITY OF ROCHESTER MEDICAL CENTER, ROCHESTER, NY, USA, [3]DEPARTMENT OF ORTHOPAEDICS AND REHABILITATION, UNIVERSITY OF ROCHESTER MEDICAL CENTER, ROCHESTER, NY, USA

## CHAPTER OUTLINE

| | |
|---|---|
| Introduction | 369 |
| Mechanism of microbial infection in the pathogenesis of osteomyelitis | 370 |
| The host response to osteomyelitis | 372 |
| Osteoblasts: the last line of defense and a novel substrate for bone infections | 374 |
| Osteoclasts modulate immune responses and functions of various cells during infections in bone | 375 |
|     Cellular responses to acute and chronic osteomyelitis | 375 |
|     Osteoclast mobilization | 376 |
|     Osteoclasts as immune cells | 377 |
| Osteoclasts and osteoclast precursors affect osteoblast functions | 378 |
| Osteoclasts as regulators of hematopoietic stem cell functions | 379 |
| The role of dendritic cells and cellular immunity in bone infection | 380 |
| DCs in adaptive immunity and infection | 380 |
| DCs in innate immunity and infection | 381 |
| DCs and osteoclasts in infection | 382 |
| The role of B cells in bone infection and the potential of passive immunization | 383 |
| References | 384 |

## Introduction

Although humans have been combating microbial pathogens since prehistoric times, infectious disease remains one of our greatest public health challenges. As testimony to the importance of this issue, antibiotic drugs and prophylactic vaccines are considered the first and third greatest scientific achievements of the twentieth century. Despite these major medical advances, one needs to look no further than the recent worldwide pandemic of H1N1 influenza virus infection to understand how vulnerable we remain to pathogenic organisms [1]. We are also vulnerable to infections of bone (osteomyelitis), which in many cases remain incurable. Despite the development of superior orthopedic surgical techniques, newer diagnostic tools, and innovative treatments, new problems including multidrug resistance and evolving bacterial virulence allow osteomyelitis to remain a significant challenge in orthopedics. Most notable is the resurgence of meticillin-resistant *Staphylococcus aureus* (MRSA) in healthy patients undergoing elective arthroplasty for degenerative arthritis [2]. Furthermore, the outcome after currently available treatment is often unsatisfactory, with a therapy failure rate greater than 20% [3]. Military conflicts in the Middle-East have seen the emergence of new musculoskeletal infections, such as *Acinetobacter baumannii*, in war-wounded soldiers at unprecedented levels [4]. The incidence of an enigmatic condition known as osteonecrosis of the jaw (ONJ) has increased significantly in recent years, particularly in patients with oral infections and multiple myeloma or metastatic bone disease treated with bisphosphonates to inhibit osteoclastic bone resorption. The precise etiology of ONJ in this setting of bisphosphonate treatment remains undetermined, but the association raises the possibility that osteoclasts derived from bone marrow myeloid precursors may play important roles in inflammatory responses [5]. Recent data suggesting that osteoclastic cells play important roles in immune responses in bone have spawned the new field of osteoimmunology [6]. Thus, a new emphasis has been placed on osteoimmunology to better explain the pathogenesis of bone infections and to derive novel interventions for osteomyelitis. Here, we provide an update on the mechanisms of microbial pathogenesis and the host response to bone infections, with a focus on immune and bone cell interactions during this process.

# Mechanism of microbial infection in the pathogenesis of osteomyelitis

Although many microbes have the ability to invade musculoskeletal tissue, the most common microorganisms that cause infections of bone (osteomyelitis) requiring medical attention are bacteria. Osteomyelitis is characterized by early progressive inflammatory destruction of bone with later apposition of new bone as part of the reparative response [7]. Many confounding variables, including the relatively low incidence of osteomyelitis, have made clinical research on the pathogenesis of acute and chronic osteomyelitis very difficult. Thus, much of our knowledge of the pathogenesis of osteomyelitis has come from animal models [8], which have been developed for chickens, mice, rats, guinea pigs, rabbits, dogs, and sheep. From these studies, it is clear that normal bone is highly resistant to infection and that osteomyelitis can occur only as a result of very large bacterial inocula, trauma, necrosis, and/or ischemia of tissue, the presence of a foreign body, or combinations of these [7]. Orthopedic surgery can require the implantation of plates, screws, prostheses, or cement, and can result in soft tissue damage, compromised blood flow, and hematoma formation. Thus, it is important to understand the susceptibility to infection that occurs as a direct consequence of orthopedic procedures.

The first step that bacterial pathogens must take to establish chronic infection of bone is entry into the host. Next, individual bacteria must adhere to and colonize host tissue, which is most commonly achieved via production of a glycocalyx, known as a biofilm. This biofilm is impenetrable by antibiotics or immune molecules, and represents a disease state that can only be managed, but not cured. In the case of osteomyelitis, the bacteria can enter the bone via three different routes: (1) Inoculation from the blood, known as hematogenous osteomyelitis, is the typical route in children [9]. This form of infection is also associated with a primary source of infection elsewhere in the body. Total joint replacements can also become infected by hematogenous spread [1]. (2) Fracture is known to facilitate the development of osteomyelitis from hematogenous seeding [11]. However, tissue damage is not a prerequisite for bacteria to lodge in bones. (3) Spread from adjacent tissues occurs commonly in the feet of patients with diabetic ulcers and fistulae, and in patients with peripheral vascular disease [12].

A common cause of acute osteomyelitis is trauma, which typically is not associated with bacteremia. Osteomyelitis is a very unusual complication of closed fractures, but approximately 60–70% of open fractures are contaminated by bacteria. Thus, infection of these fractures should be anticipated [13]. Specific high-risk environmental exposures at the time of injury are well recognized [13]. These include *Clostridium perfringens* associated with farm injuries, *Pseudomonas aeruginosa* and *Aeromonas hydrophila* after injury in freshwater, and Vibrio and Erysipelothrix infections with salt-water exposures. However, even with direct contamination of bone, most open fractures do not develop chronic osteomyelitis. In fact, most cases of post-traumatic osteomyelitis are caused by hospital-acquired pathogens, such as coagulase-positive staphylococci or enteric Gram-negative bacilli, like *P. aeruginosa* [14]. Furthermore, swabs of ulcers or fistulae that often lead to growth of coagulase-negative staphylococci can provide misleading information to suggest that these are the commonest causative microorganisms in cases of acute osteomyelitis [15].

One of the rate-limiting steps in the establishment of chronic osteomyelitis is bacterial attachment to host tissues. Intact host tissue is a poor substrate for bacterial adherence, and animal models typically require some form of bone damage to reliably establish infection. In humans, the risk of infection subsequent to trauma correlates with the degree of soft tissue injury associated with an open fracture [16]. For example, the infection rates of type 3B open fractures, which have extensive soft tissue stripping and insufficient soft tissue coverage over the fracture, are associated with a 15–40% risk of infection [14]. Besides compromising the host response, the traumatic event exposes sites for bacterial attachment and binding to bone. Once inside the host, pathogenic bacteria express a unique set of virulence genes and proteins that enable them to establish infection. Several of these mechanisms of pathogenesis for *Staphylococcus aureus* have been elucidated in elegant models of osteomyelitis in which bacteria express receptors (adhesins) for numerous host proteins, including injury-exposed collagen, laminin, bone sialoglycoprotein, and fibronectin, which covers damaged tissue and orthopedic implants [17]. As an example, Greene et al. demonstrated that the *S. aureus fnb* gene functions as an adhesin by binding to fibronectin [18], and studies from Fischer et al. showed the importance of fibronectin expression by *S. aureus* for its adhesion to titanium implants in vivo [19].

One salient feature of polymicrobial osteomyelitis is the presence of necrotic tissue and/or foreign material, which facilitate bacterial attachment. This is followed immediately by the production of biofilm that firmly establishes the infection [20]. Biofilm forms strong bonds with glycoproteins of the host tissue and is considered to be the factor most responsible for the difficulty in eradicating bacteria from bone. This is largely because biofilm acts as a dominant barrier to protect bacteria from the action of antibiotics, phagocytic cells, and antibodies, and impairs lymphocyte functions [21].

# The host response to osteomyelitis

The immediate host response to most bacterial infections is an acute inflammatory reaction [22]. Both tissue injury and bacteria trigger activation of the complement cascade, which leads to local vasodilation, tissue edema, migration of polymorphonuclear leukocytes to the lesion, as well as coating of bacteria with osponins to facilitate their intracellular uptake. Cytokines, including interleukin-1 (IL-1), IL-6, and tumor necrosis factor (TNF), are released from the site of injury and act as chemotactic factors and activators of phagocytic cells (polymorphonuclear leukocytes and macrophages) to recruit them and stimulate their production of cytolytic free radicals ($O_2^-$ and $NO^-$). This pathway, referred to as the innate immune response, is the most essential part of host defense against infection. Animals that are depleted of granulocytes rapidly succumb to non-infectious inocula of *S. aureus* and die from sepsis hours after inoculation [23]. While the innate response is highly effective under normal conditions, its power to induce host tissue damage also needs to be better understood. For example, purified bacterial components such as lipopolysaccharides (LPS) from the membrane of Gram-negative bacteria, or similar complex carbohydrates from Gram-positive bacteria, bind to specific receptors on host cells (Toll receptors) and can trigger toxic shock syndrome and death through this pathway even faster than infection itself.

Acquired immunity functions primarily to clear low numbers of persistent bacteria and to prevent recurrence of the same infection. This adaptive host response involves two important mechanisms: the cellular response in which cytotoxic or CD8$^+$ T lymphocytes lyse host cells that are infected with bacteria; and the humoral response in which B lymphocytes produce antibodies against the bacteria. Central to this acquired immunity are macrophages that phagocytose the bacteria and present bacterial antigens to helper or CD4$^+$ T lymphocytes. Macrophages orchestrate the cellular response by producing a specific set of lymphokines termed T$_H$1 (i.e. IL-12 and interferon-$\gamma$) or mediate humoral immunity by producing T$_H$2 lymphokines (i.e. IL-4 and IL-13) [24]. This process takes at least 2 weeks to reach peak effectiveness. This response is negatively regulated by suppressor T lymphocytes through poorly defined mechanisms, whose functions are to limit the extent of host tissue damage and prevent autoimmunity. However, the response can be used by the bacteria to establish disease [25]. As the bacteria are cleared, the antigenic challenge diminishes until all that remains are memory T and B lymphocytes, but these cells can participate with the innate immune response to react immediately to bacterial reactivation or reinfection.

Neutrophils and macrophages are the main phagocytic cells that participate in acute and chronic infections in bone and most other tissues. They are derived from myeloid precursors in the bone marrow. Neutrophils leave the marrow cavity as fully differentiated cells ready to participate in acute infections. They are mobilized as a consequence of the effects of chemokines, factors released from sites of acute inflammation. They are attracted to acute inflammatory foci by other chemokines. There they engulf bacteria and dead cellular debris. They are short-lived cells and many of them die at sites of acute inflammation where they comprise much of the material seen in pus, the yellow cellular debris seen in the centers of pustules and abscesses. Macrophages are also critically responsible for the removal of planktonic bacteria, apoptotic cells harboring pathogens, and necrotic tissue caused by the infections. This process is largely dependent on antibody binding to bacterial antigens, or opsonization, which has two important functions. First, it anchors the bacteria to Fc-receptors on the phagocytic cells [26]. Secondly, Fc-receptor-mediated phagocytosis activates intracellular signaling pathways that stimulate the production of free radicals necessary to kill the pathogen (i.e. superoxide and nitric oxide) [26].

An effective host response is the most important determinant of the ultimate outcome of osteomyelitis. This is reflected in the Cierny-Mader classification of long-bone osteomyelitis, which incorporates an assessment of local or systemic host response into the classification algorithm [27]. Studies using this classification system have found that the strongest predictor of failure is the health state of the host and not the extent of the bone involvement [3]. It is for this reason that patients with specific defects in host defense are particularly susceptible to osteomyelitis. Patients with chronic granulomatous disease have phagocytes that are unable to produce superoxide, the reactive oxygen species, which is required for normal bactericidal activity. Chronic osteomyelitis caused by *S. aureus* has been observed in almost 33% of these patients [14]. Abnormalities in immunoglobulin production or complement in these patients lead to a high risk of infection by encapsulated bacteria, like *Streptococcus pneumoniae*, *Haemophilus influenzae*, and *Neisseria*. These patients are also susceptible to *Mycoplasma* and *Ureaplasma* infections [14]. Patients with defects in cell-mediated immunity are at greatest risk for infections with intracellular pathogens, like *Mycobacteria*,

*Salmonella* and *Nocardia*. General host defense abnormalities that increase the risk for osteomyelitis are found in patients with sickle cell disease or diabetes, intravenous drug users, and the elderly [14].

## Osteoblasts: the last line of defense and a novel substrate for bone infections

Osteoblasts are specialized cells derived from mesenchymal precursors and are responsible for new bone formation. They make new bone by secreting matrix proteins, most prominently type I collagen, directly on the surface of bone, which has been resorbed previously by osteoclasts, and to which they are tightly attached. Although modern mouse genetic studies have been able to distinguish different kinds of osteoblastic vs. mesenchymal stromal cells [28], collectively, these bone cells are the last line of defense for the host to ward off bacterial colonization and biofilm formation in bone. To assist them in this effort, osteoblastic cells express an array of immunostimulatory molecules that are triggered by their Toll-like receptors (TLR) following ligation by bacterial factors. Osteoblasts express TLR-2, 4, and 9 [29, 30], which enable them to respond to bacterial cell surface lipopolysaccharides and DNA (CpG oligonucleotides). Following activation of these receptors by bacterial ligands, osteoblasts produce antimicrobial peptides (e.g. beta defensin-3) [29], chemokines (e.g. CCL2, CCL5, CXCL8, CXCL10) [31–33], inflammatory cytokines (e.g. IL-6) [32], co-stimulatory molecules (e.g. CD40) [34], and MHC II [34], presumably in an effort to limit bacterial infections in bone. The findings that osteoblasts participate in the innate immune response to infection are somewhat surprising. However, the early observations that bacterial ligands induce co-stimulatory molecules that are normally restricted to lymphocytes, and MHC II molecules that are normally restricted to phagocytic antigen-presenting cells, suggests that osteoblasts play an unexpected central role in the cellular immune response to bacterial infections.

Another implication of osteoblast expression of MHC II, whose function is to present exogenous antigens to helper T-cells, is that osteoblasts would internalize bacteria during infection. This theory was supported by early reports demonstrating that the *S. aureus*, which is known to be an exclusive extracellular pathogen, is internalized by osteoblasts in vitro [35], and may promote infection by inducing osteoblast apoptosis [36]. While many regarded these findings as in vitro artifacts, Reilly et al. went on to demonstrate internalization of bacteria by embryonic chick osteoblasts in vivo [37], and Nair et al. recently identified the *S. aureus* sigma B regulon as a key mediator of bacterial internalization by osteoblasts [38]. Understanding of the roles of osteoblastic cells in immune responses is at an early stage. Thus, the mechanisms by which bacteria infect osteoblasts in vivo and how bone formation is inhibited in acute osteomyelitis (see below) are important areas of future research.

## Osteoclasts modulate immune responses and functions of various cells during infections in bone

### Cellular responses to acute and chronic osteomyelitis

Osteoclasts, derived from mononuclear precursors in the myelomonocytic lineage, have long been recognized as the cells that erode bone at sites of acute and chronic infections in bone. This lineage also gives rise to polymorphonuclear leukocytes, the most abundant phagocytic immune cells recruited to sites of acute inflammation, and to macrophages, the commonest phagocytes in chronic infections. Bacteria and bacterial products, such as LPS, endotoxin, and surface-associated material (SAM), induce osteoclast formation and activation [39–42]. LPS induces the expression of osteoclastogenic cytokines, including RANKL, TNF, IL-1, and IL-6 by osteoblastic and other cells [43], leading to an increase in the RANKL/OPG ratio in favor of resorption. SAM appears to induce osteoclastogenesis from human monocytes independent of RANKL [42], a characteristic it shares with TNF [44, 45]. Interestingly, TNF also limits osteoclast formation and inflammation in inflammatory arthritis by increasing expression and preventing processing of the inhibitory NF-κB protein, p100, to its active form, p52 [45]. Whether TNF limits osteoclastogenesis in acute or chronic osteomyelitis has not been investigated, but it is likely that unprocessed p100 would limit osteoclast formation in most pathologic states.

TNF, IL-1, IL-6, and other factors induced in response to bacterial inflammation are likely to be responsible for the inhibition of bone formation seen at sites of acute inflammation in bone to account for the failure of coupling of formation to resorption that is seen during normal bone remodeling. These cytokines can inhibit bone formation through a variety of mechanisms, including direct inhibition of osteoblast precursor proliferation and differentiation, and of osteoblast activity and survival [46]. Surprisingly, although bacteria have been found inside osteoblasts, few studies have been done to determine if the bacteria responsible for acute and chronic osteomyelitis or their products have direct inhibitory effects on osteoblasts of their precursors, or on known stimulators or recently identified inhibitors of bone formation, such as sclerostin, DKK1, and noggin.

There is a well-established and accepted vicious cycle hypothesis that was developed in metastatic bone disease to explain the rapid osteolysis seen around deposits of tumor cells after they metastasize to bone [47]. This hypothesis is supported by evidence that tumor cells are attracted to sites of increased bone resorption where factors, such as TGFβ released from bone matrix during resorption, enhance tumor cell growth and more bone resorption. TGFβ can stimulate release of factors, such as PTHrP, from metastatic breast cancer cells. This, in turn, stimulates the release of RANKL from osteoblastic cells in marrow, which then enhances osteoclastogenesis and bone resorption to perpetuate the vicious cycle [48]. In murine models of metastatic bone disease, tumor cells injected into the left cardiac ventricle typically are attracted to the metaphyses of long bones of growing mice where the vicious cycle is initiated. Similarly, in children who develop acute osteomyelitis by hematogenous spread, bacteria colonize the metaphyses of their long bones. This raises the possibility that circulating bacteria are attracted to sites of bone resorption by factors released from bone, similar to that seen in metastatic bone disease. Surprisingly, however, a vicious cycle hypothesis has never been tested in animal models. Thus, the mechanisms whereby bacteria colonize particular sites in bone during septicemia remain to be fully determined.

Chronic osteomyelitis develops in a number of different clinical settings. These include: development from acute osteomyelitis at sites where infection has been treated inadequately; de novo infection as a result of an inadequate immune response to infection; and direct extension from adjacent infection, such as infected gangrene in diabetic atherosclerotic patients. Chronic osteomyelitis is characterized by a central focus of osteolysis in which plasma cells, lymphocytes, and macrophages predominate. This is associated with a reparative osteosclerotic response in which bone formation and fibrosis are stimulated presumably to help confine the inflammation. The mechanisms whereby this new bone formation is stimulated have not been investigated and remain undetermined.

## Osteoclast mobilization

Osteoclast precursors (OCPs) are more abundant than osteoclasts, but they are less easy to identify in H&E-stained tissue sections because they cannot be distinguished readily from other mononuclear cells with similar morphology. They arise in the bone marrow under the influence of M-CSF and RANKL, which are expressed predominantly by cells in the osteoblastic lineage [49]. Rather than moving directly from their site of origin in the marrow to immediately adjacent sites where they are required for bone resorption, OCPs appear to circulate in the bloodstream to which they appear to be attracted by a number of influences. These influences include changes in the concentration locally of the chemokines, such as SDF-1 [50, 51], which regulates hematopoietic cell egress and retention in bone marrow. Marrow levels of SDF-1 are reduced in response to high concentrations of TNF in a TNF-transgenic mouse model of RA [52]. This drop in the concentration of SDF-1 allows OCPs and other cells to leave the marrow and go to sites of inflammation.

OCPs are attracted into the bloodstream by another chemokine, sphingosine-1 phosphate (S1P) [53]. S1P is a bioactive lipid that regulates proliferation, survival, and migration of many cell types, especially immune cells, and plays an essential role in angiogenesis and lymphocyte trafficking [54]. It is released by red blood cells and endothelial cells to maintain a high blood/tissue gradient and is thought to protect bone from resorption by keeping OCPs in the blood and away from bone surfaces [53]. It is possible that S1P concentrations are increased at sites of infection following its release there by endothelial cells and platelets. In this way it could attract OCPs from the blood back to these sites to differentiate into osteoclasts. Osteoclasts, but not OCPs, express S1P on their cell membranes [55], and so could attract OCPs to sites of active resorption to increase mature osteoclast number and size. Interestingly, this may also be a mechanism whereby circulating or local osteoblast precursors are attracted to sites of resorption where they can function to fill in resorption lacunae. Bone formation typically is inhibited at sites of acute inflammation, and factors such as TNF inhibit bone formation by a variety of mechanisms, including suppression of precursor differentiation.

## Osteoclasts as immune cells

Immune cells can be defined in a variety of ways, based on their involvement in innate and acquired immune responses. For example, macrophages, dendritic cells, and T and B lymphocytes have been recognized as immune cells for many years. This reflects their involvement in immune responses and functions, such as expression of Fcγ receptors, cytokines, and chemokines, phagocytosis, antigen presentation, and differentiation into activated or antibody-producing cells. It has also been recognized that OCPs are involved in the immune responses of inflammatory processes affecting bone, but they have rarely been considered as immune cells. For example, early studies reported that osteoclasts produce pro-inflammatory cytokines, including IL-6, IL-1, IL-8, and TNF, in common diseases, such as RA, acute osteomyelitis, giant cell tumor of bone, and Paget's disease, and that they express receptors for cytokines [56]. Later studies showed that osteoclasts also express chemokines, including MIP-1α, RANTES, and monocyte chemo-attractant protein-3 (MCP-3), which stimulated chemotaxis of OCPs and osteoclastogenesis [57]. RANTES and especially MIP-1α increased the motility of osteoclasts, but not their resorptive activity, adhesion, or survival [57]. On the basis of these observations, the authors of these studies suggested that osteoclasts could participate in autocrine and paracrine regulation of their resorptive functions in inflammatory conditions affecting the skeleton. However, they did not conclude that osteoclasts themselves might be immune cells. They were not alone. Indeed, immunologists and hematologists have not included osteoclasts in cartoons of cells derived from HSCs until recently, and in many reviews, osteoclasts are still excluded.

We recently have pursued the notion that osteoclasts and OCPs are involved in self-amplifying cycles of autocrine stimulation in inflammatory arthritis and have proposed that they also function as immune cells [58]. The release of cytokines by OCPs and osteoclasts can lead to an increase in their numbers and activity through paracrine and autocrine actions. Specifically, in addition to induction of expression of RANKL by stromal cells, some of these cytokines such as TNF can also induce differentiation of OCPs directly and increase their expression of cytokines [59]. Furthermore, TNF and RANKL induce c-Fos expression in OCPs and osteoclasts [59]. c-Fos in turn increases the resorptive activity of osteoclasts in vitro in response to RANKL, TNF, and IL-1 [60]. Thus, cytokines can initiate and maintain self-amplifying autocrine cycles to increase not only osteoclast numbers, but also osteoclast activity and thus more bone resorption. In addition, OCP interaction with bone matrix can increase their resorptive activity by a number of mechanisms. These include integrin-mediated anti-apoptotic signaling [61], and secretion of IL-1 [59], which can increase osteoclast activity and survival. Thus there are several mechanisms whereby OCPs and osteoclasts can influence their functions and those of other cells at sites of inflammation in bone.

Like macrophages and dendritic cells, osteoclasts not only secrete cytokines and chemokines, they also express Fcγ receptors [62], and thus could function as immune cells and augment or diminish immune responses. OCPs can be detected in the blood of mice using antibodies to CD11b, c-kit, Gr-1, c-fms (the receptor for M-CSF), and RANK and in humans [52], and their numbers are increased in the blood of patients with multiple myeloma and psoriatic arthritis [63]. The definition of what constitutes an immune cell is controversial and poorly defined. Characteristics include presentation of antigens, expression of Fcγ and MHC II proteins, and secretion of and responsiveness to cytokines and chemokines. OCPs and osteoclasts possess most of these characteristics. They are abundant in most inflammatory conditions in bone and likely play more roles in these conditions than we currently are aware of.

## Osteoclasts and osteoclast precursors affect osteoblast functions

OCPs and osteoclasts can also affect osteoblastic cell functions and regulate bone mass. The first direct evidence came from a study reporting an unexpected role for EphrinB2, a ligand expressed by OCPs, and its receptor, EphB4, expressed by osteoblast precursors [64]. Direct interaction between these two cell types resulted in so-called reverse signaling through EphrinB2 into OCPs to inhibit RANKL-induced activation of c-Fos and NFATc1. This reduces osteoclastogenesis because these are two transcription factors activated by NF-κB downstream from RANK and essential for OCP differentiation [60]. More intriguing is that forward signaling through EphB4 promotes osteoblast precursor differentiation. A recent study has reported that the ephrinB2/EphB4 axis is dysregulated in multiple myeloma, which is characterized by marked osteolysis and inhibition of bone formation [64]. These investigators found that EphB4-Fc treatment inhibited myeloma growth, osteoclastogenesis, and angiogenesis in SCID-hu mice, and stimulated osteoblastogenesis and bone formation. In contrast, ephrinB2-Fc stimulated angiogenesis, osteoblastogenesis, and bone formation, but had no effect on osteoclastogenesis and myeloma growth in these mice.

These recent studies suggest that OCPs can positively regulate bone formation, but the pathway stimulated in osteoblastic cells has yet to be identified, as has the exact site in bone remodeling units where these interactions take place. They likely occur near the cutting resorbing edges of bone remodeling units, where OCPs are actively recruited to maintain adequate numbers of osteoclasts, or in the centers of these units, where osteoblast precursors differentiate before moving to the bone surface to lay down bone matrix. Recently, ephrinB2/EphB4 signaling between cells within the osteoblast lineage has been shown to have a paracrine role in osteoblast differentiation [64]. Specifically, a synthetic peptide antagonist of ephrinB2/EphB4 receptor interaction and a recombinant soluble extracellular domain of EphB4, which antagonizes both forward and reverse EphB4 signaling, were both able to inhibit mineralization and the expression of several osteoblast genes involved in the late stages of osteoblast differentiation.

In contrast to these ligand–receptor interactions, another study by Lee et al. has suggested that osteoclasts can also negatively regulate bone formation by secreting a factor, which inhibits osteoblast precursor differentiation and osteoblast matrix production [65]. Expression of this factor is regulated negatively by Atp6v0d2, a component of the V-type $H^+$ ATP6i proton pump complex in osteoclasts that secretes $H^+$ and is required for production of the acid that dissolves the mineral component of bone. Interestingly, Atp6v0d2 also positively regulates fusion of OCPs, apparently following activation of it and DC-STAMP by NFATc1 [66]. In addition to this early role as a regulator of OCP fusion during osteoclast differentiation Atp6v0d2 appears to mediate extracellular acidification in mature osteoclasts [67]. Exactly how Atp6v0d2 regulates osteoblast functions has yet to be determined. Whatever the mechanisms involved, these and the above findings suggest that OCPs are involved intimately in the regulation of bone mass in normal and disease states, such as osteoporosis and osteomyelitis in which there is a generalized or localized increase in bone resorption and a decrease in bone formation.

## Osteoclasts as regulators of hematopoietic stem cell functions

Hematopoietic stem cells (HSCs) give rise to all the hematopoietic cells in bone marrow under the influence of osteoblastic cells in the marrow and on endosteal bone surfaces. Osteoblastic cells also regulate the egress of hematopoietic stem and progenitor cells from the bone marrow into the bloodstream, a process that is in part mediated by PTH and Notch signaling through a cyclic AMP-Jagged-1-dependent mechanism [68].

Kollet et al. found unexpectedly that osteoclasts play a role in HSC mobilization from the marrow, a process that is activated by stress-inducing conditions, such as chemotherapy and inflammation. They noted that osteoclast numbers were increased on bone surfaces during HSC mobilization, associated with activation of proteolytic enzymes, which release adhesions between stem cells and the bone marrow microenvironment. When Kollet et al. treated mice with RANKL to increase osteoclast formation they found that stem and progenitor cell mobilization was increased also, and this response was associated with increased osteoclast expression of MMP-9 and cathepsin K [69]. In addition to degrading bone matrix proteins, these enzymes also cleave membrane-bound kit ligand, a growth and adhesion factor for HSCs. RANKL also decreased the expression by osteoblasts of kit ligand and osteopontin, which also affect stem cell numbers. Kollet et al. then treated protein tyrosine phosphatase-epsilon (PTP $\epsilon$)-deficient mice with RANKL. These mice have osteoclasts with defective resorbing activity and mild osteopetrosis and they did not mobilize HSCs and progenitor cells from their marrow in response to RANKL, further linking this mobilization to osteoclasts [69]. These investigators subsequently showed that CD45 expression by cells in metaphyseal bone of mice is necessary for full osteoclast differentiation and retention of and repopulation of the marrow by hematopoietic precursors [70]. Collectively, these data suggest that osteoblasts and osteoclasts play important roles in HSC mobilization and likely also hematopoietic cell numbers and functions in infections in bone.

## The role of dendritic cells and cellular immunity in bone infection

Dendritic cells (DCs) are specialized antigen-presenting cells that play an important role in defense mechanisms against pathogens, including viruses, bacteria, and parasites by affecting both innate and adaptive immunity. DCs transport antigens from sites of infection to secondary lymphoid organs, such as draining lymph nodes and spleen to initiate T-cell responses. They express a variety of cytokines, which influence T-cell function and target pathogens through a direct killing effect [71]. DCs also affect osteoclast formation by differentiating to osteoclasts directly or stimulating receptor activator of NF-κB ligand (RANKL) production by T cells [72]. Thus DCs may contribute to bone infection through multiple mechanisms.

## DCs in adaptive immunity and infection

DCs comprise migrating and resident cells. Skin DCs, such as epidermal Langerhans cells, and dermal DCs are migrating cells, which are able to move to sites where infection or inflammation occurs, and then to draining lymph nodes and spleen. CD8-expressing cells ($CD8^+$) are resident DCs, which are only present in lymphoid organs, such as lymph nodes and spleen. $CD8^+$ DCs appear to be differentiated from marrow-derived DC precursors in the blood, which have not trafficked through peripheral tissues [73]. In response to infection or inflammation, resident DCs execute their defense functions within different compartments of lymph nodes or the spleen and do not appear to migrate to other organs through the blood or lymphatic system [74].

Recent studies using mouse models of bacterial infections have demonstrated a dramatic increase in the number of DCs present both at infection sites and associated lymphoid organs [75–78]. These inflammatory DCs are derived from Ly-$6C^+$ monocytes that are also called mo-DCs. During infection, Ly-$6C^+$ monocytes are recruited to infection sites and differentiate into mo-DCs, which subsequently migrate into draining lymph nodes through lymphatic vessels. The recruitment of inflammatory Ly-$6C^+$ monocytes is mediated through the interaction of P-selectin glycoprotein ligand-1 with endothelial P- and E-selectins and the interaction of L-selectin with endothelial peripheral lymph nodes [75]. A subset of DCs, which have a phenotype similar to that of $CD8^+$ DCs, is also increased in infection sites [75, 76]. Since $CD8^+$ DCs reside only within lymphoid organs, these inflammatory $CD8^+$ DCs are considered to be formed de novo in inflammatory foci during an infection [71, 79]. Whether or not the inflammatory $CD8^+$ DCs are derived also from $Ly-6C^+$ monocytes and whether they have different function from mo-DCs is not clear.

Inflammatory DCs participate in the induction and regulation of immune responses against pathogens through mechanisms involving both adaptive and innate immunity. During infections, mo-DCs migrate to lymphoid organs and differentiate to mature DCs, which activate $CD4^+$ T cells and induce Th1 cell responses against pathogens [79, 80]. For example, in an experimental model of subcutaneous *Salmonella* infection, mature mo-DCs promote the proliferation of $CD4^+$ T cells in lymph nodes or spleen through a CCR6-dependent mechanism [80]. Apart from direct priming of $CD4^+$ T cells, DCs also affect adaptive immunity through the capture and transport of antigens to $CD8^+$ DCs in lymph nodes for cross-presentation, and the induction of $CD4^+$ T-cell responses by cross-priming $CD8^+$ T cells.

## DCs in innate immunity and infection

A subset of lymph node mo-DCs distinct from mature DCs exhibits an immature phenotype during *Leishmania* infection [79]. These immature DCs are derived from monocytes recruited to lymph nodes and they do not appear to contribute substantially to T-cell immunity against *Leishmania*. It is speculated that immature lymph node DCs may participate in innate-immunity defense mechanisms. This hypothesis is based on findings from spleen mo-DCs during *Listeria* infection [77] in which monocytes are recruited to the spleen and differentiate to mo-DCs. Interestingly, this subset of DCs has highly effective microbicidal potential because they produce TNF-alpha and inducible nitric oxide synthase and have been called Tip DCs (TNF-alpha- and iNOS-producing DCs). Tip DCs are essential for defense against *Listeria* infection, which is mediated by chemokine receptor 2 (CCR2). $CCR2^{-/-}$ mice are markedly more susceptible to *Listeria* infection than wild-type mice and have a decreased number of Tip DCs. The T- and B-cell responses to the bacteria in $CCR2^{-/-}$ mice are normal, indicating that Tip DCs are not essential for T-cell priming [77]. MCP-1 and MCP-3 are the ligands for CCR2. Both $MCP-1^{-/-}$ and $MCP-3^{-/-}$ mice have fewer Tip DCs in the spleen following *Listeria* infection [81]. DCs newly formed from migrated monocytes in the spleen during *Streptococcus pneumoniae* infection also have a Tip DC phenotype [82]. These Tip DCs produce B-cell factors, such as the TNF superfamily molecule, B-lymphocyte stimulator, to induce B-cell responses and differentiate to IgM-producing plasma cells against the bacteria.

Although Tip DCs have been shown to play a beneficial role in infection by eliminating intracellular pathogens, they could also contribute to tissue damage as a major source of TNF and inducible nitric oxide synthase in inflamed organs. During the chronic stage of *T. brucei* infection, the absence of IL-10 leads to enhanced differentiation of monocytes to Tip DCs, resulting in exacerbated pathogenicity and early death of the host [83]. In the skin lesions of psoriasis patients, a subset of DCs that has a phenotype similar to murine

Tip DCs is significantly increased. These cells infiltrate the skin tissues where the levels of TNF and inducible nitric oxide synthase are high. Anti-CD11a could reduce skin inflammation as well as infiltration of Tip DCs [84].

Little is known about the role of DCs in chronic bone infections. However, several studies have demonstrated that *Staphylococcus aureus* or its active components affect DC functions, such as activation and recruitment, and thereby modify the host immune response. Leukocidin, an exotoxin of *Staphylococcus aureus*, activates mouse bone-marrow-derived myeloid dendritic cells to generate and secrete IL-12p40, and TNF-alpha, and induce CD40 expression on their surface. Leukocidin-induced IL-12p40 production is abolished in cells from TLR4-deficient mice, indicating that Leukocidin causes the activation of DCs by triggering a TLR4-dependent signaling pathway [85]. *Staphylococcus aureus* enterotoxin B induces maturation of human monocyte-derived DCs and stimulates them to secrete high levels of IL-2. These activated DCs are able to drive polarization of naive T cells into the Th2 subset, which can be blocked by anti-TLR2 signaling [86]. Chemerin is an attractant for cells that express CMKLR1, the serpentine G protein-coupled receptor for chemerin. Chemerin is cleaved of its C terminus and is converted to a potent chemoattractant by staphopain B, a cysteine protease secreted by *Staphylococcus aureus*. The cleaved form of chemerin could recruit specialized host cells, including immunoregulatory DCs and/or macrophages, to the infection site [87]. Taken together, *Staphylococcus aureus* and its toxins may serve as new pathogen-associated molecules, which could play a major role in bacterial pathologies. Since *Staphylococcus aureus* is responsible for 90% of cases of bacterial osteomyelitis, it is speculated that *Staphylococcus aureus*-mediated changes of DC phenotypes described above may also occur in sites of bone infection.

## DCs and osteoclasts in infection

Bone infection is often associated with abnormal bone remodeling due to the massive bone destruction and repair processes at the infection site in response to bacterial toxins. Bacterial toxins have a strong stimulatory effect on osteoclasts by directly affecting osteoclast generation, survival, and activation and by indirectly promoting the production of RANKL and other osteoclastogenic factors [88, 89]. Osteoclasts are generated in the bone marrow from hematopoietic stem cells. Under the influence of c-kit ligand and stem cell factor, hematopoietic stem cells give rise to committed common myeloid precursors, which are precursors for monocytes, DCs, and osteoclasts. In the presence of macrophage colony-stimulating factor (M-CSF) and RANKL, these precursors differentiate to bone-resorbing osteoclasts [51].

RANKL is produced by bone marrow stromal cells under normal conditions. However, at active disease sites in response to pathologic stimuli, such as the bacterial component lipopolysaccharide, many cell types including activated T cells produce RANKL and result in abnormal bone loss [90]. Interestingly, DCs are responsible for T-cell activation and share common precursors with osteoclasts. Thus, DCs may participate in pathogen-induced osteoclastogenesis.

CD11c is a pan marker for DCs. An early study reported that purified bone marrow CD11c$^+$ DCs form functional osteoclasts in the presence of M-CSF and RANK in vitro, which is greatly enhanced by synovial fluid from patients with rheumatoid arthritis and involves the pro-inflammatory cytokines IL-1 or TNF, and components of the extracellular matrix, such as hyaluronic acid [91]. Recent reports demonstrated that upon activation by microbial or protein antigens and during immune interactions with CD4$^+$ T cells, CD11c$^+$ DCs from both bone marrow and spleen develop into functional osteoclasts in the presence of M-CSF and RANKL. More importantly, these CD11c$^+$ DCs can differentiate into osteoclasts and induce bone loss after adoptive transfer in vivo [92, 93]. It appears that not all CD11c$^+$ DCs have similar osteoclastogenic potential. Among the splenic DC subsets, conventional DCs have higher osteoclastogenic potential in vitro. These convensional DCs are CD11c$^+$MHC-II$^+$CD11b$^+$ and they are negative for F4/80, Ly-6C, and B220, and most of them express the co-stimulatory molecules CD80, CD86, and CD40. They also differentiate into functional osteoclasts in vivo when injected into osteopetrotic mice. This process involves the presence of activated CD4$^+$ T cells inducing high RANKL expression by bone marrow stromal cells [94]. These results indicate that DC-derived osteoclasts may be directly involved in the osteolytic lesions observed in human inflammatory and infectious bone diseases during immune interactions with CD4$^+$ T cells.

In summary, DCs are antigen-presenting cells that participate in the body's immune surveillance. In response to bacterial infection, different subsets of DCs execute their effects on cellular immunity through different mechanisms, including T-cell regulation or cytokine production. DCs also contribute to osteoclast-mediated pathologic bone loss. Although little is known about the role of DCs specifically in the ostemyelitis setting, it is conceivable that DCs may actively participate in this process, as they do in response to bacterial infections at other sites. Investigation of the role of DCs and cellular immunity in bone infection will offer a new basis for understanding the relationship between bone and immune systems and for developing specific DC-based therapy.

# The role of B cells in bone infection and the potential of passive immunization

As most bone infections are caused by extracellular pathogens, and the primary mechanism by which the host clears them is via phagocytosis of antibody-opsonized bacteria, the humoral immune response mediated by B cells is critical. In recognition of this, several research groups have performed experiments to characterize the antibody-mediated immune response and identify those proteins that are immunogenic in *S. aureus* biofilm infections [95, 96]. In these animal studies, sera were collected prior to infection and at subsequent time points thereafter, and used to characterize the Ig response and identify novel immunodominant antigens. Infection of naive mice results in an initial IgM response in the first week, followed by a very specific IgG2b response at

2 weeks [96]. Western blot assays on bacterial extracts separated by two-dimensional gel electrophoresis revealed humoral immunity against several bacterial antigens, including cell-surface-associated beta-lactamase, lipoprotein, lipase, autolysin, and an ABC transporter lipoprotein [95]. As a result of this work, several vaccine trials are currently under way.

One of the most challenging forms of osteomyelitis to treat is MRSA infection of total joint replacements [2]. As this patient population is typically >65 years of age, and are poor candidates for active immunization with purified bacterial products, several groups are proposing a potential passive immunization strategy to prevent reinfection following revision surgery. The attraction of this approach is that protective humoral immunity could be achieved 2 weeks prior to surgery by infusing the patient with protective monoclonal antibodies (mAbs). Our efforts in this regard have focused on anti-autolysin (Alt) mAb, which targets both the aminidase (Amd) and the glucosaminidase (Gmd) subunits of the *S. aureus* enzyme. Alt is one of the catalytically distinct peptidoglycan hydrolases in *S. aureus* that is required to digest the cell wall during mitosis [97]. Scanning electron microscopy studies have demonstrated that in addition to being an essential gene for growth, anti-Alt antibodies bound to *S. aureus* during binary fission localize to regions of the bacteria that are not covered by the cell wall [98]. Furthermore, Alt is also known to be an adhesin [99, 100], thus anti-Alt mAb may have protective effects via multiple mechanisms of action including: (1) inhibition of mitosis; (2) mAb-mediated serum lysis via complement binding to naked periplasm; (3) opsonization; and (4) inhibition of bacterial attachment to host tissue. Future studies designed to evaluate these mechanisms and the protection of these mAb in vivo will determine the potential of passive immunization as a therapeutic option to prevent bone infection in orthopedic patients.

# References

[1] Fraser C, Donnelly CA, Cauchemez S, Hanage WP, Van Kerkhove MD, Hollingsworth TD, Griffin J, Baggaley RF, Jenkins HE, Lyons EJ, Jombart T, Hinsley WR, Grassly NC, Balloux F, Ghani AC, Ferguson NM, Rambaut A, Pybus OG, Lopez-Gatell H, Alpuche-Aranda CM, Chapela IB, Zavala EP, Guevara DM, Checchi F, Garcia E, Hugonnet S, Roth C. Pandemic potential of a strain of influenza A (H1N1): early findings. Science 2009;324:1557–61.

[2] Darouiche RO. Treatment of infections associated with surgical implants. N Engl J Med 2004;350:1422–9.

[3] Haas DW, McAndrew MP. Bacterial osteomyelitis in adults: evolving considerations in diagnosis and treatment. Am J Med 1996;101:550–61.

[4] Crane DP, Gromov K, Li D, Soballe K, Wahnes C, Buchner H, Hilton MJ, O'Keefe RJ, Murray CK, Schwarz EM. Efficacy of colistin-impregnated beads to prevent multidrug-resistant A. baumannii implant-associated osteomyelitis. J Orthop Res 2009;27:1008–15.

[5] Khosla S, Burr D, Cauley J, Dempster DW, Ebeling PR, Felsenberg D, Gagel RF, Gilsanz V, Guise T, Koka S, McCauley LK, McGowan J, McKee MD, Mohla S, Pendrys DG, Raisz LG, Ruggiero SL, Shafer DM, Shum L, Silverman SL, Van Poznak CH, Watts N, Woo SB, Shane E. Bisphosphonate-associated osteonecrosis of the jaw: report of a task force of the American Society for Bone and Mineral Research. J Bone Miner Res 2007;22:1479–91.

[6] Mensah KA, Li J, Schwarz EM. The emerging field of osteoimmunology. Immunol Res 2009.

[7] Lew DP, Waldvogel FA. Osteomyelitis. Lancet 2004;364:369–79.

[8] Norden CW. Lessons learned from animal models of osteomyelitis. Rev Infect Dis 1988;10:103–10.

[9] Morrissy RT. Bone and joint infection in the neonate. Pediatr Ann 1989;18:33–4, 36–38, 40–34.

[10] Tsukayama DT, Goldberg VM, Kyle R. Diagnosis and management of infection after total knee arthroplasty. J Bone Joint Surg Am 2003;85-A:S75–80.

[11] Morrissy RT, Haynes DW. Acute hematogenous osteomyelitis: a model with trauma as an etiology. J Pediatr Orthop 1989;9:447–56.

[12] Caputo GM, Cavanagh PR, Ulbrecht JS, Gibbons GW, Karchmer AW. Assessment and management of foot disease in patients with diabetes. N Engl J Med 1994;331:854–60.

[13] Tsukayama DT, Gustilo RB. Antibiotic management of open fractures. In: Greene WB, 9th ed. Instructional course lectures. AAOS, Park Ridge, IL 1990;39:487–90.

[14] Tsukayama DT. Pathophysiology of posttraumatic osteomyelitis. Clin Orthop 1999:22–9.

[15] Mackowiak PA, Jones SR, Smith JW. Diagnostic value of sinus-tract cultures in chronic osteomyelitis. Jama 1978;239:2772–5.

[16] Gustilo RB, Anderson JT. Prevention of infection in the treatment of one thousand and twenty-five open fractures of long bones: retrospective and prospective analyses. J Bone Joint Surg Am 1976;58:453–8.

[17] Chuard C, Vaudaux P, Waldvogel FA, Lew DP. Susceptibility of *Staphylococcus aureus* growing on fibronectin-coated surfaces to bactericidal antibiotics. Antimicrob Agents Chemother 1993;37:625–32.

[18] Greene C, Vaudaux PE, Francois P, Proctor RA, McDevitt D, Foster TJ. A low-fibronectin-binding mutant of Staphylococcus aureus 879R4S has Tn918 inserted into its single fnb gene. Microbiology 1996;142:2153–60.

[19] Fischer B, Vaudaux P, Magnin M, el Mestikawy Y, Proctor RA, Lew DP, Vasey H. Novel animal model for studying the molecular mechanisms of bacterial adhesion to bone-implanted metallic devices: role of fibronectin in *Staphylococcus aureus* adhesion. J Orthop Res 1996;14:914–20.

[20] Gristina AG, Oga M, Webb LX, Hobgood CD. Adherent bacterial colonization in the pathogenesis of osteomyelitis. Science 1985;228:990–3.

[21] Naylor PT, Myrvik QN, Gristina A. Antibiotic resistance of biomaterial-adherent coagulase-negative and coagulase-positive staphylococci. Clin Orthop 1990:126–33.

[22] McGuire MH. The pathogenesis of adult osteomyelitis. Orthop Rev 1989;18:564–70.

[23] Verdrengh M, Tarkowski A. Role of neutrophils in experimental septicemia and septic arthritis induced by Staphylococcus aureus. Infect Immun 1997;65:2517–21.

[24] Mosmann TR, Cherwinski H, Bond MW, Giedlin MA, Coffman RL. Two types of murine helper T cell clone. I. Definition according to profiles of lymphokine activities and secreted proteins. J Immunol 1986;136:2348–57.

[25] Salgame P, Abrams JS, Clayberger C, Goldstein H, Convit J, Modlin RL, Bloom BR. Differing lymphokine profiles of functional subsets of human CD4 and CD8 T cell clones. Science 1991;254:279–82.

[26] Daeron M. Fc receptor biology. Annu Rev Immunol 1997;15:203–34.

[27] Cierny 3rd G, Mader JT. Approach to adult osteomyelitis. Orthop Rev 1987;16:259–70.

[28] Karsenty G, Kronenberg HM, Settembre C. Genetic control of bone formation. Annu Rev Cell Dev Biol 2009;25:629–48.

[29] Varoga D, Wruck CJ, Tohidnezhad M, Brandenburg L, Paulsen F, Mentlein R, Seekamp A, Besch L, Pufe T. Osteoblasts participate in the innate immunity of the bone by producing human beta defensin-3. Histochem Cell Biol 2009;131:207–18.

[30] Amcheslavsky A, Hemmi H, Akira S, Bar-Shavit Z. Differential contribution of osteoclast- and osteoblast-lineage cells to CpG-oligodeoxynucleotide (CpG-ODN) modulation of osteoclastogenesis. J Bone Miner Res 2005;20:1692–9.

[31] Marriott I, Gray DL, Rati DM, Fowler Jr VG, Stryjewski ME, Levin LS, Hudson MC, Bost KL. Osteoblasts produce monocyte chemoattractant protein-1 in a murine model of Staphylococcus aureus osteomyelitis and infected human bone tissue. Bone 2005;37:504–12.

[32] Marriott I, Gray DL, Tranguch SL, Fowler Jr VG, Stryjewski M, Scott Levin L, Hudson MC, Bost KL. Osteoblasts express the inflammatory cytokine interleukin-6 in a murine model of Staphylococcus aureus osteomyelitis and infected human bone tissue. Am J Pathol 2004;164:1399–406.

[33] Wright KM, Friedland JS. Regulation of chemokine gene expression and secretion in Staphylococcus aureus-infected osteoblasts. Microbes Infect 2004;6:844–52.

[34] Schrum LW, Bost KL, Hudson MC, Marriott I. Bacterial infection induces expression of functional MHC class II molecules in murine and human osteoblasts. Bone 2003;33:812–21.

[35] Jevon M, Guo C, Ma B, Mordan N, Nair SP, Harris M, Henderson B, Bentley G, Meghji S. Mechanisms of internalization of Staphylococcus aureus by cultured human osteoblasts. Infect Immun 1999;67:2677–81.

[36] Tucker KA, Reilly SS, Leslie CS, Hudson MC. Intracellular Staphylococcus aureus induces apoptosis in mouse osteoblasts. FEMS Microbiol Lett 2000;186:151–6.

[37] Reilly SS, Hudson MC, Kellam JF, Ramp WK. In vivo internalization of Staphylococcus aureus by embryonic chick osteoblasts. Bone 2000;26:63–70.

[38] Nair SP, Bischoff M, Senn MM, Berger-Bachi B. The sigma B regulon influences internalization of Staphylococcus aureus by osteoblasts. Infect Immun 2003;71:4167–70.

[39] Nair SP, Meghji S, Wilson M, Reddi K, White P, Henderson B. Bacterially induced bone destruction: mechanisms and misconceptions. Infect Immun 1996;64:2371–80.

[40] Itoh K, Udagawa N, Kobayashi K, Suda K, Li X, Takami M, Okahashi N, Nishihara T, Takahashi N. Lipopolysaccharide promotes the survival of osteoclasts via Toll-like receptor 4, but cytokine production of osteoclasts in response to lipopolysaccharide is different from that of macrophages. J Immunol 2003;170:3688–95.

[41] Abu-Amer Y, Ross FP, Edwards J, Teitelbaum SL. Lipopolysaccharide-stimulated osteoclastogenesis is mediated by tumor necrosis factor via its P55 receptor. J Clin Invest 1997;100:1557–65.

[42] Lau YS, Wang W, Sabokbar A, Simpson H, Nair S, Henderson B, Berendt A, Athanasou NA. Staphylococcus aureus capsular material promotes osteoclast formation. Injury 2006;37(Suppl. 2): S41–8.

[43] Henderson B, Nair SP. Hard labour: bacterial infection of the skeleton. Trends Microbiol 2003;11:570–7.

[44] Kim N, Kadono Y, Takami M, Lee J, Lee SH, Okada F, Kim JH, Kobayashi T, Odgren PR, Nakano H, Yeh WC, Lee SK, Lorenzo JA, Choi Y. Osteoclast differentiation independent of the TRANCE-RANK-TRAF6 axis. J Exp Med 2005;202:589–95.

[45] Yao Z, Xing L, Boyce BF. NF-kappaB p100 limits TNF-induced bone resorption in mice by a TRAF3-dependent mechanism. J Clin Invest 2009;119:3024–34.

[46] Kusu N, Laurikkala J, Imanishi M, Usui H, Konishi M, Miyake A, Thesleff I, Itoh N. Sclerostin is a novel secreted osteoclast-derived bone morphogenetic protein (BMP) antagonist with unique ligand specificity. J Biol Chem 2003;17:17.

[47] Mundy GR. Metastasis to bone: causes, consequences and therapeutic opportunities. Nat Rev Cancer 2002;2:584–93.

[48] Chirgwin JM, Guise TA. Skeletal metastases: decreasing tumor burden by targeting the bone microenvironment. J Cell Biochem 2007;102:1333–42.

[49] Boyce BF, Xing L. The RANKL/RANK/OPG pathway. Curr Osteoporos Rep 2007;5:98–104.

[50] Hattori K, Heissig B, Tashiro K, Honjo T, Tateno M, Shieh JH, Hackett NR, Quitoriano MS, Crystal RG, Rafii S, Moore MA. Plasma elevation of stromal cell-derived factor-1 induces mobilization of mature and immature hematopoietic progenitor and stem cells. Blood 2001;97:3354–60.

[51] Xing L, Schwarz EM, Boyce BF. Osteoclast precursors, RANKL/RANK, and immunology. Immunol Rev 2005;208:19–29.

[52] Li P, Schwarz EM, O'Keefe RJ, Ma L, Looney RJ, Ritchlin CT, Boyce BF, Xing L. Systemic tumor necrosis factor alpha mediates an increase in peripheral CD11b high osteoclast precursors in tumor necrosis factor alpha-transgenic mice. Arthritis Rheum 2004;50:265–76.

[53] Ishii M, Egen JG, Klauschen F, Meier-Schellersheim M, Saeki Y, Vacher J, Proia RL, Germain RN. Sphingosine-1-phosphate mobilizes osteoclast precursors and regulates bone homeostasis. Nature 2009;458:524–8.

[54] Rosen H, Goetzl EJ. Sphingosine 1-phosphate and its receptors: an autocrine and paracrine network. Nat Rev Immunol 2005;5:560–70.

[55] Pederson L, Ruan M, Westendorf JJ, Khosla S, Oursler MJ. Regulation of bone formation by osteoclasts involves Wnt/BMP signaling and the chemokine sphingosine-1-phosphate. Proc Natl Acad Sci U S A 2008;105:20764–9.

[56] O'Keefe RJ, Teot LA, Singh D, Puzas JE, Rosier RN, Hicks DG. Osteoclasts constitutively express regulators of bone resorption: an immunohistochemical and in situ hybridization study. Lab Invest 1997;76:457–65.

[57] Yu X, Huang Y, Collin-Osdoby P, Osdoby P. CCR1 chemokines promote the chemotactic recruitment, RANKL development, and motility of osteoclasts and are induced by inflammatory cytokines in osteoblasts. J Bone Miner Res 2004;19:2065–77.

[58] Boyce BF, Schwarz EM, Xing L. Osteoclast precursors: cytokine-stimulated immunomodulators of inflammatory bone disease. Curr Opin Rheumatol 2006;18:427–32.

[59] Yao Z, Xing L, Qin C, Schwarz EM, Boyce BF. Osteoclast precursor interaction with bone matrix induces osteoclast formation directly by an interleukin-1-mediated autocrine mechanism. J Biol Chem 2008;283:9917–24.

[60] Yamashita T, Yao Z, Li F, Zhang Q, Badell IR, Schwarz EM, Takeshita S, Wagner EF, Noda M, Matsuo K, Xing L, Boyce BF. NF-kappaB p50 and p52 regulate receptor activator of NF-kappaB ligand (RANKL) and tumor necrosis factor-induced osteoclast precursor differentiation by activating c-Fos and NFATc1. J Biol Chem 2007;282:18245–53.

[61] Teitelbaum SL, Ross FP. Genetic regulation of osteoclast development and function. Nat Rev Genet 2003;4:638–49.

[62] Shinohara M, Takayanagi H. Novel osteoclast signaling mechanisms. Curr Osteoporos Rep 2007;5:67–72.

[63] Ritchlin CT, Haas-Smith SA, Li P, Hicks DG, Schwarz EM. Mechanisms of TNF-alpha- and RANKL-mediated osteoclastogenesis and bone resorption in psoriatic arthritis. J Clin Invest 2003;111:821–31.

[64] Pennisi A, Ling W, Li X, Khan S, Shaughnessy Jr JD, Barlogie B, Yaccoby S. The ephrinB2/EphB4 axis is dysregulated in osteoprogenitors from myeloma patients and its activation affects myeloma bone disease and tumor growth. Blood 2009;114:1803–12.

[65] Lee SH, Rho J, Jeong D, Sul JY, Kim T, Kim N, Kang JS, Miyamoto T, Suda T, Lee SK, Pignolo RJ, Koczon-Jaremko B, Lorenzo J, Choi Y. v-ATPase V0 subunit d2-deficient mice exhibit impaired osteoclast fusion and increased bone formation. Nat Med 2006;12:1403–9.

[66] Kim K, Lee SH, Ha Kim J, Choi Y, Kim N. NFATc1 induces osteoclast fusion via up-regulation of Atp6v0d2 and the dendritic cell-specific transmembrane protein (DC-STAMP). Mol Endocrinol 2008;22:176–85.

[67] Wu H, Xu G, Li YP. Atp6v0d2 is an essential component of the osteoclast-specific proton pump that mediates extracellular acidification in bone resorption. J Bone Miner Res 2009;24:871–85.

[68] Calvi LM, Adams GB, Weibrecht KW, Weber JM, Olson DP, Knight MC, Martin RP, Schipani E, Divieti P, Bringhurst FR, Milner LA, Kronenberg HM, Scadden DT. Osteoblastic cells regulate the haematopoietic stem cell niche. Nature 2003;425:841–6.

[69] Kollet O, Dar A, Shivtiel S, Kalinkovich A, Lapid K, Sztainberg Y, Tesio M, Samstein RM, Goichberg P, Spiegel A, Elson A, Lapidot T. Osteoclasts degrade endosteal components and promote mobilization of hematopoietic progenitor cells. Nat Med 2006;12:657–64.

[70] Shivtiel S, Kollet O, Lapid K, Schajnovitz A, Goichberg P, Kalinkovich A, Shezen E, Tesio M, Netzer N, Petit I, Sharir A, Lapidot T. CD45 regulates retention, motility, and numbers of hematopoietic progenitors, and affects osteoclast remodeling of metaphyseal trabecules. J Exp Med 2008;205:2381–95.

[71] Leon B, Ardavin C. Monocyte-derived dendritic cells in innate and adaptive immunity. Immunol Cell Biol 2008;86:320–4.

[72] Alnaeeli M, Teng YT. Dendritic cells: a new player in osteoimmunology. Curr Mol Med 2009;9:893–910.

[73] Carbone FR, Belz GT, Heath WR. Transfer of antigen between migrating and lymph node-resident DCs in peripheral T-cell tolerance and immunity. Trends Immunol 2004;25:655–8.

[74] Reis e Sousa, Hieny CS, Scharton-Kersten T, Jankovic D, Charest H, Germain RN, Sher A. In vivo microbial stimulation induces rapid CD40 ligand-independent production of interleukin 12 by dendritic cells and their redistribution to T cell areas. J Exp Med 1997;186:1819–29.

[75] Leon B, Ardavin C. Monocyte migration to inflamed skin and lymph nodes is differentially controlled by L-selectin and PSGL-1. Blood 2008;111:3126–30.

[76] Martin P, Ruiz SR, del Hoyo GM, Anjuere F, Vargas HH, Lopez-Bravo M, Ardavin C. Dramatic increase in lymph node dendritic cell number during infection by the mouse mammary tumor virus occurs by a CD62L-dependent blood-borne DC recruitment. Blood 2002;99:1282–8.

[77] Serbina NV, Salazar-Mather TP, Biron CA, Kuziel WA, Pamer EG. TNF/iNOS-producing dendritic cells mediate innate immune defense against bacterial infection. Immunity 2003;19:59–70.

[78] Yoneyama H, Matsuno K, Zhang Y, Murai M, Itakura M, Ishikawa S, Hasegawa G, Naito M, Asakura H, Matsushima K. Regulation by chemokines of circulating dendritic cell precursors, and the formation of portal tract-associated lymphoid tissue, in a granulomatous liver disease. J Exp Med 2001;193:35–49.

[79] Leon B, Lopez-Bravo M, Ardavin C. Monocyte-derived dendritic cells formed at the infection site control the induction of protective T helper 1 responses against Leishmania. Immunity 2007;26:519–31.

[80] Ravindran R, Rusch L, Itano A, Jenkins MK, McSorley SJ. CCR6-dependent recruitment of blood phagocytes is necessary for rapid CD4 T cell responses to local bacterial infection. Proc Natl Acad Sci U S A 2007;104:12075–80.

[81] Jia T, Serbina NV, Brandl K, Zhong MX, Leiner IM, Charo IF, Pamer EG. Additive roles for MCP-1 and MCP-3 in CCR2-mediated recruitment of inflammatory monocytes during Listeria monocytogenes infection. J Immunol 2008;180:6846–53.

[82] Balazs M, Martin F, Zhou T, Kearney J. Blood dendritic cells interact with splenic marginal zone B cells to initiate T-independent immune responses. Immunity 2002;17:341–52.

[83] Guilliams M, Movahedi K, Bosschaerts T, VandenDriessche T, Chuah MK, Herin M, Acosta-Sanchez A, Ma L, Moser M, Van Ginderachter JA, Brys L, De Baetselier P, Beschin A. IL-10 dampens TNF/inducible nitric oxide synthase-producing dendritic cell-mediated pathogenicity during parasitic infection. J Immunol 2009;182:1107–18.

[84] Lowes MA, Chamian F, Abello MV, Fuentes-Duculan J, Lin SL, Nussbaum R, Novitskaya I, Carbonaro H, Cardinale I, Kikuchi T, Gilleaudeau P, Sullivan-Whalen M, Wittkowski KM, Papp K, Garovoy M, Dummer W, Steinman RM, Krueger JG. Increase in TNF-alpha and inducible nitric oxide synthase-expressing dendritic cells in psoriasis and reduction with efalizumab (anti-CD11a). Proc Natl Acad Sci U S A 2005;102:19057–62.

[85] Inden K, Kaneko J, Miyazato A, Yamamoto N, Mouri S, Shibuya Y, Nakamura K, Aoyagi T, Hatta M, Kunishima H, Hirakata Y, Itoh Y, Kaku M, Kawakami K. Toll-like receptor 4-dependent activation of myeloid dendritic cells by leukocidin of *Staphylococcus aureus*. Microbes Infect 2009;11:245–53.

[86] Mandron M, Aries MF, Brehm RD, Tranter HS, Acharya KR, Charveron M, Davrinche C. Human dendritic cells conditioned with Staphylococcus aureus enterotoxin B promote TH2 cell polarization. J Allergy Clin Immunol 2006;117:1141–7.

[87] Kulig P, Zabel BA, Dubin G, Allen SJ, Ohyama T, Potempa J, Handel TM, Butcher EC, Cichy J. *Staphylococcus aureus*-derived staphopain B, a potent cysteine protease activator of plasma chemerin. J Immunol 2007;178:3713–20.

[88] Puzas JE, Hicks DG, Reynolds SD, O'Keefe RJ. Regulation of osteoclastic activity in infection. Methods Enzymol 1994;236:47–58.

[89] Montonen M, Li TF, Lukinmaa PL, Sakai E, Hukkanen M, Sukura A, Konttinen YT. RANKL and cathepsin K in diffuse sclerosing osteomyelitis of the mandible. J Oral Pathol Med 2006;35:620–5.

[90] Ozaki Y, Ukai T, Yamaguchi M, Yokoyama M, Haro ER, Yoshimoto M, Kaneko T, Yoshinaga M, Nakamura H, Shiraishi C, Hara Y. Locally administered T cells from mice immunized with lipopolysaccharide (LPS) accelerate LPS-induced bone resorption. Bone 2009;44:1169–76.

[91] Miyamoto T, Ohneda O, Arai F, Iwamoto K, Okada S, Takagi K, Anderson DM, Suda T. Bifurcation of osteoclasts and dendritic cells from common progenitors. Blood 2001;98:2544–54.

[92] Alnaeeli M, Penninger JM, Teng YT. Immune interactions with CD4+ T cells promote the development of functional osteoclasts from murine CD11c+ dendritic cells. J Immunol 2006;177:3314–26.

[93] Rivollier A, Mazzorana M, Tebib J, Piperno M, Aitsiselmi T, Rabourdin-Combe C, Jurdic P, Servet-Delprat C. Immature dendritic cell transdifferentiation into osteoclasts: a novel pathway sustained by the rheumatoid arthritis microenvironment. Blood 2004;104:4029–37.

[94] Wakkach A, Mansour A, Dacquin R, Coste E, Jurdic P, Carle GF, Blin-Wakkach C. Bone marrow microenvironment controls the in vivo differentiation of murine dendritic cells into osteoclasts. Blood 2008;112:5074–83.

[95] Brady RA, Leid JG, Camper AK, Costerton JW, Shirtliff ME. Identification of *Staphylococcus aureus* proteins recognized by the antibody-mediated immune response to a biofilm infection. Infect Immun 2006;74:3415–26.

[96] Li D, Gromov K, Soballe K, Puzas JE, O'Keefe RJ, Awad H, Drissi H, Schwarz EM. Quantitative mouse model of implant-associated osteomyelitis and the kinetics of microbial growth, osteolysis, and humoral immunity. J Orthop Res 2008;26:96–105.

[97] Baba T, Schneewind O. Targeting of muralytic enzymes to the cell division site of Gram-positive bacteria: repeat domains direct autolysin to the equatorial surface ring of Staphylococcus aureus. Embo J 1998;17:4639–46.

[98] Yamada S, Sugai M, Komatsuzawa H, Nakashima S, Oshida T, Matsumoto A, Suginaka H. An autolysin ring associated with cell separation of Staphylococcus aureus. J Bacteriol 1996;178:1565–71.

[99] Heilmann C, Hartleib J, Hussain MS, Peters G. The multifunctional *Staphylococcus aureus* autolysin aaa mediates adherence to immobilized fibrinogen and fibronectin. Infect Immun 2005;73:4793–802.

[100] Heilmann C, Hussain M, Peters G, Gotz F. Evidence for autolysin-mediated primary attachment of *Staphylococcus epidermidis* to a polystyrene surface. Mol Microbiol 1997;24:1013–24.

# 14

# The Role of the Immune System in Hematologic Malignancies that Affect Bone

Jessica A. Fowler[1], Claire M. Edwards[1], Gregory R. Mundy[2]✠

[1]VANDERBILT CENTER FOR BONE BIOLOGY, DEPARTMENT OF CANCER BIOLOGY, VANDERBILT UNIVERSITY, NASHVILLE, TN, USA, [2]VANDERBILT CENTER FOR BONE BIOLOGY, DEPARTMENT OF MEDICINE/CLINICAL PHARMACOLOGY, VANDERBILT UNIVERSITY, NASHVILLE, TN, USA

**CHAPTER OUTLINE**

- **Introduction** ........................................................................................................... 391
- **Innate immunity – macrophages, dendritic cells, and natural killer cells** ........................ 392
  - Tumor-associated macrophages ........................................................................... 393
  - Dendritic cells .................................................................................................... 394
  - Natural killer cells ............................................................................................... 396
- **Adaptive immunity** ................................................................................................. 397
  - B cells ............................................................................................................... 397
  - Helper T cells ..................................................................................................... 398
- **Hedgehog signaling** ................................................................................................ 398
- **TNF receptor superfamily and NF-κB** ...................................................................... 398
- **Therapeutic approaches and effects on immune cells** ................................................. 400
  - Myeloma-specific cytotoxic T lymphocytes ........................................................... 400
  - Gamma delta T cells ........................................................................................... 400
  - Tumor-specific antigens ...................................................................................... 402
  - Thalidomide and lenalidomide ............................................................................ 402
- **Conclusions** ........................................................................................................... 403
- **References** ............................................................................................................ 404

## Introduction

Hematological malignancies, such as multiple myeloma, are unique in that the tumor microenvironment consists of a complex network of cellular interactions present within the bone marrow cavity. Multiple myeloma is the second most common hematological malignancy. The American Cancer Society estimated that approximately 20,000 new multiple myeloma diagnoses and 10,800 myeloma deaths occurred in 2007 in the United States alone [1]. Myeloma is characterized by the uncontrolled clonal proliferation of malignant plasma cells within the bone marrow. In addition to the associated complications that arise from the expansion of these plasma cells, patients also develop a destructive osteolytic bone disease that often results in lytic lesions, bone pain, and pathological fractures. Of all hematological malignancies, multiple myeloma is the only one in which the majority of patients develop bone disease, and the bone disease is an integral and contributing mechanism in tumor progression.

Osteolytic bone disease is a major clinical complication that arises in patients with myeloma. In the normal bone marrow microenvironment, bone is constantly undergoing remodeling with a delicate balance between osteoclastic bone resorption and osteoblastic bone formation. Within the myeloma microenvironment, there is a dysregulation in normal remodeling that results in decreased bone formation and enhanced osteoclast function. Myeloma cells play an active role in creating this dysfunctional process. Studies have shown myeloma cells in close proximity to sites of bone resorption [2–4], therefore supporting their role in stimulating osteoclast formation and activity. Myeloma cells produce various factors, including tumor necrosis factor-$\alpha$ (TNF-$\alpha$) and interleukin-6 (IL-6) that promote osteoclast differentiation and function.

The bone marrow microenvironment consists of a complex network of cellular interactions and exposure of numerous secreted factors. Cell types within this microenvironment include stromal cells, osteoclasts, osteoblasts, hematopoietic stem cells, B and T lymphocytes, macrophages, and various other immune cells. The immune system is comprised of two types of immunity: innate and adaptive. The innate immune system is generally thought of as the body's first line of defense against pathogens; however, the response to pathogens is non-specific [5]. Adaptive immunity allows for a robust immune response through specific recognition of "non-self" antigens and immunological memory of pathogens of previous exposure [6].

In addition to myeloma cells directly stimulating osteoclasts, the interaction between myeloma cells and the bone marrow microenvironment forms the vicious cycle that also contributes to the destructive bone disease. Myeloma bone disease can be perpetuated by alterations to the immune cells and surrounding stroma, found within the bone marrow. Despite extensive research in the field of multiple myeloma, there are limited studies focused on the role of other immune cells in myeloma pathogenesis.

## Innate immunity – macrophages, dendritic cells, and natural killer cells

The role of macrophages and dendritic cells in multiple myeloma draws particular interest in terms of examining immunosuppression in myeloma pathogenesis. Patients with later stages of multiple myeloma develop cumulative immunosuppression resulting from both the expansion of tumor growth and the myeloma-directed therapies. However, more recent research in myeloma has focused on the mechanisms by which immunosuppression occurs during the development of multiple myeloma.

## Tumor-associated macrophages

Cancer research has begun to investigate the role of macrophages in tumor progression. Macrophages are divided into two categories: M1 and M2 macrophages. M1 macrophages are antigen-presenting cells that promote differentiation of naïve CD4+ T cells into Th1 cells [7]. M1 macrophages are induced by "classical activators", such as lipopolysaccharide (LPS) and interferon γ (IFNγ) [7]. Tumor-associated macrophages (TAMs) are a dominant inflammatory cell population in a majority of tumor sites and often display characteristics of M2 macrophages. M2 macrophages are activated by interleukins (IL)-4, -10, and/or -13 and then subsequently stimulate differentiation of CD4+ Th2 and regulatory T (Treg) cells [8, 9]. Zheng et al. demonstrated that co-culture with TAMs protected myeloma cells against chemotherapy drug-induced apoptosis, in contrast to co-culture with normal macrophages [10]. These studies determined that this protection against apoptosis was cell-contact-dependent by inhibiting the activation and cleavage of caspase-3 and poly(ADP-ribose) polymerization (Figure 14-1). Additionally, myeloma patients had significant numbers of CD68+ macrophages present in bone marrow biopsies in comparison to control samples [10].

Another mechanism by which TAMs contribute to cancer progression is by stimulating angiogenesis. Various studies in solid tumors have demonstrated that TAMs are a rich source of proangiogenic factors and cytokines, such as vascular endothelial growth factor (VEGF). In addition to the production of angiogenic factors, TAMs also produce

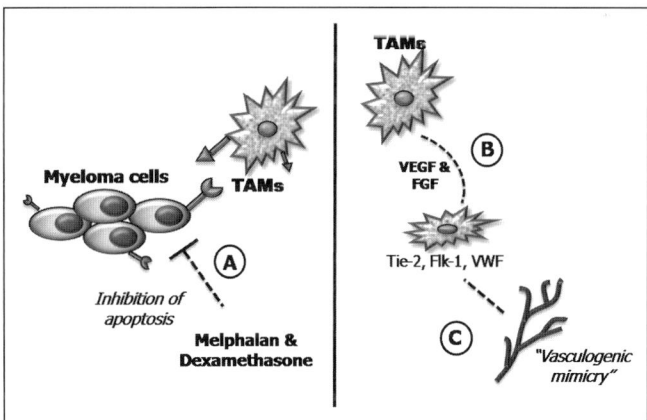

FIGURE 14-1 Tumor-associated macrophages (TAMs) promote myeloma pathogenesis. (A) TAMs within the myeloma bone marrow microenvironment are important in protecting myeloma cells from chemotherapy-induced apoptosis. (B) TAMs from myeloma patients begin expressing endothelial-cell-specific markers when stimulated with angiogenic growth factors. (C) TAMs from myeloma patients have the unique ability to undergo "vasculogenic mimicry" to form capillary-like structures.

matrix-degrading enzymes, including various members of the matrix-metalloproteinase family of enzymes. In the field of multiple myeloma, numerous groups have demonstrated the angiogenic potential of macrophages isolated from myeloma patients. Neovascularization within the bone marrow cavity is one of the hallmark features in patients with myeloma. Scavelli and colleagues have shown that macrophages from multiple myeloma patients can display "vasculogenic mimicry" (Figure 14-1) [11]. In these studies the authors demonstrated that VEGF- and basic fibroblast growth factor (bFGF)-stimulated macrophages from myeloma patients resulted in the expression of endothelial cell markers, at both the mRNA and protein level. Additionally, the authors found myeloma-associated macrophages formed capillary-like structures in Matrigel following 24 hours of exposure to VEGF and bFGF. Macrophages isolated from control patients and patients with the pre-malignant stage of monoclonal gammopathy of undetermined significance (MGUS) were less capable of forming these capillary-like structures; therefore demonstrating the unique features of TAMs in multiple myeloma.

More recently, Chen et al. demonstrated that the angiogenic factor pleiotrophin (PTN) combined with macrophage colony-stimulating factor (M-CSF) induces the expression of vascular endothelial cell genes Tie-2, Flk-1, von Willebrand factor (VWF), and VE-cadherin in CD14+ monocyte cell populations [12]. PTN was expressed in the bone marrow of myeloma patients. Culturing of monocytes with myeloma bone marrow not only induced protein expression of Flk-1, Tie-2 and VWF but also stimulated the formation Flk-1+ tube-like structures in vitro. Further in vitro investigation revealed that addition of PTN and M-CSF to monocytes promoted Flk-1+ tube formation. Treatment with both these cytokines induced expression of the endothelial precursor cell marker, CD133, on the surface of CD14+ monocytes. Co-inoculation of human monocyte and myeloma cell lines in vivo resulted in the monocytes incorporating into the tumor vasculature while also expressing vascular endothelial cell genes. These results further support the concept of "vascular mimicry" in the context of myeloma.

## Dendritic cells

Cells within the bone marrow can alter this microenvironment in favor of myeloma cell survival and growth; however, myeloma cells can also affect the resident cells within the bone marrow cavity. Dhodapkar and colleagues demonstrated that myeloma cells have effects on dendritic cells (DCs) and have the ability to promote differentiation of these cells into osteoclasts [13]. They found that co-culture of immature DCs with myeloma cells led to the formation of functional multinucleated bone-resorbing giant cells (MGCs). This effect was specific to the interactions between DCs and myeloma cells, both primary and cell line, as no other monocyte population or cancer cell type had this ability. The formation of functional osteoclast-like cells was dependent on cell–cell contact and on short-range cytokines, which resulted in fusion of DCs with other DCs to form these multinucleated cells. Microarray analysis comparing co-cultured DCs and DCs alone identified thrombospondin-1 (TSP-1) as being significantly upregulated following co-culture. The formation of MGCs and osteoclasts was significantly reduced with the addition of anti-TSP-1 antibody even in the presence of osteoclastogenic factors, RANKL and M-CSF. The authors also found that TSP-1 on the DCs interacts with CD47 present on the myeloma cells and this interaction is largely responsible for MGC formation. In addition to the interaction between TSP-1 and CD47 present on the myeloma cells, treatment of these co-cultures with OPG indicated a role for RANK–RANKL interactions in the formation of MGC. These studies also suggest that the RANKL signal from myeloma cells may not be the only source stimulating osteoclastogenesis. Finally, anti-TSP-1 treatment in vivo reduced serum calcium concentrations in mice with PTH-induced hypercalcemia compared to isotypic control antibody-treated mice. These results demonstrate that TSP-1 promotes osteoclastogenesis in vivo.

Antigen-presenting cells produce various cytokines that are responsible for induction and differentiation of T cells. A specific population of T helper cells known as Th17 cells has particular importance in the fields of pathology and autoimmune disease. Th17 cells are defined by the co-expression of IFN-γ and IL-17. Various studies investigating Th17 cells have demonstrated the ability of DCs to activate Th17 cells [14, 15] and this has significance in myeloma patients as the tumors are often infiltrated with DCs. Assessment of healthy donor peripheral blood mononuclear cells (PBMCs) found that there was only a small proportion of Th17 cells present in the entire population [16]. The Dhodapkar group also determined that DCs matured with inflammatory cytokines were more capable of activating polyfunctional Th17 cells, compared to immature DCs. The authors demonstrated that matured DCs were more potent antigen-presenting cells to induce Th17 cell expansion in comparison to monocytes. The ability of matured DCs to stimulate Th17 cells was dependent on cell–cell contact and partly on the production of IL-1β, IL-6, and IL-23. Finally, the authors found that myeloma patients had significant proportions of polyfunctional Th17 cells present in their bone marrow in comparison to patients with MGUS. The high prevalence of these cells could account for the elevated serum concentrations of IL-17 that are also seen in myeloma patients [17]. IL-17 is a known stimulator of osteoclasts [18, 19] and interestingly the proportion of Th17 cells present in the bone marrow of myeloma patients correlates with lytic bone disease [16].

Recent findings by Anderson and colleagues have shown a role for a specific population of DCs in immunosuppression, often associated with myeloma. Plasmacytoid dendritic cells (pDCs) lack lineage cell-surface markers specific for T, B, NK cells, and monocytes. These studies demonstrated pDCs from patients with multiple myeloma had impaired ability to stimulate allogeneic T-cell response despite elevated numbers of these cells in their bone marrow, in comparison to pDCs from normal donors [20]. Further examination revealed these pDCs promoted cell growth as well as the clonogenic potential of myeloma cell lines both in vitro and in vivo. The pDCs promoted drug-resistance in the presence of a proteasome inhibitor by inducing NF-κB activation in myeloma cells. The localization of pDCs also plays a role in chemotaxis of migrating MM cells and pDCs. Supernatants from pDCs co-cultured with myeloma cells greatly enhanced myeloma cell migration. Analysis of these co-culture supernatants indicated that these interactions stimulated the production of chemotactic cytokines IL-3 and SDF-1α, which are important in the migration of both tumor cells [21] and pDCs to the tumor bed. Examination of bone marrow biopsies from multiple myeloma patients demonstrated direct contact between pDCs and myeloma cells. The use of CpG-containing oligodeoxynucletides to activate pDCs via the Toll-like receptors had the ability to restore APC-stimulated T-cell proliferation and inhibited their ability to promote myeloma cell growth.

$β_2$-microglobulin ($β_2$M) plays a role in normal immune function as a component of major histocompatibility (MHC) class I molecules [22]. Normally, serum concentrations of $β_2$M are relatively low; however, elevated serum concentrations are found in patients with multiple myeloma and these levels often correlate with a poor prognosis [23–28]. In studies when monocyte-derived dendritic cells (MoDCs) were cultured with $β_2$M, the cultures remained poorly differentiated while MoDCs cultured alone developed into the enlarged cellular morphology similar to DCs [29]. $β_2$M treatment also inhibited cell yield in the cultures; however, this reduction in number was not due to induction of apoptosis. Following culture with $β_2$M, MoDCs had significantly lower expression of MHC class I and co-stimulatory molecules, including CD1a, CD40, CD54, CD80, and HLA-ABC molecules. Treating DCs with $β_2$M caused these cells to secrete high concentrations of immunosuppressive cytokines and IL-6 and IL-8, known stimulators of osteoclast formation and resorption [30–33]. Additionally, $β_2$M treatment decreased the ability of MoDCs to activate a T-cell response and the cytokine production by these activated T cells was greatly reduced.

## Natural killer cells

Dhodapkar and colleagues have provided some pivotal work examining differences between immune cells in patients with MGUS versus those with multiple myeloma. These studies found a lack of differences in natural killer T (NKT) cell (Vα24+Vβ11+), T (CD3+) and NK cell (CD3–, CD56+) populations between patients with progressive multiple myeloma, non-progressive myeloma, MGUS, and healthy individuals; however, patients with progressive myeloma had a significant decline in NKT cell function [34]. Functional NKT cells can be quantified by measuring cytokine production [35], in this specific case IFN-γ, by NKT cells in response to NKT ligand, alpha galactosyl ceramide. Using this method of detection, they found that progressive myeloma patients had significantly lower IFN-γ-producing NKT cells in both the peripheral blood and in the tumor bed, in comparison to patients in the other groups of non-progressive, MGUS, and healthy controls. Despite the unresponsiveness of fresh NKT cells from progressive patients, these cells could be restored to cytokine-producing cells in response to ligand presentation in vitro by dendritic cells.

Heat shock proteins, specifically 90 kDa (HSP-90), are overexpressed in many types of cancer [36]. The overexpression of these proteins is thought to be a requirement for cancer cell survival; therefore the inhibition of HSP-90 is thought to be a viable therapeutic option for the treatment of various cancers. Ansamycin-based compounds, such as 17-allylamino-17-demethoxygeldanamycin (17-AAG), are used for the inhibition of

HSP-90s and are currently being assessed in clinical trials for multiple myeloma. Cippitelli and colleagues investigated the mechanism of action by which HSP-90 inhibitors induce myeloma cell death [37]. Considering the immunomodulatory capabilities of the multiple myeloma therapies thalidomide and lenalidomide, these authors examined whether HSP-90 inhibitors had the ability to stimulate NK cell cytotoxic function by regulating the expression of NK-activating ligands present on myeloma cells. The authors found HSP-90 inhibitors, 17-AAG and radicicol, upregulated the expression of NK-activating ligands MICA/B, at both the protein and mRNA level. Their studies showed the unfolded protein response (UPR) was not responsible for the upregulation of these ligands. The activation and binding of the transcription factor HSP-1 to MICA and MICB promoters results from HSP-90 inhibition. Treatment with HSP-90 inhibitors also increased NK degranulation in response to enhanced NK-activating ligands.

## Adaptive immunity

### B cells

Clonotypic B lymphocytes (CBLs) are considered the stem-cell-like precursors to malignant plasma cells seen in multiple myeloma [38]. Studies have identified that these cell populations are present in both patients with multiple myeloma and MGUS [39]. Matsui et al. have detected a small sub-population of myeloma "stem cells" present in multiple myeloma patients (CD138-negative B cells), also considered CBLs [40]. This cell population has clonogenic potential [40] and chemo-resistance [41] not possessed by the myeloma cells themselves. More recent studies found that myeloma cells, which express markers CD20 and CD27, typical of memory B cells, not only have clonogenic potential in vitro but can also propagate myeloma in vivo following transplantation [41]. In a study characterizing CBLs from a cohort of multiple myeloma patients, the results indicated that this cell population expressed markers characteristic of both lymphoid lineage (CD19–, CD38 dim to moderate, CD138–) and stem cell phenotypic cells (CD34 expression) [42]. Despite the lack of CD34+ cells present in the studies performed by Matsui et al., these studies by Conway et al. suggest the presence of CD19-CD34+ cells as a possible myeloma cell precursor population.

B-lymphocyte stimulator, or BlyS, is a known activator of both the classical and alternative NF-κB signaling pathways and important for B-lymphocyte survival and differentiation. Serum concentrations of BLyS are higher in patients with multiple myeloma in comparison to normal serum samples [43]. Additionally, monocytes isolated from the bone marrow of myeloma patients expressed transmembrane activator and CAML interactor (TACI), one of the three receptors for BLyS, on their cell surface [44]. This group also found that BLyS-activated monocytes secreted high concentrations of IL-6, TNF-α, and IL-1β. Previous studies have shown that osteoclasts express high levels of BLys and APRIL and this expression contributes to the vicious cycle of cancer in bone by promoting myeloma cell growth and survival [45, 46].

### Helper T cells

Monoclonal immunoglobulin (mIg) production of specific isotypes is one of the defining features of clonal neoplasmic plasma cells in multiple myeloma patients. Given the therapeutic potential of specifically targeting malignant plasma cells, examination of monoclonal immunoglobulin-reactive lymphocytes in myeloma patients could be extremely beneficial. Ostad and colleagues co-cultured mature, mIg antigen-stimulated DCs with peripheral blood mononuclear cells from myeloma patients [47]. Co-culture resulted in a marked expansion and cell proliferation of the CD4+ T-cell population in response to mIg antigen, as opposed to CD8+ T-cell populations. Presence of MHC class II blocking antibodies in these co-cultures indicated that MHC class II molecules are involved in Ig antigen presentation to CD4+ T cells. In addition to measuring proliferation of these cells, T cells from patients demonstrated low concentrations of IFN-γ in response to mIg-pulsed DCs. Despite limited cytokine production in response to mIg, these cells were responsive to T-cell receptor stimulation and responsive to tetanus toxoid, indicative of response to exogeneous antigen presentation.

## Hedgehog signaling

The identification of antigens that are unique to tumor cells would provide many therapeutic opportunities for the treatment of cancer. Blotta et al. used the method of serologic analysis of recombinant cDNA expression library (SEREX) to identify unique tumor antigens present in MGUS patients that induced an immune response in multiple myeloma patients [48]. The authors identified ten different genes in the sera of MGUS patients, of these OFD1 induced an immune response in both MGUS and multiple myeloma patients. OFD1 encodes for a basal body/centrosome-related protein that plays a critical role in developmental processes [49, 50]. Knockdown of Ofd1 by siRNA in myeloma cells that overexpress OFD1 resulted in the inhibition of hedgehog (Hh) signaling as indicated by downregulation of *GLI1* and *PTCH1*. Wnt signaling was also altered suggesting an increase in β-catenin-dependent signaling following OFD1 knockdown. OFD1-derived peptides stimulated a CD8+ T-cell response in both MGUS and myeloma patients; therefore, this suggests OFD1 is a tumor-specific antigen important in the transition from MGUS to multiple myeloma.

## TNF receptor superfamily and NF-κB

Three proteins that belong to the tumor necrosis factor receptor family were identified as critical components for osteoclastogenesis and bone remodeling. These factors are receptor activator of NF-κB (RANK) [51], its ligand RANKL [51–54], and the RANKL decoy receptor osteoprotegerin (OPG) [55–57]. The interaction between the transmembrane receptor RANK and the membrane-bound protein RANK ligand is part of normal and pathologic bone remodeling. RANKL is upregulated in T cells following

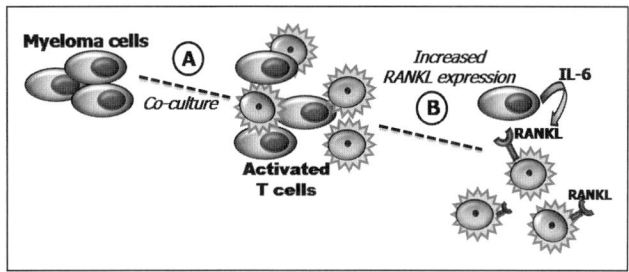

**FIGURE 14-2 Myeloma cells promote RANKL expression. (A)** Human myeloma cells co-cultured with activated T cells. **(B)** IL-6 secreted by the myeloma cells causes increased RANKL expression in activated T cells.

antigen stimulation [51, 53], which can occur from exposure to numerous infectious agents. Any activation of T cells under inflammatory or even pathologic conditions could potentially augment osteoclast formation and activity [58–60]. Giuliani et al. showed that myeloma cells have effects on T cells and their function [61]. The authors demonstrated that RANKL expression in activated T cells was increased following co-culture with human myeloma cell lines (Figure 14-2). Following co-culture, T cells produced increased amounts of soluble RANKL and the conditioned media had enhanced osteoclastogenesis capability. These studies found that IL-6 secretion by the myeloma cells was largely responsible for the increased RANKL expression. T-cell function was also suppressed as indicated by a reduction in IFN-γ production. These authors, as well as others [62], have shown T cells from multiple myeloma patients with osteolytic bone disease expressed RANKL.

NF-κB activation is evident in both myeloma cells and the bone marrow microenvironment that surrounds these cells. Recent gene expression profiling studies by Annunziata and colleagues demonstrated specific signatures indicative of dependence on NF-κB signaling [63]. Among the genes found within the expression signature were *NIK* and *TRAF3*, which encode for key regulators of both alternative and classical signaling of NF-κB [64–71]. Patients with multiple myeloma had high expression of the NF-κB activating kinase, NIK, and low expression of TRAF3, a known negative regulator of NIK, in their plasma cells in comparison to normal patients [63]. Studies with NIK overexpression or disrupted NIK–TRAF3 interactions can cause B-cell hyperplasia by either amplifying B-cell-activating factor of the TNF family (BAFF or BLyS)-induced alternative NF-κB signaling or inducing independent mechanisms [72].

TLR signaling has a role in both immunity and cancer development. Chiron et al. demonstrated that TLR3 is expressed on various human myeloma cell lines and stimulation with TLR3 ligand results in NF-κB activation in these cells [73]. Ligand stimulation resulting in NF-κB activation in the myeloma cells affected proliferation in a heterogeneous manner. Myeloma cells that showed a decrease in proliferation upon ligand stimulation were found to have an increase in caspase-dependent apoptosis. The induction of myeloma cell apoptosis was due to the production of IFN-α, which was present in the culture supernatant following stimulation. The authors determined the IFN-α production resulted from p38 MAPK activation upon stimulation.

TNF-related apoptosis-inducing ligand (TRAIL) is known to be a pro-apoptotic protein for tumor cells; however, through interaction with OPG, TRAIL can potentially promote myeloma-associated bone disease. Colucci et al. determined that T cells from multiple myeloma patients with osteolysis have the ability to promote osteoclastogenesis [62]. The authors found that T cells from these patients overexpressed RANKL, at both the mRNA and protein level. They demonstrated that T cells from patients with osteolysis were the only source of the cytokines RANKL and OPG and they also were a major source of TRAIL. The authors also found that osteoclasts derived from PBMC cultures taken from multiple myeloma patients had prolonged survival. They determined that the enhanced survival was mediated by the downregulation of the death receptor and upregulation of a decoy receptor on the osteoclasts, both of which are involved in the induction of apoptosis. Additionally, osteoclasts in the presence of T cells upregulated the expression of the anti-apoptotic protein, Bcl-2. These data further support the concept that the surrounding bone marrow microenvironment is altered by the myeloma, which facilitates cancer progression.

## Therapeutic approaches and effects on immune cells

### Myeloma-specific cytotoxic T lymphocytes

In recent years the use of a patient's own immune system has been examined as a potential anti-tumor therapeutic option. The use of dendritic cells in cancer immunotherapy has received the most attention [74, 75]; however, there are limitations to using these cells, as they are difficult to expand in vitro once they are removed from the patient and there is a need for multiple administration. An alternative approach for immunotherapy is the use of CD40-activated B cells, which activate cytotoxic T cells to produce an anti-tumor response (Figure 14-3, Panel 1). Examination of CD40 B cells as antigen-presenting cells (APCs), capable of inducing T-cell immune responses, has also been studied in the context of anti-myeloma therapy. Kim et al. demonstrated that CD40-activated B cells loaded with myeloma cell lysates had the ability to activate naïve T cells through the production of the cytokines IFN-γ, IL-6, and IL-12 [76]. The CD40 B cells have a preference to induce Th1 polarization through the secretion of high amounts of IL-12p40 and low amounts of IL-12p70. Theses studies demonstrate the possibility of using activated CD40 as APCs as opposed to dendritic cell vaccines.

### Gamma delta T cells

Maekawa et al. and Takahashi et al. investigated whether gamma delta (γδ) T cells, derived from multiple myeloma patients, can be a viable option as an antitumor immunotherapy [77–79]. Maekawa and colleagues examined the mechanism by which γδ T cells recognize myeloma cells [78]. The authors demonstrated that γδ T cells kill

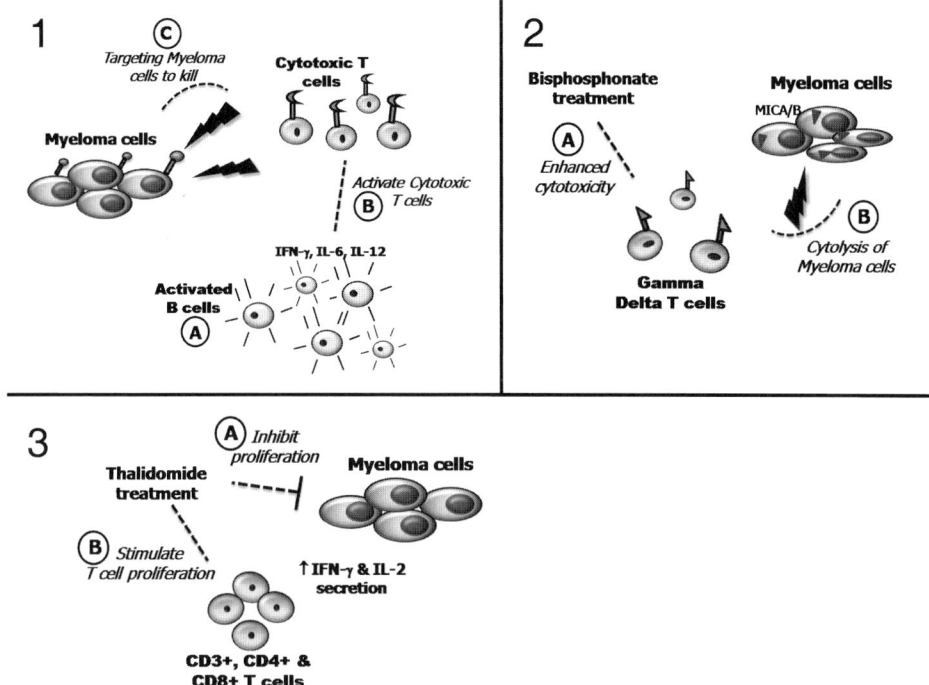

FIGURE 14-3 Immunotherapy for myeloma. Panel 1: (A) Activated B cells loaded with myeloma cell lysates produce cytokines, which (B) can activate T cells. (C) Cytotoxic T cells can then target myeloma cells as a potential anticancer vaccine. Panel 2: (A) Bisphosphonate treatment can enhance cytotoxicity of gamma delta T cells. (B) Cytolysis of myeloma cells is dependent on expression of stress-inducible ligands MICA/B present on myeloma cells. Panel 3: (A) Thalidomide treatment can inhibit myeloma cells directly and (B) stimulate T-cell proliferation and cytokine production that is important in anticancer responses.

myeloma cells through close interactions in which the γδ T cell attacks the myeloma cell, resulting in cytolysis (Figure 14-3, Panel 2). They also found that bisphosphonate treatment enhanced the cytotoxicity of γδ T cells. The ability of γδ T cells to kill myeloma cells was dependent on the expression of the adhesion molecule, ICAM-1, present on the myeloma cells. Takahashi and colleagues examined the feasibility of expanding γδ T cells from individual patients to use clinically [77]. The percentage of γδ T cells present in peripheral blood (PB) of multiple myeloma patients was significantly lower than normal donor PB. The expression and secretion of mevalonate intermediates IPP and DMAPP was increased in tumor cells. The ability to expand γδ T cells from multiple myeloma patients in vitro with the addition of zoledronate and IL2 varied among the patients; however, the level of expansion was not correlated with the stage of disease. The cytotoxicity levels of the γδ T cells from multiple myeloma patients were dependent on the expression of the stress-inducible ligand MICA/B on myeloma cells, which is recognized by NK cell activating receptors present on γδ T cells. By culturing γδ T cells with bisphosphonate ex vivo, there is an increase in cytotoxic specificity. Overall the γδ T cells were able to specifically kill a significant amount of fresh myeloma cells while having no cytotoxic effect on autologous lymphocytes.

## Tumor-specific antigens

In the context of multiple myeloma, utilizing the adaptive immune system against tumor growth may be beneficial to advance clinical treatments. In immunosurveillance, CD4+ T cells recognize tumor-specific antigens that are secreted by myeloma cells [80, 81]. In vivo studies demonstrated that some myeloma cell lines secrete high levels of antigen that can activate an immune response [82]. Myeloma cell variants that produce tumor antigen have delayed growth in vivo and the potential to be rejected. When myeloma cells do not express MHC class II molecules, they cannot directly stimulate CD4+ T-cell proliferation. However, in the presence of tumor-filtering macrophages, myeloma cell growth is suppressed in response to tumor-antigen secretion by the myeloma cells. These macrophages are recruited to the tumor site where they can become activated by the secreted antigen, resulting in MHC class II expression. MHC class II-expressing macrophages then stimulate a T-cell response. The authors finally tested the ability to deliver purified tumor-specific antigen in vivo, which resulted in reduced tumor growth. The concept of cancer vaccines has long had exciting potential; however, discovering tumor antigens that will allow for specific immune responses against tumors has been problematic [83].

## Thalidomide and lenalidomide

In recent years, thalidomide (Thal) has proven to have strong antimyeloma capabilities [84]. Despite the use of this therapeutic agent to inhibit angiogenesis, clinical evidence suggests that the antimyeloma effects of thalidomide are not due to decreases in bone marrow angiogenesis in multiple myeloma patients. Examination of Thal and Thal analogs has demonstrated immunomodulatory effects, including direct inhibition of myeloma cell proliferation [85] and the ability to act as a co-stimulatory signal to T cells [86] (Figure 14-3, Panel 3). In previous studies, Thal could only increase T-cell proliferation in the presence of anti-CD3 or dendritic cells [86]. Anderson and colleagues further examined Thal action on the immune system and identified specific effects on NK-cell function [87]. In multiple myeloma patients and in healthy individuals, there was an increase in CD3+ T-cell proliferation together with IFN-γ and IL-2 secretion. Additional T-cell subsets, including CD4+ and CD8+ T cells, also demonstrated an increase in proliferation in response to Thal and Thal analog treatment. Thal treatment of PBMCs in combination with IL-2 resulted in increased myeloma cell lysis. Additionally, depleted PBMC cultures containing CD56+ NK cells still had enhanced myeloma cell lysis in the presence of Thal; however, CD56+ depletion showed reduced lysis. The increase in NK-mediated lysis was not dependent on the upregulation of NK activation markers. In PBMCs from multiple myeloma patients, Thal or Thal analog treatment, following IL-2 treatment, resulted in enhanced lysis of myeloma cells. In treated multiple myeloma patients, there were increases in NK function and number; however, there were no differences in other T-cell subsets. The increase in CD56+ NK cells correlated with patients who responded to Thal treatment.

Ideally, the therapies for the treatment of multiple myeloma should inhibit tumor growth and improve the associated osteolytic bone disease. Lenalidomide, a thalidomide analog, not only has immunomodulatory effects, but studies have also shown it to have specific effects on osteoclasts. In vitro studies revealed that lenalidomide could dose-dependently decrease osteoclast formation and resorption activity [88]. The inhibition of osteoclast differentiation and function were due to a decrease in $\alpha V\beta 3$ integrin and cathepsin K expression following lenalidomide treatment. The authors determined that lenalidomide inhibits osteoclastogenesis during stages of differentiation, indicated by the inhibition of both ERK activation and the transcription factor, PU.1. Lenalidomide treatment of osteoclast cultures resulted in decreased secretion of the cytokines MIP-1$\alpha$, BAFF, and APRIL, which are all important for osteoclast and myeloma cell survival and growth. Additionally, bone marrow stromal cells treated with lenalidomide showed a reduction in RANKL secretion, further enhancing the inhibition of osteoclastogenesis. Multiple myeloma patients with osteolytic bone disease often have a high ratio of RANKL/OPG concentrations in their serum. Myeloma patients treated with lenalidomide had an increase in OPG serum concentrations while RANKL concentrations were significantly reduced.

## Conclusions

Osteoimmunology in the context of multiple myeloma is a newly emerging field, and it is clear that interactions between myeloma cells and cells of the immune system are important both in terms of tumor growth and the development of the osteolytic bone disease. Increasing our understanding of the role of the immune system in myeloma bone disease, for example, the effect of the immunosuppression that is found in patients with myeloma will ultimately identify new therapeutic targets for the treatment of myeloma. Our current knowledge is limited due to the difficulty of verifying intriguing in vitro observations in appropriate in vivo systems. The tumor microenvironment in the context of myeloma is rich with cells of the immune system. Immune cells can have both a deleterious and advantageous role in myeloma pathogenesis. Macrophages and T cells, for example, can become altered by the presence of myeloma cells within the bone marrow cavity and act to support further cancer progression; however, the immune system can also be utilized for anticancer therapies. Some of the ways these altered immune cells enhance cancer progression are by stimulating angiogenesis to provide nutrients to the growing tumor, inhibiting chemotherapy-induced apoptosis, and contributing to osteoclast formation through increased RANKL expression. The vicious cycle between myeloma cells and these cells within the bone microenvironment contributes to the destructive bone disease, which is one of the defining features of multiple myeloma. Within a healthy individual, the immune system provides defense mechanisms critical for protecting the body against foreign pathogens as well as tumor formation. The manipulation or stimulation of the body's natural immune system to fight cancer provides an attractive area for therapeutic potential [89–91]. Emerging cancer research of recent years not only demonstrates genetic alterations to the cancer cells but also to the surrounding microenvironment. The future of effective cancer therapeutics will have a dual focus; treating both the tumor cells and the altered microenvironment.

# References

[1] Jemal A, Tiwari RC, Murray T, Ghafoor A, Samuels A, Ward E, et al. Cancer statistics. CA Cancer J Clin 2004 Jan-Feb 2004;54(1):8–29.

[2] Mundy GR, Luben RA, Raisz LG, Oppenheim JJ, Buell DN. Bone-resorbing activity in supernatants from lymphoid cell lines. N Engl J Med 1974 Apr 18;290(16):867–71.

[3] Mundy GR, Raisz LG, Cooper RA, Schechter GP, Salmon SE. Evidence for the secretion of an osteoclast stimulating factor in myeloma. N Engl J Med 1974 Nov 14;291(20):1041–6.

[4] Valentin-Opran A, Charhon SA, Meunier PJ, Edouard CM, Arlot ME. Quantitative histology of myeloma-induced bone changes. Br J Haematol 1982 Dec;52(4):601–10.

[5] Litman GW, Cannon JP, Dishaw LJ. Reconstructing immune phylogeny: new perspectives. Nat Rev Immunol 2005 Nov;5(11):866–79.

[6] Pancer Z, Cooper MD. The evolution of adaptive immunity. Annu Rev Immunol 2006;24:497–518.

[7] Mills CD, Kincaid K, Alt JM, Heilman MJ, Hill AM. M-1/M-2 macrophages and the Th1/Th2 paradigm. J Immunol 2000 Jun 15;164(12):6166–73.

[8] Savage ND, de Boer T, Walburg KV, Joosten SA, van Meijgaarden K, Geluk A, et al. Human anti-inflammatory macrophages induce Foxp3+ GITR+ CD25+ regulatory T cells, which suppress via membrane-bound TGFbeta-1. J Immunol 2008 Aug 1;181(3):2220–6.

[9] Cua DJ, Stohlman SA. In vivo effects of T helper cell type 2 cytokines on macrophage antigen-presenting cell induction of T helper subsets. J Immunol 1997 Dec 15;159(12):5834–40.

[10] Zheng Y, Cai Z, Wang S, Zhang X, Qian J, Hong S, et al. Macrophages are an abundant component of myeloma microenvironment and protect myeloma cells from chemotherapy drug-induced apoptosis. Blood 2009 Oct 22;114(17):3625–8.

[11] Scavelli C, Nico B, Cirulli T, Ria R, Di Pietro G, Mangieri D, et al. Vasculogenic mimicry by bone marrow macrophages in patients with multiple myeloma. Oncogene 2008 Jan 24;27(5):663–74.

[12] Chen H, Campbell RA, Chang Y, Li M, Wang CS, Li J, et al. Pleiotrophin produced by multiple myeloma induces transdifferentiation of monocytes into vascular endothelial cells: a novel mechanism of tumor-induced vasculogenesis. Blood 2009 Feb 26;113(9):1992–2002.

[13] Kukreja A, Radfar S, Sun BH, Insogna K, Dhodapkar MV. Dominant role of CD47-thrombospondin-1 interactions in myeloma-induced fusion of human dendritic cells: implications for bone disease. Blood 2009 Oct 15;114(16):3413–21.

[14] LeibundGut-Landmann S, Gross O, Robinson MJ, Osorio F, Slack EC, Tsoni SV, et al. Syk- and CARD9-dependent coupling of innate immunity to the induction of T helper cells that produce interleukin 17. Nat Immunol 2007 Jun;8(6):630–8.

[15] Bailey SL, Schreiner B, McMahon EJ, Miller SD. CNS myeloid DCs presenting endogenous myelin peptides "preferentially" polarize CD4+ T(H)-17 cells in relapsing EAE. Nat Immunol 2007 Feb;8(2):172–80.

[16] Dhodapkar KM, Barbuto S, Matthews P, Kukreja A, Mazumder A, Vesole D, et al. Dendritic cells mediate the induction of polyfunctional human IL17-producing cells (Th17-1 cells) enriched in the bone marrow of patients with myeloma. Blood 2008 Oct 1;112(7):2878–85.

[17] Alexandrakis MG, Pappa CA, Miyakis S, Sfiridaki A, Kafousi M, Alegakis A, et al. Serum interleukin-17 and its relationship to angiogenic factors in multiple myeloma. Eur J Intern Med 2006 Oct;17(6):412–16.

[18] Sato K, Suematsu A, Okamoto K, Yamaguchi A, Morishita Y, Kadono Y, et al. Th17 functions as an osteoclastogenic helper T cell subset that links T cell activation and bone destruction. J Exp Med 2006 Nov 27;203(12):2673–82.

[19] Kotake S, Udagawa N, Takahashi N, Matsuzaki K, Itoh K, Ishiyama S, et al. IL-17 in synovial fluids from patients with rheumatoid arthritis is a potent stimulator of osteoclastogenesis. J Clin Invest 1999 May;103(9):1345–52.

[20] Chauhan D, Singh AV, Brahmandam M, Carrasco R, Bandi M, Hideshima T, et al. Functional interaction of plasmacytoid dendritic cells with multiple myeloma cells: a therapeutic target. Cancer Cell 2009 Oct 6;16(4):309–23.

[21] Parmo-Cabanas M, Bartolome RA, Wright N, Hidalgo A, Drager AM, Teixido J. Integrin alpha4beta1 involvement in stromal cell-derived factor-1alpha-promoted myeloma cell transendothelial migration and adhesion: role of cAMP and the actin cytoskeleton in adhesion. Exp Cell Res 2004 Apr 1;294(2):571–80.

[22] Bjorkman PJ, Parham P. Structure, function, and diversity of class I major histocompatibility complex molecules. Annu Rev Biochem 1990;59:253–88.

[23] Schardijn GH, Statius van Eps LW. Beta 2-microglobulin: its significance in the evaluation of renal function. Kidney Int 1987 Nov;32(5):635–41.

[24] Revillard JP, Vincent C. Structure and metabolism of beta-2-microglobulin. Contrib Nephrol 1988;62:44–53.

[25] Campos L, Vu Van H, Ville D, Imbert C, Gentilhomme O, Luo YC, et al. Serum beta 2 microglobulin in adult myeloid acute leukemias. Blut 1984 Apr;48(4):221–6.

[26] Nakao Y, Matsumoto H, Miyazaki T, Watanabe S, Masaoka T, Takatsuki K, et al. Genetic and clinical studies of serum beta 2-microglobulin levels in haematological malignancies. Clin Exp Immunol 1981 Oct;46(1):134–41.

[27] Bataille R, Grenier J, Sany J. Beta-2-microglobulin in myeloma: optimal use for staging, prognosis, and treatment–a prospective study of 160 patients. Blood 1984 Feb;63(2):468–76.

[28] Barlogie B, Jagannath S, Desikan KR, Mattox S, Vesole D, Siegel D, et al. Total therapy with tandem transplants for newly diagnosed multiple myeloma. Blood 1999 Jan 1;93(1):55–65.

[29] Xie J, Wang Y, Freeman 3rd ME, Barlogie B, Yi Q. Beta 2-microglobulin as a negative regulator of the immune system: high concentrations of the protein inhibit in vitro generation of functional dendritic cells. Blood 2003 May 15;101(10):4005–12.

[30] Black K, Garrett IR, Mundy GR. Chinese hamster ovarian cells transfected with the murine interleukin-6 gene cause hypercalcemia as well as cachexia, leukocytosis and thrombocytosis in tumor-bearing nude mice. Endocrinology 1991 May;128(5):2657–9.

[31] Kurihara N, Bertolini D, Suda T, Akiyama Y, Roodman GD. IL-6 stimulates osteoclast-like multinucleated cell formation in long term human marrow cultures by inducing IL-1 release. J Immunol 1990 Jun 1;144(11):4226–30.

[32] Bendre MS, Margulies AG, Walser B, Akel NS, Bhattacharrya S, Skinner RA, et al. Tumor-derived interleukin-8 stimulates osteolysis independent of the receptor activator of nuclear factor-kappaB ligand pathway. Cancer Res 2005 Dec 1;65(23):11001–9.

[33] Bendre MS, Montague DC, Peery T, Akel NS, Gaddy D, Suva LJ. Interleukin-8 stimulation of osteoclastogenesis and bone resorption is a mechanism for the increased osteolysis of metastatic bone disease. Bone 2003 Jul;33(1):28–37.

[34] Dhodapkar MV, Geller MD, Chang DH, Shimizu K, Fujii S, Dhodapkar KM, et al. A reversible defect in natural killer T cell function characterizes the progression of premalignant to malignant multiple myeloma. J Exp Med 2003 Jun 16;197(12):1667–76.

[35] Fujii S, Shimizu K, Steinman RM, Dhodapkar MV. Detection and activation of human Valpha24+ natural killer T cells using alpha-galactosyl ceramide-pulsed dendritic cells. J Immunol Methods 2003 Jan 15;272(1-2):147–59.

[36] Whitesell L, Lindquist SL. HSP90 and the chaperoning of cancer. Nat Rev Cancer 2005 Oct;5(10):761–72.

[37] Fionda C, Soriani A, Malgarini G, Iannitto ML, Santoni A, Cippitelli M. Heat shock protein-90 inhibitors increase MHC class I-related chain A and B ligand expression on multiple myeloma cells and their ability to trigger NK cell degranulation. J Immunol 2009 Oct 1;183(7):4385–94.

[38] Szczepek AJ, Bergsagel PL, Axelsson L, Brown CB, Belch AR, Pilarski LM. CD34+ cells in the blood of patients with multiple myeloma express CD19 and IgH mRNA and have patient-specific IgH VDJ gene rearrangements. Blood 1997 Mar 1;89(5):1824–33.

[39] Jensen GS, Mant MJ, Belch AJ, Berenson JR, Ruether BA, Pilarski LM. Selective expression of CD45 isoforms defines CALLA+ monoclonal B-lineage cells in peripheral blood from myeloma patients as late stage B cells. Blood 1991 Aug 1;78(3):711–19.

[40] Matsui W, Huff CA, Wang Q, Malehorn MT, Barber J, Tanhehco Y, et al. Characterization of clonogenic multiple myeloma cells. Blood 2004 Mar 15;103(6):2332–6.

[41] Matsui W, Wang Q, Barber JP, Brennan S, Smith BD, Borrello I, et al. Clonogenic multiple myeloma progenitors, stem cell properties, and drug resistance. Cancer Res 2008 Jan 1;68(1):190–7.

[42] Conway EJ, Wen J, Feng Y, Mo A, Huang WT, Keever-Taylor CA, et al. Phenotyping studies of clonotypic B lymphocytes from patients with multiple myeloma by flow cytometry. Arch Pathol Lab Med 2009 Oct;133(10):1594–9.

[43] Novak AJ, Darce JR, Arendt BK, Harder B, Henderson K, Kindsvogel W, et al. Expression of BCMA, TACI, and BAFF-R in multiple myeloma: a mechanism for growth and survival. Blood 2004 Jan 15;103(2):689–94.

[44] Chang SK, Arendt BK, Darce JR, Wu X, Jelinek DF. A role for BLyS in the activation of innate immune cells. Blood 2006 Oct 15;108(8):2687–94.

[45] Moreaux J, Cremer FW, Reme T, Raab M, Mahtouk K, Kaukel P, et al. The level of TACI gene expression in myeloma cells is associated with a signature of microenvironment dependence versus a plasmablastic signature. Blood 2005 Aug 1;106(3):1021–30.

[46] Yaccoby S, Pennisi A, Li X, Dillon SR, Zhan F, Barlogie B, et al. Atacicept (TACI-Ig) inhibits growth of TACI(high) primary myeloma cells in SCID-hu mice and in coculture with osteoclasts. Leukemia 2008 Feb;22(2):406–13.

[47] Ostad M, Andersson M, Gruber A, Sundblad A. Expansion of immunoglobulin autoreactive T-helper cells in multiple myeloma. Blood 2008 Mar 1;111(5):2725–32.

[48] Blotta S, Tassone P, Prabhala RH, Tagliaferri P, Cervi D, Amin S, et al. Identification of novel antigens with induced immune response in monoclonal gammopathy of undetermined significance. Blood 2009 Oct 8;114(15):3276–84.

[49] Ferrante MI, Zullo A, Barra A, Bimonte S, Messaddeq N, Studer M, et al. Oral-facial-digital type I protein is required for primary cilia formation and left-right axis specification. Nat Genet 2006 Jan;38(1):112–17.

[50] Romio L, Fry AM, Winyard PJ, Malcolm S, Woolf AS, Feather SA. OFD1 is a centrosomal/basal body protein expressed during mesenchymal-epithelial transition in human nephrogenesis. J Am Soc Nephrol 2004 Oct;15(10):2556–68.

[51] Anderson DM, Maraskovsky E, Billingsley WL, Dougall WC, Tometsko ME, Roux ER, et al. A homologue of the TNF receptor and its ligand enhance T-cell growth and dendritic-cell function. Nature 1997 Nov 13;390(6656):175–9.

[52] Lacey DL, Timms E, Tan HL, Kelley MJ, Dunstan CR, Burgess T, et al. Osteoprotegerin ligand is a cytokine that regulates osteoclast differentiation and activation. Cell 1998 Apr 17;93(2):165–76.

[53] Wong BR, Rho J, Arron J, Robinson E, Orlinick J, Chao M, et al. TRANCE is a novel ligand of the tumor necrosis factor receptor family that activates c-Jun N-terminal kinase in T cells. J Biol Chem 1997 Oct 3;272(40):25190–4.

[54] Yasuda H, Shima N, Nakagawa N, Yamaguchi K, Kinosaki M, Mochizuki S, et al. Osteoclast differentiation factor is a ligand for osteoprotegerin/osteoclastogenesis-inhibitory factor and is identical to TRANCE/RANKL. Proc Natl Acad Sci U S A 1998 Mar 31;95(7):3597–602.

[55] Yasuda H, Shima N, Nakagawa N, Mochizuki SI, Yano K, Fujise N, et al. Identity of osteoclastogenesis inhibitory factor (OCIF) and osteoprotegerin (OPG): a mechanism by which OPG/OCIF inhibits osteoclastogenesis in vitro. Endocrinology 1998 Mar;139(3):1329–37.

[56] Tsuda E, Goto M, Mochizuki S, Yano K, Kobayashi F, Morinaga T, et al. Isolation of a novel cytokine from human fibroblasts that specifically inhibits osteoclastogenesis. Biochem Biophys Res Commun 1997 May 8;234(1):137–42.

[57] Simonet WS, Lacey DL, Dunstan CR, Kelley M, Chang MS, Luthy R, et al. Osteoprotegerin: a novel secreted protein involved in the regulation of bone density. Cell 1997 Apr 18;89(2):309–19.

[58] Rho J, Takami M, Choi Y. Osteoimmunology: interactions of the immune and skeletal systems. Mol Cells 2004 Feb 29;17(1):1–9.

[59] Walsh MC, Choi Y. Biology of the TRANCE axis. Cytokine Growth Factor Rev 2003 Jun-Aug;14(3-4):251–63.

[60] Takayanagi H. Mechanistic insight into osteoclast differentiation in osteoimmunology. J Mol Med 2005 Mar;83(3):170–9.

[61] Giuliani N, Colla S, Sala R, Moroni M, Lazzaretti M, La Monica S, et al. Human myeloma cells stimulate the receptor activator of nuclear factor-kappa B ligand (RANKL) in T lymphocytes: a potential role in multiple myeloma bone disease. Blood 2002 Dec 15;100(13):4615–21.

[62] Colucci S, Brunetti G, Rizzi R, Zonno A, Mori G, Colaianni G, et al. T cells support osteoclastogenesis in an in vitro model derived from human multiple myeloma bone disease: the role of the OPG/TRAIL interaction. Blood 2004 Dec 1;104(12):3722–30.

[63] Annunziata CM, Davis RE, Demchenko Y, Bellamy W, Gabrea A, Zhan F, et al. Frequent engagement of the classical and alternative NF-kappaB pathways by diverse genetic abnormalities in multiple myeloma. Cancer Cell 2007 Aug;12(2):115–30.

[64] Liao G, Zhang M, Harhaj EW, Sun SC. Regulation of the NF-kappaB-inducing kinase by tumor necrosis factor receptor-associated factor 3-induced degradation. J Biol Chem 2004 Jun 18;279(25):26243–50.

[65] Xiao G, Sun SC. Negative regulation of the nuclear factor kappa B-inducing kinase by a cis-acting domain. J Biol Chem 2000 Jul 14;275(28):21081–5.

[66] Claudio E, Brown K, Park S, Wang H, Siebenlist U. BAFF-induced NEMO-independent processing of NF-kappa B2 in maturing B cells. Nat Immunol 2002 Oct;3(10):958–65.

[67] Coope HJ, Atkinson PG, Huhse B, Belich M, Janzen J, Holman MJ, et al. CD40 regulates the processing of NF-kappaB2 p100 to p52. EMBO J 2002 Oct 15;21(20):5375–85.

[68] O'Mahony A, Lin X, Geleziunas R, Greene WC. Activation of the heterodimeric IkappaB kinase alpha (IKKalpha)-IKKbeta complex is directional: IKKalpha regulates IKKbeta under both basal and stimulated conditions. Mol Cell Biol 2000 Feb;20(4):1170–8.

[69] Ramakrishnan P, Wang W, Wallach D. Receptor-specific signaling for both the alternative and the canonical NF-kappaB activation pathways by NF-kappaB-inducing kinase. Immunity 2004 Oct; 21(4):477–89.

[70] Woronicz JD, Gao X, Cao Z, Rothe M, Goeddel DV. IkappaB kinase-beta: NF-kappaB activation and complex formation with IkappaB kinase-alpha and NIK. Science 1997 Oct 31;278(5339): 866–9.

[71] Yin L, Wu L, Wesche H, Arthur CD, White JM, Goeddel DV, et al. Defective lymphotoxin-beta receptor-induced NF-kappaB transcriptional activity in NIK-deficient mice. Science 2001 Mar 16; 291(5511):2162–5.

[72] Sasaki Y, Calado DP, Derudder E, Zhang B, Shimizu Y, Mackay F, et al. NIK overexpression amplifies, whereas ablation of its TRAF3-binding domain replaces BAFF: BAFF-R-mediated survival signals in B cells. Proc Natl Acad Sci U S A 2008 Aug 5;105(31):10883–8.

[73] Chiron D, Pellat-Deceunynck C, Amiot M, Bataille R, Jego G. TLR3 ligand induces NF-{kappa}B activation and various fates of multiple myeloma cells depending on IFN-{alpha} production. J Immunol 2009 Apr 1;182(7):4471–8.

[74] Romani N, Gruner S, Brang D, Kampgen E, Lenz A, Trockenbacher B, et al. Proliferating dendritic cell progenitors in human blood. J Exp Med 1994 Jul 1;180(1):83–93.

[75] Romani N, Reider D, Heuer M, Ebner S, Kampgen E, Eibl B, et al. Generation of mature dendritic cells from human blood. An improved method with special regard to clinical applicability. J Immunol Methods 1996 Sep 27;196(2):137–51.

[76] Kim SK, Nguyen Pham TN, Nguyen Hoang TM, Kang HK, Jin CJ, Nam JH, et al. Induction of myeloma-specific cytotoxic T lymphocytes ex vivo by CD40-activated B cells loaded with myeloma tumor antigens. Ann Hematol 2009 Nov;88(11):1113–23.

[77] Saitoh A, Narita M, Watanabe N, Tochiki N, Satoh N, Takizawa J, et al. Anti-tumor cytotoxicity of gammadelta T cells expanded from peripheral blood cells of patients with myeloma and lymphoma. Med Oncol 2008;25(2):137–47.

[78] Uchida R, Ashihara E, Sato K, Kimura S, Kuroda J, Takeuchi M, et al. Gamma delta T cells kill myeloma cells by sensing mevalonate metabolites and ICAM-1 molecules on cell surface. Biochem Biophys Res Commun 2007 Mar 9;354(2):613–18.

[79] Sato K, Kimura S, Segawa H, Yokota A, Matsumoto S, Kuroda J, et al. Cytotoxic effects of gammadelta T cells expanded ex vivo by a third generation bisphosphonate for cancer immunotherapy. Int J Cancer 2005 Aug 10;116(1):94–9.

[80] Ostrand-Rosenberg S. CD4+ T lymphocytes: a critical component of antitumor immunity. Cancer Invest 2005;23(5):413–19.

[81] Lauritzsen GF, Weiss S, Dembic Z, Bogen B. Naive idiotype-specific CD4+ T cells and immunosurveillance of B-cell tumors. Proc Natl Acad Sci U S A 1994 Jun 7;91(12):5700–4.

[82] Corthay A, Lundin KU, Lorvik KB, Hofgaard PO, Bogen B. Secretion of tumor-specific antigen by myeloma cells is required for cancer immunosurveillance by CD4+ T cells. Cancer Res 2009 Jul 15; 69(14):5901–7.

[83] Houot R, Levy R. Vaccines for lymphomas: idiotype vaccines and beyond. Blood Rev 2009 May; 23(3):137–42.

[84] Mitsiades CS, Hideshima T, Chauhan D, McMillin DW, Klippel S, Laubach JP, et al. Emerging treatments for multiple myeloma: beyond immunomodulatory drugs and bortezomib. Semin Hematol 2009 Apr;46(2):166–75.

[85] Hideshima T, Chauhan D, Shima Y, Raje N, Davies FE, Tai YT, et al. Thalidomide and its analogs overcome drug resistance of human multiple myeloma cells to conventional therapy. Blood 2000 Nov 1;96(9):2943–50.

[86] Haslett PA, Corral LG, Albert M, Kaplan G. Thalidomide costimulates primary human T lymphocytes, preferentially inducing proliferation, cytokine production, and cytotoxic responses in the CD8+ subset. J Exp Med 1998 Jun 1;187(11):1885–92.

[87] Davies FE, Raje N, Hideshima T, Lentzsch S, Young G, Tai YT, et al. Thalidomide and immunomodulatory derivatives augment natural killer cell cytotoxicity in multiple myeloma. Blood 2001 Jul 1;98(1):210–16.

[88] Breitkreutz I, Raab MS, Vallet S, Hideshima T, Raje N, Mitsiades C, et al. Lenalidomide inhibits osteoclastogenesis, survival factors and bone-remodeling markers in multiple myeloma. Leukemia 2008 Oct;22(10):1925–32.

[89] Schmitt M, Casalegno-Garduno R, Xu X, Schmitt A. Peptide vaccines for patients with acute myeloid leukemia. Expert Rev Vaccines 2009 Oct;8(10):1415–25.

[90] de Souza AP, Bonorino C. Tumor immunosuppressive environment: effects on tumor-specific and nontumor antigen immune responses. Expert Rev Anticancer Ther 2009 Sep;9(9):1317–32.

[91] Bergman PJ. Cancer immunotherapy. Top Companion Anim Med 2009 Aug;24(3):130–6.

# 15

# Osteoimmunology in the Oral Cavity (Periodontal Disease, Lesions of Endodontic Origin and Orthodontic Tooth Movement)

Dana T. Graves[1], Rayyan A. Kayal[2], Thomas Oates[3], Gustavo P. Garlet[4]

[1]DEPARTMENT OF PERIODONTICS, UMDNJ, NEWARK, NJ, USA,
[2]DIVISION OF PERIODONTICS, DEPARTMENT OF ORAL BASIC SCIENCE,
FACULTY OF DENTISTRY, KING ABDULAZIZ UNIVERSITY, JEDDAH, SAUDI ARABIA,
[3]DEPARTMENT OF PERIODONTICS, UTHSC, SAN ANTONIO, TX, USA,
[4]DEPARTMENT OF BIOLOGICAL SCIENCES, SCHOOL OF DENTISTRY OF BAURU,
SAO PAULO UNIVERSITY, BAURU, BRAZIL

## CHAPTER OUTLINE

**Introduction** .................................................................................................................................. 412
**Periodontal diseases** .................................................................................................................... 412
    Gingivitis .................................................................................................................................. 414
    Periodontal disease ................................................................................................................. 415
    Prostaglandins and periodontal disease ................................................................................ 416
    Chemokines and periodontal disease .................................................................................... 417
        *Innate immune response* .................................................................................................. 418
        *Innate immune response – cell types* .............................................................................. 418
        *Cytokines of the innate immune response and periodontal disease* .......................... 419
        *Adaptive immune response* ............................................................................................. 421
    The Th17/Tregs archetype ...................................................................................................... 422
        *Cytokine control of RANKL/OPG balance in the periodontal environment* ................. 423
        *Uncoupled bone formation and periodontal disease* .................................................... 424
**Lesions of endodontic origin** ...................................................................................................... 424
    Innate immune response and lesions of endodontic origin ................................................. 426
    Cytokines of the innate immune response ........................................................................... 427
    Lesions of endodontic origin and the adaptive immune response ..................................... 428

Orthodontic tooth movement .................................................................................................. 429
Conclusions .............................................................................................................................. 432
Acknowledgments .................................................................................................................. 433
References................................................................................................................................ 433

# Introduction

There are many pathologic conditions in the oral cavity in which there is an interaction between immune cells and bone. Several of these are of systemic origin such as Paget's disease and other lytic or sclerosing lesions that can be found in the jaws. Since these lesions are covered in other chapters, we will focus on interactions that are of local origin and particularly important in dental diseases. There are two common bone-resorption lesions that are found in the oral cavity. Both have a common pathologic mechanism involving the host response. The first is periodontal disease, which is initiated by bacterial plaque attached to the tooth surface. Periodontal diseases occur in four stages, bacterial accumulation on the tooth surface (colonization), bacterial penetration of epithelium and connective tissue in the gingiva adjacent to tooth surface (invasion), stimulation of a host response that involves activation of the acquired and innate immune response (inflammation), and destruction of connective tissue attachment to the tooth surface and bone that is irreversible (irreversible tissue loss). The second is lesions of endodontic origin that are associated with bacterial contamination and necrosis of the dental pulp. These lesions occur in four stages: exposure of the dental pulp to the oral cavity and bacterial colonization, inflammation and necrosis of the dental pulp, the development of inflammation in the periapical area, and periapical resorption of bone and formation of granulomas or cysts. Both oral diseases involve bone resorption associated with bacteria that adhere to and invade soft tissue and stimulate an inflammatory response with subsequent osteoclastogenesis. Bacteria initiated resorption around dental implants, known as implantitis, is thought to involve the same sequence of events that occurs in periodontal disease. A third common interaction between immune cells and bone in the oral cavity occurs with orthodontic tooth movement. After the application of an orthodontic force, distinct areas of compression and tension are created between the tooth and supporting alveolar bone, leading to a transitory aseptic inflammation that involves a series of cytokines, chemokines, and neuropeptides. Interestingly, the nature of local immune reaction at the tension side may significantly differ from the compression side, as opposing bone resorption and formation take place in these areas, respectively.

# Periodontal diseases

The periodontium is a complex set of tissues that are challenged by large numbers of bacteria [1, 2]. Periodontal diseases are stimulated by bacterial adherence to the tooth surface. However, there is controversy about which bacteria stimulate irreversible breakdown of periodontal tissues [2, 3]. Recent advances in bacterial detection, which do not rely upon bacterial culture techniques, suggest that there are approximately 700 bacterial species in the oral cavity [1]. The bacteria that are thought to cause periodontal disease have classically been identified as Gram-negative anaerobic bacteria that survive in the gingival sulcus, a space between the tooth surface and the adjacent gingival epithelium [4]. Much attention has been spent on *Porphyromonas gingivalis* and *Aggregatibacter actinomycetemcomitans*, which have been linked to "adult periodontitis" and localized aggressive "juvenile periodontitis", respectively [4, 5]. However, newer approaches to bacterial identification have suggested that a re-evaluation of pathogenic species is warranted.

While the presence of periodontal pathogens is required, but not sufficient for disease initiation, the host response plays a critical role in periodontal tissue breakdown [6–9]. The infection starts in the gingival epithelium leading to an initial host response, whose clinical outcome is the onset of gingivitis. The host inflammatory immune response confers an efficient protection against bacteria, i.e. the systemic consequences of acute infection are very rare. Because the gingival epithelium and underlying connective tissue are chronically exposed to bacteria or their products, both the innate and acquired immune response are chronically activated in connective tissue adjacent to epithelium lining the gingiva. In most cases tissue destruction caused by activation of the host response is reversible and associated with gingivitis. However, under certain conditions the disease will progress and cause destruction of the underlying connective tissue attachment of the gingiva to the tooth surface and the tooth to bone. Indeed, periodontal diseases are distinguished from gingivitis by the irreversible nature of the attachment loss. A description of the periodontium and the changes that occur with periodontal disease are shown in Figure 15-1.

One of the most important uncertainties regarding periodontal disease is its chronic nature. Periodontal diseases may represent a series of brief insults that accumulates so that it appears to be chronic over time. Breakdown may occur in "bursts" with extended periods of remission. Alternatively, there may be relatively constant stimulation over time. There is evidence for both models [10–12]. In the burst model the length of time of the "burst" is not known. Similarly, it is not known how long the chronic destructive period lasts in the chronic model. This problem has plagued human studies since it is difficult to know whether an individual is undergoing active periodontal breakdown at any given point in time. Furthermore, the relative absence of longitudinal studies has made interpretation of results with human patients more difficult since the relationship between a given variable and the pathologic process, irreversible periodontal breakdown, is difficult to ascertain.

The strongest data demonstrating a link between the host response and periodontal bone loss come from animal studies. When the host response is altered by genetic manipulation or treatment with specific inflammatory inhibitors the severity of periodontal connective tissue and bone loss stimulated by periodontal bacteria are reduced. The first concrete evidence that inhibition of an inflammatory response reduced periodontal diseases was carried out in a dog model [13]. In these experiments inhibition of prostaglandins effectively reduced periodontal bone loss. In another animal model, non-human primates treated with

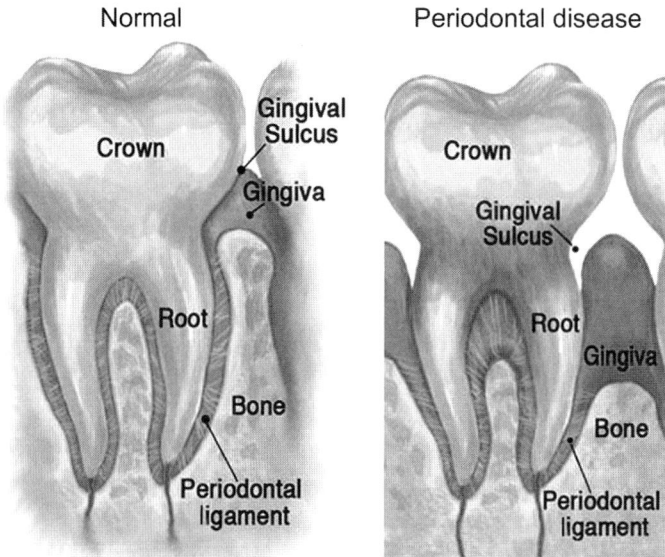

FIGURE 15-1 **Periodontal bone loss.** In periodontal disease bacteria attach to the tooth surface and invade the adjacent epithelium and connective tissue to cause inflammation and bone loss. The changes in bone level are shown in the diagram on the right. The resorbed bone is replaced with gingiva, which does not provide strong support for the tooth and can lead to tooth loss. Periodontal lesions are typically not painful unless they occur as an acute abscess. Please refer to color plate section.

inhibitors to two prominent pro-inflammatory cytokines, interleukin-1 (IL-1) and tumor necrosis factor alpha (TNF-α), had a reduction of periodontal bone loss and attachment compared to animals treated with vehicle alone [6, 7, 14]. Similarly, inhibition of RANKL decreases alveolar bone loss in several models of periodontal disease [8, 15, 16].

## Gingivitis

Classic studies on experimental gingivitis were carried out in humans [17]. Gingival inflammation develops soon after plaque accumulation, being associated with the classic vascular and cellular inflammatory events. The initial lesion appears 2–4 days after plaque accumulation. At this stage there is marked dilation of the vasculature, elevated hydrostatic pressure in the microcirculation and increased gaps between endothelial cells in the capillaries, which lead to increased permeability in microvascular beds and increased flow of gingival crevicular fluid (a transudate that forms in the space between the gingiva and tooth surface). Initially, polymorphonuclear cells (PMNs) migrate into the area under the influence of chemo-attractant gradients formed by substances from bacterial products or generated by the host response [17, 18]. After several days of plaque accumulation, the blood vessels in the dentogingival plexus increase in number and lead to clinical signs of increased redness of the marginal gingiva. In addition, the abundant vascular network in inflamed periodontal tissues facilitates subsequent leukocyte diapedesis. The loss of fibroblasts occurs through inflammation-induced apoptosis, in addition to increased proteolytic activity (mainly mediated by matrix metalloproteinases; MMPs) leading to breakdown of the collagen fiber network that makes up the gingiva. As the exposure to plaque continues, the inflammatory response in the gingival tissue continues to be enhanced. There is increased and sequential infiltration of diverse leukocyte subsets into the junctional epithelium, which is adjacent to the tooth surface, and connective tissue, which include macrophages and T and B lymphocytes. Collagen degradation increases while the inflammatory cell infiltrate expands deeper into the tissue. This lesion may remain stable for months or years and may progress to a destructive lesion, which distinguishes gingivitis from periodontitis [17, 18]. A possible explanation for the stability of gingivitis is that efficient regulatory mechanisms may restrain the local inflammatory immune response intensity, determining therefore the non-progressive nature of these lesions. Another possibility is that the inflammatory response in gingivitis is far from bone while in periodontal disease it moves in closer proximity to the bone surface [6]. This inflammation may cause irreversible damage by inducing higher levels of cell death in bone-lining cells, thereby reducing the capacity to form new bone [19]. However, the factors that are responsible for transition to irreversible loss of connective tissue attachment and bone require further investigation.

## Periodontal disease

It is hypothesized that periodontal disease progression is due to a combination of several factors, including the presence of periodontopathic bacteria, high levels of pro-inflammatory cytokines and prostaglandins, the production and activation of MMPs and RANKL, and relatively low levels of interleukin-10 (IL-10), transforming growth factor-β (TGF-β), tissue inhibitors of metalloproteinase (TIMPs) and OPG [20].

Animal models have established a causal relationship between bacteria and periodontal disease. In one animal model a ligature is tied around the teeth. This then causes plaque accumulation and facilitates bacterial penetration, which leads to inflammation and alveolar bone resorption [21]. Gnotobiotic rats treated identically do not exhibit periodontal bone loss [22], demonstrating the role of bacteria in initiating the inflammatory and bone-resorbing process in this model. Likewise, treatment with antibiotics or topical application of antimicrobial agents reduces bone resorption in this model while increasing colonization with Gram-negative bacteria enhances bone resorption [21]. In other animal models the inoculation of periodontal pathogens into the oral cavity of rodents induces bone loss. In several studies, the introduction of *P. gingivalis* by oral inoculation stimulated alveolar bone resorption in the mouse [21]. Similarly, introduction of *A. actinomycetemcomitans* in mice and rats leads to infection and the loss of alveolar bone [5, 21]. Thus, experiments with animal models support human studies, implicating bacteria in the initiation of inflammation and periodontal disease.

An essential step in the inflammatory immune response triggered by periodontal pathogens is the recognition of microbial components as "danger signals" by host cells.

Indeed, one of the critical components of the host response to bacteria or their products is a family of receptors called the Toll-like receptors (TLRs). TLRs activate the innate immune response by binding various bacterial components (i.e. diacyl lipopeptides, peptidoglycan, LPS, flagellin, bacterial DNA, etc.) [23]. Once TLRs are activated, an intracellular signaling cascade is stimulated. These lead to the activation of transcription factors such as nuclear factor-κB (NF-κB), activator protein-1 (AP-1), and C/EBP and the production of various cytokines and chemokines, many of which directly or indirectly stimulate osteoclast formation [23]. However, TLR signaling has also been shown to directly inhibit osteoclastogenesis [24]. Recent studies describe a role for both TLR-2 and TLR-4 in the recognition of *A. actinomycetemcomitans*, in the subsequent production of inflammatory cytokines and inflammatory cell migration, and in the control of alveolar bone loss [25, 26]. In addition to the TLRs, recent studies point to the nucleotide-binding oligomerization domain (NOD) receptors and the inflammasome system as potential accessory molecules that are involved in triggering host response to periodontal pathogens [27].

After host response triggering by the microbial recognition, the spatial orientation of the subsequent leukocyte infiltration into periodontal tissues is likely a contributor to periodontal disease. Human gingiva is characterized by the presence of an infiltrate in the gingiva adjacent to a tooth surface [17]. The inflammatory infiltrate is observed histologically even in the presence of minimal clinical signs of inflammation. In animal models periodontal bone loss is induced when the inflammatory infiltrate moves closer to bone [6]. Thus, when inflammation is restricted to the connective tissue closest to the gingival epithelium, gingivitis is present. When the inflammatory infiltrate moves closer to bone, osteoclastogenesis is induced. Hence, the location of the inflammation and its proximity to bone likely determine whether it induces osteoclastogenic bone loss. This suggests that the progression of the inflammatory infiltrate toward bone is a critical factor in determining whether periodontal disease occurs. This, in turn, is likely to be dependent upon the depth to which bacteria or their products have penetrated the connective tissue. Thus, the spatial aspect of the host response to the bacterial challenge may contribute to the clinical manifestation of disease (gingivitis vs. periodontitis) and the extent of tissue loss. Among the mediators potentially involved in leukocyte diapedesis and subsequent spatial localization in periodontal environment, prostaglandins and chemokines have been investigated in the last decade with special interest.

## Prostaglandins and periodontal disease

The cyclooxygenase enzymes COX-1 and COX-2 catalyze the conversion of arachidonic acid to prostaglandins (PGs). COX-1 is constitutively expressed and leads to the formation of PGs while COX-2 is induced and leads to the formation of PGs involved in inflammatory processes. PGs are potent stimulators of bone formation and resorption and are produced by bone cells, fibroblasts, gingival epithelial cells, endothelial cells, and inflammatory mononuclear cells. PGs also have inhibitory effects on fully differentiated osteoblasts and osteoclasts [28]. Prostaglandin E2 (PGE-2) production is elevated in individuals with periodontitis compared with healthy subjects [29]. PGE-2 applied topically to the gingival sulcus induces a marked increase in osteoclasts and the combined application of PGE-2 and lipopolysaccharide induces more osteoclasts than PGE-2 alone [30].

The role of PGs in the pathogenesis of periodontitis has been under investigation. PGE-2 and leukotriene B-4 were found in gingival crevicular fluid of individuals with localized aggressive periodontitis. Furthermore, *P. gingivalis* stimulates leukocyte infiltration concomitant with elevated PGE-2 levels and increased COX-2 expression by infiltrated leukocytes in vivo [31]. In an animal ligature-induced periodontitis model, both a non-selective COX inhibitor and a selective COX-2 inhibitor reduced osteoclast numbers and alveolar bone loss compared to non-treatment [32]. Many clinical trials have explored the use of a COX-2 inhibitor as an adjunct to periodontal therapy. These inhibitors improved the clinical outcome after periodontal therapy compared to periodontal therapy alone [33]. However, they are not clinically used in the treatment of human patients due to side effects. More recent evidence indicates that lipoxins and resolvins, products of omega-3 fatty acids, stimulate resolution of inflammation and protect against bacteria-induced periodontal bone loss in animal models [34]. In addition to the lipid mediators involved in the development and resolution of the inflammatory process, small chemotactic cytokines (chemokines) have been described as important regulators of inflammatory and immune functions.

## Chemokines and periodontal disease

Chemokines are a large family of chemotactic cytokines that stimulate the recruitment of inflammatory cells [35, 36]. They are divided into two major families based on their structure, CC, and CXC chemokines. There are two major classes of receptors, CC chemokine receptors (CCR) and CXC chemokine receptors (CXCR), that may bind multiple chemokine ligands. Chemokines are produced by several cell types in the periodontium, such as fibroblasts, endothelial cells, epithelial cells, osteoclasts, polymorphonuclear leukocytes, monocytes/macrophages, lymphocytes, and mast cells [35]. Some chemokines can stimulate one or more steps of bone resorption, including the recruitment, differentiation, or fusion of precursor cells to form osteoclasts or enhance osteoclast survival [35, 36]. They could also affect periodontal bone loss by recruiting cells, such as neutrophils, which protect against bacterial invasion.

Chemokines are found in both gingival tissue and crevicular fluid during the immunopathogenesis of periodontal diseases. IL-8/CXCL8, a chemoattractant of polymorphonuclear leukocytes, showed a rapid increase in gingival crevicular fluid, preceding the clinical signs of periodontal disease, following cessation of toothbrushing. In persons with periodontitis, the levels of IL-8/CXCL8 in both periodontal tissue and gingival crevicular fluid are increased and correlated with disease severity [37].

One of the most abundantly expressed chemokines in periodontitis tissues is macrophage inflammatory protein-1α (MIP-1α/CCL3), which is localized to the connective tissue subjacent to the pocket epithelium of inflamed gingival tissues [38]. It has also been shown that MIP-1α/CCL3-positive cells increase in number with increasing severity of periodontal disease. This chemokine is a ligand for the chemokine receptors CCR1 and CCR5, being primarily associated with the chemoattraction of monocytes/macrophages and dendritic cells (through the binding to CCR1), and lymphocytes polarized into Th1 phenotype (through CCR5) [39]. Since macrophages and Th1 cells are characteristic sources of bone-resorptive cytokines, MIP-1α/CCL3 has a potential role in the inflammatory bone resorption in periodontal environment. Furthermore, it is important to consider that CCR1+ and CCR5+ cell populations potentially include osteoclast precursors from monocytic lineage and RANKL+ lymphocytes [40, 41], which can directly upregulate bone resorption activity.

Upon activation, normal T cells express and secrete RANTES/CCL5. This chemokine has been detected in both the periodontal tissue and gingival crevicular fluid of persons with periodontitis, and expression levels have been directly correlated with the activity of the disease [38, 42]. Since both RANTES/CCL5 and MIP-1α/CCL3 bind CCR1 and CCR5, a similar role for these chemokines in the pathogenesis of periodontitis is inferred [40, 43]. Another chemokine that could contribute to the enhanced severity of periodontal disease is monocyte chemoattractant protein-1 (MCP-1/CCL2), which acts primarily as a monocyte/macrophage and potentially an osteoclast precursor chemoattractant [40, 43]. This is supported by analysis of data showing that MCP-1/CCL2 activity in gingiva or gingival crevicular fluid increases with the severity of the disease [44, 45].

Besides their classic role as chemoattractants, chemokines can directly influence osteoclast biology. MIP-1α/CCL3 directly stimulates osteoclast differentiation, but not activation in vitro [46]. Complementarily, stromal-cell derived factor-1 (SDF1/CXCL12) has been described as a positive regulator of osteoclast function, and was recently identified in diseased periodontium [47].

Thus, the presence of a series of chemokines, which are implicated in the chemoattraction and activity of osteoclasts, are potential factors in the disruption of alveolar bone homeostasis in periodontal lesions. In additional, sequential chemokine expression supports the interaction between the innate and adaptive immunity and may play a key role in determining the development of periodontal lesions.

*Innate immune response*

After the recognition of microbial components by host cells, the innate immunity process is immediately triggered. The innate immune response comprises the action of phagocytes and the production of cytokines that generate a loop of amplification of innate mechanisms and also provide a link to the subsequent adaptive immunity development.

*Innate immune response – cell types*

Among innate immunity cells, PMNs have been described as playing a major role in periodontal disease. They have been shown to have both protective and destructive influences [48]. Evidence that proposes a protective function is based on the observation that individuals who have neutrophil disorders such as cyclic neutropenia, Chédiak-Higashi syndrome, and leukocyte adhesion deficiency syndrome have an increased susceptibility to periodontal destruction [48]. Monocytes and macrophages are important mediators of inflammation in the connective tissue infiltrate. They produce several cytokines and lytic enzymes, present antigens to T cells, and enhance antibacterial defenses [20].

A new concept has emerged in which PMN and monocyte hyperactivity, associated with the release of destructive or inflammatory products, contributes to periodontal tissue destruction [49, 50]. These cells produce a respiratory burst in microbial killing that generates superoxides, hydrogen peroxide, hydroxyl radicals, hypochlorous acid, and chloramines. These products are responsible for oxidative killing inside the phagosome and in the extracellular microenvironment. In fact, impaired production of phagocyte reactives such as iNOS and MPO are associated with an increased load of periodontopathogens [51, 52]. While it is protective against infectious agents, the release of these products can damage adjacent cells and tissues. Moreover, PMNs and monocytes/macrophages release elastases and collagenases, which hydrolyze several extracellular proteins such as elastin, fibronectin, and collagen. Most of the collagen substrate degraded as a result of bacterial stimulation in the periodontium is derived from host leukocytes rather than directly from bacteria [53]. Moreover, greater collagenase activity is associated with progressive periodontal lesion formation [53]. In addition, PMNs and monocytes/macrophages release proinflammatory mediators that have numerous biological activities including the stimulation of osteoclastogenesis [20, 54].

Acting initially as innate immunity sensors, dendritic cells are thought to be a bridge between the innate and adaptive immune responses. Dendritic cells are derived from the monocytic lineage and present antigen to initiate primary and secondary T-cell responses and also to promote inflammation by producing chemokines and inflammatory cytokines [55]. Dendritic cells are stimulated by bacteria or their products through TLR and NOD signaling and have been implicated in periodontal disease [55–58]. Periodontal pathogens stimulate dendritic cells to express a number of cytokines that participate in the immune response such as IL-1β, IL-12, IFN-γ, TNF-α, and TNF-β [56, 57]. Moreover, recent evidence indicates that dendritic cells can, themselves, form osteoclasts [59]. Stimulation of CD4+ T cells and splenic dendritic cells in vitro with the periodontal pathogen *A. actinomycetemcomitans* results in the production of osteoclasts in a RANKL-dependent manner [60]. Thus, dendritic cells are thought to upregulate immune responses in the periodontium, to stimulate an adaptive immune response. Conversely, the adaptive immune response, when activated by periodontal pathogens, appears to stimulate dendritic cells to form osteoclasts.

### Cytokines of the innate immune response and periodontal disease

IL-1β, IL-1α, TNF-α, and IL-6 are cytokines typically associated with the innate immune response, although they are produced by multiple cell types including lymphocytes. The induction of experimental periodontitis is associated with the expression of innate immune cytokines and alveolar bone loss [61]. Inhibition of IL-1 and TNF-α together significantly reduces the progression of inflammation toward bone, osteoclastogenesis, and periodontal tissue destruction [6]. Thus, innate immune cytokines are strongly implicated in the progression of periodontal disease.

IL-1 stimulates expression of pro-resorptive cytokines such as RANKL and TNF-α and proteinases that participate in the destruction of periodontal connective tissue and bone resorption. In the periodontium IL-1 is produced by several types of cells including PMNs, monocytes, and macrophages [62, 63]. In patients with periodontal disease, IL-1β expression is elevated in gingival crevicular fluid at sites of recent bone and attachment loss [63, 64]. IL-1β is also found to be higher in gingiva from individuals with a history of periodontal disease compared to samples from healthy individuals [64]. Investigators have shown that inhibiting IL-1 causes a decrease in bone loss, while increasing IL-1 enhances bone resorption. Using a non-human primate model, Delima et al. showed that inhibition of IL-1, using human soluble IL-1 receptor type I, significantly reduced inflammation, connective tissue attachment loss, and bone resorption induced by periodontal pathogens compared to controls [7]. In other studies, IL-1-receptor-deficient mice had less *P. gingivalis* LPS-induced osteoclastogenesis compared to similarly treated wild-type mice [65]. In a different approach the exogenous application of recombinant human IL-1β in a rat model of experimental periodontitis accelerated alveolar bone destruction and inflammation over a 2-week period [66]. In addition, transgenic mice overexpressing IL-1α in gingival epithelium developed a periodontitis-like syndrome, leading to loss of attachment and destruction of periodontal bone [67]. Taken together, these studies strongly support the role of IL-1 in promoting alveolar bone destruction in periodontal disease.

TNF refers to two associated proteins, TNF-α and TNF-β. TNF-α levels are elevated in gingival crevicular fluid at sites where bone and attachment loss have recently occurred [63, 68–70]. They are also found to be higher in diseased periodontal tissue samples compared to tissue samples from healthy individuals [64]. A cause-and-effect relationship between TNF-α and periodontal bone loss has been demonstrated. Administration of recombinant TNF-α accelerates periodontal destruction in a rat periodontitis model [71]. On the other hand, *P. gingivalis*-induced osteoclastogenesis is reduced in TNF-receptor-deficient mice compared to wild-type controls, indicating that osteoclast formation is dependent on a TNF-α-regulated pathway as part of the host response to bacterial challenge [72]. In a recent study by Garlet et al., TNFR-1 knockout mice developed significantly less inflammation, indicated by decreased expression of chemokine and chemokine receptors, and less alveolar bone loss in association with decreased expression of RANKL in response to *A. actinomycetemcomitans* oral inoculation [52]. Furthermore, mRNA levels of IL-1β, IFN-γ, and RANKL in gingival tissues were significantly lower in TNFR-1 knockout mice compared to wild-type infected mice. In contrast, the level of *A. actinomycetemcomitans* quantified by real-time PCR was significantly greater in TNF receptor ablated mice compared to wild-type controls and was associated with lower levels of the neutrophilic antimicrobial myeloperoxidase [52]. Thus, the absence of TNFR-1 resulted in a lower production of cytokines in response to *A. actinomycetemcomitans* infection even in the presence of higher levels of the periodontal pathogen. Based on studies like these, it can be projected that the local production of TNF-α plays a role in bone resorption during periodontal disease. In addition to locally produced cytokines circulating inflammatory cytokines can increase bone loss in periodontal environment. It is possible that systemic cytokines could contribute to periodontal disease based on evidence that systemic administration of TNF-α enhances the severity of ligature-induced periodontitis in rats [71]. These studies may provide a possible mechanistic explanation for the concurrence of systemic conditions, such as diabetes and rheumatoid arthritis with a greater severity or incidence of periodontal disease [73, 74].

Expression of IL-6 is found at higher levels in gingival crevicular fluid and in gingival mononuclear and T cells from periodontitis patients than from healthy controls [75]. In addition, LPS from the periodontal pathogen *A. actinomycetemcomitans* induces IL-6 expression, osteoclastogenesis, and bone loss [76]. Oral inoculation of *P. gingivalis* in mice with genetically deleted IL-6 have decreased bone loss compared to wild-type mice, indicating that IL-6 contributes to the progression of bacteria-induced bone loss [9].

*Adaptive immune response*

Both innate and adaptive immunity are important in the response to the biofilm that is present on teeth. T and B lymphocytes are activated in periodontal disease tissues [20]. Lymphocytes produce cytokines associated with bone resorption, particularly receptor activator of NF-κB ligand (RANKL). There is evidence to suggest that lymphocytes are involved in mediating bacteria-stimulated periodontal bone resorption. When severe combined immunodeficient (SCID) mice, which lack B and T lymphocytes, are challenged with an oral pathogen, *P. gingivalis*, there is considerably less bone loss than in immunocompetent mice, suggesting that B and T lymphocytes contribute to bone loss [77]. When SCID mice are engrafted with human CD4(+) T cells from individuals with periodontal disease, oral inoculation with *A. actinomycetemcomitans* results in periodontal bone destruction [8]. This bone loss is inhibited by treatment with osteoprotegerin (OPG), demonstrating the importance of RANKL in the T-cell-mediated bone loss of this experimental periodontitis. Han et al. used adoptive transfer of B cells from *A. actinomycetemcomitans*-immunized rats to demonstrate the role of these cells in periodontal bone resorption. Transfer of B cells from immunized rats to a naïve host, which was subsequently challenged by injection of *A. actinomycetemcomitans* into the gingiva, increased the level of bone loss compared to transfer of B cells from non-immunized rats. This effect was antagonized by injection of an osteoprotegerin fusion protein, suggesting that bone loss was in part due to enhanced RANKL expression by B cells [15]. In other experiments oral inoculation of mice with *A. actinomycetemcomitans* stimulated formation of an inflammatory infiltrate in gingival connective tissue consisting of T and B lymphocytes and CD14+ monocytes/macrophages [8, 78]. The inoculation induced a mixed Th1 and Th2 expression profile in agreement with studies in human showing both Th1 and Th2 cytokine expression rather than polarization toward one or the other [75]. Overall, these data suggest that the adaptive immune cells are important effectors of the bone loss that is seen with bacterial infection.

CD4 T cells can be subdivided into two subsets, designated Th1 and Th2, on the basis of their pattern of cytokine production. IFN-γ is a lymphokine produced by activated Th1-type lymphocytes and natural killer cells. It plays an important role in host defense mechanisms by exerting pleiotropic activities on a wide range of cell types. IFN-γ knockout mice exhibit decreased bone loss in response to *P. gingivalis* infection compared to wild-type controls [9]. In addition, T cells extracted from periodontitis patients express more IFN-γ compared to T cells from healthy controls [75]. In an animal study, expression of IFN-γ was associated with enhanced alveolar bone loss mediated by RANKL-expressing CD4(+) T-cells in response to *A. actinomycetemcomitans* [77]. Furthermore, IFN-γ was co-expressed with RANKL in CD4(+) T cells [79]. Although IFN-γ is directly anti-osteoclastogenic, it appears to have indirect effects that promote bone resorption in animal models of periodontal disease. Thus, there is some evidence that CD4(+) T cells are important in periodontal disease progression. Th2 lymphocytes are associated with cytokines that are anti-inflammatory, such as IL-4 [80]. It has been suggested that a Th2 response attenuates the severity of experimental periodontitis [81]. However, the association of IL-4 with periodontitis outcome remains controversial, since studies suggest that a Th2-type response in periodontal lesions leads to the accumulation and activation of B cells, and consequently, to tissue destruction [82]. Therefore, while Th1 cytokines are described to upregulate pro-inflammatory response, the Th2-cytokines are potentially anti-inflammatory.

## The Th17/Tregs archetype

In addition to Th1 and Th2 polarized CD4 lymphocytes, recent studies describe two new well-defined T-cell subsets, namely Th17 and Tregs (regulatory T cells). Th17 and Tregs subsets are described as effector cells. Th17 lymphocytes secrete IL-17, which stimulates osteoclastogenesis through induction of RANKL [83]. Through its effects on the RANKL–RANK system, IL-17 may have a role in rheumatoid arthritis, periodontal disease, and the loosening of joint prostheses [83]. In experimental periodontitis in animals the predominant role of IL-17 appears to be protective since genetic deletion of IL-17 enhances periodontal bone loss, induced by the oral pathogen *P. gingivalis* [84]. This protection was associated with the production of chemokines, which induced the recruitment of neutrophils as part of a critical antibacterial defense response. In contrast, gingiva from individuals with periodontal disease exhibit elevated levels of Th17 cells and IL-17 mRNA when compared to healthy tissues, similar to increases in the Th17-related cytokines, TGF-β, IL-6, and IL-23p19. Moreover, IL-17 and RANKL are abundantly expressed in the periodontium of individuals with periodontal disease, in contrast to the detection of low levels in controls, suggesting a possible role for Th17 cells in periodontitis progression [85]. Further studies are needed to better characterize the role of Th17 cells in the destructive process.

T-regulatory cells (Tregs) have been described as potential negative regulators of host response in periodontal environment. Tregs are CD4+CD25+ T cells that specifically regulate the activation, proliferation, and effector function of activated conventional T cells. They determine the outcome of several immunological settings, ranging from infectious diseases to immunopathology and autoimmunity [86]. The presence of Tregs in periodontal tissues has been demonstrated by the expression of the phenotypic markers: Foxp3, CTLA-4, IL-10, GITR, CD103, and CD45RO [87–89]. Interestingly, Tregs-associated cytokines IL-10 and TGF-β, and the inhibitory molecule CTLA-4 have been proposed to attenuate periodontal disease progression [87]. It is important to consider that IL-10 can be produced by both Th2 and Treg subsets, and the contribution of each cell type to IL-10 production in periodontal tissues remains to be established. Thus, taken as a whole it is not clear whether specific lymphocyte subsets are protective or destructive. It is possible that the roles of specific lymphocyte subsets are dependent upon the specific microenvironment and context. Thus, additional studies are needed to determine the ultimate role of Th1, Th2, and Th17 cells as well as Tregs in the periodontal environment under specific conditions.

### Cytokine control of RANKL/OPG balance in the periodontal environment

The relative concentrations of RANKL and OPG are altered during the progression of periodontal disease in an animal model [61]. During bacteria-stimulated periodontal bone loss there is an initial increase in TNF-α, IL-1β, and RANKL and the ratio of RANKL to OPG is high. Later, IL-4, IL-10, and OPG increase and TNF-α, IL-1β and RANKL decrease. Thus, when the rate of bone loss is high the expression of pro-resorptive cytokines is high. In contrast, when the rate of bone loss slows there is a marked increase in antiresorptive cytokines [61]. Accordingly, a number of animal studies have demonstrated that inhibition of RANKL decreases periodontal bone loss, establishing a cause-and-effect relationship [8, 15, 16]. These findings correspond well with the critical role of RANKL in driving osteoclastogenesis and bone loss in periodontal disease. It is now generally accepted that RANKL plays a prominent role in periodontal bone loss. Moreover, an important component in the clinical outcome may be the RANKL and OPG balance.

A number of clinical studies have investigated the RANKL/OPG ratio in gingival tissues or gingival crevicular fluid. These studies show an increase in soluble RANKL, with or without a decrease in OPG in individuals with periodontitis compared to healthy controls [90, 91]. Interestingly, a high individual variation in RANKL and OPG mRNA levels was verified in chronic gingivitis and periodontitis, suggesting that distinct expression patterns could be associated with disease progression. A RANKL>OPG ratio predominates (70%) in chronic periodontitis lesions, being similar to that seen in sites of active bone remodeling throughout orthodontic tooth movement. Conversely, RANKL<OPG (56%) and RANKL≈OPG (35%) patterns were found to predominate in chronic gingivitis lesions, and parallel the pattern identified in healthy periodontal tissues [92].

In an attempt to identify the cellular source of RANKL in the bone-resorptive lesions of periodontal disease, Kawai et al. measured the concentrations of soluble RANKL in gingiva from individuals with a history of periodontal disease. RANKL concentrations were significantly higher in diseased gingival tissues compared to healthy tissues. RANKL was expressed at the highest level by B cells followed by T cells and to a lesser extent by monocytes. Moreover, lymphocytes isolated from gingival tissues of patients induced differentiation of mature osteoclast cells in a RANKL-dependent manner in vitro [90]. This indicates that activated T and B cells can be the cellular source of the RANKL that is stimulating bone resorption in diseased gingival tissue.

*Uncoupled bone formation and periodontal disease*

In addition to driving active bone resorption, the host response against periodontal pathogens also impairs bone formation, which can result in increased periodontitis severity. In a healthy adult with physiologic tissue turnover, an episode of bone resorption is followed by an equivalent amount of bone formation, a well-accepted process referred to as coupling [93]. Likewise, it would be expected that bacteria-induced bone resorption should be followed by an equivalent amount of bone formation in a healthy adult. However, in periodontitis there is a failure to form bone following resorption, resulting in net bone loss. Thus, a component of the pathologic process that causes periodontal bone loss is likely the failure to form new bone, a process caused by a failure in the coupling of bone resorption and formation.

It is possible that the same inflammation, which stimulates bone resorption, contributes to uncoupling. We have proposed that if inflammation is prolonged and in close proximity to bone, it will interfere with the coupling process. For example, injection of *P. gingivalis* into the connective tissue stimulates an episode of bone resorption followed by bone formation [19]. If the inflammation is prolonged by a condition such as diabetes or by activation of the acquired immune response there is a diminished capacity to form new bone and uncoupling [19, 94]. There is evidence suggesting that the prolonged inflammation in diabetic animals interferes with bone formation in the periodontium following an episode of bone resorption [95]. This interpretation is additionally supported by evidence that the application of cytokines in vivo stimulates bone resorption but also limits bone formation. Therefore, several lines of animal experimentation support the concept that inflammation uncouples bone formation from bone resorption. Thus, inflammation may not only stimulate the formation of osteoclasts and bone resorption but also affect bone by altering the function of osteoblasts and limiting reparative bone formation.

# Lesions of endodontic origin

Lesions of endodontic origin typically develop from exposure of the pulp to oral bacteria as a result of compromises of the integrity of a tooth. This may result from carious lesions that penetrate the tooth, iatrogenic causes including mechanical exposure, fracture of the

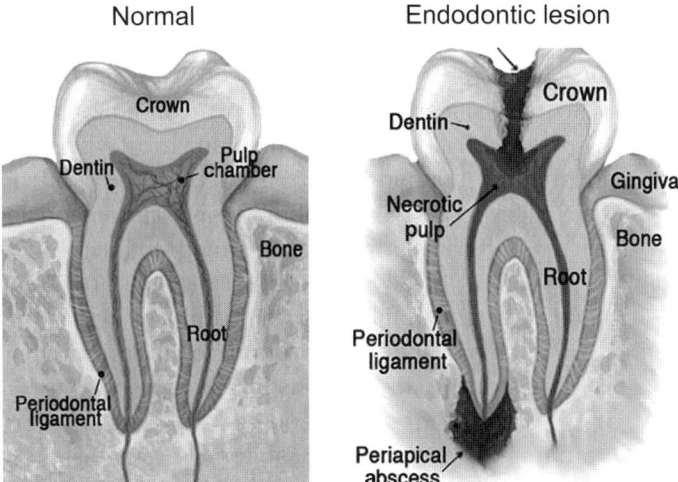

FIGURE 15-2 **Lesion of endodontic origin.** In lesions of endodontic origin disruption of the integrity of the crown typically occurs either through trauma or through the effect of cariogenic bacteria. Bacteria then contaminate the dental pulp, which becomes necrotic and forms a reservoir of bacteria that cause inflammation in the periapical area and bone resorption. These lesions are frequently painful. Please refer to color plate section.

tooth, etc. In most cases these events lead to infection, which causes formation of an inflammatory front that moves away from the exposed area, eventually leading to inflammation at the tooth apex. This inflammation involves the recruitment and activation of leukocytes of the innate and adaptive immune response, osteoclastogenesis, and bone resorption at the apex of the tooth. A description of the formation of lesions of endodontic origin is shown in Figure 15-2.

Inflammation and resorption of bone at the tooth apex, in most cases, is a consequence of the interplay between a polymicrobial infection and the host response. The role of bacteria in the development of periapical lesions was demonstrated by exposing the dental pulp to the oral cavity in germ-free animals. Pulp exposure in germ-free animals heals with transitory inflammation and partial repair of the damaged tooth by formation of a new dentin-like matrix at the exposed site. In contrast, mechanical pulp exposure in animals with a normal oral flora lead to an infection of the dental pulp, chronic inflammation, necrosis, the absence of repair, and bone resorption at the tooth apex [96]. The infection persists because the necrotic tissue of the dental pulp is inaccessible to leukocytes and hence represents a protected reservoir of bacteria [97, 98]. Bacteria and their products then chronically stimulate inflammation in the periapical area and prevent repair of the resorbed bone, i.e. cause uncoupling to occur. Unless the necrotic tissue is removed, this infection continues and eventually leads to the formation of granulomas or cysts in the apical tissues, with associated resorption of the surrounding bone [99].

In some respects it is more difficult to identify cell types and cytokines that play an essential role in osteoclastogenesis and bone resorption in periapical lesions compared to periodontal lesions. This is due to the "bacteria reservoir" that exists in the necrotic dental pulp of endodontic infections. The bacterial insult stimulates an innate vascular inflammatory response, promoting the extravasation of PMNs to fight the infection. During this process, the PMNs release leukotrienes and prostaglandins into the area. These inflammatory mediators reinforce the recruitment of additional PMNs and other leukocytes [98]. The role of the host response in lesions of endodontic origin is critical in limiting the spread of infection into the fascial planes. Even though inflammatory cytokines play an important role in osteoclastogenesis, specific inhibitors of these cytokines tend to cause the formation of larger osteolytic lesions since they compromise the ability of the host to protect itself from the reservoir of bacteria in the necrotic pulp. Thus, the use of inhibitors or mice with targeted genetic deletions may not necessarily reveal whether a particular cytokine or cell type participates in bone resorption. For example, an enhanced host response with increased numbers of PMNs and monocytes reduces the extent of apical bone resorption even though the host response is thought to be responsible for the bone resorption [100]. In another example the ablation of IL-1 receptor signaling in endodontic lesions allows the spread of infection from the surrounding bone into fascial planes and significant morbidity and mortality [101]. Accordingly, the control of periapical infection seems to be a critical issue, since the absence of the pleiotropic enzyme-inducible nitric oxide synthase (iNOS) also results in larger lesions, frequently associated with periapical abscess development [102]. This contrasts with periodontal disease in which a protected reservoir does not exist and the use of inhibitors or mice with targeted deletions of the host response typically does not sufficiently compromise the anti-bacterial defenses to complicate the analysis. Thus, in periodontal disease inhibition of a component of the host response does not seem to increase susceptibility to bacterial infection while it may with lesions of endodontic origin.

## Innate immune response and lesions of endodontic origin

Lesions of endodontic origin are initiated by an inflammatory cascade of events that includes activation of endothelial cells, PMNs, macrophages, and osteoclasts leading to rapid bone destruction. Bone resorption is typically stimulated by multiple bacteria or their products including lipolysaccharides (LPS) [98]. Bacteria are thought to initiate resorption through the stimulation of monocytes and macrophages, which then produce pro-inflammatory cytokines such as IL-1$\beta$, IL-1$\alpha$, or TNF-$\alpha$ [103]. Introduction of pro-inflammatory molecules, interleukin-1, and LPS into the root canal using a rat model increases the size of apical lesions [104]. As was noted for periodontal disease, activation of the innate immune response by microbial challenge in lesions of endodontic origin involves Toll-like receptors (TLRs) and nucleotide-binding oligomerization domain (NOD) receptors [105]. TLRs and NOD receptors are highly expressed on multiple cell types associated with the periapical lesion including pulp fibroblasts, granulocytes, monocytes/macrophages, mesenchymal cells, and osteoclast precursors, etc. [24, 105]. In addition to simulating pro-inflammatory cytokines such as IL-1, TNF-$\alpha$, and IL-6, the activation of TLRs and NOD signaling may also directly or indirectly enhance RANKL production with subsequent osteoclastogenesis and bone resorption [24, 105].

As in periodontal diseases, the initial rapid destruction of bone in the apical area of the root has been related to the generation of prostaglandins through the cyclooxygenase pathway. PGE-2 levels are significantly enhanced in periapical lesions [106]. These findings are consistent with a report that indometacin reduces the extent of bone resorption in lesions of endodontic origin [107].

## Cytokines of the innate immune response

Multiple pro-inflammatory cytokines and chemokines have been associated with the formation of periapical lesions. IL-1α or IL-1β is produced in periapical lesions by several types of cells including fibroblasts, PMNs, macrophages, and osteoclasts [108, 109]. The role for IL-1 in periapical bone destruction was demonstrated using interleukin-1 receptor antagonists to show a 60% reduction in lesion development [110]. Moreover, much of the osteoclastogenic activity in periapical lesions is related to the formation of interleukin-1α [111]. A critical role of TNF-α in modulating bone-resorptive activity in periapical lesions is less well established compared to that of IL-1α [119]. However, TNF-α expression is found in lesions of endodontic origin by cells such as fibroblasts, monocytes/macrophages, and PMNs [117, 121]. IL-6 has been demonstrated in exudates from human periapical lesions. PMNs, macrophages, T lymphocytes, osteoblasts, and fibroblasts have been identified as expressing IL-6 protein [112, 113]. The use of IL-6-deficient mice with a periapical lesion model demonstrated a protective effect for IL-6, since IL-6-deficient mice developed larger periapical lesions than did control mice [114].

In addition to the classic cytokines, chemokines also have been implicated in the pathogenesis of periapical lesions. IL-8/CXCL8 is thought to play an active role in the inflammatory events that occur in lesions of endodontic origin. It is predominantly expressed in areas with heavy infiltration of PMNs, while healthy tissue showed negative or weak IL-8/CXCL8 staining [115]. In fact, neutrophils are active in bone loss associated with endodontic lesions, since neutropenic animals demonstrate a considerable decrease in periapical lesion formation [116].

MCP-1/CCL2 is expressed by monocytes/macrophages and bone-lining cells in lesions of endodontic origin [117]. Genetic deletion of MCP-1/CCL2 results in significantly reduced numbers of monocytes and enhanced rates of apical bone resorption, suggesting that recruitment of monocytes is critical in the antimicrobial defense [118]. Complementarily, the absence of the MCP-1 receptor (CCR2) also results in larger bone resorption at the tooth apex, associated with higher levels of the osteoclastogenic and osteolytic factors, such as RANKL and cathepsin K [119]. Interestingly, the MCP-1/CCL2-CCR2 axis seems to play an active role in the acute to chronic inflammation transition, mediating the migration of monocytes/macrophages and limiting the infiltration of PMNs [119]. Likewise, genetic ablation of CC chemokine receptor five (CCR5) results in the formation of larger periapical lesions [120]. Chemokines and chemokine receptors may also play active roles in the transition from innate to adaptive immunity in periapical environment and in the development of granulomas associated with these lesions [35].

## Lesions of endodontic origin and the adaptive immune response

In addition to PMNs and monocytes/macrophages the adaptive immune response has been implicated in the development of bone resorption in lesions of endodontic origin [121]. At the height of bone resorption in apical lesions in a rat model the predominant cell types are T cells followed by B cells and monocytes/macrophages [121]. Th1 (IL-2 and IFN-γ), Th2 (IL-4 and IL-5), Th17 (IL-17A), and Treg (IL-10 and TGF-β) lymphocytes are found in periapical lesions [122–124]. Indeed, the key transcription factors T-bet, GATA-3, and FOXp2 (essential for Th1, Th2, and Tregs differentiation, respectively) have been found in periapical lesions [123] as well as IL-17A, the prototypical cytokine produced by Th17 cells [122].

A number of studies have shown that the adaptive immune response is important in protecting the host during the formation of lesions of endodontic origin. When the dental pulp is exposed in either severe combined immunodeficient (SCID) or normal control mice, apical lesions of similar size develop in both groups [125]. However, approximately one-third of the immunodeficient mice develop orofacial abscesses of endodontic origin compared with none in the immunocompetent mice. Another study showed that *nu/nu* rats, which have deficient T-cell response, had greater bone resorption following endodontic infections, suggesting a critical protective role. However, other studies showed no difference [126, 127]. The absence of IFN-γ, the prototypical Th1-cytokine resulted in increased bone resorption compared to wild-type mice, suggesting that IFN-γ played a protective role [120]. Interestingly, the majority of Th17 cells were found to also express IFN-γ, supporting a role for both pro-inflammatory responses in the pathogenesis of periapical periodontitis [122]. In contrast, genetic deletion of IL-4 had no effect, suggesting that Th2 responses did not play a critical role in protection or bone resorption [120]. However, the anti-inflammatory cytokine IL-10 is described as a protective factor against periapical bone resorption. Genetic ablation of IL-10 resulted in larger periapical lesions compared to that of wild-type mice, suggesting that IL-10 limited the size of lesions or had an essential protective role [120]. In human periapical granulomas, IL-10 mRNA levels were positively correlated with the expression of SOCS1 and SOCS3, proteins that act as negative regulators of inflammatory signaling [128]. The protective role of IL-10 and SOCS is reinforced by their inverse correlation with periapical lesions size. Recently, Tregs, a potential source of IL-10, were found in the periapical lesions following endodontic infection [124, 129].

Periapical granulomas present with heterogeneous patterns of RANKL and OPG expression, ranging from samples with RANKL/OPG ratio similar to that seen in sites with minimal or absent bone resorption [92]. However, a RANKL/OPG ratio may be indicative of whether the lesion is becoming larger (actively resorbing) or has reached a stable size [92, 130].

# Orthodontic tooth movement

Orthodontic tooth movement is achieved by the remodeling of alveolar bone in response to mechanical loading [131, 132]. When orthodontic forces are placed on the tooth, load transfer occurs from the tooth through the periodontal ligament to the alveolar bone, and causes minor reversible injury to the periodontium that supports the tooth. This minor injury plus the physical forces placed on the periodontal ligament lead to selective bone remodeling in opposing sites around the teeth. These distinct sites are referred to as "tension" and "compression", based on the predominant forces and the overall tissue response. In the tension side the displacement of the dental root creates a strain force in the PDL fibers that attach tooth to bone. On the other side of the tooth, in the direction of the force, the periodontal ligament has a compressive force as it is squeezed by the movement of the tooth toward bone. In this area the periodontal ligament fibers attaching the tooth to bone are unloaded. Therefore, the transduction of mechanical forces to the cells triggers a biologic response characterized by bone resorption at the compression site and bone formation at the tension site. Interestingly, this contrasts with bone in which the application of pressure stimulates bone formation. There are two possible explanations for why the application of pressure on the tooth leads to bone resorption whereas the application of pressure on bone leads to bone formation. One is that the application of pressure on the tooth will reduce normal tension between the periodontal ligament and adjoining bone. Thus, orthodontic tooth movement is thought to lessen mechanical forces on bone surfaces on the pressure side [131]. The second is aseptic injury to the periodontal ligament at the compression site, which is pro-inflammatory. A description of the compression and tension sites in the periodontal ligament in response to orthodontic forces is described in Figure 15-3.

The biologic response to orthodontic forces has been described as an aseptic inflammation, mediated by a variety of inflammatory cytokines, neuropeptides, and vasoactive molecules [133]. Indeed, the transitory expression of inflammatory mediators after orthodontic force application is essential for orthodontic movement, since anti-inflammatory drugs are capable of blocking tooth movement [134]. The inflammatory events at compression sites result in focal injury to the periodontal ligament, which stimulates the production of chemoattractants and the recruitment of leukocytes. The leukocytes remove the injured tissue in the periodontal ligament space. Simultaneously, osteoclast precursors are recruited to stimulate bone resorption. Conversely, the nature of inflammatory reaction at tension side may significantly differ from compression side, and the differential production of cytokines and growth factors would support the predominant bone-formation activity.

Similar to what is described with periodontal and periapical lesions, throughout orthodontic tooth movement inflammatory cytokines are associated with bone-resorbing activity [132]. TNF-α is present in high levels in gingival crevicular fluid during the orthodontic tooth movement [135], and plays an important role in response to orthodontic forces, since orthodontic tooth movement is significantly impaired in TNF receptor

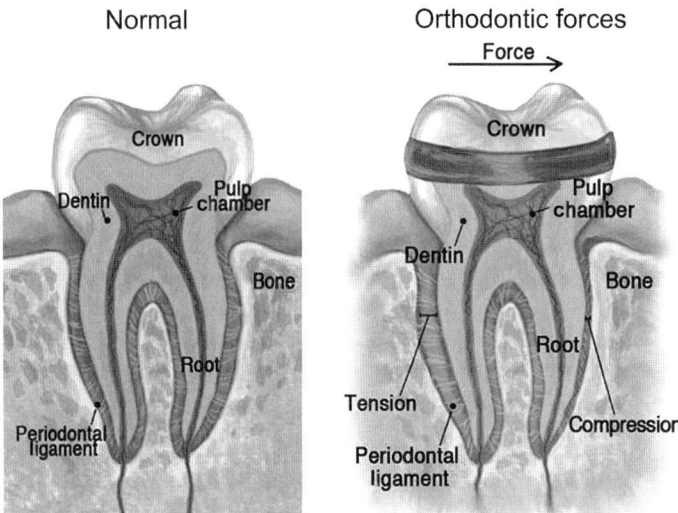

FIGURE 15-3 **Orthodontic tooth movement.** Mechanical force on the tooth causes compression of the periodontal ligament on one side and tension on the other side. The compression side is associated with aseptic injury and bone resorption and the tension side with bone formation. Please refer to color plate section.

p55-deficient mice [136]. Indeed, in the area undergoing compression periodontal ligament cells have been found to express increased levels of TNF-α, enhanced production of MMPs, and increased RANKL. Accordingly, inhibition of MMPs by chemically modified tetracyclines and inhibition of RANKL by gene transfer of OPG, significantly reduced osteoclastogenesis in response to orthodontic forces and tooth movement [137, 138]. When the RANKL/OPG expression ratio was investigated on the compression side, approximately 70% of the samples presented a RANKL>OPG ratio, suggesting that the RANKL/OPG balance is critical for bone-resorption activity in response to orthodontic forces [92]. Similarly, a differential MMP-1/TIMP-1 expression ratio induced by tensile and compressive strains may also influence tooth movement [139].

Pro-inflammatory factors that stimulate osteoclastogenesis such as IL-1β and IL-6 are produced in response to orthodontic forces [135, 140], and are thought to contribute to the predominant catabolic activity through the upregulation of RANKL and MMPs. However, inflammatory cytokines are produced with distinct kinetics in response to orthodontic force, suggesting a coordinated and complementary role for these mediators in the tooth movement process [141].

While the differentiation and activation of osteoclasts is thought to be largely controlled by RANK system, chemokines play an essential role in the recruitment of osteoclast precursors to the bone resorption site, and also can contribute to their differentiation and activation in response to orthodontic forces. Chemokines such as CCL3/MIP-1α, CXCL12/SDF-1, MCP-1/CCL2, CCL5/RANTES, and MIP-2 are intensely expressed on the compression side [142]. In agreement, MCP-1 is associated with osteoclast chemotaxis and differentiation, probably mediated by interaction with the receptor CCR2 [35]. MIP-1α stimulates osteoclast precursor chemotaxis and presumably guides them to sites where they will fuse, being also associated with osteoclast differentiation and resorption activity [35]. Similarly, SDF-1 can act on osteoclast precursors, inducing their chemotaxis, differentiation into osteoclasts, and promoting their survival [35]. Taken together, these results suggest that chemokines are potentially involved in the migration and differentiation of osteoclasts on the tension side during orthodontic tooth movement.

Interestingly, the local activation of the nervous and vascular system in the compressed PDL seems to play an active role in tissue remodeling. Blood vessels produce a series of bone active molecules (such as IL-1β, IL-6, IL-8, IL-11, and TNF-α) in the compressed periodontal ligament. PDL cells adjacent to hyalinized tissue and alveolar bone on the compressive side present VEGF immunoreactivity [143]. Local administration of VEGF enhances osteoclastogenesis at pressure sites during orthodontic tooth movement [144]. Tooth movement results in local increased production of neuropeptides, such as calcitonin gene-related peptide (CGRP) and substance P (SP), which further affect inflammation [132]. Monocytes, lymphocytes, and mast cells express receptors for neuropeptides, which stimulate cytokine production and the release of other inflammatory mediators. Indeed, sensory nerve transection impairs osteoclast formation during experimental tooth movement in rats [145].

While a series of host mediators have been associated with the tissue response to compressive forces, the nature of host response to tensile strain is less clear. Expression of the anti-inflammatory cytokine IL-10 is significantly higher on the tension side compared to the compression side, and tensile strain induces IL-10 synthesis in PDL [146, 147]. IL-10 can have a broad role in the bone environment, since this cytokine can upregulate osteoblast generation and bone formation while it inhibits bone resorption by upregulating OPG and downregulating RANKL expression [148, 149]. When the RANKL/OPG expression ratio was investigated on the tension side, the majority of the samples presented RANKL<OPG and RANKL≈OPG expression ratios, reinforcing that distinct patterns of RANKL/OPG expression may account for the distinct bone response on the tension and compression sides [92].

In addition to cytokines, chemokines produced by mechanical stimulation can also affect osteoblast behavior. SDF-1, found to be expressed on the tension side, can induce both proliferation and collagen type I mRNA expression in osteoblasts [150]. In addition, RANTES/CCL5, also expressed in PDL under tensile strain, can drive chemotaxis of osteoblasts and promotes their survival [151]. Mice lacking CCR5, one of the receptors for RANTES/CCL5, exhibited a greater amount of tooth movement after mechanical loading, associated with an increased expression of osteoclast markers (cathepsin K and RANKL) and decreased osteoblastic markers (RUNX2 and osteocalcin) [152]. However, no correlations were found between the levels of chemokines and the osteoblast and bone formation markers CBFA-1 and OCN. Besides being a target for chemokines, osteoblasts also express chemokines such as MCP-1, SDF-1, KC/CXCL1, LIX/CXCL5, CINC-1/CXCL1, and BCA-1/CXCL13 [35], and consequently may play an active role in bone remodeling associated with orthodontic forces.

While a clear distinction of pro- and anti-inflammatory cytokine expression in the tension and compression sides is described, the expression of TGF-β is increased during orthodontic tooth movement, being observed in osteoblasts in the tension zone, and in bone-resorbing osteoclasts in the compression zone [61, 153]. These results suggest a broad role for this cytokine in the tooth movement process. Irrespective of the controversies cited and discussed above, the current knowledge supports the hypothesis that the overall balance of cytokines expressed in the tension and compression sides determines the outcome of tissue response to tensile and compressive orthodontic forces.

However, it is also important to mention that the magnitude and diversity of biological response to orthodontic forces is underestimated. Variations in orthodontic force pattern (dynamic vs. static) and magnitude (light vs. heavy) have been described to result in different tissue response, reinforcing the complexity of the tooth movement process [139]. Furthermore, microarray analysis of the gene expression profiles in PDL cells, compressed by a static force using an in vitro three-dimensional culture system, demonstrates that more than a hundred genes were significantly modulated, involving multiple biological processes including cell communication, cell signaling, cell cycle, stress response, and calcium release [154].

Finally, it is important to consider that the individual effects of cytokines, chemokines, and neuropeptides on bone cells are usually investigated in highly controlled systems (i.e. in vitro or in knockout mouse). When interpreting in vivo data, their putative function must be estimated in view of the presence of other relevant factors (additional cytokines, growth factors, hormones, etc.), which can also modulate bone cell metabolism in a number of ways that affect the overall outcome. In fact, the study of the simultaneous action of multiple cytokines over PDL fibroblasts is incipient [155], and further studies are required to understand the exact coupling of force application and biological responses during tooth movement.

# Conclusions

Periodontal disease is a pathological condition, which is associated with inflammation of the tooth-supporting structures. It occurs in response to the presence of bacterial plaque on the tooth surface. The host defense system, including innate and adaptive immunity, is responsible for combating bacteria invading the periodontal tissue. In humans there is sufficient plaque accumulation even in health so that there is a continuous state of inflammation in tissue adjacent to teeth. By use of animal models and specific inhibitors both the innate and adaptive immune responses have been conclusively shown to participate in the formation of periodontal lesions.

The polymicrobial infection in lesions of endodontic origin stimulates bone resorption by interacting with the leukocytes of the host response. This is likely to involve both the innate and adaptive immune responses. However, the essential factors that regulate this response are not completely established.

The mechanotransduction of orthodontic force generates distinct areas of compression and tension in the tooth-supporting tissues, which include the alveolar bone. The nature of the transitory inflammation in these opposing areas involves differential production of cytokines, chemokines, and neuropeptides, and result in selective bone resorption and formation in these areas. However, further studies are required to understand fully the complex interaction of orthodontic force and biological responses throughout the tooth movement process.

## Acknowledgments

The authors would like to thank Allison Ryan for creating the illustrations and Juanita Hobson for help in preparing the manuscript.

## References

[1] Paster BJ, Olsen I, Aas JA, Dewhirst FE. The breadth of bacterial diversity in the human periodontal pocket and other oral sites. Periodontol 2000 2006;42:80–7.

[2] Pihlstrom BL, Michalowicz BS, Johnson NW. Periodontal diseases. Lancet 2005;366:1809–20.

[3] Sakamoto M, Umeda M, Benno Y. Molecular analysis of human oral microbiota. J Periodontal Res 2005;40:277–85.

[4] Haffajee AD, Teles RP, Socransky SS. The effect of periodontal therapy on the composition of the subgingival microbiota. Periodontol 2000 2006;42:219–58.

[5] Fine DH, Kaplan JB, Kachlany SC, Schreiner HC. How we got attached to Actinobacillus actinomycetemcomitans: a model for infectious diseases. Periodontol 2000 2006;42:114–57.

[6] Assuma R, Oates T, Cochran D, Amar S, Graves D. IL-1 and TNF antagonists inhibit the inflammatory response and bone loss in experimental periodontitis. J Immun 1998;160:403–9.

[7] Delima A, Spyros K, Amar S, Graves DT. Inflammation and tissue loss caused by periodontal pathogens is reduced by IL-1 antagonists. J Infect Dis 2002;186:511–16.

[8] Teng Y, Nguyen H, Gao X, Kong Y, Gorczynski R, Singh B, Ellen R, Penninger J. Functional human T-cell immunity and osteoprotegerin ligand control alveolar bone destruction in periodontal infection. J Clin Invest 2000;106:R59–67.

[9] Baker P, Dixon M, Evans R, Dufour L, Johnson E, Roopenian D. CD4(+) T cells and the proinflammatory cytokines gamma interferon and interleukin-6 contribute to alveolar bone loss in mice. Infect Immun 1999;67:2804–9.

[10] Socransky SS, Haffajee AD, Goodson JM, Lindhe J. New concepts of destructive periodontal disease. J Clin Periodontol 1984;11:21–32.

[11] Cohen ME, Ralls SA. Distributions of periodontal attachment levels. Mathematical models and implications. J Periodontol 1988;59:254–8.

[12] Gilthorpe MS, Zamzuri AT, Griffiths GS, Maddick IH, Eaton KA, Johnson NW. Unification of the "burst" and "linear" theories of periodontal disease progression: a multilevel manifestation of the same phenomenon. J Dent Res 2003;82:200–5.

[13] Williams R, Jeffcoat M, Kaplan M, Goldhaber P, Johnson H, Wechter W. Flurbiprofen: a potent inhibitor of alveolar bone resorption in beagles. Science 1985;227:640–2.

[14] Delima A, Oates T, Assuma R, Schwartz Z, Cochran D, Amar S, Graves D. Soluble antagonists to interleukin-1 (IL-1) and tumor necrosis factor (TNF) inhibits loss of tissue attachment in experimental periodontitis. J Clin Periodontol 2001;28:233–40.

[15] Han X, Kawai T, Eastcott JW, Taubman MA. Bacterial-responsive B lymphocytes induce periodontal bone resorption. J Immunol 2006;176:625–31.

[16] Jin Q, Cirelli JA, Park CH, Sugai JV, Taba Jr M, Kostenuik PJ, Giannobile WV. RANKL inhibition through osteoprotegerin blocks bone loss in experimental periodontitis. J Periodontol 2007;78:1300–8.

[17] Page R, Schroeder H. Pathogenesis of inflammatory periodontal disease. A summary of current work. Lab Invest 1976;34:235–49.

[18] Kinane D, Lappin D. Clinical, pathological and immunological aspects of periodontal disease. Acta Odontol Scand 2001;59:154–60.

[19] Behl Y, Siqueira M, Ortiz J, Li J, Desta T, Faibish D, Graves DT. Activation of the acquired immune response reduces coupled bone formation in response to a periodontal pathogen. J Immunol 2008;181:8711–18.

[20] Graves D. Cytokines that promote periodontal tissue destruction. J Periodontol 2008;79:1585–91.

[21] Graves DT, Fine D, Teng YT, Van Dyke TE, Hajishengallis G. The use of rodent models to investigate host-bacteria interactions related to periodontal diseases. J Clin Periodontol 2008;35:89–105.

[22] Rovin S, Costich ER, Gordon HA. The influence of bacteria and irritation in the initiation of periodontal disease in germfree and conventional rats. J Periodontal Res 1966;1:193–204.

[23] Mahanonda R, Pichyangkul S. Toll-like receptors and their role in periodontal health and disease. Periodontol 2000 2007;43:41–55.

[24] Bar-Shavit Z. Taking a toll on the bones: regulation of bone metabolism by innate immune regulators. Autoimmunity 2008;41:195–203.

[25] Gelani V, Fernandes AP, Gasparoto TH, Garlet TP, Cestari TM, Lima HR, Ramos ES, de Souza Malaspina TS, Santos CF, Garlet GP, da Silva JS, Campanelli AP. The role of toll-like receptor 2 in the recognition of *Aggregatibacter actinomycetemcomitans*. J Periodontol 2009; 80:2010–19.

[26] Lima HR, Gelani V, Fernandes AP, Gasparoto TH, Torres SA, Santos CF, Garlet GP, Silva JSd, Campanelli AP. The essential role of TLR4 in the control of *Aggregatibacter actinomycetemcomitans* infection in mice. J Clin Periodontal 2010;37:248–54.

[27] Uehara A, Takada H. Functional TLRs and NODs in human gingival fibroblasts. J Dent Res 2007;86:249–54.

[28] Fracon RN, Teofilo JM, Satin RB, Lamano T. Prostaglandins and bone: potential risks and benefits related to the use of nonsteroidal anti-inflammatory drugs in clinical dentistry. J Oral Sci 2008;50:247–52.

[29] Noguchi K, Ishikawa I. The roles of cyclooxygenase-2 and prostaglandin E2 in periodontal disease. Periodontol 2000 2007;43:85–101.

[30] Miyauchi M, Ijuhin N, Nikai H, Takata T, Ito H, Ogawa I. Effect of exogenously applied prostaglandin E2 on alveolar bone loss – histometric analysis. J Periodontol 1992;63:405–11.

[31] Pouliot M, Clish CB, Petasis NA, Van Dyke TE, Serhan CN. Lipoxin A(4) analogues inhibit leukocyte recruitment to Porphyromonas gingivalis: a role for cyclooxygenase-2 and lipoxins in periodontal disease. Biochemistry 2000;39:4761–8.

[32] Bezerra MM, de Lima V, Alencar VB, Vieira IB, Brito GA, Ribeiro RA, Rocha FA. Selective cyclooxygenase-2 inhibition prevents alveolar bone loss in experimental periodontitis in rats. J Periodontol 2000;71:1009–14.

[33] Pinho Mde N, Pereira LB, de Souza SL, Palioto DB, Grisi MF, Novaes Jr AB, Taba Jr M. Short-term effect of COX-2 selective inhibitor as an adjunct for the treatment of periodontal disease: a clinical double-blind study in humans. Braz Dent J 2008;19:323–8.

[34] Hasturk H, Kantarci A, Ohira T, Arita M, Ebrahimi N, Chiang N, Petasis NA, Levy BD, Serhan CN, Van Dyke TE. RvE1 protects from local inflammation and osteoclast-mediated bone destruction in periodontitis. FASEB J 2006;20:401–3.

[35] Silva TA, Garlet GP, Fukada SY, Silva JS, Cunha FQ. Chemokines in oral inflammatory diseases: apical periodontitis and periodontal disease. J Dent Res 2007;86:306–19.

[36] Aggarwal R, Ghobrial IM, Roodman GD. Chemokines in multiple myeloma. Exp Hematol 2006;34:1289–95.

[37] Tsai CC, Ho YP, Chen CC. Levels of interleukin-1 beta and interleukin-8 in gingival crevicular fluids in adult periodontitis. J Periodontol 1995;66:852–9.

[38] Gemmell E, Carter CL, Seymour GJ. Chemokines in human periodontal disease tissues. Clin Exp Immunol 2001;125:134–41.

[39] Alnaeeli M, Park J, Mahamed D, Penninger JM, Teng YT. Dendritic cells at the osteo-immune interface: implications for inflammation-induced bone loss. J Bone Miner Res 2007;22:775–80.

[40] Bonecchi R, Galliera E, Borroni EM, Corsi MM, Locati M, Mantovani A. Chemokines and chemokine receptors: an overview. Front Biosci 2009;14:540–51.

[41] Zhang X, Teng YT. Interleukin-10 inhibits gram-negative-microbe-specific human receptor activator of NF-kappaB ligand-positive CD4+-Th1-cell-associated alveolar bone loss in vivo. Infect Immun 2006;74:4927–31.

[42] Gamonal J, Acevedo A, Bascones A, Jorge O, Silva A. Characterization of cellular infiltrate, detection of chemokine receptor CCR5 and interleukin-8 and RANTES chemokines in adult periodontitis. J Periodontal Res 2001;36:194–203.

[43] Garlet G, Martins Jr W, Ferreira B, Milanezi C, Silva J. Patterns of chemokine and chemokine receptors expression in different forms of human periodontal disease. J Periodontal Res 2003;38:210–17.

[44] Hanazawa S, Kawata Y, Takeshita A, Kumada H, Okithu M, Tanaka S, Yamamoto Y, Masuda T, Umemoto T, Kitano S. Expression of monocyte chemoattractant protein 1 (MCP-1) in adult periodontal disease: increased monocyte chemotactic activity in crevicular fluids and induction of MCP-1 expression in gingival tissues. Infect Immun 1993;61:5219–24.

[45] Yu X, Graves D. Fibroblasts, mononuclear phagocytes and endothelial cells express MCP-1 in inflamed human gingiva. J Perio 1995;66:80–8.

[46] Watanabe T, Kukita T, Kukita A, Wada N, Toh K, Nagata K, Nomiyama H, Iijima T. Direct stimulation of osteoclastogenesis by MIP-1alpha: evidence obtained from studies using RAW264 cell clone highly responsive to RANKL. J Endocrinol 2004;180:193–201.

[47] Hosokawa Y, Hosokawa I, Ozaki K, Nakae H, Murakami K, Miyake Y, Matsuo T. CXCL12 and CXCR4 expression by human gingival fibroblasts in periodontal disease. Clin Exp Immunol 2005;141:467–74.

[48] Dennison DK, Van Dyke TE. The acute inflammatory response and the role of phagocytic cells in periodontal health and disease. Periodontol 1997;2000(14):54–78.

[49] Graves DT, Kayal RA. Diabetic complications and dysregulated innate immunity. Front Biosci 2008;13:1227–39.

[50] Gyurko R, Siqueira CC, Caldon N, Gao L, Kantarci A, Van Dyke TE. Chronic hyperglycemia predisposes to exaggerated inflammatory response and leukocyte dysfunction in Akita mice. J Immunol 2006;177:7250–6.

[51] Garlet GP, Cardoso CR, Campanelli AP, Garlet TP, Avila-Campos MJ, Cunha FQ, Silva JS. The essential role of IFN-gamma in the control of lethal *Aggregatibacter actinomycetemcomitans* infection in mice. Microbes Infect 2008;10:489–96.

[52] Garlet GP, Cardoso CR, Campanelli AP, Ferreira BR, Avila-Campos MJ, Cunha FQ, Silva JS. The dual role of p55 tumour necrosis factor-alpha receptor in *Actinobacillus actinomycetemcomitans*-induced experimental periodontitis: host protection and tissue destruction. Clin Exp Immunol 2007;147:128–38.

[53] Lee W, Aitken S, Sodek J, McCulloch CA. Evidence of a direct relationship between neutrophil collagenase activity and periodontal tissue destruction in vivo: role of active enzyme in human periodontitis. J Periodontal Res 1995;30:23–33.

[54] Kantarci A, Oyaizu K, Van Dyke TE. Neutrophil-mediated tissue injury in periodontal disease pathogenesis: findings from localized aggressive periodontitis. J Periodontol 2003;74:66–75.

[55] Strober W. The multifaceted influence of the mucosal microflora on mucosal dendritic cell responses. Immunity 2009;31:377–88.

[56] Vernal R, Leon R, Herrera D, Garcia-Sanz, Silva JA, Sanz M. Variability in the response of human dendritic cells stimulated with Porphyromonas gingivalis or *Aggregatibacter actinomycetemcomitans*. J Periodontal Res 2008;43:689–97.

[57] Jotwani R, Cutler CW. Fimbriated Porphyromonas gingivalis is more efficient than fimbria-deficient P. gingivalis in entering human dendritic cells in vitro and induces an inflammatory Th1 effector response. Infect Immun 2004;72:1725–32.

[58] Madianos PN, Bobetsis YA, Kinane DF. Generation of inflammatory stimuli: how bacteria set up inflammatory responses in the gingiva. J Clin Periodontol 2005;32(Suppl. 6):57–71.

[59] Leibbrandt A, Penninger JM. RANK/RANKL: regulators of immune responses and bone physiology. Ann N Y Acad Sci 2008;1143:123–50.

[60] Alnaeeli M, Penninger JM, Teng YT. Immune interactions with CD4+ T cells promote the development of functional osteoclasts from murine CD11c+ dendritic cells. J Immunol 2006;177:3314–26.

[61] Garlet GP, Cardoso CR, Silva TA, Ferreira BR, Avila-Campos MJ, Cunha FQ, Silva JS. Cytokine pattern determines the progression of experimental periodontal disease induced by *Actinobacillus actinomycetemcomitans* through the modulation of MMPs, RANKL, and their physiological inhibitors. Oral Microbiol Immunol 2006;21:12–20.

[62] Hou LT, Liu CM, Liu BY, Lin SJ, Liao CS, Rossomando EF. Interleukin-1beta, clinical parameters and matched cellular-histopathologic changes of biopsied gingival tissue from periodontitis patients. J Periodontal Res 2003;38:247–54.

[63] Salvi GE, Brown CE, Fujihashi K, Kiyono H, Smith FW, Beck JD, Offenbacher S. Inflammatory mediators of the terminal dentition in adult and early onset periodontitis. J Periodontal Res 1998;33:212–25.

[64] Stashenko P, Jandinski J, Fujiyoshi P, Rynar J, Socransky S. Tissue levels of bone resorptive cytokines in periodontal disease. J Perio 1991;62:504–9.

[65] Chiang C, Kyritsis G, Graves D, Amar S. Interleukin-1 and tumor necrosis factor activities partially account for calvarial bone resorption induced by local injection of lipopolysaccharide. Infection and Immunity 1999;67:4231–6.

[66] Koide M, Suda S, Saitoh S, Ofuji Y, Suzuki T, Yoshie H, Takai M, Ono Y, Taniguchi Y, Hara K. In vivo administration of IL-1 beta accelerates silk ligature-induced alveolar bone resorption in rats. J Oral Pathol Med 1995;24:420–34.

[67] Dayan S, Stashenko P, Niederman R, Kupper TS. Oral epithelial overexpression of IL-1alpha causes periodontal disease. J Dent Res 2004;83:786–90.

[68] Kurtis B, Tuter G, Serdar M, Akdemir P, Uygur C, Firatli E, Bal B. Gingival crevicular fluid levels of monocyte chemoattractant protein-1 and tumor necrosis factor-alpha in patients with chronic and aggressive periodontitis. J Periodontol 2005;76:1849–55.

[69] Bostrom L, Linder L, Bergstrom J. Clinical expression of TNF-alpha in smoking-associated periodontal disease. J Clin Periodontol 1998;25:767–73.

[70] Lee HJ, Kang IK, Chung CP, Choi SM. The subgingival microflora and gingival crevicular fluid cytokines in refractory periodontitis. J Clin Periodontol 1995;22:885–90.

[71] Gaspersic R, Stiblar-Martincic D, Osredkar J, Skaleric U. Influence of subcutaneous administration of recombinant TNF-alpha on ligature-induced periodontitis in rats. J Periodontal Res 2003;38:198–203.

[72] Graves D, Oskoui M, Volejnikova S, Naguib G, Cai S, Desta T, Kakouras A, Jiang Y. Tumor necrosis factor modulates fibroblast apoptosis, PMN recruitment, and osteoclast formation in response to P. gingivalis infection. J Dent Res 2001;80:1875–9.

[73] de Pablo P, Chapple IL, Buckley CD, Dietrich T. Periodontitis in systemic rheumatic diseases. Nat Rev Rheumatol 2009;5:218–24.

[74] Graves D, Liu R, Alikhani M, Al-Mashat H, Trackman P. Diabetes-enhanced inflammation and apoptosis – impact on periodontal pathology. J Dent Res 2006;85f:15–21.

[75] Takeichi O, Haber J, Kawai T, Smith DJ, Moro I, Taubman MA. Cytokine profiles of T-lymphocytes from gingival tissues with pathological pocketing. J Dent Res 2000;79:1548–55.

[76] Rogers JE, Li F, Coatney DD, Rossa C, Bronson P, Krieder JM, Giannobile WV, Kirkwood KL. Actinobacillus actinomycetemcomitans lipopolysaccharide-mediated experimental bone loss model for aggressive periodontitis. J Periodontol 2007;78:550–8.

[77] Baker P, Evans R, Roopenian D. Oral infection with Porphyromonas gingivalis and induced alveolar bone loss in immunocompetent and severe combined immunodeficient mice. Arch Oral Biol 1994;39:1035–40.

[78] Teng YT, Nguyen H, Hassanloo A, Ellen RP, Hozumi N, Gorczynski RM. Periodontal immune responses of human lymphocytes in Actinobacillus actinomycetemcomitans-inoculated NOD/SCID mice engrafted with peripheral blood leukocytes of periodontitis patients. J Periodontal Res 1999;34:54–61.

[79] Teng YT, Mahamed D, Singh B. Gamma interferon positively modulates Actinobacillus actinomycetemcomitans-specific RANKL+ CD4+ Th-cell-mediated alveolar bone destruction in vivo. Infect Immun 2005;73:3453–61.

[80] Agnello D, Lankford CS, Bream J, Morinobu A, Gadina M, O'Shea JJ, Frucht DM. Cytokines and transcription factors that regulate T helper cell differentiation: new players and new insights. J Clin Immunol 2003;23:147–61.

[81] Eastcott JW, Yamashita K, Taubman MA, Harada Y, Smith DJ. Adoptive transfer of cloned T helper cells ameliorates periodontal disease in nude rats. Oral Microbiol Immunol 1994;9:284–9.

[82] Gemmell E, Seymour GJ. Immunoregulatory control of Th1/Th2 cytokine profiles in periodontal disease. 2004;35(Periodontol 2000):21–41.

[83] Miossec P, Korn T, Kuchroo VK. Interleukin-17 and type 17 helper T cells. N Engl J Med 2009;361: 888–98.

[84] Yu JJ, Ruddy MJ, Wong GC, Sfintescu C, Baker PJ, Smith JB, Evans RT, Gaffen SL. An essential role for IL-17 in preventing pathogen-initiated bone destruction: recruitment of neutrophils to inflamed bone requires IL-17 receptor-dependent signals. Blood 2007.

[85] Cardoso CR, Garlet GP, Crippa GE, Rosa AL, Junior WM, Rossi MA, Silva JS. Evidence of the presence of T helper type 17 cells in chronic lesions of human periodontal disease. Oral Microbiol Immunol 2009;24:1–6.

[86] Vignali DA, Collison LW, Workman CJ. How regulatory T cells work. Nat Rev Immunol 2008; 8:523–32.

[87] Cardoso CR, Garlet GP, Moreira AP, Junior WM, Rossi MA, Silva JS. Characterization of CD4+CD25+ natural regulatory T cells in the inflammatory infiltrate of human chronic periodontitis. J Leukoc Biol 2008;84:311–18.

[88] Ernst CW, Lee JE, Nakanishi T, Karimbux NY, Rezende TM, Stashenko P, Seki M, Taubman MA, Kawai T. Diminished forkhead box P3/CD25 double-positive T regulatory cells are associated with the increased nuclear factor-kappaB ligand (RANKL+) T cells in bone resorption lesion of periodontal disease. Clin Exp Immunol 2007;148:271–80.

[89] Nakajima T, Ueki-Maruyama K, Oda T, Ohsawa Y, Ito H, Seymour GJ, Yamazaki K. Regulatory T-cells infiltrate periodontal disease tissues. J Dent Res 2005;84:639–43.

[90] Kawai T, Matsuyama T, Hosokawa Y, Makihira S, Seki M, Karimbux NY, Goncalves RB, Valverde P, Dibart S, Li YP, Miranda LA, Ernst CW, Izumi Y, Taubman MA. B and T lymphocytes are the primary sources of RANKL in the bone resorptive lesion of periodontal disease. Am J Pathol 2006;169:987–98.

[91] Crotti T, Smith MD, Hirsch R, Soukoulis S, Weedon H, Capone M, Ahern MJ, Haynes D. Receptor activator NF kappaB ligand (RANKL) and osteoprotegerin (OPG) protein expression in periodontitis. J Periodontal Res 2003;38:380–7.

[92] Menezes R, Garlet TP, Letra A, Bramante CM, Campanelli AP, Figueira Rde C, Sogayar MC, Granjeiro JM, Garlet GP. Differential patterns of receptor activator of nuclear factor kappa B ligand/osteoprotegerin expression in human periapical granulomas: possible association with progressive or stable nature of the lesions. J Endod 2008;34:932–8.

[93] Parfitt A. The coupling of bone formation to bone resorption: a critical analysis of the concept and of its relevance to the pathogenesis of osteoporosis. Metab Bone Dis Relat Res 1982;4:1–6.

[94] Al-Mashat HA, Kandru S, Liu R, Behl Y, Desta T, Graves DT. Diabetes enhances mRNA levels of proapoptotic genes and caspase activity, which contribute to impaired healing. Diabetes 2006;55:487–95.

[95] Liu R, Bal HS, Desta T, Krothapalli N, Alyassi M, Luan Q, Graves D. Diabetes enhances periodontal bone loss through enhanced resorption and diminished bone formation. J Dent Res 2006;85:510–14.

[96] Kakehashi S, Stanley HR, Fitzgerald RJ. The effects of surgical exposures of dental pulps in germ-free and conventional laboratory rats. Oral Surg Oral Med Oral Pathol 1965;20:340–9.

[97] Ricucci D, Siqueira Jr JF, Bate AL, Pitt Ford TR. Histologic investigation of root canal-treated teeth with apical periodontitis: a retrospective study from twenty-four patients. J Endod 2009;35:493–502.

[98] Nair PN. Pathogenesis of apical periodontitis and the causes of endodontic failures. Crit Rev Oral Biol Med 2004;15:348–81.

[99] Liapatas S, Nakou M, Rontogianni D. Inflammatory infiltrate of chronic periradicular lesions: an immunohistochemical study. Int Endod J 2003;36:464–71.

[100] Stashenko P, Wang C, Riley E, Wu Y, Ostroff G, Niederman R. Reduction of infection-stimulated peripical bone resorption by the biological response modifier PGG glucan. J Dent Res 1995;74:323–30.

[101] Graves D, Chen C, Douville C, Jiang Y. Interleukin-1 receptor signaling rather than that of tumor necrosis factor is critical in protecting the host from the severe consequences of a polymicrobe anaerobic infection. Infect Immun 2000;8:4746–51.

[102] Fukada SY, Silva TA, Saconato IF, Garlet GP, Avila-Campos MJ, Silva JS, Cunha FQ. iNOS-derived nitric oxide modulates infection-stimulated bone loss. J Dent Res 2008;87:1155–9.

[103] Jiang Y, Mehta C, Hsu T, Alsulaimani F. Bacteria induce osteoclastogenesis via an osteoblast-independent pathway. Infect Immun 2002;70:3143–8.

[104] Gilles J, Carnes D, Dallas M, Holt S, Bonewald L. Oral bone loss is increased in ovariectomized rats. J Endodon 1997;23:419–22.

[105] Hirao K, Yumoto H, Takahashi K, Mukai K, Nakanishi T, Matsuo T. Roles of TLR2, TLR4, NOD2, and NOD1 in pulp fibroblasts. J Dent Res 2009;88:762–7.

[106] McNicholas S, Torabinejad M, Blankenship J, Bakland L. The concentration of prostaglandin E2 in human periradicular lesions. J Endod 1991;17:97–100.

[107] Torbinejad M, Clagett J, Engel D. A cat model for the evaluation of mechanisms of bone resorption: induction of bone loss by simulated immune complexes and inhibition by indomethacin. Calcif Tissue Int 1979;29:207–14.

[108] Fouad AF. IL-1 alpha and TNF-alpha expression in early periapical lesions of normal and immunodeficient mice. J Dent Res 1997;76:1548–54.

[109] Tani-Ishii N, Wang CY, Stashenko P. Immunolocalization of bone-resorptive cytokines in rat pulp and periapical lesions following surgical pulp exposure. Oral Microbiol Immunol 1995;10: 213–19.

[110] Stashenko P, Wang CY, Tani-Ishii N, Yu SM. Pathogenesis of induced rat periapical lesions. Oral Surg Oral Med Oral Pathol 1994;78:494–502.

[111] Wang C, Stashenko P. The role of interleukin-1 alpha in the pathogenesis of periapical bone destruction in a rat model system. Oral Microbiology & Immunology 1993;8:50–6.

[112] Takeichi O, Saito I, Tsurumachi T, Moro I, Saito T. Expression of inflammatory cytokine genes in vivo by human alveolar bone-derived polymorphonuclear leukocytes isolated from chronically inflamed sites of bone resorption. Calcif Tissue Int 1996;58:244–8.

[113] Walker KF, Lappin DF, Takahashi K, Hope J, Macdonald DG, Kinane DF. Cytokine expression in periapical granulation tissue as assessed by immunohistochemistry. Eur J Oral Sci 2000;108: 195–201.

[114] Huang GT, Do M, Wingard M, Park JS, Chugal N. Effect of interleukin-6 deficiency on the formation of periapical lesions after pulp exposure in mice. Oral Surg Oral Med Oral Pathol Oral Radiol Endod 2001;92:83–8.

[115] Huang GT, Potente AP, Kim JW, Chugal N, Zhang X. Increased interleukin-8 expression in inflamed human dental pulps. Oral Surg Oral Med Oral Pathol Oral Radiol Endod 1999;88:214–20.

[116] Yamasaki M, Kumazawa M, Kohsaka T, Nakamura H. Effect of methotrexate-induced neutropenia on rat periapical lesion. Oral Surg Oral Med Oral Pathol 1994;77:655–61.

[117] Rahimi P, Wang C, Stashenko P, Lee S, Lorenzo J, Graves D. Monocyte chemoattractant protein-1 expression and monocyte recruitment in osseous inflammation in the mouse. Endocrinology 1995;136:2752–9.

[118] Chae P, Im M, Gibson F, Jiang Y, Graves D. Mice lacking monocyte chemoattractant protein 1 have enhanced susceptibility to an interstitial polymicrobial infection due to impaired monocyte recruitment. Infect Immun 2002;70:3164–9.

[119] Garlet TP, Fukada SY, Saconato IF, Avila-Campos MJ, Silva TAD, Garlet GP and Cunha FdQ. CCR2 deficiency results in increased osteolysis in experimental periapical lesions in mice. J Endod 2010;36:244–50.

[120] De Rossi A, Rocha LB, Rossi MA. Interferon-gamma, interleukin-10, intercellular adhesion molecule-1, and chemokine receptor 5, but not interleukin-4, attenuate the development of periapical lesions. J Endod 2008;34:31–8.

[121] Stashenko P, Yu SM, Wang CY. Kinetics of immune cell and bone resorptive responses to endodontic infections. J Endod 1992;18:422–6.

[122] Colic M, Gazivoda D, Vucevic D, Vasilijic S, Rudolf R, Lukic A. Proinflammatory and immunoregulatory mechanisms in periapical lesions. Mol Immunol 2009;47:101–13.

[123] Fukada SY, Silva TA, Garlet GP, Rosa AL, da Silva JS, Cunha FQ. Factors involved in the T helper type 1 and type 2 cell commitment and osteoclast regulation in inflammatory apical diseases. Oral Microbiol Immunol 2009;24:25–31.

[124] Colic M, Gazivoda D, Vucevic D, Majstorovic I, Vasilijic S, Rudolf R, Brkic Z, Milosavljevic P. Regulatory T-cells in periapical lesions. J Dent Res 2009;88:997–1002.

[125] Onunkwo O. An outbreak of infectious bursal disease (IBD) of chickens in Nigeria. Vet Rec 1975;97:433.

[126] Tani N, Kuchiba K, Osada T, Watanabe Y, Umemoto T. Effect of T-cell deficiency on the formation of periapical lesions in mice: histological comparison between periapical lesion formation in BALB/c and BALB/c nu/nu mice. J Endod 1995;21:195–9.

[127] Wallstrom J, Torabinejad M, Kettering J, McMillan P, Loma L. Role of T cells in the pathogenesis of periapical lesions. Oral Surg Oral Med Oral Pathol 1993;76.

[128] Menezes R, Garlet TP, Trombone AP, Repeke CE, Letra A, Granjeiro JM, Campanelli AP, Garlet GP. The potential role of suppressors of cytokine signaling in the attenuation of inflammatory reaction and alveolar bone loss associated with apical periodontitis. J Endod 2008;34:1480–4.

[129] Alshwaimi E, Purcell P, Kawai T, Sasaki H, Oukka M, Campos-Neto A, Stashenko P. Regulatory T cells in mouse periapical lesions. J Endod 2009;35:1229–33.

[130] Kawashima N, Suzuki N, Yang G, Ohi C, Okuhara S, Nakano-Kawanishi H, Suda H. Kinetics of RANKL, RANK and OPG expressions in experimentally induced rat periapical lesions. Oral Surg Oral Med Oral Pathol Oral Radiol Endod 2007;103:707–11.

[131] Wise GE, King GJ. Mechanisms of tooth eruption and orthodontic tooth movement. J Dent Res 2008;87:414–34.

[132] Krishnan V, Davidovitch Z. On a path to unfolding the biological mechanisms of orthodontic tooth movement. J Dent Res 2009;88:597–608.

[133] Meikle MC. The tissue, cellular, and molecular regulation of orthodontic tooth movement: 100 years after Carl Sandstedt. Eur J Orthod 2006;28:221–40.

[134] Walker JB, Buring SM. NSAID impairment of orthodontic tooth movement. Ann Pharmacother 2001;35:113–15.

[135] Uematsu S, Mogi M, Deguchi T. Interleukin (IL)-1 beta, IL-6, tumor necrosis factor-alpha, epidermal growth factor, and beta 2-microglobulin levels are elevated in gingival crevicular fluid during human orthodontic tooth movement. J Dent Res 1996;75:562–7.

[136] Andrade Jr I, Silva TA, Silva GA, Teixeira AL, Teixeira MM. The role of tumor necrosis factor receptor type 1 in orthodontic tooth movement. J Dent Res 2007;86:1089–94.

[137] Kanzaki H, Chiba M, Takahashi I, Haruyama N, Nishimura M, Mitani H. Local OPG gene transfer to periodontal tissue inhibits orthodontic tooth movement. J Dent Res 2004;83:920–5.

[138] Bildt MM, Henneman S, Maltha JC, Kuijpers-Jagtman AM, Von den Hoff JW. CMT-3 inhibits orthodontic tooth displacement in the rat. Arch Oral Biol 2007;52:571–8.

[139] Zhao Z, Fan Y, Bai D, Wang J, Li Y. The adaptive response of periodontal ligament to orthodontic force loading – a combined biomechanical and biological study. Clin Biomech (Bristol, Avon) 2008;23(Suppl. 1):S59–66.

[140] Ren Y, Hazemeijer H, de Haan B, Qu N, de Vos P. Cytokine profiles in crevicular fluid during orthodontic tooth movement of short and long durations. J Periodontol 2007;78:453–8.

[141] Yamaguchi M, Kasai K. Inflammation in periodontal tissues in response to mechanical forces. Arch Immunol Ther Exp (Warsz) 2005;53:388–98.

[142] Garlet TP, Coelho U, Repeke CE, Silva JS, Cunha Fde Q, Garlet GP. Differential expression of osteoblast and osteoclast chemmoatractants in compression and tension sides during orthodontic movement. Cytokine 2008;42:330–5.

[143] Miyagawa A, Chiba M, Hayashi H, Igarashi K. Compressive force induces VEGF production in periodontal tissues. J Dent Res 2009;88:752–6.

[144] Kaku M, Kohno S, Kawata T, Fujita I, Tokimasa C, Tsutsui K, Tanne K. Effects of vascular endothelial growth factor on osteoclast induction during tooth movemnet in mice. J Dent Res 2001;80:1880–3.

[145] Yamashiro T, Fujiyama K, Fujiyoshi Y, Inaguma N, Takano-Yamamoto T. Inferior alveolar nerve transection inhibits increase in osteoclast appearance during experimental tooth movement. Bone 2000;26:663–9.

[146] Garlet TP, Coelho U, Silva JS, Garlet GP. Cytokine expression pattern in compression and tension sides of the periodontal ligament during orthodontic tooth movement in humans. Eur J Oral Sci 2007;115:355–62.

[147] Long P, Hu J, Piesco N, Buckley M, Agarwal S. Low magnitude of tensile strain inhibits IL-1beta-dependent induction of pro-inflammatory cytokines and induces synthesis of IL-10 in human periodontal ligament cells in vitro. J Dent Res 2001;80:1416–20.

[148] Xu LX, Kukita T, Kukita A, Otsuka T, Niho Y, Iijima T. Interleukin-10 selectively inhibits osteoclastogenesis by inhibiting differentiation of osteoclast progenitors into preosteoclast-like cells in rat bone marrow culture system. J Cell Physiol 1995;165:624–9.

[149] Dresner-Pollak R, Gelb N, Rachmilewitz D, Karmeli F, Weinreb M. Interleukin 10-deficient mice develop osteopenia, decreased bone formation, and mechanical fragility of long bones. Gastroenterology 2004;127:792–801.

[150] Lisignoli G, Toneguzzi S, Piacentini A, Cristino S, Grassi F, Cavallo C, Facchini A. CXCL12 (SDF-1) and CXCL13 (BCA-1) chemokines significantly induce proliferation and collagen type I expression in osteoblasts from osteoarthritis patients. J Cell Physiol 2006;206:78–85.

[151] Yano S, Mentaverri R, Kanuparthi D, Bandyopadhyay S, Rivera A, Brown EM, Chattopadhyay N. Functional expression of beta-chemokine receptors in osteoblasts: role of regulated upon activation, normal T cell expressed and secreted (RANTES) in osteoblasts and regulation of its secretion by osteoblasts and osteoclasts. Endocrinology 2005;146:2324–35.

[152] Andrade Jr I, Taddei SR, Garlet GP, Garlet TP, Teixeira AL, Silva TA, Teixeira MM. CCR5 downregulates osteoclast function in orthodontic tooth movement. J Dent Res 2009;88:1037–41.

[153] Nagai M, Yoshida A, Sato N, Wong DT. Messenger RNA level and protein localization of transforming growth factor-beta1 in experimental tooth movement in rats. Eur J Oral Sci 1999;107:475–81.

[154] de Araujo RM, Oba Y, Moriyama K. Identification of genes related to mechanical stress in human periodontal ligament cells using microarray analysis. J Periodontal Res 2007;42:15–22.

[155] Silverio-Ruiz KG, Martinez AE, Garlet GP, Barbosa CF, Silva JS, Cicarelli RM, Valentini SR, Abi-Rached RS, Junior CR. Opposite effects of bFGF and TGF-beta on collagen metabolism by human periodontal ligament fibroblasts. Cytokine 2007;39:130–7.

# Index

ADAM8, pre-osteoclast fusion regulation, 27–28
Ae2, deficiency and bone loss, 30
Akt, macrophage colony-stimulating factor signaling, 163–164
Alveolar macrophage, osteoclast precursors, 33
Androgens, bone response, 110
Angiopoietin-1, hematopoietic stem cell–osteoblast precursor communication, 86
Ankylosing spondylitis, bone formation regulation, 315–318
AP-1
    osteoblast differentiation role, 116–117
    osteoclast differentiation role, 148–149
Arthritis, *see also* Rheumatoid arthritis; Spondylarthritis
ATF4, osteoblast differentiation role, 117

Basic multicellular unit (BMU)
    distribution in remodeling, 239
    resorption cessation, 240–241
B cell
    activation, 59
    activation, 69, 71
    development
        extrinsic signals, 62–63
        interleukin-7 role, 63–64
        intrinsic requirements, 64–67
        RANKL pathway, 68–69
        stages, 61
    estrogen response, 286
    fracture healing role, 346
    function, 69–70
    immunoglobulin
        class switching, 59
        rearrangement, 58–60
    multiple myeloma response, 397
    osteoclast precursors, 33
    osteomyelitis response, 372–373, 383–384
    osteoprotegerin generation, 250
    RANKL generation, 250–251
    receptor autoreactivity, 61–62
Bcl2, osteoclast regulation, 240
BLIMP1, B cell function, 70
BMD, *see* Bone mineral density
BMPs, *see* Bone morphogenetic proteins
BMU, *see* Basic multicellular unit
Bone marrow niche, *see* Hematopoietic stem cell
Bone mineral density (BMD)
    age-related bone loss patterns, 270–273
    measurement, 270–273
    rheumatoid arthritis studies, 312
Bone morphogenetic proteins (BMPs), osteoblast differentiation role, 113
Bone remodeling
    estrogen regulation
        bone formation effects, 278
        bone resorption and calcium homeostasis effects, 277
        osteocyte function, 279
        overview, 276–277
    extent in humans, 8, 238
    formation coupling with resorption, 108–109, 242–247, 254–255
    initiation, 240
    overview, 238–239
    purposes, 239
    reversal phase, 241
Bone repair, *see* Fracture healing
Btk, integration of RANK and ITAM signaling, 160–161
N-Cadherin, hematopoietic stem cell–osteoblast precursor communication, 87

443

Calcineurin, osteoclast differentiation
    signaling, 21
Calcitonin receptor (CTR), pre-osteoclast
    fusion regulation, 28
Calcium/calmodulin-dependent kinase IV
    (CAMKIV), osteoclast differentiation
    signaling, 21
Calcium-sensing receptor, hematopoietic
    stem cell–osteoblast precursor
    communication, 87
CAMKIV, see Calcium/calmodulin-dependent
    kinase IV
Cannabinoid receptors, bone, 113
Carbonic anhydrase II
    deficiency and bone loss, 29–30
    osteoclast acidification, 9–10
Cardiotropin-1 (CT-1), bone remodeling role,
    245
β-Catenin signaling, see Wnt
Cathepsin K, deficiency and bone loss, 30
Cavity, see Endodontic lesion
CCL2
    endodontic lesion role, 427
    overview of bone effects, 198
CCL3
    overview of bone effects, 198
    periodontal disease role, 417–418
CCL5, orthodontal tooth movement response,
    431
CCL9, overview of bone effects, 198–199
CCR1
    overview of bone effects, 199
    periodontal disease role, 418
CCR2, overview of bone effects, 199–200
CCR5, periodontal disease role, 418
CD40L, overview of bone effects, 194
CD44, pre-osteoclast fusion regulation, 27
CD47, pre-osteoclast fusion regulation,
    24–26
Celiac disease, osteoporosis and immune
    system regulation, 290
Chondrocyte, tumor necrosis factor-α effects
    in fracture healing, 354–355
CIITP, estrogen response, 287
Crohn disease, see Inflammatory bowel disease

CT-1, see Cardiotropin-1
CTLA-4, osteoclast inhibition, 252
CTR, see Calcitonin receptor
CX3CR1
    overview of bone effects, 199
    pre-osteoclast expression, 16–17
CXCL8
    endodontic lesion role, 427
    periodontal disease role, 417
CXCL12
    osteoblast signaling in hematopoietic stem
        cell microenvironment, 89, 91–92, 105
    overview of bone effects, 199
CXCR4
    osteoblast signaling in hematopoietic stem
        cell microenvironment, 89
    overview of bone effects, 199
Cyclooxygenase inhibitors, see Nonsteroidal
    anti-inflammatory drugs

DAP10, osteoclast function, 157
DAP12
    bone protective effects, 161–162
    knockout mice, 157
    Nasu-Hakola disease mutations, 156–157
    osteoclast differentiation signaling, 20–21,
        154–156
    signaling and receptors, 157–158
    Syk signaling, 159
DC, see Dendritic cell
DC-STAMP, pre-osteoclast fusion regulation,
    26, 194
Dendritic cell (DC)
    multiple myeloma response, 394–396
    osteoclast precursors, 32–33
    osteomyelitis response
        adaptive immunity, 380–381
        innate immunity, 381–382
        osteoclastogenesis role, 382–383
Denosumab, osteoporosis trials, 290–291
DKK1
    rheumatoid arthritis role, 126–127,
        309–311
    Wnt inhibition in osteoblast function,
        111, 114

E2A, B cell development role, 65–66
Ebf1, B cell development role, 66–67
Endochondral bone, formation process, 103
Endodontic lesion
   adaptive immune response, 428
   innate immune response, 426–428
   pathogenesis, 424–426
Ephrins, bone formation coupling with resorption, 243–244
Estrogen, *see also* Menopause
   bone response, 110
   immune modulation
      B cells, 285
      deficiency effects on general immunity, 286–288
      overview, 279, 286
      pro-inflammatory cytokines, 282–284
      RANKL system, 279–282
      T cells, 284–285
   lymphocyte receptors, 250
   male loss in aging, 275–276
   osteoclast apoptosis enhancement, 240

FasL, overview of bone effects, 193–194
FcRγ
   knockout mice, 157
   OSCAR interactions, 153–154
   osteoclast differentiation signaling, 20–21, 153–156
   Syk signaling, 159
c-FMS, *see* Macrophage colony-stimulating factor
c-Fos
   osteoclast differentiation role, 23
   rheumatoid arthritis role, 304
FoxO, aging effects, 288–289
Fracture healing
   cascade, 346, 349–350
   immune function
      cytokine roles
         interleukin-1, 353, 357–358
         interleukin-6, 354, 358
         osteoprotegerin, 353, 357
         overview, 351–352
         phase-specific roles, 354–358
         RANK, 353, 356
         RANKL, 353, 356–357
         tumor necrosis factor-α, 352–356
      lymphocytes, 346, 349–350
      mesenchymal stem cell modulation, 350–351
      transcriptomics
         downregulated genes, 348
         upregulated genes, 347
   nonsteroidal anti-inflammatory drugs
      clinical effects, 360–361
      COX-2 inhibition effects on bone cells, 359–360
   postnatal regenerative process perspective, 344
   stages, 345
Ftl3L, B cell development role, 63

Gab2, RANK interactions, 145
GALT, *see* Gut-associated lymphoid tissue
G-CSF, *see* Granulocyte colony-stimulating factor
Gingivitis, *see* Periodontal disease
Glucocorticoids
   bone response, 110–111
   inflammatory bowel disease-induced bone loss, 332
GM-CSF, *see* Granulocyte–macrophage colony-stimulating factor
gp130 receptors, osteoclast induction, 235
Granulocyte colony-stimulating factor (G-CSF), bone effects, 191
Granulocyte–macrophage colony-stimulating factor (GM-CSF)
   osteoclast differentiation role, 190
   osteoclast inhibition, 251
Gut-associated lymphoid tissue (GALT), functional overview, 326

Hedgehog, multiple myeloma signaling, 398
Hematopoietic stem cell (HSC)
   adult cell features, 85
   differentiation overview, 44–45
   long-term reconstituting cells, 44, 84–85

Hematopoietic stem cell (HSC) (*Continued*)
  niche, 44, 83–85, 91–94, 104–105
  osteoblast precursors
    signaling in cell–cell communication
      angiopoietin-1/Tie2, 86
      N-cadherin, 87
      calcium-sensing receptor, 87
      Notch, 87–88
      osteopontin, 86–87
      thrombopoietin, 86
    support, 83–85
  osteoclast function modulation in osteomyelitis, 379–380
Hox, osteoblast differentiation role, 116
HSC, *see* Hematopoietic stem cell

IBD, *see* Inflammatory bowel disease
IFN-α, *see* Interferon-α
IFN-β, *see* Interferon-β
IFN-γ, *see* Interferon-γ
Ikaros, B cell development role, 64–65
IKK, *see* Inhibitor of family κB
IL-1, *see* Interleukin-1
IL-3, *see* Interleukin-3
IL-4, *see* Interleukin-4
IL-6, *see* Interleukin-6
IL-7, *see* Interleukin-7
IL-8, *see* Interleukin-8
IL-10, *see* Interleukin-10
IL-11, *see* Interleukin-11
IL-12, *see* Interleukin-12
IL-13, *see* Interleukin-13
IL-15, *see* Interleukin-15
IL-17, *see* Interleukin-17
IL-18, *see* Interleukin-18
IL-23, *see* Interleukin-23
IL-27, *see* Interleukin-27
IL-32, *see* Interleukin-32
Inflammatory bowel disease (IBD)
  adaptive immunity and pathogenesis, 328–329
  bone mass effects
    mechanisms
      general factors, 331–332
      immune factors, 332–334
      RANKL and osteoprotegerin, 334–335
    osteoporosis and immune system regulation, 290
    overview, 330–331
  clinical features, 325
  epidemiology, 325
  immunological differences between diseases, 329–330
  innate immunity and pathogenesis, 327–328
  intestinal immune system, 326–327
Inhibitor of family κB (IKK), osteoclast differentiation role, 22–23
Integrins
  associated proteins, 171–174
  DAP12 in signaling, 158
  integrin αvβ3 and osteoclast resorption, 168–171
  signaling modes, 170
Interferon-α (IFN-α), overview of bone effects, 203
Interferon-β (IFN-β)
  induction by RANKL, 241
  overview of bone effects, 203
Interferon-γ (IFN-γ)
  endodontic lesion role, 428
  inflammatory bowel disease-induced bone loss, 332, 334
  osteoclast inhibition, 251
  overview of bone effects, 202–203
Interleukin-1 (IL-1)
  discovery of osteoclast activation, 2
  endodontic lesion role, 427
  estrogen response, 282–283
  fracture healing role, 353, 357–358
  overview of bone effects, 191–192
  periodontal disease role, 419–420
  rheumatoid arthritis role, 307
Interleukin-3 (IL-3), osteoclast differentiation role, 190
Interleukin-4 (IL-4)
  osteoclast inhibition, 252
  overview of bone effects, 203–204
  rheumatoid arthritis role, 308

Interleukin-6 (IL-6)
  fracture healing role, 354, 358
  osteoclast induction, 235
  overview of bone effects, 194–195
  periodontal disease role, 421
  rheumatoid arthritis role, 307
Interleukin-7 (IL-7)
  B cell development role, 63–64, 105
  estrogen response, 284
  overview of bone effects, 196–197
  T cell function, 52
Interleukin-8 (IL-8)
  endodontic lesion role, 427
  overview of bone effects, 198
  periodontal disease role, 417
Interleukin-10 (IL-10), overview of bone effects, 200
Interleukin-11 (IL-11), overview of bone effects, 195
Interleukin-12 (IL-12)
  osteoclast inhibition, 252
  overview of bone effects, 200–201
Interleukin-13 (IL-13), overview of bone effects, 203–204
Interleukin-15 (IL-15), overview of bone effects, 201
Interleukin-17 (IL-17)
  overview of bone effects, 201
  rheumatoid arthritis role, 306
Interleukin-18 (IL-18)
  osteoclast inhibition, 251–252
  overview of bone effects, 202
Interleukin-23 (IL-23)
  osteoclast inhibition, 252
  overview of bone effects, 202
Interleukin-27 (IL-27), overview of bone effects, 202
Interleukin-32 (IL-32), overview of bone effects, 204
ITAM signalingDAP12; FcRγ
  Nasu-Hakola disease defects, 156
  osteoclast differentiation, 154–156
  phospholipase Cγ2, 159–160
  Src kinases in activation, 158–159
  Syk signaling, 159
  Tek kinases and integration of RANK signaling, 160–161

Lenalidomide, multiple myeloma management, 403
Leptin, bone response, 110
Leukemia inhibitory factor (LIF), overview of bone effects, 196
Leukemia, hematopoietic stem cell microenvironment, 93
LIF, see Leukemia inhibitory factor

Macrophage
  osteoclast ontogenic relationship, 232–233
  tumor-associated macrophages in multiple myeloma, 393–394
Macrophage colony-stimulating factor (M-CSF)
  bone cell precursor proliferation and survival role, 162
  c-FMS receptor signaling, 162–164
  DAP12 in receptor signaling, 158
  osteoclast cytoskeleton function, 174–175
  osteoclast differentiation role, 18, 189–190, 236
Macrophage inflammatory protein-1α (MIP-1α), periodontal disease role, 417–418
Macrophage migration inhibitory factor (MIF), overview of bone effects, 204
MAPK, see Mitogen-activated protein kinase
Matrix metalloproteinases (MMPs)
  MMP9 and osteoclast function, 30–31
  orthodontal tooth movement response, 430
  rheumatoid arthritis role, 315
MCP-1, see Monocyte chemoattractant protein-1
Membranous bone, formation process, 103
Menopause
  bone loss patterns, 274
  cytokines in bone loss, 2, 250
Mesenchymal stem cell (MSC), fracture healing and immune function modulation, 350–351

Methotrexate, rheumatoid arthritis studies, 313
Microphthalmia transcription factor (MITF), osteoclast differentiation role, 24
MicroRNA, osteoblast differentiation role, 121–123
MIF, see Macrophage migration inhibitory factor
MIP-1α, see Macrophage inflammatory protein-1α
MITF, see Microphthalmia transcription factor
Mitogen-activated protein kinase (MAPK)
   activation by RANKL, 149
   macrophage colony-stimulating factor signaling, 163–164
MMPs, see Matrix metalloproteinases
Monocyte, osteoclast differentiation, 11–17
Monocyte chemoattractant protein-1 (MCP-1)
   endodontic lesion role, 427
   orthodontal tooth movement response, 430
MSC, see Mesenchymal stem cell
Multiple myeloma
   adaptive immunity
      B cells, 397
      T helper cells, 398
   epidemiology, 391–392
   Hedgehog signaling, 398
   immunotherapy
      γδ-T cells, 400–402
      myeloma-specific cytotoxic T cells, 400
      prospects, 403–404
      tumor-specific antigens, 402
   innate immunity
      dendritic cells, 394–396
      natural killer cells, 396–397
      tumor-associated macrophages, 393–394
   lenalidomide therapy, 403
   nuclear factor-κB activation, 399
   osteoclast
      precursors, 33
      stimulation, 392
   RANK/RANKL expression, 398–399
   thalidomide therapy, 402–403
   Toll-like receptor signaling, 399
   TRAIL role, 400

Nasu-Hakola disease, gene mutations, 156–157
Natural killer cell, multiple myeloma response, 396–397
NF-κB, see Nuclear factor-κB
NFATc1, see Nuclear factor of activated T cells c1
Nonsteroidal anti-inflammatory drugs (NSAIDs)
   COX-2 inhibition effects on bone cells, 359–360
   fracture healing clinical effects, 360–361
Notch, hematopoietic stem cell–osteoblast precursor communication, 87–88
NSAIDs, see Nonsteroidal anti-inflammatory drugs
Nuclear factor of activated T cells c1 (NFATc1)
   autoamplification and epigenetic regulation, 151
   osteoclast differentiation role, 24, 149–150
   RANKL induction, 150
   regulation by other transcription factors, 152
   target genes, 152
Nuclear factor-κB (NF-κB)
   multiple myeloma activation, 399
   osteoclast differentiation role, 22, 147–148
   periodontal disease role, 416

OAF, see Osteoclast-activating factor
Oncostatin M, overview of bone effects, 196
ONJ, see Osteonecrosois of the jaw
OPG, see Osteoprotegerin
OPN, see Osteopontin
Orthodontal tooth movement, bone remodeling regulation, 429–432
OSCAR
   FcRγ interactions, 153–154
   osteoblast–osteoclast interactions, 152–153
Osteoblast
   bone formation coupling with resorption, 108–109, 242–247, 254–255
   differentiation regulation
      morphogens, 113–115
      nuclear microenvironments and specificity of osteoblast lineage determination, 118–123

overview, 229–231
transcription factors, 115–118
functional overview, 1–2
growth/differentiation stages and markers, 107
hormonal regulation, 110
life cycle, 229
osteoblast crosstalk, 124
osteomyelitis response, 374
precursors
   hematopoietic stem cell support of niche, 83–85
   heterogeneity, 82–83
   mesenchymal stem cell development into osteoprogenitor cells, 105–108
   signaling in hematopoietic stem cell communication
      angiopoietin-1/Tie2, 86
      N-cadherin, 87
      calcium-sensing receptor, 87
      Notch, 87–88
      osteopontin, 86–87
      thrombopoietin, 86
pro-osteoclastic signal regulation, 235–238
signaling pathways in microenvironment
   CXCL12/CXCR4, 89
   prostaglandin E2, 89–90
   transforming growth factor-$\beta$, 90
   Wnt, 88–89
tumor necrosis factor-$\alpha$ effects in fracture healing, 354–355
Osteoclast
acid generation for bone resorption, 9–10, 164–165
adhesion molecules, 8–9
bone formation role, 244–245
bone resorption coupling with formation, 108–109, 242–247, 254–255
bone site differences, 31–32
dendritic cell, osteoclastogenesis role, 382–383
formation control with lineage, 233–235
functional overview, 142–143, 231–232
integrin $\alpha v \beta 3$
   associated proteins, 171–174
   resorption role, 168–171
lymphocyte-mediated inhibition, 251–253
macrophage colony-stimulating factor and cytoskeleton, 174–175
markers, 10
nuclei, 8, 142
osteomyelitis response
   acute versus chronic disease, 375–376
   hematopoietic stem cell function modulation, 379–380
   immune function, 377–378
   mobilization, 376–377
   osteoblast function modulation, 378–379
precursors
   alveolar macrophages, 33
   B cells, 33
   dendritic cells, 32–33
   external signals
      macrophage colony-stimulating factor, 18
      osteoprotegerin, 19
      RANKL, 18–19, 189–190
   fusion regulation
      ADAM8, 27–28
      calcitonin receptor, 28
      CD44, 27
      CD47, 24–26
      DC-STAMP, 26, 194
      SIRP$\alpha$, 24–26
      thrombospondin-1, 24–26
      V-ATPase, 26–27
markers, 12
monocyte–macrophage lineage, 11–17
multiple myeloma cells, 33
prospects for study, 34
signal transduction
   calcineurin, 21
   calcium/calmodulin-dependent kinase IV, 21
   DAP12, 20–21, 154–156
   FcR$\gamma$, 20–21, 153–156
   TRAF6, 20, 145–146
transcription factors
   AP-1, 148–149
   c-Fos, 23

Osteoclast (*Continued*)
    inhibitor of family κB, 22–23
    microphthalmia transcription factor, 24
    nuclear factor of activated T cells c1, 24, 149–152
    nuclear factor-κB, 22, 147–148
    peroxisome proliferator-activated receptor-γ, 23
    Spi1, 21–22
  ruffled border, 166–167
  sealing zone
    microtubules, 165–166
    podosomes, 165
  subtypes, 29–31
  tumor necrosis factor-α effects in fracture healing, 355–356
Osteoclast-activating factor (OAF), *see* Interleukin-1
Osteoclast differentiation factor, *see* RANKL
Osteocyte
  estrogen regulation, 279
  functional overview, 108
  mechanotransduction and bone viability, 109–110
Osteomac
  discovery, 253
  plasticity, 253–254
Osteomyelitis
  B cell response, 372–373, 383–384
  dendritic cell response
    adaptive immunity, 380–381
    innate immunity, 381–382
    osteoclastogenesis role, 382–383
  host immune response, 372–374
  microbial infection in pathogenesis, 370–372
  osteoblast response, 374
  osteoclast response
    acute versus chronic disease, 375–376
    hematopoietic stem cell function modulation, 379–380
    immune function, 377–378
    mobilization, 376–377
    osteoblast function modulation, 378–379
Osteon, features, 103
Osteonecrosois of the jaw (ONJ), 370
Osteonectin (SPARC), microRNA regulation, 122
Osteopetrosis, bone marrow transplantation therapy, 11
Osteopontin (OPN), hematopoietic stem cell–osteoblast precursor communication, 86–87
Osteoporosis
  age-related bone loss patterns, 270–273, 288–289
  cytokine therapeutic targeting, 290–291
  fracture incidence in aging, 273
  secondary causes and immune system regulation, 289–290
  sex steroids
    aging and bone loss
      men, 274–276
      women, 274
    bone remodeling regulation
      bone formation effects, 278
      bone resorption and calcium homeostasis effects, 277
      osteocyte function, 279
      overview, 276–277
    estrogen and immune modulation
      B cells, 285
      deficiency effects on general immunity, 286–288
      overview, 279, 286
      pro-inflammatory cytokines, 282–284
      RANKL system, 279–282
      T cells, 284–285
Osteoprotegerin (OPG)
  B cell
    development role, 68–69
    expression, 250
  endodontic lesion role, 428
  estrogen response, 281
  fracture healing role, 353, 357
  functional overview, 188–189
  inflammatory bowel disease-induced bone loss, 334–335
  orthodontal tooth movement response, 430
  osteoclast differentiation role, 19, 236

periodontal disease role, 421, 423
rheumatoid arthritis role, 314
T cell generation, 249
Osterix
nonsteroidal anti-inflammatory drug effects, 359
osteoblast differentiation role, 118, 230

p62, TRAF6 regulation, 146
p100, osteoclastogenesis effects, 189
Parathyroid hormone (PTH)
anabolic effects, 245–246
bone formation coupling with resorption, 243, 245–246
bone response, 110
T cells in catabolic action, 250
Parathyroid hormone receptor
B cell development role, 63–64
hematopoietic stem cell–osteoblast precursor communication, 87
parathyroid hormone-related peptide as ligand and actions, 246–247
Parathyroid hormone-related peptide (PTHrP)
bone effects, 246–247
rheumatoid arthritis role, 305
Pax5, B cell development role, 67
Periodontal disease
bacteria in oral cavity, 412–413
bone loss, 413–414
gingivitis pathology, 414–415
immune response on progression, 415–416
prostaglandins in pathogenesis, 416–417
Periosteum, features, 103
Peroxisome proliferator-activated receptor-γ (PPAR-γ), osteoclast differentiation role, 23
PGE2, see Prostaglandin E2
Phospholipase Cγ2, ITAM signaling, 159–160
PPAR-γ, see Peroxisome proliferator-activated receptor-γ
Prostaglandin E2 (PGE2)
adaptive immune response, 421–422
chemokine modulation, 417–419
cytokine modulation, 419–421
EP4 receptor, 359

osteoblast signaling in hematopoietic stem cell microenvironment, 89–90
osteoclast induction, 235
periodontal disease pathogenesis, 416–417
PTH, see Parathyroid hormone
PTHrP, see Parathyroid hormone-related peptide
PU.1, see Spi1
Pyk2, osteoclast function, 172

RA, see Rheumatoid arthritis
Rac, osteoclast function, 173–174
RANK
co-stimulatory receptors, 152–156
estrogen response, 282
fracture healing role, 353, 356
Gab2 interactions, 145
history of study, 145
multiple myeloma expression, 398–399
mutations and disease, 189
Tek kinases and integration of ITAM signaling, 160–161
TRAF6 activation, 144–146
RANKL
B cell
development role, 68–69
expression, 250–251
cytokine enhancement of osteoclast response, 2
endodontic lesion role, 428
estrogen response, 280–281
fracture healing role, 353, 356–357
history of study, 143, 236–237
inflammatory bowel disease-induced bone loss, 334–335
knockout mice, 188
mitogen-activated protein kinase activation, 149
multiple myeloma expression, 398–399
mutation and disease, 144
nuclear factor of activated T cells c1 induction, 150
orthodontal tooth movement response, 430
osteoclast differentiation role, 18–19, 144, 236–238

RANKL (*Continued*)
  osteoporosis therapeutic targeting, 290–291
  periodontal disease role, 421–424
  rheumatoid arthritis role, 125, 304–308, 314
  T cell expression, 3, 249–250
RANTES, orthodontal tooth movement response, 431
Rheumatoid arthritis (RA)
  generalized bone loss, 312–315
  inflammatory cytokines, 125
  joint margin and subchondral bone loss, 303–311
  pathogenesis, 301–302
  peri-articular bone loss, 311–312
  remodeling alterations, 302–303
  repair mechanisms, 125–126
  Wnt signaling regulation, 126–127
RhoA, osteoclast function, 173
Ruffled border, osteoclasts, 166–167, 232
Runx
  osteoblast–T cell crosstalk, 124
  Runx2
    nonsteroidal anti-inflammatory drug effects, 359
    osteoblast differentiation role, 117–120, 229–230

Schnurri, osteoblast differentiation role, 118
Sex hormone-binding globulin (SHBG), aging effects, 274–275
SHBG, *see* Sex hormone-binding globulin
SIRPα, pre-osteoclast fusion regulation, 24–26
Sirtuins, bone mass effects, 389
SOST
  mutation and disease, 244
  osteoblast differentiation role, 115
SPARC, *see* Osteonectin
Sphingosine 1-phosphate, T cell periphery migration signal, 52–53
Spi1
  B cell development role, 64
  osteoclast differentiation role, 21–22
Spondylarthritis, bone regulation in seronegative spondyloarthropathies, 315–318

Src kinase, osteoclast function, 172–173
Syk, osteoclast function, 173

TAK1, TRAF6 activation, 146
TAM, *see* Tumor-associated macrophage
Tartrate-resistant acid phosphatase (TRAP), osteoclast expression, 232, 234
T cell
  activation, 53–56
  aging effects, 289
  bone resorption stimulation, 249–250
  cytokine generation, 248
  cytotoxic T cell function, 56
  estrogen response, 284–286
  fracture healing role, 349–350
  functional heterogeneity, 48–49
  inflammatory bowel disease pathogenesis, 328–329
  maintenance, 51–52
  memory T cell generation, 57–58
  multiple myeloma
    immunotherapy
      γδ-T cells, 400–402
      myeloma-specific cytotoxic T cells, 400
      prospects, 403–404
      tumor-specific antigens, 402
    natural killer cell, 396–397
    T helper cell response, 398
  osteoblast crosstalk, 124
  osteoclast inhibition, 251–253
  osteomyelitis response, 372–373
  periodontal disease role, 422–424
  periphery migration signals, 52–53
  selection, 50
  T helper cell polarization, 54–55
  thymopoiesis, 49–53
  types, 46, 48–49
T cell receptor (TCR)
  ligands, 45, 47
  priming, 54
  rearrangement, 47, 50
  structure, 46
  types, 46–47
TCR, *see* T cell receptor
TEC, *see* Thymic epithelial cell

Teeth, *see* Endodontic lesion; Orthodontal tooth movement; Periodontal disease
Tek, integration of RANK and ITAM signaling, 160–161
Testosterone, male loss in aging, 275–276
TGF-β, *see* Transforming growth factor-β
Thalidomide, multiple myeloma management, 402–403
Thrombopoietin (TPO), hematopoietic stem cell–osteoblast precursor communication, 86
Thrombospondin-1 (TSP1), pre-osteoclast fusion regulation, 24–26
Thymic epithelial cell (TEC), T cell development role, 51
Tie2, hematopoietic stem cell–osteoblast precursor communication, 86
TLRs, *see* Toll-like receptors
TNF-α, *see* Tumor necrosis factor-α
Toll-like receptors (TLRs)
  endodontic lesions, 426–427
  inflammatory bowel disease pathogenesis, 327–329
  multiple myeloma signaling, 399
  periodontal disease, 416
TPO, *see* Thrombopoietin
TRAF6
  activation and signaling, 145–146
  osteoclast differentiation signaling, 20, 145
  RANK activation, 144–146
  TAK1 activation, 146
TRAIL
  multiple myeloma role, 400
  overview of bone effects, 194
TRANCE, *see* RANKL
Transforming growth factor-β (TGF-β)
  bone formation coupling with resorption, 242
  osteoblast differentiation role, 113
  osteoblast signaling in hematopoietic stem cell microenvironment, 90
TRAP, *see* Tartrate-resistant acid phosphatase

TREM-2
  Nasu-Hakola disease mutations, 156
TSP-1, *see* Thrombospondin-1
Tumor-associated macrophage (TAM), multiple myeloma, 393–394
Tumor necrosis factor-α (TNF-α)
  estrogen response, 282–284
  fracture healing role, 352–356
  inflammatory bowel disease-induced bone loss, 332, 334
  orthodontal tooth movement response, 429–430
  overview of bone effects, 192–193
  periodontal disease role, 420–421
  rheumatoid arthritis role, 304, 306–307
  spondylarthritis therapeutic targeting, 318
Tumor-specific antigen, multiple myeloma immunotherapy, 402
Twist, osteoblast differentiation role, 116

Ulcerative colitis, *see* Inflammatory bowel disease

Vascular endothelial growth factor (VEGF), orthodontal tooth movement response, 431
V-ATPase, pre-osteoclast fusion regulation, 26–27
VEGF, *see* Vascular endothelial growth factor
Vitamin D3
  bone response, 111
  osteoclast induction, 235

Wnt
  osteoblast differentiation role, 113–114, 230–231
  osteoblast signaling in hematopoietic stem cell microenvironment, 88–89
  osteoblast–T cell crosstalk, 124
  rheumatoid arthritis and signaling regulation, 126–127

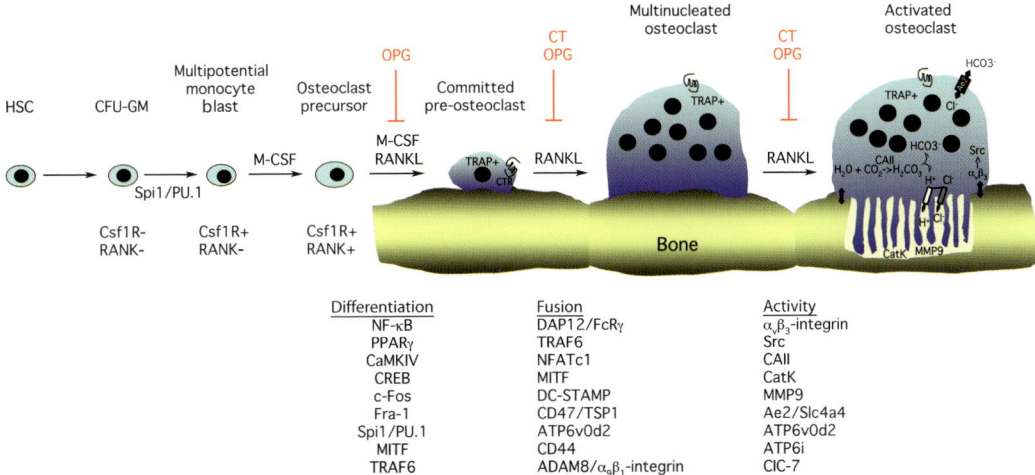

**FIGURE 2-1 Stages of osteoclast differentiation.** In basal osteoclastogenesis, membrane-bound forms of M-CSF and RANKL produced and presented largely by osteoblasts induce the differentiation, activation, and survival of osteoclasts from osteoclast precursors through a series of steps. Depicted are some of the key transcription factors regulating the steps of osteoclastogenesis (Spi1/PU.1, MITF, NF-κB, PPARγ, CREB, c-Fos, Fra-1, NFATc1), signal transduction molecules (TRAF6, CaMKIV, DAP12/FcRγ, Src), fusion regulators (transmembrane proteins DC-STAMP, CTR, CD47, and CD44, the CD47 ligand TSP1, the vacuolar-ATPase subunit ATP6v0d2, and the disintegrin and metalloproteinase ADAM8 and its cognate receptor $\alpha_9\beta_1$-integrin), and genes important for osteoclast function ($\alpha_v\beta_3$-integrin, CAII, proteinases CatK and MMP9, anion exchangers Ae2 and Slc4a4, vacuolar-ATPase subunit ATP6i, the chloride channel ClC-7). Only a few of the regulatory cytokines are denoted (M-CSF, RANKL, OPG, and CT). (Please refer to Chapter 2, page 9).

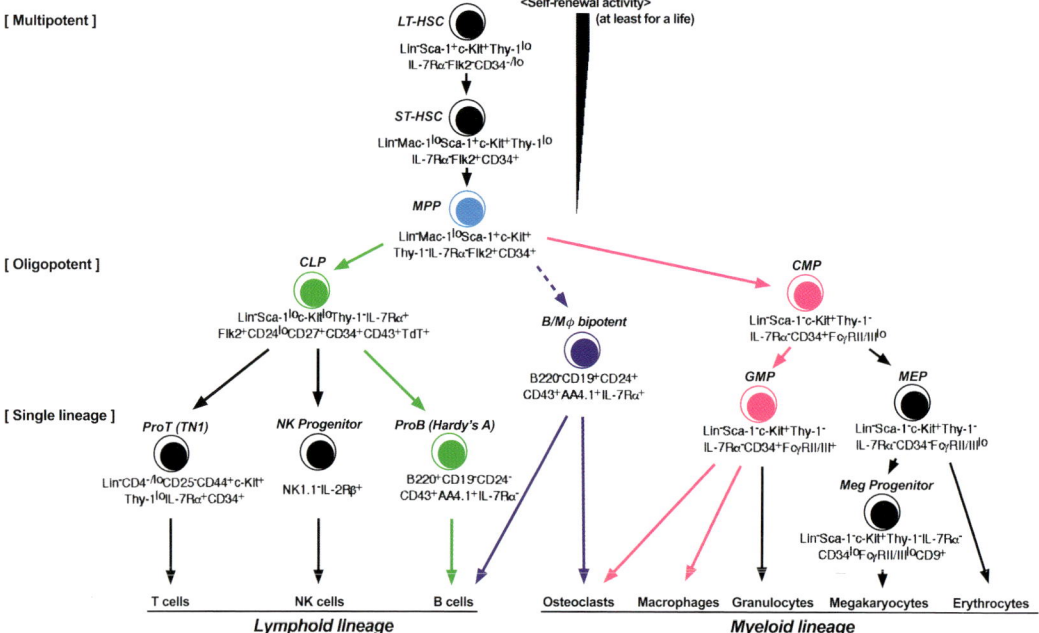

**FIGURE 3-1 Hematopoietic cell differentiation.** All hematopoietic cells arise from hematopoietic stem cells (HSC) that give rise to multipotential progenitor cells (MPP, blue). B cells differentiate from the common lymphoid progenitor (CLP, green) and osteoclasts arise from the common myeloid progenitor (CMP, pink). There appears to exist a distinct lineage derived from the MPP that expresses both macrophage and B-cell characteristics (B/Mϕ bipotent), which can differentiate to B cells, macrophages and possibly osteoclasts (purple). *(Adapted from Kondo M, Wagers AJ, Manz MG, Prohaska SS, Scherer DC, Beilhack GF, Shizuru JA, Weissman IL. Ann. Rev. Immunol. 2003; 21:759–806).* (Please refer to Chapter 3, page 45).

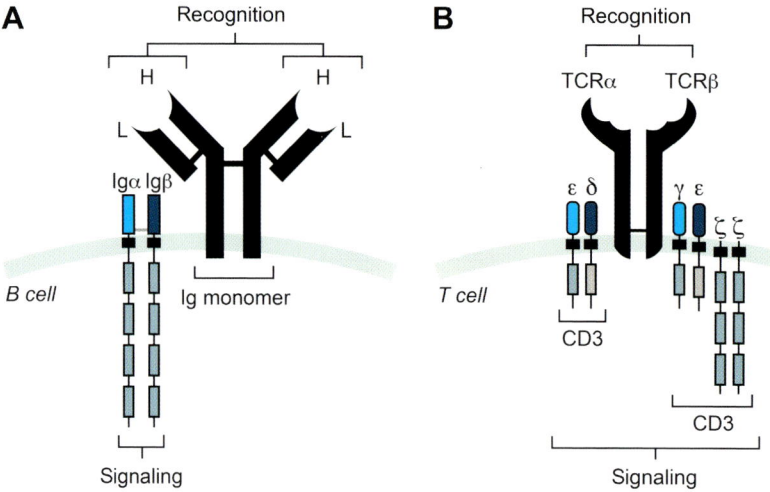

**FIGURE 3-2 Structures of the complete BCR and TCR complexes. (A)** The BCR complex is composed of an Ig monomer associated with an Igα and an Igβ chain. An Ig monomer is composed of two identical heavy (H) chains and two identical light (L) chains. These chains combine to form two identical antigen-binding sites that can reorganize unprocessed antigen. Igα and Igβ transduce intracellular signaling initiated by the binding of antigen to the Ig monomer. **(B)** The TCR is composed of a heterodimeric receptor protein and the CD3 complex. In a majority of T cells, the receptor protein is composed of a TCRα and a TCRβ chain; a minority of T cells bears a receptor protein composed of a TCRγ and a TCRδ chain (not shown). The CD3 complex transduces intracellular signaling initiated by the engagement of the receptor protein. (Please refer to Chapter 3, page 48).

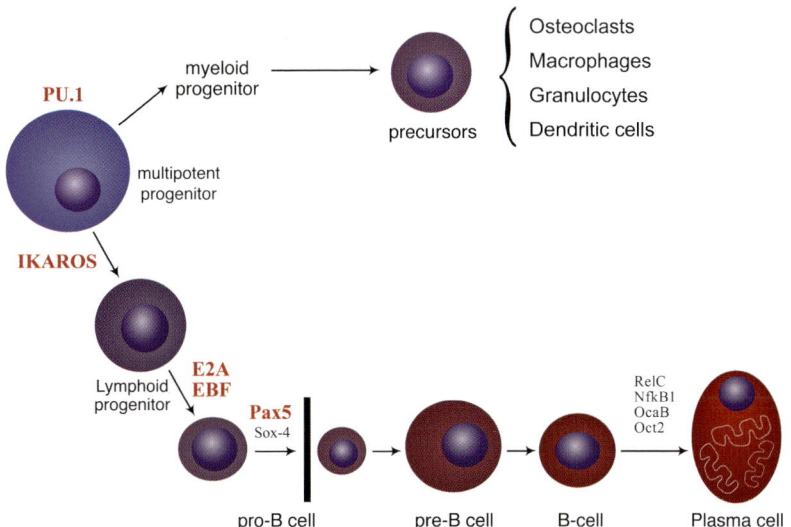

**FIGURE 3-3 Transcriptional regulation of B-cell differentiation.** B-cell differentiation is regulated, in part, by the expression of a series of transcription factors that function in a temporal manner. These transcription factors include PU.1, Ikaros, E2A, Ebf1, and Pax5. Loss of these specific factors precludes the cells from continued maturation, and results in a developmental block of cells at the latest stage of differentiation prior to the arrest. In addition to the absence of B cells, mice deficient in PU.1, Ebf1, and Pax5 have profound changes to their skeletons. No data are available on the bone phenotype in mice deficient in Ikaros or E2A. (Please refer to Chapter 3, page 65).

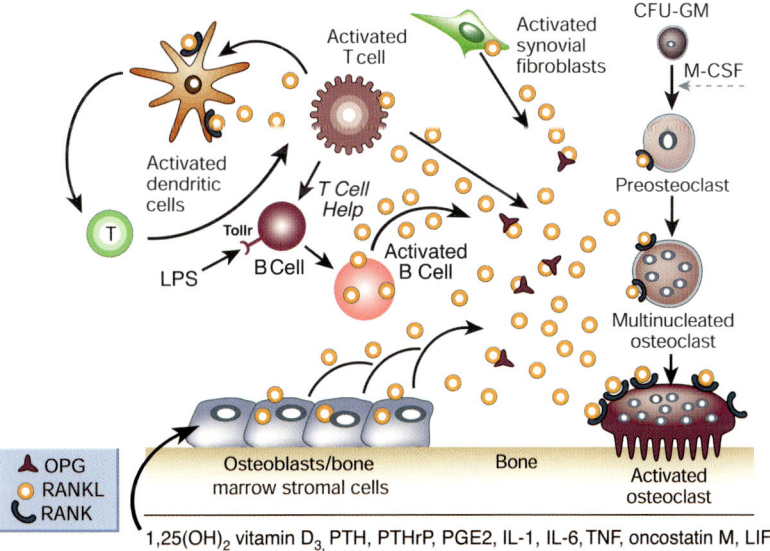

**FIGURE 3-4 Activated B cells induce osteoclast differentiation.** B cells activated by the adaptive immune system (antigen specific) or through the innate immune system (LPS-Toll receptor) result in B cells that secrete or express on their cell surface molecules like RANKL that induce osteoclastogenesis. *(Adapted from Boyle WJ, Simonet WS, Lacey DL. Nature 2003;**423**:337–42).* (Please refer to Chapter 3, page 71).

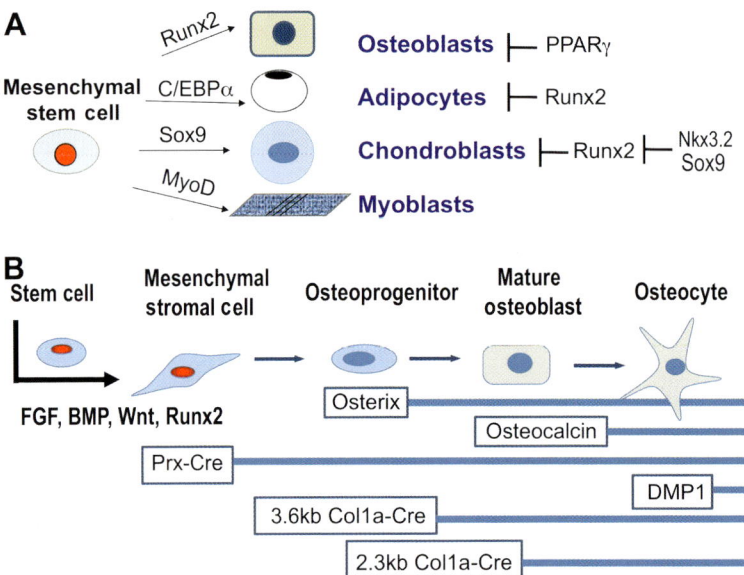

**FIGURE 5-1 Lineage determination from mesenchymal stem cells. (A)** Master transcription factors (TFs) essential for commitment to the indicated phenotypes are shown. In vitro overexpression of the PPARγ will block the osteoblast phenotype and Runx2 prevent adipocyte differentiation. Stromal cells isolated from marrow pluripotent mesenchymal cells (MSCs) are Runx2-positive and become downregulated by increases in TFs required for other lineages, e.g., Nkx3.2 and Sox9 in chondroblasts. **(B)** Illustrates stages in the osteogenic lineage in which genes can be excised in mice using Cre drivers with characterized promoters expressing at the indicated stage (reviewed in [218]). (Please refer to Chapter 5, page 106).

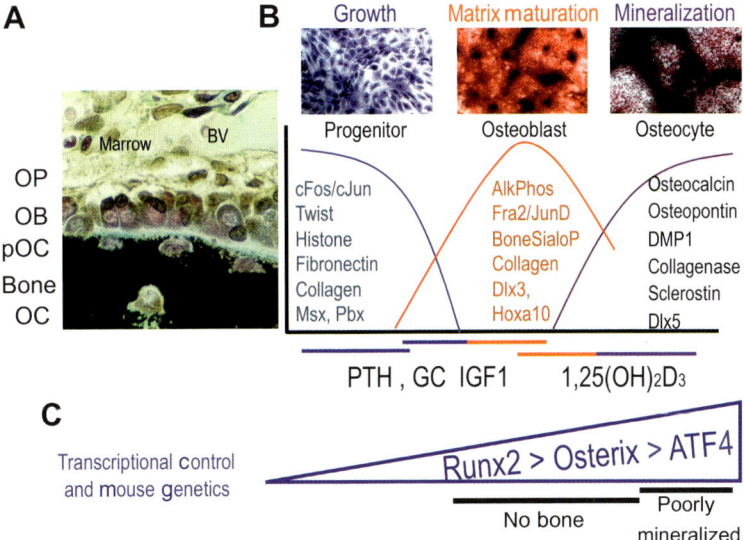

**FIGURE 5-2 Stages and markers of osteoblast growth and differentiation. (A)** Histologic section of the surface of bone showing organization of osteoprogenitor (OP) cells behind a row of mature osteoblasts (OB) on the bone surface secreting an osteoid non-mineralized matrix (pale blue color). As hydroxyapatite deposits in the osteoid, OBs become preosteocytes (pOC) until fully surrounded by mineral (bone, black, von Kossa stain) and differentiated to osteocytes (OC). **(B)** Isolation of osteoblasts from perinatal calvarial tissue can recapitulate the stages of osteoblast maturation from the proliferating progenitor to the postproliferative osteoblast and osteocyte. Growth stage (Toluidine-blue-stained cells) are characterized by proliferation markers (histone) and transcription factors that suppress genes expressed in OBs. As cells multilayer, they become alkaline-phosphatase-positive, designated as the matrix maturation stage, with maximal secretion, collagens and non-collagenous proteins such as bone sialoprotein, osteocalcin, osteopontin. This allows the matrix to mineralize, designated mineralization stage. In the osteocyte E11, MEPE, Dentin matrix protein 1 and Sclerostin, a Wnt pathway inhibitor, serve as identification markers for this population of bone cells. Hormones that have a major effect at each stage are indicated. **(C)** Transcription factors in hierarchical order are shown in the triangle as these have been proven in mouse models to be essential for initiation of bone formation and mineralization. Lines indicate the phenotype in the null mouse for each gene. (Please refer to Chapter 5, page 107).

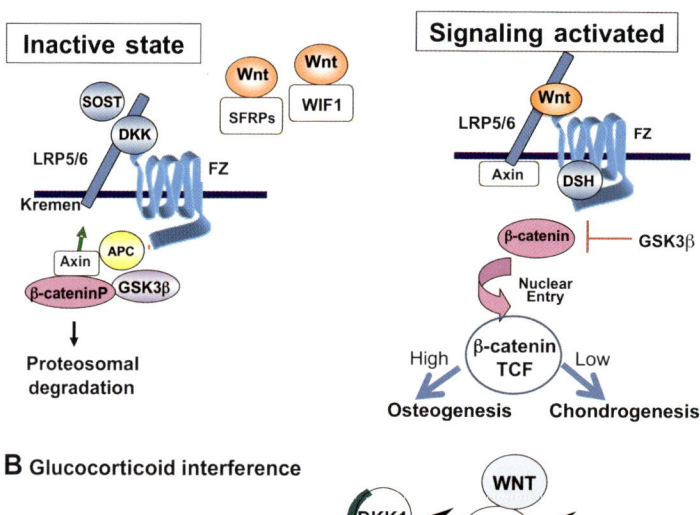
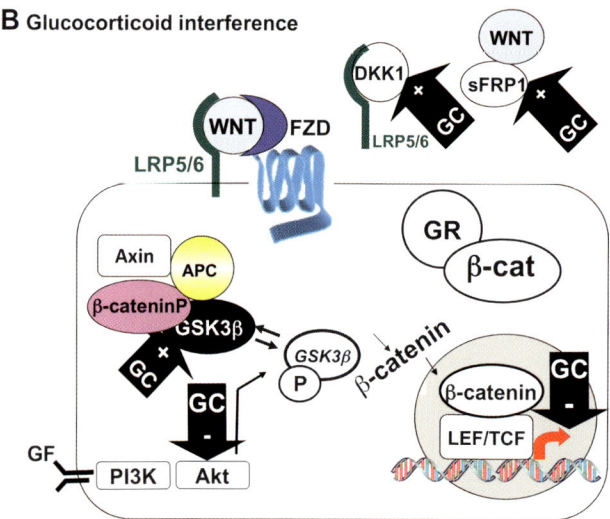

**FIGURE 5-3 The canonical Wnt pathway (A) and disruption by glucocorticoids (B). (A)** The inactive state is shown when the frizzled–LRP5/6 receptor complex is bound by multiple inhibitors, DKK, SOST, and Kremen. In addition Wnt signaling can be regulated by soluble inhibitors, WIF1 (Wnt inhibitory factor), and SFRPs (secreted frizzled related proteins) that bind Wnt ligands. In the absence of ligand binding, β-catenin, the intracellular mediator of Wnt signaling, is held in a complex with Axin, APC, and GSK3β, which phosphorylates β-catenin, resulting in its proteosomal degradation. With activation of the Wnt receptor, the intracellular β-catenin complex components are reorganized, releasing stabilized β-catenin (in the absence of phosphorylation) and allowing β-catenin nuclear translocation and complexing with the canonical transcription factor TCF/Lef. Several mouse genetic models have established that high cellular levels of β-catenin will drive an MSC to osteogenesis while lower levels can promote chondrogenesis (see text for references). **(B)** Multiple effects of glucocorticoids (GC) on Wnt signaling. Suppression of bone formation by dexamethasone is in part contributed by direct effects of glucocorticoids that inhibit Wnt signaling at multiple levels of the pathway as illustrated. GCs increases expression of Wnt inhibitors DKK1 and sFRP that target the receptor β-catenin degradation. The glucocorticoid receptor can bind to β-catenin reducing effectiveness of its activity for gene transcription. Together these multiple "hits" on canonical Wnt signaling decrease transcription of genes essential for bone formation. (Please refer to Chapter 5, page 112).

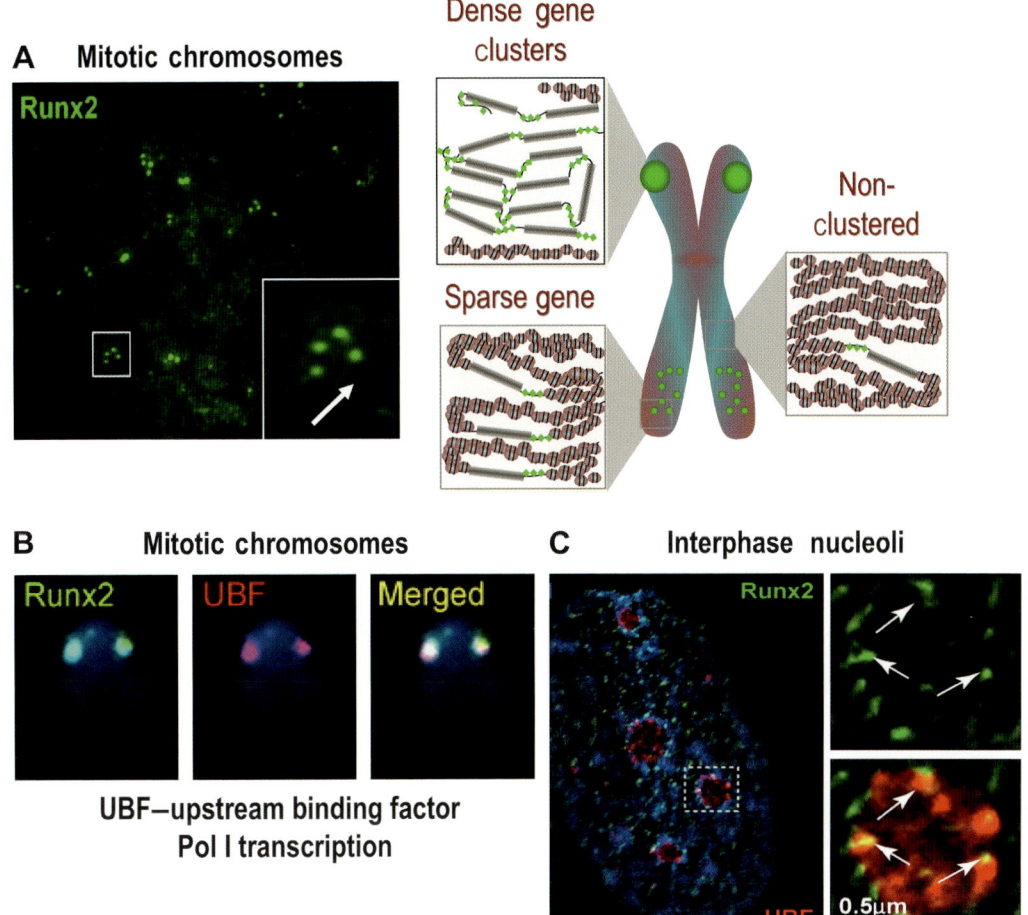

**FIGURE 5-4** Novel epigenetic functions of Runx2. **(A)** Runx2 is associated with genes on mitotic chromosomes, but the genes are not transcribed, indicating a novel level of epigenetic control [172, 219]. The Runx2 foci appear as large spots (dense clusters) and small foci illustrated in the right panel. Analyses show that the associated genes include Smads, Runx2, VEGF, and ribosomal genes to name a few [172]. These findings suggest that these genes on mitotic chromosomes with Runx2 are being "book marked" for stability of the osteoblast phenotype when the cell division is completed.
**(B)** Pairs of large Runx2 foci on eccentric chromosomes in regions are associated with ribosomal genes (UBF is a marker). Runx2 is also found to downregulate protein synthesis by binding to ribosomal genes that contain from 40 to 60 Runx2 sites. **(C)** Runx is also found in interphase nucleus in the periphery of the nucleolus where protein synthesis occurs. (Please refer to Chapter 5, page 120).

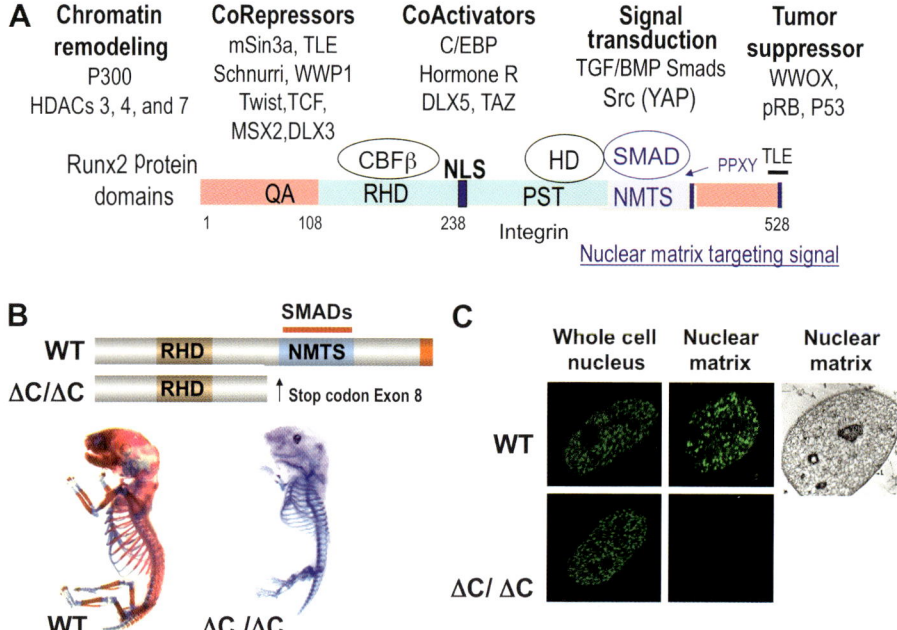

**FIGURE 5-5** Runx2 is a scaffolding protein with regulatory complexes organized in subnuclear domains. **(A)** Classes of co-repressors and co-activators forming protein–protein interactions with Runx2 reflect mechanisms by which Runx2 can both activate and repress genes during development of the osteoblast phenotype. These co-regulatory proteins have been mapped to either the N- or C-terminus. The N-terminus contains a unique QA stretch found only in Runx2 and the DNA-binding module, the runt homology domain (RHD), that requires the CBFβ partner protein. In the C-terminus is located a region of phosphorylation sites (PST), activator (NMTS, nuclear matrix targeting signal), and repressor (PPXY, TLE) domains. Note the interaction site for SMAD lies within the NMTS and requires three amino acids (HTY). **(B)** The biological significance of the NMTS has been established by a knock-in mutation at exon 8. The loss of the segment of the C-terminus results in a phenotype analogous to the complete null mouse. Runx2 DC mutant protein enters the nucleus (whole cell nucleus stained for Runx2, IHC) but is not bound to the nuclear matrix (NM scaffold remains following high salt extraction of soluble chromatin). **(C)** Loss of Runx2 binding to the nuclear matrix scaffold in Runx2$^{\Delta c/\Delta c}$ mice. An electron micrograph of the filamentous nuclear is shown. These studies have established a critical role in Runx2 functional activity mediated by forming co-regulatory protein complexes on the nuclear matrix for specificity of bone formation. (Please refer to Chapter 5, page 121).

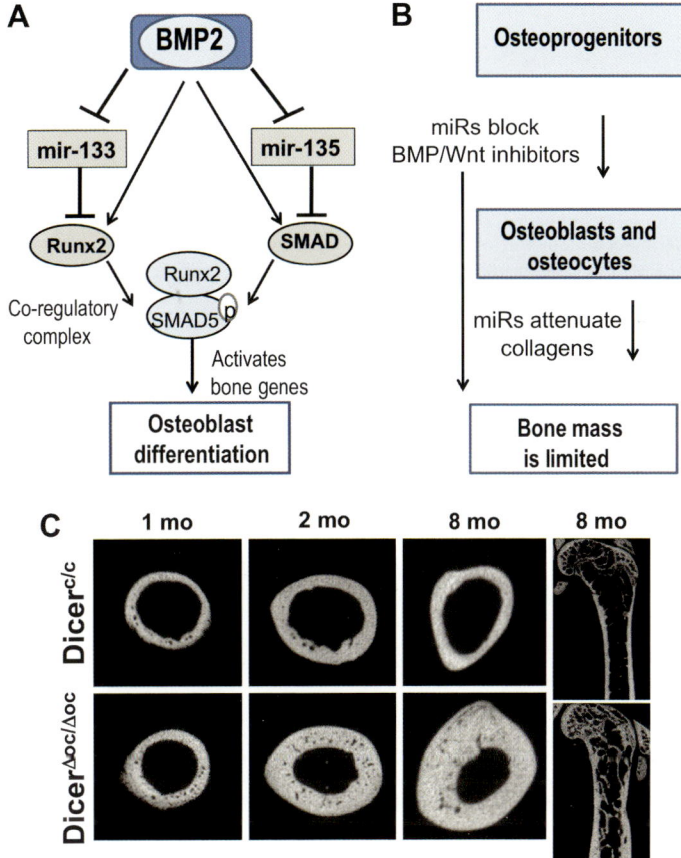

**FIGURE 5-6 MicroRNA regulation of osteogenesis. (A, B)** Schematic illustrations of characterized mature miRs. **(A)** Induction of C2C12 premyogenic cells into the osteoblast lineage by BMP2 results in downregulation of miRs that negatively regulate Runx2 (miR133) and BMP2 SMAD5 (miR135). BMP2 therefore relieves repression of these two factors essential for osteoblastogenesis. **(B)** During MC3T3 osteoblast differentiation (days 0–28 in differentiation medium), miRs are continuously upregulated and have selective functions as illustrated to either downregulate inhibitors of bone formation from osteoprogenitor proliferation stage to the osteoblast/osteocyte mineralization stage when the synthesis of bone matrix proteins is attenuated to prevent fibrosis. **(C)** Excision in osteoblast lineage cells by osteocalcin-Cre of the Dicer enzyme which processes precursor miRs to mature functional miRs, results in a high bone mass phenotype. Beginning from 1 month, cortical expansion and trabecular bone are increased at indicated ages. Single slice images from μCT three-dimensional data acquisition are shown [104]. (Please refer to Chapter 5, page 123).

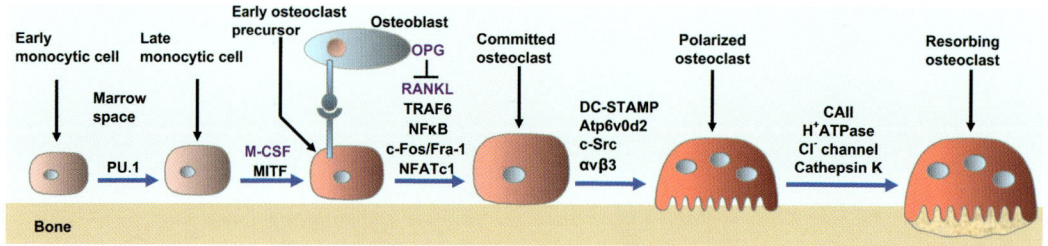

**FIGURE 6-1 Regulation of osteoclast formation and function.** The osteoclast is a member of the monocyte/macrophage family. Early non-specific differentiation along the osteoclast pathway is dependent on PU.1 and the MITF family of transcription factors, as well as the macrophage proliferation and survival cytokine M-CSF. Activation of RANK by osteoblast-expressed RANK ligand (RANKL) commits the cell to the osteoclast fate, which is mediated by signaling molecules such as AP-1 transcription factors, tumor necrosis factor receptor associated factor 6 (TRAF6), nuclear factor κB (NFκB), c-Fos, and Fra-1. RANKL-stimulated osteoclastogenesis is inhibited by the RANKL decoy receptor osteoprotegerin (OPG). Committed osteoclasts express the fusogenic genes DC-STAMP and Atp6v0d2, allowing formation of the multinucleated cell. The initial event in the development of the resorptive capacity of the mature osteoclast is its polarization, which requires c-Src and the $\alpha v\beta 3$ integrin. Once polarized, the osteoclast mobilizes the mineralized component of bone. Bone mobilization is achieved through the acidifying molecules, carbonic anhydrase II (CAII), an electrogenic H+ATPase and a charge-coupled $Cl^-$ channel. Cathepsin K mediates bone organic matrix degradation. (Please refer to Chapter 6, page 143).

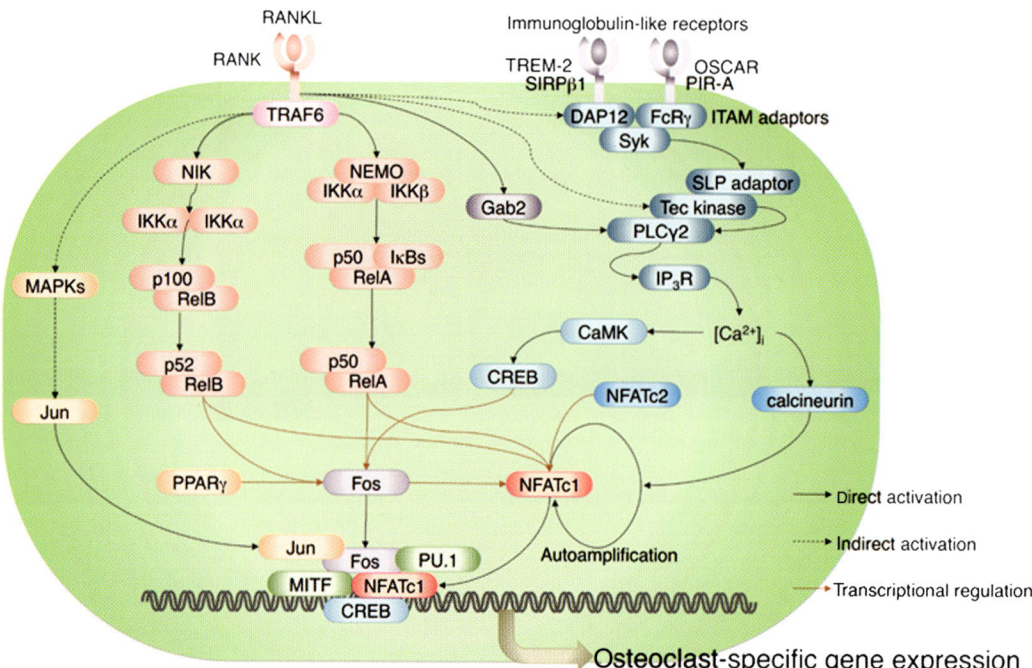

**FIGURE 6-2 RANK signaling in osteoclast differentiation.** RANKL stimulation leads to activation of several signaling pathways including MAPK and the canonical/noncanonical NF-κB pathways through TRAF6 as well as a $Ca^{2+}$ pathway through the ITAM adaptors for immunoglobulin-like receptors such as OSCAR, TREM-2, SIRPβ1, and PIR-A. NF-κB pathways contribute to induction of Fos and NFATc1, which are essential transcription factors for osteoclast differentiation. The Fos induction is also mediated by CREB and PPARγ. NFATc1 is activated by a $Ca^{2+}$ signal downstream of Ig-like receptors through the tyrosine phosphorylation of signaling molecules such as Syk, Tec kinases, SLP adaptors, and PLCγ2. Finally, NFATc1 orchestrates the transcription of osteoclast-specific genes together with AP-1, CREB, PU.1, and MITF. (Please refer to Chapter 6, page 144).

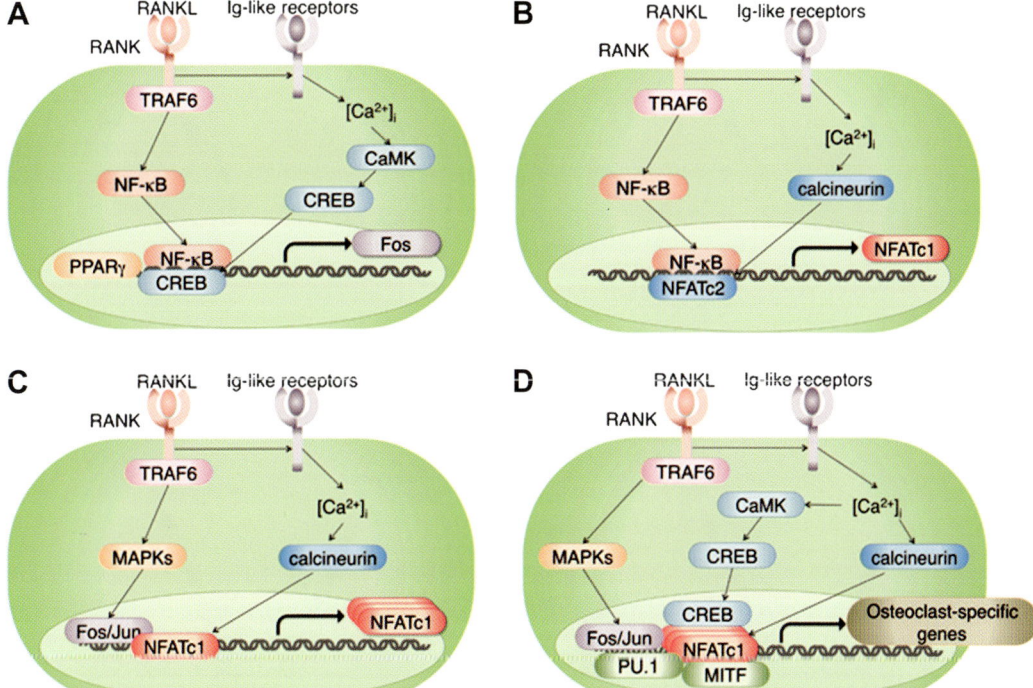

**FIGURE 6-3 Temporal regulation of gene expression by transcription factors during osteoclast differentiation.**
**(A) Induction of Fos**. RANKL binding to RANK results in the activation of NF-κB. At the same time, RANK activation results in the phosphorylation of ITAM adaptors, DAP12 and FcRγ. Activation of NF-κB and CREB in the downstream of TRAF6 and ITAM adaptors, respectively, leads to induction of c-Fos at the early stage of osteoclast differentiation. The induction of c-Fos also requires PPARγ activity. **(B) Initial induction of NFATc1.** NFATc1 is initially induced by NF-κB and NFATc2. **(C) Autoamplification of NFATc1.** NFATc1 and AP-1 transcription factors are essential for the robust induction of NFATc1. The NFATc1 promoter is epigenetically activated through histone acetylation and NFATc1 binds to an NFAT-binding site in its own promoter. **(D) Terminal differentiation of osteoclasts.** NFATc1 works together with other transcription factors, such as AP1, PU.1, MITF, and CREB, to induce various osteoclast-specific genes, including TRAP, cathepsin K, and the calcitonin receptor. (Please refer to Chapter 6, page 153).

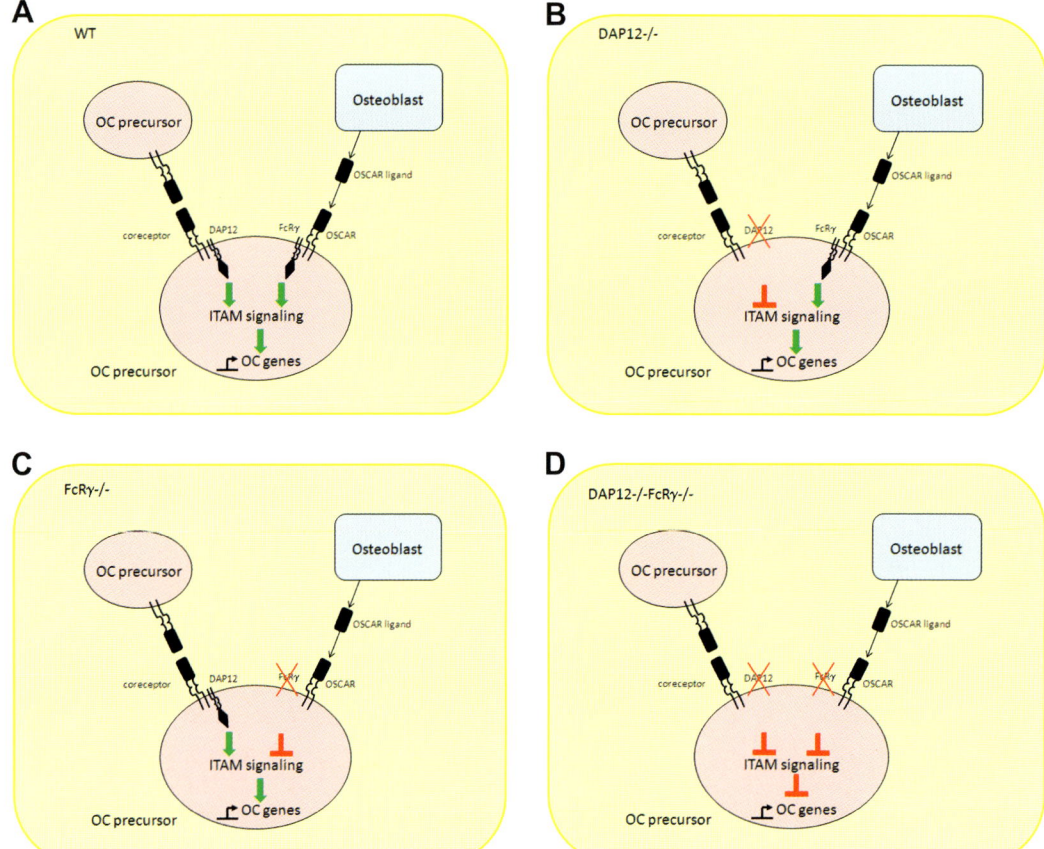

**FIGURE 6-4 Role of DAP12 and FcRγ in osteoclastogenesis. (A)** In the presence of exogenous RANKL, the ITAM co-stimulatory signals are provided by DAP12, its co-receptors and as yet unidentified ligands, expressed on osteoclast precursor cells. Under this circumstance, activation of FcRγ is not required for osteoclast formation. **(B)** In the absence of DAP12, ITAM co-stimulatory signals are not activated by exogenous RANKL, resulting in defective upregulation of osteoclastogenic genes. However, in the co-culture system, osteoblasts provide OSCAR-L and activate the OSCAR/FcRγ pair in $DAP12^{-/-}$ osteoclasts. $DAP12^{-/-}$ osteoclasts can differentiate but remain dysfunctional. **(C)** In the absence of FcRγ, DAP12, its associated receptors and their ligands can transmit ITAM-dependent co-stimulatory signals from adjacent osteoclast precursors, whether in co-culture or in the presence of exogenous RANKL, allowing their differentiation. **(D)** When both DAP12 and FcRγ are deleted, the two ITAM pathways are blocked, and osteoclastogenesis is not observed in vivo or under any in vitro condition since co-stimulatory signals cannot be delivered. (Please refer to Chapter 6, page 155).

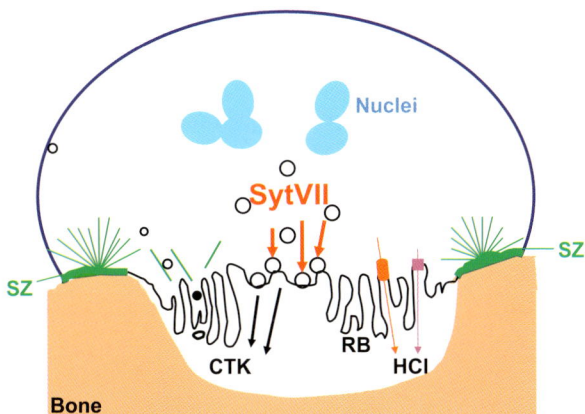

**FIGURE 6-7 Generation of the osteoclast ruffled border.** Lysosomal enzyme, vacuolar $H^+$ATPase (proton pump) and chloride channel-containing vesicles are inserted into the bone-apposed plasma membrane, under the aegis of synaptotagmin VII (SytVII), permitting delivery of HCl and cathepsin K (CTK) into the resorptive microenvironment. (Courtesy of Dr. Haibo Zhao.) Ruffled border (RB); sealing zone (SZ). (Please refer to Chapter 6, page 168).

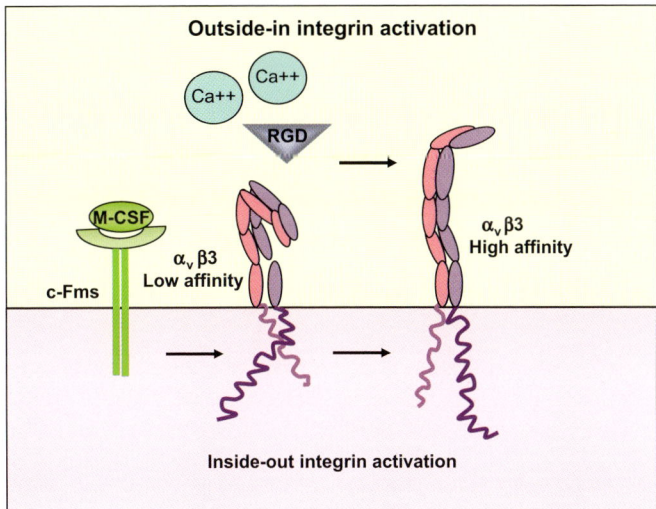

**FIGURE 6-8 Outside-in and Inside-out $\alpha v\beta 3$ integrin activation.** $\alpha v\beta 3$ integrin can exist in a low-affinity/low-binding conformation (inactive) or in a high-affinity/high-ligand-binding conformation (active state). Signals from outside the cells (e.g. integrin ligands and calcium cations) or inside the cells (e.g. signals from the M-CSF receptor, c-Fms) can both induce conformational changes allowing integrin activation. (Please refer to Chapter 6, page 170).

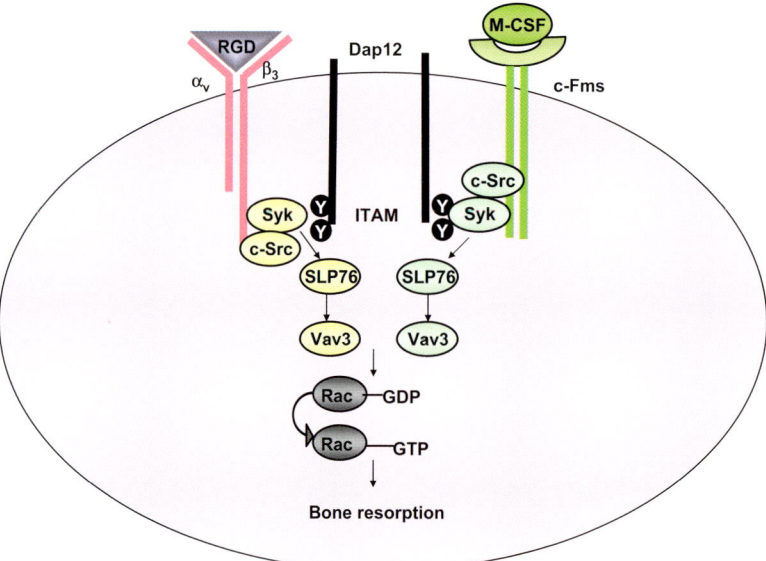

**FIGURE 6-9 αvβ3 integrin and c-Fms collaborate to organize the osteoclast cytoskeleton.** Activation of the αvβ3 integrin or the M-CSF receptor, c-Fms, organizes the osteoclast cytoskeleton by a mechanism involving stimulation of the tyrosine kinases, c-Src and Syk, phosphorylation of the ITAM-containing protein, DAP12, the adaptor, SLP-76, and the guanine nucleotide exchange factor, Vav3. This series of events activates the Rho GTPase family of transcription factors, particularly Rac, ultimately permitting the osteoclast to restructure its cytoskeleton and resorb bone. Whether this series of events represents independent activation of intracellular signals by αvβ3 and c-Fms or inside-out activation of the integrin by the cytokine receptor is unknown. (Please refer to Chapter 6, page 175).

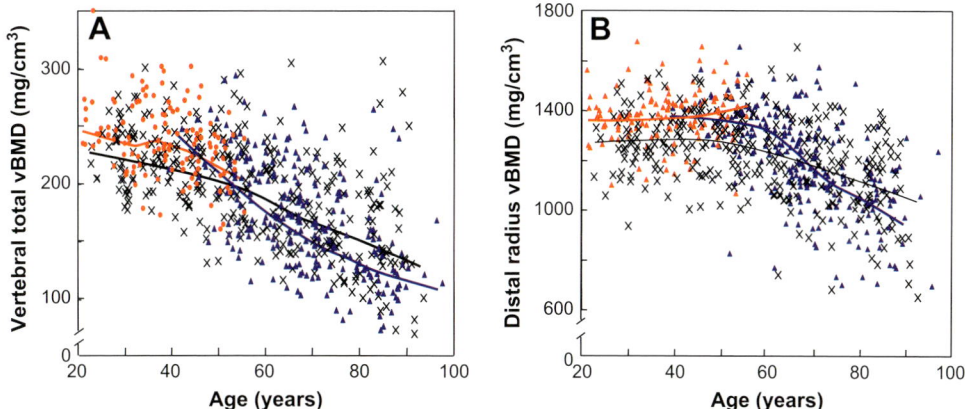

**FIGURE 9-2 (A)** Values for vBMD (mg/cm$^3$) of the total vertebral body in a population sample of Rochester, Minnesota, women and men between the ages of 20 and 97 years. Individual values and smoother lines are given for premenopausal women in red, for postmenopausal women in blue, and for men in black. **(B)** Values for cortical vBMD at the distal radius in the same cohort. Color code is as in Panel A. All changes with age were significant ($P < 0.05$). Reproduced from Riggs et al. [2], with permission. (Please refer to Chapter 9, page 271).

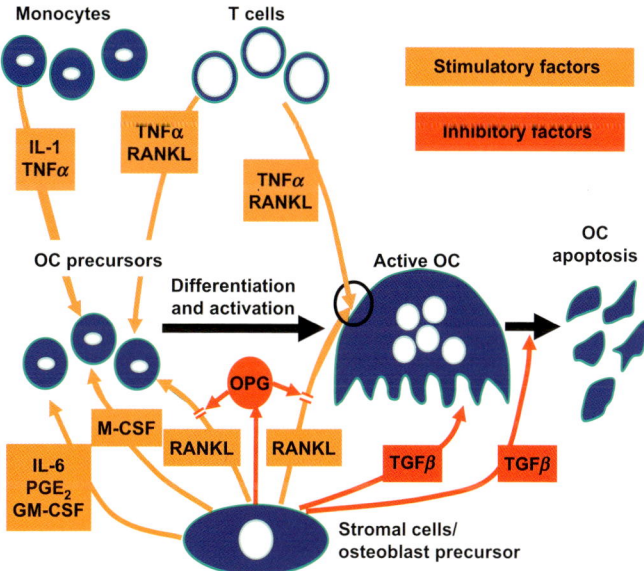

**FIGURE 9-7** Model for effects of estrogen on osteoclast (OC) formation and function by cytokines in the bone marrow microenvironment. The positive (E+) and negative (E−) effects of estrogen on stimulatory and inhibitory factors are shown. Modified from Riggs et al. [136], with permission. (Please refer to Chapter 9, page 287).

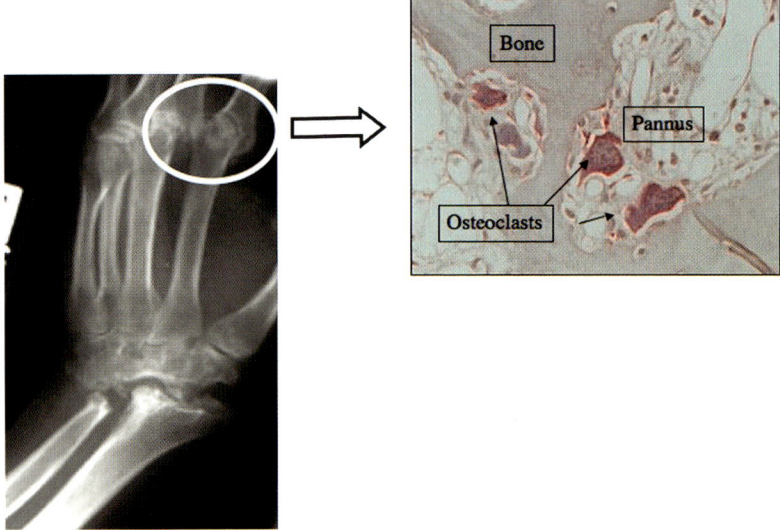

**FIGURE 10-1** Radiographic and histopathologic features of RA joint erosions. Hand radiograph from a patient with advanced RA demonstrating extensive articular bone destruction and accompanying joint erosion (highlighted by the circle). A representative histopathologic section from the bone pannus interface with osteoclasts in resorption lacunae on the bone surface. (Please refer to Chapter 10, page 303).

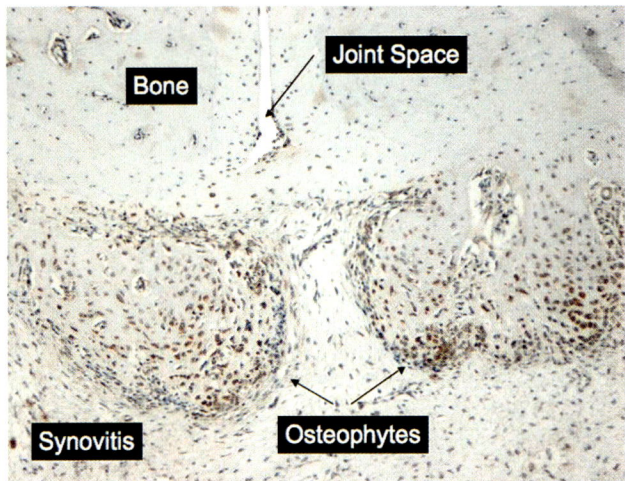

**FIGURE 10-5** Osteophytes emerging from the peri-articular periosteum of a rat induced for adjuvant arthritis. The two lesions are surrounded by inflammatory tissue (synovitis). Brown-stained cells show expression of the proliferation marker Ki-67. (Please refer to Chapter 10, page 316).

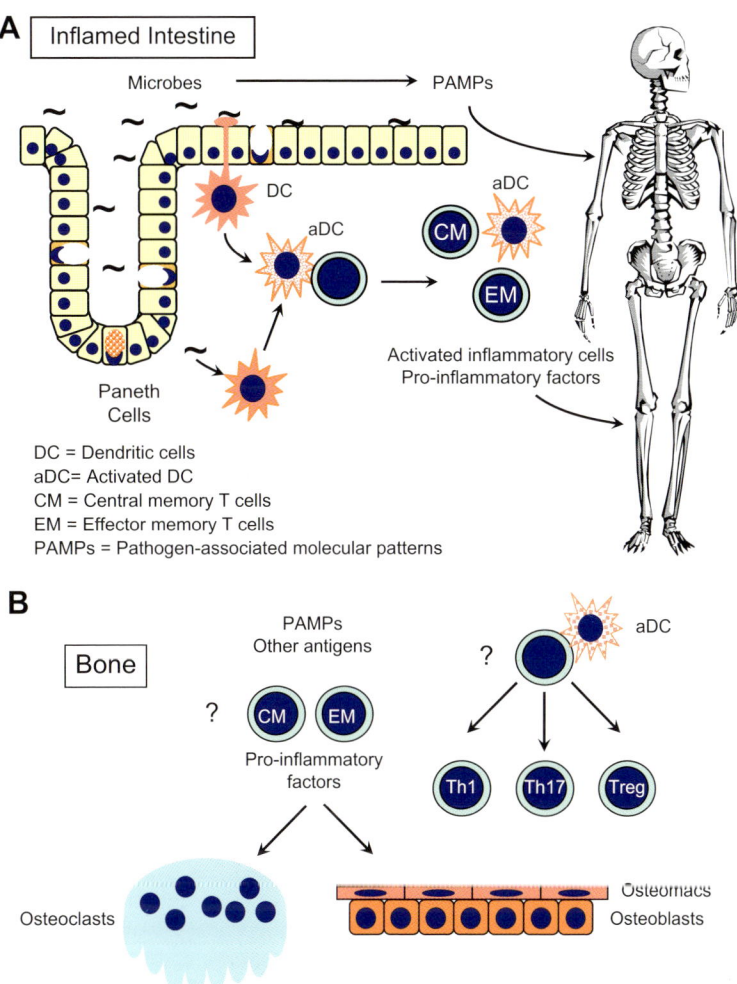

**FIGURE 11-1 (A)** The intestine is colonized with over 100 trillion microorganisms, which are contained in its lumen by a single layer of epithelial cells. In the steady state, the host tolerates this enormous antigenic load due to special adaptations of dendritic cells (DC), macrophages, and the presence of regulatory T cells. In IBD, these immune-suppressive mechanisms are lost, and dendritic cells, macrophages, and T cells become activated. T cells acquire immunological memory for their cognate antigen (central memory and effector memory cells), and intestinal mucosal addressins. T memory cells and dendritic cells leave mesenteric lymph nodes, Peyer's patches and intestinal lymphoid follicles, enter the circulation and home back to the intestine or establish residence in the bone marrow. In addition, pro-inflammatory factors, endotoxin, flagellin, and other bacterial products may enter the circulation and gain access to the bone microenvironment. **(B)** In bone, dendritic cells that carry antigen may activate the differentiation of T cells into effector cell lineages. The commitment to Th1, Th17, or Th2 cells will depend on the cytokine milieu. In addition, microbial antigens may be taken up by resident bone cells like osteomacs and activate immune responses that affect bone cell function. IBD is associated with innate immune defects that may facilitate the survival of gut-derived microorganisms, and generate vigorous adaptive immune response. (Please refer to Chapter 11, page 333).

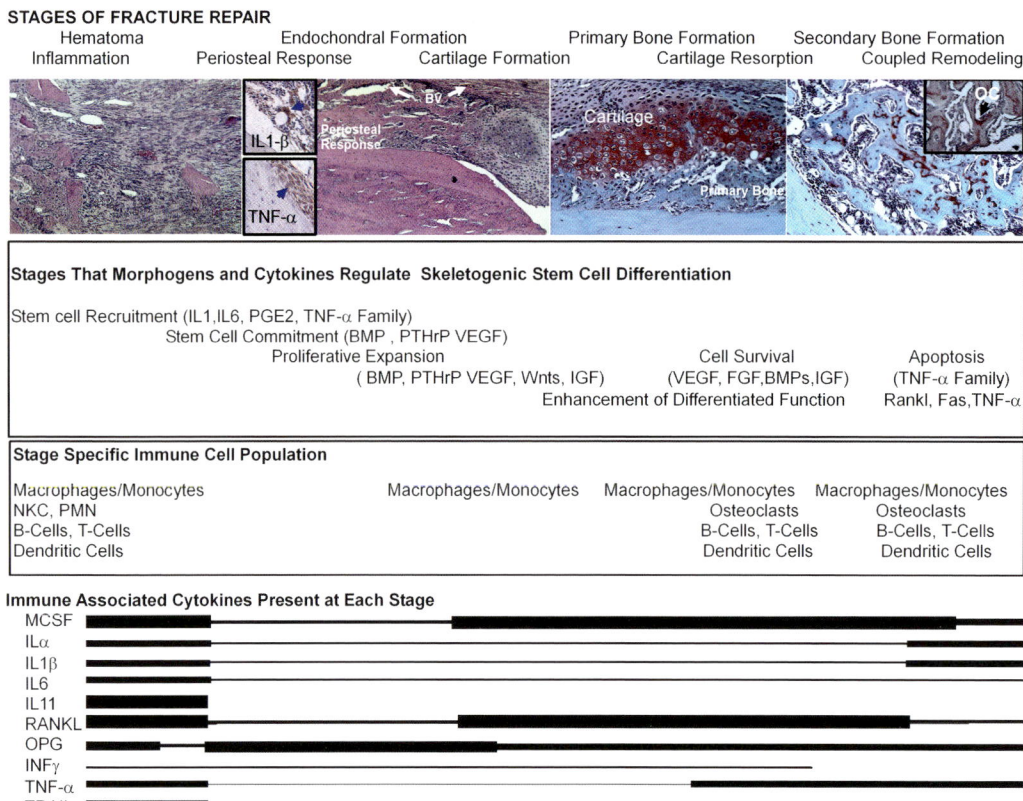

FIGURE 12-1 Summary of the multiple stages of fracture healing. Histological sections are presented for each stage and a summary of the various processes that are associated with each stage is presented. All histological specimens are from sagittal sections of mouse tibia transverse fractures and were stained with H&E or safranin O and fast green, micrographic images are at 200× magnification. Section for the initial injury was taken from the fracture site 24 hours post injury (far left). Two small inserts depict immunohistological identification of the predominant cell types in the hematoma expressing IL1 (macrophages and monocytes) and TNF-α (periosteal-lining cells). Sections depicting the initial periosteal response and endochondral formation are from 7 days post injury (left middle). Arrows denote blood vessels (BV) of the vascular in-growth from the peripheral areas of the periosteum. Sections depicting the period of primary bone formation and cartilage tissue resorption are from 14 days post injury (middle right). Sections depicting the period of secondary bone formation are from 21 days post injury (far right). Callus sites. Insert depicts 400× images of an osteoclast (∗chondroclast) resorbing an area of calcified cartilage. Summary of the stages of fracture repair, the major immune cells associated with each stage and their associated molecular processes and regulators are summarized in the three boxes below the histological sections. Bottom box depicts the relative levels of expression of various mRNAs that have been examined in our laboratories and are denoted by three line widths. The levels of expression are by percent over baseline for each and are not comparable between individual mRNAs. Data for expression levels for the pro-inflammatory cytokines and the ECM mRNAs were from Kon et al. [20], Lehmann et al. [77], and Gerstenfeld et al. [4]. (Please refer to Chapter 12, page 345).

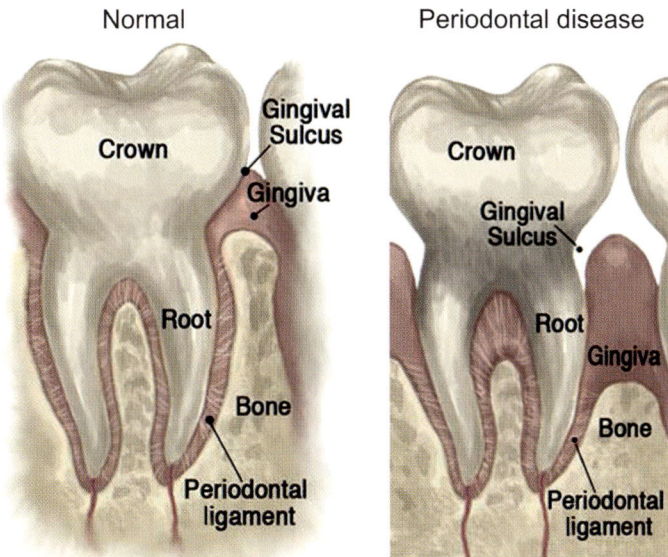

**FIGURE 15-1 Periodontal bone loss.** In periodontal disease bacteria attach to the tooth surface and invade the adjacent epithelium and connective tissue to cause inflammation and bone loss. The changes in bone level are shown in the diagram on the right. The resorbed bone is replaced with gingiva, which does not provide strong support for the tooth and can lead to tooth loss. Periodontal lesions are typically not painful unless they occur as an acute abscess. (Please refer to Chapter 15, page 414).

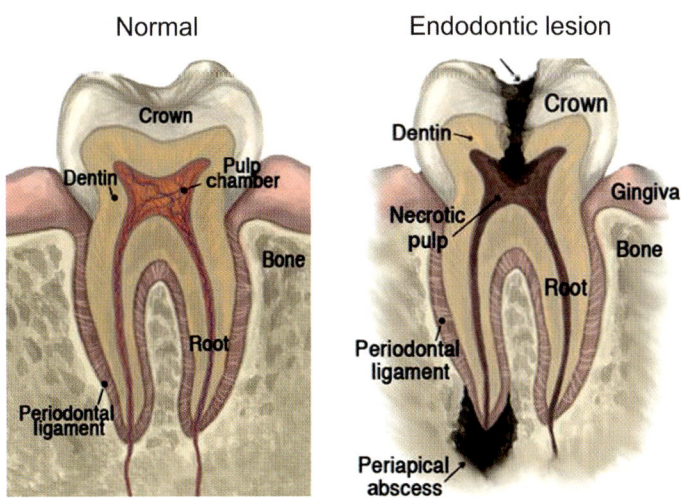

**FIGURE 15-2 Lesion of endodontic origin.** In lesions of endodontic origin disruption of the integrity of the crown typically occurs either through trauma or through the effect of cariogenic bacteria. Bacteria then contaminate the dental pulp, which becomes necrotic and forms a reservoir of bacteria that cause inflammation in the periapical area and bone resorption. These lesions are frequently painful. (Please refer to Chapter 15, page 425).

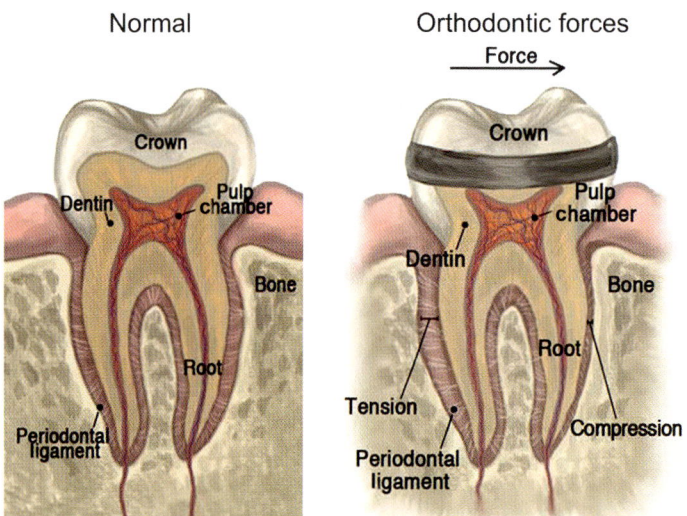

**FIGURE 15-3 Orthodontic tooth movement.** Mechanical force on the tooth causes compression of the periodontal ligament on one side and tension on the other side. The compression side is associated with aseptic injury and bone resorption and the tension side with bone formation. (Please refer to Chapter 15, page 430).